Thoroughly Revised and Updated

Reasoning & Aptitude *for* GATE 2025 and ESE Pre 2025

Comprehensive Theory *with* **Examples**
and **Solved Questions of**
GATE *and* **ESE Prelims**

Also useful for
UPSC (CSAT), MBA Entrance, Wipro, SSC, Bank (PO), TCS , Railways, Infosys, various Public Sector Units and other Competitive Exams conducted by UPSC

MADE EASY PUBLICATIONS

MADE EASY Publications Pvt. Ltd.
Corporate Office: 44-A/4, Kalu Sarai (Near Hauz Khas Metro Station), New Delhi-110016
E-mail: infomep@madeeasy.in
Contact: 9021300500, 8860378007
Visit us at: www.madeeasypublications.org

Reasoning & Aptitude for GATE 2025 & ESE Prelims 2025

1st Edition : 2008
2nd Edition : 2009
3rd Edition : 2010
4th Edition : 2011
5th Edition : 2011
6th Edition : 2012
7th Edition : 2013
8th Edition : 2014
9th Edition : 2015
10th Edition : 2016
11th Edition : 2017
12th Edition : 2018
13th Edition : 2019
14th Edition : 2020
15th Edition : 2021
16th Edition : 2022
17th Edition : 2023
18th Edition : 2024

Director's Message

B. Singh (Ex. IES)

Engineering is one of the most chosen graduation fields, choosing to become an engineer after high school is usually a matter of interest but this eventually develops into "the purpose of being an engineer" and then a student thinks of cracking various competitive exams like ESE, GATE, PSUs exams, and other state engineering services exams. With the objective nature of these competitive exams and with increasing competition, it becomes necessary for the student to study and practice every topic and also get acclimatize with the style of questions asked in the exam.

Studying engineering in university is one aspect but studying to crack different prestigious competitive exams requires altogether different strategies, crystal clear concepts and rigorous practice of previous years' questions. Every student can achieve great results through proper guidance and exam-oriented study material, and hence we have come up with this book covering all the previous years' questions. This book will help aspirants to develop an understanding of important and frequently asked areas in the exam and will also help in strengthening concepts. MADE EASY Team has put sincere efforts in framing accurate and detailed explanations for all the previous years' questions. The explanation provided for each question is not only question specific but it will also give insight on the concept as a whole which will beneficial for the student from the exam point of view to handle similar questions.

All the previous years' questions are segregated subject wise and further, they have been categorized topic-wise for easy learning and this certainly assists aspirants to solve all previous years' questions of a particular area in one place. I would like to acknowledge the efforts of the entire MADE EASY team who worked hard to solve previous years ' questions with accuracy. I hope this book will stand up to the expectations of aspirants and my desire to serve the student community by providing the best study material will get accomplished.

B. Singh (Ex. IES)
CMD, MADE EASY Group

Contents

A

Section

Arithmetic

1.1
CHAPTER

Number System

In Quantitative Aptitude (QA), Number System is one of the modules which is of critical importance. We can consider this module as the back bone as well as basic foundation and building block for QA as well as for reasoning. Applications of concepts of numbers can be easily found in puzzles, reasoning based questions, number series and many more reasoning areas. This is why it is our suggestion to students to understand the concepts discussed in the module thoroughly alongwith understanding of applications.

Classifications of Numbers

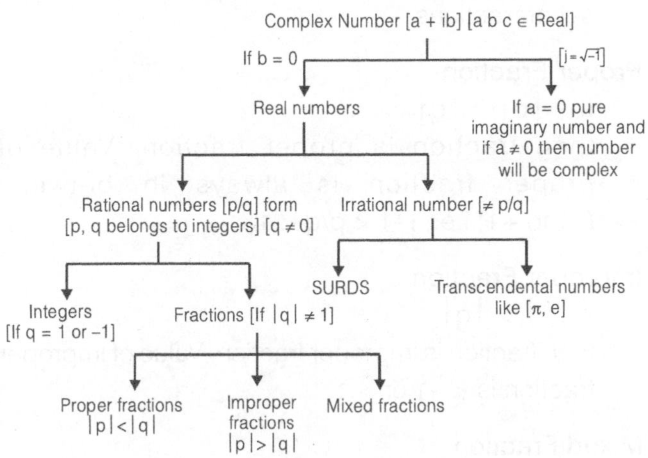

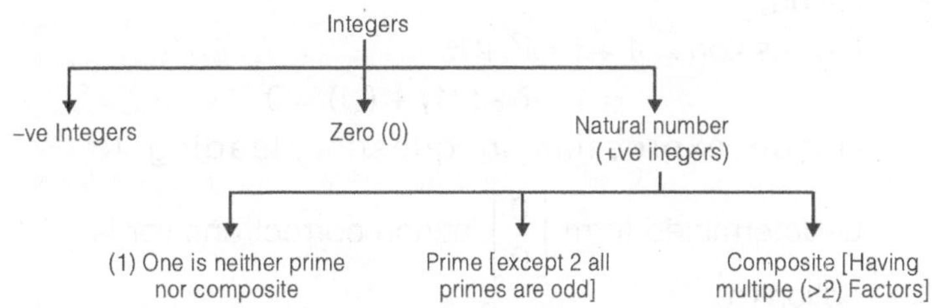

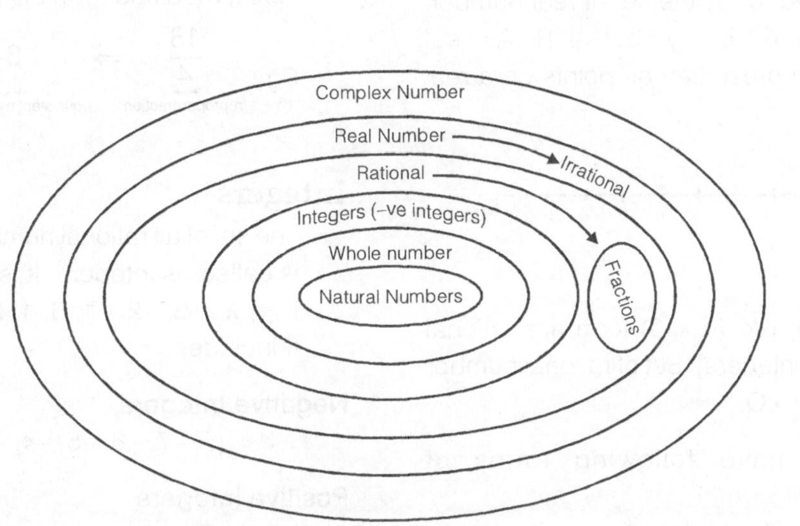

Our main focus in this module of numbers in on **real number system**. However in context of imaginary numbers only following property is important.

Imaginary Numbers

$i = \sqrt{-1} \Rightarrow i^{4K+1} \equiv \sqrt{-1} \equiv i$

$i^2 = -1 \Rightarrow i^{4K+2} \equiv -1 \equiv i^2$

$i^3 = -i \Rightarrow i^{4K+3} \equiv -i \equiv i^3$

$i^4 = 1 \Rightarrow i^{4K} \equiv 1 \equiv i^4$

Ex.1

What is the value of expression

$$\frac{i^{12} + i^{13} + i^{14} + i^{15}}{i^{18} + i^{19} + i^{20} + i^{21}} ?$$

(a) i^2 (b) -1
(c) $1/i^2$ (d) None of these

Ans. (d)

$$\frac{i^{12}\left(1+i+i^2+i^3\right)}{i^{18}\left(1+i+i^2+i^3\right)}$$

If we commit a mistake of cancelling out common terms in numerator and denominator options a, b, c all one correct hence my answer should be (d) but

Expression $1 + i + i^2 + i^3$
$= 1 + i + (-1) + (-i) = 0$

Hence expression in question leading to undetermined form $\left[\frac{0}{0}\right]$ hence correct answer is option (d).

Real Number System

Entire real numbers group of rational and irrational numbers combined forms the set of real number, which is represented by symbol → R. All real numbers can be represented as points on a real number line.

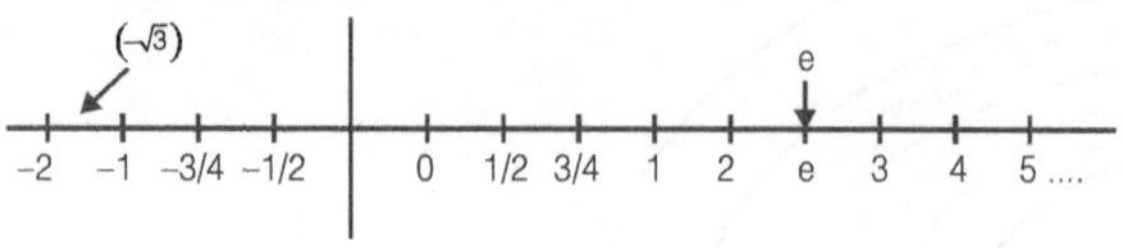

Rational Number

All the numbers in p/q ($q \neq 0$) form are rational numbers [p, q are integers]. Set of rational number is represented by → Q.

Rational Numbers have following forms of representations.

(a) Terminating decimal forms
for example 0.125

$\Rightarrow 0.125 = \frac{125}{1000} \Rightarrow$ Rational

(b) Nonterminating but recurring decimal forms.

(i) For example
Q = 0.37373737 ...
100 Q = 37.373737...
99Q = 37 ⇒ Q = 37/99 ⇒ rational

(ii) For example
Q = 0.37292929 ...
100Q = 37.292929 ...
10000Q = 3729.292929 ...
9900Q = (3729 − 37)

$Q = \left(\frac{3729 - 37}{9900}\right)$

$= \frac{p}{q}$ form ⇒ rational

Fraction

All rational numbers in which $|q| \neq 1$ comprise the set of fractions.

Proper Fraction

If $|p| < |q|$

then fraction is proper fraction. Value of proper fraction is always in between (−1 to +1) i.e., $[-1 < p/q < 1]$

Improper Fraction

If $|p| > |q|$

than fraction is improper fraction. Value of improper fraction is < −1 or > 1.

Mixed Fraction

Just a modified form of improper fraction.

Eg. $\underbrace{\frac{13}{4}}_{\text{Improper fraction}} \Rightarrow \underbrace{3\frac{1}{4}}_{\text{equivalent mixed fraction}}$

Integers

The set of all rational numbers in p/q form $[|q| = 1]$ is called as integers. It is denoted by

I = { .... −3, −2, −1, 0, 1, 2, 3, ...}

It includes.

Negative Integers

$I^- = \{... -7, -6, -5, -4, -3, -2, -1\}$

Positive Integers

$I^+ = \{1, 2, 3 .... \}$

Note: Status of 0 (zero) is neutral neither positive nor negative.

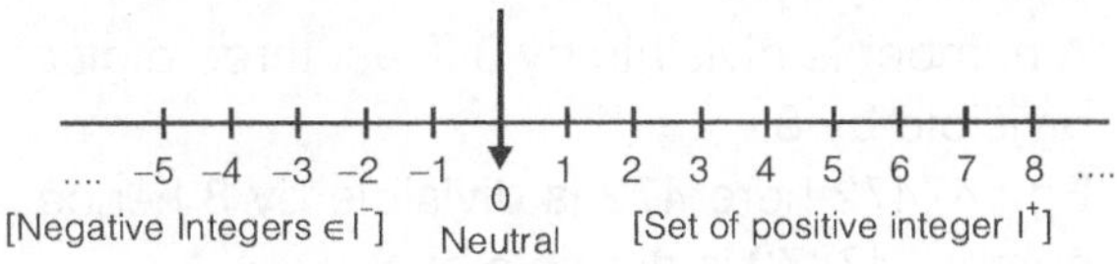

Natural Numbers

All counting numbers or set of positive integers is considered as set of natural numbers. It denoted by set [N or I^+]

N = {1, 2, 3, 4, ...}

Whole Number

Set of all nonnegative integers are considered as whole number; it is denoted by set W = {0, 1, 2, 3, 4 ......}

Note: If terms "numbers" is used without any qualifier than it means natural number henceforth.

Even Numbers & Odd Numbers

1. **Even Numbers**

   All numbers divisible by 2 are considered as even numbers.

   Note: Property evenness is applicable in entire integral number line. Hence [–2, –4, –6, ...] are even integers but they are not even numbers.

2. **Odd Number**

   All numbers not divisible by 2 are odd.

   [1, 3, 5, 7, ...] are odd numbers.

   [.......... –5, –3, –1 ....] are odd integers.

Properties of numbers based on even & odd

Even + Even = Even

Odd + Odd = Even

Odd + Odd + Odd = Odd

Odd × Odd = Odd

Odd × Even = Even

Even × Even = Even

$(\text{Even})^{\text{Odd}} \Rightarrow$ Even

$(\text{Odd})^{\text{Even}} \Rightarrow$ Odd

$(\text{Even})^{\text{Odd}} \Rightarrow$ Even

$(\text{Odd})^{\text{Odd}} \Rightarrow$ Odd

These properties can be used extensively to find out alternative method to get answers quickly with the help of options. Here are few examples.

Ex. 1

There are two, 2-digit numbers ab and cd, ba is the another two digit number prepared by reversing the digits of ab, if ab × cd = 493, ba × cd = 2059, what is value 'g' sum of (ab + cd) = ?

(a) 43 (b) 45

(c) 47 (d) 46

Ans: (d)

Value 'g' = ab × cd is odd.

It means ab and cd both are odd.

Hence there sum must be even, only one option is there which is even. Hence answer is option d.

Ex. 2

I have multiple gift vouchers of value, Rs. 101, 107, 111, 121, 131, 141, 151, 171. I have to pick exactly 10 vouchers to make payment of Rs. 1121. In how many ways I can do that?

(a) one (b) two

(c) more than two (d) none of these

Ans. (d)

Reasoning is very simple, if I'll add 10 odd numbers their sum will be always even. Hence there is no way to accomplish this.

Prime Number & Composite Numbers

Prime Numbers

Number which are perfectly divisible either by 1 or by itself only are called prime numbers. 25 prime number are there which are less than 100. 2 is the only even prime number. All prime numbers greater than 5 can be expressed as (6K ± 1) (K ∈ N) form but all the numbers in form of (6K ± 1) form are not necessarily prime.

Composite Numbers

All the numbers which can be factorized into multiple prime numbers are called composite number.

Number (1) one is neither prime nor composite.

How to check whether given number is prime or not?

1. Take the square root of number
2. Consider the prime numbers, starting from 2 till the number. Take all prime numbers upto this square root value or nearest higher integer.

3. If number is divisible by any of these prime numbers, then number is composite.

Learn it by example:

Suppose we want to check, is 629 prime or not? Square root of 627 is just more than 25. Then prime no. till 25 are 2, 3, 7, 5, 11, 13, 17, 19, 23, 29. 629 is not divisible by 2, 3, 5, 7, 11, 13 but is divisible by 17.

Hence it is not prime number

One more example: 179

Square root of 179 is more than 13. Hence we need to check divisibility of 179 against 2, 3, 5, 7, 11, 13, 17

179 is not divisible by either of these hence it is a prime number.

Test of Divisibility

1. **Divisibility by 2**

   A number is divisible by 2 if the unit digit is zero or divisible by 2.

   Eg.: 22, 42, 84, 3872 etc.

2. **Divisibility by 3**

   A number is divisible by 3 if the sum of digit in the number is divisible by 3.

   Eg.: 2553

   Here 2 + 5 + 5 + 3 = 15, which is divisible by 3 hence 2553 is divisible by 3.

3. **Divisibility by 4**

   A number is divisible by 4 if its last two digit are divisible by 4.

   Eg.: 2652, here 52 is divisible by 4 so 2652 is divisible by 4.

   Eg.: 3772, 584, 904 etc.

4. **Divisibility by 5**

   A number is divisible by 5 if the units digit in number is 0 or 5.

   Eg.: 50, 505, 405 etc.

5. **Divisibility by 6**

   A number is divisible by 6 if the number is even and sum of digits is divisible by 3.

   Eg.: 4536 is an even number also sum of digit 4 + 5 + 3 + 6 = 18 is divisible by 3.

   Eg: 72, 8448, 3972 etc.

6. **Divisibility by 8**

   A number is divisible by 8 if last three digit of it is divisible by 8.

   Eg.: 47472 here 472 is divisible by 8 hence this number 47472 is divisible by 8.

7. **Divisibility by 9**

   A number is divisible by 9 if the sum of its digit is divisible by 9.

   Eg.: 108936 here 1+0+8+9+3+6 is 27 which is divisible by 9 and hence 108936 is divisible by 9.

8. **Divisibility by 10**

   A number is divisible by 10 if its unit digit is 0.

   Eg.: 90, 900, 740, 34920 etc.

9. **Divisibility by 11**

   A number is divisible by 11 if the difference of sum of digit at odd places and sum of digit at even places is either 0 or divisible by 11.

   Eg.: 1331, the sum of digits at odd place is 1+3 and sum of digit at even places is 3+1 and their difference is 4 – 4 = 0. so 1331 is divisible by 11.

HCF and LCM of Numbers

H.C.F.

(Highest Common Factor) of two or more number is the greatest number that divides each one of them exactly. For example 8 is the highest common factor of 16 and 40.

HCF is also called greatest common divisior (G.C.D.)

L.C.M.

(Least Common Multiple) of two or more number is the least or a lowest number which is exactly divisible by each of them.

For example LCM of 8 and 12 is 24, because it is the first number which is multiple of both 8 and 12.

LCM and HCF of Fractions

Fractions are written in form of $\frac{\text{Numerator}}{\text{Denominator}}$. Where denominator is not equal to zero.

$$\text{H.C.F of Fraction} = \frac{\text{(H.C.F. of Numerators)}}{\text{(LCM of Denominators)}}$$

$$\text{L.C.M of Fraction} = \frac{\text{(LCM of Numerators)}}{\text{(HCF of Denominators)}}$$

All Fractions have to be in their simplest form:

Example: Find HCF & LCM of $\frac{1}{2}, \frac{2}{3}$ and $\frac{3}{7}$

$$\text{H.C.F.} = \frac{\text{H.C.F. of } (1,2,3)}{\text{L.C.M } (2, 3, 7)} = \frac{1}{42}$$

$$\text{L.C.M} = \frac{\text{L.C.M of } (1,2,3)}{\text{H.C.F. of } (2, 3, 7)} = \frac{6}{1} = 6$$

Important Algebraic Formulae

1. $(a+b)^2 = a^2 + 2ab + b^2$
2. $(a-b)^2 = a^2 - 2ab + b^2$
3. $(a-b)(a+b) = a^2 - b^2$
4. $(a+b)^2 + (a-b)^2 = 2(a^2 + b^2)$
5. $(a+b)^2 - (a-b)^2 = 4ab$
6. $(a+b)^3 = a^3 + 3a^2b + 3ab^2 + b^3$
   $= a^3 + b^3 + 3ab(a+b)$
7. $(a-b)^3 = a^3 - 3a^2b + 3ab^2 - b^3$
   $= a^3 - b^3 - 3ab(a-b)$
8. $a^3 + b^3 = (a+b)(a^2 - ab + b^2)$
9. $a^3 - b^3 = (a-b)(a^2 + ab + b^2)$
10. $\dfrac{a^3 + b^3 + c^3 - 3abc}{a^2 + b^2 + c^2 - ab - bc - ca} = (a+b+c)$
11. $a^4 - b^4 = (a^2)^2 - (b^2)^2 = (a^2 + b^2)(a^2 - b^2)$
    $= (a^2 + b^2)(a+b)(a-b)$

[Condition of Divisibility for Algebric Function

1. $a^n + b^n$ is exactly divisible by a+b only when n is odd
   Ex.: $a^3 + b^3 = (a+b)(a^2 - ab + b^2)$ is divisible by a+b, also $a^5 + b^5$ is divisible by a+b
2. $a^n + b^n$ is never divisible by a-b (whether n is odd or even)
   Ex.: $a^3 + b^3 = (a + b)(a^2 - ab + b^2)$ is not divisible by (a-b)
   $a^7 + b^7$ is also not divisible by $(a - b)$
3. $a^n - b^n$ is always divisible by $(a - b)$ (whether n is odd or even)
   Ex.: $a^9 - b^9$ is exactly divisible by $(a - b)$ also $a^{12} - b^{12}$ is also exactly divisible by $(a - b)$.
4. $a^n - b^n$ is divisible by a + b only when 'n' is even natural number.
   Ex. : $a^4 - b^4 = (a^2 - b^2)(a^2 + b^2) = (a - b)(a + b)(a^2 + b^2)$. Hence $a^4 - b^4$ is always divisible by (a + b) but $a^3 - b^3$ will not be.]

Factors of Composite Number

Composite numbers are the numbers which can be factorised into prime factors, or simply we can say that composite number are those numbers which are not prime.

For eg.: 8 is a composite number since it can be factorised into

$8 = 2 \times 2 \times 2$

Similarly 9 is also a composite number, i e

$9 = 3 \times 3$

Composite number $= P_1^{\lambda_1} \times P_2^{\lambda_2} \times P_3^{\lambda_3} \ldots\ldots P_n^{\lambda_n}$ here, P_1, P_2, P_3 ..... P_n are distinct prime numbers and λ_1, λ_2, ... λ_n are their respective powers.

Factors of composite number =

$(\lambda_1 + 1).(\lambda_2 + 1)\ldots(\lambda_n + 1)$

For eg.: $18 = 2 \times 3 \times 3 = 2^1 \times 3^2$

Factors of 18 = $(1 + 1) \times (2 + 1) = 2 \times 3 = 6$

Clearly it contains six factors 1, 2, 3, 6, 9 and 18

Factors of other Composite numbers $6 = 2^1 \times 3^1$

Factors = $(1 + 1) \times (1 + 1) = 4$ = 1, 2, 3 and 6

$72 = 2 \times 2 \times 2 \times 3 \times 3 = 2^3 \times 3^2$

Factors = $(3 + 1) \times (2 + 1) = 12$

Ex.1 Find the factors of composite number 360

Sol.: $360 = 2 \times 2 \times 2 \times 3 \times 3 \times 5$
$= 2^3 \times 3^2 \times 5^1$

Factors $= (3 + 1)(2 + 1)(1 + 1) = 24$.

□□□□

Counting Number of Trailing Zeros

Sometimes we come across problems in which we have to count number of zeros at the end of factorial of any number. For example

Number of zero at the end of 10!

$$10! = 10 \times 9 \times 8 \times 7 \times 6 \times 5 \times 4 \times 3 \times 2 \times 1$$

Here basically we have to count number of fives, because multiplication of five by any even number will result in 0 at the end of final product. In 10! we have 2 fives thus total number of zeros are 2.

Short Cut:

Counting number of zeros at the end of n!

Value will be $\frac{n}{5}+\frac{n}{5^2}+\frac{n}{5^3}+\frac{n}{5^4}...$

The integral value of this sum will be the total number of zeros.

Ex. 1 Number of zeros at the end of 100!

Sol.: $\frac{100}{5}+\frac{100}{5^2}+\frac{100}{5^3}+$

integral value will be

20 + 4 = 24 zeros

Ex.2 Number of zeros at the end of 126!

Sol.: $\frac{126}{5}+\frac{126}{5^2}+\frac{126}{5^3}+\frac{126}{5^4}$

integral value will be

25 + 5 + 1 = 31 zeros.

□□□□

Cyclicity

Cyclicity of a number is used mainly for the calculation of unit digits.

1. **Cyclicity of 1.**
   In 1^n, unit digit will always be 1.

2. **Cyclicity of 2.**
   $2^1 = 2$
   $2^2 = 4$
   $2^3 = 8$
   $2^4 = 16$
   $2^5 = 32$
   $2^6 = 64$
   $2^7 = 128$
   $2^8 = 256$

After every four intervals it repeats so cycle of 2 is 2, 4, 8, 6.

Ex.1 Unit digit of 2^{323}

Sol.: Here 2, 4, 8, 6 will repeat after every four interval till 320 next digit will be 2, 4, $\boxed{8}$. So unit digit of 2^{323} will be 8.

Ex.2 Find unit digit of $12^{12} \times 22^{22}$

Sol.: Unit digit of 12^{12} will be 6 and 22^{22} will be 4. So unit digit of $12^{12} \times 22^{22}$ will be

$6 \times 4 = 2\boxed{4}$; 4 Ans.

3. **Cyclicity of 3.**
   $3^1 = 3$
   $3^2 = 9$
   $3^3 = 27$
   $3^4 = 81$
   $3^5 = 243$
   $3^6 = 729$
   $3^7 = 2187$
   $3^8 = 6561$

After every four intervals 3,9, 7 and 1 are repeated. So cycle of 3 is 3, 9, 7, 1.

Ex.1 Find unit digit of 133^{133}.

Sol.: Cycle of 3 is 3, 9, 7, 1 which repeats after every four intervals till 133^{132}. So next unit digit will be 3.

4. **Cyclicity of 4.**
   $4^1 = 4$
   $4^2 = 16$
   $4^3 = 64$
   $4^4 = 256$
   Cycle is 4, 6, i.e.
   Unit digit of 4^n depends on value of n.
   If n is odd unit digit is 4 and if n is even digit is 6.

Ex.1 Find unit digit of 4^{1024}.

Sol.: Since 1024 is even number unit digit will be 6.

Ex.2 Find unit digit of $133^{63} \times 4^{49}$.

Sol.: Unit digit of 133^{63} is 7 and unit digit of 4^{49} is 4 so unit digit of $133^{63} \times 4^{49}$ will be $7 \times 4 = 28$ i.e. 8.

5. **Cyclicity of 5.**
   $5^1 = 5$
   $5^2 = 25$
   $5^3 = 125$
   $5^4 = 625$
   Unit digit will always be 5.

6. **Cyclicity of 6.**
   $6^1 = 6$
   $6^2 = 36$
   $6^3 = 216$
   $6^4 = 1296$
   Unit digit will always be 6.

Ex.1 Find unit digit of $4^{69} \times 6^5$

Sol.: Unit digit of 4^{69} is 4 and unit digit of 6^5 is 6 so unit digit of $4^{69} \times 6^5$ will be $4 \times 6 = 24$ i.e. 4.

7. **Cyclicity of 7.**

$7^1 = 7$
$7^2 = 49$
$7^3 = 343$
$7^4 = 2401$
$7^5 = 16807$
$7^6 = 117649$
$7^7 = 823543$
$7^8 = 5764801$

Cycle of 7 is 7, 9, 3, 1

Ex.1 Find unit digit of $17^{17} \times 27^{27}$

Sol.: Unit digit of 17^{17} is 7 and unit digit of 27^{27} is 3. So unit digit of $17^{17} \times 27^{27}$ will be $7 \times 3 = 21$ i.e. 1.

8. **Cyclicity of 8.**

$8^1 = 8$
$8^2 = 64$
$8^3 = 512$
$8^4 = 4096$
$8^5 = 32768$

So cycle of 8 is 8, 4, 2, 6.

Ex. 1 Find unit digit of $18^{18} \times 28^{28} \times 288^{288}$.

Sol.: Unit digit of 18^{18} is 4, unit digit of 28^{28} is 6, unit digit of 288^{288} is 6. So unit digit of $18^{18} \times 28^{28} \times 288^{288}$ will be $4 \times 6 \times 6 = 144$ i.e. 4.

9. **Cyclicity of 9.**

$9^1 = 9$
$9^2 = 81$
$9^3 = 729$
$9^4 = 6561$

Cycle of 9 is 9, 1.

In 9^n unit digit will be 9 if n is odd and unit digit will be 1 if n is even.

Ex. 1 Find unit digit of

$11^{11} + 12^{12} + 13^{13} + 14^{14} + 15^{15}$

Sol.: Unit digit of 11^{11} is 1

Unit digit of 12^{12} is 6

Unit digit of 13^{13} is 3

Unit digit of 14^{14} is 6

Unit digit of 15^{15} is 5

So unit digit of given sum will be

$1 + 6 + 3 + 6 + 5 = 21$ i.e. 1.

Remember

Cyclicity table

1 : 1
2 : 2, 4, 6, 8
3 : 3, 9, 7, 1
4 : 4, 6
5 : 5
6 : 6
7 : 7, 9, 3, 1
8 : 8, 4, 2, 6
9 : 9, 1
0 : 0

□□□□

Remainder Theorem

Remainder of expression $\frac{a \times b \times c}{n}$ [i.e. $a \times b \times c$ when divided by n] is equal to the remainder of expression $\frac{a_n \times b_n \times c_n}{n}$ [i.e. $a_n \times b_n \times c_n$ when divided by n], where

a_n is remainder when a is divided by n,
b_n is remainder when b is divided by n, and
c_n is remainder when c is divided by n.

Ex.1 Find the remainder of $15 \times 17 \times 19$ when divided by 7.

Sol.: Remainder of expression $\frac{15 \times 17 \times 19}{7}$ will be equal to $\frac{1 \times 3 \times 5}{7} = \frac{15}{7} = \frac{1}{7}$.

i.e. 1.

On dividing 15 by 7 we get 1 as remainder
On dividing 17 by 7 we get 3 as remainder
On dividing 19 by 7 we get 5 as remainder and combined remainder will be equal to remainder of $\frac{15}{7}$ i.e. 1.

□□□□

Polynomial Theorem

This is very powerful theorem to find the remainder.

According to polynomial theorem.

$$(x + a)^n = x^n + {}^nC_1 x^{n-1} + {}^nC_2 x^{n-2} a^2 + {}^nC_3 x^{n-3} a^3 \dots {}^nC_{n-1} x^1 a^{n-1} + a^n \quad \dots(1)$$

$$\therefore \frac{(x+a)^n}{x} = \frac{\begin{pmatrix} {}^nC_0 x^n + {}^nC_1 x^{n-1} a^1 + {}^nC_2 x^{n-2} a^2 + \\ {}^nC_3 x^{n-3} + \dots {}^nC_{n-1} x^1 a^{n-1} + {}^nC_n a^n \end{pmatrix}}{x} \quad \dots(2)$$

remainder of expression (2) will be equal to remainder of $\frac{a^n}{x}$ because rest of the term contains x are completely divisible by x.

Ex.1 Find the remainder of $\frac{9^{99}}{8}$.

Sol.: $\frac{9^{99}}{8} = \frac{(8+1)^{99}}{8}$

According to polynomial theorem remainder will be equal to remainder of the expression $\frac{1^{99}}{8}$ which is equal to 1.

Ex.2 Find remainder of $\frac{5^{100}}{7}$

Sol.: $\frac{5^{100}}{7} = \left[\frac{3\times7+4}{7}\right]^{50} \Rightarrow \frac{(4)^{50}}{7}$

$\Rightarrow \frac{2^{100}}{7} \Rightarrow \frac{(2^3)^{33}\times2}{7} \Rightarrow \frac{(7+1)^{33}}{7}\times2 \Rightarrow \frac{1\times2}{7}$

$\Rightarrow$ Remainder is 2.

More on Remainders

Case-I

On dividing a number by a, b & c if we get a-k, b-k and c-k as remainder respectively then that number will be **n × LCM of [a, b, c]-k.**

For ex (I): On dividing a number by 4, 5 & 6 we get 3, 4, & 5 as remainder. Find the number.

Sol.:

	4,	5,	6
Remainder	3,	4,	5,

which is equal to (4 – 1), (5 – 1), (6 – 1), so that number will be:

n × LCM of (4, 5, 6) – 1, = 60n – 1

If n = 1, 60 – 1 = 59 is smallest such natural number.

Note: n such numbers are possible. Here we have taken n as 1. Other numbers are 119, 179, 239, etc. Where value of n is 2, 3, & 4 respectively.

Ex.1 On dividing a number by 5, 6 and 7 we get 3, 4 and 5 as remainder. Find the number.

Sol.:

	5,	6,	7
Remainder	3,	4,	5

which is equal to (5 –2), (6–2), (7–2)

that number will be:

n × LCM of (5, 6, 7) – 2 = 210 –2 = 208.

Note: Here we have taken value of n as 1.

Ex.2 On dividing a number by 4, 5 and 6 we get 2, 3 and 4 as remainder find highest possible three digit such number.

Sol.:

	4,	5,	6
Remainder	2,	3,	4

which is equal to (4 –2), (5 – 2),(6 – 2), that number will be:

n × LCM of [4, 5, 6]–2 = n × 60 – 2

When n = 1 we get 58. Highest possible three digit such number will be 958.

Ex.3 On dividing a number by 5, 6 and 7 we get 3, 4 and 5 as remainder. Find highest possible three digit such number.

Sol.:

	5,	6,	7
Remainder	3,	4,	5

which is equal to (5 –2), (6 – 2),(7 – 2) that number will be:

n ∗ LCM (5, 6, 7)–2= n × 210 – 2

Highest possible three digit number will be 838.

Case-II

On dividing a number a, b and c if we get k as remainder always, then that number will be (n – 1) LCM of (a, b, c)+ k.

Ex.1 On dividing a number by 5, 6 and 7 if we get 2 as remainder always, find that number

Sol.: That number will be

(n – 1) × LCM of [5, 6, 7]+ 2

$\Rightarrow$ 2 is such smallest number

next number will be = 210 + 2 = 212

Case-III

If a number after adding k is exactly divisible by a , b and c then that number will be.

$n \times \text{LCM}(a, b, c) - k$

Ex.1 Find a number which after adding 7 is divisible by 10, 11 and 12.

Sol.: That number will be

$n \times \text{LCM of } [10, 11, 12] - 7$

if n = 1 then

660 – 7 = 653 Ans.

□□□□

Squares of Numbers

Squares of numbers are frequently used for calculations on various types of problems. It is advisable to remember square of at least first thirty numbers.

$1^2 = 1$	$11^2 = 121$
$2^2 = 4$	$12^2 = 144$
$3^2 = 9$	$13^2 = 169$
$4^2 = 16$	$14^2 = 196$
$5^2 = 25$	$15^2 = 225$
$6^2 = 36$	$16^2 = 256$
$7^2 = 49$	$17^2 = 289$
$8^2 = 64$	$18^2 = 324$
$9^2 = 81$	$19^2 = 361$
$10^2 = 100$	$20^2 = 400$

From following table we come to know that square of a number always ends with 0, 1, 4, 5, 6 & 9 as unit digit. Square of a number can never have 2, 3, 7 & 8 in its unit place.

On observing squares of numbers between 21 to 29 we get following pattern.

$21^2 = 4\,41$	$29^2 = 8\,41$
$22^2 = 4\,84$	$28^2 = 7\,84$
$23^2 = 5\,29$	$27^2 = 7\,29$
$24^2 = 5\,76$	$26^2 = 6\,76$
$25^2 = 6\,25$	

Last two digits are common.

Observation

Square of two digit number having 5 in unit places can be calculated very easily

n5 here n may 1 to 9.

$(n5)^2 = [n * (n + 1)]25$

Ex.1 $65^2 = ?$

Sol.: $[6 \times (6 + 1)]25 = 4225$

Ex.2 $85^2 = ?$

Sol.: $[8 \times (8 + 1)]25 \Rightarrow 7225$

Ex.3 $95^2 = ?$

Sol.: $[9 \times (9 + 1)]25 \Rightarrow 9025$

Base System

The Number system is used to represent any number using a set of symbols (digits /letters). The base defines the number of symbols in particular base system. We generally work in Decimal system as there are 10 digits (0, 1, 2, ...9). Some others systems are;

Binary base system: 2 symbols: 0, 1

Octal base system: 8 symbols: 0,1,2,3,4,5,6,7

Hexadecimal system: 16 symbols:

0,1,2,3,4,5,6,7,8,9, A = 10, B = 11, C = 12, D = 13, E = 14, F = 15

Converting any number from any Base system to Decimal number system:

$$abcd.efg_B = a \times B^3 + b \times B^2 + c \times B^1 + d \times B^0 + e \times B^{-1} + f \times B^{-2} + g \times B^{-3}$$

Example:

$$1234.56_8 = 1 \times 8^3 + 2 \times 8^2 + 3 \times 8^1 + 4 \times 8^0 + 5 \times 8^{-1} + 6 \times 8^{-2}$$

$= 512 + 128 + 24 + 4 + 0.625 + 0.093750$

$= 668.718750$

Converting any number from Decimal to other Base system:

Divide the number by base and get the first remainder r_1 and Quotient q_1.

Now divided q_1 by base and get remainder r_2 and Quotient q_2.

Repeat the following process till we get the quotient $q_n = 0$.

Now the decimal number in base b is $r_n r_{n-1} \ldots r_3 r_2 r_1$.

Example 1:

1. $(149)_{10} = (\ )_7$

7	149	Remainder
7	21	2
7	3	0
	0	3

$(149)_{10} = (302)_7$

2. Add $(432)_7 + (355)_7$

$$\begin{array}{r} 11 \leftarrow \text{carry} \\ (432)_7 \\ \underline{(355)_7} \\ 1120 \end{array}$$

as $2 + 5 = (7)_{10} = (10)_7$

$3 + 5 + 1 = (9)_{10} = (12)_7$

$1 + 4 + 3 = (8)_{10} = (11)_7$

Solved Examples

1. The sum of the digits of a two-digit number is 10, while when the digits are reversed, the number decreases by 54. Find the the changed number.

(a) 28 (b) 19
(c) 37 (d) 46

Ans: (a)

Going through options we get 82 – 28 = 54

2. The sum of two numbers is 15 and their geometric mean is 20% lower than their arithmetic mean. Find the numbers.

(a) 11, 4 (b) 12, 3
(c) 13, 2 (d) 10, 5

Ans: (b)

Going through options only 12 and 3 satisfices the conditon

$$AM = \frac{12+3}{2} = 7.5$$

$GM = \sqrt{12} \times 3 = 6\sqrt{3}$ which is 20% less than 7.5.

3. If A381 is divisible by 11, find the value of the smallest natural number A?

(a) 5 (b) 6
(c) 7 (d) 9

Ans. (c)

A 381 is divisible by 11 if and only if (A + 8) – (3 + 1) is divisible by 11.

So, A=7 Satisfies the condition

4. Find the LCM of 5/2, 8/9, 11/14.

(a) 280 (b) 360
(c) 420 (d) None of these

Ans: (d)

$$\text{LCM of fraction} = \frac{\text{LCM of numerators}}{\text{H. C. F of Denominators}}$$

Here, 5/2, 8/9, 11/14, so

$$\text{LCM} = \frac{\text{LCM of } (5, 8, 11)}{\text{HCF of } (2, 9, 14)} = \frac{440}{1} = 440$$

5. Find the number of divisors of 1420.

(a) 14 (b) 15
(c) 13 (d) 12

Ans: (d)

$1420 = 142 \times 10 = 71 \times 2 \times 2 \times 5 = 2^2 \times 5^1 \times 71^1$

No. of divisor = (2+1) (1+1) (1+1) = 12

6. A milkman has three different qualities of milk. 403 gallons of 1st quality, 465 gallons of 2nd quality and 496 gallons of 3rd quality. Find the least possible number of bottles of equal size in which different milk of different qualities can be filled without mixing?

(a) 34 (b) 46
(c) 26 (d) 44

Ans: (d)

It is given that gallons of

1st quality : 403

2nd quality : 465

3rd quality : 496

least number of bottles will be in size of HCF (403, 465 and 496)

$403 = 13 \times 31$

$465 = 15 \times 31$

$496 = 16 \times 31$

HCF = 31. So we required 13+15+16 = 44 bottles.

7. What is the greatest number of 4 digits that when divided by any of the numbers 6, 9, 12, 17 leaves a remainder of 1?

(a) 9997 (b) 9793
(c) 9895 (d) 9487

Ans: (b)

LCM of 6, 9, 12, 17 = 612

greatest number of 4 digit divisible by 612 is 9792, to get remainder 1 number should be 9792+1

8. Which of the following is not a perfect square?

(a) 100858 (b) 3, 25, 137
(c) 945723 (d) All of these

Ans: (d)

Square of number never ends up with 2, 3, 7, 8

9. The LCM of $(16 - x^2)$ and $(x^2 + x - 6)$ is
(a) $(x - 3)(x + 3)(4 - x^2)$
(b) $4(4 - x^2)(x + 3)$
(c) $(4 - x^2)(x - 3)$
(d) None of these
Ans: (d)

$$16 - x^2 = (4 - x)(4 + x)$$

$$(x^2 + x - 6) = (x + 3)(x - 2)$$

LCM will $(16 - x^2)(x^2 + x - 6)$

10. GCD of $x^2 - 4$ and $x^2 + x - 6$ is
(a) $x + 2$ (b) $x - 2$
(c) $x^2 - 2$ (d) $x^2 + 2$
Ans: (b)

$$x^2 - 4 = (x - 2)(x + 2)$$

$$(x^2 + x - 6) = (x + 3)(x - 2)$$

$$\text{GCD} = (x - 2)$$

11. Decompose the number 20 into two terms such that their product is the greatest.
(a) $x_1 = x_2 = 10$ (b) $x_1 = 5, x_2 = 15$
(c) $x_1 = 8, x_2 = 12$ (d) None of these
Ans: (a)
If $x + y$ = constant then xy will be maximum when $x = y$
here, $x_1 + x_2 = 20$
$x_1 = x_2 = 10$

12. Find the GCD of the polynomials $(x + 3)^2 (x - 2)(x + 1)^2$ and $(x + 1)^3 (x + 3)(x + 4)$.
(a) $(x + 3)^3 (x + 1)^2 (x - 2)(x + 4)$
(b) $(x + 3)(x - 2)(x + 1)(x + 4)$
(c) $(x + 3)(x + 1)^2$
(d) None of these
Ans: (c)
GCD of $(x+3)(x-2)(x+1)^2$ and

$(x+1)^3(x+3)(x+4)$ will be $(x+3)(x+1)^2$

13. Find the LCM of $(x + 3)(6x^2 + 5x - 4)$ and $(2x^2 + 7x + 3)(x + 3)$
(a) $(2x + 1)(x + 3)(3x + 4)$
(b) $(4x^2 - 1)(x + 3)^2 (3x + 4)$
(c) $(4x^2 - 1)(x + 3)(3x + 4)$
(d) $(2x - 1)(x + 3)(3x + 4)$
Ans: (b)

$$(x+3)(6x^2 + 5x - 4) = (x+3)(2x-1)(3x+4)$$

$$(2x^2 + 7x + 3)(x+3) = (2x+1)(x+3)(x+3)$$

$$\text{LCM} = (2x+1)(2x-1)(x+3)^2(3x+4)$$

$$= (4x^2 - 1)(x+3)^2(3x+4)$$

14. The product of three consecutive natural numbers, the first of which is an even number, is always divisible by
(a) 12 (b) 24
(c) 8 (d) All of these
Ans: (d)
Three consecutive number will be n(n + 1) (n + 2) if n is even number then (n + 2) will also be an even number and one of them will be divisible by 3. Hence number is always divisible by 12.

15. Find the pairs of natural numbers whose least common multiple is 78 and the greatest common divisor is 13.
(a) 58 and 13 or 16 and 29
(b) 38 and 23 or 36 and 49
(c) 18 and 73 or 56 and 93
(d) 78 and 13 or 26 and 39
Ans: (d)
LCM = 78 and GCD = 13
Clearly 13, 78 and 26, 39 are the two numbers

16. Fill in the blank indicated by a star in the number 4* 56 so as to make it divisible by 33
(a) 3 (b) 4
(c) 5 (d) None of these
Ans. (a)
4*56 is divisible by 33 if and only if it is divisible by 3 and 11.
4* 56 will be divisible by 3 if * will be equal to 0, 3, 6, 9
4*56 is divisible by 11 if (4 + 5) – (* + 6) will be divisible by 11 so * should be 3.

17. Find the least number which being divided by 9, 12, 16 and 30 leaves in each case a remainder 3?
(a) 623 (b) 723
(c) 728 (d) None of these

Ans. (b)
LCM of 9, 12, 16 and 30 is 720 so required number is LCM + 3 = 723

18. Find the greatest number less than 10000 which is divisible by 48, 60 and 64
(a) 9600 (b) 8500
(c) 7600 (d) None of these
Ans. (a)
The required number will be the largest four digit number in form of n*(LCM) of 48, 60 and 64 LCM of 48, 60 and 64 is 960
So the largest four digit number will be 9600

19. Find the least multiple of 11 which when divided by 8, 9, 12, 14 leaves 4 as remainder in each case.
(a) 1012 (b) 1037
(c) 1090 (d) None of these
Ans. (a)
The number is divisible by 11 and can be written in form n(LCM) +4, LCM of 8, 9, 12, 14 is 504
So the number may be 508 & 1012 but 508 is not divisible by 11 so it is 1012

20. The LCM of two number is 12 times their HCF. The sum of HCF and LCM is 403. If one number is 93 find the other.
(a) 134 (b) 124
(c) 128 (d) None of these
Ans. (b)
It is given that LCM=12 times HCF
also LCM + HCF = 403
So, 13 HCF = 403, ⇒ HCF = 31
LCM = 372 also we know that HCF

$$\text{HCF} \times \text{LCM} = \text{Number}(1) \times \text{Number}(2)$$

$$31 \times 372 = 93 \times \text{N2} \quad \therefore \text{N2} = 124$$

21. I have to spend 1/10 of my income on house rent, 1/10 of remainder on conveyance 1/3 of further remainder on children's education after which I have Rs. 648 left over. What is my income?
(a) Rs. 1200 (b) Rs. 1400
(c) Rs. 1700 (d) None of these
Ans. (a)
One alternate method
Let I have x rupees

After spending $\frac{1}{10}$ of it on house rent I have $\frac{9x}{10}$.

Now out of $\frac{9x}{10}$ I spent $\frac{1}{10}$ of it i.e., $\frac{9}{100}x$ on conveyance so remainder will be

$$\frac{9}{10}x - \frac{9}{100}x = \frac{81x}{100}.$$

Further I spent $\frac{1}{3}$ of $\frac{81x}{100}$ i.e. $\frac{27x}{100}$ into childrens education now I have $\frac{54x}{100}$,

So, $\frac{54x}{100} = 648, \; x = 1200$

22. A man had two sons. To the elder he gave $\frac{5}{11}$ of his property, to the younger $\frac{5}{11}$ of remainder, the rest to the widow. Find the Share of the sons if the widow gets Rs. 3600.
(a) Rs. 1200, 1000 (b) Rs. 6000, 2000
(c) Rs. 7500, 1000 (d) None of these
Ans. (d)
Younger son gets

$$3600 \times \left(\frac{1}{1-\frac{5}{11}}\right) \times \frac{5}{11} = \text{Rs. } 3000$$

Elder son gets

$$3000 \times \left(\frac{1}{1-\frac{5}{11}}\right) = \text{Rs. } 5500$$

□□□□

I Practice Exercise

1. $\sqrt{3\sqrt{80}+\dfrac{3}{9+4\sqrt{5}}}=?$
(a) $\sqrt{3\sqrt{5}}$ (b) 3
(c) $3\sqrt{3}$ (d) $3+2\sqrt{5}$

2. x and y are integers and If $\dfrac{x^2}{y^3}$ is even integer then which of the following must be an even integer?
(a) $x - y$ (b) y + 1
(c) $\dfrac{x^2}{y^4}$ (d) xy

3. What is the tens' digit of the sum of the first 50 terms of 1, 11, 111, 1111, 11111, 111111,.......................?
(a) 2 (b) 4
(c) 5 (d) 8

4. If $81^y = \dfrac{1}{27^x}$, in terms of y, x=?
(a) $\dfrac{3y}{4}$ (b) $-\dfrac{3y}{4}$
(c) $\dfrac{4y}{3}$ (d) $-\dfrac{4y}{3}$

5. If $\dfrac{1}{n+1}<\dfrac{1}{31}+\dfrac{1}{32}+\dfrac{1}{33}<\dfrac{1}{n}$; then n ?
(a) 9 (b) 10
(c) 11 (d) 12

6. If one integer is greater than another integer by 3, and the difference of their cubes is 117, what could be their sum?
(a) 11 (b) 7
(c) 8 (d) 9

7. Which of these has total 24 positive factors?
(a) $21^5 \times 2^3$ (b) $2^7 \times 12^3$
(c) $2^6 \times 3^4$ (d) 63×55

8. Two numbers, x and y are such that when divided by 6, they leave remainder 4 and 5 respectively. Find the remainder when $x^3 + y^3$ is divided by 6?
(a) 2 (b) 3
(c) 4 (d) 5

9. What is the remainder when N = $(1! + 2! + 3! + ...1000!)^{40}$ is divided by 10?
(a) 1 (b) 3
(c) 7 (d) 8

10. Set A is formed by selecting some of the numbers from the first 100 natural numbers such that the HCF of any two numbers in the set A is 5, what is the maximum number elements that set A can have?
(a) 7 (b) 8
(c) 9 (d) 10

11. Let x and y be positive integers such that x is prime and y is composite. Then,
(a) $y - x$ cannot be an even integer
(b) $\dfrac{x+y}{x}$ cannot be an even integer
(c) $(x + y)$ cannot be even.
(d) None of the above statements are true

12. Let N = 1421 × 1423 × 1425. What is the remainder when N is divided by 12?
(a) 0 (b) 9
(c) 3 (d) 6

13. When a four digit number is divided by 85 it leaves a remainder of 39. If the same number is divided by 17 the remainder would be ?
(a) 2 (b) 5
(c) 7 (d) 9

14. Integers 34041 and 32506 when divided by a three-digit integer n leave the same remainder. What is n?
(a) 289 (b) 367
(c) 453 (d) 307

15. A box contains 100 tickets, numbered from 1 to 100. A person picks out three tickets from the box, such that the product of the numbers on two of the tickets yields the number on the third ticket. Which of the following tickets can never be picked as third ticket?
(a) 10 (b) 12
(c) 25 (d) 26

16. N is a natural number , then how many values of N are possible such that $\dfrac{6N^3+3N^2+N+24}{N}$ is also a Natural Number?
(a) 6 (b) 7
(c) 8 (d) 9

17. What is the unit digit of $39^{53} \times 27^{23} \times 36^{12}$?
(a) 2 (b) 4
(c) 6 (d) 8

18. How many number of zeros are there if we multiply all the prime numbers between 0 and 200.
(a) 1 (b) 2
(c) 3 (d) 4

19. A man wrote all the natural numbers starting from 1 in a series. What will be the 50th digit of the number?
(a) 1 (b) 2
(c) 3 (d) 4

20. N = n(n + 1) (n + 2) (n + 3) (n + 4); where n is a natural number. Which of the following statement/s is/are true?
1. Unit digit of N is 0.
2. N is perfectly divisible by 24.
3. N is perfect square.
4. N is odd.
(a) 3 only (b) 3 and 4 only
(c) 1 only (d) 1 and 2 only

21. How many factors of
$N = 12^{12} \times 14^{14} \times 15^{15}$ are multiple of
$K = 12^{10} \times 14^{10} \times 15^{10}$
(a) $2 \times 4 \times 5$ (b) $3 \times 5 \times 6$
(c) $8 \times 7 \times 4 \times 5$ (d) $9 \times 8 \times 6 \times 5$

22. In a certain base
137 + 254 = 402 then
What is the sum of 342 + 562 in that base
(a) 904 (b) 1014
(c) 1104 (d) 1024

Answers

1. (c)	2. (d)	3. (b)	4. (d)	5. (b)
6. (b)	7. (d)	8. (b)	9. (a)	10. (c)
11. (d)	12. (c)	13. (b)	14. (d)	15. (c)
16. (c)	17. (a)	18. (a)	19. (c)	20. (d)
21. (d)	22. (b)			

Solutions

1. (c)

Method (i) $\sqrt{3\sqrt{80}+\frac{3}{9+4\sqrt{5}}}$ using rationalization

$$= \sqrt{3\sqrt{80}+\frac{3}{9+4\sqrt{5}}\times\left(\frac{9-4\sqrt{5}}{9-4\sqrt{5}}\right)}$$

$$= \sqrt{3\sqrt{80}+\frac{(3\times 9-3\times 4\sqrt{5})}{9^2-(4\sqrt{5})^2}}$$

$$= \sqrt{3\sqrt{80}+\frac{27-12\sqrt{5}}{81-80}}$$

$$= \sqrt{3\sqrt{16\times 5}+27-12\sqrt{5}}$$

$$= \sqrt{3\times 4\times\sqrt{5}+27-12\sqrt{5}}$$

$$= \sqrt{12\sqrt{5}+27-12\sqrt{5}}$$

$$= \sqrt{27} = 3\sqrt{3}$$

Alternative Method

$$\sqrt{\left(3\sqrt{80}+\frac{3}{9+4\sqrt{5}}\right)}$$

$$3\sqrt{80} \cong 3\sqrt{81} \cong 27$$

and $\frac{3}{9+4\sqrt{5}} < 1$

Thus, $\sqrt{3\sqrt{80}+\frac{3}{9+4\sqrt{5}}} \cong \sqrt{3\sqrt{81}}$

$$\cong \sqrt{3\times 9} = 3\sqrt{3}$$

2. (d)

if $\frac{x^2}{y^3}$ = even

$x^2 = y^3$ even

$\Rightarrow$ $x^2 \Rightarrow$ even

and x is integer

$\Rightarrow$ x = even

so only xy must be even.

3. (b)

```
                 1
                1 1
              1 1 1
              : : :
50 terms..... 1 1 1
             ------
                40
```

unit digit (1 + 1...... 50 times)= 0
and carry = 5
tens digit (1 + 1 + ... 49 times)+ carry 5 = 4

4. **(d)**

$$81^y = \frac{1}{27^x}$$

$\Rightarrow$ $(3)^{4y} = (3^{-3})^x$

$\Rightarrow$ $(3)^{4y} = (3)^{-3x}$

as $a^m = a^n$

and $a \neq -1, 0, +1$

$\Rightarrow$ $m = n$

so $4y = -3x$

$$x = -\frac{4}{3}y$$

5. **(b)**

$\frac{1}{30} > \frac{1}{33}; \frac{1}{30} > \frac{1}{31}; \frac{1}{30} > \frac{1}{31};$

$\Rightarrow \frac{1}{30} + \frac{1}{30} + \frac{1}{30} > \frac{1}{31} + \frac{1}{32} + \frac{1}{33}$

$\Rightarrow \frac{3}{30} > \frac{1}{31} + \frac{1}{32} + \frac{1}{33}$

$\Rightarrow \frac{1}{10} > \frac{1}{31} + \frac{1}{32} + \frac{1}{33}$

$\Rightarrow \frac{1}{n} = \frac{1}{10} \Rightarrow n = 10$

and

OR $\frac{1}{31} > \frac{1}{33}; \frac{1}{32} > \frac{1}{33}; \frac{1}{33} = \frac{1}{33}$

$\Rightarrow \frac{1}{31} + \frac{1}{32} + \frac{1}{33} > \frac{3}{33}$

$\Rightarrow \frac{1}{31} + \frac{1}{32} + \frac{1}{33} > \frac{1}{11} = \frac{1}{n+1}$

n = 10

6. **(b)**

Use plugging in

as $(x + 3)^3 - x^3 = 117$

$x + 3 \geq 5$ as $(x + 3)^3 \geq 117$

put $x = 2$

$\Rightarrow$ $5^3 - 2^2 = 125 - 8 = 117$

$\Rightarrow$ $x = 2, \; x + 3 = 5$

So, sum of both numbers = 7

Alternative Method

$$(x + 3)^3 - x^3 = 117$$

$x^3 + (3)^3 + 3(x)(3)(x + 3) - x^3 = 117$

$x^3 + 27 + 27x + 9x^2 - x^3 = 117$

$\Rightarrow 9x^2 + 27x - 90 = 0$

$\Rightarrow$ $x^2 + 3x - 10 = 0$

$\Rightarrow$ $(x + 5)(x - 2) = 0$

$x = 2, -5$

$\Rightarrow$ so either 2, 5, or –5, –2

Thus sum = 7, OR = –7

7. **(d)**

Put in prime factorization theorem.

(a) $21^5 \times 2^3 = 3^5 \times 7^5 \times 2^3$

Total factors = $6 \times 6 \times 4 \neq 24$

(b) $2^7 \times (2^2 \times 3)^3 = 2^{13} \times 3^3$

Total factors = $14 \times 4 \neq 24$

(c) $2^6 \times 3^4$ = factors $7 \times 5 \neq 24$

(d) $63 \times 55 = 3^2 \times 7^1 \times {}^15 \times 11^1$

Factors = $3 \times 2 \times 2 \times 2 = 24$

8. **(b)**

$$\text{Rem}\frac{[x]}{6} = 4, \quad \text{Rem}\frac{[y]}{6} = 5$$

$$\text{Rem} = \frac{\left[x^3 + y^3\right]}{6} = \frac{\left[4^3 + 5^3\right]}{6}$$

$$= \frac{[6 + 125]}{6}$$

$$\Rightarrow \text{Rem}\left[\frac{4+5}{6}\right] = \text{Rem}\left(\frac{9}{6}\right) = 3$$

9. **(a)**

N = (1! + 2! + 3! + 4! +....1000!)

Now we have to check only

1! + 2! + 3! + 4! as after that will factorial has unit digit as 0

5! = 120

6! = 720 and so on

Thus unit digit of (1! + 2! + 3! + 4! + 0) for all other

$\Rightarrow$ $(1 + 2 + 6 + 4 + 0)^{40}$

Unit digits of all factorial

$\Rightarrow$ $(3)^{40} = (3)^{4k}$

$\Rightarrow 3 \times 3 \times 3 \times 3 \Rightarrow 1$ unit digit. Hence when $(3)^{40} \div 10$, Remainder will be 1.

10. **(c)**

As there should be only 5 which should be common between any two number

so number may be $(5x_1, 5x_2, 5x_3.....5x_n)$

and HFC $(xi, xj) = 1$

Thus, number are

$(5 \times 1, 5 \times 2, 5 \times 3, 5 \times 5......)$

because $(xi, xj) = 1$

Then only prime number will work.

so $5 \times 19 = 95$ is biggest number.

so $5 \times 1, 5 \times 2, 5 \times 3, 5 \times 5, 5 \times 7, 5 \times 11, 5 \times 13, 5 \times 17, 5 \times 19$, total 9 number.

11. (d)

Plugging in

(a) $y = 10, x = 2$

$= 10 - 2 = 8$ X

(b) $\frac{x+y}{x} = 1 + \frac{y}{x}$

$y = 9, x = 3$

$1 + \frac{9}{3} = 1 + 3 = 4$ X

(c) $(x + y) = 10 + 2 = 12$ X

(d) None of above is true.

12. (c)

$$\text{Rem}\left[\frac{1421 \times 1423 \times 1425}{12}\right]$$

$$\Rightarrow \text{Rem}\frac{[5 \times 7 \times 9]}{12} = \text{Rem}\frac{[35 \times 9]}{17}$$

$$\text{Rem}\frac{[11 \times 9]}{12} = \text{Rem}\frac{[99]}{12} = 3$$

13. (b)

$n = 85K + 39$

Now $\frac{n}{17} = \frac{85K+39}{17} = 5K + \frac{34+5}{17}$

$= 5K + 2 + \frac{5}{17}$

$\Rightarrow$ Remainder is 5.

14. (d)

Suppose division 'd' and remainder is 'r'.

$\Rightarrow$ $34041 = q_1 d + r$

and $32506 = q_2 d + r$

$\Rightarrow$ $15365 = (q_1 - q_2)d$

$\Rightarrow$ d should be factor 1535

So, only 307 is factor of 1535 in given option.

15. (c)

(a) $100 = 5 \times 2$ possible

(b) $12 = 6 \times 2$ possible

(c) $25 = 5 \times 5$ not possible as only one 5 Number ticket

(d) $30 = 15 \times 2$ possible

16. (c)

Suppose

$$\frac{6N^3 + 3N^2 + N + 24}{N} = M = \text{Natural Number}$$

So, $M = 6N^2 + 3N + 1 + \frac{24}{N}$

Thus to be M as an natural number $\frac{24}{N}$ should be Natural number thus N should be factor of 24.

so $24 = 2^3 \times 3^1$

Factors $= 4 \times 2 = 8$

8 values can be taken by N.

17. (a)

Unit digit of

$39^{53} \times 27^{23} \times 36^{12}$

Unit digit of $(9^{53} \times 7^{23} \times 6^{12})$

$= (9^{4K+1} \times 7^{4K+3} \times 6^{4K})$

$\Rightarrow$ $(9 \times 3 \times 6) = 2$ = units digit

18. (a)

Number of zero's can be calculated by getting power of 10 in number or minimum among powers of 2 or 5 possible in n.

Thus $2 \times 3 \times 5 \times$ 97

so only $1 \rightarrow 5$'s and $1 \rightarrow 2$'s

Thus only one zero as both 2 and five will occur only once.

19. (c)

123.....910........

One digit numbers (from 1 to 9) = 9

Now 2 digit number till 50 are 50 – 9 = 41 thus total digits used to make pair.

10 11 12.........29.

So total $20 \times 2 = 40$ digits would be required to write numbers from 10 to 29.

Total digits consumed so far till we write 29 will be $9 + 40 = 49$.

50th digit which we will write is 3 of 30.

Hence 50th digit will be 3.

49 digits are used. So far till 29.

Now, 50th digit would be first digit of 30 thus 3.

20. (d)

$N = n(n+1)(n+2)(n+3)(n+4)$ is product of 5 consecutive number.

Thus divisible by 5! = 120

So, (i) unit digit is 0 TRUE.

(ii) Perfectly divisible by 24 TRUE.

(iii) N is perfect square not TRUE.

(iv) N is odd X not TRUE.

Thus (d) (i) and (ii) only.

21. (d)

$N = 12^{12} \times 14^{14} \times 15^{15}$

$= (2^2 \times 3)^{12} \times (7 \times 2)^{14} \times (5 \times 3)^{15}$

$= 2^{24} \times 3^{12} \times 7^{14} \times 2^{14} \times 5^{15} \times 3^{15}$

$= 2^{38} \times 3^{27} \times 5^{15} \times 7^{14}$

$K = (12)^{10} \times (14)^{10} \times (15)^{10}$

$= 2^{20} \times 3^{10} \times 7^{10} \times 2^{10} \times 3^{10} \times 5^{10}$

$= 2^{30} \times 3^{20} \times 5^{10} \times 7^{10}$

Thus factors of N which are multiple of k are

$2^{30} \times 3^{20} \times 5^{10} \times 7^{10} \, (2^8 \times 3^7 \times 5^5 \times 7^4)$

will be $(8 + 1) \times (7 + 1) \times (5 + 1) \times (4 + 1)$

$9 \times 8 \times 6 \times 5$

22. (b)

137_9

254_9

402_9

$(7 + 4) = (11)_{10} = (2)$

unit digit is 2 thus 9 is possible base

Thus we have to add using base 9

so 342_9

562_9

1014_9

■■■■

1.2 CHAPTER

Percentages

The term 'percent' indicates the value out of hundred. This concept of percentage is developed to make the comparison of ratio easier by taking the denominator value as 100.

The concept of percentage is very useful in reasoning & aptitude and specially for data interpretation section, where every logic that has to be used, is based on percentage.

Calculation of Percentage

As we know that percent value is the value out of hundred. The percent value is calculated as

$$\frac{\text{Value}}{\text{Total Value}} \times 100$$

The basic thing that has to be kept in mind is that the value in the base must be taken care of.

Percentage & Fraction

The value percentage can be represented in three different forms. Each form is important for making calculation similar. Here are some percentage values given to understand different form.

% form	Fraction form	Decimal form
100%	1	1
50%	$\frac{1}{2}$	0.5
$33.\overline{33}\%$	$\frac{1}{3}$	$0.\overline{33}$
25%	$\frac{1}{4}$	0.25
20%	$\frac{1}{5}$	0.20
$16.\overline{66}\%$	$\frac{1}{6}$	$0.1\overline{6}$

Multiplication factor

Multiplication factor is the value by which a particular quantity has to be multiplied to show the final changed value. It is nothing but fractional or decimal form of percentage.

It has to be understood in this manner, that a quantity is 200. Its 60% value will be $\frac{60}{100}, \frac{3}{5}$, 0.6 times of 200. Here all these values are the multiplication factor for the quantity 200.

Multiplication factors are very important when we talk about the percentage increment or decrement of a quantity. Let us understand with some examples.

Example 1.

A quantity 200 is increased by 75% then what will be the new quantity?

Solution.

One thing we can do is to find out 75% of 200 i.e.

$$\frac{75}{100} \times 200 = 150$$

and then the new quantity will be 200 + 150 = 350

But we can move in other way as the quantity initially was 100% i.e. 1 as a multiplying factor. Now it is increased by 75% i.e. 3/4 or 0.75 as a multiplying factor.

So the new quantity becomes

$$\left(1+\frac{3}{4}\right) \times 200 = 350$$

or $(1 + 0.75) \times 200 = 350$

These $\left(1+\frac{3}{4}\right)$ and (1 + 0.75) are the multiplying factors. Let us see one more example to understand it clearly.

Example 2.

A quantity 300 when increased by some percentage becomes 360. Find out by what percentage it has increased?

Solution.

Now we know that

300 × Multiplying Factor (MF) = 360

So $MF = \frac{360}{300} = 1.2$ or $\frac{6}{5}$

This shows that the multiplying factor 1 is increased by 0.2 or $\frac{1}{5}$ which is equal to 20%.

so net increment percentage is 20%.

Successive Percentage Change

Successive percentage change is one percentage change over and above another percentage change. Let us take one example to understand it.

Example 1.

A quantity 300 is increased by 20%. Then it was decreased by 10%. Find out the new quantity is how much percentage more/less than 300?

Solution.

300 when increased by 20% becomes 300(1.2) or

$300\left(1+\frac{1}{5}\right)$. Now this new quantity is decreased by 10%. So the final quantity will be

$$[300(1.2)](1-0.1)$$

or $\left[300\left(\frac{6}{5}\right)\left(1-\frac{1}{10}\right)\right]$

$= 300(1.08)$

or $300\left(1+\frac{4}{50}\right)$

So we can understand the net result is 0.08 or $\frac{4}{50}$ increment in MF. Which is equal to net 8% increment in the quantity 300.

So it will be very clearly understood from the example that **in successive percentage changes, the multiplying factors are multiplied directly**. So the point that has to be understood is "In case of successive percentage changes the net multiplying factor is the product of all the corresponding multiplying factors.

Let us take another example as:

Comparison Leading to the Base Change

Sometimes whenever the percentage comparison occurs between two different values, the base change occurs. Multiplying factors are again very useful here. Let us understand with some examples.

Example 1.

The salary of Ram was increased by 40%. But Ram's performance kept on declining. Because of which his employer decreased his salary to the salary before the increment. Find out what percentage deduction was provided by his employer?

Solution.

Let us assume in starting salary of Ram was x and after the increment it became y.

Now his salary was again decreased from y to x and we need to find out this percentage change, we can say it as

$$y = \left(1+\frac{40}{100}\right)x$$

$$\Rightarrow y = \left(1+\frac{2}{5}\right)x$$

$$\Rightarrow y = \frac{7}{5}x$$

So $x = \frac{5}{7}y$

or $x = \left(1-\frac{2}{7}\right)y$

So we can say the % decrement in the salary of Ram is the percentage equivalent of $\frac{2}{7}$ which is

$\frac{2}{7}\times 100\%$ i.e. 28.56%.

Example 2.

Rama uses rice as his daily meal. But because of inflation the price on rice were increased by 10%. Find out by what percentage he has to decrease his consumption to keep the expenditure same as before?

Solution.

Let us assume initial price was P and the expenditure was 'x' then,

Price × consumption = Expenditure

$$\Rightarrow \text{initial consumption} = \frac{x}{P}$$

Now price became $\left(1+\frac{10}{100}\right)P = \frac{11}{10}P$

Then final consumption $= \frac{x}{\frac{11}{10}P} = \frac{10}{11}\left(\frac{x}{P}\right)$

So final consumption = $\frac{10}{11}$ (initial consumption)

or we can say the net percent decrement in consumption will be $\left(1-\frac{10}{11}\right)$ i.e. $\frac{1}{11}$ part.

The percentage equivalent will be $\frac{1}{11}\times 100 = 9.09\%$

Alternatively: The net percent change

$$= \frac{100\times 10}{100+10}\% = 9.09\%$$

Example 3.

The price of a toy was increased by 20% & then it was sold at 20% discount. Find out the net change in the price of the toy.

Solution.

The net multiplying factor

$$= \left(1+\frac{20}{100}\right)\left(1-\frac{20}{100}\right)$$

$= 1.2 \times 0.8 = 0.96$

$= (1 - 0.04)$

So net percentage change

$= 0.04 \times 100 = 4\%$ decrease.

Example 4.

The price of a ball pen & a gel pen is same. Now the price of ball pen is decreased by 30%. While of gel pen is increased by 20%. Find out the net percentage change in the net total price of both pens.

Solution.

Let the price of individual pens be x

Total price = $2x$

New price of ball pen = $(1 - 0.3)x = 0.7x$

New price of gel pen = $1.2x$

New total is = $1.9x = 0.95(2x)$

So change = 5% decrement

Example 5.

A shop provides flat 50% discount on one shirt. While another shop provides the successive discount of 30% and 30%. If the difference of the bills is Rs. 43. Find out the net cost of shirt.

Solution.

Let the price of shirt be x

For the first shop the bill is

$= (1 - 0.5)x = 0.5x$

For the second shop the bill is

$= 0.7 \times 0.7x = 0.49x$

The difference in the bills

$= 0.01x = 43$

$\Rightarrow x =$ Rs. 4300

Example 6.

Radius of a sphere is increased by 10%. Find out the net percentage increment in the volume.

Solution.

Net percentage increment in radius = 10%

New radius = 1.1 r

So the new volume = $\frac{4}{3}\pi(1.1\,r)^3$

$= (1.1)^3$ (original volume)

$= 1.331$ (original volume)

So net percent increment = 33.1%

Example 7.

In an office 40% of the employee are males while 70% of the employees are married. If 80% of male are married find out the net percent of unmarried female in the office.

Solution.

Let us assume total employees is 100, then

	Married	Unmarried	Total
Male	80% of 40 = 32	8	40
Female	70 – 32 = 38	60 – 38 = 22	60
Total	70% of 100 = 70		100

So net percent $\frac{22}{100}\times 100 = 22\%$

□□□□

Solved Examples

1. Which of the following is the largest number?

(a) 20% of 200 (b) 7% of 500
(c) 1300% of 3 (d) 600% of 7

Ans. (d)

20% of 200 = 40
7% of 500 = 35
1300% of 3 = 39
600% of 7 = 42

2. Mr. Rajesh is worried about the balance of his monthly budget. The price of petrol has increased by 40%. By what percent should he reduce the consumption of petrol so that he is able to balance his budget?

(a) 33.33 (b) 28.56
(c) 25 (d) None of these

Ans. (b)

We know that % reduction required is

$$= \frac{x}{100+x} \times 100, \text{ here } x = 40$$

$$= \frac{40}{140} \times 100 = 28.56\%$$

3. In an election between 2 candidates, Ravikant gets 65% of the total valid votes. If the total votes were 6000, what is the number of valid votes that the other candidate Shailendra gets if 25% of the total votes were declared invalid?

(a) 1625 (b) 1575
(c) 1675 (d) 1525

Ans. (b)

Total votes 6000
Invalid votes = 25% of 6000 = 1500
Total valid votes = 4500
Ravikant gets 65%
So other candidate gets 35%
35% of 4500 = 1575

4. In a medical certificate, by mistake a candidate gave his height as 25% more than normal. In the interview panel, he clarified that his height was 5 feet 5 inches. Find the percentage correction made by the candidate from his stated height to his actual height.

(a) 20 (b) 28.56
(c) 25 (d) None of these

Ans. (a)

Height increased = 25%
here x = 25 to reduce to normal % correction required is

$$= \frac{25}{100+25} \times 100 = 20\%$$

5. A number is mistakenly divided by 5 instead of being multiplied by 5. Find the percentage change in the result due to this mistake.

(a) 96% (b) 95%
(c) 2400% (d) None of these

Ans. (a)

Let number is 100
It is divided by 5 we get 20
Now actual result should be $5 \times 100 = 500$
So % change in result

$$= \frac{500-20}{500} \times 100 = 96\%$$

6. In a mixture of 80 litres of milk and water, 25% of the mixture is milk. How much water should be added to the mixture so that milk becomes 20% of the mixture?

(a) 20 litres (b) 15 litres
(c) 25 litres (d) None of these

Ans. (a)

Total mixture 80 litre

Milk is 25% i.e. $\frac{25}{100} \times 80 = 20$ litre

to make it 20%, amount of water required to

$$\text{add} = \frac{60+x}{80+x} \times 100 = 80\%$$

$6000 + 100x = 6400 + 80x$
$20x = 400, x = 20$

7. A landowner increased the length and the breadth of a rectangular plot by 10% and 20% respectively. Find the percentage change in the cost of the plot assuming land prices are uniform throughout his plot.

(a) 33% (b) 35%
(c) 22.22% (d) None of these

Ans. (d)

here $x = 10$ or $y = 20$

$$\% \text{ change in area} = \left(x + y + \frac{xy}{100}\right)$$

$$= 10 + 20 + \frac{10 \times 20}{100} = 32$$

8. The length, breadth and height of a room in the shape of a cuboid are increased by 10%, 20% and 50% respectively. Find the percentage change in the volume of the cuboid.

(a) 77% (b) 75%
(c) 88% (d) 98%

Ans. (d)

Let l, b, h be length, breadth and height of the cuboid

Volume v = l.b.h

Now, l, b, h are increased by 10%, 20% 50% respectively

$$v = l\left[1+\frac{10}{100}\right]\times b\left[1+\frac{20}{100}\right]+h\left[1+\frac{50}{100}\right]$$

= 1.98 l.b.h

% change = 98%

9. The price of sugar is reduced by 25% but inspite of the decrease, Aayush ends up increasing his expenditure on sugar by 20%. What is the percentage change in his monthly consumption of sugar?

(a) +60% (b) –10%
(c) +33.33% (d) 50%

Ans. (a)

Let price of sugar be x & expenditure E

Now it is reduced by 25%. So it is $\frac{3}{4}x$ now expenditure of sugar is also increased by 20% i.e.

1.2 E or $\frac{6}{5}$E

So quantity of sugar that can be purchased

$$= \frac{\frac{6}{5}E}{3/4x} = \frac{8}{5}\frac{E}{x} = 160\% \text{ of } \frac{E}{x}$$

Hence consumption increased by 60%.

10. 30% of a number when subtracted from 91, gives the number itself. Find the number.

(a) 60 (b) 65
(c) 70 (d) None of these

Ans. (c)

$\frac{30}{100}x = 30\%$ of number x

Now $91 - \frac{30}{100}x = x$

$91 = \frac{130}{100}x,\ x = 70$

11. The population of the village of Rampur is 10,000 at this moment. It increases by 10% in the first year. However, in the second year, due to immigration, the population drops by 5%. Find the population at the end of the third year if in the third the population increases by 20%.

(a) 12,340 (b) 12,540
(c) 1,27,540 (d) 12,440

Ans. (b)

P = 10000, x = +10%, y = –5%, z = 20%

Population at the end of third year

$$= P\left(1+\frac{x}{100}\right)\left(1+\frac{y}{100}\right)\left(1+\frac{2}{100}\right)$$

$$= 10000\left(1+\frac{10}{100}\right)\left(1-\frac{5}{100}\right)\left(1+\frac{20}{100}\right)$$

= 12540

12. In an examination, Mohit obtained 20% more marks than Sushant but 10% less than Rajesh. If the marks obtained by Sushant is 1080, find the percentage marks obtained by Rajesh if the full marks is 2000.

(a) 86.66% (b) 72%
(c) 78.33% (d) None of these

Ans. (b)

Marks obtained by Shushant is 1080

Mohit's marks $= \frac{120}{100}\times 1080 = 1296$

Rajesh Marks, R $\Rightarrow \frac{90}{100}\times R = 1296$

R = 1440

% of Rajesh Marks $= \frac{1440}{2000}\times 100 = 72\%$

13. The population of a village is 5500. If the number of males increases by 11% and the number of females increases by 20%, then the population becomes 6330. Find the population of females in the town.

(a) 2500 (b) 3000
(c) 2000 (d) 3500

Ans.(a)

x is population of male

$\therefore$ (5500 – x) is female population

$$\frac{x\times 111}{100}+\frac{(5500-x)\times 120}{100} = 6330$$

On calculating we get x = 3000

So female population = 2500.

14. Vicky's salary is 75% more than Ashu's. Vicky got a raise of 40% on his salary while Ashu got a raise of 25% on his salary. By what percent is Vicky's salary more than Ashu's now?

(a) 96% (b) 51.1%
(c) 90% (d) 52.1%

Ans. (a)

Lets Ashu's salary = Rs. 100
Vicky's salary = Rs. 175
Vicky's salary increased by 40% i.e.

$$= \frac{140}{100} \times 175 = \text{Rs. } 245$$

Ashu's salary increased by 25% ie 125
Vicky's Salary is 120 more that Ashu's in % term

$$\frac{120}{125} \times 100 = 96\%$$

15. During winters, an athlete can run 'x' metres on one bottle of Glucose. But in the summer, he can only run 0.5 x metres on one bottle of Glucose. How many bottles of Glucose are required to run 400 metres during summer?

(a) 800/x (b) 890/x
(c) 96 (d) 454/x

Ans. (a)

During summer to run –5x one bottle of glucose is required

then to run 1 km $\frac{1}{0.5x}$ bottles

$\therefore$ to run 400m $\frac{1}{0.5x} \times 400 = \frac{800}{x}$ bottle

16. In an examination, 80% students passed in Physics, 70% in Chemistry while 15% failed in both the subjects. If 325 students passed in both the subjects. Find the total number of students who appeared in the examination.

(a) 500 (b) 400
(c) 300 (d) 600

Ans. (a)

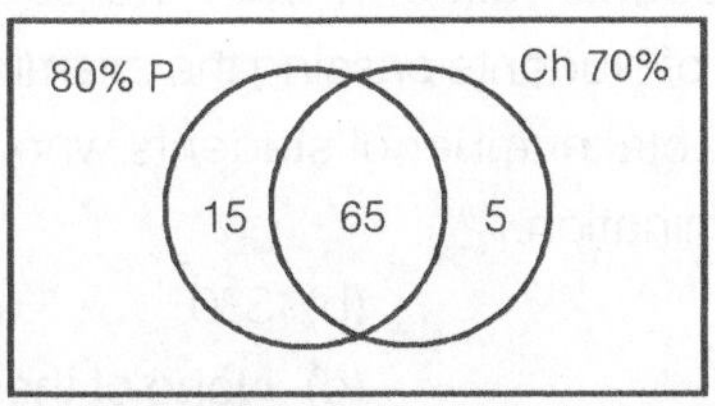

15% student failed in both subjects
So total passed = 85%
80 + 70 = 85+x, x = 65%
65% of total = 325
total = 500.

17. 40% of 20% + 30% of 25% + 50% of 28% is equivalent to

(a) 29.5% (b) 28.5%
(c) 30.5% (d) None of these

Ans. (a)

$$40\% \text{ of } 20\% = \frac{40}{100} \times \frac{20}{100} = \frac{8}{100} = 8\%$$

$$30\% \text{ of } 25\% = \frac{30}{100} \times \frac{25}{100} = \frac{75}{100} = 7.5\%$$

$$\text{and, } 50\% \text{ of } 28\% = \frac{50}{100} \times \frac{28}{100}$$

$$= \frac{14}{100} = 14\%$$

$\therefore$ (40% of 20%+ 30% of 25% + 50% of 28%)
= 8% + 7.5% + 14% = 29.5%.

18. A man's wages were decreased by 50%. The reduced wages were increased by 50%. He had a loss of

(a) 35% (b) 25%
(c) 20% (d) None of these

Ans. (b)

Here, x = –50 and y = 50.
$\therefore$ The net % change in wages

$$= \left(x + y + \frac{xy}{100}\right)\%$$

$$= \left(-50 + 50 - \frac{50 \times 50}{100}\right)\% \text{ or } -25\%.$$

Since the sign is –ve, he has a loss of 25%.

19. The radius of a sphere is increased 10%. The surface area increases by

(a) 21% (b) 31%
(c) 41% (d) None of these

Ans. (a)

Since 4π × radius × radius = surface area
$\therefore$ Net % change in area

$$= \left(x + y + \frac{xy}{100}\right)\%$$

$$= \left(10 + 10 + \frac{10 \times 10}{100}\right)\% = 21\%.$$

□□□□

Percentage

I Practice Exercise: I

1. If a number is 20% more than the other, how much percent is the second number less than the first?

(a) $12\frac{1}{3}\%$ (b) $16\frac{2}{3}\%$
(c) $16\frac{1}{3}\%$ (d) None of these

2. If the given two numbers are respectively 7% and 28% of a third number, then what percentage is the first of the second?

(a) 20% (b) 25%
(c) 18% (d) None of these

3. The price of cooking oil has increased by 15%. The percentage of reduction that a family should effect in the use of cooking oil so as not to increase the expenditure on this account is

(a) $15\frac{2}{23}\%$ (b) $13\frac{1}{23}\%$
(c) $17\frac{1}{23}\%$ (d) None of these

4. The difference between a discount of 35% and two successive discounts of 20% and 20% on a certain bill was Rs. 22. Find the amount of the bill.

(a) Rs. 3200 (b) Rs. 2200
(c) Rs. 1800 (d) None of these

5. Two shopkeepers sell a radio of similar brand and type at the same list price of Rs. 1000. The first allows two successive discounts of 20% and 10% and the second allows the successive discounts of 15% and 15%. Find the difference in discounts offered by the two shopkeepers.

(a) Rs. 3.50 (b) Rs. 1.50
(c) Rs. 2.50 (d) None of these

6. The tax on a commodity is diminished by 10% and its consumption increases by 10%. Find the effects on revenue.

(a) 1% increase (b) 2% increase
(c) 3% decrease (d) None of these

7. If the side of a square is increased by 30%, its area is increased by

(a) 49% (b) 69%
(c) 79% (d) None of these

8. In measuring the sides of a rectangle, one side is taken 10% in excess and the other 20% in deficit. Find the error percent in area calculated from the measurement.

(a) 12% deficit (b) 10% deficit
(c) 12% excess (d) None of these

9. Water tax is increased by 20% but its consumption is decreased by 20%. The increase or decrease in the expenditure is

(a) 4% decrease (b) 4% increase
(c) 8% decrease (d) 8% increase

10. The population of a city increases at the rate of 10% annually. Its present population is 90.51 lacs. The population 3 years ago was nearly.

(a) 72 lacs (b) 68 lacs
(c) 80 lacs (d) None of these

11. The value of a machine is Rs. 6250. It decreases by 10% during the first year, 20% during the second year and 30% during the third year. What will be the value of the machine after 3 years?

(a) Rs. 2650 (b) Rs. 3050
(c) Rs. 3150 (d) None of these

12. An army lost 10% its men in war, 10% of the remaining due to diseases and 10% of the rest were disabled. Thus, the strength was reduced to 729000 active men. Find the original strength.

(a) 1000000 (b) 1200000
(c) 1500000 (d) None of these

13. In an examination, 30% and 35% students respectively failed in History and Geography while 27% students failed in both the subjects. If the number of students passing the examination is 248, find the total number of students who appeared in the examination.

(a) 425 (b) 380
(c) 400 (d) None of these

14. In an examination, there were 2000 candidates, out of which 900 candidates were boys and rest were girls. If 32% of the boys 38% of the girls passed, then the total percentage of failed candidates is

(a) 35.3% (b) 64.7%
(c) 68.5% (d) 70%

15. If the price of gold increases by 30%, find by how much the quantity of ornaments must be reduced so that the expenditure may remain same as before?

(a) $27\frac{2}{3}\%$ (b) $23\frac{1}{13}\%$
(c) 30% (d) 19%

16. The price of an article is cut by 20%. To restore it to its original price, the new price must be increased by
(a) 20% (b) 22.5%
(c) 25% (d) 40%

17. In a fraction, numerator is increased by 25% and the denominator is diminished by 10%. The new fraction obtained is $\frac{5}{9}$. The original fraction is

(a) $\frac{2}{5}$ (b) $\frac{5}{9}$

(c) $\frac{3}{5}$ (d) None of these

Answers

1. (b)	2. (b)	3. (b)	4. (b)	5. (c)
6. (d)	7. (b)	8. (a)	9. (a)	10. (b)
11. (c)	12. (a)	13. (c)	14. (b)	15. (b)
16. (c)	17. (a)			

Solutions

1. **Ans.(b)**
Here, $x = 20$

$\therefore$ Required answer $= \left(\frac{x}{100+x} \times 100\right)\%$

$$= \left(\frac{20}{100+20} \times 100\right)\% = 16\frac{2}{3}\%.$$

2. **Ans. (b)**
Here, $l = 7$, and m = 28.
$\therefore$ First number

$$= \frac{l}{m} \times 100\% \text{ of second number}$$

$$= \frac{7}{28} \times 100\% \text{ of second number}$$

Or 25% of second number.

3. **Ans. (b)**
Reduction in consumption

$$= \left(\frac{P}{100+P} \times 100\right)\%$$

$$= \left(\frac{15}{100+15} \times 100\right)\% \text{ or } 13\frac{1}{23}\%.$$

4. **Ans. (b)**
The equivalent discount of two successive discounts of 20% and 20%.

$$= \left(x + y + \frac{xy}{100}\right)\%$$

$$= \left(-20 - 20 + \frac{20 \times 20}{100}\right)\% \text{ or } -36\%.$$

Given: 36% – 35% = Rs. 22.
$\therefore$ Amount of the bill = 22 × 100 = Rs. 2200.

5. **Ans. (c)**
The equivalent discount of two successive discounts of 20% and 10%

$$= \left(x + y + \frac{xy}{100}\right)\%$$

$$= \left(-20 - 10 + \frac{20 \times 10}{100}\right)\% \text{ or } 28\%$$

$\therefore$ Discount on the list price of radio offered by the first shopkeeper

$$= 28\% \text{ of } 1000 = \frac{28}{100} \times 1000 = \text{Rs. } 280.$$

Also, the equivalent discount of two successive discounts of 15% and 15%

$$= \left(x + y + \frac{xy}{100}\right)\%$$

$$= \left(-15 - 15 + \frac{15 \times 15}{100}\right)\% \text{ or } 27\frac{3}{4}\%$$

$\therefore$ Discount on the list price of radio offered by the second shopkeeper.

$$= 27\frac{3}{4}\% \text{ of } 1000 = \frac{111}{400} \times 1000$$

= Rs. 277.50

∴ Difference in discounts offered by the two shopkeepers = Rs. 280 – Rs. 277.50 = Rs. 2.50.

6. Ans. (d)

Since tax × consumption = revenue

∴ Net % change in revenue

$$= \left(x + y + \frac{xy}{100}\right)\%$$

$$= \left(-10 + 10 + \frac{10 \times 10}{100}\right)\%$$

(here $x = -10$ and $y = 10$) = –1%

∴ The revenue decreases by 1%.

7. Ans. (b)

Since side × side = area

∴ Net % change in area

$$= \left(x + y + \frac{xy}{100}\right)\%$$

$$= \left(30 + 30 + \frac{30 \times 30}{100}\right)\% = 69\%.$$

∴ The area is increased by 69%.

8. Ans. (a)

Since $side_1 \times side_2$ = area

∴ Error % in area $= \left(x + y + \frac{xy}{100}\right)\%$

$$= \left(10 - 20 - \frac{10 \times 20}{100}\right)\%$$

(Here, $x = 10$ and $y = -20$)

= –12%, i.e. 12% deficit.

9. Ans. (a)

Since tax × consumption = expenditure

∴ Net % change in expenditure

$$= \left(x + y + \frac{xy}{100}\right)\%$$

$$= \left(20 - 20 - \frac{20 \times 20}{100}\right)\% \quad [x = 20 \,\&\, y = -20]$$

= –4%.

∴ Expenditure decreases by 4%.

10. Ans. (b)

We have, P = 90.51, r = 10 and n = 3.

∴ The population 3 years ago

$$= \frac{P}{\left(1 + \frac{r}{100}\right)^n} = \frac{90.51}{\left(1 + \frac{10}{100}\right)^3}$$

$$= \frac{9051}{100} \times \frac{100}{110} \times \frac{100}{110} \times \frac{100}{110} = 68 \text{ lacs.}$$

11. Ans. (c)

Here, A = 6250, $x = -10$, y= –20 & z= –30.

∴ Value of the machine after 3 years

$$= A\left(1 + \frac{x}{100}\right)\left(1 + \frac{y}{100}\right)\left(1 + \frac{z}{100}\right)$$

$$= 6250\left(1 - \frac{10}{100}\right)\left(1 - \frac{20}{100}\right)\left(1 - \frac{30}{100}\right)$$

$$= \frac{6250 \times 90 \times 80 \times 70}{100 \times 100 \times 100} = \text{Rs. } 3150.$$

12. Ans. (a)

Let A be the original strength.

Then, $A\left(1 + \frac{x}{100}\right)\left(1 + \frac{y}{100}\right)\left(1 + \frac{z}{100}\right)$

= 729000 (Given)

Here, $x = -10$, y = –10 and z = –10.

$$\therefore \quad A\left(1 - \frac{10}{100}\right)\left(1 - \frac{10}{100}\right)\left(1 - \frac{10}{100}\right)$$

= 729000

$$\Rightarrow \quad A = \frac{729000 \times 100 \times 100 \times 100}{90 \times 90 \times 90}$$

= 1000000 men.

13. Ans. (c)

Percentage of students passing the examination

= (100 – (30 + 35 – 27))%

[here, $x = 30$, y = 35 and z = 27]

= (100 – 38)% = 62%.

Let the total number of students appearing in the examination x.

Given : 62% of x = 248

or, $\frac{62}{100} \times x = 248$ or $x = \frac{248 \times 100}{62} = 400.$

Therefore, 400 students appeared in the examination.

14. Ans. (b)

Boys = 900, Girls = 1100

Passed = (32 % of 900) + (38% of 1100)

= 288 + 418 = 706

Failed = 2000 – 706 = 1294

$$\text{Failed \%} = \left(\frac{1294}{2000}\times 100\right)\% = 64.7\%$$

15. Ans. (b)

$$\text{Reduction} = \frac{30}{100+30}\times 100\% = 23\frac{1}{13}\%$$

16. Ans. (c)

New price must be increased by

$$\left(\frac{20}{100-20}\times 100\right)\% = 25\%.$$

17. Ans. (a)

Let the fraction be $\frac{x}{y}$

Then, $\frac{x+0.25x}{y-0.10y} = \frac{5}{9}, \frac{x(1.25)}{y(0.9)} = \frac{5}{9}$

$\frac{x}{y} = \frac{5}{9}\times\frac{90}{125}, \Rightarrow \frac{x}{y} = \frac{2}{5}$

□□□□

II Practice Exercise: II

1. Of the total amount received by Prerna, 20% was spent on purchases and 5% of the remaining on transportation. If he is left with Rs. 1520, the initial amount was:
   (a) Rs. 2800 (b) Rs. 2000
   (c) Rs. 2400 (d) Rs. 1600

2. The price of jute has been reduced by 20%. if the reduced price is Rs. 800 per quintal, the original price per quintal was
   (a) Rs. 900 (b) Rs. 640
   (c) Rs. 960 (d) Rs. 1000

3. 5/9 part of the population in a village are males. If 30% of the males are married, the percentage of unmarried females in the total population is:
   (a) 70% (b) 40%
   (c) $27\frac{7}{9}\%$ (d) 20%

4. A school has only three classes which contain 40, 50 and 60 students respectively. The pass percentages of these classes are 10, 20 and 10 respectively. The pass percentage of the school is:
   (a) $13\frac{1}{3}$ (b) 15
   (c) 20 (d) $16\frac{2}{3}$

5. If x% of a is the same as y% of b, then z% of b is
   (a) $\frac{yz}{x}$% of a (b) $\frac{xy}{z}$% of a
   (c) $\frac{xz}{y}$% of a (d) None of these

6. From a container having pure milk, 20% is replaced by water and the process is repeated thrice. At the end of the third operation, the milk is:
   (a) 40% pure (b) 50% pure
   (c) 51.2% pure (d) 58.8% pure

7. The salaries of A and B together amount to Rs. 2000. A spends 95% of his salary and B, 85% of his. If now, their savings are same, what is A's salary?
   (a) Rs. 1500 (b) Rs. 1250
   (c) Rs. 750 (d) Rs. 1600

8. 300 grams of sugar solution has 40% sugar in it. How much sugar should be added to make it 50% in the solution?
   (a) 10 gms (b) 40 gms
   (c) 60 gms (d) 80 gms

9. In an examination, there are 3 papers of Mathematics of 100 marks each. A boy secures 60% in the first paper and 70% in the second paper. In order to secure 70% in the aggregate the percentage of marks he should secure in third paper will be:
   (a) 90% (b) 80%
   (c) 75% (d) 70%

10. Two numbers are less than a third number by 30% and 37% respectively. How much percent is the second number less than the first?
    (a) 10% (b) 7%
    (c) 4% (d) 3%

11. In an examination, A got 10% marks less than B, B got 25% marks more than C and C got 20% less

than D. If A got 360 marks out of 500, the percentage of marks obtained by D was
(a) 70 (b) 75
(c) 80 (d) 85

12. If the numerator of a fraction be increased by 15% and its denominator be dimimished by 8%, the value of the fraction becomes $\frac{15}{16}$. The original fraction is
(a) $\frac{3}{5}$ (b) $\frac{3}{4}$
(c) $\frac{3}{7}$ (d) $\frac{2}{3}$

13. In an examination, 35% candidates failed in one subject and 42% failed in another subject while 15% failed in both the subjects. If 2500 candidates appeared at the examiniation, how many passed in either subject but not in both?
(a) 325 (b) 1175
(c) 2125 (d) None of these

14. The boys and girls in a college are in the ratio 3 : 2. If 20% of the boys and 25% of the girls are adults, the percentage of students who are not adults is:
(a) 58% (b) 67.5%
(c) 78% (d) 82.5%

15. The price of sugar is increased by 20%. As a result, a family decreases its consumption by 25%. The expenditure of the family on sugar will be decreased by:
(a) 10% (b) 5%
(c) 14% (d) 15%

16. A building worth Rs. 133,100 is constructed on land worth Rs. 72,900. After how many years will the value of both be the same if land appreciates at 10% p.a. and building depreciates at 10% p.a.?
(a) 2.5 (b) 2
(c) 1.5 (d) 3

17. A reduction of 21% in the price of wheat enables a person to buy 10.5 kg more for Rs. 100. What is the reduced price per kg?
(a) Rs. 2 (b) Rs. 2.25
(c) Rs. 2.30 (d) Rs. 2.50

18. The length of a rectangle is increased by 60%. By what percent would the width have to be decreased to maintain the same area?
(a) $37\frac{1}{2}\%$ (b) 60%
(c) 75 % (d) None

19. Ram sells his goods 25% cheaper than Shyam and 25% dearer than Bram. How much percentage is Bram's goods cheaper than Shyam's?
(a) 33.33% (b) 50%
(c) 66.66% (d) 40%

20. Rajiv wanted to subtract 5 from a number. Unfortunately, he added 5 instead of subtracting. Find the percentage change in the result.
(a) 300%
(b) 66.66%
(c) 50%
(d) Cannot be determined

21. The salary of Amit is 30% more than that of Varun. Find by what percentage is the salary of Varun less than that of Amit?
(a) 26.12% (b) 23.07%
(c) 21.23% (d) None of these

22. Ram spends 20% of his monthly income on his house hold expenditure, 15% of the rest of books, 30% of the rest on clothes and saves the rest. On counting, he comes to know that he has finally saved Rs. 9520. Find his monthly income.
(a) 10000 (b) 15000
(c) 20000 (d) None of these

23. An ore contains 25% of an alloy that has 90% iron. Other than this, in the remaining 75% of the ore, there is no iron. How many kilograms of the ore are needed to obtain 60 kg of pure iron?
(a) 250 kg (b) 275 kg
(c) 300 kg (d) 266.66 kg

24. Ram sells his goods 20% cheaper than Bobby and 20 dearer than Chandilya. How much percentage is Chandilya's goods cheaper/dearer than Bobby's.
(a) 33.33% (b) 50%
(c) 42.85% (d) None of these

25. Out of the total production of iron from hematite, an ore of iron, 20% of the ore gets wasted, and out of the remaining ore, only 25% is pure iron. If the pure iron obtained in a year from a mine of hematite was 80,000 kg, then the quantity of hematite mined from that mine in the year is

(a) 5,00,000 kg (b) 4,00,000 kg
(c) 4,50,000 kg (d) None of these

26. Ram spends 30% of his salary on house rent, 30% of the rest he spends on his children's education and 24% of the rest salary he spends on clothes. After his expenditure, he is left with Rs. 2500. What is Ram's salary?
(a) Rs. 6713.2 (b) Rs. 20,000
(c) Rs. 10,000 (d) Rs. 15,000

Answers

1. (b)	2. (d)	3. (c)	4. (a)	5. (c)
6. (c)	7. (a)	8. (c)	9. (b)	10. (a)
11. (c)	12. (b)	13. (b)	14. (c)	15. (a)
16. (d)	17. (a)	18. (a)	19. (d)	20. (d)
21. (b)	22. (c)	23. (d)	24. (a)	25. (b)
26. (a)				

Solutions

1. **Ans. (b)**

Purchases = 20% of $x = \frac{x}{5}$.

Balance $= x - \frac{x}{5} = \frac{4x}{5}$.

Transportation = 5% of

$\frac{4x}{5} = \frac{5}{100} \times \frac{4x}{5} = \frac{x}{25}$.

Balance $= \frac{4x}{5} - \frac{x}{25} = \frac{19x}{25}$.

$\therefore \frac{19x}{25} = 1520 \Rightarrow x = \frac{1520 \times 25}{19} = 2000.$

2. **Ans. (d)**

80% of $x = 800 \Rightarrow \frac{80}{100}x = 800$

$\Rightarrow x = 800 \times \frac{100}{80} = 1000.$

3. **Ans. (c)**

Let, total population = x. Males $= \frac{5}{9}x$.

Married males = 30% of

$\frac{5}{9}x = \frac{30}{100} \times \frac{5}{9}x = \frac{x}{6}$.

Married females $= \frac{x}{6}$.

Total females $= \left(x - \frac{5}{9}x\right) = \frac{4x}{9}$.

Unmarried females $= \left(\frac{4x}{9} - \frac{x}{6}\right) = \frac{5x}{18}$.

$\therefore$ Required percentage

$= \left(\frac{5x}{18} \times \frac{1}{x} \times 100\right)\% = 27\frac{7}{9}\%.$

4. **Ans. (a)**

Number of passed candidates

$= \left(\frac{10}{100} \times 40 + \frac{20}{100} \times 50 + \frac{10}{100} \times 60\right)$

$= (4 + 10 + 6) = 20$

Passed percentage $= \frac{20}{(40+50+60)} \times 100$

$= \left(\frac{20}{150} \times 100\right)\% = 13\frac{1}{3}\%.$

5. **Ans. (c)**

$x\%$ of a = y% of b $\Rightarrow \frac{x}{100}a = \frac{y}{100}b \Rightarrow b$

$= \left(\frac{x}{100} \times \frac{100}{y}\right)a = \left(\frac{x}{y}\right)a$

$\therefore$ z% of b $= \left(z\% \text{ of } \frac{x}{y}\right)a$

$= \left(\frac{xz}{y \times 100}\right)a = \left(\frac{xz}{y}\right)\%$ of a.

6. **Ans. (c)**
Let total quantity of original milk=1000 gm.
Milk after first operation
= 80% of 1000 = 800 gm.
Milk after second operation
= 80% of 800 = 640 gm.
Milk after third operation
= 80% of 640 = 512 gm.
$\therefore$ Strength of final mixture = 51.2%.

7. **Ans. (a)**

Let A's salary = x, Then, B's = $(2000 - x)$
5% of A = 15% of B, i.e.

$\frac{5}{100}x = \frac{15}{100}(2000 - x)$ or $x = 1500$.

8. **Ans. (c)**

Sugar $= \left(\frac{40}{100} \times 300\right)$ gms = 120 gms, water = 180 gms.

Let x gm sugar may be added.

Then, $\frac{120+x}{300+x}\times 100 = 50 \Rightarrow 240+2x$

$= 300 + x \Rightarrow x = 60$

9. **Ans. (b)**

$60+70+x = \frac{70}{100}\times 300$ or $x = 80\%$.

10. **Ans. (a)**

Let, third number be x. Then,

First number = 70% of $x = \frac{7x}{10}$,

Second number = 63% of $x = \frac{63x}{100}$.

Required Percentage

$= \left(\frac{7x}{100}\times\frac{10}{7x}\times 100\right)\% = 10\%$.

11. **Ans. (c)**

$A = \frac{90}{100}B$, $B=\frac{125}{100}C$ and $C=\frac{80}{100}D$.

$\therefore\ B = \frac{10}{9}A, C = \frac{4}{5}B$ and $D = \frac{5}{4}C$.

$B = \frac{10}{9}\times 360 = 400,\ C = \frac{4}{5}\times 400 = 320$

and $D = \frac{5}{4}\times 320 = 400$.

Percentage of $D = \left(\frac{400}{500}\times 100\right)\% = 80\%$.

12. **Ans. (b)**

Let the given fraction be x/y.

Then, $\frac{115\% \text{ of } x}{92\% \text{ of } y} = \frac{15}{16} \Rightarrow \frac{115x}{92y} = \frac{15}{16}$

$\Rightarrow\ \frac{x}{y} = \left(\frac{15}{16}\times\frac{92}{115}\right) = \frac{3}{4}$.

13. **Ans. (b)**

Failed in 1st subject $= \left(\frac{35}{100}\times 2500\right) = 875$.

Failed in 2nd subject $= \left(\frac{42}{100}\times 2500\right) = 1050$.

Failed in both $= \left(\frac{15}{100}\times 2500\right) = 375$.

Failed in 1st subject only = (875 – 375) = 500.

Failed in 2nd subject only = (1050 – 375) = 675.

Passed in 2nd only + Passed in 1st only

= (675 + 500) = 1175.

14. **Ans. (c)**

Suppose boys = $3x$ and girls = $2x$.

Not adults $= \left(\frac{80}{100}\times 3x\right) + \left(\frac{75}{100}\times 2x\right)$

$= \left(\frac{12x}{5}+\frac{3x}{2}\right) = \frac{39x}{10}$.

Required percentage

$= \left(\frac{39x}{10}\times\frac{1}{5x}\times 100\right)\% = 78\%$.

15. **Ans. (a)**

Let original consumption = 100 units & original price = Rs. 100/unit.

Original expenditure = Rs. (100 × 100)

= Rs. 10000.

New expenditure = Rs. (120 × 75) = Rs. 9000.

$\therefore$ Decrease in expenditure

$= \left(\frac{1000}{10000}\times 100\right)\% = 10\%$.

16. **Ans. (d)**

$72900\left(1+\frac{10}{100}\right)^n = 133100\times\left(1-\frac{10}{100}\right)^n$

$\therefore\ \left(\frac{11}{10}\right)^n \times \left(\frac{10}{9}\right)^n = \frac{133100}{72900} = \frac{1331}{729}$

$\therefore\ \left(\frac{11}{9}\right)^n = \left(\frac{11}{9}\right)^3 \Rightarrow n = 3$.

17. **Ans. (a)**

Let original price = Rs. x/kg. Reduced price

$= \left(\frac{79}{100}x\right)$/kg.

$\frac{100}{\frac{79x}{100}} - \frac{100}{x} = 10.5 \Rightarrow \frac{10000}{79x} - \frac{100}{x} = 10.5$

$10000 - 7900 = 10.5 \times 79x$

or $x = \frac{2100}{10.5\times 79}$

$\therefore$ Reduced price

$= \text{Rs.}\left(\frac{79}{100}\times\frac{2100}{10.5\times 79}\right)/\text{kg} = \text{Rs. } 2/\text{kg}$

18. Ans. (a)

Let length = l and breadth = b.

Let the required decrease in breadth be x%.

$$\text{Then, } \frac{160}{100}l \times \frac{(100-x)}{100} \times b = lb$$

$$\Rightarrow 160(100-x) = 100 \times 100$$

$$\text{or } 100 - x = \frac{10000}{160} = \frac{125}{2}$$

$$\Rightarrow x = \left(100 - \frac{125}{2}\right) = 37\frac{1}{2}.$$

19. Ans. (d)

Let Shyam sells good at Rs. 100

So Ram sells good at Rs. 75

Now this Rs. 75 is 25% dearer than Bram i.e.

$$\text{Bram} \times \frac{125}{100} = 75$$

Bram's price = 60

Bram's good is Rs. 40 cheaper than Shyam in %

$$\text{term } \frac{40}{100} \times 100 \text{ i.e. } 40\%$$

20. Ans. (d)

This cannot be determined because after adding or substracting 5 to different numbers variable results are obtained.

21. Ans. (b)

Let Varun's salary is 100

Amit salary = 30% more that Varun's Salary = 130 in % salary of Varun less than Amit

$$= \frac{30}{130} \times 100 = 23.07\%$$

22. Ans. (c)

Let he earns Rs. P

$x = -20$, y= –15, z = –30

$$P\left[1+\frac{x}{100}\right]\left[1+\frac{y}{100}\right]\left[1+\frac{z}{100}\right] = 9520$$

$$P\left[\frac{80}{100}\right] \times \left[\frac{85}{100}\right]\left[\frac{70}{100}\right] = 9520$$

Px. 476 = 9520, P = 20000

23. Ans. (d)

Let amount of ore is 100 kg

25% or 25 kg has 90% iron

$$= \frac{25 \times 90}{100} = 22.5 \text{ kg iron}$$

75% or 75 kg has no iron = 0 kg

To obtain 22.5 kg 100 kg ore is required

So to obtain 60 kg

$$\frac{100}{22.5} \times 60 = 266.66 \text{ kg}$$

24. Ans. (a)

Let Bobby's sale price is Rs. 100

Ram sale price = Rs. 80

Ram sale price is 20% dearer than chandilya So, Chandilya's price = CP

$$CP \times \frac{120}{100} = 80, \; CP = 66.66$$

Now Chandilya good is Rs. 33.33 cheaper than Bobby's i.e.

$$= \frac{33.63}{100} \times 100 = 33.33\%$$

25. Ans. (b)

Let x kg ore is there

20% washed away so remaining is

80% i.e. 4/5x

out of $\frac{4}{5}x$, 25% is pure iron $\frac{25}{100} \times \frac{4}{5}x = \frac{1}{5}x$

i.e. $\frac{1}{5}x$ kg is obtained from x kg then 1 kg is obtained from 5 kg

∴ 80000 kg is obtained from

5 × 80000 = 400000 kg

26. Ans. (a)

Let total salary is Rs. x

$$P\left[1+\frac{x}{100}\right]\left[1+\frac{y}{100}\right]\left[1+\frac{z}{100}\right] = 2500$$

$$P\left[1-\frac{30}{100}\right]\left[1-\frac{30}{100}\right]\left[1-\frac{24}{100}\right] = 2500$$

[(–)ve sign because of spending]

$$P\left[\frac{70}{100}\right]\left[\frac{70}{100}\right]\frac{76}{100} = 2500$$

P = Rs. 6713.21

■■■■

1.3 CHAPTER

Profit and Loss

In any organisation, the most important aspect is profit and loss calculation of any transaction conducted. So the profit/loss calculation is having a very important impact from business point of view.

While dealing with the profit/loss calculation, the following terminologies have to be understood.

Terminologies in Profit & Loss

Cost price: Cost Price (CP) of an article is the expenditure incurred on the purchase or the production of the article. It can be taken as the 'INVESTMENT' on the article.

Selling price: Selling price of an article is the revenue generated by selling that particular article. So it is the 'REVENUE'.

Both CP & SP are respective quantities. They are not absolute quantities i.e. the value keeps on changing with respect to the person.

If A is selling a quantity to B. In that case selling price for A is the cost price for B.

Profit/Loss: The difference of SP & CP is the profit value or loss value. If SP > CP, the value is profit. If SP < CP, the value is loss.

Percent Profit/Loss: The profit or loss as a percentage of the 'INVESTMENT' is profit or loss percentage.

$$\% \text{ Profit/loss} = \frac{\text{Profit/Loss Value}}{\text{Investment}} \times 100$$

Margin: Margin is generally used in terms of percentage. It is the profit or loss as percentage of the 'REVENUE' or selling price.

$$\% \text{ Margin} = \frac{\text{Profit/Loss value}}{\text{Revenue}} \times 100$$

Marked Price: It is the price of the product being displayed on the label. Generally it is taken as MRP.

Discount: Discount is the rebate given to the customer before the selling of the article. Generally discount is given on marked price but to have a loss the discount can be given on the cost price as well.

Mark up: This is the increment on the cost price before the article is sold to the customer.

From these terminologies it has to be understood that the relation of selling price and cost price is nothing but a percentage increment or decrement situation based on profit or loss respectively. So if profit or loss is P% we can use the multiplying factor

i.e. $\left(1 \pm \frac{P}{100}\right)$

So $SP = CP\left(1 \pm \frac{P}{100}\right)$

Now let us understand it with the help of some examples.

Example 1.

Ram bought a scooter for Rs. 10000 and sold it for 20% loss to Rahul. Rahul invested Rs. 500 in it and sold at 10% profit to Sohan. Find out the amount paid by Sohan.

Solution.

The CP for Rahul

$$= 10000\left(1 - \frac{20}{100}\right)$$

= Rs. 8000

Net investment of Rahul on the scooter

= 8000 + 500 = Rs. 8500

Rahul sold it for 10% profit

So amount paid by Sohan

$$= 8500\left(1 + \frac{10}{100}\right) = \text{Rs. } 9350$$

Example 2.

Komal sold two articles at same selling price one at 20% profit while another at 20% loss. Find out his net profit or loss percent?

Solution.

Let the SP of one article be x

One article is sold at 20% profit so

$$1.2 \times (CP)_1 = x$$

or $(CP)_1 = \frac{x}{1.2}$

Similarly $0.8 \times (CP)_2 = x$

or $(CP)_2 = \frac{x}{0.8}$

Net $CP = \frac{x}{1.2} + \frac{x}{0.8} = \frac{2x}{0.96}$

So loss $= \frac{2x}{0.96} - 2x$

Loss% $= \frac{2x(0.04)/0.96}{2x/0.96} \times 100$

= 4% loss

Example 3.

Rose bought a bullock and a cart in rupees 6000. She sold bullock, at 20% profit and cart at 10% profit and got Rs. 7000. Find out the cost price of bullock.

Solution.

Let the CP of bullock be 'b' & of cart be 'c'

Then b + c = 6000

and 1.2 b + 1.1c = 7000

By solving both the equation we get

b = 4000 and c = 2000

So CP of bullock is Rs. 4000.

Example 4.

A shopkeeper gives a discount of 10% on a article but his net profit becomes 8% still. Find out by what percent he mark up the article.

Solution.

Let the CP be x & the mark up be P%

Then SP $= \left(1+\frac{8}{100}\right)x$ (As per the profit)

SP $= \left(1+\frac{P}{100}\right)x\left(1-\frac{10}{100}\right)$ As per markup

(Since mark up & discounts are successive percent changes on CP)

So $\left(1+\frac{P}{100}\right)0.9x = 1.08x$

$\Rightarrow \left(1+\frac{P}{100}\right) = 1.2$

$\Rightarrow$ P = 20%

False Weight

Sometimes the shopkeeper keeps the CP & SP of the articles same. But instead of giving us the indicated quantity he supplies less quantity & makes profit. So for the calculation of profit percentage we have to see his investment & the investment is the quantity he is providing to the customer. So profit or loss percent is calculated on the amount being supplied to the customer.

Example 1.

A shopkeeper sells the quantity in the same price rate for which he has bought. But he gives 20% less quantity to the customer. Find his profit percent.

Solution.

Let the indicated weight be 100 gm & CP & SP be 1 Rs./gm

Now quantity given to the customer

= (1 – 0.2) × 100

So investment of shopkeeper

= 80 × 1 = Rs. 80

The amount gained from the customer

= 100 × 1 = Rs. 100

Profit % $= \frac{100-80}{80} \times 100 = 25\%$ profit

Example 2.

Ramesh purchased a radioset at Rs. 1500 and sold it at Rs. 1200. Find loss incurred by him?

Loss= C.P. – SP

= 1500 – 1200 = 300.

Also in this case we can calculate

Loss% $= \frac{300}{1500} \times 100 = 20\%$

Thus he incurred 20% loss.

Basic Formulae

1. When SP and Gain% are Given then

$$CP = \left(\frac{100}{100+\text{Gain\%}}\right) \times S.P.$$

2. When the C.P. and Gain % are given then

$$S.P. = \frac{100+\text{Gain\%}}{100} \times C.P.$$

3. When C.P. and loss% are given then

$$SP = \frac{100-\text{Loss\%}}{100} \times C.P.$$

4. When S.P. and loss percentages are given

$$CP = \left(\frac{100}{100 - \text{Loss}\%}\right) \times 100$$

5. If the cost price (C.P.) of m articles is equal to selling price of n article, then

$$\% \text{ gain or loss } = \left[\frac{m-n}{n}\right] \times 100$$

If m >n, it is % gain and if m<n, it is % loss

6. When two different articles are sold at the same S.P., getting gain/loss of x% on the first and gain/loss of y% on the second, the overall % gain or % loss in the transaction is given by

$$\left[\frac{100(x+y)+2xy}{(100+x)+(100+y)}\right]\%$$

The above expression represent overall gain or loss according to its given (+)ve or (–ve).

7. When two different articles are sold at the same selling price getting gain of x% on the first and loss of x% on the second, then the overall % loss in the transaction is given by $\left(\frac{x}{10}\right)^2\%$

8. A merchant uses faulty measure and sells his goods at gain/loss of x%. The overall % gain or loss (g) is given by

$$\frac{100+g}{100+x} = \frac{\text{True measure}}{\text{Faulty measure}}$$

Note: If merchant sells his goods at cost price then $x = 0$.

Example 1.

A dishonest shopkeeper professes to sell his goods at the cost price but uses faulty measure. His 1 kg weight measures 950 gms only. Find his gain percent.

Solution:

Here, True measure = 1000 gms

False measure = 950 gms

Since the Shopkeeper sells the goods at cost price.

$\therefore \quad x = 0,$

$\therefore$ overall gain % is given by

$$\frac{\text{True measure}}{\text{Faulty measure}} = \frac{100+g}{100+x}$$

$$\Rightarrow \frac{1000}{950} = \frac{1000 \times g}{950}$$

So, $100 + g = \frac{1000 \times 100}{950}$, $g = 5\frac{5}{19}\%$.

□□□□

Discount

9. If two successive discount of an article are m% and n% respectively, then a single discount equivalent to the successive discount will be $\left(m+n-\frac{mn}{100}\right)\%$

It can also be calculated as

$$\left[100 - 100 \times \frac{(100-m)}{100} \times \frac{(100-n)}{100}\right]\%$$

□□□□

Example 1.

Find the single discount which is equivalent to a successive discounts of 50% and 40%.

Solution:

Single discount will be equal to

$$\left(m+n-\frac{mn}{100}\right)\%$$

$$\Rightarrow \left(50+40-\frac{50 \times 40}{100}\right)\%$$

$\Rightarrow$ 70%

Example 2.

Find a single discount which is equivalent to three successive discounts of 20%, 30% and 20%.

Solution:

Single discount equivalent to 20% and 30% is

$$\Rightarrow \left[20+30-\frac{20 \times 30}{100}\right]\%$$

$\Rightarrow$ 44%

Now, we will find single discount which is equivalent to two successive discounts of 44% and 20%.

$$\Rightarrow \left[44+20-\frac{44 \times 20}{100}\right]\%$$

$\Rightarrow$ 55.2%

Alternative:

$x \times 0.8x \times 7x \times 8 = 0.448x$

So the final value is $0.448x$

Which is reduced by 55.2%

□□□□

Solved Examples

1. By selling a watch for Rs. 495, a shopkeeper incurs a loss of 10%. Find the cost price of the watch for the shopkeeper.
(a) Rs. 545 (b) Rs. 550
(c) Rs. 555 (d) None of these

Ans. (b)
Here S.P. = 495
Loss = 10%

$$C.P. = \frac{SP}{(100 - Loss\%)} \times 100$$

$$CP = \frac{495}{90} \times 100 = Rs.\ 550$$

2. A shopkeeper sold goods for Rs. 2400 and made a profit of 25% in the process. Find his profit percent if he had sold his goods for Rs. 2040.
(a) 6.25% (b) 7%
(c) 6.20% (d) 6.5%

Ans. (a)
SP = 2400, Profit% = 25

$$C.P. = \frac{SP}{(100 + P\%)} \times 100 = \frac{2400}{125} \times 100 = 1920$$

If sold at 2040, profit = Rs. 120

$$\text{Profit \%} \Rightarrow \frac{120}{1920} \times 100 = 6.25$$

3. A digital diary is sold for Rs. 935 at a profit of 10%. What would have been the actual profit or loss on it, if it had been sold for Rs. 810?
(a) Rs. 45 (b) Rs. 40
(c) Rs. 48 (d) Rs. 50

Ans. (b)
SP = 935, Profit %= 10%

$$CP = \frac{935}{110} \times 100 = 850$$

Diary if sold at 810, Loss = Rs. 40

4. A shopkeeper bought 240 chocolates at Rs. 9 per dozen. If he sold all of them at Rs. 1 each, what was his profit percent?
(a) 66(1/6)% (b) 33(1/3)%
(c) 24% (d) 27%

Ans. (b)
CP of 12 chocolate = Rs. 9

$$\text{CP of 1 chocolate} = \frac{9}{12} = Rs.\ 0.75$$

Now SP = Rs. 1 Proift = Rs. 0.25

$$\text{Profit \%} = \frac{0.25}{0.75} \times 100 = 33\frac{1}{3}\%$$

5. A coal merchant makes a profit of 20% by selling coal at Rs. 25 per quintal. If he sells the coal at Rs. 22.50 per quintal, what is his profit percent on the whole investment?
(a) 6% (b) 6.66%
(c) 7.5% (d) 8%

Ans. (d)
Profit % = 20%, SP=25

$$CP = \frac{25}{120} \times 100 = \frac{125}{6}\ Rs. = 20.83$$

Profit if SP = 22.50
= 22.50 – 20.83 = 1.667

$$\text{Profit \%} = \frac{1.667}{20.83} \times 100 = 8\%$$

[in fractional term this can be solved very easily]

6. The cost price of a shirt and trouser is Rs. 371. If the shirt costs 12% more than the trousers, find the cost price of the trouser.
(a) Rs. 125 (b) Rs. 150
(c) Rs. 175 (d) Rs. 200

Ans. (c)
Let CP of trouser be Rs. x

$$\text{Now CP of shirt} = Rs. \frac{112x}{100}$$

$$\text{According to given condition } x + \frac{112x}{100} = 371,$$

$$\frac{212}{100}x = 371,$$

x = Rs. 175

7. The marked price of a table is Rs. 1200, which is 20% above the cost price. It is sold at a discount of 10% on the marked price. Find the profit percent.
(a) 10% (b) 8%
(c) 7.5% (d) 6%

Ans. (b)
MP = 1200

$$C.P. = \frac{1200}{120} \times 100 = 1000$$

If sold at 10% discount

$$S.P. = 1200 \times \frac{90}{100} = 1080$$

Profit on CP = Rs. 80

$$\text{Profit \%} = \frac{80}{1000} \times 100 = 8\%$$

8. A dozen pairs of gloves quoted at Rs. 80 are available at a discount of 10%. Find how many pairs of golves can be bought for Rs. 24?
(a) 4 (b) 5
(c) 6 (d) 8

Ans. (a)
12 pairs of gloves are available at Rs. 80 (with 10% discount) $80 \times \frac{90}{100} = \text{Rs.}72$

$\because$ In Rs. 72, 12 pairs are available then in Rs. 1 $\frac{12}{72}$ pairs

$\therefore$ In 24 Rs. $\frac{12}{72} \times 24 = 4$

9. How much percent more than the cost price should a shopkeeper mark his goods, so that after allowing a discount of 12.5% he should have a gain of 5% on his outlay?
(a) 9.375 (b) 16.66%
(c) 20% (d) 25%

Ans. (c)
Let the cost price be Rs. x and printed price be Rs. y.
Hence, price after giving a discount of 12.5%

$$= y - y \times \frac{12.5}{100}$$

$$\text{Given, } = y - y \times \frac{12.5}{100} = x + x \times \frac{5}{100}$$

$$\Rightarrow y \times \frac{87.5}{100} = x \times \frac{105}{100}$$

$$\therefore y = \frac{105}{87.5}x \Rightarrow y - x = \frac{105}{87.5}x - x = \frac{17.5}{87.5}x$$

Required Percentage

$$= \frac{y-x}{x} \times 100 = \frac{17.5}{87.5} \times 100 = 20\%$$

10. In order to maintain the price line, a trader allows a discount of 10% on the marked price of goods in his shop. However, he still makes a gross profit of 17% on the cost price. Find the profit percent he would have made on the selling price had he sold at the marked price.
(a) 23.07 (b) 30%
(c) 21.21% (d) 25%

Ans. (b)
Let the cost price be Rs. x and marked price be Rs. y.

$$\text{Given, } y - y \times \frac{10}{100} = x + x \times \frac{17}{100}$$

$$\Rightarrow y \times \frac{90}{100} = x \times \frac{117}{100} \Rightarrow y = \frac{117}{90}x$$

$$\therefore \quad y - x = \frac{27}{90}x$$

$\therefore$ Required profit percent

$$= \frac{y-x}{x} \times 100 = \frac{27}{90} \times 100 = 30\%$$

11. A reduction of 10% in the price of sugar enables a housewife to buy 6.2 kg more for Rs. 279. Find the reduced price per kilogram.
(a) Rs. 5 (b) Rs. 4.5
(c) Rs. 4.05 (d) None of these

Ans. (b)
Let original rate = Rs. x per kg
New rate = 90% of

$$x = \text{Rs.}\left(\frac{90}{100}x\right) = \text{Rs.}\frac{9x}{10}$$

$$\text{Original quantity for Rs. 279} = \frac{279}{x}$$

$$\text{New quantity} = 279 \times \frac{10}{9x} = \frac{310}{x}$$

$$\therefore \frac{310}{x} - \frac{279}{x} = 6.2 \quad \Rightarrow \frac{31}{x} = 6.2$$

$$\Rightarrow x = \frac{31}{6.2} = 5$$

$$\therefore \quad \text{Reduced Price} = \frac{9 \times 5}{10} = \text{Rs. 4.5 per kg}$$

12. A man sells an article at 5% above its cost price. If he had bought it at 5% less than what he paid for it and sold it for Rs. 2 less, he would have gained 10%. Find the cost price of the article.
(a) Rs. 500 (b) Rs. 360
(d) Rs. 425 (d) Rs. 400

Ans. (d)
Let the cost price of the article = Rs. x
Price when it is bought at 5% less than cost price

$$= x - x \times \frac{5}{100} = \frac{95}{100}x$$

Selling price when it is sold for Rs. 2 less.

$$= x + x \times \frac{5}{100} - 2 = \frac{105}{100}x - 2$$

Given, $\frac{105}{100}x - 2 - \frac{95}{100}x = \frac{95}{100}x \times \frac{10}{100}$

$\Rightarrow \frac{10}{100}x - 2 = \frac{95}{1000}x$

$\therefore \quad x = \frac{2 \times 1000}{5} = \text{Rs. } 400$

13. A briefcase was sold at a profit of 10%. If its cost price was 5% less and it was sold for Rs. 7 more, the gain would have been 20%. Find the cost price of the briefcase.
(a) Rs. 175 (b) Rs. 200
(c) Rs. 225 (d) Rs. 160

Ans. (a)
Let the cost price = Rs. x

Price 5% less than cost price $= \frac{95}{100}x$

Selling price when sold for Rs. 7 more

$= \frac{110}{100}x + 7$

Given, $\frac{110}{100}x + 7 - \frac{95}{100}x = \frac{95x}{100} \times \frac{20}{100}$

$\Rightarrow \frac{20}{500}x = 7 \quad \therefore x = \frac{7 \times 500}{20} = \text{Rs.} 175$

14. A dishonest dealer professes to sell at cost price but uses a 900 gram weight instead of a 1 kilogram weight. Find the percent profit to the dealer.
(a) 10% (b) 11.11%
(c) 12.5% (d) None of these

Ans. (b)
Let the cost price be Rs. x per kg.

Then cost price of 900 gm $= \frac{9}{10}x$

Hence % profit

$$= \frac{x - \frac{9}{10}x}{\frac{9}{10}x} \times 100 = \frac{100}{9}\% = 11.11\%$$

□□□□

I Practice Exercise

1. Cost of 3 cricket balls = cost of 2 pairs of leg pads.
Cost of 3 pairs of leg pads = cost of 2 pairs of gloves.
Cost of 3 pairs of gloves = cost of 2 cricket bats.
If a cricket bat costs Rs. 54, what is the cost of a cricket ball?
(a) Rs. 12 (b) Rs. 14
(c) Rs. 16 (d) Rs. 18

2. There would be 10% loss if a toy is sold at Rs. 10.80 per piece. At what price should it be sold to earn a profit of 20%?
(a) Rs. 12 (b) Rs. 12.96
(c) Rs. 14.40 (d) None of these

3. If books bought at prices ranging from Rs. 200 to Rs. 350 are sold at prices ranging from Rs. 300 to Rs. 425, what is the greatest possible profit that might be made in selling eight books?
(a) Rs. 400
(b) Rs. 600
(c) Cannot be determined
(d) None of these

4. If the selling price of 18 articles is equal to the C.P. of 21 articles, the loss or gain percent is:
(a) $16\frac{2}{3}\%$ gain (b) $14\frac{2}{7}\%$ gain
(c) $16\frac{2}{3}\%$ loss (d) $14\frac{2}{7}\%$ loss

5. A man sold 250 chairs and had a gain equal to selling price of 50 chairs. His profit percent is:
(a) 5% (b) 10%
(c) 25% (d) 50%

6. If I purchased 11 books for Rs. 10 and sold all the books at the rate of 10 books for Rs. 11, the profit percent is
(a) 10% (b) 11%
(c) 21% (d) 100%

7. Ajay bought 15 kg of dal at the rate of Rs. 14.50 per kg and 10 kg at the rate of Rs. 13 per kg. He mixed the two and sold the mixture at the rate of Rs. 15 per kg. What was his total gain in this transaction?
(a) Rs. 1.10 (b) Rs. 16.50
(c) Rs. 11 (d) Rs. 27.50

8. Pure ghee costs Rs. 100 per kg. After adulterating it with vegetable oil costing Rs. 50 per kg, a shopkeeper sells the mixture at the rate of Rs. 96 per kg, thereby making a profit of 20%. In what ratio does he mix the two?
(a) 1 : 2 (b) 3 : 2
(c) 3 : 1 (d) None of these

9. A dealer professing to sell his goods at cost price, uses a 900 gm weight for a kilogram. His gain percent is
(a) 9 (b) 10
(c) 11 (d) $11\frac{1}{9}$

10. Toffees are bought at the rate of 3 for a rupee. To gain 50%, they must be sold at:
(a) 2 for a rupees (b) 1 for a rupees
(c) 4 for a rupee (d) 5 for a rupees

11. By selling 45 lemons for Rs. 40, a man loses 20%. How many should he sell for Rs. 24 to gain 20% in the transaction?
(a) 16 (b) 18
(c) 20 (d) 22

12. A man gains 10% by selling a certain article for a certain price. If he sells it at double the price, the profit made is
(a) 20% (b) 120%
(c) 100% (d) 140%

13. A sells a bicycle to B at a profit of 20% and B sells it to C at a profit of 25%. If C pays Rs. 1500, what did A pay for it?
(a) Rs. 825 (b) Rs. 1000
(c) Rs. 1100 (d) Rs. 1125

14. If the manufacturer gains 10%, the wholesale dealer 15% and the retailer 25%, then the cost of production of a table, whose retail price is Rs. 1265 is
(a) Rs. 632.50 (b) Rs. 800
(c) Rs. 814 (d) Rs. 834.34

15. Two mixers and one T.V. cost Rs. 7000, while two T.V.'s and a mixer cost Rs. 9800. The value of one T.V. is:
(a) Rs. 2800 (b) Rs. 2100
(c) Rs. 4200 (d) Rs. 8400

16. A horse and a cow were sold for Rs. 12000 each. The horse was sold at a loss of 20% and the cow at a gain of 20%. The entire transaction resulted in
(a) no loss or no gain
(b) loss of Rs. 1000
(c) gain of Rs. 1000
(d) gain of Rs. 2000

17. An Article is sold at a certain price. By selling it at $\frac{2}{3}$ of that price one loses 10%. The gain percent at original price is
(a) 20% (b) $33\frac{1}{3}\%$
(c) 35% (d) 40%

18. A grocer sells rice at a profit of 10% and uses weights which are 20% less than the market weight. The total gain earned by him will be
(a) 30% (b) 35%
(c) 37.5% (d) None of these

19. Two-third of a consignment was sold at a profit of 5% and the remainder at a loss of 2%. If the total profit was Rs. 400, the value of the consignment (in rupees) was:
(a) 20000 (b) 15000
(c) 12000 (d) 10000

20. A fruit seller has 24 kg of apples. He sells a part of these at 20% gain and the balance at a loss of 5%. If on the whole he earns a profit of 10%, the amount of apples sold at loss is:
(a) 6 kg (b) 4.6 kg
(c) 9.6 kg (d) 11.4 kg

21. The C. P. of an article is 40% of the S. P. The percent that the S.P. is of C.P. is:
(a) 250 (b) 240
(c) 60 (d) 40

22. If an article is sold at 5% gain instead of 5% loss, the seller gets Rs. 6.72 more. The C.P. of the article is
(a) Rs. 67.20 (b) Rs. 120
(c) Rs. 134.40 (d) Rs. 240

23. A man bought an article and sold it at a gain of 5%. If he had bought it at 5% less and sold it for Rs. 1 less, he would have made a profit of 10%. The C.P. of the article was:
(a) Rs. 100 (b) Rs. 150
(c) Rs. 200 (d) Rs. 500

24. Raghu bought 4 dozen oranges at Rs. 12 per dozen and 2 dozen oranges at Rs. 16 per dozen. He sold them all to earn 20% profit. At what price per dozen did he sell the oranges?

(a) Rs. 14.40 (b) Rs. 16
(c) Rs. 16.80 (d) Rs. 19.20

25. The profit earned by selling an article for Rs. 900 is double the loss incurred when the same article is sold for Rs. 450. At what price should the article be sold to make 25% profit?
(a) Rs. 600 (b) Rs. 750
(c) Rs. 800 (d) Data inadequate

26. A man sold an article for Rs. 75 and lost something. Had he sold it for Rs. 96, his gain would have been double the former loss. The C.P. of the article is:
(a) Rs. 81 (b) Rs. 82
(c) Rs. 83 (d) Rs. 85.50

27. A table is offered for Rs. 300 with 20% and 10% off. If in addition, a discount of 5% is offered on cash payment, then the cash price of the table is:
(a) Rs. 240 (b) Rs. 216
(c) Rs. 210 (d) Rs. 205.20

28. A tradesman marks his goods 30% above the C.P. If he allows a discount of $6\frac{1}{4}\%$, then his gain percent is:
(a) $23\frac{3}{4}\%$ (b) 22%
(c) $21\frac{7}{8}\%$ (d) None

29. What price should a shopkeeper mark on an article, costing him Rs. 153, to gain 20% after allowing a discount of 15%?
(a) Rs. 224 (b) Rs.216
(c) Rs. 184 (d) Rs.162

30. If the S.P. of Rs. 24 results in a 20% discount on list price, what S.P. would result in a 30% discount on list price?
(a) Rs. 27 (b) Rs. 21
(c) Rs. 20 (d) Rs. 9

31. A shopkeeper earns a profit of 12% on selling a book at 10% discount on the printed price. The ratio of the cost price to the printed price of the book is:
(a) 50 : 61 (b) 45 : 56
(c) 99 : 125 (d) 55 : 69

Answers

1. (c)	2. (c)	3. (d)	4. (c)	5. (c)
6. (c)	7. (d)	8. (b)	9. (d)	10. (a)
11. (b)	12. (b)	13. (b)	14. (b)	15. (c)
16. (b)	17. (c)	18. (c)	19. (b)	20. (c)
21. (a)	22. (a)	23. (c)	24. (b)	25. (b)
26. (b)	27. (d)	28. (c)	29. (b)	30. (b)
31. (b)				

Solutions

1. Ans. (c)

$3G = 54 \times 2 = 108 \Rightarrow G = 36.$

$3P = 36 \times 2 = 72 \Rightarrow P = 24$

$3C = 24 \times 2 = 48 \Rightarrow C = 16$

$\therefore$ Cost of a cricket ball = Rs. 16.

2. Ans. (c)

$90 : 10.80 = 120 : x$ or $\frac{90}{10.80} = \frac{120}{x}.$

$\therefore \quad x = \frac{120 \times 10.80}{90} = 14.40$

Hence, S.P. = Rs. 14.40

3. Ans. (d)

Least C.P. = Rs. (200 × 8)= Rs. 1600.
Greatest S.P. = Rs. (425 × 8) = Rs. 3400.
Required profit = Rs. (3400 –1600)
= Rs. 1800.

4. Ans. (c)

Let C.P. of each article be Rs. 1.
C.P. of 18 articles = Rs. 18.
S.P. of 18 articles = Rs. 21.

$\therefore \quad \text{Gain\%} = \left(\frac{3}{18} \times 100\right)\% = 16\frac{2}{3}\%.$

5. Ans. (c)

Gain = (S.P. of 250 chairs) – (C.P. of 250 chairs).
$\therefore$ (S.P. of 250 chairs) – (C.P. of 250 chairs) = S.P. of 50 chairs.
S.P. of 200 chairs = C.P. of 250 chairs.
Let C.P. of each chair be Rs. 1
C.P. of 200 chairs = Rs. 200.
S.P. of 200 chairs = Rs. 250.

$\text{Gain\%} = \left(\frac{50}{200} \times 100\right)\% = 25\%.$

6. Ans. (c)

Suppose, the number of books purchased is 110

C.P. of 110 books $= \text{Rs.}\left(\frac{10}{11} \times 110\right)$

= Rs. 100

S.P. of 110 books $= \text{Rs.}\left(\frac{11}{10}\times 110\right)$

= Rs. 121, Profit % = 21%

7. **Ans. (d)**

C.P. of 25 kg = Rs. (15 × 14.50 + 10 × 13)
= Rs. 347.50
S.P. of 25 kg = Rs. (25 × 15) = Rs. 375.
∴ Gain = Rs. (375 – 347.50) = Rs. 27.50

8. **Ans. (b)**

Mean price $= \text{Rs.}\left(\frac{100}{120}\times 96\right)$ = Rs. 80/kg.

By the rule of alligation:

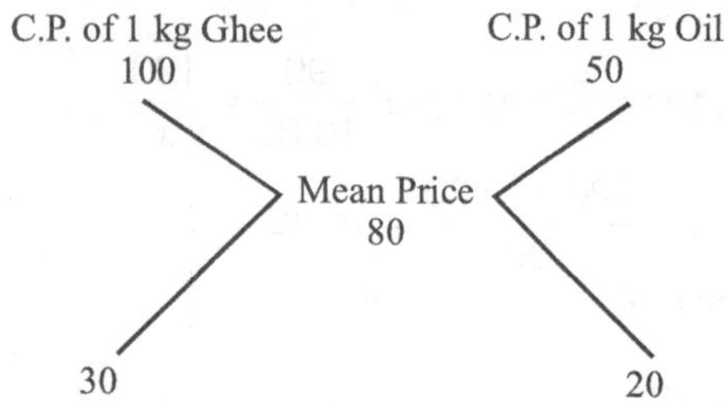

∴ Required ratio = 30 : 20 = 3 : 2.

9. **Ans. (d)**

Gain % $= \left(\frac{100}{900}\times 100\right)\% = 11\frac{1}{9}\%.$

10. **Ans. (a)**

C. P. of 3 toffees = Rs. 1

S. P. of 3 toffees = 150% of Re 1 = Rs. $\frac{3}{2}$

For Rs. $\frac{3}{2}$, toffees sold = 3.

For Rs. 1, toffees sold $= \left(3\times\frac{2}{3}\right) = 2.$

11. **Ans. (b)**

Let S.P. of 45 lemons be Rs. x

$80 : 40 = 120 : x$ or $\frac{80}{40} = \frac{120}{x}$

or $x = \frac{40\times 120}{80} = 60$

For Rs. 60, lemons sold = 45
For Rs. 24,

lemons sold $= \left(\frac{45}{60}\times 24\right) = 18.$

12. **Ans. (b)**

$110 : x = (100 + P) : 2x$ or

$\frac{110}{x} = \frac{100+p}{2x}$ or $100 + P = 220$

∴ p = 120%.

13. **Ans. (b)**

125% of 120% of A = 1500

$\Rightarrow \frac{125}{100}\times\frac{120}{100}A = 1500$

∴ $A = \left(1500\times\frac{2}{3}\right) = 1000$

14. **Ans. (b)**

125% of 115% of 110% of p = 1265.

∴ $\frac{125}{100}\times\frac{115}{100}\times\frac{110}{100}p = 1265$

or $\frac{253}{160}p = 1265.$

∴ $p = \left(\frac{1265\times 160}{253}\right) = \text{Rs. } 800$

15. **Ans. (c)**

Let C.P. of a mixer be Rs. x and that of a T.V. be Rs. y.

Then, $2x + y = 7000$ and $2y + x = 9800$.

Multiplying 2nd equation by 2 and subtracting first from it, we get

3y = 19600 – 7000 = 12600 or y = 4200

∴ C. P. of a T. V. = Rs. 4200

16. **Ans. (b)**

Total S.P. = Rs. 24000

C.P. of horse

$= \text{Rs.}\left(\frac{100}{80}\times 12000\right) = \text{Rs. } 15000.$

C.P. of a cow

$= \text{Rs.}\left(\frac{100}{120}\times 12000\right) = \text{Rs. } 10000.$

Total C.P. = Rs. 25000.

∴ Loss = 24000 – 25000 = Rs. 1000

17. **Ans. (c)**

Let original S.P. be Rs. x,

New S.P. $= \frac{2}{3}x$, loss = 10%

$\therefore$ C.P. $= \left(\frac{100}{90} \times \frac{2}{3}x\right) = \frac{20x}{27}$

New, C.P. $= \frac{20x}{27}$, S.P. = Rs. x,

Gain $= \left(x - \frac{20x}{27}\right) = \frac{7x}{27}$

$\therefore$ Gain% $= \left(\frac{7x}{27} \times \frac{27}{20x} \times 100\right)\% = 35\%$.

18. **Ans. (c)**

Let us consider a packet of rice marked 1 kg.

Its actual weight is 80% of 1000 gm = 800 gm.

Let C.P. of each gm be Rs. 1.

Then, C. P. of this packet = Rs. 800.

S.P. of this packet = 110% of C.P. of 1 kg.

$$= \left(\frac{110}{100} \times 1000\right) = \text{Rs. } 1100$$

$\therefore$ Gain $= \left(\frac{300}{800} \times 100\right)\% = 37.5\%$

19. **Ans. (b)**

Let the total value be Rs. x

Value of $\frac{2}{3}$rd $= \frac{2x}{3}$, Value of $\frac{1}{3}$rd $= \frac{x}{3}$.

Total S.P. $= \left(105\% \text{ of } \frac{2x}{3} + 98\% \text{ of } \frac{x}{3}\right)$

$$= \left(\frac{210x}{300} + \frac{98x}{300}\right) = \frac{308x}{300}$$

$$\frac{308x}{300} - x = 400 \Rightarrow \frac{308x - 300x}{300} = 400$$

$\therefore$ $x = \frac{300 \times 400}{8} = 15000$

20. **Ans. (c)**

Suppose, the quantity sold at a loss be x kg and let C.P. per kg be Rs. 1

Total C. P. = Rs. 24.

Total S. P. = 120% of $(24 - x)$ + 95% of x.

$$= \frac{6}{5}(24 - x) + \frac{19x}{20} = \frac{576 - 24x + 19x}{20}$$

$$= \frac{576 - 5x}{20}$$

$\therefore$ $\frac{576 - 5x}{20} = 110\%$ of 24

or $\frac{576 - 5x}{20} = \frac{264}{10}$

or $576 - 5x = 528$ or $5x = 48$

or $x = 9.6$ kg

21. **Ans. (a)**

C.P. $= \frac{40}{100} \times$ S.P. i.e. S.P. $= \frac{5}{2}$C.P.

$= \left(\frac{5}{2} \times 100\right)\%$ of C.P.

$\therefore$ S.P. = 250% of C.P.

22. **Ans. (a)**

Let C.P. be Rs. x

(105% of x) – (95% of x) = 6.72

or 10% of x = 6.72.

23. **Ans. (c)**

Let original C.P. be Rs. x/Its

S.P. $= \frac{105}{100}x = \frac{21x}{20}$

New C.P. $= \frac{95}{100}x = \frac{19x}{20}$

New S.P. $= \frac{110}{100} \times \frac{19x}{20} = \frac{209x}{200}$

$\frac{21x}{20} - \frac{209x}{200} = 1$ or $x = 200$

24. **Ans. (b)**

Total C.P. = Rs. (12 × 4 + 16 × 2) = Rs. 80.

S.P. of 6 dozen oranges

$$= \text{Rs.}\left(\frac{120}{100} \times 80\right) = \text{Rs. } 96.$$

S. P. per dozen = Rs. 16

25. **Ans. (b)**

Let C.P. be Rs. x

$900 - x = 2(x - 450)$

$\Rightarrow 3x = 1800 \Rightarrow x = 600$.

C.P. = Rs. 600, gain required = 25%.

$\therefore$ S.P. $= \text{Rs.}\left(\frac{125}{100}\times 600\right) = \text{Rs. } 750$

26. Ans. (b)

Let C.P. be Rs. x.

$2(x-75) = (96-x)$

$\Rightarrow 3x = 246$

$\Rightarrow x = 82$

27. Ans. (d)

Cost price = 95% of 90% of 80% of Rs. 300.

$= \left(\frac{95}{100}\times\frac{90}{100}\times\frac{80}{100}\times 300\right) = \text{Rs. } 205.20$

28. Ans. (c)

Let C.P. be Rs. 100. Then, marked price = Rs. 130.

S.P. $= \left(100-\frac{25}{4}\right)\%$ of Rs. 130

$= \left(\frac{375}{400}\times 130\right) = 121.875$

$\therefore$ Profit% $= 21.875 = \frac{21875}{1000} = 21\frac{7}{8}\%$

29. Ans. (b)

C.P. = Rs. 153, Gain = 20%.

$\therefore$ S.P. $= \left(\frac{120}{100}\times 153\right) = \text{Rs. } 183.60$

Let, the marked price be Rs. x

$\frac{85}{100}x = 183.60$

$\Rightarrow x = \frac{183.60\times 100}{85} = 216$

30. Ans. (b)

Let, the list price be Rs. x

$\frac{80}{100}x = 24 \Rightarrow x = \frac{24\times 100}{80} = 30$

Required S.P. = 70% of Rs.30 = Rs. 21

31. Ans. (b)

Let the C.P. be Rs. 100.

Then, S.P. = Rs. 112.

Let the printed price be Rs. x

Then, 90% of $x = 112 \Rightarrow \frac{90}{100}x = 112$

$\therefore x = \left(\frac{112\times 100}{90}\right) = \frac{1120}{9}$

$\therefore$ (C.P.) : (Printed price) $= 100 : \frac{1120}{9}$

$= 900 : 1120 = 45 : 56$

1.4 CHAPTER

Simple Interest & Compound Interest

Simple Interest

It is calculated on the basis of a basic amount borrowed for the entire period at a particular rate of interest.

The amount borrowed is the principal for the entire period of borrowing.

$$SI = \frac{P \times R \times T}{100}$$

Where P is principal amount

R is % rate of interest

T is the time duration

Compound Interest

The interest of the previous years are added to the principal for the calculation of the compound interest.

$$C.I. = P\left[1 + \frac{R}{100}\right]^T - p$$

Where P is principal amount

R is Rate of interest

T is time duration

Case1: When interest is compounded annually then amount A will be

$$A = \left[1 + \frac{R}{100}\right]^T$$

Case 2: When interest is compounded half yearly then

$$A = \left[1 + \frac{R/2}{100}\right]^{2T}$$

Case 3: When interest is compounded quarterly

$$A = \left[1 + \frac{R/4}{100}\right]^{4T}$$

Case 4: When differential rate of interest is charged i.e. if rate of interest is

R_1% for first year.

R_2% for second year and

R_3% for third year then

$$\text{Amount} = P\left[1 + \frac{R_1}{100}\right] \times \left[1 + \frac{R_2}{100}\right]\left[1 + \frac{R_3}{100}\right]$$

□□□□

Solved Examples-1

Example 1.

A sum of 12000 Rs. is deposited into SBI Bank for 3 years. If the bank is providing 5% rate then calculate the amount after the maturity period.

Solution:

$$SI = \frac{PRT}{100}$$

$$SI = \frac{12000 \times 5 \times 3}{100} = \text{Rs. } 1800$$

Amount = Principal + SI

13800 = 12000 + 1800

Example 2.

An amount become double in 8 years calculate the rate of interest.

Solution:

Let the amount be Rs. x

Since it become double in 8 years so SI in 8 years = Rs. x

Now $SI = \frac{P \times R \times T}{100}$, $x = \frac{x \times R \times 8}{100}$

$$\therefore \quad R = \frac{100}{8} = 12\frac{1}{2}\%$$

Example 3.

An amount become 5 times in 20 years at simple interest. Calculate the rate of interest given.

Solution:

Principal = Rs. x

Total Amount = Rs. $5x$. So

SI = Rs. $4x$

Now $SI = \frac{PRT}{100}$, $4x = \frac{x \times R \times 20}{100}$

$$R = \frac{400}{20} = 20\%$$

Example: 4.

An amount of 25000 is deposited in to Citi bank for 2 years. Calculate the interest incurred if rate of interest is 10% compounded annually.

Solution:

Here P= Rs. 25000

R= 10%

R = 2 years

$$CI = P\left[1+\frac{R}{100}\right]^T - P$$

$$= 25000\left[1+\frac{10}{100}\right]^2 - 25000$$

= 30250 – 25000 = Rs. 5250

□□□□

Solved Examples-2

1. Rs. 1200 is lent out at 5% per annum simple interest for 3 years. Find the amount after 3 years.
   (a) Rs. 1380 (b) Rs. 1290
   (c) Rs. 1470 (d) Rs. 1200

   Ans. (a)

   Here P = Rs. 1200

   R = 5%, T = 3 years

   $$SI = \frac{PRT}{100} = \frac{1200\times5\times3}{100} = \text{Rs. } 180$$

   Amount = P + SI = 1200 + 180 = 1380

2. Rs. 2100 is lent at compound interest of 5% per annum for 2 years. Find the amount after two years.
   (a) Rs. 2300 (b) Rs. 2315.25
   (c) Rs. 2310 (d) None of these

   Ans. (b)

   Here P = 2100, R = 5%, T = 2 years.

   $$A = P\left[1+\frac{R}{100}\right]^T = 2100\left[1+\frac{5}{100}\right]^2$$

   $$= 2100\times\frac{21}{20}\times\frac{21}{20} = \text{Rs. } 2315.25$$

3. Find the difference between the simple and the compound interest at 5% per annum for 2 years on a principal of Rs. 2000.
   (a) 5 (b) 105
   (c) 4.5 (d) None of these

   Ans. (a)

   Here P = Rs. 2000, R = 5%, T = 2 years

   $$CI - SI = \left[P\left[1+\frac{R}{100}\right]^T - P\right] - \frac{PRT}{100}$$

   $$= \left\{\left[2000\times\frac{105}{100}\times\frac{105}{100}\right]^2 - 2000\right\}$$

   $$-\frac{2000\times5\times2}{100}$$

   = 205 – 200 = Rs. 5

   Alternative:

   Difference between CI and SI for two years

   $$(CI - SI)_{2Y} = P\left(\frac{R}{100}\right)^2 = (200)\left(\frac{5}{100}\right)^2 = 5$$

4. What is the difference between the simple interest on a principal of Rs. 500 being calculated at 5% per annum for 3 years and 4% per annum for 4 years?
   (a) Rs. 5 (b) Rs. 10
   (c) Rs. 20 (d) Rs. 40

   Ans. (a)

   P = Rs. 500, R_1 = 5%, R_2 = 4%,

   T_1 = 3 years, T_2 = 4 years

   $$\text{Difference} = \frac{PR_1T_1}{100} - \frac{PR_2T_2}{100} = \text{Rs. } 5.$$

5. What is the simple interest for five years on a sum of Rs. 700 if the rate of interest for the first 3 years is 8% per annum and for another 2 years is 7% per annum?
   (a) 400 (b) 392
   (c) 352 (d) 266

   Ans. (d)

   $$\frac{700\times3\times8}{100} + \frac{700\times7\times2}{100} = 266$$

6. Find the compound interest on Rs. 1000 at the rate of 20% per annum for 18 months when interest is compounded half-yearly.
   (a) Rs. 331 (b) Rs. 1331
   (c) Rs. 320 (d) None of these

   Ans. (a)

   Here P = Rs. 1000, R = 20%,

   T = 18 months or $1\frac{1}{2}$ years.

Compound interest when interest is compounded annually

$$CI = 1000\left[1+\frac{R/2}{100}\right]^{2T}$$

$$= 1000\left[1+\frac{10}{100}\right]^{3} = \text{Rs. } 1331$$

7. Find the principal if the interest compounded at the rate of 10% per annum for two years is Rs. 420.
(a) Rs. 2000 (b) Rs. 2200
(c) Rs. 1000 (d) Rs. 1100

Ans. (a)

P = ?, CI = Rs. 420, R = 10%, T = 2 years.

$$CI = P\left[1+\frac{R}{100}\right]^{T} - P$$

$$420 = P\left[\left(1+\frac{10}{100}\right)^{2} - 1\right]$$

$$= P\left[\frac{21}{100}\right]$$

$$\therefore \quad P = \frac{420\times 100}{21} = \text{Rs. } 2000$$

8. What is the rate of simple interest for the first 4 years if the sum of Rs. 360 becomes Rs. 540 in 9 years and the rate of interest for the last 5 years is 6%?
(a) 4% (b) 5%
(c) 3% (d) 6%

Ans. (b)

P = Rs. 360, A = Rs. 540, SI = Rs. 180.

$$SI = \frac{PRT}{100}$$

$$180 = \frac{360\times R\times 4}{100} + \frac{360\times 6\times 5}{100}$$

$$72 = \frac{360\times R\times 4}{100}, \ R = 5\%$$

9. Ajay borrows Rs. 1500 from two money-lenders. He pays interest at the rate of 12% per annum for one loan and at the rate of 14% per annum for the other. The total interest he pays for the entire year is Rs. 186. How much does he borrow at the rate of 12%?
(a) Rs. 1200 (b) Rs. 1300
(c) Rs. 1400 (d) Rs. 300

Ans. (a)

P = 1500, R_1 = 12%, R_2 = 14%, T = 1 year

SI = 186

Let Rs. x be amount invested at 12% Rate

$$= \frac{x\times 12\times 1}{100} + \frac{(1500-x)\times 14}{100} = 186$$

$$\frac{21000}{100} - \frac{2x}{100} = 186, \ x = 1200$$

10. Two equal sums were borrowed at 8% simple interest per annum for 2 years and 3 years respectively. The difference on the interest was Rs. 56. The sum borrowed were
(a) Rs. 690 (b) Rs. 700
(c) Rs. 740 (d) Rs. 780

Ans. (b)

$$\frac{PRT_1}{100} - \frac{PRT_2}{100}$$

$$P\left[\frac{8\times 3 - 8\times 2}{100}\right] = 56, \ P = 700$$

11. If the difference between the simple interest and compound interest on some principal amount at 20% per annum for 3 years in Rs. 48, then the principle amount must be
(a) Rs. 550 (b) Rs. 500
(c) Rs. 375 (d) Rs. 400

Ans. (c)

Here P = ?, R = 20%, T = 3 year

Difference Rs. 48

$$48 = P\left[1+\frac{R}{100}\right]^{T} - P - \frac{PRT}{100}$$

$$48 = P\left[\left(1+\frac{20}{100}\right)^{3} - 1 - \frac{20\times 3}{100}\right]$$

On solving P = 375.

12. Raju lent Rs. 400 to Ajay for 2 years, and Rs. 100 to Manoj for 4 years and received together from both Rs. 60 as interest. Find the rate of interest, simple interest being calculated.
(a) 5% (b) 6%
(c) 8% (d) 9%

Ans. (a)

$$\frac{P_1R_1T_1}{100} + \frac{P_2R_2T_2}{100} = 60$$

$$\frac{400\times 2\times R}{100} + \frac{100\times 4\times R}{100} = 60$$

12R = 60, R = 5%

13. A sum of money doubles itself in 5 years. In how many years will it become four fold (if interest is compounded)?
(a) 15 (b) 10
(c) 20 (d) 12
Ans. (b)
Let sum = x

Then, $2x = x\left[1+\frac{R}{100}\right]^5$

$\Rightarrow \left[1+\frac{R}{100}\right] = 2^{1/5}$ ...(i)

$4x = x\left[1+\frac{R}{100}\right]^T$

$\Rightarrow 4 = \left[2^{1/5}\right]^T$

$\Rightarrow 2^2 = 2^{T/5}$

$\Rightarrow \frac{T}{5} = 2$

$\therefore T = 10$ yrs.

14. A sum of money placed at compound interest doubles itself in 3 years. In how many years will it amount to 8 times itself?
(a) 9 years (b) 8 years
(c) 27 years (d) 7 years
Ans. (a)
Let sum = x

Then, $2x = x\left[1+\frac{R}{100}\right]^3$

$\Rightarrow \left[1+\frac{R}{100}\right] = 2^{1/3}$ ...(i)

$8x = x\left[1+\frac{R}{100}\right]^T$

$\Rightarrow 8 = \left[2^{1/3}\right]^T$

$\Rightarrow 2^3 = 2^{T/3}$

$\Rightarrow \frac{T}{3} = 3$

$\therefore T = 9$ yrs.

Alternative:
If certain sum of money becomes 'm' times in 'y' years. Then it will become (m^n) times in 'n × y' years. Hence 2^3 in 3 × 3 = 9 years.

15. Divided Rs. 6000 into two parts so that simple interest on the first part for 2 years at 6% p.a. may be equal to the simple interest on the second part for 3 years at 8% p.a.
(a) Rs. 4000, Rs. 2000
(b) Rs. 5000, Rs. 1000
(c) Rs. 3000, Rs. 3000
(d) None of these
Ans. (a)
Let one part be Rs. x then

$\frac{x\times 2\times 6}{100} = \frac{(6000-x)\times 3\times 8}{100}$

$12x = 144000 - 24x$
$36x = 144000$
$x = 4000$

16. A sum of money becomes 7/4 of itself in 6 years at a certain rate of simple interest. Find the rate of interest.
(a) 12% (b) 12.5%
(c) 8% (d) 14%
Ans. (b)
Let sum be x

$A = \frac{7}{4}x$, So, $SI = \frac{3}{4}x$

T = 6 years, R = ?

$R = \frac{SI\times 100}{P\times T} = \frac{\frac{3}{4}x\times 100}{x\times 6} = 12.5\%$

17. Sanjay borrowed Rs. 900 at 4% p.a. and Rs. 1100 at 5% p.a. for the same duration. He had to pay Rs. 364 in all as interest. What is the time period in years?
(a) 5 years (b) 3 years
(c) 2 years (d) 4 years
Ans. (d)

$\frac{P_1R_1T_1}{100} + \frac{P_2R_2T_2}{100} = 364$

$\left[\frac{900\times 4T}{100} + \frac{1100\times 5T}{100}\right] = 364$

36 T + 55 T = 364
T = 4 years

18. If a certain sum of money becomes double at simple interest in 12 years, what would be the rate of interest per annum?
(a) $8\frac{1}{3}$ (b) 10
(c) 12 (d) 14

Ans. (a)

$x = \frac{x \times 12 \times R}{100}$

$R = \frac{100}{12} = 8\frac{1}{3}$

19. A sum of Rs. 600 amounts to Rs. 720 in 4 years at simple Interest. What will it amount to if the rate of interest is increased by 2%?
(a) Rs. 648 (b) Rs. 768
(c) Rs. 726 (d) Rs. 792

Ans. (b)

P = Rs. 600, A = Rs. 720
T = 4 years, R = ?

$R = \frac{S.I. \times 100}{P \times T} = \frac{120 \times 100}{600 \times 4} = 5\%$

at 7% Rate

$SI = \frac{600 \times 7 \times 4}{100} = 168$

A = 600 + 168 = 768

□□□□

Simple Interest

I Practice Exercise

1. Avinash borrowed Rs. 5000 from Sanjay at simple Interest. After 3years, Sanjay got Rs. 300 more than what he had given to Avinash. What was the rate of interest per annum?
(a) 2% (b) 5%
(c) 8% (d) 10%

2. Rs. 800 amounts to Rs. 920 in 3 years at simple interest. If the interest rate is increased by 3%, it would amount to how much?
(a) Rs. 992 (b) Rs. 1056
(c) Rs. 1112 (d) Rs. 1182

3. The simple interest on a sum of money at 5% is Rs. 48 for 4 years. The simple interest on the same sum for 5 years at 4% will be.
(a) Rs. 40 (b) Rs. 48
(c) Rs. 50 (d) Rs. 60

4. A sum of money at simple interest amounts to Rs. 2240 in 2 years and to Rs. 2600 in 5 years. What is the principal amount?
(a) Rs. 1520 (b) Rs. 1880
(c) Rs. 2120 (d) None

5. The simple interest on a certain sum of money at the rate of 5% p.a. for 8 years in Rs. 840. At what rate of interest the same amount of interest can be received on the same sum after 5 years?
(a) 6% (b) 8%
(c) 9% (d) 10%

6. A sum of money was lent at simple interest at 11% p.a. for $3\frac{1}{2}$ years and $4\frac{1}{2}$ years respectively. If the difference in interests for two period was Rs. 412.50, the sum is:
(a) Rs. 3250 (b) Rs. 3500
(c) Rs. 3750 (d) Rs. 4250

7. Prabhat took a certain amount as a loan from a bank at the rate of 8% p.a. S.I. and gave the same amount to Ashish as a loan at the rate of 12% p.a. If at the end of 12 years, he made a profit of Rs. 320 in the deal, what was the original amount.
(a) Rs. 2000 (b) Rs. 3000
(c) Rs. 4000 (d) None of these

8. Rahul borrowed Rs. 830 from Mr. Lal at 12% p.a. S.I. for 3 years. He then added some more money to the borrowed sum and lent it to Shobha for the same period at 14% p.a. rate of interest. If Rahul gains Rs. 93.90 in the whole transaction, how much money did he add from his side?
(a) Rs. 35 (b) Rs. 55
(c) Rs. 80 (d) Rs. 105

9. Mr. Roopchand finds that an increase in the rate of interest from $4\frac{7}{8}\%$ to $5\frac{1}{8}\%$ per annum increases his yearly income by Rs. 25. His investment is:
(a) Rs. 10,000 (b) Rs. 12,000
(c) Rs. 15,000 (d) Rs. 20,000

10. The rate at which a sum becomes four times of itself in 15 years at S.I. will be:
(a) 15% (b) $17\frac{1}{2}\%$
(c) 20% (d) 25%

11. The simple interest on a sum of money at the end of four years is $\frac{1}{5}$th of its principal. What is the rate of interest per annum?
(a) 4% (b) 5%
(c) 6% (d) Data inadequate

12. A sum of money triples itself in 15 years 6 months. In how many years would it double itself?
(a) 6 years 3 months
(b) 7 years 9 months
(c) 8 years 3 months
(d) 9 years 6 months

13. A sum of Rs.10 is lent to be returned in 11 monthly installments of Rs. 1 each, interest being simple. The rate of interest is:
(a) $9\frac{1}{11}\%$ (b) 10%
(c) 11% (d) $21\frac{9}{11}\%$

14. The rate of simple interest on a sum of money is 6% p.a. for the first 3 years, 8% p.a. for the next 5 years and 10% p.a. for the period beyond 8 years. If the simple interest accrued by the sum for a total period of 10 years is Rs. 1560, what is the sum?
(a) Rs. 1500 (b) Rs. 2000
(c) Rs. 3000 (d) Data inadequate

15. A monthly installment of Rs. 180 is required to be paid for repayment of an interest free loan in 40 months. If it is decided to pay it in 30 months, how much will be the monthly installment in rupees?
(a) 60 (b) 198
(c) 240 (d) 330

16. A sum of Rs. 1550 is lent out into two parts, one at 8% and another one at 6%. If the total annual income is Rs. 106, the money lent at 8% is:
(a) Rs. 650 (b) Rs. 720
(c) Rs. 840 (d) Rs. 900

17. A sum of Rs. 1550 was lent partly at 5% and partly at 8% p.a. simple interest. The total interest received after 3 years was Rs. 300. The ratio of the money lent at 5% to that lent at 8%.
(a) 8 : 5 (b) 5 : 8
(c) 31 : 6 (d) 16 : 15

18. A man invests an amount of Rs. 15860 in the names of his three sons A, B and C in such a way that they get the same amount after 2, 3 and 4 years respectively. If the rate of simple interest is 5%, then the ratio of amounts invested among, A, B and C will be:
(a) 10 : 15 : 20 (b) 22 : 23 : 24
(c) 6 : 4 : 3 (d) 2 : 3 : 4

Answers

1. (a)	2. (a)	3. (b)	4. (d)	5. (b)
6. (c)	7. (d)	8. (d)	9. (a)	10. (c)
11. (b)	12. (b)	13. (d)	14. (b)	15. (c)
16. (a)	17. (d)	18. (c)		

Solutions

1. Ans. (a)

$$\text{Rate} = \left(\frac{100 \times 300}{5000 \times 3}\right)\% = 2\%.$$

2. Ans. (a)

Principal = Rs. 800, S.I. = Rs. 120,
Time = 3 years.

$$\therefore \quad \text{Rate} = \left(\frac{100 \times 120}{800 \times 3}\right)\% = 5\%.$$

New rate = 8%, Principal = Rs. 800
Time 3 years.

$$\text{S.I.} = \text{Rs.}\left(\frac{800 \times 8 \times 3}{100}\right) = \text{Rs. } 192.$$

$\therefore$ New amount = Rs. (800 + 192) = Rs. 992

3. Ans. (b)

$$\text{Principal} = \text{Rs.}\left(\frac{100 \times 48}{5 \times 4}\right) = \text{Rs.}240.$$

$$\text{SI} = \text{Rs.}\left(\frac{240 \times 5 \times 4}{100}\right) = \text{Rs. } 48$$

4. Ans. (d)

S.I. for = Rs. (2600 – 2240) = Rs. 360.

$$\text{S.I. For 2 years} = \text{Rs. } \left(\frac{360}{3} \times 2\right) = \text{Rs.}240$$

$\therefore$ Principal = Rs. (2240 – 240) = Rs. 2000

5. Ans. (b)

$$\text{Sum} = \left(\frac{100 \times 840}{5 \times 8}\right) = \text{Rs. } 2100.$$

Rate required $= \left(\frac{100\times 840}{2100\times 5}\right)\% = 8\%.$

6. **Ans. (c)**

Let the sum be Rs. x. Then,

$$\left(x\times 11\times\frac{9}{2}\times\frac{1}{100} - x\times 11\times\frac{7}{2}\times\frac{1}{100}\right)$$

$= 412.50$

or $\frac{22x}{200} = 412.50$

$\Rightarrow 11x = 41250$

$\Rightarrow x = 3750$

7. **Ans. (d)**

Let the original amounts be Rs. x. then,

$$\frac{x\times 12\times 12}{100} - \frac{x\times 8\times 12}{100} = 320$$

$\Rightarrow x = \frac{2000}{3} = \text{Rs. } 666.67$

8. **Ans. (d)**

$$\frac{(830+x)\times 14\times 3}{100} - \frac{830\times 12\times 3}{100} = 93.90$$

or $830\times 42 + 42x - 830\times 36 = 9390$

$42x = 9390 - 4980$

$\Rightarrow x = \frac{4410}{42} = 105.$

$\therefore$ Money added = Rs. 105.

9. **Ans. (a)**

Let the investment be Rs. x, Then,

$$x\times\frac{41}{8}\times\frac{1}{100} - x\times\frac{39}{8}\times\frac{1}{100} = 25$$

$\Rightarrow 2x = 20000 \quad \Rightarrow x = 10000.$

10. **Ans. (c)**

Let sum $= x$, Then, S.I. $= 3x$

$\therefore$ Rate $= \left(\frac{100\times 3x}{x\times 15}\right)\% = 20\%$

11. **Ans. (b)**

Let sum $= x$, Then, S.I. $= \frac{x}{5}$, Time = 4 years.

$\therefore$ Rate $= \left(100\times\frac{x}{5}\times\frac{1}{x\times 4}\right)\% = 5\%.$

12. **Ans. (b)**

Let sum $= x$,

Then, S.I. $= 2x$ & Time $= \frac{31}{2}$ years.

Rate $= \left(\frac{100\times 2x}{x}\times\frac{2}{31}\right)\% = \left(\frac{400}{31}\right)\%$

Now, sum $= x$, S.I.$= x$ & Rate $\left(\frac{400}{31}\right)\%$.

$\therefore$ Time $= \left(\frac{100\times x}{x}\times\frac{31}{400}\right)$ years

= 7 years 9 months

13. **Ans. (d)**

Rs. 10 + S.I. on Rs. 10 for 11 months

= Rs.11 + S. I. on Re1 for (1 + 2 + 3 + 4 + ... + 10) months

Rs. 10 + S.I. on Rs. 1. for 110 months

= Rs. 11 + S.I. on Rs. 1 for 55 months

S.I. on Rs. 1 for 55 months = Rs. 1.

$\therefore$ Rate $= \left(\frac{100\times 12}{1\times 55}\right)\% = 21\frac{9}{11}\%.$

14. **Ans. (b)**

Let the sum be Rs. x. Then

$$\frac{x\times 6\times 3}{100} + \frac{x\times 8\times 5}{100} + \frac{x\times 10\times 2}{100} = 1560$$

or $78x = 156000$

or $x = 2000$

15. **Ans. (c)**

$180\times 40 = x\times 30$ or $x = 240$

16. **Ans. (a)**

Let the money lent at 8% be Rs. x. then,

$$\frac{x\times 8\times 1}{100} + \frac{(1550-x)\times 6\times 1}{100} = 106$$

or $2x + 9300 = 10600$ or $x = 650.$

17. **Ans. (d)**

Let the sum at 5% be Rs. x. Then,

$$\frac{x\times 5\times 3}{100} + \frac{(1550-x)\times 8\times 3}{100} = 300$$

or $x = 800$

$$\frac{\text{Money at 5\%}}{\text{Money at 8\%}} = \frac{800}{(1550-800)}$$

$= \frac{800}{750} = \frac{16}{15}$.

18. Ans. (c)

Let the amounts invested be x, y, z respectively.

Then, $\frac{x \times 2 \times 5}{100} = \frac{y \times 3 \times 5}{100} = \frac{z \times 4 \times 5}{100} = k$.

$\therefore\ x = 10k,\ y = \frac{20}{3}k\ \&\ z = 5k$.

So, $x : y : z = 10k : \frac{20}{3}k : 5k$

$= 30 : 20 : 15 = 6 : 4 : 3$.

□□□□

Compound Interest

I Practice Exercise

1. The difference in compound interest, and simple interest on a certain amount at 10% per annum at the end of the third year is Rs. 620. What is the principal amount?
(a) Rs. 40000 (b) Rs. 12000
(c) Rs. 10000 (d) Rs. 20000

2. If the compound interest on a certain sum at $16\frac{2}{3}\%$ for 3 years is Rs. 1270, the simple interest on the same sum at the same rate and for the same period is:
(a) Rs. 1200 (b) Rs. 1165
(c) Rs. 1080 (d) Rs. 1220

3. What is principal amount which earns Rs. 132 as compound interest for the second year at 10% per annum?
(a) Rs. 1000 (b) Rs. 1200
(c) Rs. 1320 (d) Rs. 1188

4. A sum of money amounts to Rs. 4624 in 2 years and to Rs. 4913 in 3 years at compound interest. The sum is:
(a) Rs. 4096 (b) Rs. 4260
(c) Rs. 4335 (d) Rs. 4360

5. A sum of money at compound interest amounts to thrice itself in 3 years. In how many years will it be 9 times itself?
(a) 12 (b) 9
(c) 6 (d) 8

6. In how many years will a sum of Rs. 800 at 10% per annum compounded semiannually become Rs. 926.10?
(a) $2\frac{1}{2}$ (b) $1\frac{1}{2}$
(c) $2\frac{1}{3}$ (d) $1\frac{1}{3}$

7. To find out the total compound interest on a sum of money after 5 years, which of the following informations given in the statements P and Q will be sufficient?
P : The sum was Rs. 20000.
Q : The total amount of simple interest on the sum after 5 years was Rs. 4000.
(a) Only P is sufficient
(b) Only Q is sufficient
(c) Either P or Q is sufficient
(d) Both P & Q are needed

8. The least number of complete years in which a sum of money put out at 20% compound interest will be more than doubled is:
(a) 3 (b) 4
(c) 5 (d) 6

9. A tree increases annually by $\frac{1}{8}$th of its height. By how much will it increase after 2 years, if it stands today 64 cm high?
(a) 72 cm (b) 74 cm
(c) 75 cm (d) 81 cm

10. The compound interest on a sum for 2 years is Rs. 832 and the simple interest on the same sum for the same period is Rs. 800. The difference between the compound interest and the simple interest for 3 years will be :
(a) Rs. 48 (b) Rs. 66.56
(c) Rs. 98.56 (d) None of these

11. A sum of Rs. 1100 was taken as a loan. This is to be repaid in two equal installments. If the rate of interest be 20% compounded annually, then the

value of each installment is
(a) Rs. 842 (b) Rs. 792
(c) Rs. 720 (d) Rs. 700

12. What annual payment will discharge a debt of Rs. 7620 due in 3 years at $16\frac{2}{3}\%$ per annum compound interest?
(a) Rs. 2540 (b) Rs. 3430
(c) Rs. 3260 (d) Rs. 3380

Answers

1. (d)	2. (c)	3. (b)	4. (a)	5. (c)
6. (b)	7. (d)	8. (b)	9. (d)	10. (c)
11. (c)	12. (b)			

Solutions

1. **Ans. (d)**

Let the sum be Rs. x. Then,

$$\text{C.I.} = x\left(1+\frac{10}{100}\right)^3 - x$$

$$= \left(\frac{1331}{1000} - x\right) = \frac{331x}{1000}.$$

$$\text{S.I.} = \frac{x\times10\times3}{100} = \frac{3x}{10}$$

$$\therefore\ (\text{C.I.})-(\text{S.I.}) = \left(\frac{331x}{1000}-\frac{3x}{100}\right) = \frac{31x}{1000}.$$

$$\therefore\ \frac{31x}{1000} = 620$$

or $x = 20000$.

Hence, the principal amount is Rs. 20000.

Alternative:

$$(\text{CI}-\text{SI})_{3Y} = 3P\left(\frac{R}{100}\right)^2 + P\left(\frac{R}{100}\right)^3$$

2. **Ans. (c)**

Let the sum be Rs. x. Then,

$$\text{C.I.} = \left[x\times\left(1+\frac{50}{3\times100}\right)^3\right] - x$$

$$= \left(\frac{343x}{216} - x\right) = \frac{127x}{216}.$$

$$\therefore\ \frac{127x}{216} = 1270$$

or $x = \frac{1270\times216}{127} = 2160.$

Thus, the sum is Rs. 2160.

$$\therefore\ \text{S.I.} = \text{Rs.}\left(2160\times\frac{50}{3}\times3\times\frac{1}{100}\right)$$

= Rs. 1080.

3. **Ans. (b)**

Let the principal at the end of first year be Rs. x.

Then, $\frac{x\times10\times1}{100} = 132$ or $x = 1320$.

Now, let the original principal be Rs. P.

Then, amount after 1 year

$$= P + \frac{P\times10\times1}{100} = \frac{11P}{10}.$$

$$\therefore\ \frac{11P}{10} = 1320$$

or $P = \frac{1320\times10}{11} = \text{Rs. }1200.$

4. **Ans. (a)**

Interest on Rs. 4624 for 1 year
= Rs. (6083.50 – 5290) = Rs. 793.50

$$\therefore\ \text{Rate} = \left(\frac{100\times793.50}{4624\times1}\right)\% = 6\frac{1}{4}\%$$

$$\text{Now, } x\left(1+\frac{25}{4\times100}\right)^2 = 4624$$

or $x\times\frac{17}{16}\times\frac{17}{16} = 4624$

$$\therefore\ x = \left(4624\times\frac{16}{17}\times\frac{16}{17}\right) = \text{Rs. }4096.$$

5. **Ans. (c)**

$$P\left(1+\frac{R}{100}\right)^3 = 3P \Rightarrow \left(1+\frac{R}{100}\right)^3 = 3 \quad \text{...(i)}$$

$$\text{Let } P\left(1+\frac{R}{100}\right)^n = 9P \Rightarrow \left(1+\frac{R}{100}\right)^n = 9.$$

$$\therefore\ \left(1+\frac{R}{100}\right)^n = 3^2$$

$$= \left[\left(1+\frac{R}{100}\right)^3\right]^2 = \left(1+\frac{R}{100}\right)^6 \ \{\text{using (i)}\}$$

Hence, n = 6 years.

6. **Ans. (b)**

Rate = 5% per half year.

Let time = 2n half years = n years.

Then, $800\left(1+\frac{5}{100}\right)^{2n}=926.10$

$$\Rightarrow \left(\frac{21}{20}\right)^{2n}=\frac{926.10}{800}=\frac{9261}{8000}=\left(\frac{21}{20}\right)^{3}$$

$\therefore$ $2n=3$ or $n=\frac{3}{2}$ years.

7. **Ans. (d)**
Clearly, both P and Q together are needed. Using P and Q rate can be calculated as

$$\text{Rate}=\left(\frac{100\times 4000}{5\times 20000}\right)=4\%$$

Now, C.I. can be calculated.

8. **Ans. (b)**

$$P=\left(1+\frac{20}{100}\right)^{n}>2P \text{ or } \left(\frac{6}{5}\right)^{n}>2$$

Now $\left(\frac{6}{5}\times\frac{6}{5}\times\frac{6}{5}\times\frac{6}{5}\right)>2$
So, n = 4 years.

9. **Ans. (d)**

$$\text{Increase \%}=\left(\frac{1}{8}\times 100\right)\%=12\frac{1}{2}\%.$$

Height after 2 years

$$=\left[64\times\left(1+\frac{25}{2\times 100}\right)^{2}\right]\text{cm}$$

$$=\left(64\times\frac{9}{8}\times\frac{9}{8}\right)\text{cm}=81\,\text{cm}.$$

10. **Ans. (c)**
Diff. in C.I. & S.I. for 2 years = Rs. 32.
S.I. for 1 year = Rs. 400.
$\therefore$ S.I. on Rs. 400 for 1 year = Rs.32.

$$\therefore\ \text{Rate}=\left(\frac{100\times 32}{400\times 1}\right)\%=8\%.$$

Hence, diff. in C.I. and S.I. for 3rd year
= S.I. on Rs. 832

$$=\text{Rs.}\left(832\times\frac{8}{100}\times 1\right)=\text{Rs. }66.56.$$

Total difference = Rs. (32 + 66.56)
= Rs. 98.56

11. **Ans. (c)**
Let the value of each installment be Rs. x.
Then,
(P.W. of Rs. x due 1 year hence) + (P.W. of Rs. x due 2 years hence) = Rs. 1100.

$$=\frac{x}{\left(1+\frac{20}{100}\right)}+\frac{x}{\left(1+\frac{20}{100}\right)^{2}}=1100$$

or $\frac{5x}{6}+\frac{25x}{36}=1100$; $55x=36\times 1100$

$$\therefore\ x=\left(\frac{36\times 1100}{55}\right)=\text{Rs. }720.$$

12. **Ans. (b)**
Let each installment be Rs. x. Then,

$$\frac{x}{\left(1+\frac{50}{3\times 100}\right)}+\frac{x}{\left(1+\frac{50}{3\times 100}\right)^{2}}$$

$$+\frac{x}{\left(1+\frac{50}{3\times 100}\right)^{3}}=7620.$$

or $\frac{6x}{7}+\frac{36x}{49}+\frac{216x}{343}=7620.$
$294x+252x+216x=7620\times 343$

or $x=\frac{7620\times 343}{762}=3430.$

$\therefore$ Amount of each installment = Rs.3430.

1.5 CHAPTER Ratio, Proportion & Variation

Ratio

Ratio is the relationship between the quantities of same kind. In ratio the quantities are compared as the multiple or parts of other quantities. If ratio is a : b then a is called antecedent and b is called consequent. Ratio of the quantities is expressed after removing the common factor between the quantities.

Generally, the ratio is useful in comparison. If the quantities A & B are compared & their ratio comes out as p : q

We can say A/B = p/q

or A & B can be expressed as pK, qK respectively. This provides a scope of comparison in terms of multiples of p & q.

In Equalities in the Ratio

If a ratio is given as a : b & a quantity x is added in both antecedent & consequent then

$\rightarrow \frac{a+x}{b+x} > \frac{a}{b}$ if a < b ...(i)

$\rightarrow \frac{a+x}{b+x} < \frac{a}{b}$ if a > b ...(ii)

$\rightarrow \frac{a+x}{b+x} = \frac{a}{b}$ if a = b

(i) and (ii) was considering x = +ve

If x = –ve (i) and (ii) inequalities will reverse.

Ratio as a part of total value

If two quantities A & B are in the ratio p : q then

$$\frac{A}{B} = \frac{p}{q} \Rightarrow A = pK,\ B = qK$$

So $A + B = (p + q)K$

So $$\frac{A}{A+B} = \frac{p}{p+q}$$

or $$A = \frac{p}{p+q}(A+B)$$

So A will be $\frac{p}{p+q}$ part of total sum of the quantities

similarly B will be $\frac{q}{p+q}$ part of total sum of the quantities.

Example 1.

Given that for two quantities a & b, 2a = 3b. What will be the ratio (4a + 5b) : (2a + 3b)?

Solution:

$$2a = 3b \Rightarrow \frac{a}{b} = \frac{3}{2}$$

$$\Rightarrow a = 3K,\ b = 2K$$

so $$\frac{4a+5b}{2a+3b} = \frac{4\times 3K+5\times 2K}{2\times 3K+3\times 2K}$$

$$= \frac{22K}{12K} = \frac{11}{6}$$

So the required ratio is 11 : 6

Example 2.

If a : b = 3 : 5 & a : c = 6 : 7 then what is a : b : c ?

Solution.

a : b = 3 : 5

a : c = 6 : 7

In this case the common quantity is 'a' so we will make a's value as constant.

So we can say

a : b = 6 : 10 (multiplying by 2)

and a : c = 6 : 7

so a : b : c = 6 : 10 : 7

Example 3.

If $\frac{10a^2+ab}{3ab-b^2} = \frac{10}{1}$ what will be a : b

(a) 2 : 3 (b) 2 : 5
(c) 3 : 4 (d) 3 : 7

Solution.

$$\frac{10a^2+ab}{3ab-b^2} = \frac{10}{1}$$

or $10a^2 + ab = 30ab - 10b^2$

$$\Rightarrow \frac{10a^2 + ab}{ab} = \frac{30ab - 10b^2}{ab}$$

$$\Rightarrow 10\left(\frac{a}{b}\right) + 1 = 30 - a\left(\frac{b}{a}\right)$$

Assuming $\frac{a}{b} = x$

$$\Rightarrow 10x = 29 - 10\left(\frac{1}{x}\right)$$

or $10x^2 = 29x - 10$

$\Rightarrow 10x^2 - 29x + 10 = 0$

$$\Rightarrow x = \frac{10}{25} \text{ or } \frac{25}{10}$$

$$\Rightarrow x = \frac{2}{5} \text{ or } \frac{5}{2}$$

Hence option (b)

Types of Ratio

If two quantities A & B have their ratio as A : B = P : q then

- Duplicate ratio of A : B is $P^2 : q^2$
- Triplicate ratio of A : B is $P^3 : q^3$
- Sub-duplicate ratio of A : B is $\sqrt{P} : \sqrt{q}$
- Sub-triplicate of A : B is $\sqrt[3]{P} : \sqrt[3]{q}$

Example 1.

The ratio of two quantities A & B is 4 : 9 what is the triplicate ratio of sub-duplicate ratio of A & B.

Solution.

The sub-duplicate ratio of A & B is $\sqrt{4} : \sqrt{9}$

$= 2 : 3$

The triplicate ratio of 2 : 3 is $2^3 : 3^3$

$= 8 : 27$

Proportion

Proportion is comparison of two equal ratio. When the two ratios are equal then all the quantities comprising the ratios are called in proportion i.e.

$\frac{a}{b} = \frac{c}{d}$ then a, b, c & d are in proportion & 'd' is called as fourth proportion of a, b & c.

If three quantities a, b, c are taken then

$\frac{a}{b} = \frac{b}{c}$, here c is called as third proportion of a & b, b is called as second proportion or mean proportion of a & c.

so $b^2 = ac$

or $b = \sqrt{ac}$

so mean proportion is the geometric mean of the two quantities. Here the quantities b and c will be in 'continued proportion' similarly if a, b, c d are in continued proportion then

$$\frac{a}{b} = \frac{b}{c} = \frac{c}{d}$$

or in other way we can say that if the terms are in G.P. then they will be in continued proportion.

Example 1.

Find out the mean proportion of 3 & 27 also find out their third proportion.

Solution.

For mean proportion 3, x, 27 are in proportion

i.e. $\frac{3}{x} = \frac{x}{27}$

$\Rightarrow x^2 = 3 \times 27$ or $x = 9$

So mean proportion of 3 & 27 is 9

For third proportion 3, 27, x are in proportion

i.e. $\frac{3}{27} = \frac{27}{x}$

$\Rightarrow 3x = 27^2$

$\Rightarrow x = 243$

So third proportion of 3 & 27 is 243.

Applications of Proportion

Proportion is very much useful in removing the complexity of the problems in ratios. There are some basic rules that have to be taken care of:

if a, b, c, d are in proportion

i.e. $\frac{a}{b} = \frac{c}{d}$ then

- $\frac{a+b}{b} = \frac{c+d}{d}$ (componendo rule)
- $\frac{a-b}{b} = \frac{c-d}{d}$ (dividendo rule)
- $\frac{a+b}{a-b} = \frac{c+d}{c-d}$ (componendo-dividendo rule)
- $\frac{b}{a} = \frac{d}{c}$ (invertendo rule)
- $\frac{a}{c} = \frac{b}{d}$ (alternendo rule)

and if $\frac{a}{b} = \frac{c}{d} = \frac{e}{f} = \ldots$

then $\frac{a}{b} = \frac{c}{d} = \frac{e}{f} = \frac{a+c+e}{b+d+f}$

Variation

The basic meaning of variation is respective change. In arithmetic, use of variation makes problems very easy to approach. It is the relationship between the changes of two quantities.

Let us assume there are two quantities A & B if any change is done 'A' & that change results the change in B then the quantities are called in variation. One thing has to be remembered the change we are counting is multiple change not the differential change i.e. If a quantity is 60 and then it becomes 80 then the change will not be taken as (+20). It will be taken as 80/60 = 4/3.

- Direct variation : For two quantities a & b if the nature of changes are similar then they will be directly varying i.e.

if $a \propto b$

then if 'a' becomes twice then 'b' will also be twice. If a becomes 2/3 times then b will also be 2/3 times.

Example 1.

Two quantities A & B are varying directly. Initially, A was 30 & B was 50, what has to be the value of B if A becomes 24?

Solution.

Here $A \propto B$

Change in A is 24/30 times = 4/5 times

So the change in B will be the same

So B will be 4/5 times of 50 = 4/5 × 50 = 40

In other way we can make it as

$$A \propto B$$

or $A = BK$

or $A/B = \text{constant}$

$$\Rightarrow \quad \frac{A_1}{B_1} = \frac{A_2}{B_2}$$

$$\Rightarrow \quad \frac{30}{50} = \frac{24}{x}$$

$$\text{So} \quad x = \frac{24}{30} \times 50$$

$$\Rightarrow \quad x = 40$$

Inverse Variation

When in variation the nature of changes in the two quantities are inverse then the quantities will be in inverse variation?

i.e. $A \propto \frac{1}{B}$

then $A = \frac{K}{B}$

or $A.B = K = \text{const.}$

so $A_1B_1 = A_2B_2 = A_3B_3 = ...A_nb_n = K$

Example 1.

A car running with some speed increases its speed to 4/3 times & reaches its destination 30 minutes earlier. What time it will take if it runs with original speed?

Solution.

Here we know that speed ∝ 1/time

So we can use variation property

Let us assume the initial speed as 'S' & initial time as t

Now by the property of inverse variation

$(\text{speed})_1 \times (\text{time})_1 = (\text{speed})_2 \times (\text{time})_2$

$$S \times t = \frac{4}{3} S \times (\text{time})_2$$

$$\Rightarrow \quad (\text{time})_2 = \frac{3}{4} t$$

given that $(\text{time})_1 - (\text{time})_2 = 30$ mins.

$$t - \frac{3t}{4} = 30$$

$$\Rightarrow \quad \frac{t}{4} = 30 \text{ or } t = 120 \text{ mins.}$$

So the car will take 120 mins with original speed.

Compound Ratio & Concept of Partnership

If there are two ratio a : b and c : d then their compound ratio will be a × c : b × d.

This leads to the concept of partnerships the main part of the partnership is the division of profit between the partners. The profit among the partners is always divided in the compound ratio of investment ratio and time of investment ratio i.e.

If the partners' investment ratio is $P_1 : P_2 : P_3$ & the ratio of their investment time is $t_1 : t_2 : t_3$ then the ratio of profit = $P_1t_1 : P_2t_2 : P_3t_3$

So there may be three cases of the profit sharing.

Case 1.

If the ratio of investment is unity i.e. the invest equal amount then

$P_1 : P_2 : P_3 = 1 : 1 : 1$

So ratio of profit sharing will be $t_1 : t_2 : t_3$

Example 1.

It three persons A, B, C invests rupees 20000 in a business for 3 years, 2 years & 5 years respectively then what will be the ratio of their profits.

Solution.

Since the investment is same for all the persons so the profit will be divided in the ratio of

$1 \times 3 : 1 \times 2 : 1 \times 5$

i.e. 3 : 2 : 5 will be the profit ratio.

Case 2.

If the investment time is same for all the persons i.e. $t_1 : t_2 : t_3 = 1 : 1 : 1$, then the profit will be shared in the ratio of their investment i.e. the ratio of profit will be

$P_1 : P_2 : P_3$

Example 1.

A, B, C invest rupees 10000, 15000 & 20000 respectively in a business for 2 years each, what will be the ratio of their profit?

Solution.

Since the investment time is constant. So profit will be shared in the ratio of their investment i.e. ratio of profit

= 10 K : 15 K : 20 K

= 2 : 3 : 4

Case 3.

When the investment is variable we have to look for the effective investment value for the profit sharing.

Example 1.

A, B, C invested rupees 10000, 15000 & 20000 respectively. Next year A & B invested 5000 more while C withdrew 10000 rupees. What will be the ratio of their profit after 3 years?

Solution.

Here we will see the effective investment value of A, B & C respectively.

Effective Investment Value of A

= 10 K + 15 K + 15 K

= 40 K

EIV of B = 15 K + 20 K + 20 K

= 55 K

EIV of C = 20 K + 10 K + 10 K

= 40 K

So ratio of profit will be

40 K : 55 K : 40 K

= 8 : 11 : 8

Example 2.

A mixture contains alcohol and water in the ratio 4 : 1 by adding 10 litre of water the ratio becomes 2 : 1. Find out the amount of alcohol in the mixture

(a) 20 (b) 30

(c) 40 (d) 50

Ans. (c)

The amount of alcohol = 4 K

The amount of water = K

By adding water $\frac{4K}{K+10} = \frac{2}{1}$

$\Rightarrow$ 4 K = 2 K + 20

$\Rightarrow$ 2 K = 20 or K = 10

So amount of alcohol = 4 × 10 = 40 litre

Example 3.

The present ratio of ages of Ram & Shyam is 5 : 4. 18 years ago the ratio of their ages was 16 : 11. The present sum of their ages is

(a) 90 years (b) 105 years

(c) 80 years (d) 110 years

Ans. (a)

Let the ages be 5 K, 4 K

$$\frac{5K-18}{4K-18} = \frac{16}{11} \Rightarrow K = 10$$

So sum of the ages = 50 + 40 = 90

Hence option (a)

Example 4.

The income of Ram and Shyam are in the ratio 2 : 3 and their expenditures are in the ratio 3 : 5. If each saves Rs.1000 then Ram's income is

(a) 8000 (b) 6000

(c) 5000 (d) 4000

Ans. (d)

Let income be $2x$, $3x$

and expenditure be 3y, 5y

$2x - 3y = 1000$

$3x - 5y = 1000$

So $x = 2000$

So income is 4000

Hence option (d)

Solved Examples

1. Divide Rs. 500 among A, B, C and D so that A and B together get thrice as much as C and D together, B gets four times of what C gets and C gets 1.5 times as much as D. Now the value of what B gets is

(a) 300 (b) 75

(c) 125 (d) None of these

Ans. (a)

A + B + C + D = 500

Here A + B = 3 (C + D)

So 4 (C + D) = 500

C + D = 125

A + B = 375

also B = 4C and C = 1.5 D

C + D = 125

2.5 D = 125

D = 50, C = 75

$\therefore$ B = 300

2. If 4 examiners can examine a certain number of answer books in 8 days by working 5 hours a day, for how many hours a day would 2 examiners have to work in order to examine twice the number of answer books in 20 days.

(a) 6 (b) 7

(c) 8 (d) 9

Ans. (c)

Man. Day. Hour = constant

= 4 . 8. 5 = 160

Next ⟶ Man. day. hour = 160 × 2 (twice as earlier)

2. 20. h = 160 × 2

h = 8

3. In a mixture of 40 litres, the ratio of milk and water is 4 : 1. How much water must be added to this mixture so that the ratio of milk and water becomes 2 : 3?

(a) 20 litres (b) 32 litres

(c) 40 litres (d) 30 litres

Ans. (c)

Let water is x

$4x : x$ = milk : water

$32 : 8 = 4x : x$ (since total 40 liters)

Now $\frac{32}{8+x} = \frac{2}{3}$, x= 40 litre

4. The ratio between two numbers is 3 : 4 and their LCM is 180. The first number is:

(a) 60 (b) 45

(c) 15 (d) 20

Ans. (b)

Let number are $3x$ and $4x$

$3x \times 4x$ = HCF × 180

Clearly HCF will be x.

$12x^2 = x \times 180$, $x = 15$

numbers will be 45 : 60

5. The incomes of A and B are in the ratio 3 : 2 and their expenditures are in the ratio 5 : 3. If each saves Rs. 1000, then, A's income is

(a) Rs. 3000 (b) Rs. 4000

(c) Rs. 6000 (d) Rs. 9000

Ans.(c)

Let incomes are $3x : 2x$

expenditures are 5y : 3y

then

$3x - 5y = 1000$ (1) also

$2x - 3y = 1000$ (2)

from (1) and (2) we get

$6x - 10y = 2000$

$+6x - 9y = 3000$

$\Rightarrow -y = -1000$

y = 1000, x= 2000

A's income is 6000

6. Divide Rs. 680 among A, B and C such that A gets 2/3 of what B gets and B gets 1/4th of what C gets. Now the share of C is?

(a) Rs. 480 (b) Rs. 300

(c) Rs. 420 (d) None of these

Ans. (a)

680 = A + B + C

$A = \frac{2}{3}B$ and $B = \frac{1}{4}C$, 4B = C

$A = \frac{2}{3}B$, C = 4B

$680 = \frac{2}{3}B + B + 4B$

$680 = \frac{17}{3}B$, B = 120, C = 480

7. The students in three batches at Made Easy are in the ratio 2 : 3 : 5. If 20 students are increased in each batch, the ratio changes to 4 : 5 : 7. The total number of students in the three batches before the increase were.

(a) 10 (b) 90

(c) 100 (d) 150

Ans. (c)

Let students are $2x : 3x : 5x$ now

According to the given condition

$2x + 20 : 3x + 20 : 5x + 20 = 4 : 5 : 7$

$x = 10$

20 : 30 : 50 = 2 : 3 : 5 and after adding 20

40 : 50 : 70

8. The speeds of three cars are in the ratio 2 : 3 : 4. The ratio between the times taken by these cars to travel the same distance is

(a) 2 : 3 : 4 (b) 4 : 3 : 2

(c) 4 : 3 : 6 (d) 6 : 4 : 3

Ans. (d)

Speeds are in the ratio 2 : 3 : 4

ratio of time taken will be $\frac{1}{2}:\frac{1}{3}:\frac{1}{4}$

$= \frac{6:4:3}{12} = 6:4:3$

9. After an increment of 7 in both the numerator and denominator, a fraction changes to 3/4. Find the original fraction.

(a) 5/12 (b) 7/9
(c) 2/5 (d) 3/8

Ans. (c)

$\frac{x+7}{y+7} = \frac{3}{4}$

$4x + 28 = 3y + 21, \quad 4x + 7 = 3y,$

only $x : y = 2/5$ satisfies the given condition.

10. If Rs. 58 is divided among 150 children such that each girl and each boy gets 25 p and 50 p respectively. Then how many girls are there?

(a) 52 (b) 54
(c) 68 (d) 62

Ans. (c)

Let the number of girls and boys are x and y then

$0.25x + 0.5y = 58$ ...(1)

also

$x + y = 150$ ...(2)

from (1) and (2) we get

$0.5x + y = 116$

$x + y = 150$

So, $0.5x = 34$, $x = 68$, $y = 82$

11. A mixture contains milk and water in the ratio 5 : 1. On adding 5 litres of water, the ratio of milk to water becomes 5 : 2. The quantity of milk in the mixture is:

(a) 16 litres (b) 25 litres
(c) 32.5 litres (d) 22.75 litres

Ans. (b)

Let milk and water are $5x : x$

Now, $\frac{5x+5}{x+5} = \frac{5}{2}$

$\Rightarrow 10x = 5x + 25$

$5x = 25, \quad x = 5$

Milk : Water = 25 : 10

12. Vijay has coins of the denomination of Re. 1, 50 p and 25 p in the ratio of 12 : 10 : 7. The total worth of the coins he has is Rs. 75. Find the number of 25 p coins that Vijay has

(a) 48 (b) 72
(c) 60 (d) None of these

Ans. (d)

Coins are $12x : 10x : 7x$

It is given that $12x + 10x \times \frac{1}{2} + \frac{7}{4}x = 75$

$\frac{48x + 20x + 7x}{4} = 75$

$75x = 4 \times 75, \quad x = 4$

13. A varies jointly with B and C directly and A = 6 when B = 3, C = 2; find A when B = 5, C = 7.

(a) 17.5 (b) 35
(c) 70 (d) 105

Ans. (b)

Let A = k (BC) then

6 = k (3.2)

So, k = 1 then

A = k (BC)

= 1 × (5 × 7) = 35

14. If x varies as y directly, and as z inversely and $x = 14$, when y = 10; find z when $x = 49$, y = 45.

(a) 14/10
(b) 10
(c) 10/14
(d) Cannot be determined

Ans. (d)

$x = k_1 y \qquad x = \frac{k_2}{z}$

According to given values

$14 = k_1.10$

So, $k_1 = \frac{14}{10} = \frac{7}{5}$

also, $14 = \frac{k_2}{z}$

So, $k_2 = 14z$,

But value of k_2 is not given.

Hence Z cannot be determined.

15. A cask contains a mixture of 49 litres of wine and water in the proportion 5 : 2. How much water must be added to it so that the ratio of wine to water may be 7 : 4?

(a) 3.5 (b) 6
(c) 7 (d) None of these

Ans. (b)

Let wine and water are $5x$ and $2x$ litres

$5x + 2x = 49$

$5x = 35, \quad 2x = 14$

Now $\frac{35}{14+w} = \frac{7}{4}$

$98 + 7w = 140$

$7w = 42, w = 6$

□□□□

Ratio & Proportion

I Practice Exercise

1. Find a fourth proportional to the numbers 60, 48, 30.
(a) 36 (b) 24
(c) 48 (d) None of these

2. Find a third proportional to the numbers 4, 42.
(a) 441 (b) 541
(c) 641 (d) None of these

3. The prices of a scooter and a television set are in the ratio of 3 : 2. If a scooter costs Rs. 600 more than the television set, then the price of television set is:
(a) Rs. 1800 (b) Rs. 1200
(c) Rs. 2400 (d) None of these

4. If A : B = 7 : 5 and B : C = 9 : 11, then A : B : C is equal to
(a) 55:45:63 (b) 63:45:55
(c) 45:63:55 (d) None of these

5. The ratio of money with Anju and Sanju is 4 : 5 and that with Sanju and Manju is 5 : 6. If Anju has Rs. 280, then the amount of money Manju has
(a) Rs. 320 (b) Rs. 420
(c) Rs. 640 (d) None of these

6. The ratio of present ages of Suresh and Mahesh is 7 : 5. If after 6 years their ages will be in the ratio of 4 : 3, the present age of Mahesh is
(a) 32 years (b) 36 years
(c) 30 years (d) None of these

7. Two numbers are in the ratio of 5 : 7. If 25 be subtracted from each, they are in the ratio of 35 : 59. Find the difference of the two numbers.
(a) 48 (b) 52
(c) 24 (d) None of these

8. The number that must be added to each of the numbers 8, 21, 13 and 31 to make the ratio of first two numbers equal to the ratio of last two number is
(a) 5 (b) 7
(c) 9 (d) None of these

9. A mixture contains alcohol and water in the ratio of 12 : 5. On adding 14 litres of water, the ratio of alcohol to water becomes 4 : 3. The quantity of alcohol in the mixture is
(a) 18 litres (b) 42 litres
(c) 26 litres (d) None of these

Answers

1. (b) 2. (a) 3. (b) 4. (b) 5. (b)
6. (c) 7. (c) 8. (a) 9. (b)

Solutions

1. **Ans. (b)**
Let x be the fourth proportion, then

$60 : 48 : 38 : x$ or, $\frac{60}{48} = \frac{38}{x}$

$\therefore \quad x = \frac{30 \times 38}{60} = 24.$

2. **Ans. (a)**
Let x be the third proportion, then

$4 : 42 : : 42 : x$ or, $\frac{4}{42} = \frac{42}{x}$.

$\therefore \quad x = \frac{42 \times 42}{4} = 441.$

3. **Ans. (b)**
Here, a = 3, b = 2, and x = 600.
∴ The price of a television set

$= \frac{bx}{a-b} = \frac{2 \times 600}{3-2} = \text{Rs. } 1200.$

4. **Ans. (b)**
Here, $n_1 = 7$, $n_2 = 9$, $d_1 = 5$ and $d_2 = 11$.
$\therefore$ A : B : C = $(n_1 \times n_2) : (d_1 \times n_2) : (d_1 \times d_2)$
$= (7 \times 9) : (5 \times 9) : (5 \times 11)$
$= 63 : 45 : 55.$

5. **Ans. (b)**
We have, A : B = 4 : 5 and B : C = 5 : 6.
Here, $n_1 = 4$, $n_2 = 5$, $d_1 = 5$ and $d_2 = 6$.
$\therefore$ A : B : C = $(n_1 \times n_2) : (d_1 \times n_2) : (d_1 \times d_2)$
$= (4 \times 5) : (5 \times 5) : (5 \times 6)$
$= 20 : 25 : 30$ or, $4 : 5 : 6$.

Thus, ratio of money with Anju, Sanju and Manju is 4 : 5 : 6. Since Anju has Rs. 280, the amount of money Manju has

$= \frac{280}{4} \times 6 = \text{Rs. } 420.$

6. **Ans. (c)**
We have, a : b = 7 : 5, c : d = 4 : 3 and x=6.

∴ The present age of Mahesh $= \frac{bx(c-d)}{ad-bc}$

$= \frac{5 \times 6 \times (4-3)}{(7 \times 3 - 5 \times 4)} = 30$ years.

7. **Ans. (c)**
We have, a : b = 5 : 7, c : d = 35 : 59 and x = 25.

∴ The first number $= \frac{ax(d-c)}{ad-bc}$

$= \frac{5 \times 25 \times (59-35)}{(5 \times 59 - 7 \times 35)} = 60.$

and, the second number $= \frac{bx(d-c)}{ad-bc}$

$= \frac{7 \times 25 \times (59-35)}{(5 \times 59 - 7 \times 35)} = 84$

∴ The difference of two numbers
= 84 – 60 = 24

8. **Ans. (a)**
Here, a = 8, b = 21, c = 13 and d = 31.
∴ The required number

$= \frac{bc - ad}{(a+d)-(b+c)}$

$= \frac{21 \times 13 - 8 \times 31}{(8+31)-(21+13)} = 5.$

9. **Ans. (b)**
Ratio of alcohol and water 12 : 5
Let their qualities be 12x and 5x respectively.
After adding the litres of water ratio becomes 4 : 3

$$\frac{12x}{5x+14} = \frac{4}{3}$$

$$x = \frac{7}{2}$$

Quantity of alcohol $= 12 \times \frac{7}{2} = 42$ litres.

□□□□

Partnership

I Practice Exercise

1. Amit, Nitin and Ravindra entered into a partnership. Amit invested Rs. 16000 for 9 months. Nitin invested Rs. 12000 for 6 months and Ravindra invested Rs. 8000 for 12 months. At the end of a year there was a profit of Rs. 26000. Find the share of Nitin in the profit.
(a) Rs. 8000 (b) Rs. 7500
(c) Rs. 6000 (d) None of these

2. A starts business with Rs. 3500 and 5 months after B joins A as his partner. After a year the profits are divided in the ratio of 2 : 3. How much did B contribute?
(a) Rs. 7000 (b) Rs. 11000
(c) Rs. 9000 (d) None of these

3. A, B and C invested capitals in the ratio 3 : 5 : 9; the timing of their investments being in the ratio 2 : 3 : 1. In what ratio would their profits be distributed?
(a) 2 : 5 : 3 (b) 3 : 2 : 5
(c) 7 : 5 : 3 (d) None of these

4. A, B and C start a business. If the ratio of their periods of investments are 2 : 3 : 6 and their profits are in the ratio of 4 : 5 : 6, then the ratio of capitals of A, B and C is
(a) 6 : 8 : 10 (b) 12 : 10: 6
(c) 10 : 12 : 6 (d) None of these

5. A and B started a business with initial investments in the ratio 5 : 7. If after one year their profits were in the ratio 1 : 2 and the period for A's investment was 7 months, B invested the money for
(a) 6 months (b) 12
(c) 10 months (d) 4 months

6. A, B, C enter into a partnership with shares in the ratio $\frac{7}{2} : \frac{4}{3} : \frac{6}{5}$. After 4 months, A increase his share by 50%. If the total profit at the end of one year be Rs. 21600, then B's share in the profit is
(a) Rs. 2100 (b) Rs. 2400
(c) Rs. 3600 (d) Rs. 4000

7. B is a sleeping partner and A working. A puts in Rs. 5000 and B puts in 6000. A received 12.5% of profit for managing the business and rest is divided in proportion to their capitals. A's share of profit in a total profit of Rs. 880 is
(a) Rs. 350 (b) Rs. 400
(c) Rs. 420 (d) Rs. 460

8. A starts business with a capital of Rs. 1200. B and C join with some investments after 3 and 6 months, respectively. If at the end of a year, the profit is divided in the ratio 2 : 3 : 5 respectively, what is B's investment in the business?
(a) Rs. 2400 (b) Rs. 1800
(c) Rs. 3600 (d) Rs. 6000

Answers

1. (c)	2. (c)	3. (a)	4. (b)	5. (c)
6. (d)	7. (d)	8. (a)		

Solutions

1. **Ans. (c)**
Here, $C_1 = 16000$, $C_2 = 12000$, $C_3 = 8000$, $t_1 = 9$, $t_2 = 6$, $t_3 = 12$ and $P = 26000$.
Nitin's share in the profit

$$= \frac{C_2 \times t_2 \times P}{C_1t_1 + C_2t_2 + C_3t_3}$$

$$= \frac{12000 \times 6 \times 26000}{16000 \times 9 + 12000 \times 6 + 8000 \times 12}$$

$$= \frac{1872000000}{312000} = \text{Rs. } 6000.$$

2. **Ans. (c)**
We have, $C_1 \times t_1 = 3500 \times 12 = 42000$ and $C_2 \times t_2 = x \times 7 = 7x$.

Then, $\frac{\text{Profit for A}}{\text{Profit for B}} = \frac{C_1 \times t_1}{C_2 \times t_2}$

$$\Rightarrow \frac{2}{3} = \frac{42000}{7x}$$

or, $x = \frac{42000 \times 3}{2 \times 7} = \text{Rs. } 9000.$

3. **Ans. (a)**
Ratio of capitals of A, B and C are 3 : 5 : 9. Let the capitals of A, B and C be $3x$, $5x$ and $9x$, respectively. Ratio of timing of their investments are 2 : 3 : 1. Let A, B and C invest their capitals for 2y, 3y and y months, respectively.
Then, profit of A : profit of B : Profit of C
$= C_1 \times t_1 : C_2 \times t_2 : C_3 \times t_3$
$= 3x \times 2y : 5x \times 3y : 9x \times y$
$= 6 : 15 : 9$ or, $2 : 5 : 3$.

4. **Ans. (b)**
We have, $P_1 : P_2 : P_3 = 4 : 5 : 6$ and $t_1 : t_2 : t_3 = 2 : 3 : 6$.

$\therefore$ Required ratio $= \frac{P_1}{t_1} : \frac{P_2}{t_2} : \frac{P_3}{t_3} = \frac{4}{2} : \frac{5}{3} : \frac{6}{6}$

or, 12 : 10 : 6
Thus, A, B and C invested their capitals in the ratio 12 : 10 : 6.

5. **Ans. (c)**
Let investments of A and B respectively be $5x$ and $7x$ and period of B's investment be y months.

Then, $\frac{(5x) \times 7}{(7x) \times y} = \frac{1}{2} \Rightarrow y = 10.$

6. **Ans. (d)**
Given ratio $= \frac{7}{2} : \frac{4}{3} : \frac{6}{5} = 105 : 40 : 36$
Let the initially invest Rs. 105, Rs. 40 and Rs. 36, respectively.
Ratio of investments
$= [105 \times 4 + (150\% \text{ of } 105) \times 8] : (40 \times 12) : (36 \times 12)$
$= 1680 : 480 : 432 = 35 : 10 : 9$

B's share $= \text{Rs.}\left(21600 \times \frac{10}{54}\right) = \text{Rs. } 4000.$

7. **Ans. (d)**
A's share for managing the business
= 12.5% of Rs. 880 = Rs. 110
Remaining proift = Rs. 770.
Profit ratio of A and B = 5 : 6

A's share $= \frac{5}{11}$ of Rs. 770 = Rs. 350.

A's total profit= Rs. 350 + Rs.110 = Rs. 460.

8. **Ans. (a)**
Profit ratio of A, B and C is
$(1200 \times 12) : (x \times 9) : (y \times 6) = 2 : 3 : 5$

$$\Rightarrow \frac{1200 \times 12}{2} = \frac{9x}{3}$$

$\therefore$ $x = \text{Rs. } 2400.$

CHAPTER 1.6 Averages, Mixture & Alligation

Average

Average is a number which represents the general characteristics of a set of numbers. It is a very effective way of representing the entire group by a single value. Average of the group is defined as:

$$\text{Average} = \frac{\text{Sum of all items in the group}}{\text{No. of items}}$$

Basic Points related with Average

- If the value of each item is increased by the some value P, then the average of the group or items will also increase by P.

Example 1.

Let us take a group of 5 numbers 25, 30, 15, 35, 40 and each number is increased by 3, then find out the new average.

Solution:

$$\text{Initial average} = \frac{25+30+15+35+40}{5}$$

$$= \frac{145}{5} = 29$$

$$\text{Final average} = \frac{20+33+18+38+43}{5}$$

$$= \frac{160}{5} = 32$$

So final average = initial average +3

- If the group items are decreased by any common value then the average value will also decrease by same.
- If the group items are multiplied or divided by any constant value then the average value will also be multiplied or divided by the same.
- The average value will always lie between the smallest value of the group and the largest value of the group.

Alternate way of Calculation of Average

Let us say that a group contains n numbers P_1, P_2, P_3 .... Then take a value which can be assumed as the mean value. Take the deviations of mean value 'x' from the given values. If deviations are Q_1, Q_2, Q_3 .... Q_n. Then the average will be

$$x + \frac{Q_1 + Q_2 + Q_3 + ... + Q_n}{n}$$

Example 1.

6 Students have their weights as 60, 62, 63, 67, 69, 74 kg. Find their average weight.

Solution:

Let us assume the mean value as 67, then average will be

$$67 + \frac{-7-5-4+2+7}{6}$$

$$= 67 - \frac{7}{6} = 65.83 \text{ kg}$$

Concept of Weighted Average

When we talk about the average of two different groups or more than two different groups. The average value of all groups taken together will not be just depending upon the average of groups. For example if we take two different groups of average weights as 60 & 70 kg respectively then the average of all the groups element taken together need not be $\frac{60+70}{2} = 65$ kg always. It can vary based on the number of elements in the group. So in this case the numbers of group elements plays a very important role, let the first group has 20 persons while the second group has 30 persons. Then the average will be

$$\frac{\text{Total weight}}{\text{Total numbers}} = \frac{20\times60+30\times70}{20+30}$$

$$= \frac{1200+2100}{50} = \frac{3300}{50} = 66 \text{ kg}$$

This sort of average is called as weighted average and weighted average is calculated as

$$\text{average} = \frac{n_1A_1 + n_1A_2 + n_3A_3 + \ldots}{n_1 + n_2 + n_3 + \ldots}$$

where n_1, n_2, n_3... are the number of elements in the groups. While A_1, A_2, A_3.... are the average of the groups.

Instead of exact value of numbers of elements we can also take the ratio of the number of elements.

Example 1.

The average marks of students of class A is 37, while that of class B is 23. If the number of students in class A & B are 42 & 63 respectively then find out the average marks of all the students taken together.

Solution:

$$\text{Average} = \frac{2 \times 37 + 3 \times 23}{2+3}$$

$$(\because \text{since } 42{:}63 = 2{:}3)$$

$$= \frac{74+69}{5} = \frac{143}{5} = 28.6$$

The weighted average has wide range of applications. The weightage need not be necessarily the number of elements. It can be any characteristic of group that affect the group's standing in the total. We will see some examples here.

Example 2.

If two alloys having the ratio of copper & gold in the ratio 3 : 4 and 2 : 5 are mixed and 14 kg of first alloy is mixed with 21 kg of the second alloy. Find out the new ratio of gold and copper in the mixture.

Solution:

The ratio of mixing = 14 : 21 = 2 : 3

So this will be the weightage for the mixing.

Now if we talk, about the quantity of gold in both the alloy.

It is $\frac{4}{7}$ and $\frac{5}{7}$.

So the amount of gold in the mixture

$$= \frac{\frac{2 \times 4}{7} + \frac{3 \times 5}{7}}{2+3} = \frac{23}{35}$$

So the ratio of gold to copper is 23 : 12.

Mixture & Alligation technique

When it comes to the mixing of two different quantities. Weighted average comes into the picture. If two solutions of concentration c_1 and c_2 are mixed in the ratio m : n. Then the weightage concentration c will be

$$c = \frac{m \times c_1 + n \times c_2}{m+n}$$

or we can rearrange the formula as

$$\frac{m}{n} = \frac{c_2 - c}{c - c_1}$$

This is called as alligation formula. It can also be understood with pictorial representation.

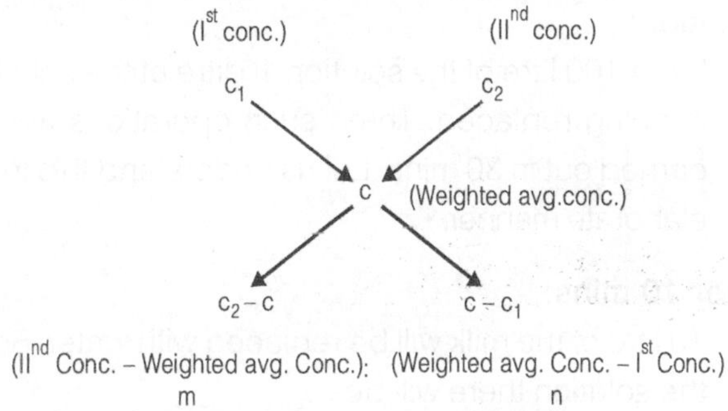

So it can be understood as the alligation technique provides us the ratio of mixing directly. Alligation technique has to be used when the weighted average is given & weighted average formula has to be used when weighted average is not given. Let us understand it better with some examples.

Example 1.

How much water has to be added in 60 litre of 70% milk solution. To make it as 40% milk solution.

Solution:

The concentration of water will be 30% and 60% respectively.

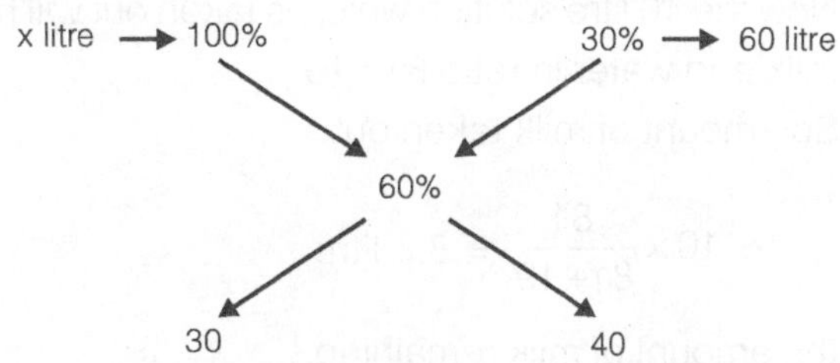

So ratio of mixing is 3 : 4

$$\Rightarrow \frac{x}{60} = \frac{3}{4}$$

$$\Rightarrow x = 45 \text{ litres}$$

Replacement

In case of replacement problem, one part of a solution is taken out and is replaced with another ingredient. We have to look for the ingredient having constant trend and the ingredient which is continuously decreasing will keep on decreasing with the same percentage value. Let us understand this with one example.

Example 1.

A container has 100 litre of pure milk, after every 10 minutes 10 litre of the solution is taken out and is replaced with water find out what will be the net amount of milk, after 30 mins. in the solution?

Solution:

From 100 litre of the solution 10 litre of the solution is being replaced. Three such operations will be carried out in 30 mins. Let us understand this in an elaborate manner.

After 10 mins:

10 litre of the milk will be replaced with water. So in the solution there will be

Milk = 90 litre

Water = 10 litre

After 20 minutes:

Now when the 10 litre solution will be taken out, it will be having milk and water in ratio 9 : 1.

So out of 10 litre solution milk taken out is = 9 litre

remaining amount of milk = 90 – 9 = 81 litre

the remaining portion will be water so the amount of water = 100 – 81 = 19 litre

Now ratio of milk and water = 81 : 19

After 30 minutes:

Now the 10 litre solution which is taken out will have milk and water in ratio 81 : 19

So amount of milk taken out

$$= 10\times\frac{81}{81+19} = 8.1 \text{ litre}$$

the amount of milk remaining

= 81 – 8.1 = 72.9 litre

the amount of water = 100 – 72.9 = 27.1 litre

Now let us take the amount of milk and water for the previous example.

Milk	Water	
100	0	
90	10	→ After Ist replacement
81	19	→ After 2nd replacement
72.9	27.1	→ After 3rd replacement

Total volume of the solution = 100 litre

total replaced volume = 10 litre

Now we can understand in every cycle 10% of the total solution is being replaced.

Talking about the flow of ingredients, milk has on outward flow (only being taken out not being added) While water has both inward & outward flow. So it we see the trend of the single flow ingredient, the value of milk is continuously decreasing by 10% successively

90 = 100 (1 – 0.1)

$81 = 100 (1 - 0.1)^2$

$72.9 = 100 (1 - 0.1)^3$

Thus, it has to be kept in mind that

"During replacement if the solution that is being replaced is P% of the total solution, then the ingredient having single-directional flow will decrease by P% after every cycle."

This is the basic idea behind replacement.

Example 2.

A person makes a profit of 15% on 20% of the quantity sold while on the rest he makes a profit of 5%. Find out the net profit percentage?

Solution:

Let us say the total quantity is x

20% of the quantity = $\frac{x}{5}$

80% of the quantity = $\frac{4x}{5}$

on $\frac{x}{5}$ the profit made = $\frac{15}{100}\times\frac{x}{5} = \frac{3x}{100}$

on $\frac{4x}{5}$ the profit made = $\frac{5}{100}\times\frac{4x}{5} = \frac{4x}{100}$

Net profit = $\frac{7x}{100}$ = 7% of x.

Alternate

The same problem can be taken as the two parts are 20% & 80% i.e. they are in the ratio 1 : 4 and he is making different profit so net profit percent

$$= \frac{1\times15+4\times5}{1+4} = \frac{35}{5} = 7\%$$

Example 3.

Ramesh deposits some amount in BIS bank at 7% interest rate p.a. & some part of amount in IBU bank, at 9% p.a. & he gets Rs. 850 on Rs. 10000 yearly. Find out the amount invested in BIS bank.

Solution:

850 out of 100000 means the net interest rate

$$= \frac{850}{10000}\times100 = 8.5\%$$

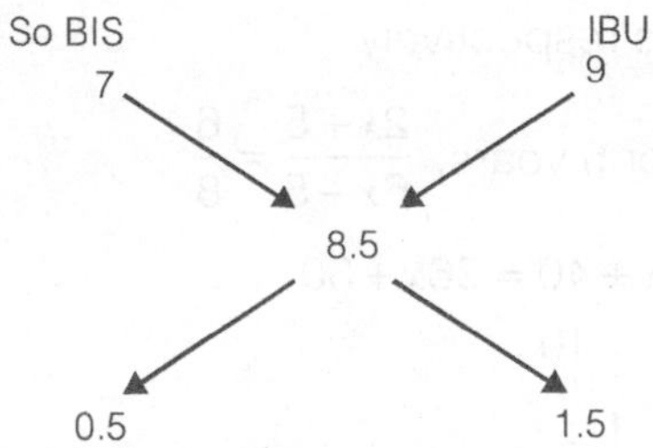

So the amount invested are in the ratio 1 : 3 for BIS & IBU

So the amount in BIS = $\frac{1}{1+3}\times10000$

= Rs. 2500

Example 4.

A Jar contains solution having milk and water in the ratio 3 : 2. If out of 100 litres, 10 litre is taken out & is replaced with water in one cycle. Find out the amount of water after 2 cycles.

Solution:

Out of 100 litre solution,

$$\text{Milk} = \frac{3}{5}\times100 = 60 \text{ litre}$$

$$\text{Water} = \frac{2}{5}\times100 = 40 \text{ litre}$$

Now percent of solution which is being replaced with water

$$= \frac{10}{100}\times100 = 10\%$$

so amount of milk after 2 such cycles

$= 60 \times (1 - 0.1)(1 - 0.1)$

$= 48.6$ litre

Net amount of water

$= 100 - 48.6 = 51.4$ litre

□□□□

Solved Examples

1. The average age of 24 students and the principal is 15 years. When the principal's age is excluded, the average age decreases by 1 year. What is the age of principal?

(a) 38 (b) 40
(c) 39 (d) Data inadequate

Ans. (c)

Average age of 24 students

$$= \frac{\text{sum of ages of 24 students}}{24}$$

$$\Rightarrow 14 = \frac{\text{sum of ages of 24 students}}{24}$$

$\therefore$ sum of ages of 24 students = 24×14

= 336 years

Let the Age of principal be x year

Then, average age of 24 students and the principal

$$= \frac{\text{sum of ages of 24 students \& principal}}{25}$$

$$\Rightarrow 15 = \frac{336+x}{25}$$

$$\Rightarrow x = 25\times15 - 336 = 39 \text{ years}$$

2. The average weight of 3 men A, B and C is 84 kg. Another man D joins the group and the average now becomes 80 kg. If another man E, whose weight is 3 kg more than that of D, replaces A then the average weight of B,C,D and E becomes 78 kg. The weight of A is

(a) 70kg (b) 72 kg
(c) 79 kg (d) 78 kg

Ans. (c)

Sum of weights of A,B and C

= 3 × 84 = 252 kg

sum of weights of A,B, C and D

= 4 × 80 = 320 kg

weight of D = 320 – 252 = 68 kg

$\therefore$ weight of E = 68 + 3 = 71 kg

Now, sum of weights of A,B, C, D and E

= 320 + 71 = 391 kg

sum of weights of B,C, D and E

= 4 × 78 = 312 kg

$\therefore$ weight of A = 391 – 312 = 79 kg

3. Vijay Tendulkar has certain average for 9 innings. In the tenth inning, he scores 100 runs thereby increasing his average by 8 runs. His new average is

(a) 20 (b) 24
(c) 28 (d) 32

Ans. (c)

Let the old average and new average by x and y respectively

Then, $y = x + 8$ ...(i)

and $y = \frac{9x+100}{10}$ ...(ii)

From (i) and (ii) we get

$10(x + 8) = 9x + 100$

$\Rightarrow x = 20$

$\therefore y = 20 + 8 = 28$

4. The average age of 8 persons in a committee is increased by 2 years when two men aged 35 years and 45 years are substituted by two women. Find the average age of the two women.

(a) 48 (b) 45
(c) 51 (d) 42

Ans. (a)

Let the average age of 8 persons in the committee be x years

Let the sum of ages of two women be y years.

Then sum of ages of 6 persons excluding two men aged 35 and 45 years

$= 8x - (35 + 45) = 8x - 80$

Again, $x + 2 = \frac{8x - 80 + y}{8}$

$\Rightarrow 8x + 16 = 8x - 80 + y$

$\Rightarrow y = 96$

$\therefore$ average age of two women

$= \frac{96}{2} = 48$ years

5. The speed of the train in going from Nagpur to Allahabad is 100 km/hr while when coming back from Allahabad to Nagpur, its speed is 150 km/hr. Find the average speed during the whole journey.

(a) 125 (b) 75
(c) 135 (d) 120

Ans. (d)

Let the distance between Allahabad and Nagpur be x km.

The average speed $= \frac{2x}{\frac{x}{100} + \frac{x}{150}}$

$= \frac{2 \times 100 \times 150}{250} = 120$ km/hr.

6. The age of Mitali and Pooja is in the ratio 2 : 6. After 5 years, the ratio of their ages will become 6 : 8. Find the average of their ages after 10 years.

(a) 12 (b) 13
(c) 17 (d) 24

Ans. (a)

Let the present age of Mitali and Pooja be $2x$ and $6x$ years respectively.

After 5 years, $\frac{2x+5}{6x+5} = \frac{6}{8}$

$\Rightarrow 16x + 40 = 36x + 30$

$\Rightarrow 20x = 10$

$\therefore x = \frac{1}{2}$

Hence Present ages are 1 and 3 years

After ten years, their ages will be 11 and 13 years.

$\therefore$ Average of their ages

$= \frac{11+13}{2} = 12$ years

7. Find the average of the first 97 natural numbers.

(a) 47 (b) 37
(c) 48 (d) 49

Ans. (d)

Sum of the first 97 natural members

$= \frac{97(97+1)}{2} = 4753$

$\therefore$ Their average $= \frac{4753}{97} = 49$

8. Find the average of all prime numbers between 30 and 50.

(a) 39.8 (b) 38.8
(c) 37.8 (d) 41.8

Ans. (a)

The prime number between 30 and 50 are 31, 37, 41, 43, 47

$\therefore$ Their average

$= \frac{31+37+41+43+47}{5} = 39.8$

9. The average of 5 consecutive numbers is n. If the next two numbers are also included, the average will.

(a) increase by 1 (b) remain the same
(c) increase by 1.4 (d) increase by 2

Ans. (a)

Let the five consecutive numbers be x, $x + 1$, $x + 2$, $x + 3$, and $x + 4$.

Given, $n = \dfrac{x+x+1+x+2+x+3+x+4}{5}$

$\Rightarrow n = x+2$

The average after including the next two numbers

$= \dfrac{x+x+1+x+2+x+3+x+4+x+5+x+6}{7}$

$= x + 3 = n - 2 + 3$

$= n + 1$

Hence the average increases by 1.

10. The average of 50 number is 38. If two numbers, namely, 45 and 55 are discarded, the average of the remaining numbers is

(a) 36.5 (b) 37
(c) 37.6 (d) 37.5

Ans. (d)

Sum of 50 numbers

$= 38 \times 50 = 1900$

Sum of remaining numbers

$= 1900 - (45 + 55) = 1800$

average of remaining numbers

$= \dfrac{1800}{48} = 37.5$

11. The average weight of 5 men is decreased by 3 kg when one of them weighing 150 kg is replaced by another person. Find the weight of the new person?

(a) 165 (b) 135
(c) 138 (d) 162

Ans. (b)

Let the average weight of 5 men be x kg before replacement of one person.

Also, Let the weight of new person be y kg.

Then, $x - 3 = \dfrac{5x - 150 + y}{5}$

$\Rightarrow 5x - 15 = 5x - 150 + y$

$\Rightarrow y = 135$ kg

12. The average age of the Indian cricket team playing the Nagpur test is 30. The average age of 5 of the players is 27 and that of another set of 5 players, totally different from the first five, is 29. If it is the captain who was not included in either of these two groups, then find the age of the captain.

(a) 75
(b) 55
(c) 50
(d) Cannot be determined

Ans. (c)

Let the age of the captain be x years

Then, $30 = \dfrac{27\times 5 + x + 29\times 5}{11}$

$\Rightarrow 330 = 135 + x + 145$

$\therefore x = 330 - 280 = 50$ years

13. Out of three numbers, the first is twice the second and three times the third. The average of the three numbers is 88. The smallest number is

(a) 72 (b) 36
(c) 42 (d) 48

Ans.(d)

Let the first number be n. then the other two numbers will be $\dfrac{n}{2}$ and $\dfrac{n}{3}$

Again, $88 = \dfrac{n+\frac{n}{2}+\frac{n}{3}}{3}$

$\Rightarrow 88 = \dfrac{n}{3}\left(1+\dfrac{1}{2}+\dfrac{1}{3}\right)$

$\Rightarrow 88 = \dfrac{n}{3}\times\dfrac{11}{6}$

$\therefore$ The smallest number $= \dfrac{n}{3} = \dfrac{144}{3} = 48$

14. The average of first five multiples of 3 is

(a) 3 (b) 9
(c) 12 (d) 15

Ans. (b)

$\text{Average} = \dfrac{3(1+2+3+4+5)}{5} = \dfrac{3\times 15}{5} = 9$

15. The average temperature of 1st, 2nd 3rd December was 24.4°C. The average temperature of the first two days was 24°C. The temperature on the 3rd of December was

(a) 20°C (b) 25°C
(c) 25.2°C (d) None of these

Ans. (c)

Let the temp. on 3rd of December be x C. Then

$24.4 = \dfrac{2\times 24 + x}{3} = 25.2$ °C

16. The average of 20 results is 30 and that of 30 more results is 20. For all the results taken together, the average is
(a) 25 (b) 50
(c) 12 (d) 24

Ans. (d)

The required average

$$= \frac{20 \times 30 + 30 \times 20}{50} = \frac{1200}{50} = 24$$

17. The average temperature on Monday, Tuesday and Wednesday was 41°C and on Tuesday, Wednesday and Thursday it was 40°C. If on Thursday it was exactly 39°C, then on Monday, the temperature was
(a) 42°C (b) 46°C
(c) 23°C (d) 26°C

Ans. (a)

The sum of temp. on Tuesday and Wednesday

$= 3 \times 40 - 39 = 81°C$

The sum of temp. on Monday, Tuesday, and Wednesday $= 3 \times 41 - 81 = 42°C$

□□□□

Average

I Practice Exercise

1. There are 35 students in a hostel. If the number of students increased by 7, the expenses of the mess were increased by Rs. 42 per day while the average expenditure per head diminished by Rs. 1. Find the original expenditure of the mess.
(a) Rs. 480 (b) Rs. 440
(c) Rs. 520 (d) Rs. 420

2. An aeroplane travels distances 2500 km, 1200 km and 500 km at the rate of 500 km/hr, 400 km, and 250 km/hr, respectively. The average is
(a) 420 km/hr (b) 410 km/hr
(c) 405 km/hr (d) 575 km/hr

3. The average weight of 24 students of section A of a class is 58 kg whereas the average weight of 26 students of section B of the same class is 60.5 kg. Find the average weight of all the 50 students of the class.
(a) 57.4 Kg. (b) 59.3 kg
(c) 58.9 kg (d) 59.7 kg

4. The average monthly salary of a staff of 9 persons is Rs. 2450. One member of the staff whose monthly salary is Rs. 2650. Find the average salary of the remaining 8 persons of the staff.
(a) Rs. 2425 (b) Rs. 2625
(c) Rs. 3025 (d) Rs. 2825

5. 3 years ago the average age of a family of 5 members was 17 years. With the birth of a new baby the average remains the same even today. Find the age of the baby.
(a) 1 years (b) 3 years
(c) $2\frac{1}{2}$ years (d) 2 years

6. A batsman in his 17th innings makes a score of 85, and thereby increases his average by 3. What is his average after the 17th innings? He had never been 'not out'.
(a) 47 (b) 37
(c) 39 (d) 43

7. The sum of three numbers is 98. If the ratio between first and second be 2 : 3 and that between second and third be 5 : 8, then the second number is
(a) 30 (b) 20
(c) 58 (d) 48

8. A number x equals 80% of the average of 5, 7, 14 and a number y. If the average of x and y is 26, the value of y is
(a) 13 (b) 26
(c) 39 (d) None of these

9. The average age of 8 men is increased by 2 years. When two of them, whose ages are 20 years and 24 years respectively, are replaced by two women. What is the average age of these women?
(a) 36 years (b) 30 years
(c) 40 years (d) 42 years

10. The average speed of a train running at a speed of 30 km/hr during the first 100 kilometers, at 40 km/hr during the second 100 kilometers and at 50 km/hr during the last 100 kilometers is nearly
(a) 38.5 km/hr (b) 38.3 km/hr
(c) 40.0 km/hr (d) 39.2 km/hr

11. The average weight of 3 men A, B and C is 84 kg. Another men D joins the group and the average now becomes 80 kg. If another men E, whose weight is

3 kg more than that of D, replaces A, then average weight of B, C, D and E becomes 79 kg. The weight of A is
(a) 70 kg (b) 72 kg
(c) 75 kg (d) 80 kg

12. The average of marks obtained by 120 candidates was 35. If the average of marks of passed candidates was 39 and that of failed candidates was 15, the number of candidates who passed the examination is
(a) 100 (b) 110
(c) 120 (d) 150

13. In a class, there are 20 boys whose average age is decreased by 2 months, when one boy aged 18 years in replaced by a new boy. The age of the new boy is
(a) 14 years 8 months
(b) 15 years
(c) 16 years 4 months
(d) 17 years 10 months

14. The average daily wages of A, B and C is Rs. 120. If B earns Rs. 40 more than C per day and A earns double of what C earns per day, the wages of A per day is
(a) Rs. 80 (b) Rs. 120
(c) Rs. 160 (d) Rs. 100

15. With an average speed of 40km/hr a train reaches its destination on time. If it goes with an average speed of 35 km/hr, it is late by 15 minutes. The total journey is
(a) 30 km (b) 40 km
(c) 70 km (d) 80 km

16. In a competitive examination, the average marks obtained was 45. It was later discovered that there was some error in computerisation and the marks of 90 candidates had to be changed from 80 to 50 and the average came down to 40 marks. The total no. of candidates who appeared in examination is
(a) 520 (b) 550
(c) 540 (d) 525

17. A man whose bowling average is 12.4 takes 5 wickets for 26 runs and thereby decreases his average by 0.4. The number of wickets, taken by him before his last match is
(a) 85 (b) 78
(c) 72 (d) 64

Answers

1. (d)	2. (a)	3. (b)	4. (a)	5. (d)
6. (b)	7. (a)	8. (c)	9. (b)	10. (b)
11. (c)	12. (a)	13. (a)	14. (c)	15. (c)
16. (c)	17. (a)			

Solutions

1. **Ans. (d)**
Let the original expenditure = Rs. x

Original average expenditure = $\frac{x}{35}$

New average expenditure = $\frac{x+42}{42}$

$\Rightarrow \frac{x}{35} - \frac{x+42}{42} = 1 \Rightarrow x = 420$

$\therefore$ Original expenditure = Rs. 420

2. **Ans. (a)**
The total time taken can be calculated as shown below :

Distance	Speed	Time
25 km	500 km/hr	5 hrs.
1200 km	400 km/hr	3hrs.
500 km	250 km/hr	2hrs.
Total 4200 km		10hrs.

Average speed = $\frac{\text{Total Distance}}{\text{Total Time}}$

Average speed = $\frac{4200}{10}$ = 420 km/hr.

3. **Ans. (b)**
Average weight of 24 students of section A = 58 kg
Total weight of 24 students of section A = 58 × 24 = 1392 kg
Total weight of 26 students of section B(a) an Average of = 60.5 kg = 60.5 × 26 = 1573
Total weight of 50 student = (1392 +1573) kg = 2965 kg
Average weight of students of students in the class

$= \frac{2965}{50} = 59.3$ kg.

4. **Ans. (a)**
Average salary of 9 persons = Rs. 2450
Total salary of 9 persons

= Rs. 2450 × 9 = Rs. 22050

Salary of the person who is transferred

= Rs. 2650

Thus, the total salary of remaining 8 persons

= Rs. 22050 – 2650 = Rs. 19400

The average salary of the remaining 8 persons

$= \text{Rs.} \dfrac{19400}{8} = \text{Rs. } 2425.$

5. Ans. (d)

Present age of 5 members

= (5 × 17 + 3 × 5) years.

= 100 years.

Present age of 5 members and a baby

= 17 × 6 = 102 yrs.

∴ Age of the baby = (102 – 100)yrs = 2 years.

6. Ans. (b)

Average score before 17th innings

= 85 – 3 × 17 = 34

∴ Average score after 17th innings

= 34 + 3 = 37.

7. Ans. (a)

Let the number be x, y and z. Then,

$x + y + z = 98, \dfrac{x}{y} = \dfrac{2}{3}$ and $\dfrac{y}{z} = \dfrac{5}{8}$

$\therefore\ x = \dfrac{2y}{3}$ and $z = \dfrac{8y}{5}$

So $\dfrac{2y}{3} + y + \dfrac{8y}{5} = 98$

or, $\dfrac{49y}{15} = 98$ or, y = 30.

8. Ans. (c)

Average of 5, 7, 14 and $y = \dfrac{5+7+14+y}{4}$

Therefore, x = 80% of

$\dfrac{5+7+14+y}{4} = \dfrac{80}{100} \times \dfrac{26+y}{4}$

$\Rightarrow x = \dfrac{26+y}{5}$ ...(i)

Also $\dfrac{x+y}{2} = 26$ ...(ii)

From (i) and ((ii), we get

$52 - y = \dfrac{26+y}{5} \Rightarrow y = 39.$

9. Ans. (b)

Let the average age of 8 men be x years

∴ Sum of the ages of 8 men = $8x$ years

Now, according to the condition of the question, average age of (6 men + 2 women)

= $(x + 2)$ years

∴ Sum of the ages of (6 men + 2 women)

= $8(x + 2) = 8x + 16$ years

Hence, it is clear that on replacing two men by two women, sum of their ages increased by 16 years

Therefore, sum of the ages of two women

= (20 + 24) + 16 = 60 yrs

∴ Average age of the women $\dfrac{60}{2} = 30$ yrs

10. Ans. (b)

Time taken to cover first 100 kilometres

$= \dfrac{100}{30} = 3\dfrac{1}{3}$hrs

Time taken to cover second 100 kilometres

$= \dfrac{100}{40} = 2\dfrac{1}{2}$hrs

Time taken to cover last 100 kilometres

$= \dfrac{100}{5} = 2$ hrs

Total time taken

$3\dfrac{1}{3} + 2\dfrac{1}{2} + 2 = \dfrac{10}{3} + \dfrac{5}{2} + 2 = \dfrac{47}{6}$ hrs.

Total distance covered = 300 km

∴ Average speed $= \dfrac{300}{47/6}$

$= \dfrac{300 \times 6}{47} = 38.3$ km/hr

11. Ans. (c)

Weight of D = (80 × 4 – 84 × 3) kg = 68 kg

Weight of E = (68 + 3) kg = 71 kg

(B + C + D + E)'s

Weight = (79 × 4) kg = 316 kg

∴ (B+C)'s weight = [316 –(68 + 71)] kg

= 177 kg

Hence, A's weight = [(84 × 3)–177] kg = 75 kg

12. Ans. (a)

Let the number of candidates who passed = x.

Then, $39 \times x + 15 \times (120 - x) = 120 \times 35$

$\therefore\ 24x = 4200 - 1800$

or, $x = \dfrac{2400}{24}$, $x = 100$.

13. Ans. (a)

Total decrease = (20×2) months

= 3 years 4 months

$\therefore$ Age of the new boy = 18 years – 3 years 4 months

= 14 years 8 months.

14. Ans. (c)

Let daily wages of C = x

Then, daily wages of A = $2x$

and, daily wages of B = $x + 40$

Hence, average daily wages of A, B and C

$= \dfrac{x + 2x + x + 40}{3} = \dfrac{4x + 40}{3}$

$\therefore\ \dfrac{4x + 40}{3} = 120$ or, $4x + 40 = 360$

$\Rightarrow 4x = 320$ or, $x = 80$

$\therefore$ Wages of A per day = 2×80 = Rs. 160.

15. Ans. (c)

$\dfrac{x}{35} - \dfrac{x}{40} = \dfrac{15}{60}$ or, $\dfrac{5x}{35 \times 40} = \dfrac{1}{4}$

or, $x = \dfrac{35 \times 40}{5 \times 4} = 70$

16. Ans. (c)

Let the number of candidates be x, Then, total marks obtained by all the candidates = $45x$

Total change in marks for 90 candidates

$= 30 \times 90 = 2700$

Total reduced marks = $45x - 2700$

Reduced average = $\dfrac{45x - 2700}{x}$

$\therefore\ 40 = \dfrac{45x - 2700}{x}$

or, $40x = 45x - 2700$

$\Rightarrow\ 5x = 2700$ or, $x = 540$

17. Ans. (a)

Let the number of wickets taken before the last match = x

Then, $\dfrac{12.4x + 26}{x + 5} = 12 \Rightarrow x = 85$

■■■■

1.7
CHAPTER

Time & Work

In real life situations we come across practical problems of accomplishing the given project in prescribed time limit. Since efficiency of different person is different, the management has to take proper note of it before distributing the task to subordinates or executives. We discuss all such problems under the heading 'Time and work'.

In Time & Work, the problems are based on real life situations to allot the work between the subordinates based on their capacities.

The basic understanding of the problems comes by quantizing a work. We will start understanding the quantization with the following example.

Let us assume a person finishes 2 units of the work in one day. So for finishing 60 units he will take 30 days. So to find out the net time we should know the capacity of the person & the net amount of work to be finished.

All time & work problems depends upon one particular relation i.e.

$$\text{Capacity} = \frac{\text{Work}}{\text{Time}}$$

So to find out any variables we need to know the other two variables. If they are not given they can be assumed based on the situation. Let us start with a simple example.

Example 1.

5 Persons finish one work in 7 days. In how many days 7 persons will finish the same amount of work.

Solution.

Here we don't know how much amount of work, is to be finished & what is the capacity of the individuals. So we will assume the capacity.

Let us assume 1 person finishes 1 unit in 1 day

⇒ 5 persons will finish 5 units in 1 day

⇒ 5 persons will finish 35 units in 7 day

⇒ The net work is 35 units

Now taking 7 persons,

they will finish 7 units/day

So time taken to finish 35 units = 35/7 = 5 days.

Example 2.

A can finish a work in 10 days & B can finish the same work in 12 days. In how many days, they will finish the task working together.

Solution.

Given:

A takes 10 days

B takes 12 days

Since we can not assume the same capacity here. So we will assume the work. Let us assume the work is 60 units.

So capacity of A = $\frac{60}{10}$ = 6 units/day

Capacity of B = $\frac{60}{12}$ = 5 units/day

Combined capacity of A & B = 11 units/day

So time taken by A & B = $\frac{60}{11} = 5\frac{5}{11}$ days

Example 3.

A finishes a task in 30 days. But with the help of B, he finishes the same task in 12 days. How much time B alone will take to finish the task.

Solution.

Let the work be 60 units

A $\xrightarrow{\text{Capacity = 2unit}}$ 30 days

A+ B $\xrightarrow{\text{Capacity = 5unit}}$ 12 days

B $\xrightarrow{\text{Capacity = 3unit}}$?

So the time taken B = $\frac{60}{3}$ = 20 days

Alternate Work

Sometimes the availability of the person is limited in that case they work in a particular sequence it can be understood with the following example.

Example 1.

A can finish a task in 12 days while B can finish a task in 15 days. In how many days both of them will finish the work if they work on alternate days started by 'A'.

Solution.

Let work = 60 units

A $\xrightarrow{\text{Capacity = 5unit/day}}$ 12 days

B $\xrightarrow{\text{Capacity = 4unit/day}}$ 15 days

Now since they work on alternates day, the work distribution will be in following manner for individual day

$\underbrace{5+4}+\underbrace{5+4}+\underbrace{5+4}+ \ ..... = 60$

So we can understand that in 2 days they finish 9 unit and the work is being completed as a multiple of 9 unit.

So 9 × 6 units will be finished in 12 days.

So the 60 units of work can be allocated as

Work allocation	9×6 units +	5 units +	1 unit	= 60 units
	↓ (A+B)	↓ A	↓ B	
Time allocation	2×6 days	1 day	1/4 day	= 13¼ day

Suppose three people can finish work individually in A, B, C days respectively. If all three of them start working on some alternate pattern then if A > B > C ∈ N. $\frac{1}{A}+\frac{1}{B}+\frac{1}{C}=\frac{1}{N}$. A, B C & N all are natural number then work will get completed in integral number of days who is starting the work will not effect the total time taken.

But if $\frac{1}{A}+\frac{1}{B}+\frac{1}{C}\neq\frac{1}{N}$ then in order to get the job done in minimum no. of days, we should use most efficient person on day one, 2nd best person on day 2 and least efficient person on day 3 and so on.

Concept of Negative Work

Negative work indicates the destructing conditions. Take for example if a bridge has to be made the flow of water of river will cause destruction of the work. So the river water does the negative work. These problems have to be dealt same as in previous cases.

Example 1.

A can build a house in 10 days, B can build the same house in 12 days while C can completely destruct the house in 20 days. in how many days the work will be finished if all three are working together.

Solution.

Let the work = 60 units

A $\xrightarrow[\text{Constructs}]{\text{Cap.=+6 units/day}}$ 10 days

B $\xrightarrow[\text{Constructs}]{\text{Cap.=+5 units/day}}$ 12 days

C $\xrightarrow[\text{destructs}]{\text{Cap.=−3 units/day}}$ 20 days

A + B + C $\xrightarrow{\text{6+5−3=8 units/day}}$ 8 units/day

As all of them are working together.

So time taken = $\frac{60}{8}=7\frac{1}{2}$days.

Concept of Wages

If the work is finished by more than one persons then the wages for them will be distributed among them based on the amount of work done by them. So,

- Ratio of wages = Ratio of work finished
  Now if all the persons are working for same no. of days then the ratio of work = ratio of capacities, i.e. ratio of wages = ratio of capacities.

Example 1.

5 man or 3 women finishes a work in 10 days if 6 women finishes the work along with 7 man then find out the wages received by them if the total work costs Rs. 3000.

Solution.

Lee the work 150 units

5 man ⟶ 10 days

⇒ 1 man $\xrightarrow{\text{Cap.=3unit/day}}$ 50 days

3 women ⟶ 10 days

⇒ 1 women $\xrightarrow{\text{Cap.=5unit/day}}$ 30 days

Since all of them are working for same no. of days so the wages will be divided in the ratio of their capacity

i.e. 3000 should be divided in

6 × 5 : 7 × 3 = 30 : 21 or 10 : 7

so amount received by 6 women = $\frac{30\times3000}{51}$

amount received by 7 men = $\frac{21\times3000}{51}$

Pipes & Cisterns

The Pipes and Cisterns problems are once again similar as of the general time and work problem only thing that has to be understood is the concept of positive and negative work.

The inlet pipe always does the positive work while the outlet pipe always does the negagive work. Let us see some problems.

Example 1.
Pipe A can fill a tank in 10 hours while pipe B can empty a completely filled tank in 15 hours. How much time it will take to fill a 2/3rd filled tank completely by both of the pipes.

Solution.
Take volume = 30 litres

$A \xrightarrow[\text{Fill}]{\text{Cap.}=3\text{litre/hr}} 10 \text{ hours}$

$B \xrightarrow[\text{Empty}]{\text{Cap.}=-2\text{litre/hr}} 15 \text{ hours}$

Capacity of (A + B) = 3 – 2 = 1 litre/hr
Amount of water to be filled

$$= \left(\frac{2}{3}\right) \times 30 = 20 \text{ litre}$$

So the time taken $= \frac{20}{1} = 20$ hrs.

Efficiency Comparison
When two person's efficiencies are compared then in that case, the basic work and time calculations become much easier. In this sort of case, we will be able to assume the capacities directly. For example:

Example 1.
A is 3 times efficient than B and can finish the work 10 days earlier as compared to B. How much time both will take to finish the same amount of work?

Solution.
If A is 3 times efficient than B
Then, B will take thrice the no. of days what A takes

as $\eta \propto \frac{1}{\text{Time}}$

So if A takes x days then B takes $3x$ days given that

$$x + 10 = 3x \Rightarrow x = 5 \text{ so}$$

$A \xrightarrow{\text{Cap.3unit/day}} 5 \text{ days}$

$B \xrightarrow{\text{Cap.1unit/day}} 15 \text{ days}$

Work 15 unit
(A + B)'s capacity = 4 unit/day

Time taken $= \frac{15}{4} = 3\frac{3}{4}$ days

Example 2.
A tank can be filled completely in 6 hours but because of the leak at its bottom, it takes 2 more hours to get filled. In how much time the leak can empty a full tank?

Solution.

$\text{Pipe} \xrightarrow{\text{Cap. 4 litre/hr}} 6 \text{ hrs}$

$\text{Pipe + Leak} \xrightarrow{\text{Cap. 3 litre/hr}} 8 \text{ hrs}$

Tank 24 litre
Leak capacity = 1 litre/hour

Time $= \frac{24}{1} = 24$ hrs

Example 3.
A monkey climbs on a greased pole. In 1 minute it goes up by 3 meter while in the next 1 minute it comes down by 2 meter if the height of the pole is 50 meter. In how many minutes the monkey will be at the top?

Solution.
When monkey will reach @ the height of 47 m. In the next minute, it will reach at the top. So we will find out how much time the monkey takes to reach @ the height of 47 m.
in first minute monkey moves = +3 m
is second minute monkey moves = –2 m
so in 2 minutes monkey effectively moves = 1 m
So it will take 47 × 2 min to reach 47 m
So total time taken = 47 × 2 + 1 min = 95 mins.

Example 4.
In a hostel there was food for 100 persons for 30 days. After 10 days 100 more persons joined the hostel for how many days the food will last?

Solution.
After 10 days the situation will be as follows:
100 person can take the food for 20 days. If 100 more persons joined, total person in the hostel = 200
If 100 persons take 20 days

Then 200 persons will take $\frac{20 \times 100}{200} = 10$ days.

Example 5.
Ram can finish 1/2 of a task in 9 days while Shyam can finish 1/4th of the same task in 3 days. How many days they will take to finish the work together?

Solution.
Ram finishes 1/2 work in 9 days
He will finish full work in 18 days

Shyam finishes 1/4th work in 3 days
He will finish full work in 12 days
Let work 36 units

So Ram $\xrightarrow{2\text{ unit/day}}$ 18 days

Shyam $\xrightarrow{3\text{ unit/day}}$ 12 days

Capacity of Ram & Shyam = 5 unit/day

Time taken $= \frac{36}{5} = 7\frac{1}{5}$ days

Example 6.

Ramu is thrice as finish efficient as Rahul. If working together they finishes the work in 20 days. How many days Rahul will take to finish the work?

Solution.

Let us assume the capacity of Ramu is 3 unit/day
then capacity of Rahul = 1 unit/day
Total capacity of Ramu & Rahul = 4 unit/day
If they finishes the work in 20 days then net work $= 20 \times 4 = 80$ unit

So time taken by Rahul $= \frac{80}{1} = 80$ days.

Example 7.

Suresh can finish a task in 24 days. Along with Monika, Suresh completed the task & they were paid Rs. 108 & Rs. 162 respectively. How many days they would have taken to finish the work together?

Solution.

If they are working together for same number of days then
Ratio of wages = ratio of capacities
i.e. Monika : Suresh = 108 : 162 = 2 : 3
Let us assume
Monika's capacity = 2 unit/day
Suresh's capacity = 3 unit/day
Then net work $= 24 \times 3 = 72$ units
So time taken when working together

$$= \frac{72}{2+3} = 14\frac{2}{5} \text{ days.}$$

Example 8.

5 men & 3 boys can finish a work in 12 days and 1 man and 3 boys can finish a work in 60 days. How much time 4 boys along with 3 men will take to finish the work?

Solution.

Let the capacity of man by 'm' and capacity of boy be 'b'

$5\text{ m} + 3\text{b} \xrightarrow{5\text{ unit/day}} 12$ days

$1\text{ m} + 3\text{ b} \xrightarrow{1\text{ unit/day}} 60$ days

Work 60 unit

$\Rightarrow 5\text{ m} + 3\text{b} = 5$
$1\text{ m} + 3\text{ b} = 1$

So $m = 1$, $b = 0$

Total capacity of 4 boys and 3 men
$= 4 \times 0 + 3 \times 1$
$= 3$ units/day

Time taken $= \frac{60}{3} = 20$ days.

□□□□

Concept of Man-Day

If 100 workers can finish the task in 50 days, then 1 worker can finish the task in $100 \times 50 = 5000$ day.

This is represented in form of man-day
So, man day is nothing but the time required to finish the task by 1 person.

Three Cases:

Case-I.

Let 100 workers can finish the task in 50 days, then how many days are required to finish the task by 25 workers?

Solution:

Here man Day $= 100 \times 50 = 5000$
how according to question only 25 workers are available so

$25 \times$ day $= 5000$

Day $= \frac{5000}{25} = 200$ days are required to finished the task by 25 workers.

Case-II.

Let 100 worker can finish the task in 50 days. 100 worker started completing the task by working for 10 days. Now to finish the task as soon as possible 100 more workers are employed. Now calculate the total number of days required to complete the task.

Solution:

Here Man × Day = constant

100 × 50 = 5000

Initially 100 workers are working for 10 days So they can finish

100 × 10 = 1000 (Man.Day)

Now 4000 man - Day is still remaining which has to be finished by 200 workers

So, 200 × Day = 4000

$$\text{Day} = \frac{4000}{200} = 20 \text{ days}$$

So total it requires 10 + 20 = 30 days to complete the task.

Case-III.

Let 100 workers can finish the task in 50 days. 100 workers started the task and working for 20 days. Now because of unforeseen situations 80 workers have to left the work. Find total number of days required to complete the task.

Solution:

Here Man . Day = constant

100 . 50 = 5000

initially 100 workers are working for 20 day so, they can finish

100 × 20 = 2000 (Man . Day)

Now, 3000 (Man . Day) is still remaining which has to be finished by 20 workers

Man × Day = 3000

20 × Day = 3000

$$\text{Day} = \frac{3000}{20} = 150 \text{ days}$$

So total it requires 20 + 150 = 170 days to complete the task.

Man × Day ×Hour

$M_1 \times D_1 \times H_1$ = work, also

$M_2 \times D_2 \times H_2$ = work

Work may be anything like construction of a wall of different size.

$$\frac{M_1 D_1 H_1}{W_1} = \frac{M_2 D_2 H_2}{W_2}$$

Also if $W_1 = l_1 b_1 h_1$ where length is l_1 breadth b_1 and height h_1 and if $W_2 = l_2 b_2 h_2$ where l_2, b_2 and h_2 are length, and height respectively of second wall then

$$\frac{M_1 D_1 H_1}{l_1 b_1 h_1} = \frac{M_2 D_2 H_2}{l_2 b_2 h_2}$$

$$= \frac{l_1 b_1 h_1}{l_2 b_2 h_2} = \frac{M_1 D_1 H_1}{M_2 D_2 H_2}$$

Concept of Efficiency

Efficiency of persons performing same task may not be same. It may be different for different persons. If efficiency is high a person can complete the work much faster than stipulated time period. Similarly if efficiency is low, it requires more time to finish the same task than that of stipulated time.

We will observe following examples to have a glimpse of problems based on concept of efficiency.

Ex.1 A is three times as efficient as B and finish the task 32 days ahead of B. Find the number of days required to finish the task if both are working simultaneously.

Sol.: Let A requires x days then

B requires $3x$ days

$3x - x = 2x \Rightarrow 32$

$x = 16$ days

$x = 16 \quad 3x = 48$ days

Together they can finish

$$\frac{1}{16} + \frac{1}{48} = \frac{4}{48} \text{ part} = \frac{1}{12} \text{ part}$$

Thus 12 days are required by them to finish the task.

Ex.2 A is four times as efficient as of B and can finish a task 45 days ahead of B. Find total number of days required to finish the task if both are working simultaneously.

Sol.: Let A requires x days then

B requires $4x$ days.

According to given condition

$4x - x = 45 \Rightarrow 3x = 45 \Rightarrow x = 15.$

Together they can finish

$$\frac{1}{15} + \frac{1}{60} = \frac{5}{60} = \frac{1}{12} \text{ part in a day}$$

So, it requires 12 days to finish the task together.

□□□□

Solved Examples

1. Nishu and Archana can do a piece of work in 10 days and Nishu alone can do it in 12 days. In How many days can Archana do it alone?
(a) 60 days (b) 30 days
(c) 50 days (d) 45 days
Ans.(a)

One day work of both $= \frac{1}{10}$

and Nishu's day work $= \frac{1}{12}$

So, Archana's one day work

$= \frac{1}{10} - \frac{1}{12} = \frac{2}{120} = \frac{1}{60}$

So, 60 days required to finish the work by Archana.

2. 4 men and 3 women finish a job in 6 days. And 5 men and 7 women can do the same job in 4 days. How long will 1 man and 1 women take to do the work?
(a) $22\left(\frac{2}{7}\right)$ days (b) $25\left(\frac{1}{2}\right)$ days
(c) $5\left(\frac{1}{7}\right)$ days (d) $12\left(\frac{7}{22}\right)$ days.
And. (a)
Let man completes m part in a day and woman completes w part in a day then

$4\text{ m} + 3\text{ w} = \frac{1}{6}$ also ...(i)

$5\text{ m} + 7\text{ w} = \frac{1}{4}$ ...(ii)

After simplifying we get

$20\text{ m} + 15\text{ w} = \frac{5}{6}$

$20\text{ m} + 28\text{ w} = 1$

So $13\text{ w} = \frac{1}{6}$

$\text{w} = \frac{1}{78}$ ...(iii)

From (i) and (iii) we get

$\text{m} = \frac{5}{156}$

$\text{m} + \text{w} = \frac{5}{156} + \frac{1}{78} = \frac{7}{156}$

So it require $\frac{156}{7}$ days $= 22\frac{2}{7}$

3. A can do a piece of work in 10 days and B can do the same work in 20 days. With the help of C, they finish the work in 5 days. How long will it take for C alone to finish the work?
(a) 20 days (b) 10 days
(c) 35 days (d) 15 days
Ans. (a)

$\frac{1}{5} - \frac{1}{10} - \frac{1}{20} = \frac{1}{20}$

This is C's one day work so 20 days are required for C alone to finish task

4. A can do a piece of work in 20 days. He work at it for 5 days and then B finishes it in 10 more days. In how many days will A and B together finish the work?
(a) 8 days (b) 10 days
(c) 12 days (d) 6 days
Ans. (a)

A's one day work $= \frac{1}{20}$

in 5 day 'A' can complete $\frac{1}{20} \times 5 = \frac{1}{4}$ part

remaining work $= \frac{3}{4}$ part

$\therefore$ B finishes $\frac{3}{4}$ part in 10 days

$\therefore$ B finishes 1 part in $\frac{10}{3/4} = \frac{40}{3}$ days

together they can finish it in

$\frac{1}{\frac{1}{20} + \frac{3}{40}} = \frac{1}{\frac{1}{8}} = 8$ days

5. Twenty workers can finish a piece of work in 30 days. After how many days should 5 workers leave the job so that the work is completed in 35 days?
(a) 5 days (b) 10 days
(c) 15 days (d) 20 days
Ans. (c)

Man × Day = Man . Day

20 × 30 = 600

Let 5 workers leave after x days then

$20x + 15(35 - x) = 600$

$5x + 525 = 600$

$5x = 75,\ x = 15$

6. Subhash can copy 50 page in 10 hours; subhash and Prakash together can copy 300 pages in 40 hours. In how much time can Prakash copy 30 pages?
(a) 13 h (b) 12 h
(c) 11 h (d) 9 h

Ans. (b)

∴ Subhash in 10 hours can copy 50 pages

∴ in 1 hour $\frac{50}{10} = 5$ pages

Both can copy 300 pages in 40 hours

So, in 1 hour $\frac{300}{40} = 7.5$

Clearly prakash can copy 2.5 pages in 1 hour So 30 pages in $\frac{1}{2.5} \times 30 = 12$ hours

7. Sashi can do a piece of work in 25 days and Rishi can do it in 20 days. They work for 5 days and then Sashi goes away. In how many more days will Rishi finish the work?

(a) 10 days (b) 12 days
(c) 14 days (d) None of these

Ans. (d)

Shashi's one day work = $\frac{1}{25}$

Rishi's one day work = $\frac{1}{20}$

Together they can finish $\frac{1}{25} + \frac{1}{20}$

$= \frac{9}{100}$ part in a day

∴ Work in 5 days $\frac{9}{100} \times 5 = \frac{45}{100}$ part

$\frac{55}{100}$ part is remaining. Rishi can finsh it in $\frac{\left(\frac{55}{100}\right)}{\left(\frac{1}{20}\right)}$

= 11 days

8. Priya can do 1/2 of the work in 8 days while Preeti can do 1/3 of the work in 6 days. How long will it take for both of them to finish the work?

(a) $\frac{88}{17}$ day (b) $\frac{144}{17}$ days

(c) $\frac{72}{17}$ days (d) 8 days

Ans. (b)

Priya's days work = $\frac{(1/2)}{8} = \frac{1}{16}$

Preeti's days work

$\Rightarrow \frac{(1/3)}{6} = \frac{1}{18}$ days

Together they can finish in

$$\frac{1}{\frac{1}{16} + \frac{1}{18}} = \frac{144}{17} \text{ days}$$

9. Manoj takes twice as much time as Ajay and thrice as much as Vijay to finish a piece of work. Together they finish the work in 1 day. What is the time taken by Manoj to finish the work?

(a) 6 days (b) 3 days
(c) 2 days (d) None of these

Ans. (a)

M : A : V

2 : 1 : 2/3 or 6 : 3 : 2

Let Manoj takes $6x$ days Ajay takes $3x$ days and vijay $2x$ days then $\frac{1}{6x} + \frac{1}{3x} + \frac{1}{2x} = 1$

$$\frac{1}{x}\left[\frac{1}{6} + \frac{1}{3} + \frac{1}{2}\right] = 1$$

$\frac{1}{x} = 1$, So $x = 1$

Manoj will take 6 days

10. Apurva can do a piece of work in 12 days. Apurva and Amit complete the work together and were paid Rs. 54 and Rs. 81 respectively. How many days must they have taken to complete the work together?

(a) 4 days (b) 4.5 days
(c) 4.8 days (d) 5 days

Ans. (c)

Payment ratio = 54 : 81

days work = $\frac{1}{12} : \frac{1}{x}$

$$54 : 81 :: \frac{1}{12} : \frac{1}{x}$$

$$81 \times \frac{1}{12} = 54 \times \frac{1}{x}$$

$$x = \frac{54 \times 12}{81}, \quad x = 8 \text{ days}$$

together they will take = $\frac{12 \times 8}{12 + 8} = 4.8$ days.

11. Raju is twice as Vijay. Together, they finish the work in 14 days. In how many days can Vijay alone do the same work?

(a) 16 days (b) 21 days
(c) 32 days (d) 42 days

Ans. (d)

Let Raju can do the work in x days then vijay in $2x$ days

Together they can complete in

$$\frac{2x \times x}{3x} = 14 \text{ days}$$

$x = 21$ days, $2x = 42$ days

12. If 12 men and 16 boys can do a piece of work in 5 days and 13 men and 24 boys can do it in 4 days, compare the daily work done by a man with that done by a boy?

(a) 1 : 2 (b) 1 : 3
(c) 2 : 1 (d) 3 : 1

Ans. (c)

$$12\text{ M} + 16\text{ B} = \frac{1}{5} \quad \text{...(i)}$$

$$13\text{ M} + 24\text{ B} = \frac{1}{4} \quad \text{...(ii)}$$

Also, $36\text{ M} + 48\text{ B} = \frac{3}{5}$

$$26\text{ M} + 48\text{ B} = \frac{2}{4}$$

$$10\text{ M} = \frac{1}{10},\ \text{M} = \frac{1}{100}$$

from eq (1) $\text{B} = \frac{1}{200}$, M : B = 2 : 1

13. A can do a work in 10 days and B can do the same work in 20 days. They work together for 5 days and then A goes away. In how many more days will B finish the work?

(a) 5 days (b) 6.5 days
(c) 10 days (d) $8\frac{1}{3}$ days

Ans. (a)

In 5 days the can complete

$$\left[\frac{1}{10} + \frac{1}{20}\right] \times 5 = \frac{3}{4} \text{ part}$$

Remaining is $\frac{1}{4}$ part

B will finish it in $\frac{\frac{1}{4}}{1/20} = 5$ days

14. 15 men could finish a piece of work in 210 days. But at the end of 100 days, 15 additional men are employed. In how many more days will the work be complete?

(a) 80 days (b) 60 days
(c) 55 days (d) 50 days

Ans. (c)

Man × Day = Man . Day

15 × 210 = 3150, so after 100 days

15 × 100 = 1500 will be completed

remaining work = 1650 Man . Day

Man . Day = 1650

30. Day = 1650

Day = 55

15. In a fort there was sufficient food for 200 soldiers for 31 day. After 27 days 120 soldiers left the fort. For how many extra days will the rest of the food last for the remaining soldiers?

(a) 12 days (b) 10 days
(c) 8 days (d) 6 days

Ans. (b)

Man × Day = Man.Day

200 × 31 = 6200

After 27 days

200 × 27 = 5400 Man. Day is finished

remaining = 800

Man . Day = 800

80 . Day = 800

Day = 10

16. A cistern is normally filled in 5 hours. However, it takes 6 hours when there is leak in its bottom. If the cistern is full, in what time shall the leak empty it?

(a) 6 h (b) 5 h
(c) 30 h (d) 15 h

Ans. (c)

In one hour $\frac{1}{5}$ part is filled now leak can empty in x hour then

$$\frac{1}{5} - \frac{1}{x} = \frac{1}{6}$$

$$\Rightarrow \frac{1}{5} - \frac{1}{6} = \frac{1}{x}$$

$$\Rightarrow \frac{1}{30} = \frac{1}{x},\ x = 30 \text{ hours}$$

17. Pipe A and B running together can fill a cistern in 6 minutes. If B takes 5 minutes more than A to fill the cistern, then the time in which A and B will fill the cistern separately will be respectively?

(a) 15 min, 20 min (b) 15 min, 10 min
(c) 10 min, 15 min (d) 25 min, 20 min

Ans. (c)

Let pipe A can fill in x and pipe B in $x+5$ minutes

$$\frac{1}{x}+\frac{1}{x+5}=\frac{1}{6}$$

or $$\frac{x(x+5)}{2x+5}=6$$

$$x = 10,\ x + 5 = 15$$

18. There are two pipes in a tank. Pipe A is for filling the tank & Pipe B is for emptying the tank. If A can fill the tank in 10 hours and B can empty the tank in 15 hours then find how many hours will it take to completely fill a half empty tank?

(a) 30 hours (b) 15 hours
(c) 20 hours (d) 33.33 hours

Ans. (b)

$$\frac{1}{10}-\frac{1}{15}=\frac{5}{150}=\frac{1}{30}$$

So it takes 30 hours to fill the tank and 15 hours to fill half the tank

19. There are three Taps A, B and C in a tank. They can fill the tank in 10 hrs, 20 hrs and 25 hrs respectively. At first, all of them are opened simultaneously. Then after 2 hours tap C is closed and A and B are kept running. After the 4th hour, tap B is also closed. The remaining work is done by Tap A alone. Find the percentage of the work done by Tap A by itself.

(a) 32 % (b) 52 %
(c) 75 % (d) None of these

Ans. (d)

	A	B	C		
Ist hours	$\frac{1}{10}$	$\frac{1}{20}$	$\frac{1}{25}$	$\frac{19}{100}$ part	
IInd hours	$\frac{1}{10}$	$\frac{1}{20}$	$\frac{1}{25}$	$\frac{19}{100}$ part	
IIIrdhours	$\frac{1}{10}$	$\frac{1}{20}$	×	$\frac{3}{20}$	$\frac{15}{100}$part
IVthhours	$\frac{1}{10}$	$+\frac{1}{20}$	×	$\frac{3}{20}$	$\frac{15}{100}$ part

So remaining part after 4 hour is

$$\frac{\frac{32}{100}}{1/10}=3.2 \text{ hours}$$

% of work done by A

$$\frac{(10+10+10+10+32)}{100}\times 100=72\%$$

□□□□

Time & Work

I Practice Exercise

1. 10 men can complete a piece of work in 15 days and 15 women can complete the same work in 12 days. If all the 10 men and 15 women work together, in how many will the work get completed?

(a) $6\frac{2}{3}$ days (b) $8\frac{1}{3}$ days
(c) $7\frac{2}{3}$ days (d) None of these

2. A is thrice as good as B and is therefore able to finish a piece of work in 60 days less than B. Find the time in which they can do it, working together.

(a) $22\frac{3}{4}$ days (b) $22\frac{1}{2}$ days
(c) 24 days (d) None of these

3. Ramesh takes twice as much time as Mahesh and thrice as much time as Suresh to complete a job. If working together, they can complete the job in 4 days, then the time taken by each of them separately to complete the work is

(a) 36, 24 and 16 days
(b) 20, 16 and 12 days
(c) 24, 42 and 18 days
(d) None of these

4. 5 men can complete a work in 2 days, 4 women can complete the same work in 3 days and 5 children can do it in 3 days. 1 man, 1 women and 1 child, working together, can complete the work in

(a) 6 days (b) 4 days
(c) 8 days (d) None of these

5. Ajay and Sunil can do a piece of work in 10 days, sunil and Sanjay in 15 days and Sanjay and Ajay in 20 days. They together work at it for 6 days and then Ajay leaves and Sunil and Sanjay go on together for 4 days more. If Sunil then leaves, how long will Sanjay take to complete the work?

(a) 12 days (b) 10 days
(c) 16 days (d) None of these

6. Bansal, Gupta and Singhal together can complete a work in 4 days. If Bansal and Gupta together can complete the work in $4\frac{4}{5}$ days, Gupta and Singhal together can do it in 8 days, then Gupta alone can complete the work in

(a) 16 days (b) 12 days
(c) 20 days (d) None of these

7. Working 7 hours daily 24 men can complete a piece of work in 27 days. In how many days would 14 men complete the same piece of work working 9 hours daily?

(a) 36 days (b) 30 days
(c) 32 days (d) None of these

8. 4 men or 6 women can finish a piece of work in 20 days. In how many days can 6 men and 11 women finish the same work?

(a) 9 days (b) 6 days
(c) 7 days (d) None of these

9. A does half as much work as B in three- fourths of the time. If together they take 18 days to complete a work, how much time shall B take to do it?

(a) 30 days (b) 35 days
(c) 40 days (d) 66 days

10. Two men undertake to do a piece of work for Rs. 600. One alone could do it in 6 days and the other in 8 days. With the assistance of a boy they finish it in 3 days. Boy's share should be

(a) Rs. 300 (b) Rs. 225
(c) Rs. 75 (d) Rs. 100

11. 5 men and 2 boys working together can do four times as much work per hour as a man and a boy together. The work done by a man and a boy should be in the ratio

(a) 1 : 2 (b) 2 : 1
(c) 1 : 3 (d) 4 : 1

12. A and B working separately can do a piece of work in 9 and 12 days, respectively. If they work for a day altenately with A starting the work. In how many days the work will be completed?

(a) $10\frac{1}{2}$ days (b) $10\frac{1}{4}$ days
(c) $10\frac{2}{3}$ days (d) $10\frac{1}{2}$ days

13. A and B can do a piece of work in 45 and 40 days, respectively. They began the work together, but A leaves after some days and B finished the remaining work in 23 days. After how many days, did A leaves?

(a) 6 days (b) 8 days
(c) 9 days (d) 12 days

14. A can do a piece of work in 30 days, B in 50 days and C in 40 days. If A is assisted by B on one day and by C on the next day alternately work will be completed in

(a) $17\frac{32}{35}$ days (b) $19\frac{2}{3}$ days
(c) $16\frac{31}{37}$ days (d) $18\frac{1}{3}$ days

Answers

1. (a)	2. (b)	3. (d)	4. (b)	5. (b)
6. (b)	7. (a)	8. (b)	9. (d)	10. (c)
11. (b)	12. (b)	13. (c)	14. (a)	

Solutions

1. **Ans. (a)**

Here, $x = 15$ and $y = 12$

∴ Working together, 10 men and 15 women will complete the work in

$$= \frac{xy}{x+y} \text{ days}$$

$$= \frac{15 \times 12}{15+12}, \text{ i.e. } \frac{20}{3} \text{ or, } 6\frac{2}{3} \text{ days.}$$

2. **Ans. (b)**

Let B takes x days to do the work. Then, A takes $(x - 60)$ days to complete the work.

Since ratio of work done by A and B is 3 : 1, ratio of time taken by A and B is 1 : 3.

We have, $\frac{x-60}{x} = \frac{1}{3}$

$\Rightarrow 3(x - 60) = x$ or, $x = 90$.

∴ Time taken by B to finish the work = 90 days

and time taken by A to finish the work = $\frac{90}{3} = 30$ days.

∴ A and B, working together, will complete the work in $\frac{xy}{x+y}$ days

$= \frac{90 \times 30}{90 + 30}$ days, i.e. $\frac{45}{2}$

or, $22\frac{1}{2}$ days.

3. Ans. (d)
Let Ramesh takes x days to finish the work.
Then, Mahesh takes $\frac{x}{2}$ and Suresh takes $\frac{x}{3}$ days to finish the same work.
$\therefore$ Ramesh, Mahesh and Suresh, working together, will complete the work in

$$\frac{xyz}{xy + yz + zx} \text{ day}$$

$$= \frac{x \times \frac{x}{2} \times \frac{x}{3}}{x \times \frac{x}{2} + \frac{x}{2} \times \frac{x}{3} + x \times \frac{x}{3}} \text{ days}$$

i.e. $\frac{x^3/6}{x^2}$ or, $\frac{x}{6}$ days

Given : $\frac{x}{6} = 4$. $\therefore$ $x = 24$

$\therefore$ Ramesh takes 24 days. Mahesh takes $\frac{24}{2}$ or 12 days and Suresh takes $\frac{24}{3} = 8$ days to finish the work.

4. Ans. (b)
1 Man can complete the work in 5 × 2 = 10 days,
1 woman can complete the work in 4 × 3 = 12 days.
And 1 child can complete the work in 5 × 3 = 15 days.
$\therefore$ 1 man, 1 woman and 1 child, working together, can complete the work in

$$\frac{xyz}{xy + yz + zx} \text{ days}$$

$$= \frac{10 \times 12 \times 15}{10 \times 12 + 12 \times 15 + 15 \times 10} = 4 \text{ days}$$

5. Ans. (b)
Ajay, sunil and Sanjay, working together, can complete the work in

$$= \left(\frac{2xyz}{xy + yz + zx}\right) \text{ days}$$

$$= \left(\frac{2 \times 10 \times 15 \times 20}{10 \times 15 + 15 \times 20 + 20 \times 10}\right)$$

$$\frac{6000}{650}, \text{ i.e. } \frac{120}{13} \text{ days}$$

$\therefore$ Work done by all of them together in 6 days

$$= \frac{6 \times 13}{120} \text{ i.e. } \frac{13}{20}$$

Also, work done by Sunil and Sanjay in 4 days

$$= \frac{4}{15}$$

$\therefore$ Remaining work

$$= 1 - \left(\frac{13}{20} + \frac{4}{15}\right) = \frac{1}{12}$$

Which is to be done by Sanjay.
Now, Ajay Sunil and Sanjay, can complete the work in $\frac{120}{13}$ days and Ajay and Sunil can complete the work in 10 days.
$\therefore$ Sanjay alone can complete the work in

$$\frac{\frac{120}{13} \times 10}{10 - \frac{120}{13}} = 120 \text{ days.}$$

$\therefore$ $\frac{1}{12}$ of the work is done by Sanjay in $\frac{120}{12} = 10$ days.

6. Ans. (b)
Bansal, Guptal and Singhal together can finish the work in 4 days.

Bansal and Gupta together can do it in $\frac{24}{5}$ days.
Gupta and Singhal together can do it in 8 days.
Therefore, Bansal alone can complete the work in

$$= \frac{xy}{y - x} \text{ days} = \left(\frac{8 \times 4}{8 - 4}\right) \text{ days}$$

[Here, $x = 4$ and $y = 8$ days
Also, Singhal alone can complete the work in

$$= \left(\frac{xy}{x - y}\right) \text{days} = \left(\frac{\frac{24}{5} \times 4}{\frac{24}{5} \times 4}\right) \text{days}$$

Here, $x = 4$ and $y = \frac{24}{5} = 24$ days.
$\therefore$ Bansal and Singhal can complete the work in

$$= \left(\frac{xy}{x + y}\right) \text{ days.} = \left(\frac{24 \times 8}{24 + 8}\right) \text{ days}$$

[Here $x = 24$ and $y = 8$] = 6 days.

Gupta alone can complete the work $\frac{6 \times 4}{6-4} = 12$.

7. Ans. (a)

We have $M_1 = 24$ $D_1 = 27$, $W_1 = 1$, $t_1 = 7$,

$M_2 = 14$, $D_2 = ?$, $W_2 = 1$, $t_2 = 9$,

$M_1D_1t_1W_2 = M_2D_2t_2W_1$

$\Rightarrow 24 \times 27 \times 7 \times 1 = 14 \times D_2 \times 9 \times 1$

$\Rightarrow D_2 = 36$ days.

8. Ans. (b)

Here, a = 4, b = 6, n = 20, c = –6, and d = 11.

∴ Required number of days

$= \left(\frac{nab}{bc+ad}\right)$ days

$= \left(\frac{20 \times 4 \times 6}{6 \times 6 + 4 \times 11}\right)$ days = 6 days.

9. Ans. (d)

If B does x part of work in 1 hour

Then B does $\frac{3}{4}x$ part in $\frac{3}{4}$ hour

Now, A does $\frac{3x}{8}$ work in 1 hour

It is given that $x + \frac{3}{8}x = \frac{1}{18}$

So $x = \frac{4}{99}$

B complete work in $\frac{99}{4}$ days and A in 66 days.

10. Ans. (c)

1st man's 3 days' work = $\frac{3}{6} = \frac{1}{2}$

2nd man's 3 days' work = $\frac{3}{8}$

Boy's 3 days' work = $1 - \left(\frac{1}{2} + \frac{3}{8}\right) = \frac{1}{8}$

∴ They should get money in the ratio

$\frac{1}{2} : \frac{3}{8} : \frac{1}{8}$ i.e. 4 : 3 : 1

∴ Boy's share = $\text{Rs.}\frac{1}{8} \times 600 = \text{Rs. } 75.$

11. Ans. (b)

Obviously.

(5 M + 2B) = 4(1M + 1B)

∴ M = 2B

∴ Work done by a man and boy are in the ratio 2 : 1

12. Ans. (b)

(A + B)'s 2 days' work $\frac{1}{9} + \frac{1}{7} = \frac{7}{36}$ Evidently, the work done by A and B during 5 pairs of days

$= 5 \times \frac{7}{36} = \frac{35}{36}$

Remaining work = $1 - \frac{35}{36} = \frac{1}{36}$

Now, $\frac{1}{9}$ work is done by A in 1 day

∴ $\frac{1}{36}$ work, will be done by A in

$9 \times \frac{1}{36} = \frac{1}{4}$ day

So, total time taken = $10\frac{1}{4}$ days.

13. Ans. (c)

B's 23 days' work = $\frac{23}{40}$

Remaining work = $1 - \frac{23}{40} = \frac{17}{40}$

Now, (A+B)'s 1 day's work

$= \frac{1}{45} + \frac{1}{40} = \frac{17}{360}$

$\frac{17}{360}$ work is done by A and B in 1 day

∴ $\frac{17}{40}$ work is done by A and B in

$\frac{360 \times 17}{17 \times 40} = 9$ days

So, A left after 9 days.

14. Ans. (a)

(A + B)'s 1 day's work = $\frac{1}{30} + \frac{1}{50} = \frac{8}{150}$

∴ (A + C)'s 1 day's work

$= \frac{1}{30} + \frac{1}{40} = \frac{7}{120}$

∴ Work done in first 2 days

$= \frac{8}{150} + \frac{7}{120} = \frac{67}{600}$

Work done in 8 × 2 = 16 days

$= \frac{67 \times 8}{600} = \frac{67}{75}$

Work left $= 1 - \frac{67}{75} = \frac{8}{75}$

On 17 th day(A + B) will work and they will finish $\frac{8}{150}$

$\therefore$ Work left $= \frac{8}{75} - \frac{8}{150} = \frac{8}{150} = \frac{4}{75}$

On 18th day (A + C) will work and they will finish it in $\frac{120}{7} \times \frac{4}{75} = \frac{32}{35}$ days

$\therefore$ Whole work will be done in $17\frac{32}{35}$ days

□□□□

Pipes & Cisterns

Practice Exercise: I

1. One tap can fill a cistern in 2 hours and another can empty the cistern in 3 hours. How long will they take to fill the cistern if both the taps are opened?
(a) 6 hours (b) 7 hours
(c) 6.30 hours (d) None of these

2. A tap can fill a tank in 25 minutes and another can empty it in 50 minutes. Find whether the tank will be filled up or emptied and in how many minutes?
(a) Tank is filled up in 50 minutes
(b) Tank is emptied in 25 minutes
(c) Tank is filled up in 25 minutes
(d) None of these

3. Two taps A and B can fill a tank in 10 hours and 15 hours, respectively. If both the taps are opened together the tank will be full in:
(a) 8 hours (b) 6 hours
(c) 5 hours (d) None of these

4. Two pipes can fill a tank in 10 hours and 12 hours, respectively. While a third pipe can empty the full tank in 20 hours. If all the three pipes operate simultaneously, in how much time the tank will be filled?
(a) 7 hours 30 minutes (b) 6 hours 40 minutes
(c) 8 hours 30 minutes (d) None of these

5. A cistern is normally filled in 8 hours but takes 2 hours longer to fill because of a leak in its bottom. If the cistern is full, the leak will empty it in
(a) 35 hours (b) 45 hours
(c) 40 hours (d) None of these

6. If two pipes function simultaneously, the reservoir will be filled in 12 hours. One pipe fills the reservoir 10 hours faster than the other. How many hours does the faster pipe take to fill the reservoir?
(a) 35 hours (b) 30 hours
(c) 40 hours (d) None of these

7. One fill pipe A is 3 times faster than second fill pipe B and takes 32 minutes less than the fill pipe B. When will the cistern be full if both pipes are opened together?
(a) 28 minutes (b) 12 minutes
(c) 30 minutes (d) data inadequate

8. Two pipes A and B can fill a cistern in 4 minutes and 6 minutes, respectively. If these pipes are turned on alternately for 1 minute each how long will it take for the cistern to fill?
(a) 4 min 40 sec (b) 3 min 20 sec
(c) 4 min 50 sec (d) 3 min 30 sec

9. A cistern is provided by two taps A and B. A can fill it in 20 minutes and B in 25 taps minutes. Both the taps are kept open for 5 minutes and then the second is turned off. The cistern will be completely filled in another
(a) 11 minutes (b) 10 minutes
(c) 15 minutes (d) 12 minutes

10. A cistern has two taps which fill it in 12 minutes and 15 minutes respectively. There is also a waste pipe in the cistern. When all the pipes are opened, the empty cistern is full 20 minutes. How long will the waste pipe take to empty a full cistern?
(a) 8 minutes (b) 10 minutes
(c) 12 minutes (d) 16 minutes

Answers

1. (a)	2. (a)	3. (b)	4. (a)	5. (c)
6. (b)	7. (b)	8. (a)	9. (a)	10. (b)

Solutions

1. **Ans. (a)**
Here, $x = 2$ and $y = 3$
$\therefore$ Part of the cistern filled in 1 hour
$$= \frac{1}{x} - \frac{1}{y} = \frac{1}{2} - \frac{1}{3} = \frac{1}{6}$$
$\therefore$ Total time taken to fill the cistern = 6 hrs.

2. **Ans. (a)**
Here, $x = 25$ and $y = 50$
$\therefore$ Part of the tank filled or emptied in 1 minutes
$$= \frac{1}{x} - \frac{1}{y} = \frac{1}{25} - \frac{1}{50} = \frac{1}{50}$$
Which is positive, therefore the tank will be filled.
$\therefore$ Total time taken to fill the tank
= 50 minutes

3. **Ans. (b)**
Here, $x = 10$ and $y = 15$.
$\therefore$ The tank will be full in
$$= \left(\frac{xy}{x+y}\right) \text{ hours}$$
$$= \left(\frac{10 \times 15}{10+15}\right) \text{ hours or 6 hours.}$$

4. **Ans. (a)**
Here, $x = 10$, $y = 12$ and $z = -20$
$\therefore$ The tank will be full in
$$= \left(\frac{x \times y \times z}{xy - yz - zx}\right) \text{ hours}$$
$$= \left(\frac{10 \times 12 \times -20}{10 \times 12 - 12 \times 20 - 20 \times 10}\right) \text{ hours}$$
$$= \left(\frac{15}{2}\right) \text{ hours or, 7 hours 30 minutes}$$

5. **Ans. (c)**
Here, $x = 8$ and $y = 8 + 2 = 10$.
$\therefore$ The leak will empty the cistern in
$$= \left(\frac{xy}{y-x}\right) \text{ hours}$$
$$= \left(\frac{8 \times 10}{10-8}\right) \text{ hours or, 40 hours}$$

6. **Ans. (b)**
Let one pipe takes x hours to fill the reservoirs. Then, another pipe takes $(x - 10)$ hours.
$$\therefore \quad \frac{1}{x} + \frac{1}{x-10} = \frac{1}{12}$$
$$\Rightarrow \quad x(x-10) = 12(x + x - 10)$$
$$\Rightarrow \quad x^2 - 34x + 120 = 0$$
or $\quad (x-30)(x-4) = 0$
$\therefore \quad x = 30$ or $x = 4$
$\therefore$ The faster pipe takes 30 hours to fill the reservoir.

7. **Ans. (b)**
As pipe 'A' is 3 times faster than pipe 'B'. So if 'A' takes 'x' min, 'B' will take '$3x$' min. to fill tank. According to question $3x - x = 2x = 32$, $x = 16$,
$3x = \frac{4}{8}$.
Both pipe together can fill tank in
$$\frac{1}{\frac{1}{16} + \frac{1}{48}} = 12 \text{ min.}$$

8. **Ans. (a)**
As the pipes are operating alternately, thus their 2 minutes job is $\frac{1}{4} + \frac{1}{6} = \frac{5}{12}$
In the next 2 minutes the pipes can fill another $\frac{5}{12}$ part of cistern. Therefore, in 4 minutes the two pipes which are operating alternately will fill
$$\frac{5}{12} + \frac{5}{12} = \frac{10}{12} = \frac{5}{6} \text{ part.}$$
The part of the cistern left unfilled = $1 - \frac{5}{6} = \frac{1}{6}$
Pipe A can fill $\frac{1}{4}$ of the cistern in 1 minutes.
Pipe A can fill $\frac{1}{6}$ of the cistern in
$$4 \times \frac{1}{6} = \frac{2}{3} \text{ minutes}$$
Total time taken to fill the cistern
$$4 + \frac{2}{3} = 4\frac{2}{3} \text{ min.}$$
Or, 4 minutes 40 seconds.

9. **Ans. (a)**
Part filled in 1 minute $\frac{1}{20} + \frac{1}{25} = \frac{9}{100}$
Part filled in 5 minutes = $\frac{9}{100} \times 5 = \frac{9}{20}$

Unfilled part = $1-\frac{9}{20}=\frac{11}{20}$

This is to be filled by A alone and hence will be filled in $20\times\frac{11}{20}$ = 11 minutes.

10. Ans. (b)

Work done by waste pipe in 1 minutes

$$=\left(\frac{1}{12}+\frac{1}{15}\right)-\frac{1}{20}=\left(\frac{3}{20}-\frac{1}{20}\right)=\frac{1}{10}$$

∴ Waste pipe can empty the cistern in 10 minutes.

□□□□

Pipes & Cisterns

II Practice Exercise: II

1. Two pipes can fill a tank in 10 hours and 12 hours respectively while a third pipe empties the full tank in 20 hours. If all the three pipes operate simultaneously, in how much time the tank will be filled?
   (a) 7 hrs (b) 8 hrs
   (c) 7 hrs 30 min. (d) 8 hrs 30 min.

2. An electric pump can fill a tank in 3 hours. Because of a leak in the tank, it took 3½ hours to fill the tank. The leak can drain out all the water of the tank in:
   (a) 10:30 hrs (b) 12 hrs
   (c) 21 hrs (d) 24 hrs

3. If two pipes function simultaneously, the reservoir will be filled in 12 hours. One pipe fills the reservoir 10 hours faster than the other. How many hours it takes the second pipe to fill the reservoir?
   (a) 25 hrs (b) 28 hrs
   (c) 30 hrs (d) 35 hrs

4. 12 buckets of water fill a tank when the capacity of each bucket is 13.5 liters. How many buckets will be needed to fill the same tank, if the capacity of each bucket is 9 liters?
   (a) 8 (b) 16
   (c) 15 (d) 18

5. Bucket P has thrice the capacity as bucket Q. It takes 60 turns for bucket P to fill the empty drum. How many turns it will take for both the buckets P and Q, having each turn together to fill the empty drum?
   (a) 30 (b) 40
   (c) 45 (d) 90

6. Two pipes A and B can fill a cistern in 12 minutes and 15 minutes respectively but a third pipe C can empty the full tank in 6 minutes, A and B are kept open for 5 minutes in the begining and then C is also opened. In what time is the cistern emptied?
   (a) 30 min. (b) 33 min.
   (c) $37\frac{1}{2}$ min. (d) 45 min.

7. Three pipes A, B and C can fill a tank in 6 hours. After working at it together for 2 hours, C is closed and A and B can fill the remaining part in 7 hours. The number of hours taken by C alone to fill the cistern, is:
   (a) 10 (b) 12
   (c) 14 (d) 16

8. A leak in the bottom of tank can empty the full tank in 8 hours. An inlet pipe fills water at the rate of 6 liters per minute. When the tank is full, the inlet is opened and due to the leak, the tank is empty in 12 hours. How many liters does the cistern hold?
   (a) 7580 (b) 7960
   (c) 8290 (d) 8640

9. A cistern has two taps which fill it in 12 min. and 15 min, respectively. There is also a waste pipe in the cistern. When all the three are opened, the empty cistern is full in 20 minutes. How long will the waste pipe take to empty the full cistern?
   (a) 8 min (b) 10 min
   (c) 12 min (d) 16 min

10. Two pipes A and B can fill a cistern in 12 minutes and 16 minutes respectively. If both the pipes are opened together, then after how much time B should be closed so that the tank is full in 9 minutes?
    (a) $3\frac{1}{2}$ min. (b) 4 min.
    (c) $4\frac{1}{2}$ min. (d) $4\frac{3}{4}$ min.

Answers

1. (c)	2. (c)	3. (c)	4. (d)	5. (c)
6. (d)	7. (c)	8. (d)	9. (b)	10. (b)

Solutions

1. **Ans. (c)**
Net part filled in 1 hour
$= \left(\frac{1}{10}+\frac{1}{12}-\frac{1}{20}\right) = \frac{8}{60} = \frac{2}{15}$
$\therefore$ The tank will be full in $\frac{15}{2}$ hrs.
= 7 hrs. 30 min.

2. **Ans. (c)**
Work done by the leak in 1 hour
$= \left(\frac{1}{3}-\frac{2}{7}\right) = \frac{1}{21}$
$\therefore$ Leak will empty the tank in 21 hours.

3. **Ans. (c)**
Let the reservoir be filled by first pipe in x hours.
The second pipe will fill it in $(x+10)$ hours.
$\therefore \frac{1}{x}+\frac{1}{x+10} = \frac{1}{12} \Leftrightarrow \frac{x+10+x}{x(x+10)} = \frac{1}{12}$
$\Rightarrow x^2 - 14x - 120 = 0$
$\Rightarrow (x-20)(x+6) = 0 \Rightarrow x = 20$
$\therefore$ Second pipe takes 30 hrs to fill the reservoir.

4. **Ans. (d)**
Capacity of the tank = (12 × 13.5) litres
= 162 litres.
Capacity of each bucket = 9 litres
Number of buckets needed = $\left(\frac{162}{9}\right)$ = 18

5. **Ans. (c)**
Let capacity of P be x litres.
Then, capacity of Q = $\frac{x}{3}$ litres.
Capacity of the drum = $60x$ litres.
Required number of turns
$= \frac{60}{\left(x+\frac{x}{3}\right)} = \left(60 \times \frac{3}{4x}\right) = 45.$

6. **Ans. (d)**
Part filled in 5 min
$= 5\left(\frac{1}{12}+\frac{1}{15}\right) = \left(5 \times \frac{9}{60}\right) = \frac{3}{4}$
Part emptied in 1 min., when all the pipes are opened
$= \frac{1}{6}-\left(\frac{1}{12}+\frac{1}{15}\right) = \left(\frac{1}{6}-\frac{3}{20}\right) = \frac{1}{60}$
Now, $\frac{1}{60}$ part is emptied in 1 min
$\therefore \frac{3}{4}$ part will be emptied on $\left(60 \times \frac{3}{4}\right)$ = 45 min.

7. **Ans. (c)**
Part filled in 2 hours = $\frac{2}{6} = \frac{1}{3}$
Remaining part = $\left(1-\frac{1}{3}\right) = \frac{2}{3}$
$\therefore$ (A + B)'s 7 hour's work = 2/3
$\therefore$ (A+B)'s 1 hour's work = 2/21
$\therefore$ C's 1 hour's work
= [(A+B+C)'s 1 hour's work] – (A+B)'s 1 hour's work = $\left(\frac{1}{6}-\frac{2}{21}\right) = \frac{1}{14}$
$\therefore$ C alone can fill the tank in 14 hours.

8. **Ans. (d)**
Work done by the inlet in 1 hour
$= \left(\frac{1}{8}-\frac{1}{12}\right) = \frac{1}{24}$
Work done by the inlet in 1 min.
$= \left(\frac{1}{24} \times \frac{1}{60}\right) = \frac{1}{1440}$
$\therefore$ Volume of $\frac{1}{1440}$ part = 6 litres
$\therefore$ Volume of whole = (1440 × 6) litres
= 8640 litres

9. **Ans. (b)**
Work done by waste pipe in 1 min.
$= \frac{1}{20}-\left(\frac{1}{12}+\frac{1}{15}\right) = -\frac{1}{10}$
[-Ve sign means emptying]
$\therefore$ Waste pipe will empty the full cistern in 10 min.

10. **Ans. (b)**
Let B be closed after x minutes.
Then, Part filled by (A +B) in x min. + Part filled by A in $(9-x)$ min. = 1
$\therefore x\left(\frac{1}{12}+\frac{1}{16}\right)+(9-x).\frac{1}{12} = 1$
or $\frac{7x}{48}+\frac{9-x}{12} = 1$
or $7x + 36 - 4x = 48$ or $x = 4$.
So, B must be closed after 4 minutes.

1.8 CHAPTER

Time, Speed & Distance

Time, Speed & Distance is a very important chapter to understand the logical development and the visualisation capacity of students. The problems on TSD are always favourite to the question setters. So for this chapter we need to come back on the basics of ratio and variation to solve the problem in TSD.

In this chapter, we will study

- Relative speed
- Average speed
- Races
- Circular tracks
- Application of variation

The whole chapter is based on one relation i.e.

$$\text{Speed} = \frac{\text{Distance}}{\text{Time}}$$

The units of speed & distance have to be kept same so if speed has to be converted we can use

$$1 \text{ km/hr} = \frac{5}{18}\text{m/s}$$

or $$1 \text{ m/s} = \frac{18}{5}\text{km/hr}$$

Example 1.

A car travels 200 m with the speed of 72 km/hr. How much time it takes to travel the distance.

Solution.

$$\text{Speed} = 72 \text{ km/hr} = 72\times\frac{5}{18} = 20 \text{ m/s}$$

$$\text{Time} = \frac{200}{20} = 10 \text{ sec}$$

Average Speed

For calculating the average speed the first thing that has to be wiped out from the mind is that the averge speed is not the average value of different speeds. It is the ratio of the total distance travelled to the total time taken to cover that distance.

$$\text{Average speed} = \frac{\text{Total distance}}{\text{Total time}}$$

Example 1.

A car covers first 400 meters with the speed of 72 km/hr and next 300 meters with the speed of 36 km/hr. Find out the average speed of the car.

Solution.

Time taken to cover first 400 cm

$$= \frac{400}{72\times\frac{5}{18}} = 20 \text{ s}$$

Time taken to cover next 300 m

$$= \frac{300}{36\times\frac{5}{18}} = 30 \text{ s}$$

Total time = 50 s

Total distance = 700 m

$$\text{Average speed} = \frac{700}{50} = 14 \text{ m/s}$$

Where equal distances are covered with different uniform speed, the value of distance does not matter while calculating the average speed.

Example 2.

A train covers a distance with the speed of 90 km/hr and then double of the previous distance with the speed of 60 km/hr. Find out its average speed.

Solution.

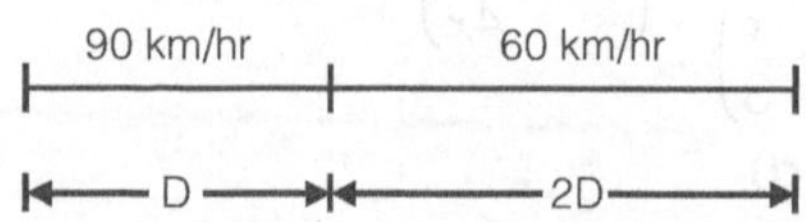

$$\text{Time taken for first distance} = \frac{D}{90}$$

$$\text{Time taken for second distance} = \frac{2D}{60} = \frac{D}{30}$$

Net distance = D + 2D = 3D

So average speed = $\frac{3D}{\frac{D}{90}+\frac{D}{30}} = \frac{3D}{\frac{D+3D}{90}}$

$= \frac{3D}{4D}\times 90 = \frac{270}{4} = 67.5$ km/hr

Relative Speed

Relative speed is very good application part of TSD. Relative speed is defined as the speed of one moving body with respect to another moving body. If two bodies are running with the speed *'a'* and *'b'* then the relative speed of one body when

- They are moving on same plane
  ($a-b$) for same direction
  ($a+b$) for opposite direction
- They are moving as one on another
  (As man on an moving escalator)
  ($a+b$) for same direction
  ($a-b$) for opposite direction

The basic funda of using relative speed in the problems is to make one body as constant and let the other bodies run with the relative speed.

Problems on Trains

When a train crosses anybody it has to cover its own length apart from the length of other object, so relative distance will be the sum of lengths of train and length of the object. If L_T is length of train, L_O is length of the object, U_T is speed of train and U_O is speed of object, then the time taken for crossing will be given by

$$t = \frac{L_T + L_O}{U_T \pm U_O}$$

For same direction we will take –ve sign while for opposite direction we will take +ve sign.

Example 1.

A train running with the speed of 90 km/hr crosses a platform in 20 secs. The length of train is 200 m. Find the length of platform.

Solution.

Here $L_T = 200$ m, $t = 20$ sec

$U_T = 90$ km/hr $= 90\times\frac{5}{18} = 25$ m/s

$U_O = 0,\ L_O = ?$

So $20 = \frac{200+L_O}{25\pm 0}$

$\Rightarrow\ 200 + L_O = 500$

or $L_O = 300$ m

Rivers & Boats

For the case of boats and river the condition is similar as of the trains problems. Only thing to understand is

- Upstream → Moving against the river flow
- Downstream → Moving along the river flow

If U_O is speed of object and U_r is the speed of river then

for upstream → relative speed is $U_O - U_r$

for downstream → relative speed is $U_O + U_r$

Example 1.

A person covers a distance downstream in 3 hrs and returns back in 9 hrs. If the speed of river is 10 m/s, find out the speed of the person.

Solution.

Upstream speed = $U_O - 10$

Downstream speed = $U_O + 10$

Since the distance covered are same

$(U_O - 10)9 = 3(U_O + 10)$

$\Rightarrow\ 9U_O = 3U_O + 30 + 90$

$\Rightarrow\ 9U_O = 120$

$U_O = 20$ m/s

Circular Tracks

When talking about circular track the first thing to be kept in mind that the time interval between any two consecutive meetings is same. There are 3 basic things that can be asked

- First ever meeting time (T_f)
- First meeting time at starting point (T_s)
- No. of distinct point on the track where the meeting occurs. (η)

There are two cases that can be taken.

Case 1.

When two persons are running on the track. If the length of the track be D and the speed of persons be a and b then

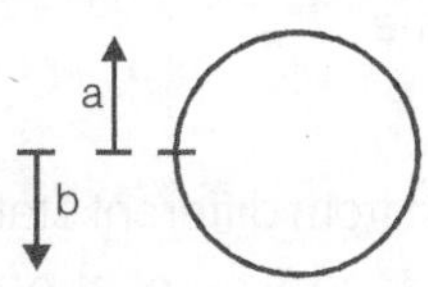

$T_f = \frac{D}{a\pm b}$ (based on direction)

$$T_s = \text{LCM of } \left\{\frac{D}{a}, \frac{D}{b}\right\}$$

$$\eta = \frac{T_s}{T_f}$$

Case 2.

When more than two persons are running then

$$T_f = \text{LCM of } \left\{\frac{D}{b-a}, \frac{D}{c-a}, \frac{D}{d-a}.....\right\}$$

$$T_s = \text{LCM of } \left\{\frac{D}{a}, \frac{D}{b}, \frac{D}{c}.....\right\}$$

$$\eta = \frac{T_s}{T_f}$$

Example 1.

Two persons start running on a circular track with speed 20 m/s & 30 m/s in opposite direction. If length of the track is 100 m, then find out at how many distinct point they will meet.

Solution.

For opposite direction $T_f = \dfrac{100}{20+30} = 2\text{s}$

$$T_s = \text{LCM of } \frac{100}{20}, \frac{100}{30} = \frac{100}{10} = 10 \text{ s}$$

$$\text{So } \eta = \frac{T_s}{T_f} = \frac{10}{2} = 5$$

Application of Variation

There are three different sort of variations possible in TSD

- When distance is constant

$$\text{Speed} \propto \frac{1}{\text{Time}} \text{ (inverse variation)}$$

As per this variation if speed becomes m/n times of previous speed then time will become n/m times of previous time.

Example 1.

Two trains start from different stations towards each other at same time & meet at point P. From point P both trains take 4 hrs & 9 hrs to reach their destination. Find out the ratio of their speeds.

Solution.

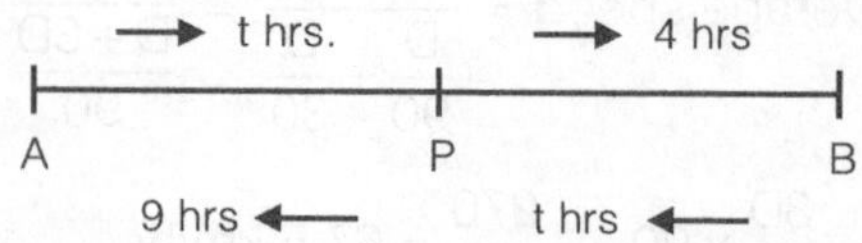

Let us assume that train takes 't' time to reach at 'P', then

$$\Rightarrow \text{ ratio of speeds} = \frac{1}{\text{ratio of times}}$$

$$\Rightarrow \frac{A}{B} = \frac{1}{t/9} = \frac{9}{t} \quad ...(i)$$

Similarly for PB

$$\frac{A}{B} = \frac{1}{4/t} = \frac{t}{4} \quad ...(ii)$$

By eq. (i) & (ii)

$$\frac{9}{t} = \frac{t}{4} \Rightarrow t = 6$$

$$\text{So } \frac{A}{B} = \frac{9}{6} = \frac{3}{2}$$

Example 2.

Two trains start from Delhi & Mumbai towards each other at 10 am in the morning & they reach their destination on same day at 4 pm & 8 pm respectively find out at what time they cross each other.

Solution

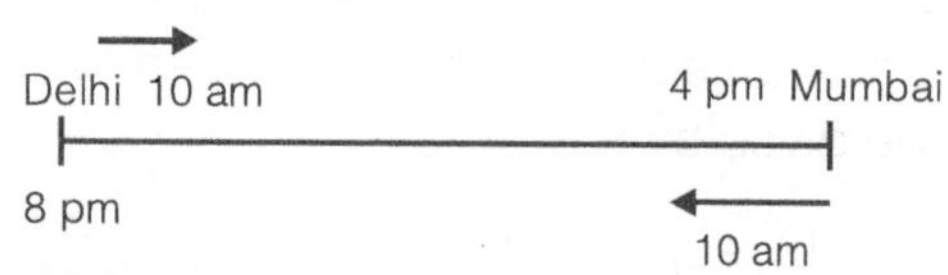

So Time taken by Delhi train = 6 hrs.

Time taken by Mumbai train = 10 hrs

$$\text{So speed ratio } \frac{D}{M} = \frac{10}{6} \text{ (inverse variation)}$$

$$= \frac{5}{3}$$

Let speed of D = 5 K

M = 3 K

The distance between Delhi & Mumbai

= 5 K × 6 = 30 K unit

or 3 K × 10 = 30 K unit

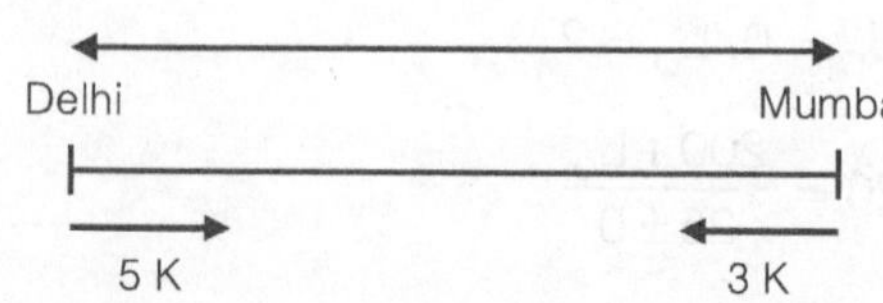

So now the problem is two trains are at a distance 30 K apart running towards each other

So time taken to meet = $\dfrac{30K}{5K+3K}=\dfrac{15}{4}$hrs

= 3 hr 45 min

So both trains will meet at 1 : 45 pm

Races

In races there are two sort of comparisons which are generally carried out.

- Comparison of distance & speed.
- Comparison of speed & time.

Let us see some examples.

Example 1.

In a 100 meter race, A beats B by 20 m & B beats C by 20 m. Find out by what distance A beats C.

Solution

A beats B by 20 m

⇒When A covers 100 m, B covers 80 m so ratio of speeds of A : B = 10 : 8 = 5 : 4

Similarly B : C = 5 : 4

So A : C = 25 : 16

So, when A covers 25 m, 'C' covers 16 m. So, when A will cover 100 m, 'C' will cover 16 × 4 = 64 m.

So 'A' beats 'C' by 36 meters.

□□□□

Solved Examples

1. Walking at 3/4 of his normal speed, Ankur is 16 minutes late in reaching his office. The usual time taken by him to cover the distance between his home and his office is

(a) 48 minutes (b) 60 minutes
(c) 42 minutes (d) 62 minutes

Ans. (a)

Speed × Time = Distance

S × T = D

Here S is changed to $\frac{3}{4}$S, so T must be changed to $\frac{4}{3}$T to keep D constant i.e.

$$\frac{3}{4}S\times\frac{4}{3}T=D$$

Now $\frac{4}{3}T=T+16$

So $\frac{1}{3}T=16$; T = 48 minutes

2. Two trains, Calcutta Mail and Bombay Mail, start at the same time from stations Calcutta and Bombay respectively towards each other. After passing each other, they take 12 hours and 3 hours to reach Bombay and Calcutta respectively. If the Calcutta Mail is moving at the speed of 48 km/h, the speed of the Bombay Mail is

(a) 24 km/h (b) 22km/h
(c) 21 km/h (d) 96 km/h

Ans. (d)

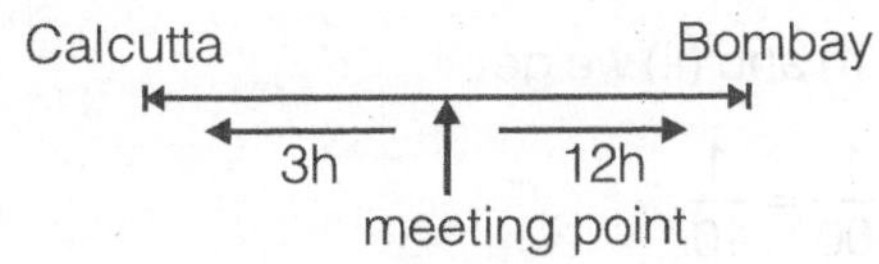

in this case $\dfrac{S_1}{S_2}=\dfrac{\sqrt{T_2}}{\sqrt{T_1}}$

$=\dfrac{48}{S_2}=\dfrac{\sqrt{3}}{\sqrt{12}}=\dfrac{1}{2}$ $S_2 = 96$ km/h

3. Rajdhani Express travels 650 km in 5 h and another 940 km in 10 h. What is the average speed of train?

(a) 159 km/h (b) 63 km/h
(c) 106 km/h (d) 126 km/h

Ans. (c)

Average speed = $\dfrac{\text{Total Distance Travelled}}{\text{Total Time taken}}$

$=\dfrac{650+940}{15}$ = 106 km/h

4. A car travels from A to B at V_1 km/h, travels back from B to A at V_2 km/h and again goes back from A to B at V_2 km/h. The average speed of the car is :

(a) $\dfrac{2V_1V_2}{V_1+2V_2}$ (b) $\dfrac{2V_1V_2}{V_2+2V_1}$

(c) $\dfrac{3V_1V_2}{V_2+2V_1}$ (d) $\dfrac{3V_1V_2}{V_1+2V_2}$

Ans. (c)

$$\text{Average speed} = \frac{\text{Total Distance}}{\text{Total Time}}$$

$$= \frac{3D}{\frac{D}{V_1}+\frac{D}{V_2}+\frac{D}{V_2}} = \frac{3}{\frac{1}{V_1}+\frac{2}{V_2}} = \frac{3V_1V_2}{2V_1+V_2}$$

5. Narayan Murthy walking at a speed of 20 km/h reaches his college 10 minutes late. Next time he increases his speed by 5 km/h, but finds that he is still late by 4 minutes. What is the distance of his college from his house.
(a) 20 km (b) 6 km
(c) 12 km (d) None of these

Ans. (d)
Let distance be D km

$$\frac{D}{20} = T + \frac{10}{60} \quad \text{...(i)}$$

$$\text{Also } \frac{D}{25} = T + \frac{4}{60} \quad \text{...(ii)}$$

From (i) and (ii) we get,

$$D \times \frac{1}{100} = \frac{1}{10}$$

$$D = \frac{100}{10} = 10 \text{ km}$$

6. A motor car does a journey in 17.5 hours, covering the first half at 30 km/h and the second half at 40 km/h. Find the distance of the journey.
(a) 684 km (b) 600 km
(c) 120 km (d) 540 km

Ans. (b)
Here Total time = 17.5 hours
let total Distance be 2 D km

$$\text{then } \frac{D}{30}+\frac{D}{40} = 17.5$$

$$D\left[\frac{1}{30}+\frac{1}{40}\right] = 17.5$$

$$D \times \frac{7}{120} = 17.5$$

$$D = \frac{17.5 \times 120}{7}$$

$$D = 300 \text{ km}$$

Total Distance = 2D = 600 km

7. Manish travels a certain distance by car at the rate of 12 km/h and walks back at the rate of 3 km/h. The whole journey took 5 hours. What is the distance he covered on the car?
(a) 12 km (b) 30 km
(c) 15 km (d) 6 km

Ana. (a)
Let Distance be D km

$$\frac{D}{12}+\frac{D}{3} = 5 \quad \Rightarrow \quad D\left(\frac{1}{12}+\frac{1}{3}\right) = 5$$

$$D = \frac{5 \times 12}{5} = 12 \text{ km}$$

8. Two trains A and B start simultaneously in the opposite direction from two points A and B and arrive at their destinations 9 and 4 hours respectively after their meeting each other. At what rate does the second train B travel if the first train travels at 80 km per hour.
(a) 60 km/h (b) 100 km/h
(c) 120 km/h (d) None of these

Ans. (c)

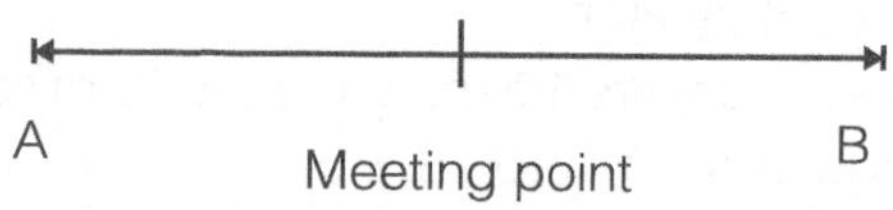

$$\text{Here } \frac{S_1}{S_2} = \frac{\sqrt{T_2}}{\sqrt{T_1}}$$

$$\frac{80}{S_2} = \frac{\sqrt{4}}{\sqrt{9}} \qquad \frac{80}{S_2} = \frac{2}{3}$$

S_2 = 120 km/hour

9. A journey of 192 km takes 2 hours less by a fast train than by a slow train. If the average speed of the slow train be 16 kmph less than that of fast train, what is the average speed of the faster train.
(a) 32 kmph (b) 16 kmph
(c) 12 kmph (d) 48 kmph

Ans. (d)
let speed of fast train be S km/hour then

$$\frac{192}{S-16} - \frac{192}{S} = 2 \quad \text{...(i)}$$

This type of questions can be directly solved by going through options.
Here using equation (i) and putting options.
We get S = 48 km/hour.

10. A passenger train takes 2 h less for a journey of 300 kilometres if its speed is increased by 5 kmph over its usual speed. Find the usual speed.

(a) 10 kmph (b) 12 kmph
(c) 20 kmph (d) 25 kmph

Ans. (d)

$$\frac{300}{S}-\frac{300}{S+5}=2 \quad \text{...(i)}$$

Going directly through options and using equation (i) we get,

S = 25 km/hour

11. A plane left half an hour later than the scheduled time and in order to reach its destination 1500 kilometre away in time, it had to increase its speed by 33.33 percent over its usual speed. Find its increased speed.

(a) 250 kmph (b) 500 kmph
(c) 750 kmph (d) None of these

Ans. (c)

$33.33\% = \frac{1}{3}$ of normal speed

$$\frac{D}{S}-\frac{D}{S+\frac{1}{3}S}=\frac{1}{2}$$

$$\frac{1500}{S}-\frac{1500}{(4/3)S}=\frac{1}{2} \quad \text{...(i)}$$

Going through options and putting in equation (i) we get,

S = 750 km/hour

12. A car travels 1/3 of the distance on a straight road with a velocity of 10 km/h, the next 1/3 with a velocity of 20 km/h and the last 1/3 with a velocity of 60 km/h. What is the average velocity of the car for the whole journey?

(a) 18 km/h (b) 10 km/h
(c) 20 km/h (d) 15 km/h

Ans. (a)

$$\text{Average speed} = \frac{\text{Distance Covered}}{\text{Time taken}}$$

$$=\frac{\frac{D}{3}+\frac{D}{3}+\frac{D}{3}}{\frac{D}{3}\times\frac{1}{20}+\frac{D}{3}\times\frac{1}{10}+\frac{D}{3}\times\frac{1}{60}}=\frac{3}{\frac{1}{20}+\frac{1}{10}+\frac{1}{60}}$$

= 18 km/hour

13. Two trains for Patna leave Delhi at 6 a.m. and 6.45 a.m. and travel at 98 kmph and 136 kmph respectively. How many kilometres from Delhi will the two trains meet?

(a) 262.4 km (b) 260 km
(c) 200 km (d) None of these

Ans.(d)

Delhi ⟵⟶ Patna

Distance travelled by 1st train in 45 minutes

$$= 98\times\frac{45}{60} = 73.5 \text{ km} = \frac{147}{2} \text{ km}$$

Relative speed = $S_2 - S_1$ = 136 − 98 = 38 km/hr

Time taken to cover 73.5 km is

$$\frac{73.5}{38}=\frac{147}{2/38}=\frac{147}{76} \text{ hour}$$

$$\text{Distance} = \frac{147}{76}\times 98 = 189.55$$

Total distance = 73.5 + 189.55
= 263 km approx.

14. A motorboat went downstream for 28 km and immediately returned. It took the boat twice as long to make the return trip. If the speed of the river flow were twice as high, the trip downstream and back would take 672 minutes. Find the speed of the boat in still water and the speed of the river flow.

(a) 9 km/h, 3 km/h (b) 9 km/h, 6 km/h
(c) 8 km/h, 2 km/h (d) 12 km/h, 3 km/h

Ans.(a)

Let speed of boat be S_B

Speed of stream = S_S

$$\text{Then } \frac{28}{S_B - S_S} = 2\times\frac{28}{S_B + S_S}$$

$$\text{Also } \frac{28}{S_B + S_S}+\frac{28}{S_B - S_S} = 672 \text{ minutes}$$

Going directly by option

S_B = 9 km

S_S = 3 km

15. A train requires 7 seconds to pass a pole while it requires 25 seconds to cross a stationary train which is 378 metres long. Find the speed of the train.

(a) 75.6 km/h (b) 75.4 km/h
(c) 76.2 km/h (d) 21 km/h

Ans. (a)

To cross a pole

$$t = \frac{L_T}{S_T} = \frac{\text{length of train}}{\text{speed of train}}$$

$$7 = \frac{L_T}{S_T} \quad ...(i)$$

Also to cross a stationary train $t = \frac{L_T + L_S}{S_T}$

L_S = length of stationary train

$$25 = \frac{L_T + 378}{S_T} \quad ...(ii)$$

From (i) and (ii)

$7 \times S_T = L_T$

$25 \times S_T = L_T + 378$

$18\, S_T = 378$

$S_T = 21 \text{ m/s} = 21 \times \frac{18}{5} = 75.6 \text{ km/h}$

16. A boat sails down the river for 10 km and then up the river for 6 km. The speed of the river flow is 1 km/h. What should be the minimum speed of the boat for the trip to take a maximum of 4 hours?

(a) 2 kmph (b) 3 kmph
(c) 4 kmph (d) 5 kmph

Ans. (c)

$$\frac{D_1}{S_B + S_S} + \frac{D_2}{S_B - S_S} = 4 \text{ hour}$$

$$\frac{10}{S_B + 1} + \frac{6}{S_B - 1} = 4 \text{ hour} \quad ...(i)$$

Going by option and (i) we get,

$S_B = 4$ km/hour

17. Two trains are running on parallel lines in the same direction at speeds of 40 kmph and 20 kmph respectively. The faster train crosses a man in the second train in 36 seconds. The length of the faster train is

(a) 200 metres (b) 185 metres
(c) 225 metres (d) 210 metres

Ans. (a)

t = time required

$$\frac{L_T + L_o}{S_T - S_o} \quad ...(i)$$

$S_T - S_O = 40 - 20 = 20$ km/h

$= 20 \times \frac{5}{18} = \frac{50}{9} \text{m/s}$

Now $t = \frac{L_T}{50/9} \Rightarrow 36 = \frac{L_T}{50/9}$

So, $L_T = 200$ m

18. The speed of the boat in still water is 12 km/h and the speed of the stream is 2 km/h. A distance of 8 km, going upstream, is covered in

(a) 1 hr (b) 1 hr 15 min
(c) 1 hr 12 min (d) None of these

Ans. (d)

$$t = \frac{D}{S_B - S_S} \text{ (upstream)}$$

$$= \frac{8}{12-2} = \frac{8}{10} = \frac{8}{10} \times 60 = 48 \text{ min}$$

19. A boat goes 15 km upstream in 80 minutes. The speed of the stream is 5 km/h. The speed of the boat in still water is

(a) 16.25 km/h (b) 16 km/h
(c) 15 km/h (d) 17 km/h

Ans. (a)

$$t = \frac{D}{S_B - S_S} \text{ (upstream)}$$

$$\frac{80}{60} = \frac{15}{S_B - 5} \Rightarrow \frac{4}{3} = \frac{15}{S_B - 5}$$

$S_B = 16.25$ km/h

20. In a stream, B lies in between A and C such that it is equidistant from both A and C. A boat can go from A to B and back in 6 hr 30 minutes while it goes from A to C in 9 hr. How long would it take to go from C to A?

(a) 3.75 hr (b) 4 hr
(c) 4.25 hr (d) 4.5 hr

Ans. (b)

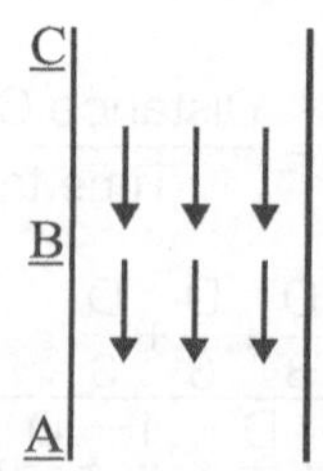

Let AB = D km then

BC = D km

AC = 2D km

$$\frac{D}{S_B - S_S} + \frac{D}{S_B + S_S} = \frac{13}{2} \quad ...(i)$$

$$\frac{2D}{S_B - S_S} = 9 \qquad ...(ii)$$

So $$\frac{D}{S_B - S_S} = \frac{9}{2} \qquad ...(iii)$$

from (i) and (iii) we get,

$$\frac{9}{2} + \frac{D}{S_B + S_S} = \frac{13}{2}$$

$$\frac{D}{S_B + S_S} = 2$$

to go from C to A it takes

$$\frac{2D}{S_B + S_S} = 4 \text{ hours}$$

21. Two trains are travelling in the same direction at 50 km/h and 30 km/h respectively. The faster train crosses a man in the slower train in 18 seconds. Find the length of the faster train.

(a) 0.1 km (b) 1 km
(c) 1.5 km (d) 1.4 km

Ans. (a)

$$t = \frac{L_T + L_o}{S_T - S_o}$$

50 – 30 = 20 km/hr = 50/9 m/s

$$\text{now } 18 = \frac{L_T}{50/9} \Rightarrow L_T = 100 \text{ m}$$

22. Without stoppage, a train travels at an average speed of 75 km/h and with stoppages it covers the same distance at an average speed of 60 km/h. How many minutes per hour does the train stop?

(a) 10 minutes (b) 12 minutes
(c) 14 minutes (d) 18 minutes

Ans. (b)

Speed of trains are 75 km/h and 60 km/h. and to cover a distance of 60 km train will take 60 minutes in second case (with stoppage) while without stoppage it will take $\frac{60}{75}$ = 48 minuts

So, stoppage = 12 minutes

23. A boat rows 16 km up the stream and 30 km down stream taking 5 h each time. The velocity of the current

(a) 1.1 km/h (b) 1.2 km/h
(c) 1.4 km/h (d) 1.5 km/h

Ans. (c)

$$\frac{16}{S_B - S_S} = 5 \text{ hour}$$

$$\frac{30}{S_B + S_S} = 5 \text{ hour}$$

$$16 = 5S_B - 5\,S_S \qquad ...(i)$$
$$30 = 5\,S_B + 5\,S_S \qquad ...(ii)$$
$$14 = 10\,S_S$$
$$S_S = 1.4 \text{ km/hr}$$

24. A man can row 30 km upstream and 44 km downstream in 10 hours. It is also known that he can row 40 km upstream and 55 km downstream in 13 hours. Find the speed of the man in still water.

(a) 4 km/h (b) 6 km/h
(c) 8 km/h (d) 12 km/h

Ans. (c)

$$\frac{30}{S_B - S_S} + \frac{44}{S_B + S_S} = 10 \text{ also}$$

$$\frac{40}{S_B - S_S} + \frac{55}{S_B + S_S} = 13$$

Let $\frac{1}{S_B - S_T} = x$ and $\frac{1}{S_B + S_S} = y$

Then

$$30x + 44y = 10 \qquad ...(i)$$
$$40x + 165 = 13 \qquad ...(ii)$$

By solving we get,

$S_B = 8$ km/hr.

25. In a stream that is running at 2 km/h, a man goes 10 km upstream and comes back to the starting point in 55 minutes. Find the speed of the man in still water.

(a) 20 km/h (b) 22 km/h
(c) 24 km/h (d) 28 km/h

Ans.(b)

$$\frac{10}{S_B - 2} + \frac{10}{S_B + 2} = \frac{55}{60} = \frac{11}{12} \qquad ...(i)$$

On solving e.g. (i) we get,

$S_B = 22$ km/hr

26. A motorboat went down the river for 14 km and then up the river for 9 km. It took a total of 5 hours for the entire journey. Find the speed of the river flow if the speed of the boat in still water is 5 km/h.

(a) 1 kmph (b) 1.5 kmph
(c) 2 kmph (d) 3 kmph

Ans.(c)

$$\frac{14}{S_B + S_S} + \frac{9}{S_B - S_S} = 5$$

$$\frac{14}{5 + S_S} + \frac{9}{5 - S_S} = 5$$

Using equation and going by option we get

$S_S = 2\text{km/hr}$

27. A motorboat whose speed in still water is 10 km/h went 91 km downstream and then returned to its starting point. Calculate the speed of the river flow if the round trip took of 20 hours.

(a) 3 km/h (b) 4 km/h
(c) 2 km/h (d) 8 km/h

Ans.(a)

$$\frac{91}{S_B + S_S} + \frac{91}{S_B - S_S} = 20$$

$$\frac{91}{10 + S_S} + \frac{91}{10 - S_S} = 20$$

$S_S = 3$ km/hr.

□□□□

Time, Speed & Distance

I Practice Exercise

1. A train passes a railway bridge 150 m long in 18 seconds. If the train is running at a speed of 60 km/hr., then the length of the train in metres is

(a) 160 m (b) 150 m
(c) 180 m (d) None of these

2. Sound travels 330 metres a second. The sound of a thunder-cloud follows the flash after 10 seconds. The thunder-could is at a distance of

(a) 3.7 km (b) 3.5 km
(c) 3.3 km (d) None of these

3. The wheel of an engine is $3\frac{3}{4}$ metres in circumference and makes 4 revolutions in 2 seconds. The speed of the train is

(a) 27 km/hr (b) 31 km/hr
(c) 35 km/hr (d) None of these

4. On a tour a man travels at the rate of 64 km an hour for the first 160 km, then travels the next 160 km at the rate of 80 km an hour. The average speed in km per hour for the first 320 km of the tour is

(a) 81.13 km/hr (b) 73.11 km/hr
(c) 71.11 km/hr (d) None of these

5. Rakesh sets out to cycle from Delhi to Mathura and at the same time Suresh starts from Mathura to Delhi. After passing each other they complete their journeys in 9 and 16 hours, respectively. At what speed does Suresh cycle if Rakesh cycles at 16 km per hour?

(a) 12 km/hr (b) 16 km/hr
(c) 14 km/hr (d) None of these

6. A car during its journey travels 30 minutes at a speed of 40 km/hr, another 45 minutes at a speed of 60 km/hr, and 2 hours at a speed of 70 km/hr. The average speed of the car is

(a) 63 km/hr (b) 65 km/hr
(c) 70 km/hr (d) None of these

7. By walking at $\frac{3}{4}$ of his usual speed, a man reaches office 20 minutes later than usual. His usual time is

(a) 65 minutes (b) 60 minutes
(c) 70 minutes (d) None of these

8. A car starts from A for B travelling 20 km an hour. $1\frac{1}{2}$ hours later another car starts from A and travelling at the rate of 30 km an hour reaches B $2\frac{1}{2}$ hours before the first car. Find the distance from A to B

(a) 280 km (b) 260 km
(c) 240 km (d) None of these

9. A train does a journey without stopping in 8 hours. If it had travelled 5 km an hour faster, it would have done the journey in 6 hours 40 min. What is its slower speed?

(a) 35 km/hr (b) 25 km/hr
(c) 40 km/hr (d) None of these

10. Without any stoppage a person travels a certain distance at an average speed of 42 km/hr and with stoppages he covers the same distance at an average speed of 28 km/hr. How many minutes per hour does he stop?
(a) 25 minutes (b) 30 minutes
(c) 20 minutes (d) None of these

11. A train 300 metres long is running at a speed of 90 km/hr. How many seconds will it take cross a 200 metres long train running in the same direction at a speed of 60 km/hr?
(a) 70 sec. (b) 60 sec.
(c) 50 sec. (d) None of these

12. Two trains are running in opposite directions with the same speed. If the length of each train is 135 metres and they cross each other in 18 seconds, the speed of each train is
(a) 29 km/hr (b) 35 km/hr
(c) 27 km/hr (d) None of these

13. Two trains are moving in the same direction at 50 km/hr and 30 km/hr. The faster train crosses a man in the slower train in18 seconds. Find the length of the faster train.
(a) 120 m (b) 110 m
(c) 100 m (d) None of these

14. Two trains, 130 m and 110 m long, while going in the same direction, the faster traintakes one minutes to pass the other completely. If they are moving in opposite direction, they pass each other completely in 3 seconds. Find the speed of each train.
(a) 42 m/sec., 38 m/sec.
(b) 38 m/sec., 36 m/sec.
(c) 36 m/sec., 42 m/sec.
(d) None of these

15. Two stations A and B are 100 km apart on a straight line. One train starts from A at 7 A.M. and travels towards B at 20 km/hr speed. another train starts from B at 8 A.M. and travels towards A at 25 km/hr. speed. At what time will they meet?
(a) 10.30 A.M. (b) 11 A.M
(c) 10 A.M (d) None of these

16. Two trains start at the same time from Mumbai and Pune and proceed towards each other at the rate of 60 km and 40 km per hour. respectively. When they meet, it is found that one train has travelled 20 km more than the other. Find the distance between Mumbai and Pune.
(a) 150 km (b) 100 km
(c) 120 km (d) None of these

17. A and B are two stations. A train goes from A to B at 64 km/hr and returns to A at a slower speed. If its average speed for the whole journey is 56 km/hr, at what speed did it return?
(a) 48 km/hr (b) 49.77 km/hr
(c) 30 km/hr (d) 47.46 km/hr

18. Ramesh sees a train passing over 1 km long bridge. The length of the train is half that of bridge. If the train clears the bridge in 2 minutes, the speed of the train is
(a) 45 km/hr (b) 43 km/hr
(c) 50 km/hr (d) None of these

19. A motor cyclist goes from Mumbai to Pune, a distance of 192 kms, at an average speed of 32 km p.h. Another man starts from Mumbai by car, $2\frac{1}{2}$ hours after the first and reaches Pune half an hour earilier. What is the ratio of the speed of the motorcycle and the car?
(a) 1 : 2 (b) 1 : 3
(c) 10 : 27 (d) 5 : 4

20. A person sets to cover a distance of 12 km in 45 minutes. If he covers $\frac{3}{4}$ of the distance in $\frac{2}{3}$rd time, what should be his speed to cover the remaining distance in the remaining time?
(a) 16 km/hr (b) 8 km/hr
(c) 12 km/hr (d) 14 km/hr

21. A train 110 metres in length passes a man walking at the speed of 6 km/hr. against it in 6 seconds. The speed of the train in km per hour is
(a) 60 km/hr (b) 45 km/hr
(c) 50 km/hr (d) 55 km/hr

Answers

1. (b)	2. (c)	3. (a)	4. (c)	5. (a)
6. (a)	7. (b)	8. (c)	9. (b)	10. (c)
11. (b)	12. (c)	13. (c)	14. (a)	15. (d)
16. (b)	17. (b)	18. (a)	19. (a)	20. (c)
21. (a)				

Solutions

1. Ans. (b)

Let the length of the train be x m.

$\therefore$ Total distance covered by the train = $(x + 150)$m

Speed of the train = 60 km/hr

$$= 60 \times \frac{5}{18} = \frac{50}{3} \text{ m/sec}$$

Since, Distance = Speed × time

$$\therefore \quad x + 150 = \frac{50}{3} \times 18 = 300$$

or, $x = 300 - 150 = 150$ m.

$\therefore$ Length of the train = 150 m.

2. Ans. (c)

Distance of thunder-cloud

= distance travelled by sound in 10 sec.

= (330 × 10) metres

= 3.3 km.

3. Ans. (a)

Distance covered in 2 seconds

$$= \frac{15}{4} \times 4 = 15 \text{ m.}$$

$$\therefore \quad \text{Speed} = \frac{\text{Distance}}{\text{Time}} = \frac{15}{2} \text{ m/sec}$$

$$= \left(\frac{15}{2} \times \frac{18}{5}\right) \text{km/hr} = 27 \text{ km/hr.}$$

4. Ans. (c)

Let the speed on the return journey be x km/hr.

$$\text{Then, } 56 = \left(\frac{2s_1s_2}{s_1 + s_2}\right) = \frac{2 \times 64 \times x}{64 + x}$$

$\therefore$ $7(64 + x) = 16x$ or $9x = 448$

$$\therefore \quad x = \frac{448}{9} = 49.78 \text{ km/hr.}$$

5. Ans. (a)

$$\frac{\text{Rakesh's speed}}{\text{Suresh's speed}} = \frac{\sqrt{T_2}}{\sqrt{T_1}} = \frac{\sqrt{16}}{\sqrt{9}} = \frac{4}{3}$$

$$\therefore \quad \text{Suresh's speed} = \frac{3}{4} \text{ Rakesh's speed}$$

$$= \frac{3}{4} \times 16 = 12 \text{km/hr}$$

6. Ans. (a)

Here, $T_1 = \frac{30}{60}$, $T_2 = \frac{45}{60}$, $T_3 = 2$, $s_1 = 40$

$s_2 = 60$ and $s_3 = 70$.

$\therefore$ The average speed of the car

$$= \frac{s_1T_1 + s_2T_2 + s_3T_3}{T_1 + T_2 + T_3}$$

$$= \frac{40 \times \frac{30}{60} + 60 \times \frac{45}{60} + 70 \times 2}{\frac{30}{60} + \frac{45}{60} + 2}$$

= 63 km/hr.

7. Ans. (b)

Here, change in time = 20 and $\frac{a}{b} = \frac{3}{4}$

We have, change in time

$$= \left(\frac{b}{a} - 1\right) \times \text{original time}$$

$$\Rightarrow \text{Original time} = \frac{\text{Change in time}}{\left(\frac{b}{a} - 1\right)}$$

$$= \frac{20}{\left(\frac{4}{3} - 1\right)} = 60 \text{ minutes}$$

8. Ans. (c)

Here, difference in speed = 30 – 20 = 10.

Difference in time = $2\frac{1}{2} + 1\frac{1}{2} = 4$.

And product of speed = 20 × 30 = 600.

We have,

$$\frac{\text{Product of speed}}{\text{d}} = \frac{\text{difference of speed}}{\text{difference of time}}$$

$$\Rightarrow \text{d} = \text{product of speed} \times \left(\frac{\text{Diff. of time}}{\text{Diff. of speed}}\right)$$

$$600 \times \frac{4}{10} = 240 \text{km}$$

9. Ans. (b)

Let the slower speed = s km/hr.

Since the distance travelled is same in both the cases therefore,

$$\frac{s_1}{T_2} = \frac{s_2}{T_1} \Rightarrow s_1 \times T_1 = s_2 \times T_2$$

$\Rightarrow \quad s \times 8 = (s+5) \times \frac{20}{3}$

$\Rightarrow \quad 24s = 20(s+5)$

$\therefore \quad s = 25$ km/hr.

10. Ans. (c)

Here, $s_1 = 42$ and $s_2 = 28$.

$\therefore$ Stoppage time/hr $= \frac{s_1 - s_2}{s_1} = \frac{42-28}{42}$

$= \frac{1}{3}$ hour = 20 minutes

11. Ans. (b)

Here, $L_1 = 300$ m, $L_2 = 200$ m,
$s_1 = 90$ km/hr and $s_2 = 60$ km/hr

$\therefore \quad s_1 - s_2 = 90 - 60 = 30 \text{ km/hr} = 30 \times \frac{5}{18}$ m/s

$\therefore$ Time taken $= \frac{L_1 + L_2}{s_1 - s_2} = \frac{300+200}{30 \times \frac{5}{18}}$

$= \frac{500 \times 18}{30 \times 5} = 60$ sec.

12. Ans. (c)

Let the speed of each train be x m/sec.

We have, $L_1 = L_2 = 135$ m

and $S_1 = S_2 = x$ m/sec.

$\therefore$ Therefore time taken $= \frac{L_1 + L_2}{S_1 + S_2}$

$\Rightarrow 18 = \frac{135+135}{x+x}$

or, $x = \frac{270}{2 \times 18}$ m/s

$= \frac{270}{2 \times 18} \times \frac{18}{5} \text{ km/hr} = 27 \text{ km/hr}.$

13. Ans. (c)

Relative speed = (50 – 30)km/hr = 20 km/hr

$= \left(20 \times \frac{5}{18}\right) = \left(\frac{50}{9}\right)$ m/sec.

Distance covered in 18 sec at this speed

$= \left(18 \times \frac{50}{9}\right)$ m = 100 m.

$\therefore$ Length of faster train = 100 m.

14. Ans. (a)

Speed of the faster train

$= \left(\frac{L_1 + L_2}{2}\right)\left(\frac{T_1 + T_2}{T_1 T_2}\right) = \left(\frac{130+110}{2}\right)\left(\frac{60+3}{60 \times 3}\right)$

= 42 m/sec.

Speed of the slower train

$\left(\frac{L_1 + L_2}{2}\right) \times \left(\frac{T_1 - T_2}{T_1 T_2}\right)$

$= \left(\frac{130+110}{2}\right)\left(\frac{60-3}{60 \times 3}\right) = 38$ m/sec.

15. Ans. (d)

Time from 7 A.M. to 8 A.M. = 1 hour.

Therefore, time of their meeting

$= \left(\frac{d + s_2 T}{s_1 + s_2}\right) \text{hr} = \left(\frac{100 + 25 \times 1}{20+25}\right)$ hr

= 2 hrs 47 min. after 7 A.M.

16. Ans. (b)

Distance between Mumbai and Pune

$= d\left(\frac{s_1 + s_2}{s_1 - s_2}\right)$ km

$= 20\left(\frac{60+40}{60-40}\right)$ km = 100 km.

17. Ans. (b)

Let the required speed by x km/hr.

Then, $\frac{2 \times 64 \times x}{64 + x} = 56$

$\therefore \quad 128x = 64 \times 56 + 56x$

$\therefore \quad x = \frac{64 \times 56}{72} = 49.77$ km/hr

18. Ans. (a)

Distance travelled in 2 minutes

$= \left(1 + \frac{1}{2}\right)$ km i.e. $\frac{3}{2}$ kms.

Distance covered in 1 hr. $= \left(\frac{3}{2} \times \frac{60}{2}\right)$ km

= 45 km

$\therefore$ Speed of the train = 45 km/hr

19. Ans. (a)

Speed of the first man = 32 km/hr.

Time taken = 192 ÷ 32 = 6 hr

Second man covers 192 km in 3 hr

∴ Speed of the second man

= 192 ÷ 3 = 64 km/hr

Ratio = 32 : 64 or 1 : 2

20. Ans. (c)

Distance already covered = $\frac{3}{4} \times 12 = 9$ km

Time spent = $\frac{2}{3} \times 45$ min = 30 min

Distance left = (12 – 9) km = 3 km

Time left = (45 – 30)min = 15 min

∴ Required speed = $\frac{3}{15/60}$ km/hr

= 12 km/hr

21. Ans. (a)

Let the speed of the train in km/hr = x

Then, relative speed = $(x + 6)$ km/hr

$= (x+6) \times \frac{5}{18}$ m/sec

∴ $(x+6) \times \frac{5}{18} \times 6 = 110$

∴ $x = 60$

∴ Speed of the train = 60 km/hr.

□□□□

Boats & Streams

I Practice Exercise

1. If a man can swim downstream at 6 kmph and upstream at 2 kmph, his speed in still water is:
   (a) 4 km/hr (b) 2 km/hr
   (c) 3 km/hr (d) 2.5 km/hr

2. If Anshul rows 15 km upstream and 21 km downstream taking 3 hours each time, then the speed of the stream is:
   (a) 1 km/hr (b) 1.5 km/hr
   (c) 2 km/hr (d) 12 km/hr

3. If a boat goes 7 km upstream in 42 minutes and the speed of the stream is 3 kmph, then the speed of the boat in still water is:
   (a) 4.2 km/hr (b) 9 km/hr
   (c) 13 km/hr (d) 21 km/hr

4. A man can row $9\frac{1}{3}$ kmph in still water and finds that it takes him thrice as much time to row up than as to row down the same distance in the river. The speed of the current is :
   (a) $3\frac{1}{3}$ km/hr (b) $3\frac{1}{9}$ km/hr
   (c) $4\frac{2}{3}$ km/hr (d) 14 km/hr

5. A man can row a boat at 10 kmph in still water. If the speed of the stream is 6 kmph, the time taken to row a distance of 80 km down the stream is :
   (a) 8 hours (b) 5 hours
   (c) 10 hours (d) 20 hours

6. A boat takes 4 hours for travelling downstream from point A to point B and coming back to point A upstream. If the velocity of the stream is 2 kmph and the speed of the boat in still water is 4 kmph, what is the distance between A and B?
   (a) 4 kms (b) 6 kms
   (c) 8 km (d) 9 km

7. Speed of a boat in standing water is 9 kmph and the speed of the stream is 1.5 kmph. A man rows to a distance of 10.5 km and comes back to the starting point. The total time taken by him is:
   (a) 16 hours (b) 18 hours
   (c) 20 hours (d) 24 hours

8. A man rows to a place 48 km distant and back in 14 hours. He finds that he can row 4 km with the steam in the same time as 3 km against the stream. The rate of the stream is :
   (a) 1 km/hr (b) 1.8 km/hr
   (c) 3.5 km/hr (d) 1.5 km/hr

9. A man can row three-quarters of a kilometre against the stream in $11\frac{1}{4}$ minutes and return in $7\frac{1}{2}$ minutes. The speed of the man in still water is:
   (a) 2 km/hr (b) 3 km/hr
   (c) 4 km/hr (d) 5 km/hr

10. A man can row 5 kmph in still water. If the river is running at 1 kmph, it takes him 75 minutes to row to a place and back. How far is the place?
(a) 3 km (b) 2.5 km
(c) 4 km (d) 5 km

Answers

1. (a)	2. (a)	3. (c)	4. (c)	5. (b)
6. (b)	7. (d)	8. (a)	9. (d)	10. (a)

Solutions

1. **Ans. (a)**

Speed in still water $= \frac{1}{2}(6+2)$ kmph
$= 4$ kmph.

2. **Ans. (a)**

Rate upstream $= \left(\frac{15}{3}\right)$ kmph $= 5$ kmph.

Rate downstream $= \left(\frac{21}{3}\right)$ kmph $= 7$ kmph.

$\therefore$ Speed of stream $= \frac{1}{2}(7-5)$ kmph
$= 1$ kmph.

3. **Ans. (c)**

Rate upstream $= \left(\frac{7}{42}\times 60\right)$ kmph $= 10$ kmph

Speed of stream = 3 kmph.
Let speed in still water be x km/hr
Then speed upstream $= (x-3)$ km/hr.
$\therefore$ $x - 3 = 10$ or $x = 13$ kmph.

4. **Ans. (c)**

Let speed upstream be x kmph.
Then, speed downstream $= 3x$ kmph.

$\therefore$ speed in still water $= \frac{1}{2}(3x + x)$ kmph
$= 2x$ kmph

$\therefore$ $2x = \frac{28}{3}$ $\Rightarrow x = \frac{14}{3}$

$\therefore$ Speed upstream $= \frac{14}{3}$ km/hr,

Speed downstream = 14 km/hr

$\therefore$ Speed of the current

$= \frac{1}{2}\left(14 - \frac{14}{3}\right)$ km/hr

$= \frac{14}{3}$ km/hr $= 4\frac{2}{3}$ km/hr

5. **Ans. (b)**

Speed downstream $= (10+6)$ km/hr
$= 16$ km/hr.
Time taken to cover 80 km downstream

$= \left(\frac{80}{16}\right)$ hrs $= 5$ hrs.

6. **Ans. (b)**

Let the distance between A and B be x km
Speed downstream = 6 kmph,
Speed upstream = 2 kmph.
$\therefore$ Distance A B = 6 km.

7. **Ans. (d)**

Speed upstream = 7.5 kmph,
Speed downstream = 10.5 kmph.

$\therefore$ Total time taken $= \left(\frac{105}{7.5} + \frac{105}{10.5}\right)$ hrs
$= 24$ hours

8. **Ans. (a)**

Suppose he moves 4 km downstream in x hours.

Then, speed downstream $= \left(\frac{4}{x}\right)$ km/hr, speed upstream $= \left(\frac{3}{x}\right)$ km/hr.

$\therefore$ $\frac{48}{(4/x)} + \frac{48}{(3/x)} = 14$ or $x = \frac{1}{2}$

$\therefore$ Speed downstream = 8 km/hr.
Speed upstream = 6 km/hr

Rate of the stream $= \frac{1}{2}(8-6)$ km/hr

$= 1$ km/hr.

9. **Ans. (d)**

Speed upstream $= \left(\frac{3}{4}\times\frac{4}{45}\times 60\right)$ kmph

$= 4$ kmph

Speed downstream $= \left(\frac{3}{4}\times\frac{2}{15}\times 60\right)$ kmph

$= 6$ kmph

Speed in still water = $\frac{1}{2}(6 + 4)$ kmph

= 5 kmph

10. Ans. (a)

Speed downstream = (5 + 1) km/hr = 6 km/hr.

Speed upstream = (5 –1) km/hr = 4 km/hr.

Let the required distance be x km

Then, $\frac{x}{6}+\frac{x}{4}=\frac{75}{60}$ or, $2x+3x = 15$ or $x = 3$.

∴ Required distance = 3 km.

□□□□

Problems on Trains

I Practice Exercise

1. A train is moving at a speed of 132 kmph. If the length of the train is 110 metres, how long will it take to cross a railway platform 165 m long?
   (a) 5 sec (b) 7.5 sec
   (c) 10 sec (d) 15 sec.

2. A train 700 m long is running at 72 kmph. If it crosses a tunnel in 1 minute, the length of the tunnel is
   (a) 700 m (b) 600 m
   (c) 550 m (d) 500 m

3. A train 300 m long crossed a platform 900 m long in 1 minute 12 seconds. The speed of the train (in km/hr) is
   (a) 45 (b) 50
   (c) 54 (d) 60

4. A train speeds past a pole in 15 seconds and a platform 100 m long in 25 seconds. Its length is
   (a) 200 m (b) 150 m
   (c) 50 m (d) Data inadequate

5. If a train 110 m long passes a telegraph pole in 3 seconds, then the time taken by it to cross a railway platform 165 m long is
   (a) 3 sec (b) 4 sec
   (c) 5 sec (d) 7.5 sec

6. A train 150 m long moving at a speed of 25 metres per second overtakes a man moving at 5 metres/sec in opposite direction. The train will pass the man in
   (a) 5 sec (b) 6 sec
   (c) $4\frac{2}{7}$ sec (d) 8 sec

7. Two trains 200 m and 150 m long are running on parallel rails at the rate of 40 kmph and 45 kmph respectively. In how much time will they cross each other, if they are running in the same direction?
   (a) 72 sec (b) 132 sec
   (c) 192 sec (d) 252 sec

8. Two train 126 m and 114 m long are running in opposite directions, one at the rate of 30 kmph and another one at 42 kmph. From the moment they meet will cross each other in
   (a) 10 sec (b) 11 sec
   (c) 12 sec (d) 13 sec

9. A train 110 m long passes a man, running at 6 kmph in the direction opposite to that of the train, in 6 seconds. The speed of the train is
   (a) 60 km/hr (b) 66 km/hr
   (c) 54 km/hr (d) 72 km/hr

10. A train 108 m long moving at a speed of 50 km/hr crosses a train 112 m long coming from opposite direction in 6 seconds. The speed of the second train is
    (a) 48 kmph (b) 54 kmph
    (c) 66 kmph (d) 82 kmph

11. A train B speeding with 120 kmph crosses another train C, running in the same direction in 2 minutes. If the lengths of the trains B and C be 100 m and 200m respectively, what is the speed of the train C?
    (a) 111 kmph (b) 127 kmph
    (c) 123 kmph (d) 129 kmph

12. A train overtakes two persons who are walking in the same direction in which the train is going, at the rate of 2 kmph and 4 kmph and passes them completely in 9 and 10 seconds respectively. The length of the train is
    (a) 72 m (b) 54 m
    (c) 50 m (d) 45 m

13. Two stations A and B are 110 km apart on a straight line. One train starts from A at 7 a.m. and travels

towards B at 20 kmph. Another train starts from B at 8 a.m. and travels towards A at a speed of 25 kmph. At what time will they meet?

(a) 9 a.m. (b) 10 a.m.
(c) 11 a.m. (d) 10.30 a.m

14. Two trains are running in opposite directions towards each other with speeds of 54 kmph and 48 kmph respectively. If the length of the train is 250 m and they cross each other in 18 seconds, the length of the other train is

(a) 145 m (b) 230m
(c) 260 m (d) 180 m

15. A train travelling at 48 kmph completely crosses another train having half its length and travelling in opposite direction at 42 kmph, in 12 seconds. It also passes a railway platform in 45 seconds. The length of the platform is

(a) 560 m (b) 400 m
(c) 600 m (d) 450 m

16. A train of length 150 m takes 10 seconds to pass over another train 100 m long coming from the opposite direction. If the speed of the first train be 30 kmph, the speed of the second train is

(a) 36 kmph (b) 54 kmph
(c) 60 kmph (d) 72 kmph

17. A man sees a train passing over a bridge 1 km long. The length of the train is half that of the bridge. If the train clears the bridge in 2 minutes, the speed of the train is

(a) 30 km/hr (b) 45 km/hr
(c) 50 km/hr (d) 60 km/hr

Answers

1. (b)	2. (d)	3. (d)	4. (b)	5. (d)
6. (a)	7. (d)	8. (c)	9. (a)	10. (d)
11. (a)	12. (c)	13. (b)	14. (c)	15. (b)
16. (c)	17. (b)			

Solutions

1. **Ans. (b)**

Speed $= \left(132 \times \frac{5}{18}\right)$ m/sec $= \frac{110}{3}$ m/sec.

Total distance covered
$= (110 + 165)\text{m} = 275$ m.

$\therefore$ Required time $= \left(275 \times \frac{3}{110}\right)$ sec.
$= 7.5$ seconds.

2. **Ans. (d)**

Speed $\left(72 \times \frac{5}{18}\right)$ m/sec $= 20$m/sec.

Time = 60 sec.

$\therefore \frac{700 + x}{20} = 60 \Leftrightarrow 700 + x = 1200$

$\Rightarrow x = 500$ m

3. **Ans. (d)**

Total distance covered = (300 +900)m
= 1200 m.

Time taken = 1 min. 12 sec = 72 sec,

$\therefore$ Speed $= \left(\frac{1200}{72}\right)$m/sec.

$= \left(\frac{1200}{72} \times \frac{18}{5}\right)$ km/hr = 60 km/hr

4. **Ans. (b)**

Let the length of train be x metres and its speed by y metres/sec.

Then, $\frac{x}{y} = 15 \Rightarrow y = \frac{x}{15}$

$\frac{x+100}{25} = \frac{x}{15} \quad \Rightarrow x = 150$ m/s

5. **Ans. (d)**

Speed $= \left(\frac{110}{3}\right)$ m/sec

Time taken to cross railway platform

$= \left[(110 + 165) \times \frac{3}{110}\right]$ sec

$= \left(275 \times \frac{3}{110}\right)$ sec = 7.5 sec

6. **Ans. (a)**

Speed of train relative to man
= (25 + 5) m/sec
= 30 m/sec

∴ Time taken to pass the man

$= \left(\frac{150}{30}\right)$ sec = 5 sec.

7. **Ans. (d)**

Relative speed = (45–40)kmph = 5 kmph.

$= \left(5 \times \frac{5}{18}\right)$ m/sec $= \left(\frac{25}{18}\right)$ m/sec

Total distance covered = Sum of length of trains = 350 m

∴ Time taken $= \left(350 \times \frac{18}{25}\right)$ sec. = 252 sec.

8. **Ans. (c)**

Relative speed = (30+42) kmph = 72 kmph

$= \left(72 \times \frac{5}{18}\right)$ m/sec = 20 m/sec

Distance covered in crossing each other = (126 + 114) m = 240 m.

Required time $= \left(\frac{240}{20}\right)$ sec = 12 sec.

9. **Ans. (a)**

Speed of the train relative to man

$= \left(\frac{110}{6}\right)$ m/sec. $= \left(\frac{110}{6} \times \frac{18}{5}\right)$ kmph

= 66 kmph.

Let the speed of the train be x kmph.

Then, relative speed = $(x + 6)$ kmph.

$x + 6 = 66$ or $x = 60$ kmph.

10. **Ans. (d)**

Let the speed of the second train be x kmph

Relative speed = $(x + 50)$ kmph

$= \left[(x+50) \times \frac{5}{18}\right]$ m/sec.

$= \left(\frac{250+5x}{18}\right)$ m/sec.

Distance covered = (108 + 112) = 220 m.

∴ $\frac{220}{\left(\frac{250+5x}{18}\right)} = 6$ or $250 + 5x = 660$

or $x = 82$ kmph

11. **Ans. (a)**

Let the speed of train C be x kmph.

Speed of B relative to C = $(120 - x)$

$= \left[(120 - x) \times \frac{5}{18}\right]$ m/sec

$= \left(\frac{600 - 5x}{18}\right)$ m/sec

Distance covered = (100 + 200)m = 300 m.

∴ $\frac{300}{\left(\frac{600-5x}{18}\right)} = 120$

$\Rightarrow 5400 = 120(600 - 5x) \quad \Rightarrow x = 111.$

12. **Ans. (c)**

2 kmph $= \left(2 \times \frac{5}{18}\right)$ m/sec

$= \frac{5}{9}$ m/sec and 4 kmph $= \frac{10}{9}$ m/sec

Let the length of the train be x metres and its speed be y m/sec.

Then, $\frac{x}{\left(y - \frac{5}{9}\right)} = 9$ and $\frac{x}{\left(y - \frac{10}{9}\right)} = 10.$

∴ $9y - 5 = x$ and $10(9y - 10) = 9x$

∴ $9y - x = 5$ and $90y - 9x = 100$

On solving we get $x = 50$.

∴ Length of the train is 50 m.

13. **Ans. (b)**

Suppose they meet x hours after 7 a.m.

Distance covered by A in x hours

$= 20x$ km

Distance covered by B in $(x - 1)$ hours

$= 25(x - 1)$km

∴ $20x + 25(x - 1) = 110$

or $45x = 135$ or $x = 3$.

So, they meet at 10 a.m.

14. **Ans. (c)**

Relative speed = (54 + 48) kmph

$= \left(102 \times \frac{5}{18}\right)$ m/sec $= \left(\frac{85}{3}\right)$ m/sec

Let the length of the other train be x metres.

Then, $(250 + x) \times \frac{3}{85} = 18$ or $750 + 3x$

$= 1530$ or $x = 260$ m

$\therefore$ The length of the other train is 260 m.

15. Ans. (b)

Let the length of first train be x metres.

Then, the length of second train is (x/2) metres.

Relative speed = (48 + 42) kmph

$$= \left(90 \times \frac{5}{18}\right) \text{m/sec} = 25 \text{ m/sec.}$$

$$\therefore \frac{\left(x + \frac{x}{2}\right)}{25} = 12 \text{ or } \frac{3x}{2} = 300$$

or $x = 200$.

$\therefore$ Length of first train = 200m.

Let the length of platform be y metres.

Speed of the first train

$$= \left(48 \times \frac{5}{18}\right) \text{m/sec} = \frac{25}{3} \text{m/sec.}$$

$$\therefore (200 + y) \times \frac{3}{40} = 45 \Rightarrow 600 + 3y$$

$$= 1800 \Rightarrow y = 400 \text{ m}$$

16. Ans. (c)

Let the speed of second train be x kmph.

Relative speed = $(30 + x)$ kmph

$$= (30 + x) \times \frac{5}{18} \text{m/sec}$$

$$(150 + 100) \times \frac{18}{5(30 + x)} = 10$$

$\Rightarrow x = 60$

17. Ans. (b)

Length of bridge = 1000 m.

Length of train = 500 m.

Toal distance covered in clearing the bridge = 1500 m.

Time taken = 120 seconds.

$$\therefore \text{Speed} = \left(\frac{1500}{120}\right) \text{m/sec}$$

$$= \left(\frac{25}{2} \times \frac{18}{5}\right) \text{kmph} = 45 \text{ kmph.}$$

■■■■

B Section

Algebra & Set Theory

2.1
CHAPTER

Surds, Indices & Logarithms

LAWS OF INDICES

(1) $a^m \times a^n = a^{m+n}$

Example:

→ $2^4 \times 2^{13} = 2^{17}$

→ $3^8 \times 3^7 = 3^{15}$

(2) $\dfrac{a^m}{a^n} = a^{m-n}$

Example:

(i) $\dfrac{7^5}{7^2} = 7^{5-2} = 7^3$

(ii) $\dfrac{27^5}{27^2} = 27^3$

(3) $(a^m)^n = a^{mn}$

Example:

(i) $(3^2)^4 = 3^{2\times4} = 3^8$

(ii) $(5^3)^9 = 5^{3\times9} = 5^{27}$

(4) $(a \times b)^n = a^n \times b^n$

Example:

(i) $(3\times5)^2 = 3^2 \times 5^2$

(ii) $(4\times5)^3 = 4^3 \times 5^3$

(5) $\left(\dfrac{a}{b}\right)^n = \dfrac{a^n}{b^n}$

Example:

(i) $\left(\dfrac{3}{4}\right)^2 = \dfrac{3^2}{4^2} = \dfrac{9}{16}$

(ii) $\left(\dfrac{5}{8}\right)^3 = \dfrac{5^3}{8^3} = \dfrac{125}{512}$

(6) $a^0 = 1$

Here a may be any number

Example:

(i) $5^0 = 1$

(ii) $1000^0 = 1$

(7) $a^x = a^y$

If and only if Condition

Example:

(i) $x = y$ or

(ii) $x = 0, y = 0$

(8) $a^{-m} = \dfrac{1}{a^m}$

Example:

(i) $4^{-2} = \dfrac{1}{4^2} = \dfrac{1}{16}$

(ii) $5^{-3} = \dfrac{1}{5^3} = \dfrac{1}{125}$

(9) $a^{-1} = \dfrac{1}{a}$

Example: $10^{-1} = \dfrac{1}{10}$

Laws of Surds

We write $\sqrt[n]{a} = a^{1/n}$ and it is called a surd of order 'n'

(1) $\left(\sqrt[n]{a}\right)^n = \left(a^{\frac{1}{n}}\right)^n = a$

(2) $\sqrt[n]{ab} = \sqrt[n]{a} \,.\, \sqrt[n]{b}$

Example:

(i) $\sqrt[3]{2\times5} = \sqrt[3]{2} \times \sqrt[3]{5}$

(ii) $\sqrt[5]{7\times13} = \sqrt[5]{7} \times \sqrt[5]{13}$

(3) $\sqrt[n]{\dfrac{a}{b}} = \dfrac{\sqrt[n]{a}}{\sqrt[n]{b}}$

Example:

(i) $\sqrt[3]{\dfrac{2}{5}} = \dfrac{\sqrt[3]{2}}{\sqrt[3]{5}}$

(ii) $\sqrt[5]{\dfrac{7}{13}} = \dfrac{\sqrt[5]{7}}{\sqrt[5]{13}}$

(4) $\left(\sqrt[n]{a}\right)^m = \sqrt[n]{a^m}$

Example:

(i) $\left(\sqrt[3]{2}\right)^5 = \sqrt[3]{2^5} = \sqrt[3]{32}$

(ii) $\left(\sqrt[5]{4}\right)^3 = \sqrt[5]{4^3} = \sqrt[5]{64}$

(5) $\sqrt[m]{\sqrt[n]{\sqrt[p]{a}}} = \sqrt[mnp]{a} = a^{\frac{1}{mnp}}$

Example:

(i) $\sqrt[2]{\sqrt[3]{\sqrt[4]{10}}} = 10^{\frac{1}{2\times3\times4}} = 10^{\frac{1}{24}}$

(ii) $\sqrt[3]{\sqrt[5]{25}} = 25^{\frac{1}{3\times5}} = 25^{\frac{1}{15}}$

□□□□

Solved Examples

1. Find the value of $\left(\frac{32}{243}\right)^{-3/5}$

Sol.: $\left(\frac{32}{243}\right)^{-3/5} = \left(\frac{2^5}{3^5}\right)^{-3/5} = \left(\frac{2}{3}\right)^{5\times\frac{-3}{5}}$

$= \left(\frac{2}{3}\right)^{-3} = \left(\frac{3}{2}\right)^{3} = \frac{27}{8}$

2. Solve $4^{3x-2} \cdot 8 = 1$

Sol.: $4^{3x-2} \times 8 = \left(2^2\right)^{3x-2} \times 2^3 = 1$

$= 2^{6x-4+3} = 1$

$= 2^{6x-1} = 2^0$ $\quad [\because 2^0 = 1]$

$= 2^{6x-1} = 2^0 \Rightarrow x = \frac{1}{6}$

3. Find the value of x if $\left(\frac{a}{b}\right)^{3x-5} = \left(\frac{b}{a}\right)^{x+3}$

Sol.: $= \left(\frac{a}{b}\right)^{3x-5} = \left(\left(\frac{a}{b}\right)^{-1}\right)^{x+3}$

$= \left(\frac{a}{b}\right)^{3x-5} = \left(\frac{a}{b}\right)^{-x-3}$

$= 3x - 5 = -x - 3$

$= 4x = 2 \Rightarrow x = \frac{1}{2}$

4. Which is greater $\sqrt[4]{4}$ or $\sqrt[5]{5}$

Sol.: $\sqrt[4]{4} = 4^{\frac{1}{4}}$ and $\sqrt[5]{5} = 5^{\frac{1}{5}}$

Now the LCM of 4 & 5 is 20, so to compare two surds raise the power by 20.

$\left(4^{1/4}\right)^{20} = 4^5 = 1024$

$\left(5^{1/5}\right)^{20} = 5^4 = 625$

Hence, $\sqrt[4]{4} > \sqrt[5]{5}$

5. Simplify, $\frac{\sqrt{8}+\sqrt{2}}{\sqrt{8}-\sqrt{2}}$

Sol.: $\frac{\sqrt{8}+\sqrt{2}}{\sqrt{8}-\sqrt{2}} \times \frac{\sqrt{8}+\sqrt{2}}{\sqrt{8}+\sqrt{2}}$

(Multiply numerator and denominator by conjugate of denominator)

$= \frac{\left(\sqrt{8}+\sqrt{2}\right)^2}{\left(\sqrt{8}\right)^2 - \left(\sqrt{2}\right)^2} = \frac{8+2+2\sqrt{16}}{8-2} = \frac{10+8}{6} = 3$ Ans.

6. Simplify $\frac{\sqrt{a^2+1}+\sqrt{a^2-1}}{\sqrt{a^2+1}-\sqrt{a^2-1}}$

Sol.:

$\frac{\sqrt{a^2+1}+\sqrt{a^2-1}}{\sqrt{a^2+1}-\sqrt{a^2-1}} \times \frac{\sqrt{a^2+1}+\sqrt{a^2-1}}{\sqrt{a^2+1}+\sqrt{a^2-1}}$

$= \frac{a^2+1+a^2-1+2\left(\sqrt{a^2+1}\right)\left(\sqrt{a^2-1}\right)}{a^2+1-\left(a^2-1\right)}$

$= \frac{2a^2+2\sqrt{a^4-1}}{2} = a^2 + \sqrt{a^4-1}$

7. Which is greater $\sqrt[3]{4}$ or $\sqrt[4]{5}$

Sol.: $\sqrt[3]{4} = 4^{\frac{1}{3}}$ and $\sqrt[4]{5} = 5^{\frac{1}{4}}$

raise the power of both the surds by 12(LCM of 3 and 4)

$\left(4^{\frac{1}{3}}\right)^{12} = 4^4 = 256$

$\left(5^{\frac{1}{4}}\right)^{12} = 5^3 = 125$

Hence, $\sqrt[3]{4} > \sqrt[4]{5}$

8. If $2^{2x-1} = \frac{1}{8^{x-3}}$ then find value of x

Sol.: $2^{2x-1} = \frac{1}{8^{x-3}}$

$= 2^{2x-1} = \frac{1}{(2^3)^{x-3}} = \frac{1}{2^{3x-9}}$

$= 2^{2x-1} = 2^{-3x+9} \quad \left[\because \frac{1}{a^m} = a^{-m}\right]$

$= 5x = 10, \ x = 2$

9. If $2^x = 3^y = 6^{-z}$ then $\left(\frac{1}{x}+\frac{1}{y}+\frac{1}{z}\right)$ is equal to?

Sol.: let $2^x = 3^y = 6^{-z} = k$

So, $2 = k^{1/x}$, $3 = k^{1/y}$, $6 = k^{-1/z}$

We know that 2 × 3 = 6 this gives

$k^{1/x} \times k^{1/y} = k^{-1/z} = \frac{1}{x}+\frac{1}{y} = -\frac{1}{z}$

$= \frac{1}{x}+\frac{1}{y}+\frac{1}{z} = 0$

10. Arrange $\sqrt{2}$, $\sqrt[3]{4}$ and $\sqrt[4]{6}$ in ascending order

Sol.: $\sqrt{2} = 2^{1/2}$

$\sqrt[3]{4} = 4^{1/3}$, $\sqrt[4]{6} = 6^{1/4}$

Now raise the power of given surds by LCM of 2, 3, 4 that is 12

$\left(2^{1/2}\right)^{12} = 2^6 = 64$

$\left(4^{1/3}\right)^{12} = 4^4 = 256$

$\left(6^{1/4}\right)^{12} = 6^3 = 216$

Ans: $\sqrt{2}$, $\sqrt[4]{6}$, $\sqrt[3]{4}$

□□□□

Logarithms

Definition

The logarithm of any number of a given base is equal to the index to which the base should be raised to obtain the given number.

For example if

$a^x = c$, then x is called logarithm of a number c to the base a. It is written as $\log_a c = x$

Similarly

$2^3 = 8$ is similar as $\log_2 8 = 3$

$4^{-2} = \frac{1}{4^2} = \frac{1}{16}$ is written

as $\log_4 \frac{1}{16} = -2$

$10^3 = 1000 \Rightarrow \log_{10} 1000 = 3$

$3^5 = 243 \Rightarrow \log_3 243 = 5$

Properties of Logarithms

(a) $\log_a (m \times n) = \log_a m + \log_a n$

Example 1.

$\log_{10}(15) = \log_{10} 3 + \log_{10} 5$

(b) $\log_a\left(\frac{m}{n}\right) = \log_a (m) - \log_a (n)$

Example 2.

$\log_2\left(\frac{6}{5}\right) = \log_2 6 - \log_2 5$

(c) $\log_a (m^n) = n\log_a m$

Example 3.

$\log_5 625 = \log_5 5^4 = 4\log_5 5$

(d) $\log_{a^n}(m) = \frac{1}{n}\log_a(m)$

Example 4.

$\log_{2^4}(8) = \frac{1}{4}\log_2 8$

$\log_{10^2} 1000 = \frac{1}{2}\log_{10} 1000$

(e) $\log_a b = \frac{\log_n b}{\log_n a}$

[here n may be any natural number)

Example 5.

$\log_2 5 = \frac{\log_{10} 5}{\log_{10} 2}$

$\log_2 16 = \frac{\log_{10} 16}{\log_{10} 2}$

(f) $\log_a x = \frac{\log_n x}{\log_n a} = \frac{1}{\log_x a}$

Example 6.

$\log_2 5 = \frac{1}{\log_5 2}$

$\log_3 8 = \frac{1}{\log_8 3}$

(g) $\log_a b \times \log_b a = \dfrac{\log_n b}{\log_n a} \times \dfrac{\log_n a}{\log_n b} = 1$

Remark:

(1) When base is not mentioned it is taken as 10.

(2) Logarithms to the base 10 is known as common logarithms.

(3) $\log_a a = 1$

Example 7.

$\log_5 5 = 1$

$\log_2 2 = 1$

(i) $\log_a 1 = 0$

[a may be any natural number.]

(j) $a^{\log_a x} = x$

$5^{\log_5 3} = 3$

$3^{\log_3 7} = 7$

□□□□

Solved Examples

1. Find the logarithm of 32 to the base 2

Sol.: $\log_2 32 = \log_2 (2^5)$

$= 5\log_2 2 = 5 \times 1 = 5$

2. Find the value of $2^{\log_2 5}$

Sol.: $a^{\log_a x} = x$

So $2^{\log_2 5} = 5$

3. Find the value of $3^{2+\log_3 5}$

Sol.: $3^{2+\log_3 5} = 3^2 \times 3^{\log_3 5}$

$= 3^2 \times 5 = 45$

4. Find the value of $3^{2-\log_3 5}$

Sol.: $3^{2-\log_3 5} = \dfrac{3^2}{3^{\log_3 5}} = \dfrac{9}{5}$

5. Find the value of $\log x^3 \div \log x$

Sol.: $\log x^3 \div \log x = 3\log x \div \log x = 3$

6. If $\log_{16} a = \dfrac{1}{2}$, Find a

Sol.: $\log_{16} a = \dfrac{1}{2}$

$a = 16^{1/2}$

$a = 4$

7. $\log_a \sqrt{3} = \dfrac{1}{6}$, find the value of a

Sol.: $\log_a \sqrt{3} = \dfrac{1}{6}$

$= \log_a 3^{1/2} = \dfrac{1}{6}$

or $a^{1/6} = 3^{1/2}$, $a = 3^3 = 27$

8. If $\log_{27} x + \log_9 x + \log_3 x = 11$, find the value of x

Sol.: $\log_{27} x + \log_9 x + \log_3 x = 11$

$= \dfrac{1}{3}\log_3 x + \dfrac{1}{2}\log_3 x + \log_3 x = 11$

$= \dfrac{11}{6}\log_3 x = 11$

$= \dfrac{1}{6}\log_3 x = 1$

$= \log_3 x = 6$

$= x = 3^6 = 729$

9. Solve $\log_3 n - \log_3 4 = 2$

Sol.: $\log_3 n - \log_3 4 = \log_3 \dfrac{n}{4} = 2$

$\log_3 \dfrac{n}{4} = 2$

or $\dfrac{n}{4} = 3^2$, $n = 3^2 \times 4 = 36$

10. Find the value of $\log\dfrac{x^2}{yz} + \log\dfrac{y^2}{zx} + \log\dfrac{z^2}{xy}$

Sol.: $\log\dfrac{x^2}{yz} + \log\dfrac{y^2}{zx} + \log\dfrac{z^2}{xy}$

$= \log\dfrac{x^2}{yz} \times \dfrac{y^2}{zx} \times \dfrac{z^2}{xy}$

$= \log\dfrac{x^2y^2z^2}{y^2z^2x^2} = \log 1 = 0$

11. Evaluate $\log_7\left(\dfrac{1}{343}\right)$

Sol.: $\log_7\left(\dfrac{1}{343}\right) = \log_7 \dfrac{1}{7^3}$

$= \log_7 7^{-3} = -3\log_7 7 = -3$

12. Evaluate = $\log_{100}(0.01)$

Sol.: $\log_{100}(100)^{-1} = -1\log_{100}100 = -1$

13. If $\log_{\sqrt{8}} x = 3\frac{1}{3}$, find the value of x

Sol.: $\log_{\sqrt{8}} x = \frac{10}{3}$, $x = \left(2^{3/2}\right)^{10/3}$

$x = 2^{3/2 \times 10/3} = 2^5 = 32$

□□□□

Surds, Indices & Logarithms

I Practice Exercise: I

1. The value of $\left(\sqrt{8}\right)^{1/3}$ is:

(a) 2 (b) 4
(c) $\sqrt{2}$ (d) 8

2. $\left(\frac{1}{216}\right)^{-2/3} \div \left(\frac{1}{27}\right)^{-4/3} = ?$

(a) $\frac{3}{4}$ (b) $\frac{2}{3}$
(c) $\frac{4}{9}$ (d) $\frac{1}{8}$

3. If $\sqrt{2^n} = 64$, then the value of n is:

(a) 8 (b) 4
(c) 6 (d) 12

4. If $\frac{9^n \times 3^5 \times (27)^3}{3 \times (81)^4} = 27$, then n equals:

(a) 0 (b) 2
(c) 3 (d) 4

5. If $\left(\sqrt{3}\right)^5 \times 9^2 = 3^\alpha \times 3\sqrt{3}$, then α equals:

(a) 2 (b) 3
(c) 4 (d) 5

6. If x, y, z are real numbers, then the value of:

$\sqrt{x^{-1}y}.\sqrt{y^{-1}z}.\sqrt{z^{-1}x}$ is

(a) xyz (b) $\sqrt{xyz}$
(c) $\frac{1}{xyz}$ (d) 1

7. $\left(\frac{x^b}{x^c}\right)^{(b+c-a)} \cdot \left(\frac{x^c}{x^a}\right)^{(c+a-b)} \cdot \left(\frac{x^a}{x^b}\right)^{(a+b-c)} = ?$

(a) x^{abc} (b) x^{a+b+c}
(c) $x^{ab+bc+ca}$ (d) 1

8. If $2^{x+4} - 2^{x+2} = 3$, then x is equal to:

(a) 0 (b) 2
(c) –1 (d) –2

9. If $2^{2x-1} = \frac{1}{8^{x-3}}$, the value of x is:

(a) 3 (b) 2
(c) 0 (d) –2

10. If $\sqrt{5 + \sqrt[3]{x}} = 3$, then x is equal to:

(a) 125 (b) 64
(c) 27 (d) 9

11. If $a^x = b^y = c^z$ and $b^2 = ac$, then y equals:

(a) $\frac{xz}{x+z}$ (b) $\frac{xz}{2(x-z)}$
(c) $\frac{xz}{2(z-x)}$ (d) $\frac{xz}{2(x+z)}$

12. If $x = y^a$, $y = z^b$ and $z = x^c$, then the value of abc is:

(a) 4 (b) 3
(c) 2 (d) 1

Answers

1. (c) 2. (c) 3. (d) 4. (c) 5. (d)
6. (d) 7. (d) 8. (d) 9. (b) 10. (b)
11. (d) 12. (d)

Solutions

1. Ans. (c)

$$\left(\sqrt{8}\right)^{1/3} = \left(8^{1/2}\right)^{1/3} = 8^{\left(\frac{1}{2}\times\frac{1}{3}\right)} = 8^{1/6}$$

$$= \left(2^3\right)^{1/6} = 2^{\left(3\times\frac{1}{6}\right)} = 2^{1/2} = \sqrt{2}.$$

2. Ans. (c)

$$\left(\frac{1}{216}\right)^{-2/3} \div \left(\frac{1}{27}\right)^{-4/3}$$

$$= (216)^{2/3} \div (27)^{4/3} = \left(6^3\right)^{2/3} \div \left(3^3\right)^{4/3}$$

$$= 6^{\left(3\times\frac{2}{3}\right)} \div 3^{\left(3\times\frac{4}{3}\right)} = 6^2 \div 3^4 = \frac{36}{81} = \frac{4}{9}.$$

3. Ans. (d)

$$\sqrt{2^n} = 64 \quad \Rightarrow 2^{n/2} = 64 = 2^6.$$

$$\therefore \quad \frac{n}{2} = 6 \text{ or } n = 12.$$

4. Ans. (c)

$$\frac{9^n \times 3^5 \times (27)^3}{3 \times (81)^4} = 27$$

$$\Rightarrow \frac{3^{2n} \times 3^5 \times \left(3^3\right)^3}{3 \times \left(3^4\right)^4} = 3^3.$$

$$\text{or} \quad \frac{3^{2n} \times 3^5 \times 3^9}{3^1 \times 3^{16}} 3^3$$

$$\text{or} \quad 3^{2n+5+9} = 3^3 \times 3^1 \times 3^{16}$$

$$\text{or} \quad 3^{2n+14} = 3^{20}$$

$$\therefore \quad 2n + 14 = 20$$

$$\text{or} \quad 2n = 6 \text{ or } n = 3.$$

5. Ans. (d)

$$\left(\sqrt{3}\right)^5 \times 9^2 = 3^\alpha \times 3\sqrt{3}$$

$$\Rightarrow \left(3^{1/2}\right)^5 \times \left(3^2\right)^2 = 3^\alpha \times 3^1 \times 3^{1/2}$$

$$\therefore \quad 3^{5/2} \times 3^4 = 3^\alpha \times 3^1 \times 3^{1/2}$$

$$\text{or} \quad 3^{\left(\frac{5}{2}+4\right)} = 3^{\left(\alpha+1+\frac{1}{2}\right)}$$

$$\text{or,} \quad 3^{13/2} = 3^{\alpha+\frac{3}{2}} \qquad \text{So, } \alpha + \frac{3}{2} = \frac{13}{2}$$

$$\text{or} \quad \alpha = \left(\frac{13}{2} - \frac{3}{2}\right) = 5.$$

6. Ans. (d)

$$\sqrt{x^{-1}y}.\sqrt{y^{-1}z}.\sqrt{z^{-1}x}$$

$$= \sqrt{\frac{y}{x}}.\sqrt{\frac{z}{y}}.\sqrt{\frac{x}{z}} = \frac{\sqrt{y}}{\sqrt{x}} \times \frac{\sqrt{z}}{\sqrt{y}} \times \frac{\sqrt{x}}{\sqrt{z}} = 1.$$

7. Ans. (d)

Given Exp.

$$= x^{(b-c)(b+c-a)}.x^{(c-a)(c+a-b)}.x^{(a-b)(a+b-c)}$$

$$= x^{(b^2-c^2)+(c^2-a^2)+(a^2-b^2)}.x^{-a(b-c)-b(c-a)-c(a-b)}$$

$$= x^0.x^0 = 1$$

8. Ans. (d)

$$2^{x+4} - 2^{x+2} = 3$$

$$\Rightarrow 2^{x+2}\left(2^2 - 1\right) = 3$$

$$\Rightarrow 2^{x+2} = 1 = 2^\circ.$$

$$\therefore \quad x + 2 = 0 \text{ or } x = -2.$$

9. Ans. (b)

$$2^{2x-1} = \frac{1}{8^{x-3}} \qquad \Rightarrow 2^{2x-1} = \frac{1}{(2^3)^{(x-3)}}$$

$$\Rightarrow 2^{2x-1} = \frac{1}{2^{3x-9}} \qquad \Rightarrow 2^{2x-1} = 2^{9-3x}$$

$$\therefore \quad 2x - 1 = 9 - 3x \text{ or } 5x = 10 \text{ or } x = 2.$$

10. Ans. (b)

On squaring both sides, we get :

$5 + \sqrt[3]{x} = 9$ or $\sqrt[3]{x} = 4$.

Cubing both sides, we get

$x = (4 \times 4 \times 4) = 64.$

11. Ans. (d)

Let $a^x = b^y = c^z = k.$

Then $a = k^{1/x}$, $b = k^{1/y}$ and $c = k^{1/z}$.

$$b^2 = ac \Rightarrow k^{2/y} = k^{1/x}.k^{1/z}$$

$$\Rightarrow k^{2/y} = k^{\left(\frac{1}{x}+\frac{1}{z}\right)}.$$

$$\Rightarrow \frac{2}{y} = \frac{1}{x} + \frac{1}{z}$$

$$\Rightarrow \frac{2}{y} = \frac{z+x}{xz} \Rightarrow y = \frac{xz}{2(x+z)}$$

12. Ans. (d)

$$x = y^a = \left(z^b\right)^a = z^{ab} = \left(x^c\right)^{ab} = x^{abc}. \therefore \quad abc = 1.$$

□□□□

Surds, Indices & Logarithms

II Practice Exercise: II

1. The value of $\log_{343} 7$ is:
   (a) $\frac{1}{3}$ (b) –3
   (c) $-\frac{1}{3}$ (d) 3

2. The value of $\log_{10}(.0001)$ is:
   (a) $\frac{1}{4}$ (b) $-\frac{1}{4}$
   (c) –4 (d) 4

3. If $\log_8 x = \frac{2}{3}$, then the value of x is :
   (a) $\frac{3}{4}$ (b) $\frac{4}{3}$
   (c) 4 (d) 3

4. $\log_{32} x = 0.8$, then x is equal to:
   (a) 25.6 (b) 16
   (c) 10 (d) 12.8

5. If $\log_4 x + \log_2 x = 6$, then x is equal to
   (a) 2 (b) 4
   (c) 8 (d) 16

6. If log 2 = 0.30103, then the number of digits in 5^{20} is:
   (a) 14 (b) 16
   (c) 18 (d) 25

7. The value of $\log_2(\log_5 625)$ is:
   (a) 2 (b) 5
   (c) 10 (d) 15

8. $(\log_b a \times \log_c b \times \log_a c)$ is equal to:
   (a) 0 (b) 1
   (c) abc (d) a+b+c

9. $\left[\frac{1}{(\log_a bc)+1} + \frac{1}{(\log_b ca)+1} + \frac{1}{(\log_c ab)+1}\right]$ is equal to:
   (a) 1 (b) 2
   (c) 3 (d) $\frac{3}{2}$

10. If $\log_2[\log_3(\log_2 x)] = 1$, then x is equal to:
    (a) 512 (b) 128
    (c) 12 (d) 0

11. If $\log_{10} 2 = 0.3010$, then $\log_2 10$ is :
    (a) .3322 (b) 3.2320
    (c) 3.3222 (d) 5

12. If $\log_{10} 2 = 0.3010$, the value of $\log_{10} 5$ is:
    (a) 0.3241 (b) 0.6911
    (c) 0.6990 (d) 0.7525

13. The value of $(\log_9 27 + \log_8 32)$ is :
    (a) 4 (b) 7
    (c) $\frac{7}{2}$ (d) $\frac{19}{6}$

14. If $\log_5(x^2 + x) - \log_5(x+1) = 2$, then the value of x is:
    (a) 5 (b) 32
    (c) 25 (d) 10

15. If $\log x + \log y = \log(x+y)$, then:
    (a) $x = y$ (b) $xy = 1$
    (c) $y = \frac{x-1}{x}$ (d) $y = \frac{x}{x-1}$

16. The value of $\left[\frac{1}{\log_{(p/q)} x} + \frac{1}{\log_{(q/r)} x} + \frac{1}{\log_{(r/p)} x}\right]$ is:
    (a) 3 (b) 2
    (c) 1 (d) 0

17. If log 2 = 0.3010 and log 3 = 0.4771 then the value of log 4.5 is:
    (a) 0.6532 (b) 0.7727
    (c) 0.3266 (d) None of these

Answers

1. (a)	2. (c)	3. (c)	4. (b)	5. (d)
6. (a)	7. (a)	8. (b)	9. (a)	10. (a)
11. (c)	12. (c)	13. (d)	14. (c)	15. (d)
16. (d)	17. (a)			

Solutions

1. Ans. (a)

Let $\log_{343} 7 = m$.

Then, $(343)^m = 7 \Rightarrow (7^3)^m = 7$

$\Rightarrow \quad 7^{3m} = 7$

$\Rightarrow 3m = 1 \Rightarrow m = \frac{1}{3}$.

$\therefore \quad \log_{343} 7 = \frac{1}{3}$.

2. Ans. (c)

Let $\log_{10}(.0001) = m$. Then,

$10^m = .0001 \Rightarrow 10^m = \frac{1}{10000} \Rightarrow 10^m$

$= \frac{1}{10^4} \Rightarrow 10^m = 10^{-4} \Rightarrow m = -4$

$\therefore \quad \log_{10}(.0001) = -4$.

3. Ans. (c)

$\log_8 x = \frac{2}{3} \Rightarrow x = 8^{2/3} = (2^3)^{2/3}$

$= 2^{\left(3 \times \frac{2}{3}\right)} = 2^2 = 4$.

4. Ans. (b)

$\log_{32} x = 0.8 \Rightarrow x = (32)^{0.8}$

$= (2^5)^{4/5} = 2^4 = 16$.

5. Ans. (d)

$\log_4 x + \log_2 x = 6 \Rightarrow \frac{\log x}{\log 4} + \frac{\log x}{\log 2} = 6$

$\therefore \quad \frac{\log x}{2 \log 2} + \frac{\log x}{\log 2} = 6 \Rightarrow 3\log x = 12\log 2$

or $\log x = 4\log 2 \Rightarrow \log x$

$= \log 2^4$ or $x = 2^4 = 16$.

6. Ans. (a)

$\log 5^{20} = 20\log 5 = 20 \times \left[\log\left(\frac{10}{2}\right)\right]$

$= 20 \times [\log 10 - \log 2] = 20 \times [1 - 0.3010]$

$= 20 \times 0.6990 = 13.9800$

$\therefore$ Characteristic = 13.

$\therefore$ Number of digits in 5^{20} is 14.

7. Ans. (a)

Let $\log_5 625 = x$. Then, 5^x

$= 625 = 5^4$ or $x = 4$

Let $\log_2(\log_5 625) = y$. Then, $\log_2(4) = y$

or $\quad 2^y = 4 = 2^2$. So, $y = 2$.

$\therefore \quad \log_2(\log_5 625) = 2$.

8. Ans. (b)

Given Exp. $= \left(\frac{\log a}{\log b} \times \frac{\log b}{\log c} \times \frac{\log c}{\log a}\right) = 1$.

9. Ans. (a)

$= \frac{1}{\log_a bc + \log_a a} + \frac{1}{\log_b ca + \log_b b}$

$+ \frac{1}{\log_c ab + \log_c c}$

$= \frac{1}{\log_a (abc)} + \frac{1}{\log_b (abc)} + \frac{1}{\log_c (abc)}$

$= \log_{abc} a + \log_{abc} b + \log_{abc} c$

$= \log_{abc}(abc) = 1$.

10. Ans. (a)

$\log_2[\log_3(\log_2 x)] = 1$

$\Rightarrow \quad \log_3(\log_2 x) = 2$

$\Rightarrow \log_2 x = 3^2 = 9$

$\Rightarrow x = 2^9 = 512$

11. Ans. (c)

$\log_2 10 = \frac{1}{\log_{10} 2} = \frac{1}{0.3010} = 3.3222$

12. Ans. (c)

$$\log_{10} 5 = \log_{10}\left(\frac{10}{2}\right) = \log_{10} 10 - \log_{10} 2$$

$$= 1 - \log_{10} 2 = (1 - 0.3010) = 0.6990.$$

13. Ans. (d)

$\log_9 27 + \log_8 32$

$\log_{3^2} 3^3 + \log_{2^3} 2^5$

$\frac{3}{2}\log_3 3 + \frac{5}{3}\log_2 2$

So, $\frac{3}{2} + \frac{5}{3} = \frac{19}{6}$

14. Ans. (c)

$\log_5 (x^2 + x) - \log_5 (x+1) = 2$

$$\Rightarrow \log_5\left(\frac{x^2 + x}{x+1}\right) = 2$$

$$\therefore \log_5\left[\frac{x(x+1)}{x+1}\right] = 2$$

or $\log_5 x = 2$ or $x = 5^2 = 25$.

15. Ans. (d)

$\log x + \log y = \log(x + y)$

$\Rightarrow \log(x + y) = \log(xy)$

$\therefore x + y = xy$ or $x = y(x - 1)$

$$\Rightarrow y = \frac{x}{x-1}$$

16. Ans. (d)

Given Expression

$$= \log_x\left(\frac{p}{q}\right) + \log_x\left(\frac{q}{r}\right) + \log_x\left(\frac{r}{p}\right)$$

$$\log_x\left(\frac{p}{q} \times \frac{q}{r} \times \frac{r}{p}\right) = \log 1 = 0.$$

17. Ans. (a)

$$\log 4.5 = \log\left(\frac{9}{2}\right) = \log 9 - \log 2$$

$$= \log(3^2) - \log 2 = 2\log 3 - \log 2$$

$$= (2 \times 0.4771 = 0.3010) = 0.6532.$$

■■■■

2.2 CHAPTER

Progressions

Talking about a series 1, 3, 5, 7, 9....... . In this series every term is made by adding a fixed value in the previous term. So it can be stated as the series is in the form of a, (a + d), (a + 2d), ... This sort of series is called as Arithmetic Series and Term 'a' is first term while 'd' is the common difference between two consecutive terms.

Basics of Arithmetic Progression (AP)

- Finding out the nth term of an AP
  $T_n = a + (n - 1)d \rightarrow n^{th}$ term,
  $T_n = a + (n - 1)d$
- Sum of n terms of an AP

$$S_n = \Sigma T_n = \frac{n}{2}\left[2a+(n-1)d\right]$$

$$= \frac{n}{2}\left[a+a+(n-1)d\right]$$

$$= \frac{n}{2}\left[a+T_n\right]$$

$$\text{So } S_n = \frac{n}{2}\left[a+T_n\right] = \frac{n}{2}\left[2a+(n-1)d\right]$$

Properties of Arithmetic Progression

(a) If three numbers are in AP then they can be assumed as a – d, a, a + d

(b) If four numbers have to be assumed then they can be a – 3d, a – d, a + d, a + 3d

(c) If five numbers have to be assumed then they will be (a – 2d), (a – d), a, (a + d), (a + 2d)

If in an AP sum of n terms is 'm' and sum of 'm' terms is 'n' then sum of (m + n) terms will be "–(m + n)"

Example 1.

For an AP sum of 2 terms is '4' and sum of '4' terms is '2'. Find out the sum of 6 term of that 'AP.

Solution.

Let the first term be 'a' and common difference be 'd' the

$$\frac{2}{2}\left[2a+(2-1)d\right] = 4$$

$$\Rightarrow \quad 2a + d = 4 \quad ...(i)$$

$$\text{and} \quad \frac{4}{2}\left[2a+(4-1)d\right] = 2$$

$$\Rightarrow \quad 4a + 6d = 2 \quad ...(ii)$$

By solving (i) and (ii) we get,

$$d = \frac{3}{2} \text{ and } a = \frac{11}{4}$$

So sum of 6 terms

$$= \frac{6}{2}\left[2\times\frac{11}{4}+(6-1)\left(\frac{-3}{2}\right)\right]$$

$$= \frac{6}{2}\left[\frac{11}{2}-\frac{15}{2}\right]$$

$$= -6$$

- If in an AP m^{th} term is 'n' and n^{th} term is 'm' then
  (a) The common difference of such AP will be '–1'
  (b) The $(m + n)^{th}$ term of the AP will be 'zero'
  (c) The first term of the AP will be '(m + n –1)'

Example 2.

Talk about any AP of common difference – 1.

7, 6, 5, 4, 3, 2, 1, 0

Its 2nd term is 6

and 6th term is 2

then 8th term will be '0'

and the first term is 6 + 2 – 1 = '7'

- If same number is added or subtracted from each term of the AP. The new series will again be an AP.

Example 3.

If 2, 6, 10, 14, 18, ... is an AP and a number 5 is either added or subtracted, then the new series formed are

7, 11, 15, 19, 23, ...
–3, 1, 5, 9, 13, ... [Both series are AP]

- If same number is multiplied in all terms of AP, the new series formed will be an AP.

Example 4.

Take series 1, 3, 5, 7, 9.......
Multiplying it by 2 we get
2, 6, 10, 14, 18, ... This series is again an AP.

- If sum of m terms of an AP is equal to sum of n terms of the same AP then the AP will have $\left(\frac{m+n+1}{2}\right)^{th}$ term as 'zero'.

It is valid only when (m + n) is odd.

- When m + n is even
There is no possibility that the term will be zero i.e. in that A.P. No term can be zero. Let us understand it with one example and discuss both the cases.

Example 5.

The sum of first 8 terms of an AP is equal to the sum of first (a) 15 terms (b) 16 terms then which term in the AP will have 'zero' value.

(a) $S_8 = S_{15}$

$$\Rightarrow \frac{8}{2}[2a+7d] = \frac{15}{2}[2a+14d]$$

$\Rightarrow$ 8a + 28d = 15a + 105d
$\Rightarrow$ 7a = –77d
$\Rightarrow$ a = –11d
So –11d + (n – 1)d = 0
$\Rightarrow$ n = 12

(b) $S_8 = S_{16}$

$$\Rightarrow \frac{8}{2}[2a+7d] = \frac{16}{2}[2a+15d]$$

$\Rightarrow$ 2a + 7d = 4a + 30d

$$\Rightarrow a = \frac{-23}{2}d$$

$$\text{So } \frac{-23}{2}d + (n-1)d = 0$$

$$\Rightarrow n = \frac{25}{2}$$

Which is not possible.

- If two APs are compared & the ratio of T_m of first AP to T_n of the second AP is p : q then they will have the ratio of sums determinable only in one case & that will be

$$\frac{S_{2m-1}}{S_{2n-1}} = \frac{2m-1}{2n-1}\left(\frac{T_m}{T_n}\right)$$

Some Important Applications

- Sum of first n natural numbers

$$S_n = \Sigma n = \frac{n(n+1)}{2}$$

- Sum of the squares of first n natural numbers

$$(S_n)^2 = \frac{n(n+1)(2n+1)}{6}$$

- Sum of the cubes of first n natural no.

$$(S_n)^3 = \left\{\frac{n(n+1)}{2}\right\}^2$$

So $(S_n)^3 = (S_n)^2$

Geometric Series

Consider a series 2, 4, 8, 16, 32... . In this series, we see every term in the series is obtained by multiplying the precedent term by a fixed value. This sort of series are called as geometric series. So a geometric series is in the form.

a, ar, ar^2, ar^3, ar^4 .... ar^n.

Here, r is the common ratio and a is the first term.

Basics of GP

- N^{th} term of the GP is = ar^{n-1}
- Sum of n terms

$$S_n = a\left(\frac{r^n-1}{r-1}\right) \quad \text{when } r > 1$$

$$S_n = a\left(\frac{1-r^n}{1-r}\right) \quad \text{when } r < 1$$

Infinite GP

This is a geometric series with its number of terms as infinite. The sum of infinite terms of GP can be calculated as

$$S_\infty = \infty \quad \text{(when } r > 1)$$

$$S_\infty = \frac{a}{1-r} \quad \text{(Wen } r < 1)$$

Basic Properties of GP

Assumption of number of terms in geometric progression is same as of the AP for three terms will be assume $\frac{a}{r}$, a, ar, for four terms $\frac{a}{r^3}$, $\frac{a}{r}$, ar, ar^3.

Harmonic Series

If all the terms of an arithmetic series are reversed then the series obtained will be harmonic series. Let us take a series 1, 3, 5, 7, 9, ... It is an arithmetic series, if we are going to reverse the term it means the series $\frac{1}{1}, \frac{1}{3}, \frac{1}{5}, \frac{1}{7}, \frac{1}{9}, \frac{1}{11}$, ... This is harmonic series.

Basics of HP

The nth term of a harmonic series can be calculated only by finding out the nth term of the arithmetic series obtained by inversing the harmonic series and then again reversing the term.

Example.

Find out the nth term for the series

$\frac{1}{5}, \frac{1}{11}, \frac{1}{17}, \frac{1}{23}, ...$

The terms are reversed then the AP obtained is 5, 11, 17, 23, ...

So n^{th} term of HP is $\frac{1}{6n-1}$

Basic Means

For the given terms there are three types of means

- Arithmetic Mean (AM)
- Geometric Mean (GM)
- Harmonic Mean (HM)

Arithmetic Mean

For n terms $a_1, a_2, a_3, a_4, ... a_n$, the AM will be

$$\frac{a_1 + a_2 + a_3 + ...a_n}{n}$$

Geometric Mean

Geometric mean is the term which when included, makes all the term in GP. Let us say there are two terms a and b then GM will be in such a way that a, GM, be will be in GP

So $GM = \sqrt{(a, b)}$

or it can be written as

- GM (for two terms = $(a.b)^{1/2}$

If there are n terms $a_1, a_2, a_3, a_4, ... a_n$ then

- $GM = (a_1.a_2.a_3, ... a_n)^{1/n}$

Harmonic Mean

Harmonic mean is the term when included between the terms make all the terms in harmonic progression. The harmonic mean for the two terms a and b is

$$HM = \frac{2ab}{a+b}$$

Relation between AM, GM, HM

When compared for two fixed term, the arithmetic mean is having the highest value.

$AM \geq GM \geq HM$

- When taken for two terms a and b, arithmetic mean, geometric mean and harmonic mean
  $HM \times AM = GM^2$

□□□□

Solved Examples

1. If in an AP 10th term is 86 and 86th term is 10, then what will be the 100th term?

(a) 1 (b) 3
(c) 4 (d) –4

Ans. (d)

$$a + 9d = 86$$
$$a + 85d = 10$$
$$\Rightarrow -76d = 76 \text{ or } d = -1$$

substituting the values

We get $a = 95, d = -1$

So 100th term $T_{100} = 95 + (100 - 1)(-1)$

$\Rightarrow T_{100} = -4$

hence option (d)

Alternate: For this series from the property of AP. We can say that 96th term is 0 and d = –1 so 100th term = –4.

2. Given that a function, S ={2, 3, 4,..... 2n + 1} and X is average of all the odd integers of S while Y is average of all the even integers of S, find out what is (X – Y)?

(a) 1 (b) n – 1
(c) n (d) None of these

Ans. (a)

$$X = \frac{3+5+7+....2n+1}{\text{No. of terms}}$$
$$= \frac{3+(2n+1)}{2} = n+2$$
$$Y = \frac{2+4+6+....2n}{\text{No. of terms}}$$
$$= \frac{2+2n}{2} = n+1$$

So X – Y = 1

Hence option (a)

Alternate: Difference of 2 consecutive terms i.e. odd term and even term = 1 so difference of avg. of odd and avg. of even = 1
i.e. $x - y = 1$.

3. The number of terms between 30 to 530, which are divisible by 11 are
(a) 32 (b) 35
(c) 36 (d) 46

Ans. (d)
The first term divisible by 11 = 33
The last term divisible by 11 = 528

No. of terms = $\frac{528-33}{11}+1=46$

Hence option (d).

4. Split 74 into four parts in such a manner that all parts are in AP and the product of first part and last part is 2 less than the product of second part and third part.

Sol.
Let the four parts be
a – 3d, a – d, a + d, a + 3d
no a – 3d + a – d + a + d + a + 3d = 74

$\Rightarrow a = \frac{37}{2}$

and (a – 3d)(a + 3d)+ 2 = (a – d)(a + d)
$\Rightarrow a^2 - 9d^2 + 2 = a^2 - d^2$
$\Rightarrow 8d^2 = 2 \Rightarrow d = \pm 1/2$
So the parts are 17, 18, 19, 20

□□□□

Progression

I Practice Exercise: I

1. How many terms are there in the AP 20, 25, 30, ...130.
(a) 22 (b) 23
(c) 21 (d) 24

2. Find the 1st term of an AP whose 8th and 12th terms are respectively 39 and 59.
(a) 5 (b) 6
(c) 4 (d) 3

3. Find the 15th term of the sequence 20, 15, 10, ...
(a) –45 (b) –55
(c) –50 (d) 0

4. A number 15 is divided into three parts which are in AP and the sum of their squares is 83. Find the smallest number.
(a) 5 (b) 3
(c) 6 (d) 8

5. The sum of the first 16 terms of an AP whose first term and third are 5 and 15 respectively is
(a) 600 (b) 765
(c) 640 (d) 680

6. A boy agrees to work at the rate of one rupee on the first day, two rupees on the second day, four rupees on the third day and so on. How much will the boy get if he starts working on the 1st of February and finishes on the 20th of February?
(a) 2^{20} (b) $2^{20} - 1$
(c) $2^{19} - 1$ (d) 2^{19}

7. The seventh term of a GP is 8 times the fourth term. What will be the first term when it's fifth term is 48?
(a) 4 (b) 3
(c) 5 (d) 2

8. The sum of three numbers on a GP is 14 and the sum of their squares is 84. Find the largest number.
(a) 8 (b) 6
(c) 4 (d) None of these

9. How many natural numbers between 300 to 500 are multiples of 7?
(a) 29 (b) 28
(c) 27 (d) None of these

10. The 4th and 10th term of an GP are 1/3 and 243 respectively. Find the 2nd term.
(a) 3 (b) 1
(c) 1/27 (d) 1/9

11. The 7th and 21st terms of an AP are 6 and –22 respectively. Find the 26th term.
(a) –34 (b) –32
(c) –12 (d) –10

12. Find the number of terms of the series 1/81, –1/27, 1/9, ... –729.
(a) 11 (b) 12
(c) 10 (d) 13

13. 'a' and 'b' are two number whose AM is 25 and GM is 7. Which of the following may be value of 'a'?
(a) 10 (b) 20
(c) 49 (d) 25

14. Two number A and B are such that their GM is 20% lower than their AM. Find the ratio between the numbers.
(a) 3 : 2 (b) 4 : 1
(c) 2 : 1 (d) 3 : 1

Answers

1. (b)	2. (c)	3. (c)	4. (b)	5. (d)
6. (b)	7. (b)	8. (a)	9. (a)	10. (c)
11. (b)	12. (a)	13. (c)	14. (b)	

Solutions

1. Ans. (b)
In series 20, 25, 30 ....130.
a = 20, d = 5
n^{th} term is 130
$\Rightarrow$ 20 + (n – 1) 5 = 130
$\Rightarrow$ 5n = 115, n = 23

2. Ans. (c)
Here a + 7d = 39 ...(i)
a + 11d = 59 ...(ii)
So, 4d = 20
d = 5
a + (7 × 5) = 39
So, a = 4

3. Ans. (c)
AP 20, 15, 10
Here a = 20
d = – 5
15th term will be a + 14d which is equal to
20 + (–5) × 14 = –50

4. Ans. (b)
Let numbers are
a – d, a and a + d
now it is given that
sum = 15 & sum of there square is 83 i.e.
3a = 15 ...(i)
$(a - d)^2 + a^2 + (a + d)^2 = 83$ ...(ii)
$3a^2 + 2d^2 = 83$
$75 + 2d^2 = 83$
$2d^2 = 8$
$d^2 = 4$, d = 2
So, least term is a – d = 3

5. Ans. (d)
a = 5, a + 2d = 15
So d = 5
now sum of 16 term will be
16/2 [10 + 15 × 5] = 680

6. Ans. (b)
This sequence is in GP 1, 2, 4, 8, ...
Now sum of first 20 term will be

$$\frac{a(r^n - 1)}{r - 1} = 2^{20} - 1$$

7. Ans. (b)
It is given that
$ar^6 = 8\,ar^3$
$r^3 = 8,$ r = 2
Now $ar^4 = 48$
$a \times 2^4 = 48,$ a = 3

8. Ans. (a)
a, ar, ar^2
now $a + ar + ar^2 = 14$ ...(i)
also $a^2 + a^2r^2 + a^2r^4 = 84$ ...(ii)
only suitable combination is 2, 4, 8
So largest term is 8

9. Ans. (a)
Between 300 and 500 the first term divisible by 7 is 301 and last term divisible by 7 is 497
So, here a = 301, d = 7
L = a + (n – 1) d
497 = 301 + (n–1) × 7
$\Rightarrow$ 7n = 203
$\Rightarrow$ n = 29

10. Ans. (c)
It is given that
$ar^3 = 1/3$ ...(i)
$ar^9 = 243$ ...(ii)
From (i) and (ii) we get
$r^6 = 729$
r = 3 ...(iii)
From (i) and (iii) we get

$$a \times 3^3 = \frac{1}{3}, \qquad a = \frac{1}{81}$$

second term will be $\frac{1}{27}$

11. Ans. (b)

It is given that

$a + 6d = 6$ and

$a + 20d = -22$

$\Rightarrow 14d = -28, \quad d = -2,$

$\Rightarrow a = 18$

26th term will be -32

12. Ans. (a)

It is given that

GP is $\frac{1}{81}, \frac{-1}{27}, \frac{7}{9}, \ldots\ldots\ldots\ldots -729$

Here $a = \frac{1}{81}, \quad r = -3$

Now $ar^{n-1} = 729$

$\frac{1}{81} \times (-3)^{n-1} = -729$

$(-3)^{n-1} = 3^{10}, n = 11$

13. Ans. (c)

$\frac{a+b}{2} = 25$

$a + b = 50$

also $\sqrt{ab} = 7$

$ab = 49$

here $a = 1$ and $b = 49$ or vice versa

14. Ans. (b)

$\frac{a+b}{2} = AM$

$\sqrt{ab} = GM$

Here it is given that

GM = 20% less than AM, that is

$GM = \frac{4}{5} AM$

$\Rightarrow \sqrt{ab} = \frac{4}{5} \times \frac{a+b}{2}$

only possibility is $a = 1$ & $b = 4$

or $a = 4, b = 1$

So, ratio is 4 : 1

□□□□

II Practice Exercise: II

1. Determine k so that $\frac{2}{3}$, k, and $\frac{5}{8}$k are three consecutive terms of an A.P.
   (a) 16/33 (b) 14/33
   (c) 12/33 (d) 18/33

2. If 7 time the 7th term of an A.P. is equal to 11 times its, 11th term, then the 18th term of the A.P. is
   (a) 1 (b) 2
   (c) 0 (d) 3

3. If the pth, q^{th} and r^{th} terms of an A.P. are a, b, c, respectively, find the value of
   $a(q - r) + b(r - p) + c(p - q)$
   (a) 2 (b) 1
   (c) 0 (d) 3

4. A ball rolling up an incline covers 36 metres during the first second, 32 metres during the second, 28 metres during the next and so on. How much distance will it travel during the 8th second?
   (a) 8 metres (b) 6 metres
   (c) 7 metres (d) 9 metres

5. The sum of a series in A.P. is 525. Its first term is 3 and last term is 39. Find the common difference.
   (a) 3/2 (b) 3/3
   (c) 2/3 (d) 1/3

6. The sum of p terms of an A.P. is $3p^2+4p$. Find the nth term?
   (a) 5n + 2 (b) 6n + 1
   (c) 8n + 3 (d) 7n + 3

7. The 5th term of a G.P. is 2, find the product of first 9 terms.
   (a) 508 (b) 512
   (c) 504 (d) 516

8. The 3rd term of a G.P. is the square of the first term. If the second term is 8, determine the 6th term.
   (a) 136 (b) 132
   (c) 128 (d) 124

9. The sum of first three terms of a G.P. is to the sum of first six terms is 125 : 152. Find the common ratio of G.P.

(a) $\frac{2}{5}$ (b) $\frac{4}{5}$

(c) $\frac{3}{5}$ (d) $\frac{1}{5}$

10. Evaluate $\sum_{j=1}^{11}(2+3^j)$

(a) $22+\frac{3}{2}(3^{11}-1)$ (b) $11+\frac{3}{2}(3^{11}-1)$

(c) $22+\frac{3}{2}(3^{10}-1)$ (d) None of these

11. The common ratio of a G.P. is $-\frac{4}{5}$ and the sum to infinity is $\frac{80}{9}$. Find the first term.

(a) 14 (b) 16
(c) 12 (d) 10

12. Sum the series to infinity

$$\frac{3}{4}-\frac{5}{4^2}+\frac{3}{4^3}-\frac{5}{4^4}+\frac{3}{4^5}-\frac{5}{4^6}+............$$

(a) $\frac{8}{15}$ (b) $\frac{7}{17}$

(c) $\frac{7}{15}$ (d) $\frac{8}{17}$

13. Find the 9th term of the H.P. 6, 4, 3, ...

(a) 7/5 (b) 6/5
(c) 5/6 (d) None of these

14. If $x > 1$, $y > 1$, $z > 1$ are in G.P., then $\frac{1}{1+\log x}, \frac{1}{1+\log y}, \frac{1}{1+\log z}$ are in

(a) A.P. (b) H.P.
(c) G.P. (d) None of these

15. If a be the first term of a G.P., l the nth term and P the product of first n terms then P =

(a) $(al)^{n/2}$ (b) $(a-l)^{n/2}$
(c) $(a+l)^{n/2}$ (d) None of these

Answers

1. (a)	2. (c)	3. (c)	4. (a)	5. (a)
6. (b)	7. (b)	8. (c)	9. (c)	10. (a)
11. (b)	12. (c)	13. (b)	14. (b)	15. (a)

Solutions

1. Ans. (a)

$\because \frac{2}{3}, k, \frac{5}{8}k$ are in A.P.

$\therefore k-\frac{2}{3}=\frac{5}{8}k-k \Rightarrow \frac{5k}{8}-2k=\frac{-2}{3}$

$\Rightarrow \frac{-11k}{8}=\frac{-2}{3} \Rightarrow k=\frac{16}{33}$.

2. Ans. (c)

Let a be the first term and d, the common difference of an A.P.

$\therefore a_7 = a + 6d$

$a_{11} = a + 10d \quad \therefore 7a_7 = 11a_{11}$

$7(a + 6d) = 11(a + 10d)$

$\Rightarrow 7a + 42d = 11a + 110d$

$\Rightarrow 4a + 68d = 0$

$\Rightarrow a + 17d = 0$

$\Rightarrow a_{18} = 0$

3. Ans. (c)

Let A be the first term and D, the common difference of A.P.

$a_p = a, \therefore A + (p-1)D = a$ ....(i)

$a_{q=b}, \therefore A+(q-1)D = b$ ....(ii)

$a_r = c, \therefore A+(r-1)D = c$ ....(iii)

$\therefore a(q-r) + b(r-p) + c(p-q)$

$= [A + (p-1)D](q-r) + [A+(q-1)D](r-p)$

$+[A+(r-1)D](p-q)$

$= (q-r+r-p+p-q)A + [(p-1)(q-r)$

$+(q-1)(r-p)+(r-1)(p-q)]D$

$= 0{\cdot}A + 0{\cdot}D = 0$

4. Ans. (a)

Distance covered during the I^{st} second = 36 m

Distance covered during the 2nd second = 32 m

Distance covered during the 3rd second = 28 m.

The distance covered form an A.P.

= 36 + 32 + 28 ..... in which

a = 36, d = –4

$\therefore$ Distance covered in 8th second

= 8th term of the A.P.
= $a + 7d = 36 + 7(-4)$
= $36 - 28 = 8$ metres.

5. Ans. (a)

If n be the number of terms, then

$a_n = a + (n-1)d$,

where a is the first term and d the common difference.

$\therefore\ 39 = 3 + (n-1)d.$

Also $S_n = \frac{n}{2}[a_1 + a_n]$

$\therefore\ 525 = \frac{n}{2}[3+39] \Rightarrow 1050 = n(42)$

or, $n = \frac{1050}{42} = 25$

putting n = 25 in (i), we get

$(25-1)d = 36 \Rightarrow d = 36 \div 24 = \frac{3}{2} = 1\frac{1}{2}$

6. Ans. (b)

Here, $S_p = 3p^2 + 4p$

Putting p = n, we have

$S_n = 3n^2 + 4n$

Changing n to (n – 1), we get

$S_{n-1} = 3(n-1)^2 + 4(n-1)$
$= 3(n^2 - 2n + 1) + 4n - 4$
$= 3n^2 - 2n - 1$

$\therefore\ a_n = S_n - S_{n-1}$
$= 3n^2 + 4n - 3n^2 + 2n + 1 = 6n + 1.$

7. Ans. (b)

Let a be the first term and r the common ratio

$\therefore\ a^5 = 2 \Rightarrow ar^4 = 2$ ...(i)

Now, product of first 9 terms
$= a \times ar \times ar^2 \times \ldots\ ar^8$
$= a^9 r^{1+2+\ldots+8} = a^9 r^{36}$
$= (ar^4)^9 = 2^9 = 512.$

8. Ans. (c)

Let a be the first term and r be the common ratio of G.P.

We have $a_3 = (a_1)^2 \Rightarrow ar^2 = a^2$

$\Rightarrow r^2 = a$ ...(i)

Also, $a_2 = 8$

$\Rightarrow ar = 8$ ...(ii)

Multiplying (i) and (ii) we get

$ar^3 = 8 \times a$

$\therefore\ r^3 = 8 \Rightarrow r = 2$

From (i) $a = (2)^2 = 4$

Hence, $a^6 = ar^5 = (4)(2)^5$
$= 4 \times 32 = 128.$

9. Ans. (c)

Here, $\frac{S_3}{S_6} = \frac{125}{152}, \frac{a(r^3-1)/(r-1)}{a(r^6-1)/(r-1)} = \frac{125}{152}$

$\Rightarrow \frac{r^3-1}{r^6-1} = \frac{125}{152}, \quad \therefore \frac{r^3-1}{(r^3-1)(r^3+1)} = \frac{125}{152}$

$\Rightarrow \frac{1}{r^3+1} = \frac{125}{152}, \quad \therefore 152 = 125r^3 + 125$

$\Rightarrow 125r^3 = 27, \Rightarrow r^3 = \frac{27}{125}.$

or, $r^3 = \left(\frac{3}{5}\right)^3, \quad \therefore r = \frac{3}{5}.$

Hence, the common ratio of G. P. is $\frac{3}{5}$.

10. Ans. (a)

$(2+3^1) + (2+3^2) + (2+3^3) + .. + (2+3^{11})$
$= 2 + 2 + 2 + \ldots$ up to 11 terms
$+(3 + 3^2 + 3^3 + \ldots$ up to 11 terms)
$= 11 \times 2 + \frac{3(3^{11}-1)}{3-1} = 22 + \frac{3}{2}(3^{11}-1)$

11. Ans. (b)

$S_\infty = \frac{a}{1-r} \Rightarrow \frac{80}{9} = \frac{a}{1-\left(\frac{-4}{5}\right)} \Rightarrow \frac{80}{9} = \frac{a}{9/5}$

$\Rightarrow a = \frac{80}{9} \times \frac{9}{5} = 16$

Hence, the first term is 16.

12. Ans. (c)

$\left(\frac{3}{4} + \frac{3}{4^3} + \frac{3}{4^5} + \ldots \text{to } \infty\right) - \left(\frac{5}{4^2} + \frac{5}{4^4} + \frac{5}{4^6} + \ldots \text{to } \infty\right)$

$= \frac{3/4}{1-\left(\frac{1}{4}\right)^2} - \frac{5/4^2}{1-\left(\frac{1}{4}\right)^2}$

$$= \frac{3}{4}\times\frac{16}{15}-\frac{5}{16}\times\frac{16}{15}=\frac{4}{5}-\frac{1}{3}$$

$$= \frac{12-5}{15}=\frac{7}{15}.$$

13. Ans.(b)

The given sequence is 6, 4, 3, ... which is H..P.
The sequence of reciprocals of its terms is

$\frac{1}{6},\frac{1}{4},\frac{1}{3}$........ which is an A,.P

Here, $a = \frac{1}{6}$, $d = \frac{1}{4}-\frac{1}{6}=\frac{1}{12}$

$\therefore$ a_9 of A.P. = a + 8d

$$= \frac{1}{6}+8\times\frac{1}{12}=\frac{1}{6}+\frac{4}{6}=\frac{5}{6}$$

$\therefore$ 9th term of H.P. is $\frac{6}{5}$.

14. Ans. (b)

$\because$ x, y, z are in G.P.

$\because$ $y^2 = xz$

Taking log on both sides

2 log y = log x + log z

$\Rightarrow$ 2 + 2 log y = (1+log x) + (1+log z)

$\Rightarrow$ 2 (1+log y) = (1+log x) + (1 + log z)

$\Rightarrow$ 1+ log x, 1+log y, 1+log z, are in A.P.

$\Rightarrow$ $\frac{1}{1+\log x},\frac{1}{1+\log y},\frac{1}{1+\log z}$ are in H.P.

15. Ans. (a)

If r is the common ratio of G.P., then

$l = a r^{n-1}$ ...(i)

The first n terms of the G.P. are

a, ar, ar^2, ar^3, ... ar^{n-1}

$P = a \times ar \times ar^2 \times ... \times ar^{n-1}$

$= a^n \times r^{1+2+3...+(n-1)}$

$$= a^n \times r^{(n-1)\times\frac{n}{2}} = (a^2)^{n/2}\times(r^{n-1})^{n/2}$$

$$= (a^2r^{n-1})^{n/2} = (a.\ ar^{n-1})^{n/2} = (al)^{n/2}.$$

□□□□

III Practice Exercise: III

1. If the sum of the 6th and the 15th elements of an arithmetic progression is equal to the sum of the 7th, 10th and 12th elements of the same progression, then which element of the series should necessary be equal to zero?
   (a) 10th (b) 8th
   (c) 1st (d) None of these

2. If p, q, r, s are in harmonic progression and p > s, then
   (a) $\frac{1}{ps}<\frac{1}{qr}$ (b) q + r = p + s
   (c) $\frac{1}{q}+\frac{1}{r}=\frac{1}{p}+\frac{1}{s}$ (d) None of these

3. If $\log_x a$, $a^{x/2}$ and $\log_b x$ are in GP, then x is
   (a) $\log_a(\log_b a)$
   (b) $\log_a(\log_e a) + \log_a(\log_e b)$
   (c) $-\log_a(\log_a b)$
   (d) $\log_a(\log_e b) - \log_a(\log_e a)$

4. A person pays Rs. 975 in monthly installments, each monthly installment being less than the former by Rs. 5. The amount of the first installment is Rs. 100. In what tune, will the entire amount be paid?
   (a) 12 months (b) 26 months
   (c) 15 months (d) 18 months

5. Let S_n denote the sum of the first 'n' terms of an A.P. $S_{2n} = 3S_n$. Then, the ratio S_{3n}/S_n is equal to
   (a) 4 (b) 6
   (c) 8 (d) 10

6. Three numbers are in G.P. Their sum is 28 and their product is 512. The numbers are
   (a) 6, 9 and 13 (b) 4, 8 and 16
   (c) 2, 8 and 18 (d) 2, 6 and 18

7. The sum of the series:
   $1^2 + 2^2 + 3^2 + 4^2 + ... + 15^2$ is
   (a) 1080 (b) 1240
   (c) 1460 (d) 1620

8. If $\frac{1}{b}-\frac{1}{c}=\frac{1}{a}-\frac{1}{b}$, then a, b, c form a/an
   (a) Arithmetic progression
   (b) Geometric Progression
   (c) Harmonic Progression
   (d) None of these

9. The value of
$(1^3 + 2^3 + 3^3 + \ldots. 15^3) - (1 + 2 + 3 + \ldots. 15)$ is
(a) 14280 (b) 14400
(c) 12280 (d) 13280

10. The mean of the cubes of the first n natural numbers is
(a) $\frac{n(n+1)^2}{4}$ (b) n^2
(c) $\frac{n(n+1)(n+2)}{8}$ (d) (n^2+n+1)

11. If $\frac{3+5+7+\ldots.+n \text{ terms}}{5+8+11+\ldots+10 \text{ terms}} = 7$, then the value of n is
(a) 35 (b) 36
(c) 37 (d) 40

12. If the sum of first n natural numbers is one-fifth of the sum of their squares, then *n* is
(a) 5 (b) 6
(c) 7 (d) 8

13. What is the least value of *n* such that $(1 + 3 + 3^2 + \ldots + 3^n)$ exceeds 2000?
(a) 7 (b) 5
(c) 8 (d) 6

14. The sum of 12 terms of an A.P., whose first term is 4, is 252. What is the last term?
(a) 35 (b) 36
(c) 37 (d) 38

15. If log 2, $\log(2^x - 1)$ and $\log(2^x + 3)$ (all to the base 10) be three consecutive terms of an Arithmetic Progression, then the value of *x* is equal to
(a) 0 (b) 1
(c) $\log_2 5$ (d) $\log_{10} 2$

16. The third term of a Geometric Progression is 4. The product of the first five terms is
(a) 4^3 (b) 4^5
(c) 4^4 (d) None of these

Answers

1. (b)	2. (c)	3. (a), (c)		4. (c)
5. (b)	6. (b)	7. (b)	8. (c)	9. (a)
10. (a)	11. (a)	12. (c)	13. (c)	14. (d)
15. (c)	16. (b)			

Solutions

1. **Ans. (b)**
Let a be the first term and d be the common difference of an A.P.
$\therefore \quad (a+5d)+(a+14d)$
$= (a + 6d) + (a + 9d) + (a + 11d)$
$\Rightarrow a + 7d = 0$
$\Rightarrow 8^{th}$ term = 0.

2. **Ans. (c)**
p, q, r, s are in harmonic progression
$\Rightarrow \frac{1}{p}, \frac{1}{q}, \frac{1}{r}$ and $\frac{1}{s}$ are in A.P.
$\Rightarrow \frac{1}{q} - \frac{1}{p} = \frac{1}{s} - \frac{1}{r}$
$\Rightarrow \frac{1}{q} + \frac{1}{r} = \frac{1}{p} + \frac{1}{s}.$

3. **Ans. (a), (c)**
Given statement
$\Rightarrow (a^{x/2})^2 = (\log_b x) \times (\log_x a)$
$= a^x = \log_b a = \log_a(\log_b(a))$
$\Rightarrow x \log a = \log_a[\log_b a]$
$\Rightarrow x = \log_a[\log_b a] = -\log_a(\log_a b)$

4. **Ans. (c)**
Let n be the number of months in which all the installments can be paid
First Installment = Rs. 100
Common Difference = –5
$\Rightarrow$ Sum of the series with *n* terms whose first term is 100 or common difference is (–5) = 975
i.e. $\frac{n}{2}[2a+(n-1)d] = 975$
i.e. $\frac{n}{2}[2\times 100+(n-1)(-5)] = 975$
i.e. $n^2 - 41n + 390 = 0$
i.e. n = 26 or n = 15
For n=15, total amount paid
$= \frac{15}{2}[2\times 100+(15-1)(-5)]$
$= \frac{15}{2}[200-70] = 975.$

5. **Ans.(b)**

$$S_n = \frac{n}{2}[a+(n-1)d]$$

[where a is the first term and d is the common difference]

$$S_{2n} = \frac{2n}{2}[a+(2n-1)d]$$

$$S_{3n} = \frac{3n}{2}[a+(3n-1)d]$$

Given, $S_{2n} = 3S_n$

$$\Rightarrow n[a+2nd-d] = 3\left[\frac{n}{2}(a+nd-d)\right]$$

$$\Rightarrow d = \frac{a}{1+n}$$

$$\therefore \frac{S_{3n}}{S_n} = \frac{\frac{3n}{2}[a+3nd-d]}{\frac{n}{2}[a+nd-d]}$$

$$= \frac{3\left[a+\frac{3na}{1+n}-\frac{a}{1+n}\right]}{a+\frac{na}{1+n}-\frac{a}{1+n}} = 6.$$

6. **Ans. (b)**

Let the three numbers be a, ar, ar^2, where r is common ratio.

$$\therefore a + ar + ar^2 = 28 \text{ and } a^3r^3 = 512$$

$$\therefore ar = 8 \Rightarrow a + ar^2 = 20$$

$$\Rightarrow 8r^2 - 20r + 8 = 0$$

$$\Rightarrow r = 2,\ r = \frac{1}{2}.$$

if r = 2, a = 4. Therefore, the three numbers are 4, 8, 16

7. **Ans. (b)**

The sum of the squares of the first n natural numbers is

$$\frac{n(n+1)(2n+1)}{6}$$

Put n = 15, we have, $1^2 + 2^2 + 3^2 + 4^2 + ... + 15^2$

$$= \frac{15(15+1)(30+1)}{6} = 1240.$$

8. **Ans. (c)**

$$\frac{2}{b} = \frac{1}{a} + \frac{1}{c}$$

$$\Rightarrow b = \frac{2ac}{a+c} \Rightarrow \frac{1}{a},\frac{1}{b},\frac{1}{c} \text{ are in A.P.}$$

$\Rightarrow$ a, b, c are in H.P.

9. **Ans. (a)**

Given expression

$$= \left[\frac{15\times16}{2}\right]^2 - \frac{15\times16}{2} = (120)^2 - 120$$

$$= 120 \times 119 = 14280.$$

10. **Ans. (a)**

Sum of the cubes of first n natural numbers

$$= \left[\frac{n(n+1)}{2}\right]^2 = \frac{n^2(n+1)^2}{4}$$

$$\therefore \text{Mean} = \frac{n(n+1)^2}{4}.$$

11. **Ans. (a)**

S_n = Sum of n terms of an A.P.

$$= \frac{n}{2}[2a+(n-1)d]$$

where a = first term,

d = common difference

$$\therefore \frac{3+5+7+...+n \text{ terms}}{5+8+11+...+10 \text{ terms}} = 7$$

$$\Rightarrow \frac{\frac{n}{2}[2\times3+(n-1)\times2]}{\frac{10}{2}[2\times5+(10-1)\times3]} = 7$$

$$\Rightarrow \frac{n(2n+4)}{370} = 7$$

$$\Rightarrow 2n^2 + 4n - 2590 = 0$$

$$\Rightarrow n^2 + 2n - 1295 = 0$$

$$\Rightarrow n^2 + 37n - 35n - 1295 = 0$$

$$\Rightarrow n(n+37) - 35(n+37) = 0$$

$$\Rightarrow (n-35)(n+37) = 0$$

$$\Rightarrow n = 35.$$

12. **Ans. (c)**

Sum of the first n natural numbers $= \frac{n(n+1)}{2}$

Sum of the squares of the first n natural numbers

$$= \frac{n(n+1)(2n+1)}{6}$$

$\therefore \quad \frac{n(n+1)}{2} = \frac{1}{5}\left(\frac{n(n+1)(2n+1)}{6}\right)$

$\Rightarrow \quad 2n+1 = 15 \Rightarrow n = 7.$

13. Ans. (c)

$\frac{3^n - 1}{3-1} > 2000 \Rightarrow 3^n > 4001 \Rightarrow n = 8.$

14. Ans. (d)

$S_n = \frac{n}{2}[\text{First term + Last term}]$

Where a = 4, n = 12, L = ?

$\therefore \quad S_{12} = 252 = \frac{12}{2}[4 + L]$

$\Rightarrow \quad 4 + L = 42$

$\Rightarrow \quad L = 38$

15. Ans. (c)

log 2, $\log(2^x-1)$,

$\log(2^x+3)$ are in A.P.

$\Rightarrow \quad 2\left[\log\left(2^x - 1\right)\right] = \log 2 + \log\left(2^x + 3\right)$

$= \log\left[2 \times \left(2^x + 3\right)\right]$

$\Rightarrow \quad \log\left(2^x - 1\right)^2 = \log\left[2^{x+1} + 6\right]$

$\Rightarrow \quad \left(2^x - 1\right)^2 = 2^{x+1} + 6 = 2^x.2 + 6$

Let $2^x = y$

$\therefore \quad (y-1)^2 = 2y + 6 \Rightarrow y^2 - 2y + 1 = 2y + 6$

$\Rightarrow \quad y^2 - 4y - 5 = 0$

$\Rightarrow \quad (y-5)(y+1) = 0$

$\Rightarrow \quad y = 5, -1$

If $y = 5 \Rightarrow 2^x = 5 \Rightarrow x\log 2 = \log 5$

$\Rightarrow \quad x = \frac{\log 5}{\log 2} \Rightarrow x = \log_2 5.$

16. Ans.(b)

Let a be the first term of a G. P. and r be the common ratio

$\therefore$ First five terms of a G.P. are a, ar, ar^2, ar^3, ar^4.

$\therefore$ Third term = $ar^2 = 4 \Rightarrow (ar^2)^5 = 4^5$.

□□□□

IV Practice Exercise: IV

1. If for an AP m^{th}, n^{th}, p^{th} terms are a, b, c respectively then c(m – n) + a(n – p) + b(p – m) will have the values as
   (a) 0 (b) $\frac{1}{2}$
   (c) –1 (d) $\frac{abc}{mnp}$

2. The sum of all the odd 3 digit integers which are divisible by 9 is
   (a) 23400 (b) 27900
   (c) 35100 (d) 55350

3. If first, third 4 last term of an AP are a, b, 2a respectively then its sum is
   (a) ab (b) $\frac{3ab}{a-b}$
   (c) $\frac{3ab}{2(b-a)}$ (d) $\frac{3ab}{2(b-a)}$

4. If $|a| < 1$ and $S = a^2 + a^4 + a^6 + ...$ then a =
   (a) 0 (b) aS
   (c) $\frac{S}{1-S}$ (d) $\left(\frac{S}{1+S}\right)^{1/2}$

5. A car moves 25 m is 3rd second, 65 meters in 7th second, 75 meters in 8th second and so on. How for it will move in 15th second
   (a) 165 (b) 145
   (c) 155 (d) 175

6. The sum of n terms of an AP is $n^2 + 2n$, the common difference will be
   (a) 1 (b) 3
   (c) 2 (d) –1

7. If a, b, c, d are four real numbers what will be the minimum possible value of $[a^2 + b^2 + c^2 + d^2]\left[\frac{1}{a^2} + \frac{1}{b^2} + \frac{1}{c^2} + \frac{1}{d^2}\right]$
   (a) 4 (b) 2
   (c) abcd (d) 16

8. Take an AP 2, 3, 4, ... and a GP 3, 6, 12 ...., if taken upto 100. How many terms will be there which are common in both.
   (a) 23 (b) 33
   (c) 6 (d) 13

9. If second term of an AP is equal to the 9th term of another AP. What will be the sum of 17 terms of second AP as the sum of first 3 terms of first AP is given as 9.
(a) 85 (b) 102
(c) 204 (d) 51

10. There are three numbers is AP. Which have their sum as 45 and product as 1875. The highest of the number is
(a) 15 (b) 25
(c) 35 (d) 21

11. A ball is dropped from a height of 50 m and it rebounces to 1/3 of the initial height. If it continues to rebound in the same manner. What is the distance travelled by the ball in the air
(a) 75 (b) 60
(c) 150 (d) 100

12. Three numbers P, q, r are in GP if P & q are increased by 4 and r is decreased by 4 then they forms and AP. If sum of the numbers is 56 then highest among P, q, r will be
(a) 24 (b) 18
(c) 36 (d) 32

13. Find the sum to infinite term of the series

$$\frac{1}{56}+\frac{1}{72}+\frac{1}{90}+\frac{1}{110}+....$$

(a) $\frac{1}{8}$ (b) $\frac{1}{7}$
(c) $\frac{1}{6}$ (d) $\frac{1}{28}$

14. Let {1, 2, 3, ...541}. Find out how many APs can be formed by taking the element of S starting from 1 and ending at 541. Which have atleast 3 elements
(a) 12 (b) 13
(c) 23 (d) 24

15. Find the sum to first 10 terms of the series
12 + 35 + 70 + 117 + ...
(a) 2595 (b) 2575
(c) 2945 (d) 2495

Answers

1. (a)	2. (d)	3. (d)	4. (d)	5. (b)
6. (c)	7. (d)	8. (b)	9. (d)	10. (b)
11. (d)	12. (d)	13. (b)	14. (c)	15. (a)

Solutions

1. Ans. (a)

Let first term be 'x'

$$x + (m-1)d = a$$

$$\Rightarrow (m-n) = a-b$$

$$x + (n-1)d = b$$

$$\Rightarrow (n-p) = b-c$$

$$x + (p-1)d = c$$

$$\Rightarrow (p-m) = c-a$$

$$\Rightarrow c(m-n) + a(n-p) + b(p-m) = c(a-b) + a(b-c) + b(c-a) = 0$$

2. Ans. (d)

First '3' digit integer divisible by '9' = 108

Last '3' digit integer divisible by '9' = 999

Let 999 be n^{th} number

$$\Rightarrow 999 = 108 + (n-1)9$$

So $\frac{100}{2}[108+999] = 50 \times 1107 = \text{sum}$

$$\Rightarrow \text{Sum} = 55350$$

3. Ans. (d)

Common difference = (b – a)

$$2a = a + \{(n-1)(b-a)\}$$

$$\Rightarrow \frac{a}{b-a} = n-1$$

$$\Rightarrow n = \frac{b}{b-a}$$

$$\frac{1}{2}\left(\frac{b}{b-a}\right)[2a+a] = \text{sum} = \frac{3ab}{2(b-a)}$$

4. Ans. (d)

$$S = \frac{a^2}{1-a^2}$$

$$\Rightarrow S(1-a^2) = a^2$$

$$\Rightarrow S - Sa^2 = a^2$$

$$\Rightarrow a^2 = \frac{S}{1+S}$$

$$\Rightarrow a = \left(\frac{S}{1+S}\right)^{1/2}$$

5. Ans. (b)

Let speed be x and the car covers distance in A.P.

Then the distance covered in

3rd sec. = $x + 2d = 25$

$x + 6d = 65$ so $d = 10$ and $x = 5$.

So the distance covered in

15th sec. = 5 + 14 × 10 = 145 m.

6. Ans. (c)

$T_n = s_n - s_{n-1} = \{n^2 + 2n\} - \{(n-1)^2 + 2(n-1)\}$

$= (2n-1) + 1 = 2n$

$T_{n-1} = 2(n-1) = 2n - 2$

Common diff. $= 2n - (2n-2) = 2$

7. Ans. (d)

Applying A.M. ≥ G.M

$$\frac{[a^2 + b^2 + c^2 + d^2]}{4} \geq (a^2.b^2.c^2.d^2)^{1/4}$$

$$\& \ \frac{\left[\frac{1}{a^2}+\frac{1}{b^2}+\frac{1}{c^2}+\frac{1}{d^2}\right]}{4} \geq \left(\frac{1}{a^2}.\frac{1}{b^2}.\frac{1}{c^2}.\frac{1}{d^2}\right)^{1/4}$$

$$\text{So } \frac{1}{16}(a^2+b^2+c^2+d^2).\left(\frac{1}{a^2}+\frac{1}{b^2}+\frac{1}{c^2}+\frac{1}{d^2}\right) \geq 1$$

or the min. possible value = 16

8. Ans. (b)

In the A.P. all the multiples of 3 will be common with G.P. and since the common diff. is '1' every term will be there in consecutive 3 terms.

So net common terms $= \frac{100}{3} = 33$.

9. Ans. (d)

$$T_2 = T_9' \Rightarrow \frac{T_2}{T_9'} = \frac{1}{1}$$

$$\text{So } \frac{s_{(2.2-1)}}{s'_{(2.9-1)}} = \frac{(2.2-1)\,T_2}{(2.9-1)\,T_9'}$$

$$\Rightarrow \frac{s_3}{s'_{17}} = \frac{3}{17} \times \frac{1}{1} = \frac{3}{17}$$

i.e. $s_3 = 9$, so $s'_{17} = 51$.

10. Ans. (b)

$a - d + a + a + d = 45 \Rightarrow a = 15$

and $15(15-d)(15+d) = 1875$

So $d = \pm 0$. Then highest number

$= 15 + 10 = 25$.

11. Ans. (d)

This will be a infinite G.P. with common ratio as $\frac{1}{3}$

So net distance covered

$$= 50 + \frac{50}{3} + \frac{50}{3} + \frac{50}{9} + \frac{50}{9} + \frac{50}{27} + \frac{50}{27} + \ldots$$

↓ up ↓ down

$$= 100\left(1 + \frac{1}{3} + \frac{1}{9} + \ldots\right) - 50$$

$$= 100\left(\frac{1}{1-\frac{1}{3}}\right) - 50 = 100\text{m}$$

12. Ans. (d)

$q^2 = pr$ ...(i)

$2(q+4) = (p+4) + (r-4)$

$\Rightarrow 2q + 8 = p + r$ ...(ii)

$p + q + r = 56$ ...(iii)

by solving the eq. we get

$p = 8, q = 16, r = 32$

13. Ans. (b)

$$\text{The series} = \frac{1}{7.8} + \frac{1}{8.9} + \frac{1}{9.10} + \ldots$$

$$= \frac{1}{7} - \frac{1}{8} + \frac{1}{8} - \frac{1}{9} + \frac{1}{9} - \frac{1}{10} + \ldots$$

So the required

$$\text{sum} = \frac{1}{7} - 0 = \frac{1}{7}$$

14. Ans. (c)

In this case $541 = 1 + (n-1)d$

So the possible value of n will be the number of factors of 540.

So no. of factors of $540 = 3 \times 4 \times 2 = 24$

in this we can not include '1' as $n - 1 = 1$.

So $n = 2$.

Which is not acceptable. So net possibilities $= 24 - 1 = 23$

15. Ans. (a)

The series

$s = 3.4 + 5.7 + 7.10 + 9.13 + \ldots$

So for n terms

$T_n = \{3 + (n-1)2\}\{4 + (n-1)3\}$

$= (2n+1)(3n+1)$

$= 6n^2 + 5n + 1$

$S_n = 6\Sigma n^2 + 5\Sigma n + \Sigma 1$

$$= 6\frac{(n)(n+1)(2n+1)}{6} + 5\frac{(n)(n+1)}{2} + n$$

For n = 10

$$S_{10} = (10 \times 11 \times 21) + \frac{5}{2} \times 10 \times 11 + 10$$

$S_{10} = 2595$

■■■■

2.3 CHAPTER Permutation & Combination

Being an area as a measure of students reasoning ability, permutations and combinations becomes one of the important chapter from exams point of view.

Fundamental Principle of Counting

Multiplication Rule

- If a task A can be done in 'm' ways and a second task B can be done in 'n' ways, the number of ways of doing both the task one after another is 'm × n' ways.

Additive Rule

- If task 'A' can be done in 'm' ways and another task 'B' can be done in 'n' ways then total no. of ways of doing either of task.

  = (m + n) ways
- These two rules are basic building block to permutations and combinations principles.

Some Basic Examples

1. In a building there are 6 floors. In how many ways 5 persons can get down at different floors such that exactly one person is allowed at any floor. (order in which person are getting down is not important)

Solution

For 1st person there are 6 choices of floor. Now for 2nd person only 5 choices are left (because one of the floors is occupied by 1st person).

Progressing with same logic.

Total no. of ways (Because all five to get down, hence multiplicative rule)

$6 \times 5 \times 4 \times 3 \times 2 = 720$ ways

2. In a party there are 20 couples, 20 boys (single) and 20 girls (single) are invited. If handshake between only two persons of opposite sex is allowed and handshake between spouses is not allowed total how many handshakes took place in the party?

Solution.

	Males	Females
Couples	20	20
Singles	20	20

For each male among couple has a choice of 39 shake hands, because he has not do handshake will his spouse and for each of single boy there is choice of 40 handshakes.

Hence total no. of handshakes

$= 20 \times 39 + 20 \times 40 = 20 \times 79 = 1580$

Permutations

- Number of ways of arranging particular things in certain order is known as permutations. For example for three things (a, b, c) can be arranged in multiple ways in a straight line, total number different arrangements is called permutations.

$$\begin{bmatrix} a, b, c & b, a, c & c, a, b \\ a, c, b & b, c, a & c, b, a \end{bmatrix} \text{Total six ways}$$

$${}^{n}P_{r} = \frac{|\underline{n}}{|\underline{n-r}}$$

- Distinct permutations of 'r' things taken out of n things taken at timings = ${}^{n}P_{r}$.
- The terms permutation and arrangement are synonymous and can be used inter-changeably.

Combinations

- Combinations is basically selection of a group of certain things out of given set. In combination order of things is not important.
- Combinations of 'r' things out of 'n' distinct things is

$${}^{n}C_{r} = \frac{{}^{n}P_{r}}{|\underline{r}} = \frac{|\underline{n}}{|\underline{n-r} \times |\underline{r}}$$

Special Notes:

At times, question may not explicitly state whether we have find out permutations or combination but nature of question can guide us which has to be use by pure examination of statement in question?

For example

"If I say how many 3 digit number can be made using digit 1, 2, 3 exactly once". I know possible name 123, 132 order is important hence permutation will be used.

If I say out of a group of five friends that I have, in how many ways I can invite three friends, now say out of friend u, v, w, x, y suppose. If I select u, v, w the w, u, v will be same group hence selection will be applicable.

In beginning we will cover all permutation combination formulae with two basic conditions

- If all things are distinct
- Each item is used exactly for once.

Linear Permutation of 'n' distinct things taken 'r' at a time without repetition

$${}^nP_r = \frac{\lfloor n}{\lfloor n-r}$$

$n! = 1 \times 2 \times 3 \times 4 \times 5 \ldots . \times n$

$0! = 1$

Number of combinations of n distinct things taken r at time

$$= {}^nC_r = \frac{\lfloor n}{\lfloor n-r \times \lfloor r}$$

$${}^nC_r = {}^nC_{n-r}$$

Number of arrangements of 'n' items of which 'P' one of one type, 'q' second type and rest are distinct.

$$= \frac{\lfloor n}{\lfloor P \times \lfloor q \times \lfloor r}$$

Number of Permutations of 'n' distinct items where each item can be used any number of time (repetition allowed) = n^r

Number of ways of selecting one of more items from n given items = $(2^n - 1)$

$$= {}^nC_1 + {}^nC_2 + {}^nC_3 \ldots\ldots {}^nC_n = (2^n - 1)$$

Division into groups

Dividing (P + q + r) things into group of P thing, q things and r things.

$$= \frac{(P+q+r)!}{P!q!r!}$$

Special Case.

(i) Dividing 3(P) things into three distinguishable grouping P items each = $\frac{(3P)!}{(P!)^3}$

(ii) Dividing 3(P) things into three identical group of P items each = $\frac{(3P)!}{(P!)^3} \times \frac{1}{3}$

Circular Permutations

Number of circular permutations of n distinct things = (n − 1)! [when clockwise and anticlockwise arrangement are different]

$= \frac{(n-1)!}{2}$ [when clockwise and anticlockwise arrangement are not different]

Partion Rule

No. of ways of distribution of n identical things among r people.

Case-1.

When few people can have zero things

$$= {}^{n+r-1}C_{r-1}$$

Case-2.

When each person has to have atleast one thing

$$= {}^{n-1}C_{r-1}$$

The total number of ways in which a selection can be made by taking some or all out P + q + r + ... things where P are of one kind, q alike of second kind, r of a third kind and so on.

$$= [\{(P + 1)(q + 1)(r + 1)\ldots.\} - 1]$$

Some Additional Points

$${}^{n+1}C_r = {}^nC_r + {}^nC_{r-1}$$

$${}^nP_r = r \times {}^{n-1}P_{r-1} + {}^{n-1}P_r$$

$${}^nC_0 + {}^nC_1 + {}^nC_2 + \ldots . {}^nC_n = 2^n$$

$${}^nC_2 + {}^nC_4 + {}^nC_6 + \ldots . = [2^{(n-1)} - 1]$$

$${}^nC_1 + {}^nC_3 + {}^nC_5 + \ldots = 2^{(n-1)}$$

□□□□

Solved Examples

Direction: (Q.1 to Q.3): How many 5 digit numbers are possible from digits 1, 2, 3, 4, 5, 6, 7.

1. When each digit is distinct?

(a) 2520 (b) 1260

(c) 630 (d) None of these

Ans. (b)

${}^7P_5$ or $7 \times 6 \times 5 \times 4 \times 3 = 2520$

2. When repetition of digits is allowed?

(a) 7^4 (b) 5^7

(c) 7^5 (d) None of these

Ans. (c)

$(7)^5 \Rightarrow 7 \times 7 \times 7 \times 7 \times 7 = 7^5$

3. What is sum of all numbers in example 1.

(a) ${}^7P_5 \times 11111 \times 28$ (b) $\frac{{}^7P_5}{7} \times 11111 \times 28$

(c) $\frac{{}^7P_5}{7} \times 11111 \times 24$ (d) None of these

Ans. (b)

Solution for Example 3.

Total no. possible = 7 × 6 × 5 × 3 × 4 = 2520

In these 1260, number at unit plate each digit out of 1, 2, 3, 4, 5, 6, 7 will occur equal no. time that is

each digit at unit place will occur $\frac{2520}{7}$ times $\frac{{}^7P_5}{7}$

same pattern will be applicable 10th place, 100th place, 1000th place, 10000th place digit

Hence sum of all numbers unit place digits will be

$$= \frac{2520}{7}[1+2+3+4+5+6+7]$$

Hence sum of all such number will be

$$= \frac{2520}{7}[1+2+3+4+5+6+7]$$

$$[10000+1000+100+1+10+1]$$

$$= \frac{2520}{7}[28][11111] = \frac{{}^7P_5}{7}[28][11111]$$

Direction: (Q.4 to Q.6): Below is a square grid of 8 × 8.

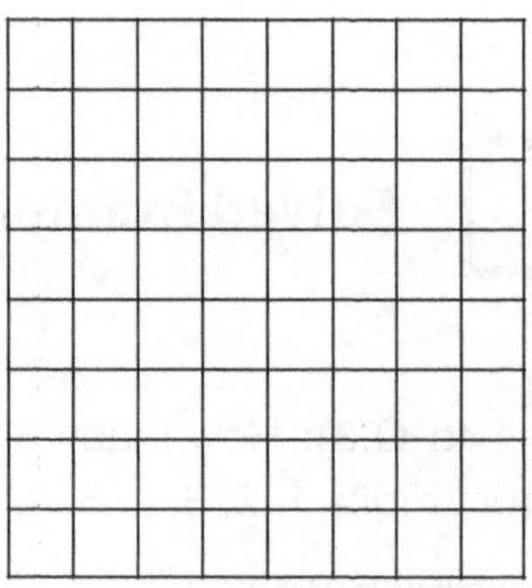

Example 4.

What is total No. of squares in the grid?

(a) 204 (b) 285

(c) 140 (d) None of these

Ans. (a)

Square of 1 × 1 unit → 8^2

Square of 2 × 2 unit → 7^2

Square of 3 × 3 unit → 6^2

....

....

Square of 7 × 7 unit → 2^2

Square of 8 × 8 unit → 1^2

Total number of square

$= 1^2 + 2^2 + 3^2 + ... + 8^2 = 204$

Hence option (a)

Example 6.

What is total number of rectangles in the grid?

(a) 36 × 36 (b) 35 × 35

(c) 34 × 34 (d) 32 × 32

Ans. (a)

To make one rectangle we need to select two horizontal lines which can selected as = ${}^9C_2$ ways and we need two vertical lines which can be selected as = ${}^9C_2$ ways.

(In a grid of 8 × 8, there will 9 horizontal lines and 9 vertical lines)

Hence total number of rectangles are

$$= {}^9C_2 \times {}^9C_2 = \left(\frac{9\times8}{2}\right)\times\left(\frac{9\times8}{2}\right) = 36 \times 36$$

Hence option (a) is correct.

Example 6.

How many rectangle of different dimensions are there?

(a) 34 (b) 35

(c) 36 (d) 33

Ans. (c)

Rectangles of

1 × 1, 1 × 2, 1 × 3, 1 × 4, 1 × 5, 1 × 6, 1 × 7, 1 × 8 ⇒ 8 rectangles

2 × 2, 2 × 3, 2 × 4, 2 × 5, 2 × 6, 2 × 7, 2 × 8 ⇒ 7 rectangles

3 × 3, 3 × 4, ... 3× 8 ⇒ 6 rectangles

4 × 4 .... 4 × 8 ⇒ 5 rectangles

5 × 5 .... 5 × 8 ⇒ 4 rectangles

6 × 6, 6 × 7, 6 × 8 ⇒ 3 rectangles

7 × 7, 7 × 8 ⇒ 2 rectangles

8 × 8 = 1 rectangles

Hence total no. of rectangles

= 8 + 7 + 6 + 5 + 4 + 3 + 2 + 1

= 36 rectangles

Direction: (Q.7 and Q.8): There are 10 points in a plane and out of which 4 are collinear.

Example 7.

How different triangles can be mode using these 10 points.

(a) ${}^{10}C_3 - {}^3C_3$ (b) ${}^9C_3 - {}^2C_2$

(c) ${}^{10}C_3 - {}^4C_3$ (d) ${}^{10}C_3 - {}^5C_3$

Ans. (c)

To make a triangle we need 3 non-collinear point

$\Rightarrow$ Out 10, 3 points can be selected in $^{10}C_3$ ways but in this selection $^4C_3$ selection are also included but these 4 points are collinear hence total number of feasible triangles.

$= {}^{10}C_3 - {}^4C_3$ Option (c)

Example 8.

How many different lines can be drawn?

(a) $^{10}C_2 - {}^4C_2$ (b) $^{10}C_2 - {}^4C_2 + 1$

(c) $^{10}C_3 - {}^4C_3 + 1$ (d) None of these

Ans. (b)

To draw a line in a plane we need two points out of 10 points we can select 2 points in $^{10}C_2$ ways, but in this selection 4 points are through which only one line passes.

Hence total number of lines

$= {}^{10}C_2 - {}^4C_2 + 1$

Hence option (b)

Example 9.

How many non-negative integral solution possible for equation below

$x + y + z + w = 20$

(a) $^{23}C_3$ (b) $^{22}C_2$

(c) $^{19}C_3$ (d) None of these

Ans. (a)

(x, y, z, w and integers can take value from 0 to 20). It is application of partion rule. 20 things to be distributed among x, y, z, w such that some values can be zero.

Hence total no. of non-negative integral solution.

$= {}^{20+4-1}C_{4-1} = {}^{23}C_3$

Hence option (a)

Example 10.

How many natural number solutions are possible for equation below?

$x + y + z + w = 20$

(a) $^{23}C_3$ (b) $^{22}C_2$

(c) $^{19}C_3$ (d) None of these

Ans. (c)

In this case again partion rule is applicable, such that each of x, y, z and w should get at least one thing.

Hence no. of ways $= {}^{20-1}C_{4-1} = {}^{19}C_3$

Hence option (c)

Example 11.

There are 15 people to be seated around a round table. In how many ways we can arrange them such that

(I) There is no restriction

(a) 14! (b) 13!

(c) 12! (d) 11!

Ans. (a)

Normal circular permutation = (15 – 1)! = 14!

Hence option (a)

(II) Such that 3 particular person A, B, C are all ways together.

(a) 12! (b) 13!

(c) $12! \times 3!$ (d) $13! \times 3!$

Ans. (c)

Consider those three persons as single group now 13 objects (12 people and a group) which can be arrange in (13 – 1)! ways but 3 persons of group can be arranged in 3! ways.

Hence total possible no. of arrangements.

$= 12! \times 3!$

Hence option (c) is correct.

(III) Such that A, B, C all three are never together.

(a) $12! \times 3!$ (b) $12! \times 176$

(c) $12! \times 182$ (d) $12! \times 142$

Ans. (b)

Total no. of required permutations

(Total circular permutations) – (Total circular permutations when all 3 are together.)

$= 14! - 12! \times 3! = 12! [14 \times 13 - 6]$

$= 12! [182 - 6] = 12! \times [176]$

Hence option (b) is correct.

□□□□

Solved Examples

1. In the above question how many numbers can be formed when repetition is allowed?

   Solution: We have to form two digit number i.e., we have to fill two place out of 4 numbers and any number can be used any number of times this can be done in $4 \times 4 = 16$ ways.

2. How many words can be formed by using letters of word MADAM?

   Solution: In MADAM we have

   2 – M's

   2 – A's

So total number of ways of forming words are

$\frac{5!}{2! \times 2!} = 30$

3. Of the different words that can be formed from letters of word MOBILE, how many begins with M and ends with E?

Solution: [M] [ ] [ ] [ ] [ ] [E]

M and E are fixed at the start and end positions, Hence, we have to arrange B, I, L, O among themselves (ie in four places), this can be done in 4! ways.

4. Six boys and 4 girls wanted to enjoy a movie. How many ways of sitting arrangement can be possible for them if girls want to sit together?

Solution: $B_1B_2B_3B_4B_5B_6, G_1G_2G_3G_4$ ...(1)

Now if all girls are sitting together then

$B_1B_2B_3B_4B_5B_6$ X ...(2)

(here X is symbolic notation that all girls are sitting together)

equation (2) Can be arranged in 7! ways

Now girls can be arranged among themselves i.e., 4! ways so total number of sitting arrangement will be 7! × 4! ways

5. In the above question find number of ways of arrangement if no two girls sit together.

Solution: This can be done in following manner

$- B_1 - B_2 - B_3 - B_4 - B_5 - B_6 -$

First of all, boys can be arranged in 6! ways, now vacant seats between them will be filled by girls. This can be done in $7P_4$ ways so total arrangement $= 7P_4 \times 6!$

6. How many 5 digit numbers can be formed by using 0, 1, 2, 3, 4, 5, 6, 7 only once?

Solution: d_1 d_2 d_3 d_4 d_5 [ ] [ ] [ ] [ ] [ ]

box d_1 cannot be filled with zero so only 7 numbers can be filled in first box rest can be done in

$7 \times 7 \times 6 \times 5 \times 4 = 5880$ ways

7. How many different sums can be formed with the following coins, selecting any number of coins from 5 rupees, 1 rupee, 50 paisa, 25 paisa, 10 paisa, 1 paisa.

Solution: A distinct sum will be formed by selecting either 1 or 2 or 3 or 4 or 5 or all 6 coins.

${}^6C_1 + {}^6C_2 + {}^6C_3 + {}^6C_4 + {}^6C_5 + {}^6C_6$ ways.

By binomial theorem this can be done in $2^6 - 1$ way.

Answer = 63.

8. Out of 15 points in a plane, 5 of which are collinear. How many straight lines can be formed?

Solution: If all 15 points were non collinear then the answer would have been ${}^{15}C_2$. However, in this case, since the 5 collinear points have also been counted. These would have been counted as ${}^5C_2$ whereas they should have been counted as 1. We need to adjust by reducing the count by $({}^5C_2 - 1)$.

Hence, the answer ${}^{15}C_2 - ({}^5C_2 - 1) = 96$

9. How many triangle can be formed by 18 points if no three among them are collinear?

Solution: To form a triangle we have to connect any three points. i.e., in this case select any three points out of 18. This can be ${}^{18}C_3$ done in ways = 816

10. In the above situation how many triangle can be formed if 5 points are collinear.

Solution: The triangles will be given by

${}^{18}C_3 - {}^5C_3 = 806$ ways

□□□□

Practice Exercise: I

1. How many numbers of 3 digits can be formed with the digits 1, 2, 3, 4, 5 (repetition of digits not allowed)?
(a) 125 (b) 120
(c) 60 (d) 150

2. How many numbers between 2000 and 3000 can be formed with the digits 0, 1, 2, 3, 4, 5, 6, 7 (repetition of digits not allowed)?
(a) 42 (b) 210
(c) 336 (d) 440

3. In how many ways can a person send invitation cards to 6 of his friends if he has four servants to distribute the cards?
(a) 6^4 (b) 4^6
(c) 24 (d) None of these

4. In how many ways can 7 Indians, 5 Pakistanis and 6 Dutch be seated in a row so that all persons of the same nationality sit together?
(a) 3! (b) 7!5!6!
(c) 3! 7! 5! 6! (d) 18!

5. How many straight lines can be formed from 8 non-collinear points on the X-Y plane?
(a) 28 (b) 56
(c) 18 (d) 19860

6. In how many ways can the letters of the word PATNA be rearranged?
(a) 60 (b) 120
(c) 119 (d) 59

7. In the above question, how many words would be there which would start with the letter P?
(a) 24 (b) 12
(c) 60 (d) 18

8. A captain and a vice-captain are to be chosen out of a team having 11 players. How many ways are there to achieve this?
(a) 10×9 (b) $^{11}C_2$
(c) 110 (d) 109!

9. In how many ways can Ram choose a vowel and a consonant from the letter ALLAHABAD?
(a) 4 (b) 6
(c) 9 (d) 5

10. In how many ways can the letters of the word EQUATION be arranged so that all the vowels come together?
(a) $^9C_4.^9C_5$ (b) 4! . 5!
(c) 9! / 5! (d) 9! – 4!5!

Direction for questions 11-12 : There are 25 points on a plane of which 7 are collinear. Now solve the following:

11. How many straight lines can be formed?
(a) 7 (b) 300
(c) 280 (d) None of these

12. How many triangles can be formed from these points?
(a) 453 (b) 2265
(c) 755 (d) None of these

13. How many batting orders are possible for the Indian cricket team if there is a squad of 15 to choose from such that Sachin Tendulkar is always chosen?
(a) $1001 \times 11!$ (b) $364 \times 11!$
(c) 11! (d) $15 \times 11!$

14. How many distinct words can be formed out of the word PROWLING that start with R and end with W?
(a) 8!/2! (b) 6!2!
(c) 6! (d) None of these

15. How many 7-digit numbers are there having the digit 3 (three times) and the digit 0 (four times)?
(a) 15 (b) $3^3 \times 4^4$
(c) 30 (d) None of these

Answers

1. (c)	2. (b)	3. (b)	4. (c)	5. (a)
6. (a)	7. (b)	8. (c)	9. (a)	10. (b)
11. (c)	12. (b)	13. (a)	14. (c)	15. (a)

Solutions

1. **Ans. (c)**
Three digits can be formed in $^5P_3$ ways that is
$$\frac{5!}{(5-3)!} = 60$$

2. **Ans. (b)**
Number between 2000 and 3000 are four digit numbers in which 2 is the first digit

d_1 d_2 d_3 d_4
[2] [ ] [ ] [ ]

So, 2 must be taken in box d_1
Now rest box can be filled in
$^7P_3$ ways
$$\Rightarrow \frac{7!}{(7-3)!} = 210 \text{ ways}$$

3. **Ans. (b)**
Person has six invitation cards 1 card can be given to any of 4 servants so six cards can be distributed in
$4 \times 4 \times 4 \times 4 \times 4 \times 4 = 4^6$ ways

4. **Ans. (c)**
We have

→ 7 Indians
→ 5 Pakistanis and
→ 6 Dutch

7 Indians can sit in 7! ways
5 Pakistanis can sit in 5! ways
6 Dutch can sit in 6! ways
Also they can sit together in 3! different ways
$7! \times 5! \times 6! \times 3!$

5. **Ans. (a)**
To draw a line we have to choose two points
So, ${}^8C_2 = 28$ ways

6. **Ans. (a)**

This can be done in $\frac{5!}{2!}$ ways

$\frac{5!}{2!} = 60$

7. **Ans. (b)**
In PATNA, if P is fixed so we have only four letter ATNA in which there are two A's.

So, $\frac{4!}{2!} = 12$

8. **Ans. (c)**
${}^{11}C_2 \times 2!$ Ways which is equal to 110.

9. **Ans. (a)**
In Allahabad we have 4 A's that is only one vowel and 4 consonants (B, D, H, L)
So any four combination.
AB, AD, AH, AL are only 4 possibilities so.

10. **Ans. (b)**
In EQUATION there are five vowels A, E, I, O and U and three consonants N,Q, T so the possible arrangements will be $4! \times 5!$ ways.

11. **Ans. (c)**
Total number of straights lines
$= {}^{25}C_2 - {}^7C_2 + 1 = 280$

12. **Ans. (b)**
Total triangle formed will be equal to ${}^{25}C_3 - {}^7C_3 = 2265$

13. **Ans. (a)**
${}^{15-1}C_{11-1} \times 11!$
$\Rightarrow {}^{14}C_{10} \times 11! = 1001 \times 11!$

14. **Ans. (c)**

l_1 l_2 l_3 l_4 l_5 l_6 l_7 l_8
□□□□□□□□

l_1 is R and l_8 is W rest boxes can be filled in 6! ways

15. **Ans. (a)**

d_1 d_2 d_3 d_4 d_5 d_6 d_7
□□□□□□□

d_1 is 3 because we have to form 7 digit number now we have rest six boxes d_2, d_3, d_4, d_5, d_6, d_7 to be filled, this can be done in

$\frac{6!}{2! \times 4!} = 15$ ways

II Practice Exercise: II

1. There are 15 buses running between Delhi and Mumbai. In how many ways can a man go to mumbai and return by a different bus?
(a) 280 (b) 310
(c) 240 (d) 210

2. A teacher of a class wants to set one question from each of two exercises in a book. If there are 15 and 12 questions in the two exercises respectively, then in how many ways can the two questions be selected?
(a) 160 (b) 140
(c) 180 (d) 120

3. A code word is to consist of two distinct English alphabets followed by two distinct numbers between 1 and 9. For example, CA23 is a code word. How many such code words are there?
(a) 615800 (b) 46800
(c) 719500 (d) 410800

4. There are 5 letters and 5 directed envelopes. Find the number of ways in which the letters can be put into the envelopes so that all are not put in directed envelopes?
(a) 129 (b) 119
(c) 109 (d) 139

5. How many different numbers of two digits can be formed with the digits 1, 2, 3, 4, 5, 6; no digits being repeated?
(a) 40 (b) 30
(c) 35 (d) 45

6. How many three-digit odd numbers can be formed from the digits 1, 2, 3, 4, 5, 6 when
(i) repitition of digits is not allowed?
(ii) repitition of digits is allowed?
(a) (i) 60, (ii) 108 (b) (i) 50, (ii) 98
(c) (i) 70, (ii) 118 (d) (i) 80, (ii) 128

7. How many numbers are there between 100 and 1000 in which all the digits are distinct?
(a) 548 (b) 648
(c) 748 (d) 756

8. If $(n + 1)! = 6 [(n - 1)!]$, find n
(a) 6 (b) 4
(c) 8 (d) 2

9. How many words, with or without meaning, can be formed using all letters of the word EQUATION, using each letter exactily once?
(a) 38320 (b) 39320
(c) 40320 (d) 38400

10. It is required to seat 5 men and 4 women in a row so that the women occupy the even places. How many such arrangements are possible?
(a) 2880 (b) 2480
(c) 3680 (c) 3280

11. Four books, one each in chemistry, Physics, Biology and Mathematics are to be arranged in a shelf. In how many ways can this be done?
(a) 12 (b) 36
(c) 24 (d) 48

12. How many different numbers of six digits can be formed with the numbers 3, 1, 7, 0, 9, 5?
(a) 500 (b) 400
(c) 400 (d) 600

13. How many three-digit numbers are there, with no digits repeated?
(a) 648 (b) 548
(c) 848 (d) 748

14. In how many different ways, the letters of the word ALGEBRA can be arranged in a row if
(i) The two As are together?
(ii) The two As are not together?
(a) (i) 720, (ii) 1800 (b) (i) 620, (ii) 1600
(c) (i) 780, (ii) 1860 (d) (i) 720, (ii) 1600

15. How many different words can be formed with the letters of the word 'BHARAT'?
In how many of these B and H are never together?
(a) 240, 180 (b) 360, 240
(c) 320, 200 (c) 380, 260

16. How many different necklaces can be formed with 6 white and 5 red beads?
(a) 18 (b) 24
(c) 21 (d) 27

17. In how many ways can 5 sportmen be selected from a group of 10?
(a) 272 (b) 282
(c) 252 (d) 242

Answers

1. (d)	2. (c)	3. (b)	4. (b)	5. (b)
6. (a)	7. (b)	8. (d)	9. (c)	10. (a)
11. (c)	12. (d)	13. (a)	14. (a)	15. (b)
16. (c)	17. (c)			

Solutions

1. Ans. (d)
The first event of going from Delhi to Mumbai can be performed in 15 ways as he can go by any of the 15 buses. But the event of coming back from Mumbai can be performed in 14 ways (a different bus is to be taken).
Hence, both the events can be performed in $15 \times 14 = 210$ ways.

2. Ans. (c)
- Since the first exercise contains 15 questions, the number of ways of choosing the first question in 15.
- Since the second exercise contains 12 questions, the number of ways of choosing the second question is 12. Hence, by the fundamental principle, two questions can be selected in $15 \times 12 = 180$ ways.

3. Ans. (b)
(i) There are in all 26 English alphabets. We have to choose 2 distinct alphabets.
First alphabet can be selected in 26 ways.
Second alphabet can be selected in 25 ways.
Again, out of 9 digits (1 to 9), first digit can be selected in 9 ways. Second digit can be selected in 8 ways.
Thus, the number of distinct codes
$= 26 \times 25 \times 9 \times 8 = 46800$

4. Ans. (b)
- Here, the first letter can be put in any one of the 5 envelopes in 5 ways. Second letter can be put in any one of the 4 remaining evelopes in 4 ways. Continuing in this way, we get the total number of ways in which 5 letters can be put into 5 envelopes
$= 5 \times 4 \times 3 \times 2 \times 1 = 120.$
- Since out of the 120 ways, there is only one way for putting each letter in the correct envelope.

Hence, the number of ways of putting letters all are not in directed envelopes = 120 – 1 = 119 ways.

5. Ans. (b)
- We have to fill up two places (since numbers are of two digits.)
- The first place can be filled up in 6 ways, as any one of the six digits can be placed in the first place. The 2nd place can be filled up in 5 ways as no digit is to be repeated. Hence, both places can be filled up in 6 × 5 = 30 ways

6. Ans. (a)
(i) When repetition of digits is not allowed:
Since we have to form a three-digit odd number, thus the digit at unit's place must be odd. Hence, the unit's place can be filled up by 1, 3 or 5, that is, in 3 ways.
Now, the ten's digit can be filled up by any of the remaining 5 digits in 5 ways and then the hundred's place can be filled up by the remaining 4 digits in 4 ways.
Hence, the number of three-digit odd numbers that can be formed 3 × 4 × 5 = 60
(ii) When repetition of digits is allowed:
Again, the unit's place can be filled up by 1, 3, 5, that is, in 3 ways. But the ten's and hundred's place can be filled up by any of the six given digits in 6 ways each. (Since repetition is allowed)
Hence, the number of three-digit odd numbers that can be formed
= 3 × 6 × 6 = 108.

7. Ans. (b)
Any number between 100 and 1000 is of three digits, Since the numbers should have distinct digits, repetition of digits 0, 1, 2, 3, 4, 5, 6, 7, 8, 9 is not allowed.
Also 0 cannot be placed on the extreme left place. Hundredth place can be filled in 9 ways. Tenth place can be filled in 9 ways. Unit's place can be filled in 8 ways.
∴ The total three-digit numbers
= 9 × 9 × 8 = 648

8. Ans. (d)

$$(n+1)! = 6[(n-1)!]$$

$$\Rightarrow (n+1).\, n.\, [(n-1)!] = 6[(n-1)!]$$

$$\Rightarrow n^2 + n = 6 \Rightarrow n^2 + n - 6 = 0$$

$$\Rightarrow (n-2)(n+3) = 0$$

∴ Either n – 2 = 0 or n + 3 = 0

⇒ n = 2 or n = –3

n being natural number, so $n \neq -3$

∴ n = 2.

9. Ans. (c)
The word EQUATION has exactly 8 letters which are all different.
∴ Number of words that can be formed = number of permutations of 8 letters taken all at a time.

$$= P(8, 8) = 8!$$

$$= 8 \times 7 \times 6 \times 5 \times 4 \times 3 \times 2 \times 1$$

$$= 40320$$

10. Ans. (a)
Total number of candidates = 5 + 2 = 9
In the row of 9 positions, the even places are 2nd, 4th, 6th and 8th. Now, number of even places = 4.
Number of women to occupy the even places = 4.
∴ Even places can be filled =P(4, 4) ways
Number of men = 5
∴ The remaining 5 places can be filled by 5 men = P(5, 5) ways
∴ By the fundamental principle of counting.
∴ The required number of seating arrangements

$$= P(4, 4) \times P(5, 5) = 4! \times 5!$$

$$= 24 \times 120 = 2880.$$

11. Ans. (c)
4 different books can be arranged among themselves, in a shelf, in P (4, 4)

$$= 4 \times 3 \times 2 \times 1 = 24 \text{ ways.}$$

12. Ans. (d)
The numbers that can be formed, by taking all six digits together.

$$= {}^6P_6 = 6!$$

But we have to neglect the numbers which begin with zero. Now the numbers in which zero comes in the 1st Place= 5!
Hence, the required number=6!-5!
= 720 – 120 = 600.

13. Ans. (a)
The required number of three-digit numbers = The permutations of the 10 objects 0, 1, 2, 3, 4, 5, 6, 7,

8, 9 take 3 at a time, with the condition that 0 is not in the hundred's place.

$= P(10, 3) - P(9, 2)$

$= \frac{10!}{7!} - \frac{9!}{7!}$

$= \frac{10\times9\times8\times7!}{7!} - \frac{9\times8\times7!}{7!}$

$= 10\times9\times8 - 9\times8$

$= 720 - 72 = 648$

14. Ans. (a)

ALGEBRA has seven letters where 2-A, 1-L, 1-G, 1-E, 1-B and 1-R.

(i) Since two A's are always together, we take both the A's as one letter.

If p is the number of arrangements, then

$p = 6! = 6\times5\times4\times3\times2\times1 = 720.$

(ii) Total number of permutations

$q = \frac{7!}{2!} = 7\times\frac{7\times6\times5\times4\times3\times2\times1}{2\times1}$

$= 2520$

In these permutations, in some permutations, two A's are together while in the rest they are not together. Hence, the number of permutations in which two A's are not together in

$q - p = 2520 - 720 = 1800.$

15. Ans. (b)

Out of letters in the word 'BHARAT' two letters, that is, A's are alike.

$\therefore$ Number of permutations $= \frac{6!}{2!} = 360.$

Number of words in which B and H are never together = Total number of words – number of words in which B and H are together

$= 360 - \frac{5!}{2!}.2! = 360 - 120 = 240.$

16. Ans. (c)

n= total no. of beads = 6 + 5 = 11

$P_1 = 6, P_2 = 5$

$\therefore$ No. of different necklaces

$= \frac{1(11-1)!}{2\times6!\,5!} = \frac{10!}{2.6!5!}$

$= \frac{10.9.8.7.6!}{2.6!5.4.3.2.1} = 3\times7 = 21.$

17. Ans. (c)

The require number of ways = C(10, 5)

$= \frac{10!}{5!5!} = \frac{10.9.8.7.6}{5.4.3.2} = 252.$

□□□□

III Practice Exercise: III

1. The letters of word PROMISE are arranged so that no two of the vowels should come together. Find the total number of arrangements.
(a) 49 (b) 1440
(c) 7 (d) 1898

2. In an examination paper there are two groups, each containing 4 questions. A candidate is required to attempt 5 questions but not more than 3 questions from any group. In how many ways can 5 questions be selected?
(a) 24 (b) 48
(c) 96 (d) None of these

3. A box contains 10 balls out of which 3 are red and the rest are blue. In how many ways can a random sample of 6 balls be drawn from the bag so that at the most 2 red balls are included in the sample and no sample has all the 6 balls of the same colour?
(a) 105 (b) 168
(c) 189 (d) 120

4. A cricket team of 11 players is to be formed from 20 players including 6 bowlers and 3 wicket keepers. The number of ways in which a team can be formed having exactly 4 bowlers and 2 wicket keepers is
(a) 20790 (b) 6930
(c) 10790 (d) 360

5. In a hockey championship, there were 153 matches played. Every two teams played one match with each other. The number of teams participating in the championship is
(a) 18 (b) 19
(c) 17 (d) 16

6. Seven points lie on a circle. How many chords can be drawn by joining these points.
(a) 22 (b) 21
(c) 23 (d) 24

7. Ten different letters of an alphabet are given. Words with 5 letters are formed from these given letters. Then the number of words which have at least one letter repeated is
(a) 69760 (b) 30240
(c) 99748 (d) None of these

8. Five persons A, B, C, D and E occupy seats in a row such that A and B sit next to each other. In how many possible ways can these five people sit?
(a) 24 (b) 48
(c) 72 (d) None of these

9. A department had 8 male and female employees each. A project team involving 3 male and 3 female members needs to be chosen from the department employees. How many different project teams can be be chosen?
(a) 112896 (b) 3136
(c) 720 (d) 112

10. From a group of 7 men and 6 women 5 people are to be selected to form a committee so that at least 3 men are there on the committee. In how many ways can it be done?
(a) 756 (b) 735
(c) 564 (d) 645

11. A polygon has 44 diagonals, then the number of its side are
(a) 11 (b) 9
(c) 7 (d) 5

12. A five digit number divisible by 3 is to be formed using the numerals 0, 1, 2, 3, 4 and 5 without repetition. The total number of ways in which this can be done is
(a) 211 (b) 216
(c) 221 (d) 311

13. How many parallelograms will be formed if 7 parallel horizontal lines intersected by 6 parallel vertical lines?
(a) 42 (b) 294
(c) 315 (d) None of these

14. A dean must select three students to serve on a committee. If she is considering five students, then from how many different possible threesomes must she choose?
(a) 2 (b) 3
(c) 10 (d) 15

Answers

1. (b)	2. (b)	3. (b)	4. (a)	5. (a)
6. (b)	7. (a)	8. (b)	9. (b)	10. (a)
11. (a)	12. (b)	13. (c)	14. (c)	

Solutions

1. **Ans. (b)**

The four consonants can be written in 4! ways i.e. 24 ways. The three vowels can be written in 3! ways, i.e. 6 ways. Since no two vowels can come together, therefore vowels can be inserted in any three places out of the five places available, such as,

ⓥP ⓥR ⓥMⓥ Sⓥ, i.e. in $^5C_3$ ways, i.e. 10 ways.

Total number of arrangements required
$= 24 \times 6 \times 10 = 1440$

2. **Ans. (b)**

$$^4C_3 \times {^4C_2} + {^4C_2} \times {^4C_3} = 4\times6+4\times6 = 48.$$

3. **Ans. (b)**

The possible ways are as follows:

(i) 1 red ball out of the three and 5 blue balls out the seven.

(ii) 2 red balls out of the three and 4 blue balls out of the seven.

∴ Total number of ways in which a random sample of six balls can be drawn.

$$^3C_1 \times {^7C_5} + {^3C_2} \times {^7C_4} = 168$$

4. **Ans. (a)**

There are 6 bowlers, 3 wicket keepers and 11 batsmen in all. The number of ways in which a team of 4 bowlers, 2 wicket keepers and 5 batsmen can be chosen.

$$= {^6C_4} \times {^3C_2} \times {^{11}C_5}$$

$$= {^6C_2} \times {^3C_1} \times {^{11}C_5}$$

$$= \frac{6\times5}{2\times1} \times \frac{3}{1} \times \frac{11\times10\times9\times8\times7}{5\times4\times3\times2\times1} = 20790.$$

5. **Ans. (a)**

$$^{18}C_2 = \frac{18\times17}{2} = 153.$$

6. **Ans. (b)**

$$^7C_2 = \frac{7!}{2!5!} = \frac{7\times6\times5!}{5!\times2!} = 21.$$

7. **Ans. (a)**

$10^5 - {}^{10}P_5 = 69760$

8. **Ans. (b)**

$4!\times2$ ways, i.e. $24\times2 = 48$.

9. **Ans. (b)**

$$^8C_3\times{}^8C_3 = \frac{8\times7\times6}{3\times2}\times\frac{8\times7\times6}{3\times2}$$

$= 56\times56 = 3136.$

10. **Ans. (a)**

$$^7C_3\times{}^6C_2 + {}^7C_4\times{}^6C_1 + {}^7C_5\times{}^6C_0$$

$$= {}^7C_3\times{}^6C_2 + {}^7C_3\times6 + {}^7C_2\times1$$

$$= \frac{7\times6\times5}{3\times2\times1}\times\frac{6\times5}{2\times1} + \frac{7\times6\times5}{3\times2\times1}\times6 + \frac{7\times6}{2\times1}$$

$= 525 + 210 + 21 = 756.$

11. **Ans. (a)**

Let the number of sides be n.

$$\therefore {}^nC_2 - n = 44, n > 0 \Rightarrow \frac{n(n-1)}{2} - n = 44$$

$$\Rightarrow n^2 - 3n - 88 = 0$$

$$\Rightarrow n^2 - 11n + 8n - 88 = 0$$

$$\Rightarrow n(n-11) + 8(n-11) = 0$$

$$\Rightarrow (n-11)(n+8) = 0$$

$$\Rightarrow n = 11.$$

12. **Ans. (b)**

All permutations formed with 1, 2, 3, 4, 5 (sum = 15) will be divided by 3.

There are 5! = 120 such permutations. Such numbers can also be formed using 0 and 1, 2, 4, 5. There are 4 × 4! such numbers, i.e. 96. (Factor of 4 for four positions of 0 and 4! for different permutations of these four numbers)

∴ Total of such numbers = 120 + 96 = 216.

13. **Ans. (c)**

$$^7C_2\times{}^6C_2 = 315.$$

14. **Ans. (c)**

$$^5C_3 = {}^5C_2 = \frac{5\times4}{2\times1} = 10.$$

■■■■

2.4 CHAPTER

Probability

Introduction

The probability has its origin in the problems dealing with games of chance such as gambling, coin tossing, die throwing and playing cards. In all these cases the outcome of a trial is uncertain. These days probability is widely used in business and economics in the field of prediction for future.

Lots of questions based on coins, die, playing cards and alphabetical arrangements are asked in various public sector examination.

Theory

Probability means the chance of occurrence of an event. In layman terms, we can say that it is likelyhood that something that is defined as an events will or will not occur mathematically,

$$\text{Probability (P)} = \frac{\text{Number of favourable events}}{\text{Total Number of events}}$$

for example probability of getting an even number after throwing a die can be calculated as.

$$(P) = \frac{\{2, 4, 6\}}{\{1, 2, 3, 4, 5, 6\}} = \frac{3}{6} = \frac{1}{2}$$

Were {2, 4, 6} are even numbers and hence favorable out comes and {1,2,3,4,5,6} etc. all 6 possible outcomes. Similarly, we can calculate probability of getting a number divisible by 3 out of 9 single digit number in following manner Probability

$$= \frac{\text{Number divisible by 3}}{\text{All nine single digit number}}$$

$$= \frac{\{3, 6, 9\}}{\{1, 2, 3, 4, 5, 6, 7, 8, 9\}} = \frac{3}{9} \Rightarrow \frac{1}{3}$$

The followings remark may be important for learning this chapter on probability

1. **Die:** A die is small cube used in games of chance. On its six faces dots are marked numbering (1, 2, 3, 4, 5, 6).
2. **Playing card:** A pack (or deck) of playing cards has 52 cards, divided into four suits:
   (1) Spades (2) Clubs
   (3) Hearts (4) Diamonds
   Each suit has 13 cards
   (a) Nine number cards 2, 3, 4, 5, 6, 7, 8, 9, 10
   (b) An Ace, a king, a Queen and a Jack are known as face cards
   Ace - A King - K
   Queen - Q Jack -J
   Spade and clubs are black faced cards while Hearts and Diamonds are red- faced cards.
   The King, Queens and Jacks are called court cards.
3. **Unbiased coin** - coin having two faces head **(H)** and Tail **(T).**

Some Basic Concepts

Random experiment

An experiment whose out come has to be among a set of events that are completely known but whose exact out come is unknown is random experiment.
Eg: (Throwing of a dice, tossing of a coin)

Sample Space

This is defined in the context of a random experiment and denotes the set representing all the possible outcomes of the random experiment.

Eg: 1. Sample Space when a coin is tossed is Head (H) or Tail (T)
2. Sample space when a dice is thrown is (1, 2, 3, 4, 5, 6)

Event

The set representing the desired out come of a random experiment is called event. Event is subset of a sample space for example: Probability of getting a number divisible by 5 in a single throw of a die if odd numbers are obtained.

Here sample space is (1, 2, 3, 4, 5, 6), event is (1, 3, 5) favourable outcome is (5)only.

$$\text{So probability} = \frac{\{5\}}{\{1, 3, 5\}} = \frac{1}{3}$$

Non Event

The outcome that is opposite of the desired event is the non-event.

Note: If the event occurs, the non-event does not ocur and vice versa.

Impossible Event

An event that can never occur is an impossible event. The probability of an impossible event is 0.

Eg. (Probability of occurance of 7 when a dice with 6 faces numbered 1-6 is thrown).

Mutually Exclusive Events

A set of event is mutually exclusive when the occurence of any one of them means that the other event cannot occur.

Eg: 1. If head appears on a coin tail will not appear and vice versa.
2. If 5 occurs in a single throw of a die then (1, 2, 3, 4, 6) will never appear.

Equally likely Events

If two events have the same probability or chance of occurence they are called equally likely events.

Example: In a throw of a dice, the chance of 1 showing on the dice is equal to 2, is equal to 3, is equal to 4, is equal to 5, is equal to 6 appearing on the dice.

Exhaustive Set of Event

A set of events that includes all the possibilities of a sample space is said to be an exhaustive set of an events.

Example: In a throw of a dice the number is less than four or more than or equal to four.

Independent Events

An event is described as such if the occurence of an event has no effect on the probability of ocurence of another event.

Example: If the first child of a couple is girt there is no effect on the chances of the second child being a girl.

Conditional Probability

Probability of the occurence of an event A given that event B has already ocured.

This is denoted by P (A/B).

Example: The probability that in two throws of a dice we get a total of 7 or more given that in the first throw of the dices number 4 had occured.

Concept of AND and OR

Whenever we use AND as the natural conjunction joining two separate parts of event definitions, we replace the AND by the multiplication sign.

Example: If probability of passing in an exam is 1/2 for A and probability of passing the same is 1/3 for B then probability of passing of A and B is P(A)× P (B)

$$\text{i.e.: } \frac{1}{2}\times\frac{1}{3}=\frac{1}{6}$$

Whenever we use OR as the natural conjunction join two separate parts of the event definition, we replace the OR by their addition sign.

Example: If we have the probability of A winning a race as 2/3 and that of B as 1/6 then probability that either A or B wins a race is given by

$$P(A) + P(B) = \frac{2}{3}+\frac{1}{6} = \frac{5}{6}.$$

□□□□

Solved Example

1. In throwing a fair dice, what is the probability of getting the number '3'?

(a) $\frac{1}{3}$ (b) $\frac{1}{6}$

(c) $\frac{1}{9}$ (d) $\frac{1}{12}$

Ans. (b)

2. Find the chance of throwing at least one ace in a simple throw with two dice.

(a) $\frac{1}{12}$ (b) $\frac{1}{3}$

(c) $\frac{1}{4}$ (d) $\frac{11}{36}$

Ans. (d)

Ace is nothing but 6. we can get it in 11 different ways (1,6), (6,1), (2, 6), (6, 2), (3, 6), (6, 3), (4, 6), (6, 4), (5, 6), (6, 5), (6, 6)

$\Rightarrow \frac{11}{36}$ (Probability)

3. Find the chance of drawing 2 blue balls in succession from a bag containing 5 red and 7 blue balls, if the balls are not being replaced.

(a) $\frac{3}{13}$ (b) $\frac{21}{64}$

(c) $\frac{7}{22}$ (d) $\frac{21}{61}$

Ans. (c)

Probability of getting blue ball in single draw is $\frac{7}{12}$ and probability of getting blue ball in second draw is $\frac{6}{11}$ so $\frac{7}{12} \times \frac{6}{11} = \frac{7}{22}$

4. If a card is picked up at random from a pack of 52 cards. Find the probability that it is

(i) a spade.

(a) $\frac{1}{9}$ (b) $\frac{1}{6}$

(c) $\frac{1}{4}$ (d) $\frac{1}{3}$

(ii) a king or queen.

(a) $\frac{3}{13}$ (b) $\frac{2}{13}$

(c) $\frac{7}{52}$ (d) $\frac{1}{169}$

(iii) 'a spade' or 'a king' or 'a queen'

(a) $\frac{21}{52}$ (b) $\frac{5}{13}$

(c) $\frac{19}{52}$ (d) None of these

(i) Ans. (c)
(ii) Ans. (b)
(iii) Ans. (c)

(i) $\frac{13}{52} = \frac{1}{4}$

(ii) 4 king or 4 Queen = 8

$\Rightarrow \frac{8}{52} = \frac{2}{13}$

(iii) Spade = 13, King = 4 Queen = 4 but there will 1 King of spade and 1 Queen of spade so we have 19 to choose from $\Rightarrow \frac{19}{52}$

5. Three coins are tossed. What is the probability of getting 2 Tails and 1 Head?

(a) $\frac{1}{2}$ (b) $\frac{3}{8}$

(c) $\frac{2}{3}$ (d) $\frac{3}{4}$

Ans. (b)

When three coins are tossed we get

HHH or (1)
HHT or (2)
HTT or (3)
HTH or (4)
THH or (5)
THT or (6)
TTT or (7)
TTH or (8)

We get 2T and one H in 3 cases so probability = $\frac{3}{8}$

6. For the above question, the probability that there is at least one tail is :

(a) $\frac{2}{3}$ (b) $\frac{7}{8}$

(c) $\frac{3}{8}$ (d) $\frac{1}{2}$

Ans. (b)

For at least one tail, we have to exclude the case of all Heads {H, H, H} and rest 7 are favourable.

So $P = \frac{7}{8}$

7. A bag contains 3 green and 7 white balls. Two balls are drawn from the bag in succession without replacement. What is the probability that both are white?

(a) $\frac{1}{7}$ (b) $\frac{5}{11}$

(c) $\frac{7}{11}$ (d) $\frac{7}{15}$

Ans. (d)

To get both white

$$\frac{^7C_2}{^{10}C_2} = \frac{21}{45} = \frac{7}{15}$$

8. What is the probability of throwing a number greater than 2 with a fair dice?

(a) $\frac{2}{3}$ (b) $\frac{2}{5}$

(c) 1 (d) $\frac{3}{5}$

Ans. (a)

Number greater than 2 = { 3, 4, 5, 6}

So $P = \frac{4}{6} = \frac{2}{3}$

9. Two fair coins are tossed. Find the probability of obtaining

(i) 2 Heads

(a) 1 (b) $\frac{2}{3}$

(c) $\frac{1}{2}$ (d) $\frac{1}{4}$

(ii) 1 Head and 1 Tail

(a) $\frac{1}{4}$ (b) 1

(c) $\frac{1}{2}$ (d) $\frac{2}{3}$

(iii) 2 Tails

(a) 1 (b) $\frac{1}{4}$

(c) $\frac{2}{3}$ (d) $\frac{1}{2}$

(i) Ans. (d)
(ii) Ans. (c)
(iii) Ans. (b)

When two coins are tossed then four out-comes are

HH, HT, TH, or TT

(i) $P = \frac{1}{4}$

(ii) $P = \frac{2}{4} = \frac{1}{2}$

(iii) $P = \frac{1}{4}$

10. In rolling two dices, find the probability that

(i) there is at least one '6'

(a) $\frac{11}{36}$ (b) $\frac{22}{36}$

(c) $\frac{15}{36}$ (d) $\frac{29}{36}$

(ii) the sum is 5

(a) $\frac{1}{4}$ (b) $\frac{1}{9}$

(c) $\frac{1}{2}$ (d) $\frac{1}{6}$

(i) Ans. (a)
(ii) Ans. (b)

Rolling two dice gives following sums

1. Two (1, 1)
2. Three (1, 2), (2, 1)
3. Four (1, 3), (3, 1), (2, 2)
4. Five (1, 4), (2, 3), (3, 2), (4, 1)
5. Six (1, 5), (5, 1), (2, 4), (4, 2), (3, 3)
6. Seven (1, 6), (6, 1), (5, 2), (2, 5), (4, 3), (3, 4),
7. Eight (6, 2), (2, 6), (5, 3), (3, 5), (4, 4)
8. Nine (6, 3), (3, 6), (5, 4), (4, 5)
9. Ten (4, 6), (4, 6), (5, 5)
10. Eleven (6, 5), (5, 6)
11. Twelve (6, 6)

(i) $P = \frac{11}{36}$

(ii) $P = \frac{4}{36} = \frac{1}{9}$

11. From a bag containing 4 white and 5 black balls a man draws 3 at random. What are the odds against these being all black?

(a) $\frac{5}{37}$ (b) $\frac{37}{5}$

(c) $\frac{11}{13}$ (d) $\frac{13}{37}$

Ans. (b)

Probability that all 3 balls are black

$\frac{^5C_3}{^9C_3} = \frac{10}{84} = \frac{5}{42}$ out of 42 cases only 5 are favourable and 37 are not favourable so odd against these being all black is $\frac{37}{5}$.

12. Two balls are to be drawn from a bag containing 8 grey and 3 blue balls. Find the chance that they will both be blue.

(a) $\frac{1}{5}$ (b) $\frac{3}{55}$

(c) $\frac{11}{15}$ (d) $\frac{14}{45}$

Ans. (b)

8 grey and 3 blue balls

probability that all are black $= \frac{^3C_2}{^{11}C_2} = \frac{3}{55}$

13. A bag contains four black and five red balls. If three balls from the bag are chosen at radom, what is the chance that they are all black?

(a) $\frac{1}{21}$ (b) $\frac{1}{20}$

(c) $\frac{2}{23}$ (d) $\frac{1}{9}$

Ans. (a)

4 black and 5 red balls to get all black

$$P = \frac{^4C_3}{^9C_3} = \frac{4}{84} = \frac{1}{21}$$

14. If a number of two digits is formed with the digits 2, 3, 5, 7, 9 without repetition of digits, what is the probability that the number formed is 35?

(a) $\frac{1}{10}$ (b) $\frac{1}{20}$

(c) $\frac{2}{11}$ (d) $\frac{1}{11}$

Ans. (b)

Total two digit numbers formed $= {}^5P_2 = 20$. So probability this number being 35 is $\frac{1}{20}$.

15. From a pack of 52 playing cards, three cards are drawn at random. Find the probability of drawing a king, a queen and jack.

(a) $\frac{16}{5525}$ (b) $\frac{1}{13^3}$

(c) $\frac{1}{14^3}$ (d) $\frac{1}{15^3}$

Ans. (a)

$$P = \frac{^4C_1 \times {}^4C_1 \times {}^4C_1}{^{52}C_3} = \frac{64}{22100} = \frac{16}{5525}$$

16. A bag contains 3 red, 6 white and 7 black balls. Two balls are drawn at random. What is the probability that both are black?

(a) $\frac{1}{8}$ (b) $\frac{7}{40}$

(c) $\frac{12}{40}$ (d) $\frac{13}{40}$

Ans. (b)

Here 3 red, 6 white and 7 black balls

$$P = \frac{^7C_2}{^{16}C_2} = \frac{21}{120} = \frac{7}{40}$$

17. A bag contains 4 white and 2 black balls. Another contains 3 white and 5 black balls. If one ball is drawn from each bag, find the probability that

(i) both are white.

(a) $\frac{1}{3}$ (b) $\frac{2}{3}$

(c) $\frac{1}{4}$ (d) $\frac{3}{4}$

(ii) both are black.

(a) $\frac{3}{24}$ (b) $\frac{1}{24}$

(c) $\frac{3}{12}$ (d) $\frac{5}{24}$

(iii) one is white and one is black.

(a) $\frac{13}{24}$ (b) $\frac{15}{24}$

(c) $\frac{11}{21}$ (d) $\frac{1}{2}$

(i) Ans. (c)

(ii) Ans. (d)

(iii) Ans. (a)

Bag 1 $\longrightarrow$ 4 white and 2 black and

Bag 2 $\longrightarrow$ 3 white and 5 black ball

(i) Probability of getting both white $\frac{4}{6} \times \frac{3}{8} = \frac{1}{4}$

(ii) Probability of geting both black

$$= \frac{2}{6} \times \frac{5}{8} = \frac{5}{24}$$

(iii) $\frac{4}{6} \times \frac{5}{8} + \frac{2}{6} \times \frac{3}{8} = \frac{13}{24}$

□□□□

I Practice Exercise: I

1. Three coins are tossed. Find the probability of no heads.

(a) $\frac{3}{8}$ (b) $\frac{1}{8}$
(c) $\frac{1}{2}$ (d) None of these

2. A coin is tossed three times. Find the chance that head and tail show alternately.

(a) $\frac{3}{8}$ (b) $\frac{1}{4}$
(c) $\frac{1}{8}$ (d) None of these

3. In a single throw of two dice, find the probability of getting a total of 3 or 5.

(a) $\frac{1}{3}$ (b) $\frac{2}{3}$
(c) $\frac{1}{6}$ (d) $\frac{5}{6}$

4. In a single throw of two dice, what is the probability of a doublet?

(a) $\frac{1}{6}$ (b) $\frac{5}{6}$
(c) $\frac{1}{9}$ (d) $\frac{1}{18}$

5. What is the chance that a leap year, selected at random will contain 53 Sunday?

(a) $\frac{1}{7}$ (b) $\frac{2}{7}$
(c) $\frac{3}{7}$ (d) $\frac{4}{7}$

6. Find the probability that in a random arrangement of letter of the words "UNIVERSITY" two 'I's do not come together.

(a) $\frac{4}{5}$ (b) $\frac{1}{5}$
(c) $\frac{3}{5}$ (d) $\frac{2}{3}$

7. An interger is chosen at random from first two hundred natural numbers. What is the probability that the integer chosen is divisible by 6 or 8?

(a) $\frac{1}{4}$ (b) $\frac{3}{4}$
(c) $\frac{1}{2}$ (d) None of these.

8. In a simultaneous throw of two dice, find P (A or B) if A denotes the event 'a total of 11 and B denotes the event' 'an odd number on each die'.

(a) $\frac{11}{36}$ (b) $\frac{1}{4}$
(c) $\frac{5}{18}$ (d) $\frac{1}{6}$

9. A and B are mutually exclusive events of an experiment. If P ('not A') = 0.65, P(A∪B) = 0.65 and P(B) = P, find the value of p.

(a) 0.70 (b) 0.30
(c) 0.63 (d) 0.35

10. The probabilities that a student will recive an A, B, C or D grade are 0.30, 0.38, 0.22 and 0.01, respectively. What is the probability that the student will receive at least B grade?

(a) 0.38 (b) 0.42
(c) 0.68 (d) None of these

11. A card is drawn from an ordinary pack and a gambler bets that it is a spade or an ace. What are the odds against his winning the bet?

(a) 9 : 4 (b) 4 : 9
(c) 5 : 9 (d) 9 : 5

12. A problem in Statistics is given to four students A, B, C and D. Their chances of solving it are $\frac{1}{3}, \frac{1}{4}, \frac{1}{5}$, and $\frac{1}{6}$, respectively. What is the probability that the problem will be solved?

(a) $\frac{1}{3}$ (b) $\frac{2}{3}$
(c) $\frac{4}{5}$ (d) None of these

Direction for Question No. 13-14: An urn contains 25 balls numbered 1 to 25. Suppose getting an odd numbered ball is considered a 'success'. If two balls are drawn from the urn with replacement.

13. Find the probability of getting two successes.

(a) $\frac{169}{625}$ (b) $\frac{312}{625}$

(c) $\frac{481}{625}$ (d) $\frac{144}{625}$

14. Find the probability of getting no success.

(a) $\frac{169}{625}$ (b) $\frac{312}{625}$

(c) $\frac{481}{625}$ (d) $\frac{144}{625}$

Direction for Question No. 15-18: A husband and wife appear in an interview for two vacancies in the same post. The probability of husband's selection is 1/7 and that of wife's is 1/5.

15. What is the probability that only one of them will be selected?

(a) $\frac{2}{7}$ (b) $\frac{1}{35}$

(c) $\frac{24}{25}$ (d) $\frac{11}{35}$

16. What is the probability that both of them will be selected?

(a) $\frac{2}{7}$ (b) $\frac{1}{35}$

(c) $\frac{24}{35}$ (d) $\frac{11}{35}$

17. What is the probability that none of them will be selected?

(a) $\frac{2}{7}$ (b) $\frac{1}{35}$

(c) $\frac{24}{35}$ (d) $\frac{11}{35}$

18. What is the probability that at least one of them will be selected?

(a) $\frac{2}{7}$ (b) $\frac{1}{35}$

(c) $\frac{24}{35}$ (d) $\frac{11}{35}$

19. A man speaks truth in 80% of the cases and another in 90% of the cases. While stating the same fact, what is the probability that they contradict?

(a) $\frac{37}{50}$ (b) $\frac{13}{50}$

(c) $\frac{16}{50}$ (d) None of these

20. A can solve 90% of the problems given in a book and B solve 70%. What is the probability that atleast one of them will solve a problem selected at random from the book?

(a) $\frac{3}{100}$ (b) $\frac{97}{100}$

(c) $\frac{83}{100}$ (d) $\frac{17}{100}$

21. Find the probability that in a random arrangement of the letters of the word DAUGHTER, the letter D occupies the first place.

(a) $\frac{1}{8}$ (b) $\frac{1}{4}$

(c) $\frac{3}{8}$ (d) $\frac{1}{2}$

Answers

1. (b)	2. (b)	3. (c)	4. (a)	5. (b)
6. (a)	7. (a)	8. (a)	9. (b)	10. (c)
11. (a)	12. (b)	13. (a)	14. (d)	15. (a)
16. (b)	17. (c)	18. (d)	19. (b)	20. (b)
21. (a)				

Solutions

1. Ans. (b)

Sample space S = {HHH, HHT, HTH, HTT, THT, TTH, THH, TTT}

Number of exhaustive cases = 8

P (no heads) = P (all tails)= $\frac{1}{8}$

($\because$ there is only favourable case TTT).

2. Ans. (b)

Sample space S = {HHH, HHT, HTH, HTT, THT, TTH, THH, TTT}

Number of exhaustive cases = 8

Favourable case are HTH, THT

Number of favourable = 2.

$\therefore$ Required probability = $\frac{2}{8} = \frac{1}{4}$.

3. Ans. (c)

A total of 3 or 5 may be obtained in 6 ways, viz., (1, 2), (2, 1), (1, 4), (2, 3), (3, 2), (4, 1).

No. of exhaustive cases = 6 × 6 = 36.

∴ Probability of getting a total of 3 or 5 = $\frac{6}{36} = \frac{1}{6}$.

4. **Ans. (a)**

A 'doublet' means that both the dice show the same number on the upper most faces. Therefore, the outcomes, favourable to this event are (1, 1), (2, 2), (3, 3), (4, 4), (5, 5), (6, 6)

Thus, the number of favourable cases = 6.

Hence, P(doublet) = $\frac{6}{36} = \frac{1}{6}$

5. **Ans. (b)**

We know that a leap year has 366 days and thus a leap year has 52 weeks and 2 days over.

The two left over (successive days have the following likely cases:

(i) Sunday and Monday

(ii) Monday and Tuesday

(iii) Tuesday and Wednesday

(iv) Wednesday and Thursday

(v) Thursday and Friday

(vi) Friday and Saturday

(vii) Saturday and Sunday

∴ Number of exhaustive cases 'n' = 7.

Out of these, the favourable cases are...(i) and (vii)

∴ Number of favourable cases 'm' = 2

∴ Probability of having 53 Sunday = $\frac{2}{7}$.

6. **Ans. (a)**

Out of the letters in the word 'UNIVERSITY' two letters 'I' are alike.

∴ Number of permutations = $\frac{10!}{2}$ ...(i)

Number of words in which two 'I' are never together = Total number of words – Number of words in which two 'I' are together

$$\frac{10!}{2} - 9! = \frac{10! - 2.9!}{2} = \frac{9![10-2]}{2} = \frac{9!.8}{2} = 9!.4$$

∴ Required probability = $\frac{9.4}{10!/2} = \frac{9!.8}{10!}$

$$= \frac{8}{10} = \frac{4}{5}.$$

7. **Ans. (a)**

A : Integer chosen is divisible by 6B : Integer chosen is divisible by 8

$n(A) = 33,\ n(B) = 25,\ n(A \cap B) = 8,\ n(S) = 200$

$$P(A) = \frac{33}{200},\ P(B) = \frac{25}{200}$$

$$P(A \cup B) = P(A) + P(B) - P(A \cap B)$$

$$= \frac{33}{200} + \frac{25}{200} - \frac{8}{200} = \frac{50}{200} = \frac{1}{4}.$$

8. **Ans. (a)**

A:Getting total of 11 B: Getting odd number one each die

A = [(6, 5), (5, 6)]

B = [(1, 1), (1, 3), (1, 5), (3, 1), (3, 3), (3, 5), (5, 1), (5 ,3), (5, 5)]

$$P(A) = \frac{2}{36}, P(B) = \frac{9}{36}, P(A \cap B) = 0$$

∴ Required probability

$$= P(A) + P(B) - P(A \cap B)$$

$$= \frac{2}{36} + \frac{9}{36} - 0 = \frac{11}{36}.$$

9. **Ans. (b)**

We know $P(A) = 1 - P(\bar{A})$

= 1 – 0.65=0.35 and

$P(A \cup B) = P(A) + P(B) - P(A \cap B)$

⇒ 0.65 = 0.35 + P(B) – 0

[∵ A and B are mutually exclusive events]

⇒ P(B) = 0.65 – 0.35 = 0.30.

10. **Ans. (c)**

P(at least B grade) = P(B grade) + P(A grade) = 0.38 + 0.30 = 0.68.

11. **Ans. (a)**

Let A : a spade is drawn and

B: an ace is drawn

Probability of winning the bet = P(A or B)

= P (A) + P(B) – P(A and B)

$$= \frac{13}{52} + \frac{4}{52} - \frac{1}{52} = \frac{16}{52} = \frac{4}{52}$$

Probability of losing the bet $= 1 - \frac{4}{13} = \frac{9}{13}$

Odds against winning the bet = 9 : 4.

12. Ans. (b)

Probability that A fails to solve the problem is

$1-\frac{1}{3}=\frac{2}{3}$

Probability that B fails to solve the problem is

$1-\frac{1}{4}=\frac{3}{4}$

Probability that C fails to solve the problem is

$1-\frac{1}{5}=\frac{4}{5}$

Probability that D fails to solve the problem is

$1-\frac{1}{6}=\frac{5}{6}$

Since the events are independent, the probability that all the four students fail to solve the problem is

$\frac{2}{3}\times\frac{3}{4}\times\frac{4}{5}\times\frac{5}{6}=\frac{1}{3}.$

∴ The probability that the problem will be solved

$=1-\frac{1}{3}=\frac{2}{3}.$

13. Ans. (a)

Success: Getting odd number $P=\frac{13}{25}$

$\Rightarrow q=1-P=1-\frac{13}{25}=\frac{12}{25}$

P(two successes) = p × p

$=\frac{13}{25}\times\frac{13}{25}=\frac{169}{625}.$

14. Ans. (d)

P(no success) $= q\times q=\frac{12}{25}\left(\frac{12}{25}\right)=\frac{144}{625}.$

15. Ans. (a)

A : Husband selected

B : Wife selected

$P(A)=\frac{1}{7}\Rightarrow P(\bar{A})=1-P(A)=1-\frac{1}{7}=\frac{6}{7}$

$P(B)=\frac{1}{5}\Rightarrow P(\bar{B})=1-P(B)=1-\frac{1}{5}=\frac{4}{5}$

P(only one of them will be selected)

$=P(A)\times P(\bar{B})+P(B)P(\bar{A})$

$=\frac{1}{7}\left(\frac{4}{5}\right)+\frac{1}{5}\left(\frac{6}{7}\right)=\frac{4+6}{35}=\frac{10}{35}=\frac{2}{7}.$

16. Ans. (b)

P(both of them will be selected)

$=P(A)\times P(B)=\frac{1}{7}\times\frac{1}{5}=\frac{1}{35}.$

17. Ans. (c)

P(none of them will be selected)

$=P(\bar{A})P(\bar{B})=\frac{6}{7}\times\frac{4}{5}=\frac{24}{35}.$

18. Ans. (d)

P(at least one of them will be selected)

$=1-\left[P(\bar{A})\times P(\bar{B})\right]$

$=1-\left(\frac{6}{7}\times\frac{4}{5}\right)=1-\frac{24}{35}=\frac{11}{35}.$

19. Ans. (b)

Let the two men be A and B.

A: A speaks truth

B: B speaks truth

$$P(A)=\frac{80}{100}\Rightarrow P(\bar{A})=1-P(A)=1-\frac{80}{100}=\frac{20}{100}$$

$$P(B)=\frac{90}{100}\Rightarrow P(\bar{B})=1-P(B)=1-\frac{90}{100}=\frac{10}{100}$$

∴ Required probability

$=P(A)P(\bar{B})+P(B)P(\bar{A})$

$=\frac{80}{100}\times\frac{10}{100}+\frac{90}{100}\times\frac{20}{100}$

$=\frac{8+18}{100}=\frac{26}{100}=\frac{13}{50}.$

20. Ans. (b)

A: a solves the problem;

B: B solves the problem.

$$P(A)=\frac{90}{100}\Rightarrow P(\bar{A})=1-P(A)=1-\frac{90}{100}=\frac{10}{100}$$

$$P(B)=\frac{70}{100}\Rightarrow P(\bar{B})=1-P(B)=1-\frac{70}{100}=\frac{30}{100}$$

Required probability $=1-P(\bar{A})P(\bar{B})$

$=1-\frac{10}{100}\times\frac{30}{100}=1-\frac{3}{100}=\frac{97}{100}$

21. Ans. (a)

If D occupies the first place

n(A) = 7!, Total words = n(S) = 8!

∴ Required probability

$=P(A)=\frac{n(A)}{n(S)}=\frac{7!}{8!}=\frac{1}{8}.$

□□□□

II Practice Exercise: II

1. The Probability of raining on day 1 is 0.2 and on day 2 is 0.3. What is the probability of raining on both the days?
(a) 0.2 (b) 0.1
(c) 0.06 (d) 0.25

2. A bag contains 5 red balls and 8 blue balls. It also contains 4 green and 7 black balls. If a ball is drawn at random, then find the probability that it is not green.
(a) 5/6 (b) 1/4
(c) 1/6 (d) 7/4

3. If the probability that A will live 15 years is $\frac{7}{8}$ and that B will live 15 years is $\frac{9}{10}$, then what is the probability that both will live after 15 years?
(a) $\frac{1}{20}$ (b) $\frac{63}{80}$
(c) $\frac{1}{5}$ (d) None of these

4. The probability that a student is not a swimmer is 1/5. Then the probability that out of the five students, exactly four are swimmers, is
(a) ${}^5C_4\left(\frac{4}{5}\right)^2\left(\frac{1}{5}\right)$ (b) $\left(\frac{4}{5}\right)^4\left(\frac{1}{5}\right)$
(c) ${}^5C_1\left(\frac{1}{5}\right)\left(\frac{4}{5}\right)^4$ (d) None of these

5. If events A and B are independent and $P(A) = 0.15$, $P(A \cup B) = 0.45$ then $P(B) =$
(a) $\frac{6}{13}$ (b) $\frac{6}{17}$
(c) $\frac{6}{19}$ (d) $\frac{6}{23}$

6. Four different objects 1, 2, 3, 4 are distributed at random in four places marked 1, 2, 3, 4. What is the pobability that none of the objects occupy the place corresponding to its number?
(a) 17/24 (b) 3/8
(c) 1/2 (d) 5/8

7. If the probability of rain on any given day in Pune city is 50%, then what is the probability that it rains on exactly 3 days in a 5-day period?
(a) 8/125 (b) 5/16
(c) 8/25 (d) 2/25

8. The probability that an event A happens in one trial of an experiment is 0.4. Three independent trial of the experiment are formed. The probability that the event A happens at least once is
(a) 0.934 (b) 0.784
(c) 0.548 (d) 0.343

9. A number is chosen at random among the first 120 natural numbers. The probability of the number chosen being a multiple of 5 or 15 is
(a) 1/5 (b) 1/6
(c) 1/7 (d) 1/9

10. From a pack of 52 playing cards, two cards are drawn together at random. Calculate the probability of both the cards being Kings.
(a) 1/15 (b) 25/57
(c) 35/256 (d) None of these

11. From a box containing 60 standard and 40 substandard article, two articles are chosen one by one. What is the probability that one of them is standard and the other substandard?
(a) $\frac{60}{100}\times\frac{40}{100}$ (b) $\frac{60}{100}\times\frac{39}{100}$
(c) $\frac{16}{33}$ (d) 24%

Answers

1. (c)	2. (a)	3. (b)	4. (c)	5. (b)
6. (c)	7. (b)	8. (b)	9. (a)	10. (d)
11. (c)				

Solutions

1. Ans. (c)
$0.2 \times 0.3 = 0.06$

2. Ans. (a)
The probability that the ball drawn is of green colour
$= \frac{4}{24} = \frac{1}{6}$
Probability that the ball drawn is not of green colour
$= 1 - \frac{1}{6} = \frac{5}{6}.$

3. **Ans. (b)**

$P(A) \times P(B) = \frac{7}{8} \times \frac{9}{10} = \frac{63}{80}.$

4. **Ans. (c)**

5. **Ans. (b)**

$P(A \cap B) = P(A) \times P(B) = 0.15 \times P(B)$

$P(A \cup B) = P(A) + P(B) - P(A \cap B)$

$0.45 = 0.15 + P(B) - (0.15) \times P(B)$

$= 0.15 + P(B)(1 - 0.15)$

$= 0.15 + 0.85P(B)$

$\therefore \quad 0.85P(B) = 0.45 - 0.15 = 0.30$

$\therefore \quad P(B) = \frac{0.30}{0.85} = \frac{30}{85} = \frac{6}{17}.$

6. **Ans. (c)**

Let the four places be 1 2 3 4

Now object 1 cannot occupy the place 1..... (A)

Suppose object 2 occpuies the place 1. Then other placements can be done in 6 ways as follows:

(1)	2	1	3	4
(2)	2	1	4	3
(3)	2	3	1	4
(4)	2	3	4	1
(5)	2	4	1	3
(6)	2	4	3	1

Here out of the six ways, only three are permissible, because (1), (3) and (6) are not permissible because of the non-fulfilment of condition (A). Hence, required probability is $\frac{3}{6} = \frac{1}{2}$. Similarly you can allow objects 3 and 4 to occupy place 1 and in each case you can find that the probability is 1/2.

7. **Ans. (b)**

Probability that it rains on any day $= \frac{1}{2}$ and it doesn't rain $= 1 - \frac{1}{2} = \frac{1}{2}$

Prob. that it it rains on exactly 3 days in a 5-day period.

$= {}^5C_3 \times \left(\frac{1}{2}\right)^3 \times \left(\frac{1}{2}\right)^2$

${}^5C_3$ for choosing which three days among given five days. $\left(\frac{1}{2}\right)^3$ for that it rains on three days and $\left(\frac{1}{2}\right)^2$ for it doesn't rains on remaining 2 days.

8. **Ans. (b)**

Required probability

$= {}^3C_1(0.4)(0.6)^2 + {}^3C_2(0.4)^2(0.6) + {}^3C_3(0.4)^3$

$= 3(0.144) + 3(0.096) + 1(0.064) = 0.784$

Alternative

Probability that A happens = 0.4

Probability that A doesn't happen

= 1 – 0.4 = 0.6

So required probability = 1 – Probability (that event does not happens i.e. not even one in three times)

= 1 – (0.6) × (0.6) (0.6)

= 1 – 0.216

= 0.784

9. **Ans. (a)**

10. **Ans. (d)**

Two cards can be drawnfrom a pack of 52 playing cards in ${}^{52}C_2$ ways,

i.e. $\frac{52 \times 51}{2} = 1326$ ways.

The event that two kings appear in a single draw can appear in ${}^4C_2$ ways, i.e. 6 ways.

$\therefore$ The probability that the two cards drawn from a pack of 52 cards are kings $= \frac{6}{1326} = \frac{1}{221}.$

11. **Ans. (c)**

Required probability

$= \frac{60}{100} \times \frac{40}{99} + \frac{40}{100} \times \frac{60}{99}$

$= \frac{4800}{100 \times 99} = \frac{16}{33}.$

■■■■

CHAPTER 2.5 Set Theory

Venn Diagrams

Venn Diagrams are pictorial ways of representation of sets of things having different attributes. Basic understanding of regions represented by a Venn diagram is the tool for solving these questions.

If a Venn diagram is representing distribution of n attributes then total 2^n regions will be created in entire Venn diagram (including a region when particular thing will not belong to any of attributes.)

Venn-diagram involving two attributes

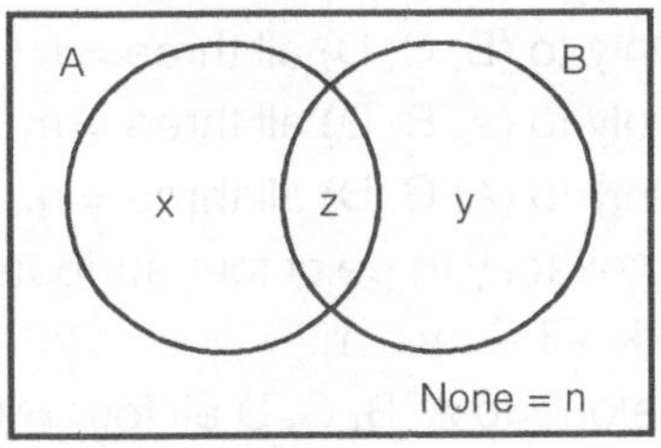

Clearly 4 regions are then (x, y, z and n)

Region x can be referred as

(a) z = things that belongs to both attribute A & B

(b) x = things that belong to A but not to B

(c) y = things that belong to B but not to A

(d) n = represent region of thing which neither belongs to A nor to B.

Further observations

(a) If things belong to exactly one attribute then region = $x + y$.

(b) If things belong to at least one of the attribute = exactly one + exactly two $n(A \cup B) = x + y + z$.

(c) Total sample (μ) : [(Things belongs to atleast one) + (things belong to none of attribute.)] $\mu = x + y + z + n$.

(d) Things belong to both attribute $n(A \cap B) = Z$

(e) Things belong to attribute A $n(A) = x + z$

(f) Things belong to attribute Y $n(B) = y + z$ $n(A \cup B) = n(A) + n(B) - n(A \cap B)$

(g) Things does not belong to attribute A = (y + n)

(h) Things does not belong to attribute B = (x + n)

Venn diagram with three attributes

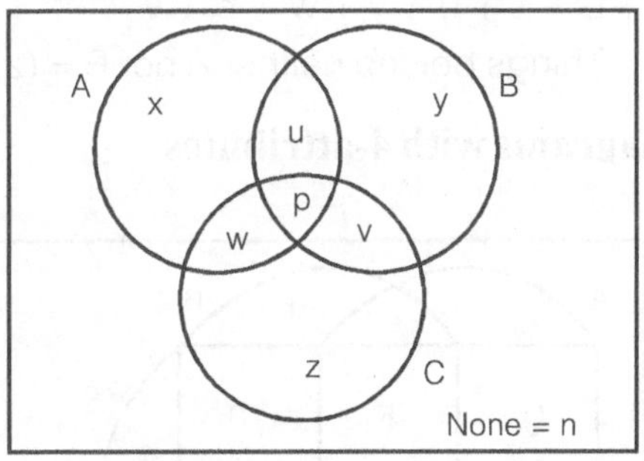

Total sample sets = 3

Here X, Y, Z are different 3 attributes hence

(a) Total number of regions = $2^3 = 8$

(b) $n(A) = x + w + u + p$, things belong to only A = x

(c) $n(B) = u + p + v + y$, things belong to only B = y

(d) $n(C) = w + p + v + z$, things belong to only C = z

(e) Things belong to exactly one of attribute $= x + y + z$

(f) Things belong to attribute (A & B) both $n(A \cap B) = u + p$

(g) Things belong to attribute (B & C) both $n(B \cap C) = v + p$

(h) Things belong to attribute (A & C) both $n(A \cap C) = w + p$

(i) Things belong to only attribute (A & B) both = u

(j) Things belong to only attribute (B & C) both = v

(k) Things belong to only attribute (A & C) both = w

(l) Things belong to all the three attributes $n(A \cap B \cap C) = p$

(m) Things belong to all the three attributes $n(A \cap B \cap C) = p$

(n) Things belong to none of the attributes = n

(o) Things belong to at least one of the attributes
Total sample – n = μ – n
= Exactly one + Exactly two + Exactly three
= x + y + z + u + v + w + p
$n(A \cup B \cup C) = n(A) + n(B) + n(C) - n(A \cap B) - n(B \cap C) - n(C \cap A) + n(A \cap B \cap C)$

(p) Things does not belong to A = n + z + u + y

(q) Things belong to A or B = x + u + y + w + p + v

(r) Things belong neither A nor B = (z + n)

Venn diagrams with 4-attributes

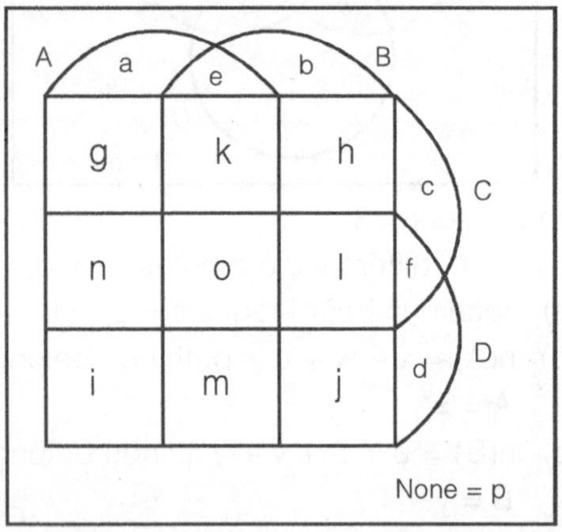

Total sample = m

1. n(A) = a + e + g + k + n + o + j + m
2. n(B) = b + e + h + k + l + o + j + m
3. n(C) = c + f + h + l + k + o + g + n
4. n(D) = d + f + j + l + m + o + j + n
5. Only A = a
6. Only B = b
7. Only C = c
8. Only D = d
9. Things belong to exactly one of attributes = a + b + c + d
10. A & B = $n(A \cap B)$ = e + k + o + m
11. A & C both = $n(A \cap C)$ = g + k + o + m
12. A & D both = $n(A \cap D)$ = n + o + i + m
13. B & C both = $n(B \cap C)$ = k + h + o + l
14. B & D both = $n(B \cap D)$ = m + j + o + l
15. C and D both = $n(C \cap D)$ = m + o + l + f
16. Only to (A & B) both = e
17. Only to (A & C) both = g
18. Only to (A & C) both = i
19. Only to (B & C) both = h
20. Only to (B & D) both = j
21. Only to (C & D) both = f
22. To exactly two out of four = e + f + g + h + i + j
23. (A, B & C) all three = k + 0 = $n(A \cap B \cap C)$
24. (B, C & D) all three = l + 0 = $n(B \cap C \cap D)$
25. (A, B & D) all three = m + 0 = $n(A \cap B \cap D)$
26. (A, C & D) all three = n + 0 = $n(A \cap C \cap D)$
27. Only to (A, B, C) all three = k
28. Only to (B, C, D) all three = l
29. Only to (A, B, D) all three = m
30. Only to (A, C, D) all three = h
31. To exactly three of four attributes = k + l + m + n
32. Belong to A, B, C, D all four = O
33. None among A, B, C, D = p

Solved Examples

Direction (Q.1 to Q. 5): These examples are based on the following diagram.

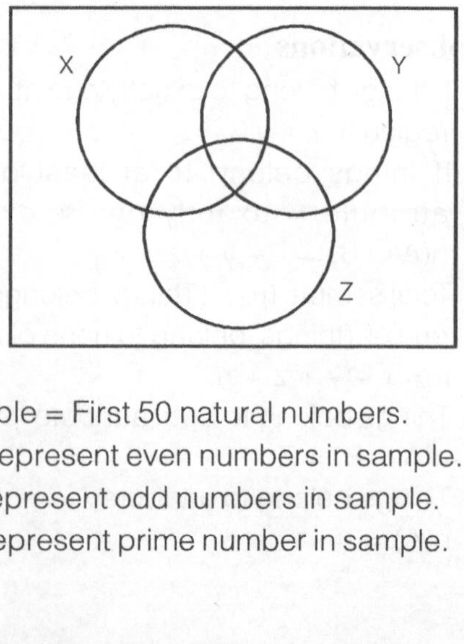

Total sample = First 50 natural numbers.
Circle X- represent even numbers in sample.
Circle Y-represent odd numbers in sample.
Circle Z-represent prime number in sample.

Solution : Following venn-diagram can be drawn.

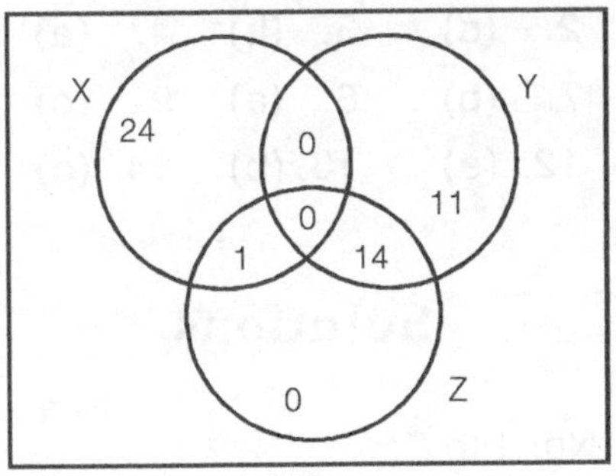

1. How many numbers belong to X only?
   (a) 25 (b) 24
   (c) 23 (d) 22

Ans. (b)
Only even number but not prime number hence 24 elements, option (b).

2. How many numbers belong to Y only

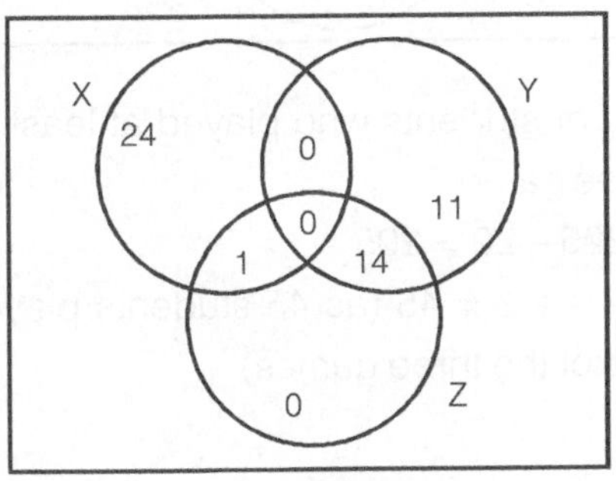

   (a) 14 (b) 25
   (c) 11 (d) 13

Ans. (c)
All odd numbers which are not prime 11 elements.

3. How many elements are these in $(Y \cap Z)$?
   (a) 14 (b) 12
   (c) 24 (d) 13

Ans. (a)
All odd prime number till 50 are 14.

4. How many elements are there in $(X \cap Z)$?
   (a) 0 (b) 1
   (c) 2 (d) 3

Ans. (b)
2 is the only even and prime number. Hence only one element.

5. How many elements are there which do not belong to Z?
   (a) 35 (b) 25
   (c) 34 (d) 14

Ans. (a)
All non prime number till 50, hence 35 numbers which are non-prime.

Practice Exercise: I

Direction (Q.1 to Q. 3): During the annual day celebrations of a college 225 students participated. Out of them 100 played Cricket, 105 played Hockey, 95 played Football, 20 played none of these three, 45 students played exactly two of the above three sports, 20 student played all three game.

1. If 45 students played only Cricket, then find the number of students who played Football & Hockey only
   (a) 15 (b) 10
   (c) 20 (d) 25

2. If 20 students played both Cricket & Football but not Hockey, find the number of students who played only Hockey.
   (a) 50 (b) 55
   (c) 45 (d) 60

3. If 50 students played Hockey but not Football, find the number of students who played Cricket or Hockey
   (a) 110 (b) 120
   (c) 135 (d) 140

Direction (Q.4 to Q.8): In an apartment, a survey was conducted among 600 people to find out readership of 3 magazines India today, Business today and Front line. It is known that 200 people read at least two of these magazines. 460 people read either India today or Front line. 360 read exactly one among three. 160 read neither India today nor Business today. 260 read the India today or Business today but not Front line.

4. How many people read at least one of the other two magazines along the Front line?
   (a) 140 (b) 180
   (c) 200 (d) 220

5. How many people read all three magazines?
   (a) 140 (b) 120
   (c) 80 (d) Can't determined

6. How many people read India today only?
   (a) 100 (b) 120
   (c) 140 (d) 160

7. How many people read none of three magazines?
   (a) 60 (b) 40
   (c) 30 (d) None of these

8. How many people read exactly one among three magazines?
(a) 360 (b) 300
(c) 200 (d) 260

Direction (Q.9 to Q. 16): In a party there were 70 people who took tea. There were 80 people who took coffee. If there were total 120 people in the party then.

9. What is the maximum possible no. of people who took none of these two?
(a) 60 (b) 30
(c) 40 (d) None of these

10. What is the maximum possible number of people who took exactly one of these two drink?
(a) 80 (b) 90
(c) 120 (d) 100

11. What is the maximum possible number of people who took both of the tea and coffee?
(a) 80 (b) 90
(c) 70 (d) 100

12. What is the maximum possible number of people who took at least one of these two drinks?
(a) 120 (b) 100
(c) 90 (d) None of these

13. What is the minimum possible number who took both the drinks?
(a) 40 (b) 30
(c) 50 (d) None of these

14. What is the minimum possible no. of people who took at least one of the two drink?
(a) 70 (b) 40
(c) 80 (d) None of these

15. What is the minimum possible number of people, who took at most one of the two drinks?
(a) 40 (b) 50
(c) 60 (d) 70

16. What is the minimum possible number of people who took none of these two drinks?
(a) 20 (b) 30
(c) 40 (d) None of these

Answers

1. (b)	2. (d)	3. (b)	4. (a)	5. (d)
6. (a)	7. (b)	8. (a)	9. (c)	10. (b)
11. (c)	12. (a)	13. (b)	14. (c)	15. (b)
16. (d)				

Solutions

Question. No. 1 to 3

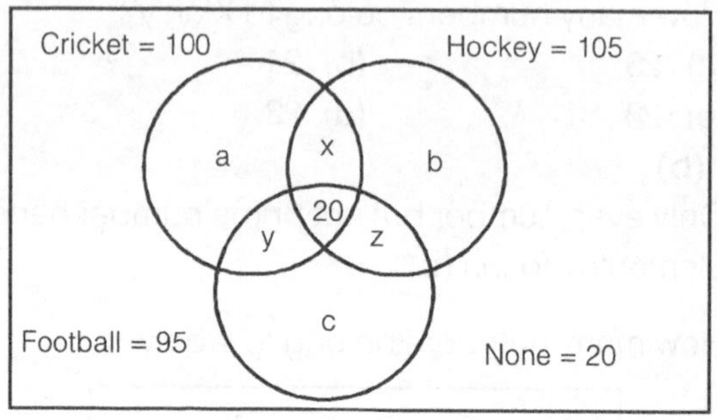

No. of students who played at least one out of three games

$= 225 - 20 = 205$

$x + y + z = 45$ (as 45 students played exactly two of the three games)

1. (b)

a = 45 (given)

$45 + x + y + 20 = 100$

$x + y = 35$

$z = 45 - (x + y) = 10$

Hockey and Football only = 10

Option (b).

2. (d)

y = 20

$x + z + b + 20 = 105$ (Total Hockey = 105)

$x + y + z = 45$

$x + z = 45 - 20 = 25$

$b + 20 + 25 = 105$

$b = 105 - 45 = 60$

Hence option (d)

3. (b)

$x + b = 50$

$x + b + z + 20 = 105$

$\Rightarrow z = 35$

$x + y + z = 45$

$\Rightarrow x + y = 10$

$a + x + y + 20 = 100$

$\Rightarrow a = 70$

$a + x + b = 70 + 50 = 120$

Hence option (b)

Question 4 to 8

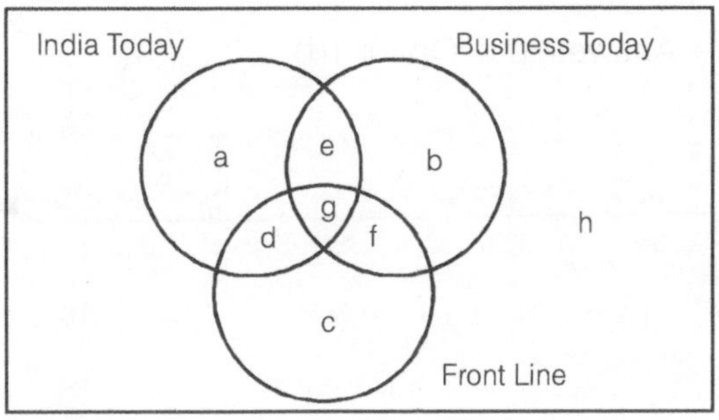

$a + b + c + d + e + f + g + h = 600$

$e + d + g + f = 200$

$a + b + c = 360 \Rightarrow h = 40$ ...(i)

$a + d + c + e + f + g = 460$

$b + h + 460 = 600$

$b = 100$ ...(ii)

$c = 160$ ...(iii)

$a = 100$ ...(iv)

$a + e + b = 260$

$e \Rightarrow 260 - (a + b) = 60$ ...(v)

$e + f + g = 460 - (a + b + c)$

$= 460 - 360 = 100$

$d = 200 - 100 = 100$ ...(vi)

4. (a)
$d + g + e + f = 200$, $e = 60$ from (v)
No. of people required
$= d + g + f = 200 - 60 = 140$.

5. (d)
Can't be determined.

6. (a)
India today only = a = 100.

7. (b)
N one of three $\Rightarrow h = 40$.

8. (a)
$a + b + c = 360$.

Question 9 to 16.

9. (c)

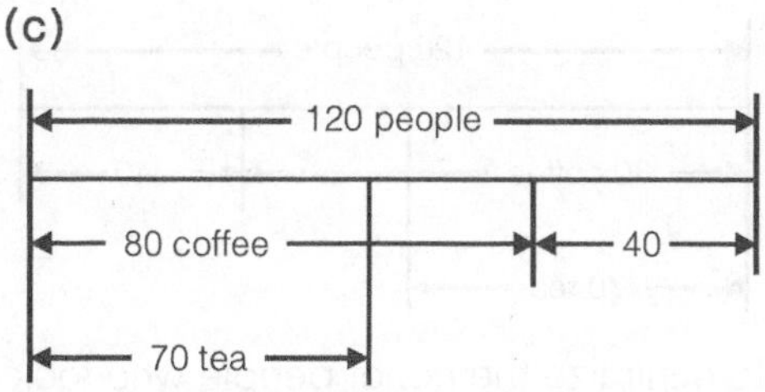

Maximum possible no. of people took none of these two.
(120 – 80) = 40. Option (c)

10. (b)
To maximize the no. of only one drink taken try to minimize both and none.

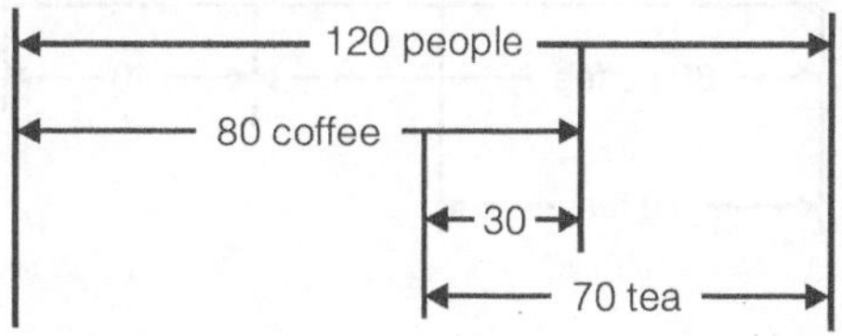

Maximum number of only one drink, taken
= (80 – 30) + (70 – 30) = 90. Option (b)

11. (c)

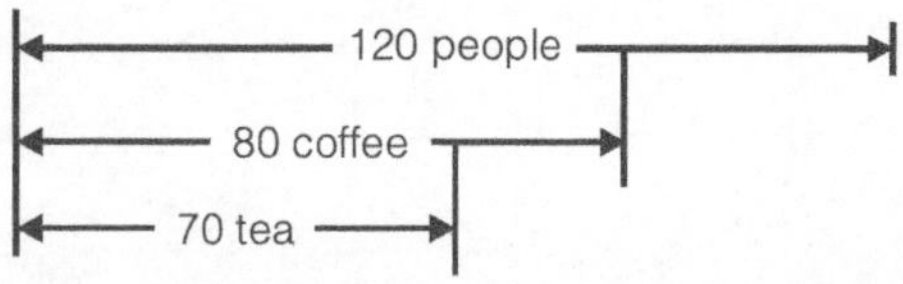

Maximum no. of people who took both
= 70. Option (c)

12. (a)

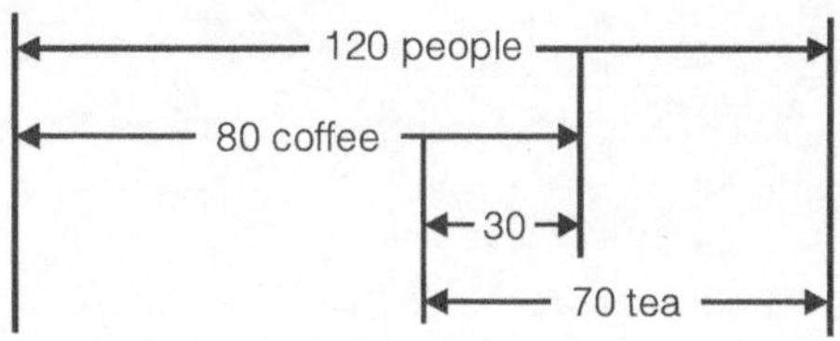

To maximize at least one of two means we have to minimize the none.
Hence 120. Option (a)

13. (b)

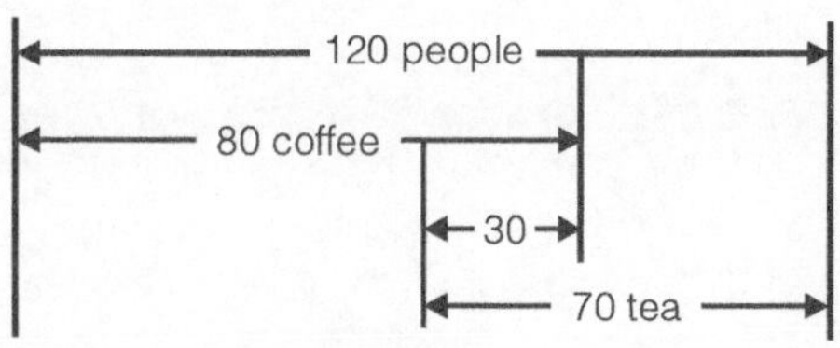

Answer 30. Option (b)

14. (c)

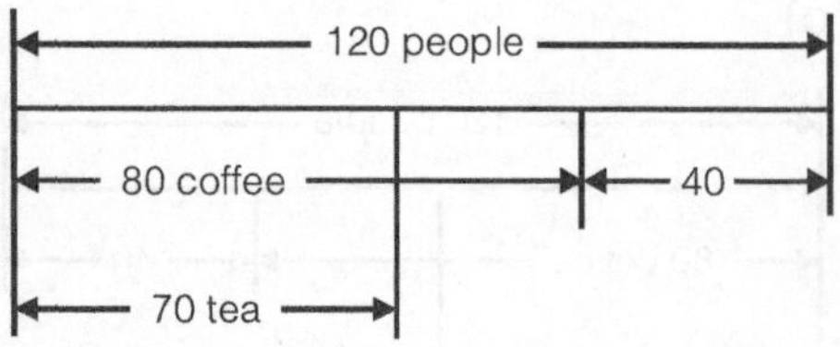

To minimize the no. of people who took at least one, we need to maximize the number of people who took none. Option (c).

15. (b)

To minimize the no. of people who took at most one. I need to maximize who took both.

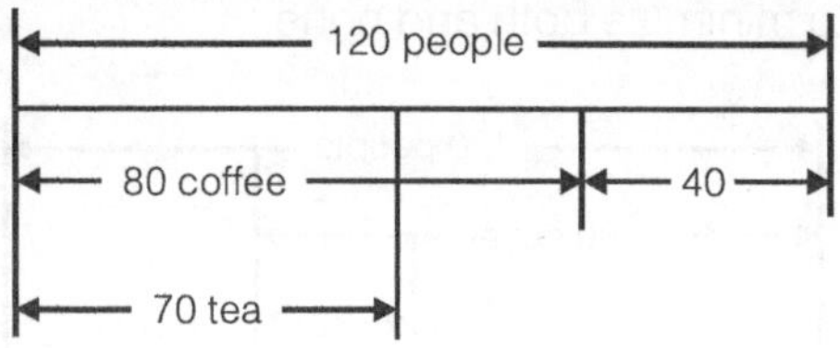

Maximum 70 people can take both drinks. Hence min. 50 people are here who will take at most one drink. Option (b)

16. (d)

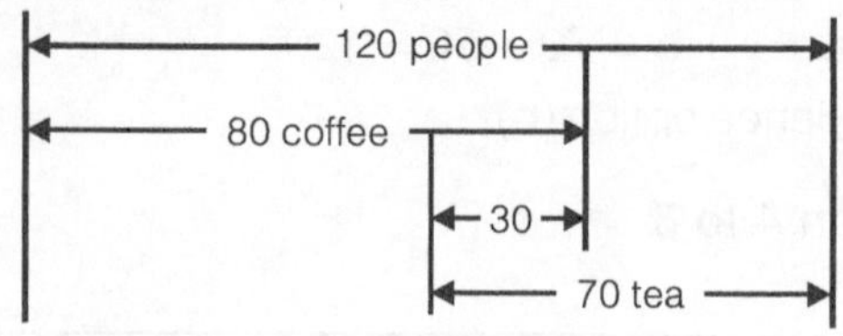

Answer zero. Option (d)

■■■■

C

Section

Reasoning & Data Interpretation

3.1

CHAPTER

Blood Relationship

In the exams, the success of a candidate in the questions on blood relations depends upon his knowledge about various blood relations. The easiest and non-confusing way to solve these problems would be to draw a family tree diagram and increase the levels in the hierarchy as shown below:

1st stage:

Grand Parents—Grand father, Grand mother, Grand uncle, Grand aunt.

2nd stage:

Parents and in laws—Father, Mother, Uncle, Aunt, Father-in-law, Mother-in-law.

3rd stage:

Siblings, spouse and in laws—Brother, Sister, Cousin, Wife, Husband, Brother-in-law, Sister-in-law.

4th stage:

Children and in laws—Son, Daughter, Niece, Nephew, Son-in-law, Daughter-in-law.

5th stage:

Grand children—Grand son, Grand daughter.

Mother's or father's brother	Uncle
Mother's or father's sister	Aunt
Mother's or father's mother	Grand mother
Mother's or father's father	Grand father
Grand father's or grand mother's brother	Grand uncle
Grand father's or grand mother's sister	Grand aunt
Children's children (Grand son, Grand daughter)	Grand children
Uncle or Aunt's son/daughter	Cousin
Son's wife	Daughter-in-law
Dauther's husband	Son-in-law
Sister's or brother's son	Nephew
Husband's or wife's brother or sister's husband	Brother-in-law
Husband's or wife's sister or brother's wife	Sister-in-law
Chindren at same parents: siblings (could be brother, all sister's or some brother's & some sister's	

Example 1.

Pointing to a man in a photograph a man said to a woman "his mother is the only daughter of your father". How is the woman related to the man in the photograph.

(a) Sister (b) Mother
(c) Wife (d) Daughter

Sol.(b) The only daughter of the woman's father is the woman herself, and hence the man in the photograph is her son. Therefore the woman is the mother of the man in the photograph.

Example 2.

Pointing to a man in photograph a woman said "The father of his brother is the only son of my grand father". How is the woman related to the man in photograph.

Sol.(d) The only son of woman's grand father is the father of woman and the father of the man's brother is the father of the man. On combining these two information together single information emerges that the man's father is the woman's father. Hence woman is sister of the man in the photograph.

Direction for Ex. 3 to 5.

A + B means A is the daughter of B

A – B means A is the husband of B

A × B means A is the brother of B

Example 3.

If P + Q – R, which of the following is true:

(a) R is the mother of P
(b) R is the sister-in-law of P
(c) R is the aunt of P
(d) R is the mother-in-law of P

Sol.(d) P + Q – R means P is the daughter at Q who is the husband of R. i.e. R is P's mother.

Example 4.

If P × Q + R, which of following is true

(a) P is the brother of R
(b) P is the uncle of R
(c) P is the son of R
(d) P is the father of R

Sol.(c) P × Q + R means P is the brother of Q who is daughter of R. i.e. P is son of R.

Example 5.

If P + Q × R, which at following is true

(a) P is the Niece of R
(b) P is Daughter of R
(c) P is Cousin of R
(d) P is the Daughter-in-law of R

Sol. (a) P + Q × R, means P is the daughter of Q who is the brother of R i.e. P is niece of R

Example 6.

A party consists of grand mother, father, mother, four sons and their wives and one son and two daughters of each of the sons. How many females are?

(a) 14 (b) 16
(c) 18 (d) 24

Sol. (a) 14

Grandmother is one female, mother is another, wives of four sons are 4 female and two daughters of all four sons are 8 female. So all are

1 + 1 + 4 + 8 = 14 female

Example 7.

Laxmi and Meena are Rohan's wives, Shalini is Meena's stepdaughter. How is Laxmi related to Shalini?

(a) Sister (b) Mother-in-law
(c) Mother (d) Stepmother

Sol. (c) Mother

Shalini is Meena's stepdaughter Means Shalini is the daughter of other wife of Rohan, so Shalini is a daughter of Laxmi or Laxmi is mother of Shalini.

Direction for Ex. 8 to 10.

A + B means A is daughter of B.
A × B means A is the son of B.
A – B means A is the wife of B.

Example 8.

If P × Q – S, which of the following is true?

(a) S is wife of Q
(b) S is father of P
(c) P is daughter of Q
(d) Q is father of P

Sol. (b) P × Q – S means P is the son of Q who is the wife of S.

Example 9.

If T – S × B – M, which of the following is not true?

(a) B is mother of S
(b) M is husband of B
(c) T is wife of S
(d) S is daughter of B

Sol. (d) T – S × B – M means T is the wife of S, who is the son of B who is the wife of M i.e. T is the wife of son at M (father) i.e. T is the daughter-in-law of M. But as this is not given in the choice. So the fact in (d) is not true.

Example 10.

If Z × T – S × U + P, what is U to Z.

(a) Mother
(b) Grand mother
(c) Father
(d) Can't be determined

Sol. (b) Z × T – S × U means Z is the son of T who is the wife of S who is the son of U. i.e. Z is the son of S who is the son of U. i.e. U is the grandson of U or U is the grand mother of Z. From the choice (b) is true.

Some Standard Coding Technique

1. **Proper noun should always be denoted by capital letters**

   E.g. John, Merry and Paul can be codes as
   John – J, Merry – M, Paul – P

2. In questions where the sex of proper noun is crucial to the solving of questions, you can denote the female by under lining the capital letter used

   ex. → Merry is a beautiful girl. Merry – $\underline{M}$
   ex. → Sheena is the only girl student in her class
   Sheena – $\underline{S}$

3. **Relationship:**

   (a) Is the Son of

   ex. → Ram is the son of Mr. Pratap → $\begin{matrix}P\\R\end{matrix}$

   (b) Is the daughter of

   ex. → Sweta is daughter of Mr. Pratap → $\begin{matrix}P\\\underline{S}\end{matrix}$

   (c) Is the Parent of

   ex. → A is the parent of B → $\begin{matrix}A\\B\end{matrix}$ or $\begin{matrix}A\\\underline{B}\end{matrix}$

   (d) Is the spouse of

   ex. → A is spouse of B → $A — \bar{B}$ or $B — \bar{A}$

[Note: Here sex of A & B is not mentioned]
ex. → Seema is wife of Rahul → R – S
ex. → Priya is wife of Rajesh R – P
[**Note:** It is better to write male on the left & female on the right hand side to avoid any confusion]

(e) Is the sibling of

ex → A is sibling of B

1. A .. B or (A & B are brothers)
2. A .. B or (A is sister of B)
3. A .. B or (B is sister of A)
4. A .. B or (A & B are sisters)

(f) Grand father, Grand son relation

ex. → Ramesh and Param are the two sons of Mr. Shiv Kumar, who is the son of Mr. Sundarlal

Su
→ SK
R, P

(**Note:** Shiv Kumar and Sundarlal are coded as SK and Su to avoid any confusion)

(g) Where the sex of a person is not confirmed write it in a box

Like Q is the only son of [S]

(Here Q is male but sex of S is not mentioned)

□□□□

Practice Exercise: I

Direction (Qs. 1 to 6): Read the following information carefully and answer the question below it.
All the six members of a family P, Q, R, S, T and U are travelling together Q is the son of R but R is not the mother of Q. P and R are a married couple. T is the brother of R. S is the daughter of P. U is the brother of Q.

1. How may male members are there in the family?
(a) 1 (b) 2
(c) 3 (d) 4

2. Who is the mother of Q?
(a) S (b) U
(c) T (d) P

3. How many children does P have?
(a) One (b) Two
(c) Three (d) Four

4. Who is the wife of T?
(a) P
(b) U
(c) Q
(d) can't be determined

5. Which of the following is a pair of females?
(a) PT (b) QS
(c) SU (d) PS

6. How is U related to S?
(a) Father
(b) Brother
(c) Uncle
(d) can't be determined

Direction (Qs. 7 to 12): Read the following information carefully and answer the question below it.
A family consists of six members P, Q, R, X, Y and Z. Q is the son of R but R is not mother of Q. P and R are a married couple. Y is the brother of R. X is the daughter of P. Z is the brother of P.

7. Who is the brother-in-law of R?
(a) P (b) Z
(c) Y (d) X

8. Who is the father of Q?
(a) R (b) P
(c) Z (d) None of these

9. How many children does P have?
(a) One (b) Two
(c) Three (d) Four

10. How many female member are there in the family?
(a) One (b) Two
(c) Three (d) Four

11. How is Q related to X?
(a) Husband (b) Father
(c) Brother (d) Uncle

12. Which is a pair of brothers?
(a) P and X (b) P and Z
(c) Q and X (d) R and Y

Direction (Qs. 13 to 15) : Read the following information carefully and answer the questions given below it
P is the son of Q. R, Q's sister has a son S and daughter T. U is the maternal uncle of S.

13. How is P related to S?
(a) Cousin (b) Nephew
(c) Uncle (d) Brother

14. How is T related to U?
(a) Sister (b) Daughter
(c) Niece (d) Wife

15. How many nephew does U have?
(a) Nill (b) One
(c) Two (d) Three

16. Ranjan is the brother of Sachin and Manick is the father of Ranjan. Jagat is the brother of Priya and Priya is the daughter of Sachin. Who is the uncle of Jagat?
(a) Ranjan (b) Sachin
(c) Manick (d) None of these

Direction (Qs. 17 to 18): Read the following infor-mation carefully and answer the questions given below it.
(i) A is the father of C.
(ii) E is the daughter of C.F is the spouse of A.
(iii) B is the brother of C.D is the son of B.
(iv) G is the spouse of B. H is the father of G.

17. Who is the grandmother of D?
(a) A (b) C
(c) F (d) H

18. Who is the son of F?
(a) B (b) C
(c) D (d) E

Answers

1. (d)	2. (d)	3. (c)	4. (d)	5. (d)
6. (b)	7. (b)	8. (a)	9. (b)	10. (b)
11. (c)	12. (d)	13. (a)	14. (c)	15. (c)
16. (a)	17. (c)	18. (a)		

Solutions

Hints (1–6)

Sol. → Q is son of R
→ R is not the mother, then R is father of Q
→ P and R are married couple R – P
→ T is brother of R T .. R
→ S is daughter of P P / S
→ U is brother of Q Q .. U

Complete Solution

T.. R – P
Q.. U .. S

1. (d)
2. (d)
3. (c)
4. (d)
5. (d)
6. (b)

Sol. (7 – 12)

Y .. R – P .. Z
Q .. X

7. (b)
8. (a)
9. (b)
10. (b)
11. (c)
12. (d)

Sol.(13 to 15)

Q .. R ..U
P S .. T

13. (a)
14. (c)
15. (c)

Sol.(16)

M
R .. S
P .. J

16. (a)

Sol.(17 – 18)

A – F H
C .. B – G
E D

17. (c)
18. (a) here sex of C is not confirmed

II Practice Exercise: II

1. Introducing Asha to guests, Bhasker said, "Her father is the only son of my father". How is Asha related to Bhasker?

(a) Daughter (b) Mother
(c) Sister (d) Niece

2. Akash said to Mohit "The boy in blue shirt is younger of the two brothers of the daughter of my father's wife". How is the boy in blue shirt related to Akash?
(a) Father (b) Uncle
(c) Brother (d) Nephew

3. Pointing to a man on the stage, Rita said, "He is the brother of the daughter of the wife of my husband". How is the man on the stage related to Rita?
(a) Son (b) Husband
(c) Cousin (d) Nephew

4. Pointing to a photograph, Vipul said, "She is the daughter of my grand father is only son". How is Vipul related to the girl in the photograph?
(a) Father (b) Brother
(c) Cousin (d) Data inadequate

5. A woman introduce a man as the son of the brother of her mother. How is man related to the woman?
(a) Nephew (b) Son
(c) Cousin (d) Uncle

6. Looking at a picture of a man, Harsh said, "His mother is the wife of my father's son. Brothers and sisters I have none". At whose picture was Harsh looking?
(a) His son (b) His cousin
(c) His uncle (d) His nephew

7. A man said to a lady "your mother's husband's sister is my aunt. How is the lady related to the man?
(a) Daughter (b) Grand daughter
(c) Mother (d) Sister

8. If Meena says, "Anita's father, Raman is the only son of my father-in-law, Mahipal", then how is Bindu, who is the sister of Anita, related to Mahipal?
(a) Grand daughter (b) Daughter
(c) Sister (d) Niece

9. Pointing to a girl in the photograph Amar said, "Her mother's brother is the only son of my mother's father". How is the girl's mother related to Amar?
(a) Mother (b) Sister
(c) Aunt (d) Grand mother

10. Daya has a brother Amit. Daya is the son of Chandra, Bimal is Chandra's father. In terms of relationship what is Amit to Bimal?
(a) Son (b) Grand son
(c) Brother (d) Grand father

11. Rahul's mother is the only daughter of Monika's father, how is Monika's husband related to Rahul
(a) Uncle (b) Father
(c) Grand father (d) Brother

12. If (i) M is brother of N
(ii) B is brother of N and
(iii) M is brother of D then which of the following statements is definitely true?
(a) N is brother of B (b) N is brother of D
(c) M is brother of B (d) D is brother of M

13. Deepak is brother of Ravi, Rekha is sister of Atul, Ravi is son of Rekha. How is Deepak related to Rekha?
(a) Son (b) Brother
(c) Nephew (d) Father

14. A is B's sister, C is B's mother, D is C's father, E is D's mother. Then, how is A related to D?
(a) Grand mother (b) Grand father
(c) Daughter (d) Grand daughter

15. Given that (i) A is brother of B (ii) C is father of A (iii) D is brother of E (iv) E is daughter of B, then uncle of D is
(a) A (b) B
(c) C (d) D

16. E is the son of A, D is the son of B, E is married to C, C is B's daughter. How is D related to E?
(a) Brother (b) Uncle
(c) Father-in-law (d) Brother-in-law

Answers

1. (a)	2. (c)	3. (a)	4. (b)	5. (c)
6. (a)	7. (d)	8. (a)	9. (c)	10. (b)
11. (b)	12. (c)	13. (a)	14. (d)	15. (a)
16. (d)				

Solutions

1. (a)

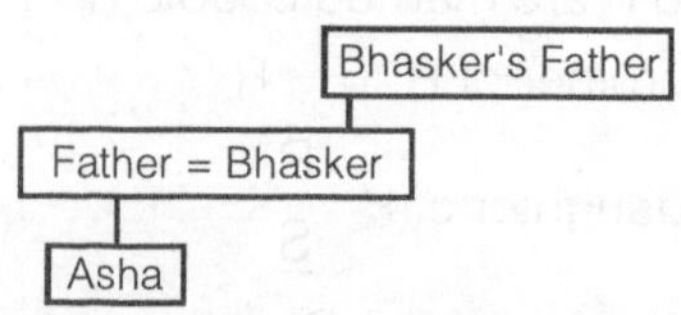

Hence, Asha is Bhasker's daughter.

2. (c)

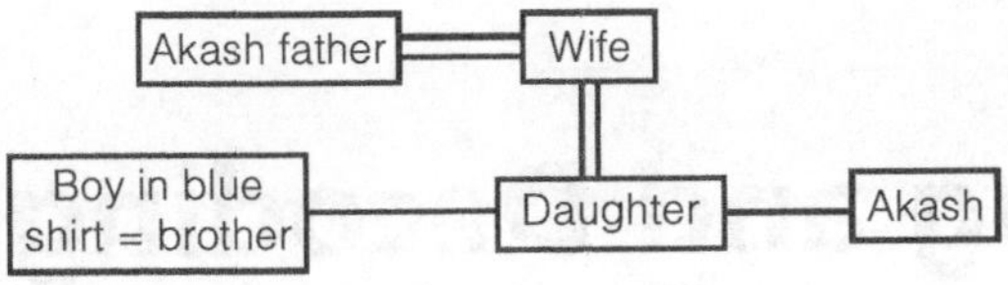

Hence brother.

3. (a)

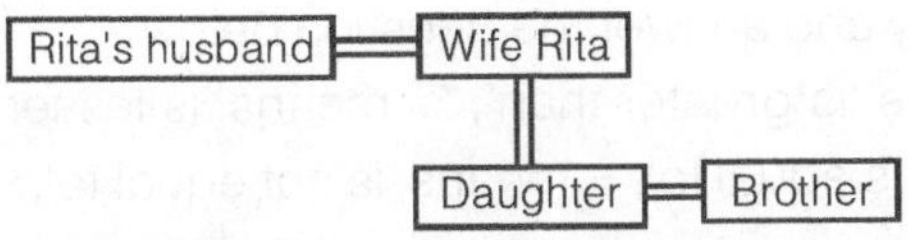

Hence person on stage is Rita's son.

4. (b)

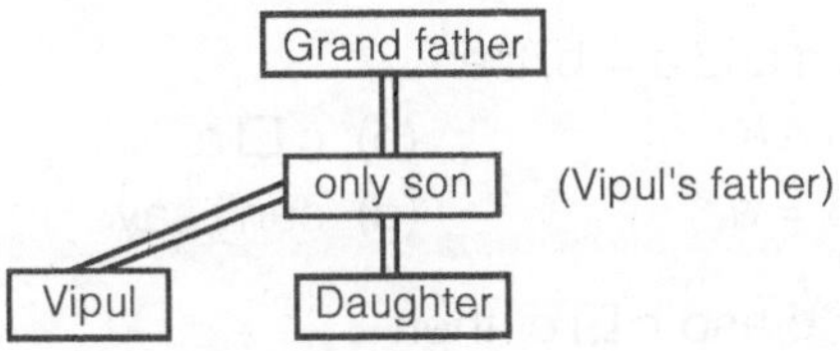

Hence Vipul is brother of girl.

5. (c)

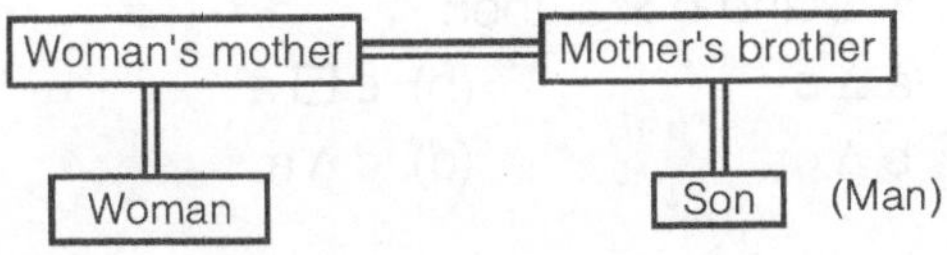

Hence man is woman's Cousin.

6. (a)

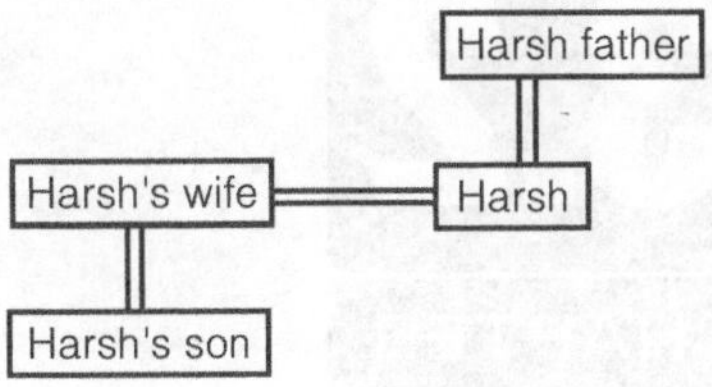

Harsh is looking at his son's picture.

7. Sister (d)

8. Grand daughter (a)

9. Aunt (c)

10. Grand son (b)

11. Father (b)

12. M is brother of B (c)

13. Son (a)

14. Grand daughter (d)

15. Uncle of D is A (a)

16. (d)

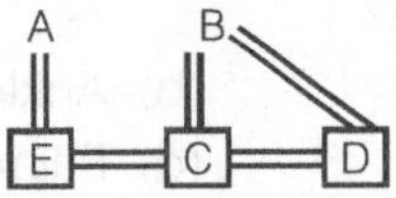

D is brother in law of E.

3.2

CHAPTER

Coding and Decoding

I Practice Exercise

Direction (Qs. 1 to 5): Read the following information carefully and answer the question below it.

1. If 'Rat is called Dog', 'Dog is called Mongoose', 'Mongoose' is called Lion;, 'Lion is called Snake' and 'Snake is called Elephant', which animal is reared as pet?
   (a) Rat (b) Dog
   (c) Mongoose (d) Lion

2. If Finger is called Toe, Toe is called Foot, Foot is called Thumb, Thumb is called Ankle, Ankle is called Palm and Palm is called Knee, which one finger has different name?
   (a) Thumb (b) Ankle
   (c) Knee (d) Palm

3. In a certain code language, 'kew xas huma deko' means 'she is eating apples', 'kew tepo qua' means 'she sells toys' and 'su lim deko' means 'I like apples'. Which word in that language means she and apples?
   (a) xas & deko (b) xas & kew
   (c) kew & deko (d) kew & xas

4. If 'gnr tag zog qmp' stands for 'Seoul Olympic Organising Committee, 'hydo gnr emf' stands for 'Summer Olympic Games' and 'esm sdr hyto' stands for 'Modern Games History', which would be the code for Summer?
   (a) hyto (b) gnr
   (c) emf (d) zog

5. In a certain code language, 'Pat Zoo Sim' means 'Eat Good Mangoes', 'Pus Sim Tim' means 'Mangoes and Sweets's and 'Tim Zoo Kit' means 'Purchase Good sweets', which word in the language means Good?
   (a) Zoo (b) Pus
   (c) Sim (d) Tim

Direction (Qs. 6 to 10): Read the following information carefully and answer the question below.

Δ means 'is greater than', % means 'is lesser than', $\square$ means 'is equal to', = means 'is not equal to', + means 'is a little more than', $\times$ means 'is a little less than'.

6. If $a \Delta b$ and $b + c$, then
   (a) a % c (b) c % a
   (c) $c + a$ (d) can't say

7. If $c = a$ and $a = b$, then
   (a) $b \Delta a$ (b) $c \square a$
   (c) $b = a$ (d) can't say

8. If $a \times b$ and $b \square c$, then
   (a) $c + a$ (b) $b \Delta c$
   (c) $a + c$ (d) $c \square a$

9. If c % b and $b \times a$, then
   (a) $a \Delta c$ (b) $c \square a$
   (c) $b \Delta c$ (d) $c \Delta a$

10. If ac + bc then
    (a) $a \square c$ (b) $b \Delta c$
    (c) $c \Delta b$ (d) b % a

Direction (Qs. 11 to 15): Read the following information carefully and answer the question below it.

If $>$ denotes $+$, $<$ denotes $-$, $+$ denotes $\div$, $\wedge$ denotes $\times$, $-$ denotes $=$, $\times$ denotes $>$ and $=$ denotes $<$, choose the correct statement in each of the following questions.

11. (a) $6 + 3 > 1 = 4 + 2 < 1$
    (b) $4 > 6 + 2 \times 32 + 4 < 1$
    (c) $8 < 4 + 2 = 6 > 3$
    (d) $14 + 7 > 3 = 6 + 3 > 2$

12. (a) $14 > 18 + 9 = 16 + 4 < 1$
    (b) $4 > 3 \wedge 8 < 1 - 6 + 2 > 24$
    (c) $3 < 6 \wedge 4 > 25 = 8 + 4 > 1$
    (d) $12 > 9 + 3 < 6 \times 25 + 5 > 6$

13. (a) $13 > 7 < 6 + 2 = 3 \wedge 4$
    (b) $9 > 5 > 4 - 18 + 9 > 16$

(c) 9 < 3 < 2 > 1 × 8 ^ 2
(d) 28 + 4 ^ 2 = 6 ^ 4 + 2

14. (a) 29 < 18 + 6 = 36 + 6 ^ 4
(b) 18 > 12 + 4 × 7 > 8 ^ 2
(c) 32 > 6 + 2 = 6 < 7 ^ 2
(d) 31 > 1 < 2 = 4 > 6 ^ 7

15. (a) 7 > 7 < 7 + 7 = 14
(b) 7 ^ 7 > 7 + 7 = 7 ^ 7 > 1
(c) 7 < 7 + 7 = 6
(d) 7 + 7 > 7 = 8

Direction (Qs. 16 to 18): Read the following information carefully and answer the question below it.
In a certain code, 'il be pee' means ' roses are blue'; 'silk hee' means 'red flowers' and 'pee mit hee' means 'flowers are vegetables'.

16. How is 'red' written in that code?
(a) hee (b) silk
(c) be (d) none of these

17. How is 'roses' written in that code?
(a) il
(b) pee
(c) be
(d) can not be determined

18. How is 'vegetables are red flowers' written in this code ?
(a) pee silk mit hee
(b) silk pee hee be
(c) il silk mit hee
(d) can not be determined

Direction (Qs. 19 to 22): Read the following information carefully and answer the question below it.
Shamita is very found of collecting greeting cards. Her collection of 211 cards has a good mix of Birthday cards, New year cards, Deepawali cards, Christmas cards and even a Marriage Anniversary card. The number of Birthday cards is equal to the sum of all other except the Marriage Anniversary card. The number of New Year card is double of Deepawali cards which in turn is double of Christmas cards.

19. The number of Birthday cards in the collection is
(a) 95 (b) 105
(c) 110 (d) 85

20. The number of Deepawali cards is
(a) 25 (b) 30
(c) 35 (d) 40

21. The number of Christmas cards is
(a) 10 (b) 15
(c) 20 (d) 25

22. The ratio of Birthday cards to Christmas cards is
(a) 7 : 1 (b) 4 : 1
(c) 2 : 1 (d) 1 : 1

Direction (Qs. 23 to 28): Read the following information carefully and answer the question below it.
Rahul's roll number in IES Exam is a number consisting of three non-zero distinct digits, such that the sum of the digits at hundred's and unit's place is equal to that half of the digit at ten's place. Also the sum of all possible three digit numbers obtained using these three digits without repetition is 2664.

23. The digits in the unit is
(a) 2 (b) 1
(c) 3 (d) can't say

24. The digits in the tens place is
(a) 6 (b) 8
(c) 9 (d) 2

25. The digit in the hundreds place is
(a) 3 (b) 6
(c) 9 (d) can't say

26. The sum of the three digits is
(a) 7 (b) 12
(c) 10 (d) 14

27. The three digit number is
(a) 138 (b) 183
(c) 381 (d) can't say

28. The three digit is number is always divisible by :
(a) 3 (b) 18
(c) 29 (d) None of these

Direction (Qs. 29 to 30): Read the following information carefully and answer the question below it
P, Q, R, S, T, U and V are seven positive integers and (P × Q × R × S × T × U × V) is odd.

29. Maximum how many of these integers can be odd?
(a) 4 (b) 5
(c) 6 (d) 7

30. Minimum how many of these integers can be even?
(a) 3 (b) 2
(c) 4 (d) 0

31. In the sequence of alphabets, which letter would be eight to the right of the letter which is sixteenth from the left ?
(a) G (b) Y
(c) Z (d) X

32. CEGJLN ____ XZB. The missing group of letters in the series are ...
(a) QSU (b) NPR
(c) PRT (d) TUX

Direction (Qs. 33 to 37): Read the following information carefully and answer the question below it.

In each of the Letter Analogy various terms of a letter series are given with one term missing as shown by (?). Choose the missing term out of the given alternatives.

33. HUA GTZ FSY ERX ?
(a) DWQ (b) DQW
(c) WDQ (d) WQD

34. DF GJ KM NQ RT ?
(a) UW (b) YZ
(c) XZ (d) UX

35. DCXW FEVU HGTS ?
(a) LKPQ (b) ABZ
(c) JIRQ (d) LMRS

36. AB DEF HIJK ? STUVWX
(a) MNOPQ (b) LMNOP
(c) LMNO (d) QRSTU

37. C G L R ?
(a) Y (b) S
(c) U (d) Z

Answers

1. (c)	2. (b)	3. (c)	4. (c)	5. (a)
6. (b)	7. (c)	8. (a)	9. (a)	10. (d)
11. (c)	12. (b)	13. (b)	14. (d)	15. (a)
16. (b)	17. (d)	18. (a)	19. (b)	20. (b)
21. (b)	22. (a)	23. (d)	24. (b)	25. (d)
26. (b)	27. (d)	28. (a)	29. (d)	30. (d)
31. (d)	32. (a)	33. (b)	34. (d)	35. (c)
36. (a)	37. (a)			

Solutions

Answer 1 to 5:

1. (c) 'Dog' is reared as pet. But 'Dog' is called 'Mongoose'. So, a 'Mongoose' is reared as pet.

2. (b) The 'Thumb' is a finger having a different name. But 'Thumb' is called 'Ankle'. So, 'Ankle' is the finger that has a different name.

3. (c) In the 1st and 2nd statements, the common code word is kew and the common word is she. So, kew means she. In the 1st and 3rd statements, the common code is deko and the common word is apples. So, deko means apples.

4. (c) In the first and second statements, the common code word is gnr and the common word is Olympic. So, gnr means Olympic. In the second and third statements, the common code is hyto and the common word is games. So, hyto means games. Thus, in the second statement, emf means summer.

5. (a) From 1st and 3rd statements. Zoo means Good.

Answer 6 to 10:

6. (b) $a \Delta b$ and $b + c \Rightarrow a > b$
and b is a little more than c
$\Rightarrow a > c$
$\Rightarrow c < a$
i.e. c % a.

7. (c) $c = a$ and $a = b \Rightarrow c \neq a$ and $a \neq b$
$\Rightarrow b \neq a$ i.e. $b = a$.

8. (a) $a \times b$ and $b \square c \Rightarrow$ a is a little less than b and $b = c$.
$\Rightarrow$ a is a little less then c.
$\Rightarrow$ c is a little more than a i.e. $c + a$.

9. (a) c % b $\Rightarrow c < b$
and $b \times a \Rightarrow$ b is a little less than a.
$\Rightarrow c < a \Rightarrow a > c$ i.e. $a \Delta c$.

10. (d) $ac + bc \Rightarrow ac > bc \Rightarrow a > b \Rightarrow b < a$ i.e. b % a.

Answer 11 to 15 :

11. (c) Using the proper notations in (3), we get the statements as $8 - 4 + 2 < 6 + 3$ or $6 < 9$, which is true.

12. (b) Using the proper notations in (2), we get the statements as $4 + 3 \times 8 - 1 = 6 \div 2 + 24$ or $27 = 27$.

13. (b) Using the proper notations in (2), we get the statements as $9 + 5 + 4 = 18 + 9 + 16$ or $18 = 18$.

14. (d) Using the proper notations in (4), we get the statements as $31 + 1 - 2 < 4 + 6 \times 7$ or $30 < 46$.

15. (a) Using the proper notations in (1), we get the statements as $7 + 7 - 7 \div 7 < 14$ or $13 < 14$.

Answer 16 to 18:

16. (b) In the second and third statements, the common code word is 'hee' and the common word is

'flowers'. So, 'hee' stands for 'flowers'. Thus, in the second statement, 'silk' stands for 'red'.

17. (d) Since from the given information, we can only find the code for 'are' in the first statement, it cannot be determined which of the remaining two codes for 'roses'.

18. (a) Clearly, the required code will consist of the same codes as in the third statements with the code for 'red' i.e. silk added to it.

Answer 19 to 22:

If, number of Christmas cards = x

Then, number of Deepawali cards = $2x$

Number of New Year cards = $4x$

and number of Birthday cards

$= x + 2x + 4x$

$= 7x$

i.e. $7x + (x + 2x + 4x) + 1 = 211$

$14x = 210$

$x = 15$

Thus, number of Birthday cards

$= 15 \times 7$

$= 105$

Number of New Year cards

$= 15 \times 4$

$= 60$

Number of Deepawali cards

$= 15 \times 2$

$= 30$

Number of Christmas cards

$= 15 \times 1$

$= 15$

Number of Marriage Anniversary card

$= 1$

Total $= 211$

19. (b)

20. (b)

21. (b)

22. (a) 105 : 15 = 7 : 1.

Answer 23 to 28:

Let the digits be x, y, z such that the number is $100x + 10y + z$

Thus, the other five 3 digit number which we may obtained using these 3 digits will be

$100x + 10z + y$

$100y + 10x + z$

$100y + 10z + x$

$100z + 10x + y$

$100z + 10y + x$

Now when we add all the six three digit numbers possible to be formed by these three digits:

$222 (x + y + z) = 2664$ or $x + y + z = 12$

Also $x + z = y/2$ thus $y = 8$

Thus, $x + z = 4$

As the digits are non zero and distinct thus x and z have to be 1 and 3 but not necessarily in the same order. Thus we cannot say whether the number is 183 or 381.

23. (d)

24. (b)

25. (d)

26. (b)

27. (d)

28. (a)

Answer 29 to 32:

29. (d) All integer should be odd to get odd result.

30. (d) None of the integer should be even.

31. (d)

32. (a)

Answer 33 to 37:

33. (b) All the letters of each term are moved one step backward to obtain the corresponding letters of the next term.

34. (d) There is a gap of one letter between both the letters of first term, a gap of two letters between both the letters of second term and again a gap of one and two letters between the letters of third and fourth terms respectively. Besides, the last letter of each term and the first letter of next term are in alphabetical.

35. (c) First two letters of each term are in reverse order. Similarly third and fourth letters are also in reverse order. Besides, the second letter of each term is the letter next to the first letter of the proceeding term.

36. (a) The number of letters in the term goes on increasing by 1 at each step. Each term consists of letters in alphabetical. The last letter of each term and the first letter of the next term are alternate.

37. (a) There is a gap of three letters between the first and the second term, four letters between the second and the third term; and five letters between the third and the fourth term. So, there should be a gap of six letters between the fourth term and the missing term.

■■■■

3.3
CHAPTER

Cubes and Dice

Cubes

A type of cuboid in which all the sides i.e. length, breadth & height are equal. All faces of cubes are of same area.

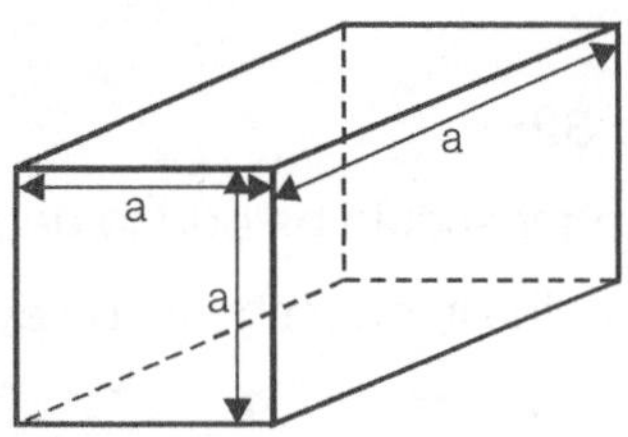

In a cube there are 8-corners/vertices
There are 6-faces [all equal in area]
There are 12-edges [all equal in length]
No. of edges = No. of vertices + No. of face – 2

Primarily question in cubes are based on painting of faces of cubes and then cutting the painted cube into identical cubelets. Because questions are based on the way cube is painted and how cuts are being made, hence we don't recommend any thumb rule. Still however because some basic rules are always applicable which we'll be using as building blocks for answering questions.

- If n equidistanced cuts are made (all parallel to same surface), cube will be divided into (n + 1) identical cuboidal pieces with each such cut there will $2a^2$ new surface area will be generated which will be unpainted.
- If I want to cut my bigger cube into identical n^3 cubelets, using minimum number cuts, I need total 3(n – 1) cuts, such that (n – 1) cuts parallel to each of these faces which are joining to corner.
- If number of cuts are not multiple of three then cube can never be cut into identical cubes but still it can be cut into maximum number of identical cuboidal pieces. To maximize such number of pieces we need to split the number of cuts into three parts which are closest. That we'll see in upcoming examples.

I Practice Exercise: I

Directions for Questions (1 - 9):
If a cube is cut into n^3 identical cubelets using minimum no. of cuts, after painting all faces of cube with white color, then answer the following questions.

1. What is minimum number cuts required?
2. What is maximum number of possible painted faces in once of such of cubelets?
3. How many cubelets will have exactly 3 faces painted?
4. How many cubelets will have exactly 2 faces painted?
5. How many cubelets will have exactly 1 face painted?
6. How many cubelets will have no faces painted?
7. How many cubelets will have at most 2 faces painted?
8. How many cubelets will have at least 1 face painted?
9. If initially (V) liters of paints was required to paint all faces of cube than how much extra paint will be required such that all the unpained faces of all cubelets will be painted.

Direction (Qs. 10 to 12) : A cube is divided into 216 identical cubelets. Each cut is made parallel to some surface of cube. But before cutting the cube is colored with green color on one set of opposite faces, red on the other set of opposite faces and blue on the third set.

10. How many cubelets are there which are painted with exactly one color?
    (a) 96 (b) 108
    (c) 124 (d) 48
11. How many cubelets are there which are painted with at least two coluor?
    (a) 96 (b) 56
    (c) 108 (d) 124

12. How many cubelets are there which are painted with at most one color?
(a) 124 (b) 108
(c) 160 (d) 96

Direction (Qs. 13 to 17): A cube is divided into 343 identical cubelets. Each cut is made parallel to some surface of the cube. But before doing that the cube is colored with green color on one set of adjacent faces, red on the second and blue on the third set.

13. How many cubelets are there which are colored with exactly are color?
(a) 180 (b) 150
(c) 165 (d) 148

14. How many minimum cuts you have made?
(a) 15 (b) 18
(c) 21 (d) 9

15. How many cubelets are colored with exactly two colors?
(a) 48 (b) 36
(c) 51 (d) 50

16. How many cubelets are thin which are colored with exactly three colors?
(a) 6 (b) 4
(c) 2 (d) 8

17. How many cubelets are painted with multiple colors?
(a) 51 (b) 53
(c) 115 (d) 120

Direction (Qs. 18 to 22): A cuboid is divided into 192 identical cubelets. This is done by making minimum no. of cuts possible. All cuts are parallel to some of the face. But before doing so. The cube is painted with green color on one set of opposite faces. Blue on other set of opposite faces and red on remaining their pair of opposite faces.

18. What is the maximum number of cubelets possible which one colored with green colored only?
(a) 48 (b) 64
(c) 96 (d) 72

19. What is the minimum no. of cuts I have made?
(a) 15 (b) 18
(c) 21 (d) 12

20. What is the minimum no. of cubelets possible which are painted with green and blue colors?
(a) 16 (b) 24
(c) 32 (d) 48

21. What is the number of cubelets which are painted with exactly 3 colors?
(a) 4 (b) 6
(c) 8 (d) 2

22. What is the number of cubelets possible which are painted with none of the color?
(a) 16 (b) 24
(c) 36 (d) 48

Answers

10. (a)	11. (b)	12. (c)	13. (c)	14. (b)
15. (c)	16. (c)	17. (b)	18. (a)	19. (a)
20. (a)	21. (c)	22. (d)		

Solutions

1. Total number cuts required in 3(n – 1). (n – 1) equidistant cuts parallel to each of 3 face which are joining to corner.

2. Maximum faces painted will be three in any such cubelets, which will be in case of cubelets coming out of corners of big cube after cutting.

3, 4, 5, 6.

Cubes painted from 3 faces = 8

Cubes painted from 2 faces = 12(n – 2)

Cubes painted from 1 face = $6(n-2)^2$

Cubes painted from none face = $(n-2)^2$

Where 'n' is number of part into which each side is divided.

7. To find out the number of cubelets with at most 2 faces painted, I need to remove all those cubes which have exactly 3 faces painted.

= Total no. of cubes – No. of cubes with 3 faces painted.

$= (n^3 - 8)$

8. To find out the number of cubelets with at least one face painted, I need to remove all the cubelets which have no face painted.

$= n^3 - (n-2)^3$

9. As with every cut unpainted area of $2a^2$ is created Total no. of cuts used is 3(n – 1).
Hence total unpainted area generated is, (which is equal to areas of all unpainted faces of all small cubelets).

$= 2a^2 \times 3(n-1) = 6a^2(n-1)$

As the total surface area of cube was $6a^2$ for which (V) liters of paint was required, hence total paint required to paint, all unpainted faces of small cubelets is $= \frac{6a^2(n-1)}{6a^2} \times V = (n-1)V$ liters

10, 11 & 12.

Because cube was painted with same color on opposite set of faces. We can still used formula.

$[(n-2)+2]^3 = {}^3C_0(n-2)^3 + {}^3C_1(n-2)^2 \times 2 + {}^3C_2(n-2) \times 2^2 + {}^3C_3 \times 2^3$

10. Exactly one color painted

$= {}^3C_1(n-2)^2 \times 2$

$= 3 \times (n-2)^2 \times 2$

$= 3 \times 4^2 \times 2 = 96$

Option (a)

11. Basically we are looking for no. of cubes which colored on two faces or three face

$= {}^3C_2(n-2) \times 2^2 + {}^3C_3 \times 2^3$

$= 3 \times (n-2) \times 4 + 8$

$= 3 \times 4 \times 4 + 8 = 56$

Option (b)

12. Basically in this question we are looking for no. of cubelet, which are uncolored and no. of cubelets which are colored on just one face.

$= {}^3C_0(n-2)^3 + {}^3C_1(n-2)^2 \times 2$

$= (n-2)^3 + 3 \times 2 \times (n-2)^2$

$= 4^3 + 6 \times 4^2$

$\Rightarrow 64 + 96 = 160$. Option (c)

Alternatively: Total no. cubelets which are colored an at most are face = total no. of cubelets – total no. of cubelets which are colored on multiple faces.

$= 216 - 56 = 160$. option (c)

13. From each face $5 \times 5 = 25$ cubelets are painted is one color. Such total $= 25 \times 6 = 150$ cubelets. Further because each face is having one similar color adjacent face and 3 other different coloured faces hence $5 \times 3 = 15$ extra cubelets will come from three edges, where adjacent faces of similar color are meeting hence total $= 150 + 15 = 165$ cube will be then colored in exactly one color?

Hence option (c)

14. $n^3 = 343 = 7^3 \Rightarrow n = 7$

Minimum number of cuts $= 3(n-1)$

$= 3(7-1) = 3 \times 6 = 18$. Option (b)

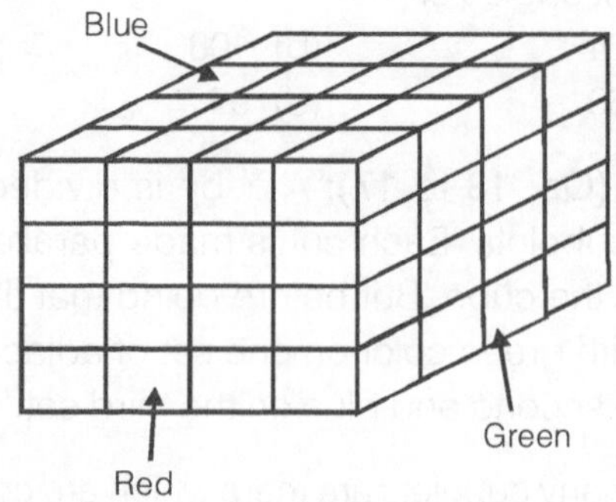

As we can see in above figure. 3 faces are visible in 3-diff. colours, out of hidden faces, bottom is red, one is green and another is blue.

15. Number of cubelet with no face painted will

$= (n-2)^3 = (7-2)^3 = 125$

No. of cubelets with 2 color

= Total no. of cubelet – [cubelets with one color + cubelets with no color + cubelets with three color]

$= 343 - [125 + 2 + 165] = 51$

Option (c)

16. Only two cubelets one is visible in figure and 2nd one exactly diagonally opposite corner of it will be colored with 3 three colors. Option (c)

17. No. of cubelets with multiple color

= No. of cubelets with three color + No. cubelets with two color

$= 2 + 51 = 53$. Option (b)

18. As we want maximize number of cubelets with green in color only cuboid has to painted with green color on set of opposite faces of 6×8. Hence no. of green colored only cubelets will $(6-2) \times (8-4) = 4 \times 6 = 24$, from face. There are two such faces hence maximum total no. of only green painted cubelets will be $= 24 \times 2 = 48$

Hence option (a)

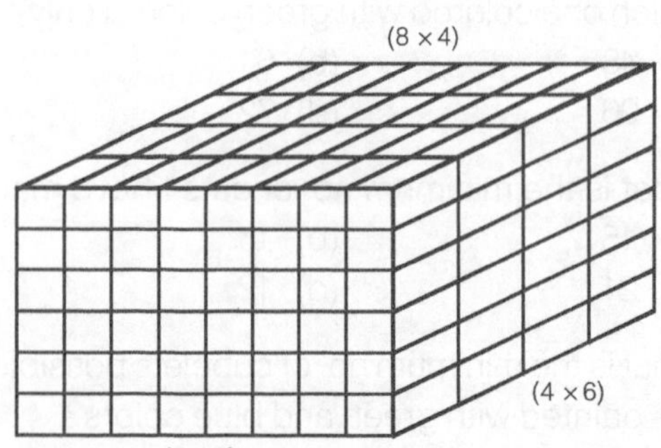

19. The no. of cuts will be minimum when 192 will be factorized into 3 closest factors. $192 = 4 \times 6 \times 8$, so we have two opposite faces of 4×6 and two opposite faces of 6×8 and two opposite faces of 4×8. Hence the minimum number of cuts will be $= (4 - 1) + (6 - 1) + (8 - 1) = 3 + 5 + 7 = 15$. Hence option (a)

20. To minimize the number of cubelets with blue and green blue should be selected on (4×6) or (6×8) and green should be colored on (6×8) or (4×6) respectively cubelets from the edges where green and blue faces will be meeting will give the number of desired cubelets $= 4 + 4 + 4 = 16$
Hence option (a)

21. Again only the cubelets from corners of cuboid will be painted with 3 colors; hence 8 cubelets will be colored in 3 colors. Hence option (c)

22. Number of unpainted cubelets will be
$(4 - 2) \times (6 - 2) \times (8 - 2) = 2 \times 4 \times 6 = 48$
Hence option (d)

❑❑❑❑

DICES

Dices are cubical structure with number of points from 1 to 6 are marked on faces. Dices are of two kinds.

1. **Symmetric:** Where sum of numbers on opposite pair of faces is same.
   Naturally the faces opposite to each other will be
   $1 \rightarrow 6$, $2 \rightarrow 5$, $3 \rightarrow 4$
2. **Asymmetric:** When sum of numbers on opposite pair of faces is different.

Example:

A dice with it's faces numbered 1 to 6 is shown in three different positions x, y, z

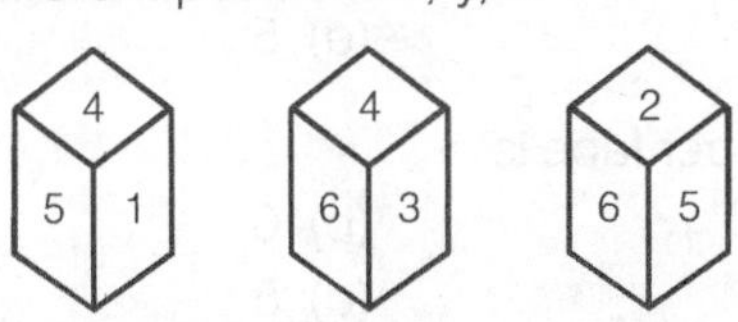

(x) (y) (z)

Find opposite faces pairs.

Solution.

Hence pair
$1 \rightarrow 6$
$2 \rightarrow 4$
$5 \rightarrow 3$

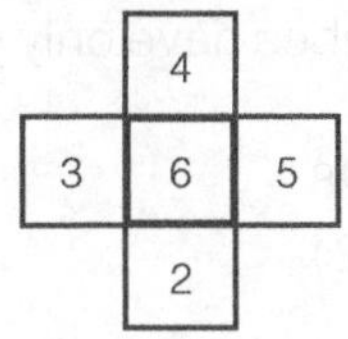

Hence 1 is marked opposite to 6

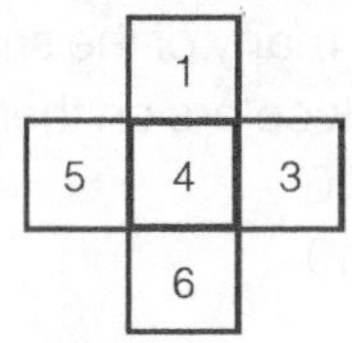

2 is marked opposite to 4

❑❑❑❑

Practice Exercise: I

Direction (Qs. 1 to 5) : A cube is painted and then divided cut into 336 smaller but identical pieces by making the minimum number of cuts possible. All cuts are parallel to some face.

1. What is the minimum no. of cuts required?
   (a) 21 (b) 18
   (c) 24 (d) None of these
2. How many smaller pieces have exactly 3 painted faces?
   (a) 20 (b) 8
   (c) 12 (d) 6
3. How many smaller pieces have at least 2 face painted?
   (a) 64 (b) 68
   (c) 72 (d) 76
4. How many smaller pieces have at most one face painted?
   (a) 248 (b) 264
   (c) 268 (d) None of these
5. How many smaller pieces have no face painted?
   (a) 144 (b) 124
   (c) 120 (d) 108

Direction (Qs. 6 to 13) : Questions are based on the following information.

Three adjacent faces of one cube is painted in pink, one adjacent pair of faces is painted in black and the remaining faces are painted in violet. The cube is the cut into 216 identical cubelets.

6. How many of cubelets will have all three colors on them?
   (a) 1 (b) 2
   (c) 3 (d) 4

7. How many of the smaller cubes have only pink and black colors on them?
(a) 16 (b) 18
(c) 19 (d) 24

8. How many smaller cubes or cubelets have exactly two colors on them?
(a) 32 (b) 36
(c) 37 (d) 42

9. How many the cubelets have exactly two painted surface in two different colors?
(a) 36 (b) 32
(c) 48 (d) 38

10. How many of the cubelets have exactly one color on them?
(a) 107 (b) 109
(c) 96 (d) 113

11. How many of the cubelets have exactly one painted surface of exactly one color?
(a) 84 (b) 108
(c) 96 (d) 102

12. How many cubelets do not have pink color on them?
(a) 120 (b) 100
(c) 150 (d) 125

13. How many cubelets have black or violet color on them but not pink color on them?
(a) 60 (b) 59
(c) 51 (d) 64

14. The six faces of a cube are marked as 1, 2, 3, 4, 5 and 6. Given below are two different view of same cube.

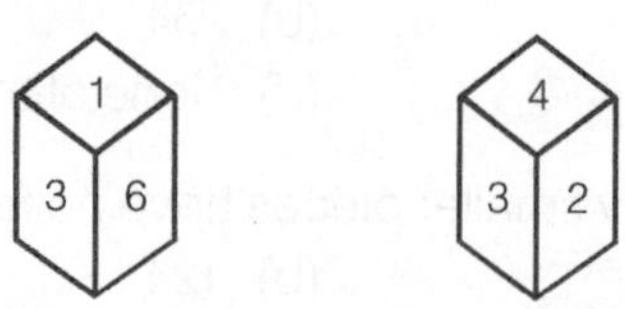

(I) (II)

Which face is opposite the face marked 5?
(a) 2 (b) 4
(c) 3 (d) 6

15. The six faces of a cube are marked as faces 1, 2, 3, 4, 5 and 6. Given below are two different views. Which face is opposite the marked 6?

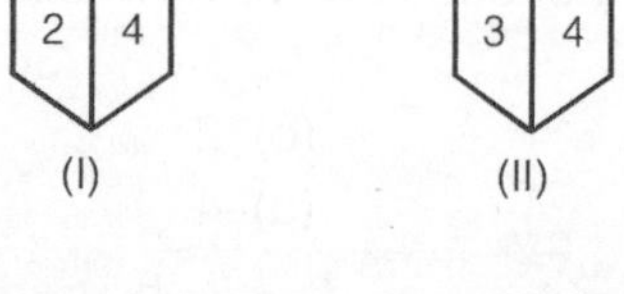

(I) (II)

(a) 4 (b) 5
(c) 1 (d) 2

16. The six faces of a cube as marked 1, 2, 3, 4, 5 and 6. Given below are two different view of same cube.

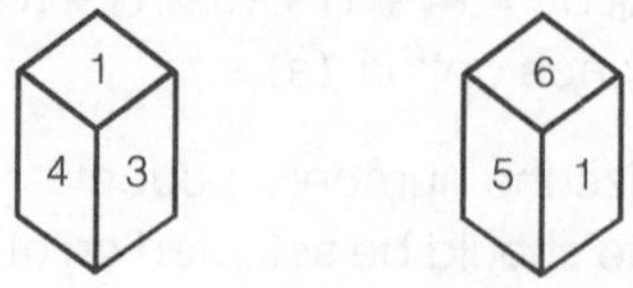

(I) (II)

Which face is opposite the face mark as 6?
(a) 3 (b) 4
(c) 2 (d) None of these

17. How many cuboids of dimensions 4 × 5 × 6 are required to form a cube of least size if cuboids have to be placed adjacent, above or below each other?
(a) 1600 (b) 1800
(c) 1200 (d) None of these

18. A cube has been cut into cuboids of size 2 × 3 × 4. What is least possible integer length of the edge of cube and how many such cuboids are obtained from this cube?
(a) 96 (b) 72
(c) 60 (d) 84

Direction (Qs. 19 to 22) : Read the following information carefully and answer the question below it.

A dice is prepared in following manner?
(i) 1 should lie between 2 and 3
(ii) 2 should lie opposite to 3
(iii) 4 should lie between 5 and 6
(iv) 5 and 6 should lie opposite to each other
(v) 4 should lie face down

19. The face opposite to 1 is
(a) 2 (b) 4
(c) 6 (d) 5

20. The upper face is
(a) 1 (b) 6
(c) 2 (d) 5

21. The face adjacent to 5 are
(a) 2, 6, 1, 4 (b) 1, 3, 4, 6
(c) 3, 4, 2, 1 (d) 2, 6, 4, 5

22. The face adjacent to 3 are
(a) 5, 4, 1, 2 (b) 1, 2, 5, 6
(c) 5, 6, 4, 1 (d) 2, 6, 4, 5

Answers

1. (b)	2. (b)	3. (b)	4. (c)	5. (c)
6. (b)	7. (c)	8. (c)	9. (b)	10. (d)
11. (c)	12. (d)	13. (c)	14. (c).	15. (d)
16. (b)	17. (b)	18. (b)	19. (b)	20. (a)
21. (c)	22. (c)			

Solutions

Answer 1 to 5: Number of identical pieces 336 = 8 × 7 × 6. Hence we need 7 cut's in Z-direction, 6 cuts in Y-direction, 5-cuts in X-directions.

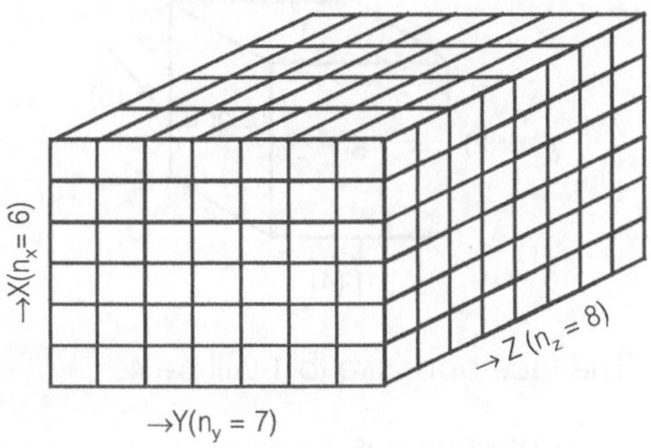

Means our cube cut into 8 parts along Z-directions say $n_Z = 8$, similarly $n_Y = 7$, $n_X = 6$.
For total number of identical pieces we can say
$n_X \times n_Y \times n_Z = 336$
$\{(n_X - 2) + 2\} \times \{(n_Y - 2) + 2\} \times \{(n_Z - 2) + 2\}$

Now look at table below:

No. of pieces	Formula	Answers for 6 × 7 × 8 cube
3-face painted (Corner pieces)	2^3	8
2-faces painted piece	$4[(n_x - 2) + (n_y - 2) + (n_z - 2)]$	$4[(8-2)+(7-2)+(6-2)] = 60$
Only one face painted pieces	$2[(n_x-2)(n_y-2)+(n_y-2)(n_z-2)+(n_z-2)(n_x-2)]$	$2[(8-2)(7-2)+(7-2)(6-2)+(8-2)\times(6-2)]$
Pieces no face painted	$(n_x-2)(n_y-2)(n_z-2)$	$(8-2)(7-2)(6-2) = 120$

1. (b) Minimum no. of cuts
$(8 - 1) + (7 - 1) + (6 - 1) = 18$

2. (b) All pieces from corners of cube = 8

3. (b) 2 faces painted + 3 faces painted
= 8 + 60 = 68

4. (c) No. faces painted + one face painted
= 120 + 148 = 268

5. (c) $(8 - 2)(7 - 2)(6 - 2) = 120$

Answer 6 to 13: Cube is cut in 6 × 6 × 6 = 216 cubelets
Pattern of painting is an follows.

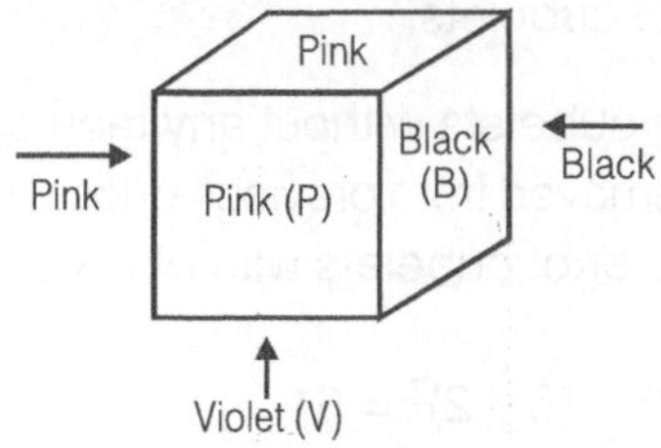

Let us analysis the color combination

Corners	Edges	Faces
PPP – 1, BBP = 1	PP = 3, PB = 4	P = 3
PPB – 2, BBV = 1	BB = 1, PV = 2	B = 2
PPV – 1, PVB = 2	BV = 2	V = 1

6. (b) The small pieces with three color will from corners. Only 2-such diagonl opposite y cubelets are possible.

7. (c) The cubelets with only pink and black are found at corners and along the edges PB
= 2 PPB cubelets + 1 BBP cubelet + 4 P edges × (6 – 2) = 19 cubelets.

8. (c) The cubelets with exactly two colors are found at corners and along the edges having different colors on either side = 5 corner pieces + 8 edges × (6 – 2) = 37 cubelets.

9. (b) In the previous question there is no. restriction on the number of painted surface. But in this question there is a restriction. The paint should be only on two faces only. Hence we'll not consider cubelets from corner. Hence only 32 cubelets.

10. (d) The cubelets with exactly one color are
(a) One exactly at corner (PPP) = 1
(b) At the three edges each of PP category = 3(6 – 2) = 12
(c) At the one edge of BB category = 1(6 – 2) = 4
(d) At the middle of each of the six faces = 6 × (6 – 2) = 96
So total cublets with exactly one colour are
1 + 12 + 4 + 96 = 113

11. (c) In this question only the cubelets from middle of faces of cube will be considered = $6 \times (6 - 2)^2 = 96$ cubelets.

12. (d) The number of cubelets with no pink color
= 216 – (36 from one pink surface + 30 from second + 25 from third)
= 125 cubelets.

13. (c) Here cubelets without any face colored has to be removed from previous question number. The number of cubelets with black or violet but not pink
= 125 – (6 – 2)^3 = 61

14. (c) As 3 is adjacent to (1 and 6) as well as (4 and 2) hence 3 is opposite to the remaining face 5.

15. (d) 2 and 6 both are adjacent to both 3 and 4. Hence 2 and 6 have to be opposite.

16. (b) From (I) & (II) we know that 1 is adjacent to 3, 4, 5, 6. Hence 1 is opposite to 2. We can find the other two pairs of opposite faces by rotating one of the view to reach another.

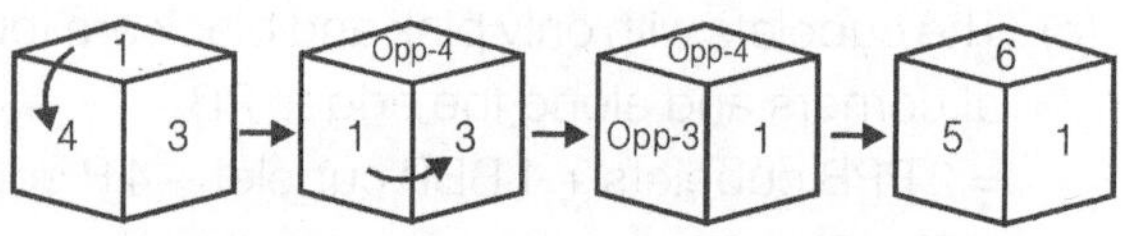

⇒ Opp to 4 is 6 and opp. to 2 is 5.

17. (b) Least possible dimension of cube is LCM of (4, 5, 6) = 60
The number of cuboids required

$$= \frac{60 \times 60 \times 60}{4 \times 5 \times 6} = 1800$$

18. (b) The least possible dimension of cube
LCM of 2, 3, 4 = 12
No. of cuboids that can be obtained

$$= \frac{12 \times 12 \times 12}{2 \times 3 \times 4} = 72$$

Answer 19 to 22:

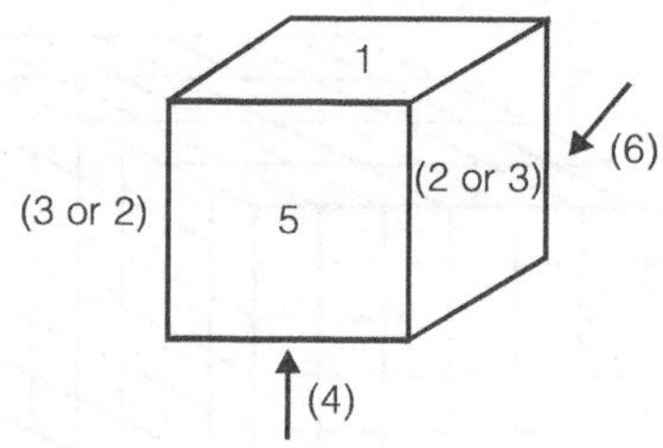

19. (b) The face opposite to 1 will be 4.

20. (a) The upper face is 1.

21. (c) 6 is opposite to 5, hence 1, 2, 3, 4 must be adjacent to 5.

22. (c) 3 is opposite to 2 hence 1, 4, 5, 6 are adjacent to 3.

■■■■

3.4 CHAPTER

Direction Sense Test

Practice Exercise: I

1. Dalbir is facing south. He turns 135° in the anticlockwise direction and then 180° in the clockwise direction. Which direction is he facing now?
   (a) North-east (b) North-west
   (c) South-east (d) South-west
2. Ram a college student is facing north-west. He turns 90° in the clockwise direction and then 135° in the anticlockwise direction. Which direction is he facing now?
   (a) East (b) West
   (c) North (d) South
3. Rakesh starts walking straight towards east. After walking 75 m, he turns to the left and walks 25 m straight. Again he turns to the left, walks a distance of 40 m straight, again he turns to the left and walks a distance of 25 m. How far is he from the starting point?
   (a) 25 m (b) 50 m
   (c) 140 m (d) none of these
4. Neelesh leaves for his office from his house. He walks towards East. After moving a distance of 20 m, he turns towards South and walks 10 m. Then he walks 35 m towards the West and further 5 m towards the North. He then turns towards East and walks 15 m. What is the straight distance in m between his initial and final positions?
   (a) 0
   (b) 5
   (c) 10
   (d) cannot be determined
5. Vinod walks 20 m towards North. He then turns left and walks 40 m. He again turns left and walks 20 m. Further, he moves 20 m after turning to the right. How far is he from his original position?
   (a) 20 m (b) 30 m
   (c) 50 m (d) 60 m
6. Kuldeep starts from his house towards West. After walking a distance of 30 m, he again turned towards right and walked 20 m. He then turned left and moving a distance of 10 m turned to his left and walked 40 m, turns to the left and walks 5 m. Finally he turns to his left. In which direction is he walking now?
   (a) North (b) South
   (c) East (d) South-West
7. A rat runs 20 m towards East and turns a right, runs 10 m and turns to right, runs 9 m and again turns to left, runs 5 m and then turns to left, runs 12 m and finally turns to left and runs 6 m. Now, which direction is the rat facing?
   (a) East (b) West
   (c) North (d) South
8. Bhawna leaves from her home. She first walks 30 m in North-west direction and then 30 m in South-West direction. Next, she walks 30 m in South-east direction. Finally, she turns towards her house. In which direction is she moving?
   (a) North-east (b) North-west
   (c) South-east (d) South-west
9. I am facing South. I turn right and walk 20 m. Then I turn right again and walk 10 m. Then I turn left and walk 10 m and then turning right walk 20 m. Then I turn right again and walk 60 m. In which direction am I from the starting point?
   (a) North (b) North-west
   (c) East (d) North-east
10. A man walks 1km towards East and then he turns to South and walks 5 km. Again he turns to East and walks 2 km, after this he turns to North and walks 9 km. Now, how far is he from his starting point?

(a) 3 km (b) 4 km
(c) 5 km (d) 7 km

11. From his house Sanjay went 15 kms to the North. Then he turned West and covered 10 kms. Then, he turned South and covered 5 kms. Finally, turning to East, he covered 10 kms. In which direction is he from his house?
(a) East (b) West
(c) North (d) South

12. Going 50 m to the South of her house, Radhika turns left and goes another 20 m. Then, turning to the North, she goes 30 m and then starts walking to her house. In which direction is she walking now?
(a) North-west (b) North
(c) South-east (d) East

13. Michael walks 20 m North. Then he turns right and walks 30 m. Then he turns right and walks 35 m. Then he turns left and walks 15 m. Then he again turns left and walks 15 m. In which direction and how many meters away is he from his original position?
(a) 15 m west (b) 30 m east
(c) 30 m west (d) 45 m east

14. A child is looking for his father. He went 90 m is the East before turning to his right. He went 20 m before turning to his right again to look for his father at his uncle's place 30 m from this point. His father was not there. From here he went 100 m to the North before meeting his father in a street. How far did the son meet his father from the starting point?
(a) 80 m (b) 100 m
(c) 140 m (d) 260 m

15. The door of Aditya's house faces the East. From the back side of his house, he walks straight 50 m, then turns to the right and walks 50 m again. Finally he turns towards left and stops after walking 25 m. Now, Aditya is in which direction from the starting point?
(a) South-east (b) North-east
(c) South-west (d) North-west

16. Two buses start from the opposite points of a main road, 150 kms apart. The first bus runs for 25 kms straight and takes a right turn and then runs for 15 kms. It then turns left and runs for another 25 kms and takes the direction back to reach the main road. In the meantime, due to a minor breakdown, the other bus has run only 35 kms along the main road. What would be the distance between the two buses at this point?
(a) 65 kms (b) 75 kms
(c) 80 kms (d) 85 kms

17. X and Y start moving towards each other from two places 200 m apart. After walking 60 m, Y turns left and goes 20 m, then he turns right and goes 40 m. He then turns right again and comes back to the road on which he had started walking. If X and Y walk with the same speed, what is the distance between them now?
(a) 20 m (b) 30 m
(c) 40 m (d) 50 m

18. Five boys are standing in a row facing East. Deepak is to the left of Sameer, Tushar and Shailendra. Sameer, Tushar and Shailendra are to the left of Sushil. Shailendra is between Sammer and Tushar. If Tushar is fourth from the left, how far is Sameer from the right?
(a) First (b) Second
(c) Third (d) Fourth

19. After walking 6 km, I turned right and covered a distance of 2 km, then turned left and covered a distance of 10 km. In the end, I was moving towards the north. From which direction did I start my journey?
(a) North (b) South
(c) East (d) West

20. A postman was returning to the post office which was in front of him to the north. When the post office was 100 m away from him, he turned to the left and moved 50 m to deliver the last letter at Shantivilla. He then moved in the same direction for 40 m, turned to his right and moved 100 m. How many metres was he away from the post office.
(a) 0 (b) 90
(c) 150 (d) 100

21. A boy rode his bicyle northwards, then turned left and rode one km and again turned left and rode 2 km. He found himself exactly one km west of his starting point. How far did he ride northwards initially?
(a) 1 km (b) 2 km
(c) 3 km (d) 5 km

22. A square field ABCD of side 90 m is so located that its diagonal AC is from north to south and the corner

B is to the west of D. Muong and Thames start walking along the sides from B and C respectively in the clockwise and anticlockwise direction with speeds of 8 m/s and 10 m/s. Where shall they cross each other the second time?

(a) On AD at a distance of 30 m from A
(b) On BC at a distance of 10 m from B
(c) On AD at a distance of 30 m from D
(d) On BC at a distance of 10 m from C

Answers

1. (d)	2. (b)	3. (d)	4. (b)	5. (d)
6. (a)	7. (c)	8. (a)	9. (d)	10. (c)
11. (c)	12. (a)	13. (d)	14. (b)	15. (d)
16. (a)	17. (c)	18. (d)	19. (b)	20. (b)
21. (b)	22. (d)			

Solutions

1. (d) As shown in Fig., Dalbir initially faces in the direction OA, On moving 135° anticlockwise, he faces in the direction OB. On further moving 180° clockwise, he faces in the direction OC, which is South-west.

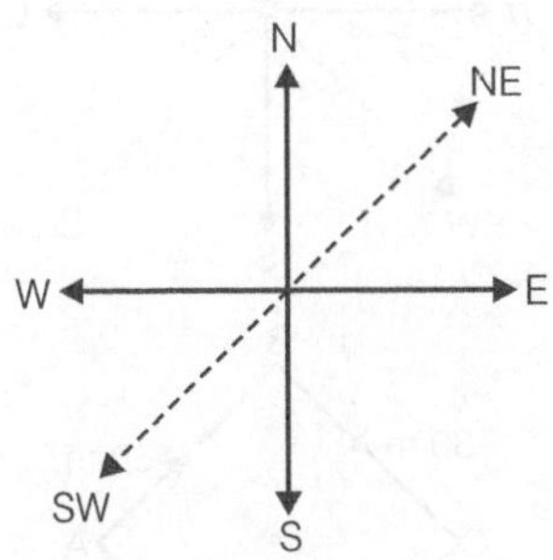

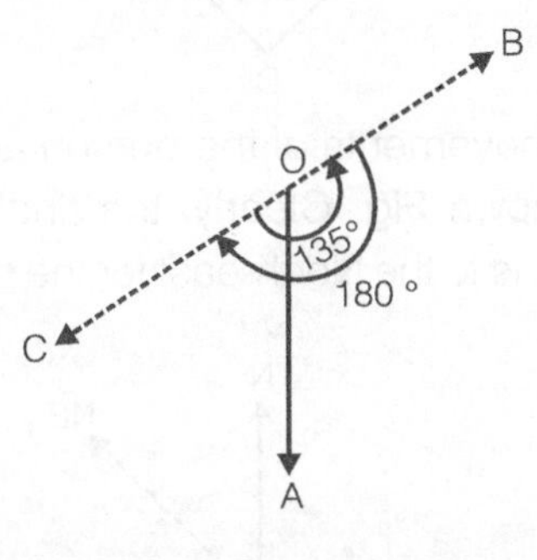

2. (b) As shown in Fig., Ram initially faces in the direction OP. On moving 90° clockwise, he faces in the direction OQ. On further moving 135° anti-clockwise, he faces in the direction OR, which is West.

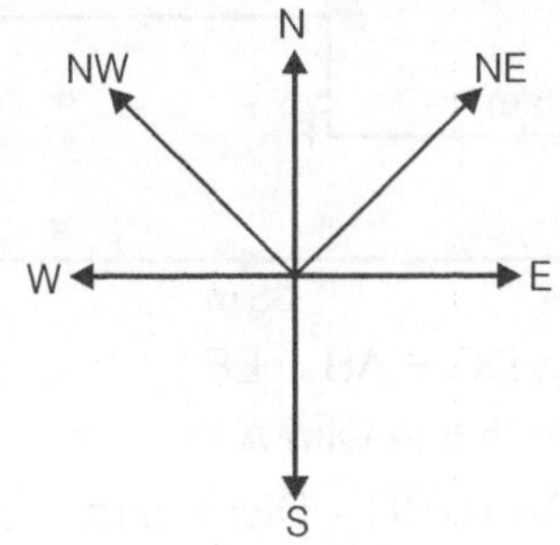

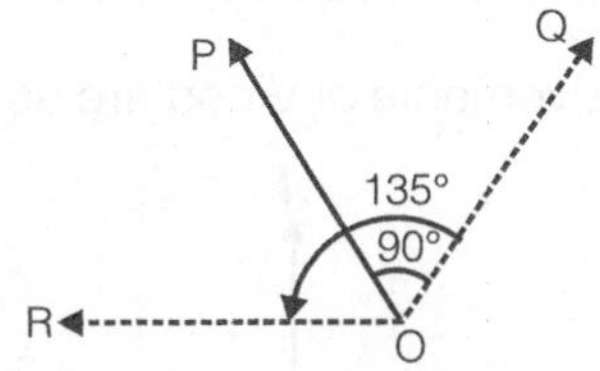

3. (d) The movement of Rakesh are as shown in Fig.

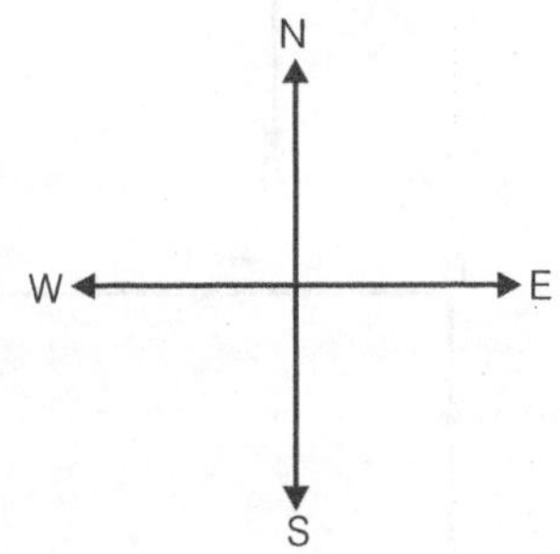

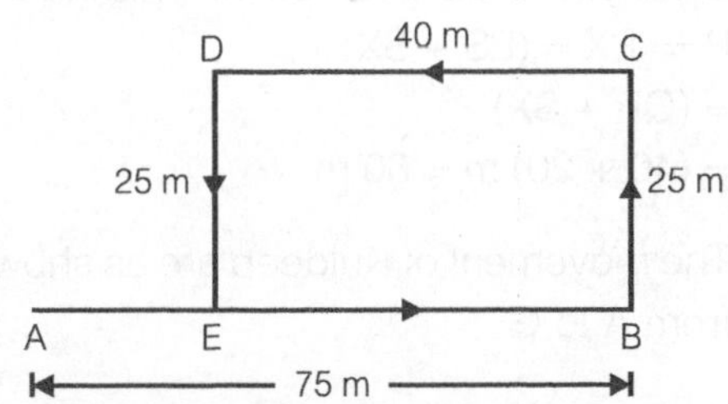

Clearly, EB = DC = 40 m.

∴ Rakesh's distance from the starting point A = (AB – EB) = (75 – 40) m = 35 m.

4. (b) The movement of Neelesh from A to F are as shown in Fig.

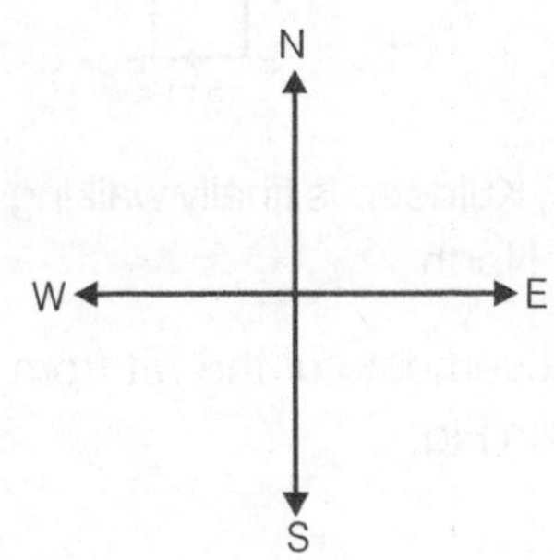

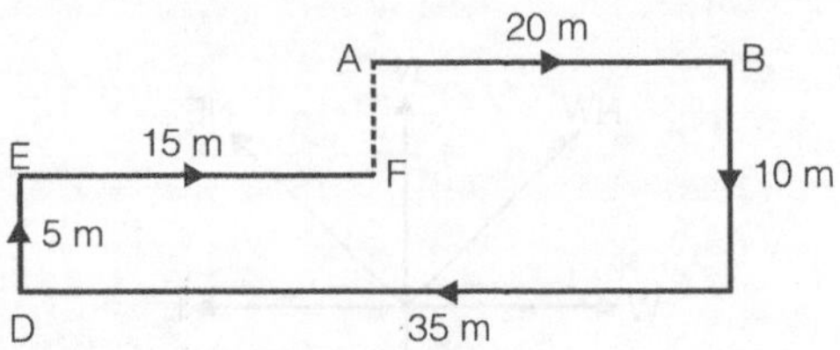

Clearly, DC = AB + EF.

∴ F is in line with A.

Also, AF = (BC – DE) = 5 m.

So, Neelesh is 5 m away from his initial position.

5. (d) The movements of Vinod are as shown in Fig.

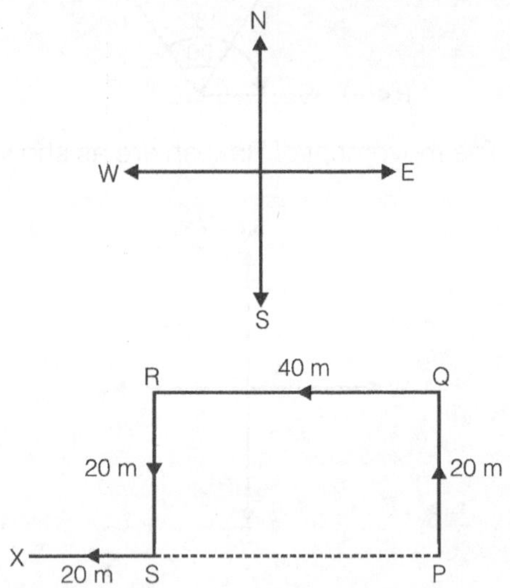

Clearly; Vinod's distance from his initial position

P ⇒ PX = (PS + SX)

= (QR + SX)

= (40 + 20) m = 60 m.

6. (a) The movement of Kuldeep are as shown in Fig. from A to G.

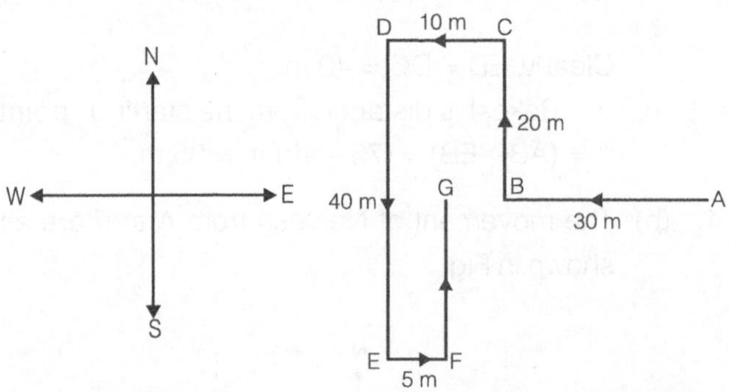

Clearly, Kuldeep is finally walking in the direction FG i.e, North

7. (c) The movements of the rat from A to G are as shown in Fig.

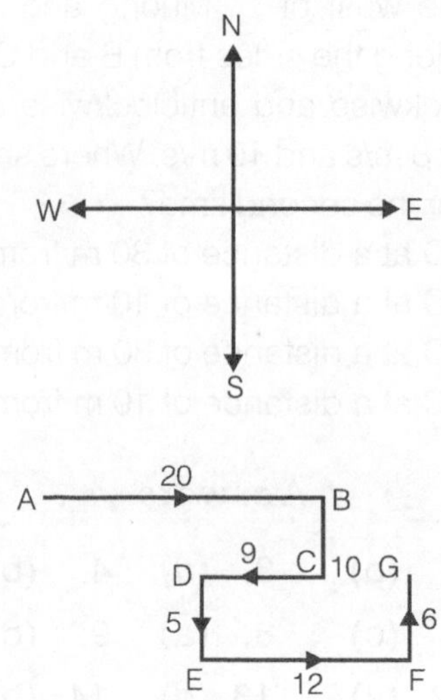

Clearly, it is finally walking in the direction FG i.e, North

8. (a) The movements of Bhawna are as shown in Fig. (A to B, B to C, C to D, D to A)

Clearly, she is finally moving in the direction DA i.e, North-east

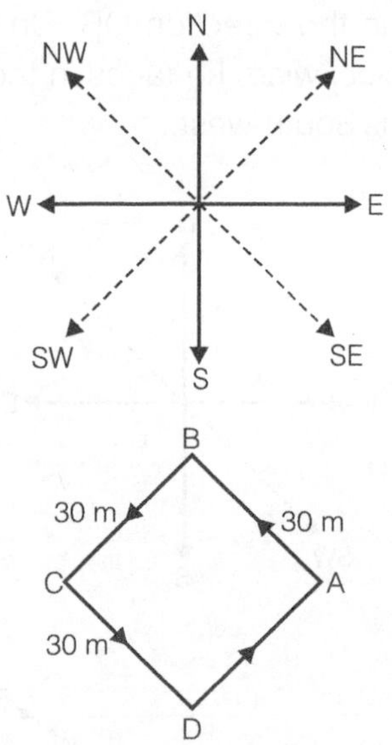

9. (d) The movements of the person are from A to F, as shown Fig. Clearly, the final position is F which is to the North-east for the starting point A

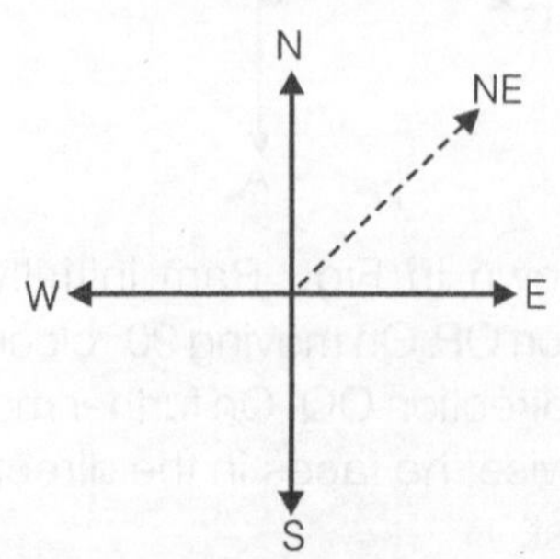

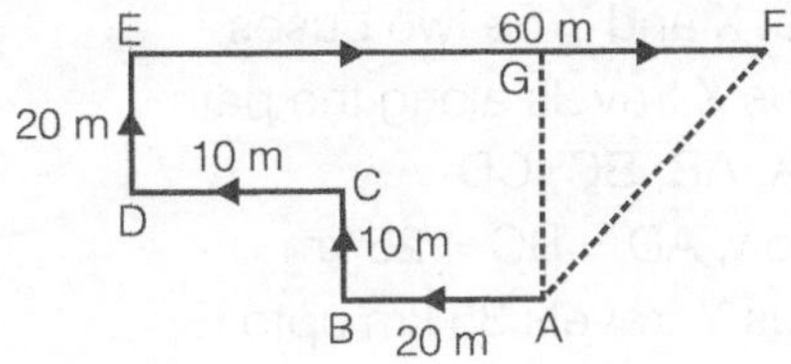

10. (c) The movements of the man are as shown in Fig. (A to B, B to C, C to D, D to E).

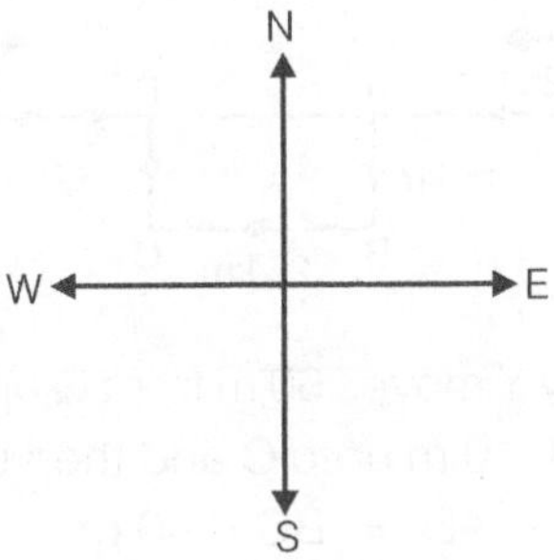

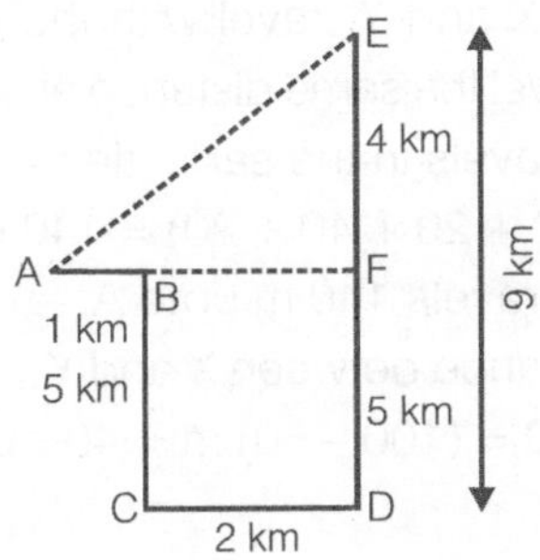

Clearly, DF = BC = 5 km.

EF = (DE – DF) = (9 – 5) km = 4 km.

BF = CD = 2 km

AF = AB + BF = AB + CD = (1 + 2) km = 3 km

$\therefore$ Man's distance from starting point A

$= AE = \sqrt{AF^2 + EF^2} = \sqrt{3^2 + 4^2}$

$= \sqrt{25} = 5$ km

11. (c) The movements of Sanjay are as shown in Fig. (A to B, B to C, C to D and D to E).

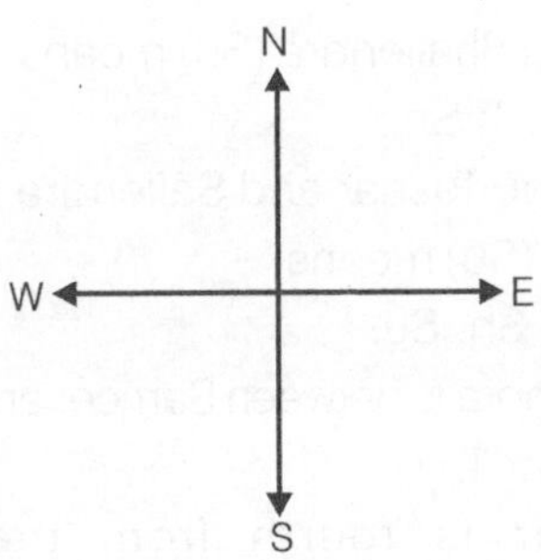

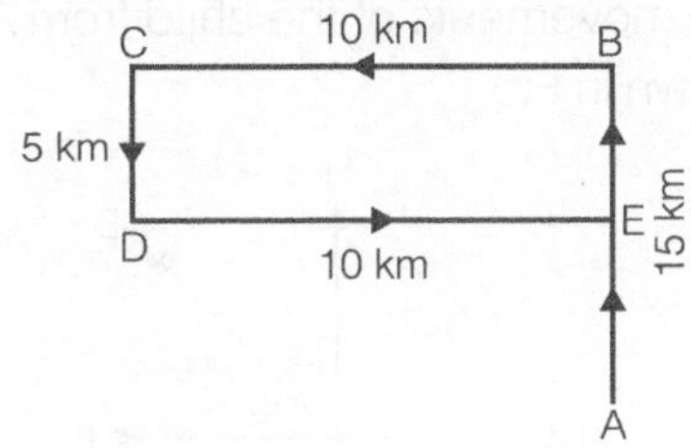

Clearly, his final position is E which is to the North of his house at A.

12. (a) The movements of Radhika are as shown in Fig. (A to B, B, to C, C to D and D to A).

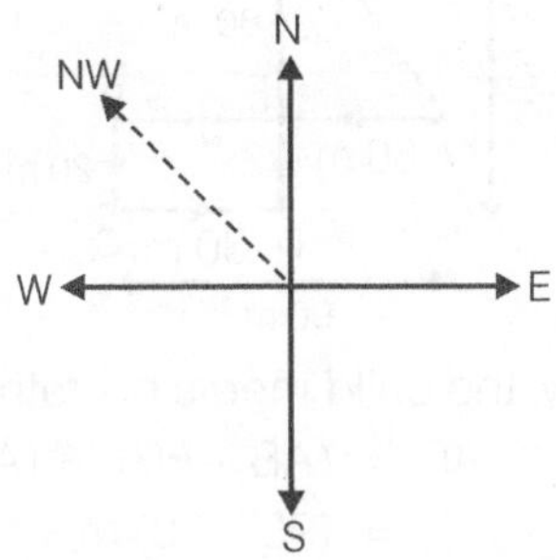

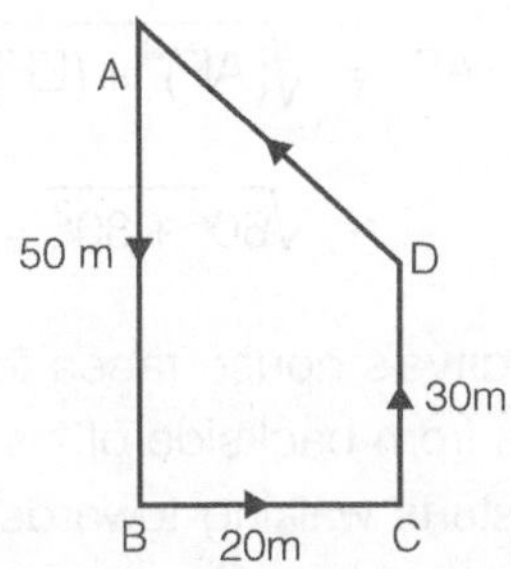

Clearly, she is finally moving in the direction DA i.e. North-west.

13. (d) The movement of Michael from A to F are as shown in Fig.

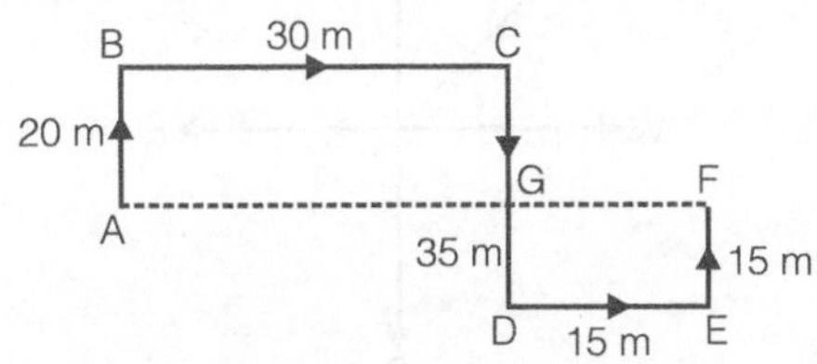

Since CD = AB + EF, so F lies in line with A

$\therefore$ Michael's distance from original position A

= AF = (AG + GF)

= (BC + DE) = (30 + 15) = 45 m.

Also, F lies to the east of A

14. (b) The movements of the child from A to E are as shown in Fig.

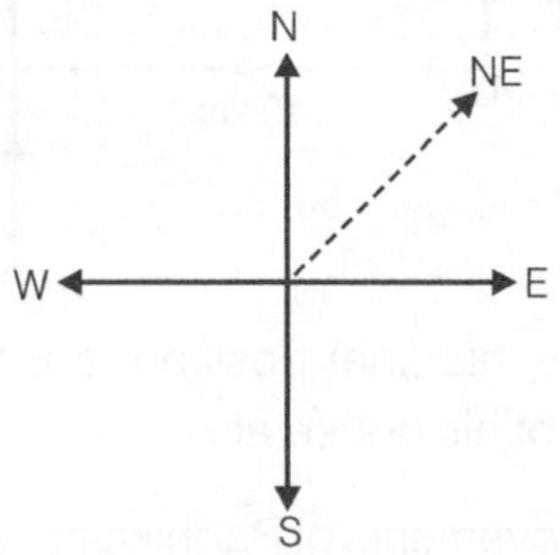

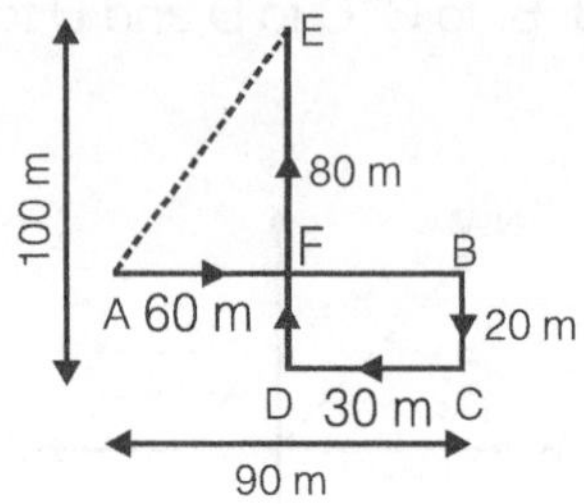

Clearly, the child meets his father at E.

Now, AF = (AB – FB) = (AB – DC)

= (90 – 30) m = 60 m.

EF = (DE – DF) = (DE – BC)

= (100 – 20) m = 80 m

$$AE = \sqrt{(AF)^2 + (EF)^2}$$

$$= \sqrt{60^2 + 80^2} = 100 \text{ m}$$

15. (d) Since Aditya's house faces towards East and he walks from backside of his house, it means that he starts walking towards West. Thus, the movements of Aditya are as shown in Fig. (A to B, B, to C, C to D).

Clearly, Aditya's final position is D which is to the North-west of the starting point A.

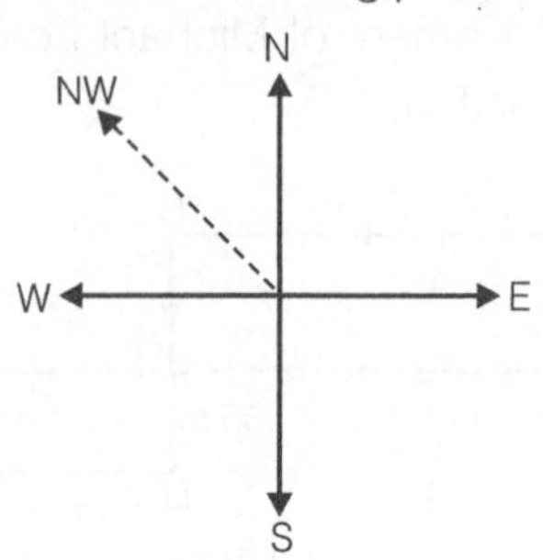

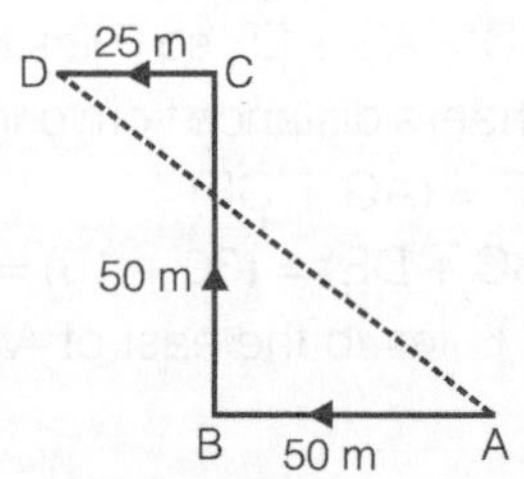

16. (a) Let X and Y be two buses.

Bus X travels along the path PA, AB, BC, CD

Now, AD = BC = 25 km

Bus Y travels 35 km upto E

∴ Distance between two buses

= PQ – (PD + QE)

= [150 – (50 + 35)] = 65 km

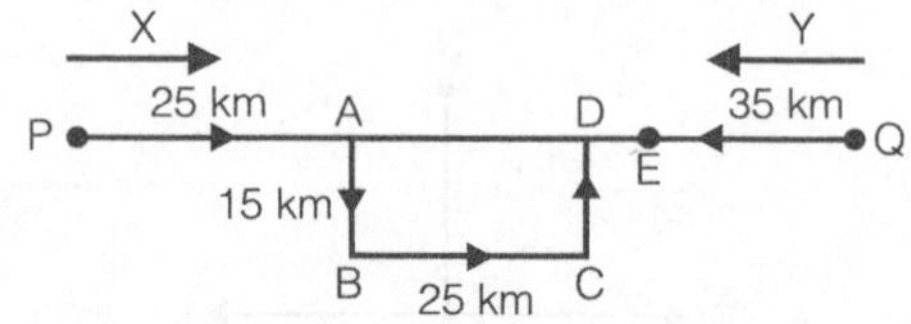

17. (c) Clearly Y moves 60 m from Q upto A, then 20 m upto B, 40 m upto C and then upto D.

So, AD = BC = 40 m.

QD = (60 + 40) m = 100m.

Since X and Y travel with the same speed, X will travel the same distance along the horizontal as Y travels in the same tiem

i.e. (60 + 20 + 40 + 20) = 140 m.

So, X travels 140 m upto A.

∴ Distance between X and Y

= AD = (100 – 60) m = 40 m.

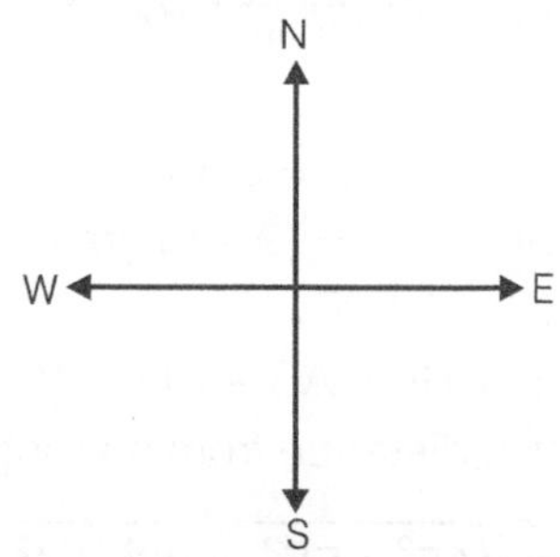

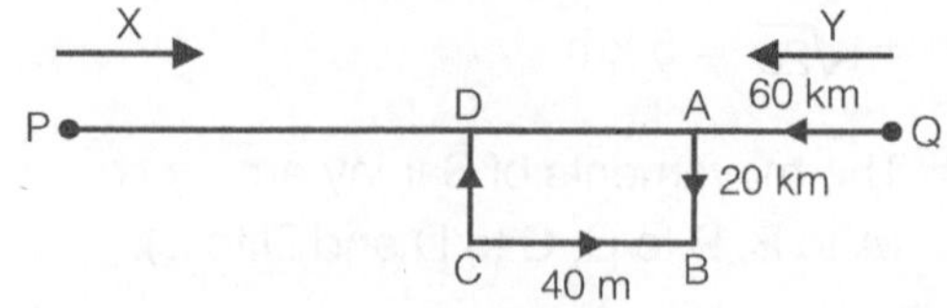

18. (d) Deepak (D) is to the left of Sameer (Sm), Tushar (T) and Shailendra (Sh) means

D, Sm, T, Sh.

Sameer, Tushar and Sailendra are to the left of Sushil (Su) means

Sm, T, Sh, Su.

Shailendra is between Sameer and Tushar means

Sm, Sh, T.

Tushar is fourth from the left means

□□□T.

Combining all the arrangements, we have D, Sm, Sh, T, Su.
So Sameer is fourth from the right.

19. (b) Clearly, the route is as shown in the adjoining diagram. Thus, I started the journey from the South and moved northwards

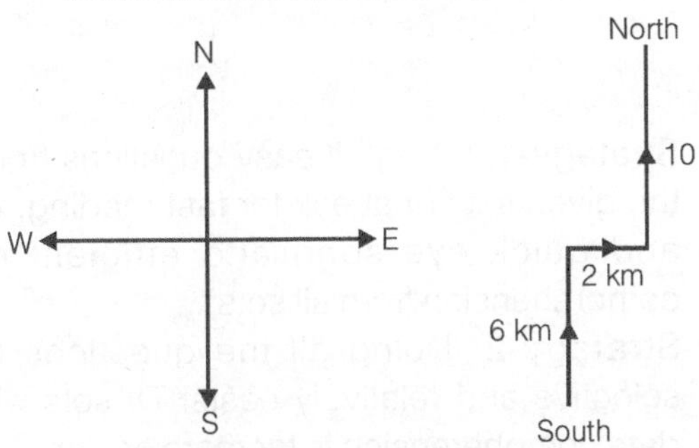

20. (b) Clearly, the route of the postman is as shown. So, at the final point the distance of postman from post office

$$= PD = AC = AB + BC$$
$$= 50 + 40 = 90 \text{ m}$$

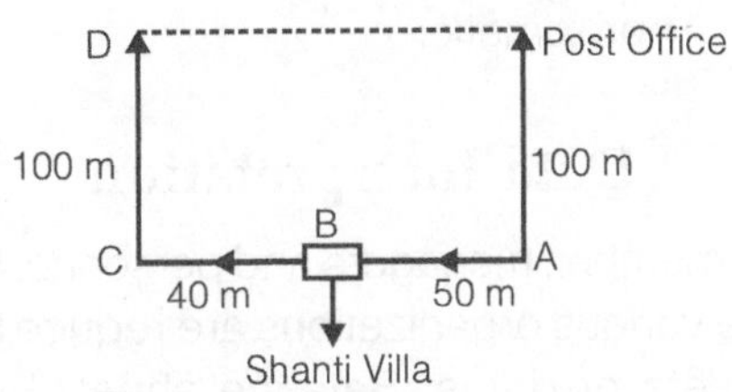

21. (b) Clearly, the boy rode from A to B, to C and finally upto D. Since D lies to the west of A, so required distance

$$= AB = CD = 2 \text{ km}$$

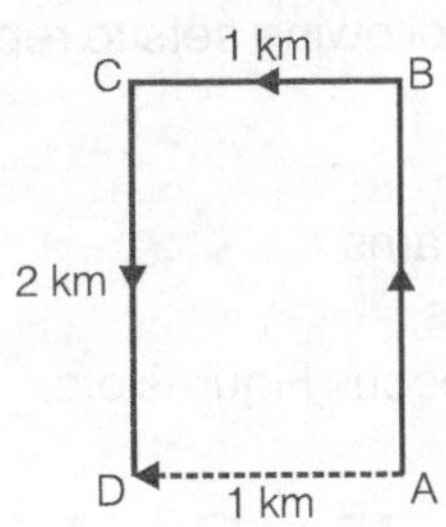

22. (d) Clearly, the arrangement is as shown in the adjoining diagram.

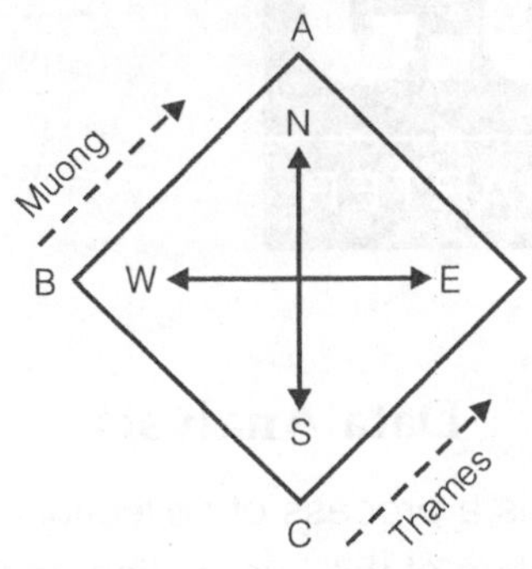

It is given that
Ram speed = 8 m/s
Thames's speed = 10 m/s
It is clearly visible from the above figure that to meet for the first time they have to cover a distance of 270 m i.e three arms length of the field.
BA + AD + CD = 90 + 90 + 90 = 270 m
Since they are running in clockwise and anticlockwise direction respectively, they have to cover a distance of 360 m (i.e. 4 arms length of the field) for their second meeting from the first meeting point.
Thus, total distance to be covered for their second meeting = 270 + 360 = 630 m

$$\text{Time required} = \frac{630}{8+10} = \frac{630}{18}$$

= 35 seconds
Distance coverd by Ram in clockwise direction in 35 seconds = 35 × 8 = 280 m
Distance covered by Thames in anticlockwise direction in 35 seconds
= 35 × 10 = 350 m
Thus, Ram and Thames will meet on BC at a point 10 m from C and 80 m from B.

■■■■

3.5

CHAPTER

Line Graphs

Data Analysis

Data analysis is a process of collection, presentation and interpretation of the information contained in the data to aid to reach some kind of conclusions. Usually data collection and presentation is further coupled with mathematical and analytical tools to react on further conclusion. Visual representation is the best representation of data because of ease of comprehension. Visual representation data can be primarily represented into following forms.

1. Line-graphs/cartesian graphs
2. Tables
3. Bar diagrams
4. Caselets
6. Some other non-conventional miscellaneous figure

Skills: That individual need to, to handle DI (Data Interpretation) efficiently from exampoint of view.

1. **Data Compresion Skills:** Ability to select relevant information from any given set of data in context of question be asked, based upon that data set.
2. **Ability to differentiate between easy & difficult questions:** It has many aspect for example, how many question in data set requires actual calculation and how many can be answered by observation which are based an patterns. How many questions which are having options can't not be determined or data insufficient and out of such actually for how many question required information is not available.
3. **Ability to apply approximation techniques:** That is based upon the scenario that how far the options are value wise.
4. **Last but not the least:** In entire DI section which strategy can help me to maximize my attempts with higher accuracy.

   Strategy-1: Doing all easy questions from all the given sets (that ask for fast reading, wide and quick eye span and efficient data comprehension from all sets).

   Strategy-2: Doing all the questions from selective and relatively easier DI sets where data compherension is far more easier.
5. Basic command over elementary mathematical calculation tool. Which comprises of following:
   - Percentage and equivalent fractional forms
   - Averages
   - Ratios
   - Approximation

Data Interpretation

Executive member, managers and personnel at different level in the various organizations are require to analyse different sets of datas. Balance sheet, Records of Annual General Meeting, Quarterly Report etc. requires interpretation of various data types. On the basis of the given sets of datas future projections can be made and corrective action can be taken at regular intervals. Mainly there are following sets to represent datas.

- Graph
- Table
- Bar Diagrams
- Pie Charts
- Miscellaneous Figures etc.

Line Graph

Line graph is the simplest way to represent data. Single set or multiple sets of datas can be shown in a graph. Normally following things are required to analyse.

- Increase in profit in absolute terms
- Increase in profit in percentage term
- Growth rate of the given duration
- Average annual growth rate
- Average profit
- Capacity utilisation
- In case of trade, Trade Deficit and Trade Surplus etc.

Advantage: Advantage with line graph is that it gives an ideal about the quantity expressed in this graph between the given time period as well.

Disadvantage: Usually exact values are not available, when we try to represent exact value on the line graph, incase of multiples line representation graph become complex.

Lets have an example of graph

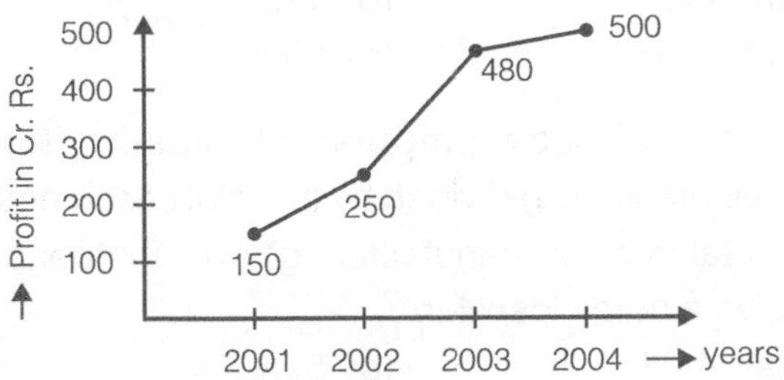

Fig. (I) Balance Sheet of ABC corporation.

On the basis of the given balance sheet of ABC corporation we can calculate following things.

1. Increase in profit in absolute terms

2001-02 250-150 = 100 Cr.

2002-03 480-250 = 230 Cr.

2003-04 500-480 = 20 Cr.

- The highest absolute increment in profit is observed during financial year 2002-03.
- The least absolute increment in profit is observed during financial year 2003-04.

2. Percent increase in profit

2001-02 $\frac{250-150}{150}\times 100 = 66.66\%$

2002-03 $\frac{480-250}{250}\times 100 = 92\%$

2003-04 $\frac{500-480}{480}\times 100 = 4.16\%$

- The highest profit increment in percentage term is for financial year 2002-03.
- The minimum profit increment in percentage term is for financial year 2003-04.

3. Growth rate for the duration 2001-04

$$\Rightarrow \frac{500-150}{150}\times 100 = 233\%$$

[Growth rate for the duration]

$$= \frac{\text{Final years Profit} - \text{Base years profit}}{\text{Base years profit}}$$

4. Average Annual Growth Rate

$$= \frac{\text{Increase in profit for the duration}}{\text{Base years profit}} \times \frac{100}{\text{Number of years}}$$

In the above example

$$\frac{500-150}{150}\times\frac{100}{3}$$

$$= \frac{350}{150}\times\frac{100}{3} = 77.77\%$$

5. Average profit $= \frac{\text{Sum of profits of the duration}}{\text{Total number of years}}$

$$\Rightarrow \frac{150+250+480+500}{4} = \frac{1380}{4} = 345.$$

Capacity Utilisation

$$\text{Capacity utilisation} = \frac{\text{Total Production}}{\text{Total Capcity}} \times 100$$

Example 1.

The total capacity of Tata Nano car plant in Gujrat is 150 car per day. In the month of April 2009 the plant manufactured at the rate of 120 car a day. Find the capacity utilisation in the month of April.

Sol.: Capacity utilisation in April 2009

$$= \frac{120}{150} \times 100 = 80\%$$

Example 2.

The manufacturing capacity of Nokia plant based in Yokohama, Japan is 3,00,000 cellphones per day. In the month of March 2009 there was slump in demand due to recession and capacity utilisation come down to 30%. Find the production of cellphones in March 2009.

Sol.: Capacity utilisation

$$= \frac{\text{Total Production}}{\text{Total Capacity}} \times 100$$

Total production in march 2009

$$= \frac{\text{Capacity Utilisation} \times \text{Total Capacity}}{100}$$

$$\Rightarrow \frac{30 \times 3,00,000}{100}$$

$\Rightarrow$ 90,000 Ans.

I Practice Exercise

Direction (Qs. 1 to 5) : Refer to the graph (Fig.) and answer the questions given below that

(Consumer price index in 1993-1994)

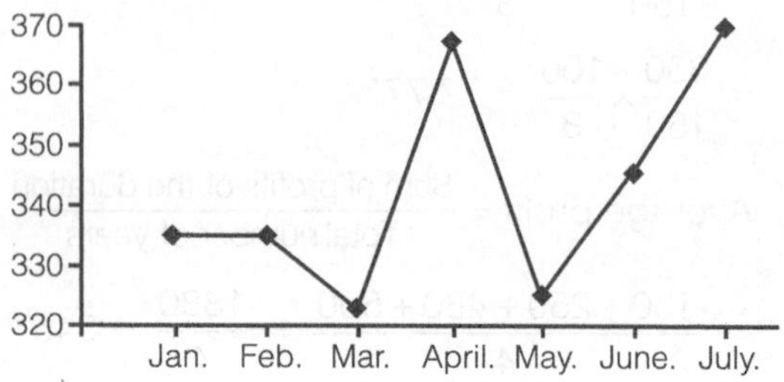

1. Which month showed the highest absolute difference in the Consumer Price Index (CPI) over the previous month?
(a) March (b) April
(c) May (d) July

2. Which month showed the highest percentage difference in the CPI over the previous month?
(a) March (b) April
(c) May (d) July

3. For how many month was the CPI greater than 350?
(a) One (b) Two
(c) Three (d) Four

4. In how many months was there a decrease in the CPI over the previous month?
(a) One (b) Two
(c) Three (d) Four

5. The difference in the number of months in which there was an increase in the CPI and the number of months in which there was a decrease was
(a) One (b) Two
(c) Three (d) Four

Direction (Qs.6 to Q.10) : Study the following graph to answer the given questions:

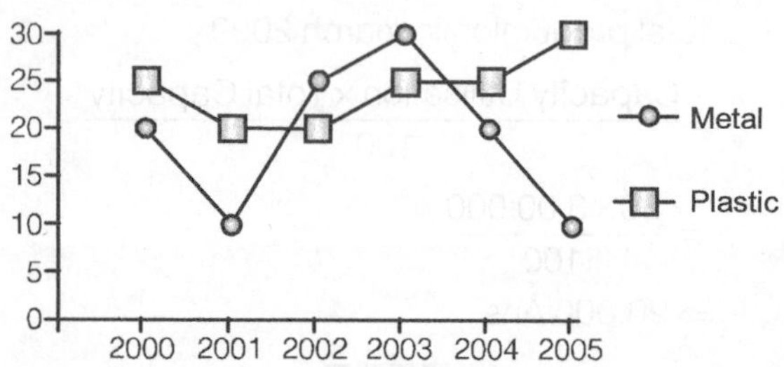

Consumption of Metals versus plastics in the given Years for Car manfucaturing (in thousands tonnes)

The graph shows the trend of consumption of metals and plastic in the production of cars between 2000-05.

6. The number of years for which the consumption of Metals was less than the consumption of Plastic over the given time period was
(a) One (b) Two
(c) Three (d) Four

7. The total consumption of plastic (for car manufacturing) divided by the total consumption of Metal (for car manufacturing) over the period will give a ratio closest to
(a) 4 : 3 (b) 5 : 4
(c) 6 : 5 (d) 7 : 4

8. Which item and for which year shows the highest percentage change in consumption over the previous year?
(a) Metal 2003 (b) Plastic 2003
(c) Metal 2002 (d) Plastic 2005

9. For the two data series shown, how many years have shown decrease in consumption over previous year for both the items?
(a) One (b) Two
(c) Three (d) Four

10. Which year showed the highest percentage decrease in the total consumption of the two?
(a) 2001 (b) 2002
(c) 2004 (d) 2005

Direction (Qs. 11 to 15): Study the following graph to answer the given questions:

Percent profit earned by two companies over the given years:

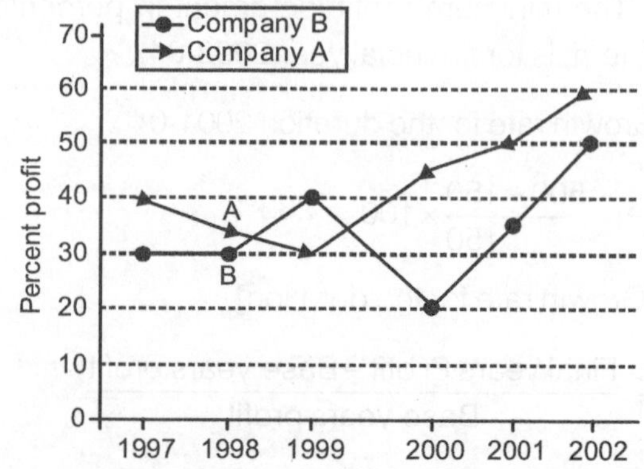

11. If the expenditure of Company B in year 2000 was Rs. 200 crore, what was its income?
(a) Rs. 240 crore
(b) Rs. 220 crore
(c) Rs. 160 crore
(d) Cannot be determined

12. If the income of Company A in year 2002 was Rs. 600 crore, what was its expenditure?
(a) Rs. 360 crore
(b) Rs. 480 crore
(c) Rs. 375 crore
(d) Can't be determined

13. If the income of Company B in 1998 was Rs. 200 crorers, what was its profit in 1999?
(a) Rs. 21.5 crore
(b) Rs. 153 crore
(c) Rs. 46.15 crore
(d) Can't be determined

14. If the income of the two companies in 1998 were equal, what was the ratio of their expenditures?
(a) 1 : 2
(b) 26 : 27
(c) 100 : 67
(d) Can't be determined

15. What is the percent increase in percent profit for company B from year 2000 to 2001?
(a) 75 (b) 175
(c) 42.86 (d) Can't be determined

Direction (Qs. 16 to 20): Study the following graph carefully and answer the questions given below it.

Imports of 3 companies over the years (Rs. in crore)

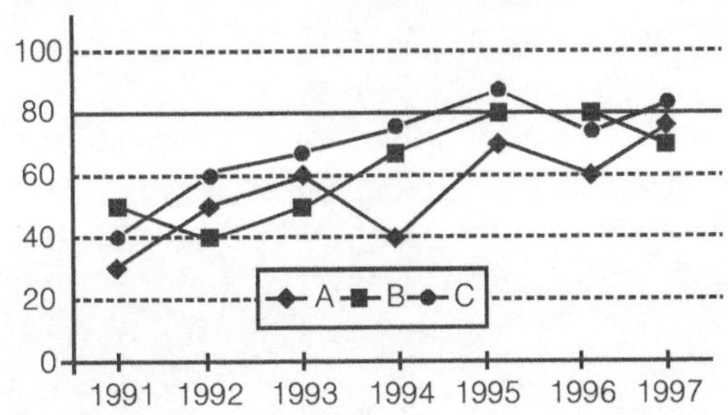

16. In which of the following years, the imports made by Company A was exactly equal to its own average imports?
(a) 1992 (b) 1993
(c) 1994 (d) None of these

17. In which of the following years was the difference between the imports made by Company B and C the maximum?
(a) 1995 (b) 1994
(c) 1991 (d) 1992

18. In which of the following years was the imports made by Company A exactly half of the total imports made by Company B and C together in that year?
(a) 1992 only (b) 1993 only
(c) 1992 and 1993 (d) 1995 only

19. What was the percentage increase in imports of Company B from 1992 to 1993?
(a) 10 (b) 25
(c) 40 (d) 20

20. In which of the following years was the total imports made of all the three companies together the maximum?
(a) 1996 only (b) 1997 only
(c) 1995 only (d) 1995 & 1997 only

Answers

1. (b)	2. (b)	3. (b)	4. (b)	5. (a)
6. (d)	7. (b)	8. (c)	9. (a)	10. (a)
11. (a)	12. (c)	13. (d)	14. (b)	15. (d)
16. (d)	17. (d)	18. (a)	19. (b)	20. (c)

Solutions

Answer 1 to 5:

1. (b)
2. (b) Visually clear that it is April
3. (b) April and July – Two.
4. (b) The CPI drecreased in March and May
5. (a) The CPI increased in three months (April , June and July) while it decreased in two months (March and May).
6. (d) Visully seen as 4. 2000, 2001, 2004 & 2005.
7. (b) 145 : 115 = 29 : 23 is closest to 5 : 4
8. (c) Metals in 2002 is more than double over it's value in 2001.
9. (a) 2001 is the only year which satisfy the condition.
10. (a) 33.33% between 200 to 2001.
11. (a) Income of Company B in 2000

$= 200 \times \frac{120}{100} =$ Rs. 240 cr.

12. (c) Expenditure of Company A in 2002

$= 600 \times \frac{100}{160} =$ Rs. 375 cr.

13. (d) We can find out the amount of profit in 1998, we do not know the income and expenditure of A and B therefore option (d) is the correct choice.

14. (b) Ratio of their expenditures

$= \frac{100}{135} \times \frac{130}{100} = 26. : 27.$

15. (d) Reqd. % increase $= \frac{35-20}{20} \times 100 = 75\%.$

16. (d) Average imports made by company A

$= \frac{30+50+60+40+70+60+75}{7}$

$= \frac{385}{7} = 55$

In none of the given years the imports is exactly equal to 55 (crore). Hence, the answer is (d)

17. (d) By visual inspection it is clear that 1992 is the desired year (as the distance between two points is the maximum in 1992).

18. (a) By observation, in year 1992

$\left(\text{as } 50 = \frac{40+60}{2}\right)$

So 1992 is the desired year. We do not need any calculation see the year where the point A lies exactly in the middle of points of B and C

19. (b) Reqd percentage increase $= \frac{50-40}{40}$

= 25%

20. (c) The total imports (in crore) made by all the three companies together : From the height of the points we observe that the total heights of three points is the maximum either in 1995 or 1997. If you observe carefully our clear answer is 1995, but to be sure we find actual values for the two years.

In 1995 = 70 + 80 + 85 = 235

In 1997 = 75 + 70 + 85 = 230

Clearly, 1995 is the desired year.

3.6 CHAPTER

Table

A table serves both needs of visual communication and arranging data.

Advantage: Huge amount of data can be put in a tabular format. As exact values are available there is no possibility of approximation error.

Disadvantage: Locating pattern, trend or visual observations are difficult.

I Practice Exercise

Direction (Qs. 1 to 5): Study the following Table carefully and answer the questions given below

Production of main crops in India (in million tonnes)

Crops	91-92	92-93	93-94	94-95	95-96	96-97
Pulses	20.5	22.4	24.6	23.5	27.8	28.2
Oilseeds	32.4	34.6	40.8	42.4	46.8	52.4
Rice	80.5	86.4	88.2	92.6	94.2	90.8
Sugarcane	140.8	150.2	152.2	160.3	156.4	172.5
Wheat	130.2	138.4	146.8	141.6	152.2	158.4
Coarse grain	45.6	52.8	60.4	62.2	58.2	62.8
Total	450	484.8	513.2	522.8	535.6	565.1

1. Production of sugarcane in 1993-94 was approximately what percentage of production of rice in 1992-93?
   (a) 50 (b) 75
   (c) 150 (d) 175
2. Production of which type of crop was continuously increasing in the given years?
   (a) Rice (b) Pulse
   (c) Sugarcane (d) Oilseeds
3. What was the average production of pulse in the given years?
   (a) 26.8 million tonnes
   (b) 20.5 million tonnes
   (c) 24.5 million tonnes
   (d) 22.5 million tonnes
4. Production of oilseeds was what percentage of the total crops produced in the year 1991-92?
   (a) 7.2 (b) 8.4
   (c) 2.7 (d) 6.4
5. In which of the following years the total production of oilseeds in the years 1994-95, 1995-96 and 1996-97 was equal to the production of wheat?
   (a) 1993-94 (b) 1994-95
   (c) 1996-97 (d) 1992-93

Direction (Qs. 6 to 9): The table given below shows production of five types of cars by a company in the years 1989 to 1994. Study the table and answer questions.

Production of cars by a company

Year→ Type ↓	1989	1990	1991	1992	1993	1994	Total
P	8	20	16	17	21	6	88
Q	16	10	14	12	12	14	78
R	21	17	16	15	13	8	90
S	4	6	10	16	20	31	87
T	25	18	19	30	14	27	133
Total	74	71	75	90	80	86	476

6. In which year the total production of cars of types P and Q together was equal to the total production of cars of types R and S together?
   (a) 1990 (b) 1991
   (c) 1994 (d) None of these
7. During the period 1989-94, in which type of cars was a continuous increase in production?
   (a) P (b) Q
   (c) R (d) S
8. The production of which type of cars was 25% of the total production of all types of cars during 1993?
   (a) S (b) R
   (c) Q (d) P

9. The per cent increase in total production of all types of cars in 1992 to total in 1991 was?
(a) 15 (b) 20
(c) 25 (d) 30

Direction (Qs. 10 to 14): The following five questions are to be answered on the basis of the following table:

Weight Distribution in the Average Adult

Organs	Weight (in grams)
Muscles	30,000
Skeleton	10,000
Blood	5,000
Gastrointestinal Tract	2,000
Lungs	1,000
Liver	1,700
Brain	1,500

10. The total body weight of the average adult is
(a) 70,000 grams
(b) More than 51 kg
(c) 50,000 grams
(d) Less than 50 kg

11. If the weight of the skeleton is represented as S, then the weight of the liver can be represented as
(a) 1.7 S (b) 0.17 S
(c) 17 S (d) 71 S

12. The ratio expressed in decimals of the weight of the blood to the weight of the gastrointestinal tract is
(a) 0.4 (b) 4.0
(c) 2.5 (d) 0.25

13. The ratio expressed in decimal for weight of the brain to the weight of the mucles is
(a) 0.50 (b) 0.15
(c) 0.20 (d) 0.005

14. The ratio decimal of the weight of the brain to the weight of the lungs is
(a) 1.5 (b) 0.15
(c) 15.0 (d) 5.1

Direction (Qs. 15 to 19): The figures for a country's Foreign Trade for the years 1990-91 to 1996-97 are given in the following table. Answer these questions on the basis of the information given:

A Country's Foreign Trade (Rupees in Crores)

Year	Exports	Imports	Trade Deficit
1990-91	6711	12549	5838
1991-92	7806	13608	5802
1992-93	8803	14293	5490
1993-94	9771	15831	6060
1994-95	11855	17173	5318
1995-96	10420	18371	7951
1996-97	12550	20063	7513

15. Which of the following showed an increase every year?
(a) Exports (b) Imports
(c) Trade deficit (d) All of these

16. The ratio of imports to exports was maximum in the year
(a) 1990-91 (b) 1996-97
(c) 1995-96 (d) 1992-93

17. The percentage increase in exports over previous year was maximum in the year
(a) 1990-91 (b) 1996-97
(c) 1994-95 (d) 1993-94

18. The total trade deficit for the last five years?
(a) Rs 28,508 crore (b) Rs. 32,332 crore
(c) Rs. 44,322 crore (d) Rs. 33,232 crore

19. The difference between imports and exports was maximum in the year
(a) 1995-96 (b) 1996-97
(c) 1994-95 (d) 1993-94

Direction (Qs. 20 to 25): Information given below shows % of students, who applied for selection in a company. For final selection one has to pass through five major stages written test, aptitude test, group task, group discussion, interview & a medical checkup in the same order. At every stage only those candidates who fulfill the required standards are qualified for next stage. Table below given the region wise details.

Task / Region	Written test	Aptitude	Group Task	Group Discussion	Inter view	Medical
West	70%	90%	85%	50%	90%	90%
North	68%	88%	90%	60%	85%	90%
South	80%	85%	80%	70%	70%	80%
North-east	75%	95%	78%	40%	65%	85%
Central	70%	80%	68%	50%	60%	70%

For Example: 70% of the people from central region who appeared for written test have qualified for aptitude test & 80% of those have taken aptitude are qualified for group test and so on.

20. If the candidate appearing for written test from Northern & southern region are in ratio of 4 : 5 what is the approximate ratio of people who will appeared for final interview from these region respectively,
(a) 13 : 19 (b) 19 : 13
(c) 16 : 19 (d) 19 : 16

21. If equal number of candidates are called for group task from each region, then find the region from where maximum candidate. Participated in selection process.
(a) North (b) South
(c) West (d) Central

22. If equal no. of candidates are called for group task from each region, then find the region from where minimum candidate will be able to clear their medical test
(a) Northeast (b) West
(c) Central (d) None of these

23. If candidates from South & North who passed the group discussion are in ratio 9 : 8 and finally from South 1000 candidates clear the medical checkup then approx. How many candidates from North have cleared their medical checkup?
(a) 1300 (b) 1200
(c) 1400 (d) 1600

24. If same number of candidates appeared for written test from all regions then from which region final selection % is best
(a) South (b) West
(c) North (d) Central

25. If finally same number of candidates have been selected from each region, then from which region success rate was best in % terms
(a) South (b) West
(c) North (d) Central

Answers

1. (d)	2. (d)	3. (c)	4. (a)	5. (b)
6. (d)	7. (d)	8. (a)	9. (b)	10. (b)
11. (b)	12. (c)	13. (d)	14. (a)	15. (b)
16. (a)	17. (c)	18. (b)	19. (a)	20. (a)
21. (d)	22. (a)	23. (b)	24. (c)	25. (c)

Solutions

Answer 1 to 5:

1. (d) Required per cent = $\frac{152.2}{86.4} \times 100 \approx 175\%$.
2. (d)
3. (c) Average production of pulse
$= \frac{20.5+22.4+24.6+23.5+27.8+28.2}{6}$
$= \frac{147.0}{6} = 24.5$ million tonne.
4. (a) Required percentage = $\frac{32.4}{450} \times 100$
$= 7.2\%$
5. (b) Total production of oilseeds in the given years
$= 42.4 + 46.8 + 52.4 = 141.6$
Which is equal to the production of wheat in 1994-95.
6. (d) In year be 1993
$P + Q = 33 = R + S$
7. (d)
8. (a) 25% of 80 = 20 = production of car S in 1993
9. (b) Required per cent increase
$= \frac{90-75}{75} \times 100 = 20\%$
10. (b) 51200 gm = 51 kg and 200 gms
11. (b) S = 10000
∴ Weight of Liver = 1700
$= \frac{S}{10000} \times 1700 = 0.17\,S$
12. (c) $\frac{5}{2} = 2.5$
13. (d) $\frac{1500}{30000} = \frac{1}{20} = 0.05$
14. (a) $\frac{1500}{1000} = \frac{3}{2} = 1.5$.
15. (b)
16. (a) The ratio of imports to exports in the year
1990-91 = $\frac{12549}{6711} = 1.87$
1996-97 = $\frac{20063}{12550} = 1.60$
1995-96 = $\frac{18311}{10420} = 1.76$

$$1992\text{-}93 = \frac{14293}{8803} = 1.62$$

17. (c) Percentage increase in exports in

$$1996\text{-}97 = \frac{12520 - 10420}{10420} \times 100 = 20.15\%$$

$$1994\text{-}95 = \frac{11855 - 9771}{9771} \times 100 = 21.33\%$$

$$1993\text{-}94 = \frac{9771 - 8803}{8803} \times 100 = 11\%$$

18. (b) 5490 + 6060 + 5318 + 7951 + 7513
= Rs. 32332 crore.

19. (a) The difference between the imports in
1995-96 = Rs. 7951 crore
1996-97 = Rs. 7513 crore
1997-98 = Rs. 5318 crore
1993-97 = Rs. 6060 crore

20. (a) No. of students from northern region = $4x$
No. of students appeared for final interview
= $4x \times 0.68 \times 0.88 \times 0.9 \times 0.6 = 1.2925x$
Similarly no. of students for final interview from southern region
= $5x \times 0.8 \times 0.85 \times 0.80 \times 0.70 = 1.904x$
Hence required ratio = 13 : 19. Option (a)

21. (d) Suppose number of candidates from west be equal to x. So $x \times 0.7 \times 0.9 = 0.63x$ candidate will appear for group test. Hence lower the multiplication factor of x we will have higher value of x. Hence maximum candidate appeared from central region. Option (d)

22. (a) From table, Smallest multiplication factor for interview will be for Northeast. Hence option (a).

23. (b) Say number of candidates from south who cleared group discussion is $9x$, the number of candidate who cleared medical test.
= $9x \times 0.7 \times 0.8 = 1000$

$$x = \frac{1000}{9 \times 0.56}$$

No. of candidates from North who cleared medical test

$$= 8 \times \frac{1000}{9 \times 0.56} \times 0.85 \times 0.9$$

= 1214 approximately option (b).

24. (c) From table it is clear final multiplication factor is highest corresponding to North.
Hence option (c).

25. (c) North, hence option (c).

■■■■

CHAPTER 3.7

Bar Diagrams

Bar Diagrams

There are mainly six types of Bar Diagrams.

- Simple Bar Diagram
- Multiple Bar Diagram
- Compound Bar Diagram
- Percent Bar Diagram
- Horizontal Bar Diagram
- Floating Bar Diagram, etc.

Advantage

Can represent many different categories. We can directly compare multiple categories.

Disadvantage

Parameters of different entity need to approximate the values.

Simple Bar Diagram

Simple Bar Diagram is similar to Line graph. Only one set of data can be represented in simple Bar Diagram. It is different to Line graph only in an aspect that in place of points there are bars which represent values of respective years.

Ex. 1

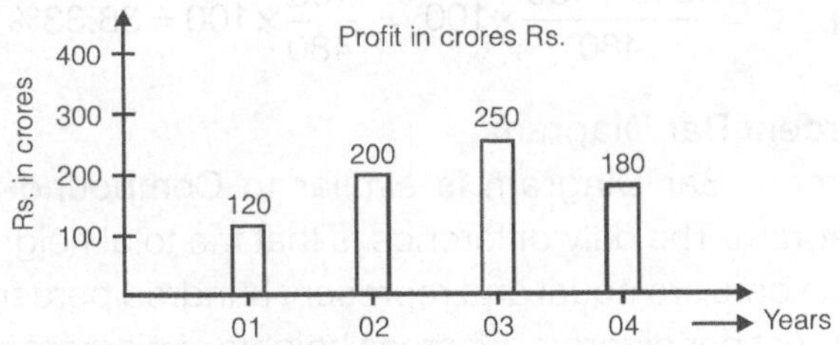

Fig. (1) Balance Sheet of MTS incorporation

Sol.: From the Bar Diagram given in Fig. (1) We can determine following things

1. Percent increment in profits

$$2001-02 \Rightarrow \frac{80}{120}\times 100 = 66.66\%$$

$$2002-03 \Rightarrow \frac{50}{200}\times 100 = 25\%$$

$$2003-04 \Rightarrow \frac{-70}{250}\times 100 = -28\%$$

(– Sign indicates decrement)

- Maximum increase in profit was observed in financial year 2001-02.
- Minimum increase in profit was observed in financial year 2003-04.

2. Average profit $= \frac{120+200+250+180}{4}$

$$= \frac{750}{4} = 187.5 \text{ crore}$$

3. Annual average growth rate from 2001-2004

$$= \frac{180-120}{120}\times\frac{100}{3}$$

$$= \frac{60}{120}\times\frac{100}{3} = 16.66\%$$

Multiple Bar Diagram

In Simple Bar Diagrams only one set of data can be represented. To represent multiple set of data for different years we use multiple bar diagrams.

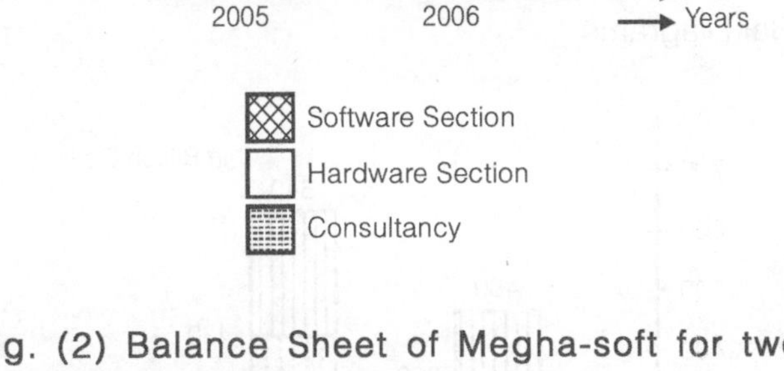

Fig. (2) Balance Sheet of Megha-soft for two financial years 2005 and 2006.

From the multiple bar diagram shown in Fig. (2) we can calculate following things

1. Sectional Growth

(a) Software section 180 – 150 = 30 Cr.

(b) Hardware section 400 – 350 = 50 Cr.

(c) Consultancy 260 – 200 = 60 Cr.

- Highest absolute increase in profit is observed by consultancy section.
- Minimum absolute increase in profit is observed by software section

2. Growth rates in percent terms.

$$\text{Software section} \Rightarrow \frac{180-150}{150}\times 100 = 20\%$$

$$\text{Hardware section} \Rightarrow \frac{400-350}{350}\times 100 = 14.28\%$$

$$\text{Consultancy} \Rightarrow \frac{260-200}{200}\times 100 = 30\%$$

- Minimum growth is observed by Hardware section
- Maximum growth is observed by consultancy section.

3. Growth rate of megha soft for the duration 2005-06.

$$\frac{\text{Profit in 2006} - \text{Profit in 2005}}{\text{Profit in 2005}}\times 100$$

$$\Rightarrow \frac{(180+400+260)-(150+350+200)}{(150+350+200)}\times 100$$

$$\Rightarrow \frac{840-700}{700}\times 100$$

$$= \frac{140\times 100}{700} = 20\%$$

Compound Bar Diagram

Compound Bar Diagrams are similar to Multiple Bar Diagrams. The only difference between these two Bar Diagrams is that in Compound Bar Diagram a single Bar Diagram is subdivided into different parts, while multiple bars are used in Multiple Bar Diagrams. Compound Bar Diagrams are also known as sub-divided Bar Diagrams.

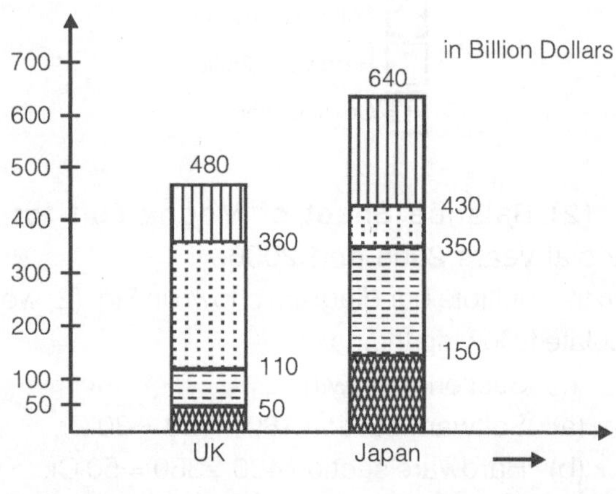

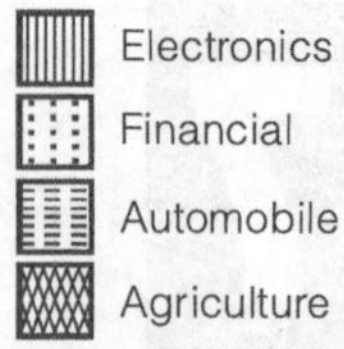

Fig. Export Basket of two countries UK and Japan for year 2005 in $ billion

From the Compound Bar Diagram we answer following set of questions.

Example 1.

Export of Agriculture product of UK is how much percent of export of Automobile product of Japan?

Sol. Export of Agriculture product of UK

= $ 50 billion

Export of Automobile product of Japan

= $ 200 billion

$$= \frac{50}{200}\times 100 = 25\%$$

Example 2.

Export of Automobile product of Japan is how many times of automobile product.

Sol. Automobile Export of Japan

= $ 200 billion

Automobile Export of UK = $60 billion

$$\frac{260}{60} = \frac{10}{3} = 3.33 \text{ times}$$

Example 3.

Total export given in Fig. of Japan is how much percentage higher than that of UK?

Sol. $= \frac{640-480}{480}\times 100 = \frac{160}{480}\times 100 = 33.33\%$

Percent Bar Diagram

Percent Bar Diagram is similar to Compound Bar Diagram. The only difference is that the total height of each bars are equal and represent hundred percent in percent bar diagram, whereas heights of different bars in compound bar diagram may or may not be equal.

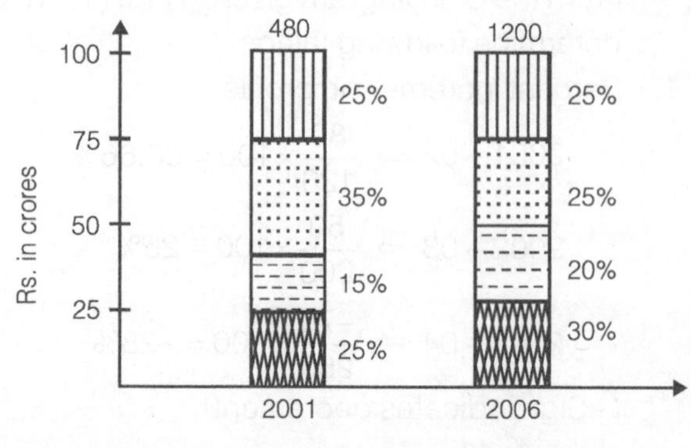

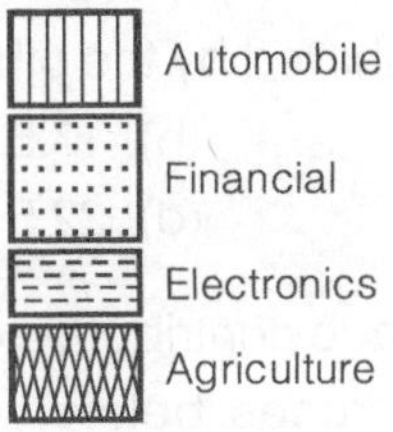

Fig. Export Basket of XYZ for financial year 2001 and 2006 in billion dollars.

From the above figure we can calculate following things.

(1) Percent increment in export for the given duration 2001-06

$$\frac{1200-480}{480}\times 100 = 150\%$$

(2) Export of Japan for financial year 2001 consist of

- \$ 120 billion for Automobiles
- \$158 billion for Financial services
- \$ 72 billion for Electronic Goods
- \$120 billion for Agriculture product

(3) Export Basket of Japan for financial year 2006 consists of

- \$ 360 billions for Automobiles
- \$ 300 billion for Financial services
- \$ 240 billions for Electronic goods
- \$ 300 billion for Agriculture product

(4) Export of financial services in year 2006 is how many times more than export of electronic goods in 2001?

Sol. Export of financial services in 2002 = 300

Export of electronic goods in 2001 = 72

$$\frac{300-72}{72} = 3.16 \text{ times approx.}$$

Horizontal Bar Diagrams

Horizontal Bars are similar to simple bar diagrams. The only difference is that in case of horizontal bar, representation is horizontal, whereas vertical bars are used for simple bar diagram. Horizontal bar diagram is used in cases when we requires to represent data sets for comparatively large number of years. All the calculation and observation of both the Bar Diagrams are similar.

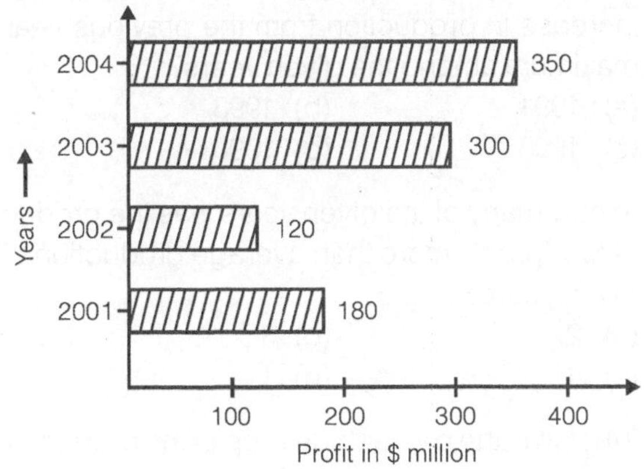

Fig. (5) Profits of Tamasha.com for four consecutive financial year.

□□□□

Practice Exercise: I

Direction (Qs. 1 to 5): Study the following graph carefully and answer the questions given below:

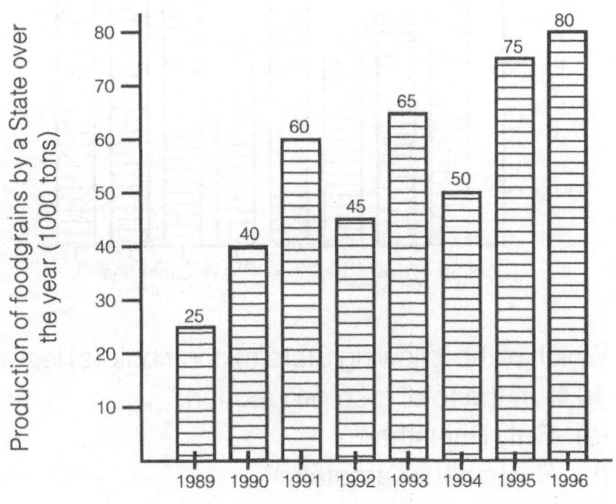

1. The average production of 1990 and 1991 was exactly equal to the average production of which of the following pairs of years?

(a) 1991 and 1992 (b) 1992 and 1994

(c) 1993 and 1994 (d) None of these

2. What was the difference in the production of foodgrains between 1991 and 1994?

(a) 10000 tons (b) 15000 tons

(c) 500 tons (d) 5000 tons

3. In which of the following years was the percentage increase in production from the previous year the maximum among the given years?
(a) 1991 (b) 1993
(c) 1995 (d) 1990

4. In how many of the given years was the production of foodgrains more than average production of the given years?
(a) 2 (b) 3
(c) 4 (d) 1

5. What was the percentage drop in the production of foodgrains from 1991 to 1992?
(a) 15 (b) 20
(c) 25 (d) 30

Direction (Qs. 6 to 10): Study the following graph carefully and answer the questions given below :

Rose Production

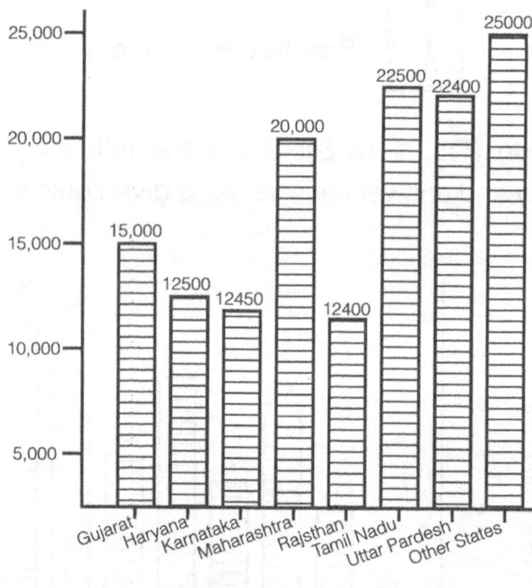

6. Which of the following State(s) contribute(s) less than 10 % in the total rose production?
(a) Only Rajasthan
(b) Rajasthan, Karnataka
(c) Rajasthan, Karnataka, Haryana
(d) Rajasthan, Karnataka, Haryana and Gujarat

7. By what percentage rose production of other States is more than that of the Maharashtra?
(a) 25 (b) 30
(c) 20 (d) 15

8. What is the approximate average production of roses (in thousands) across all the states?
(a) 21 (b) 20
(c) 19 (d) 18

9. Approximately what percentage of the total rose production is shared by the other States?
(a) 12.5 (b) 17.5
(c) 19.5 (d) 22.5

10. If total percentage contribution of the States having production of roses below twenty thousand is considered, which of the following statement is true?
(a) It is little above 40%
(b) It is approximatey 36.8%
(c) It is below 35%
(d) It is little below 30%

Direction (Qs. 11 to 15): These questions are based on the following bar graph. Read the graph and answer the questions

Finances of XYZ Railway

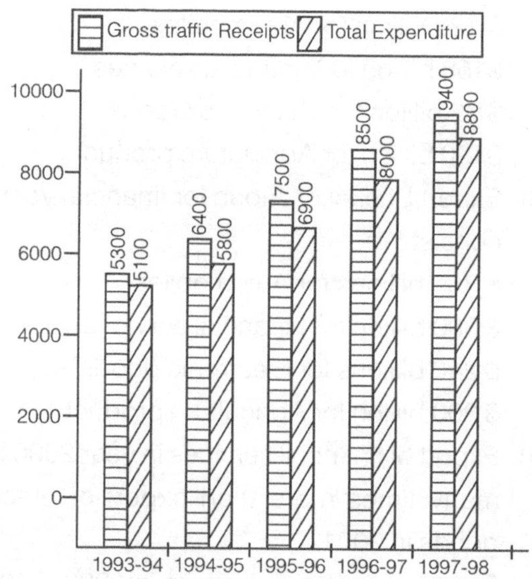

11. What is the percentage increase in the gross traffic receipts in 1995-96 as compared to 1993-94?
(a) 33.9% (b) 29.3%
(c) 20.7% (d) 17%

12. If profit = gross traffic receipts–total expenditure, then in 1996-97 what percentage of gross traffic receipts is the profit made?
(a) 5.9% (b) 6.4%
(c) 7.2% (d) 8%

13. In which year was the profit as a percentage of gross traffic receipts the highest?

(a) 1997-98 (b) 1996-97
(c) 1995-96 (d) 1994-95

14. In order to make a profit of 10%, what should have been the gross traffic receipts (in Rs. crore) in 1994-95, total expenditure remaining the same?
(a) 5667 (b) 5876
(c) 6444 (d) 7667

15. By what amount (in Rs. crore) has the expenditure increased over the period 1993-94 to 1997-98?
(a) 4100 (b) 3900
(c) 3850 (d) 3700

Direction (Qs. 16 to 20): Study the following graph to answer these questions:

Tea in India (in million kg)

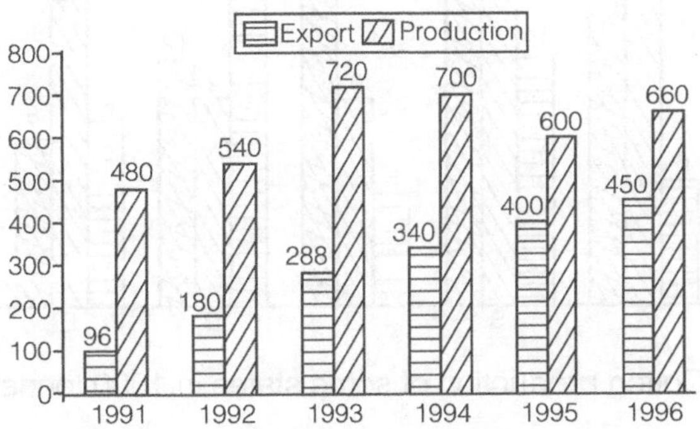

Per Capita Availablity in gm

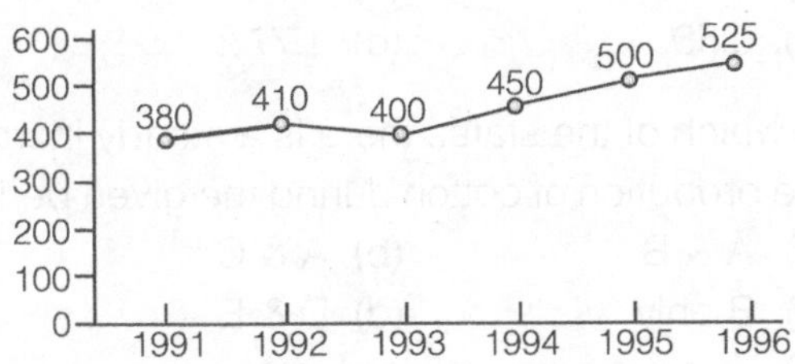

16. Which year shows the maximum percentage of export with respect to production?
(a) 1992 (b) 1993
(c) 1996 (d) 1995

17. Assuming that entire quantity is consumed after export, the population of India in 1993 was
(a) 800 million (b) 1080 million
(c) 985 million (d) 900 million

18. The proportion of tea exported to the tea production over the entire period
(a) 0.87 (b) 0.47
(c) 0.58 (d) 0.66

19. In which year, there was minimum percentage of export with respect of production?
(a) 1991 (b) 1992
(c) 1993 (d) 1994

20. In which year we had maximum quantity of tea for domestic consumption
(a) 1994 (b) 1991
(c) 1993 (d) 1996

Answers

1. (d)	2. (a)	3. (d)	4. (c)	5. (c)
6. (c)	7. (a)	8. (d)	9. (b)	10. (b)
11. (b)	12. (a)	13. (d)	14. (c)	15. (d)
16. (c)	17. (b)	18. (b)	19. (a)	20. (c)

Solutions

Answer 1 to 5:

1. (d)

2. (a) Required difference
$= 60 - 50 = 10{,}000$ tonne.

3. (d) Percentage increase in production in 190 is height given by
$$= \frac{15}{25} \times 100 = 60\%$$

4. (c) Average production
$$= \frac{25+40+60+45+65+50+75+80}{8}$$
$$= \frac{440}{8} = 55.$$
More than the average in 4 years namely 91, 93, 95, 96.

5. (c) Required percentage drop
$$= \frac{60-45}{60} \times 100 = 25\%$$

6. (c) Total rose production
$= (15 + 12.5 + 12.45 + 20 + 12.4 + 22.4 + 22.5 + 25) \times 1000 = 142250$
Percentage production of rose in the 3 States (is lower than 10%)

Rajasthan	Karnataka	Haryana
8.71%	8.75%	8.78%

7. (a) Required percentage

$$= \frac{25-20}{20} \times 100 = 25\% \text{ (more)}$$

8. (d) Total production of rose by all the states = 142250

$$\therefore \text{Average} = \frac{142250}{8 \times 1000} = 18 \text{ (thousand)}$$

9. (b) Required percentage

$$= \frac{25}{142.25} \times 100 \approx 20\%$$

10. (b) It is 36.8% approximately.

11. (b) $\frac{7500-5300}{7500} \times 100 = 29.33\%$

12. (a) $\frac{8500-8000}{8500} \times 100 = 5.88\%$

13. (d)

14. (c) Let GTR be Rs. x

$\therefore x - x \times 10\% = 5800$

$\therefore x = \frac{5800 \times 10}{9} = \text{Rs. } 6444.4$

15. (d) 8800 – 5100 = 3700.

16. (c) As per options

In 1992, $\frac{180}{540} \times 100 = 33\frac{1}{3}\%$

In 1993 $\frac{288}{720} \times 100 = 40\%$

In 1995, $\frac{400}{600} \times 100 = 66.66\%$

In 1996, $\frac{450}{660} \times 100 = 68.18\%$

17. (b) 720 – 288 = 432 million = Total consumption of tea in 1993

$$\text{Population of India} = \frac{432}{0.4} = 1080 \text{ million.}$$

18. (b) Average tea exported during 1991–1996 = 1754 million kg. Average tea produced during 1991–1996 = 3700 million kg

$$\therefore \text{ Required ratio} = \frac{1754}{3700} = 0.47.$$

19. (a)

20. (c)

In 1991, 480 – 96 = 384 million tonnes

In 1993, 720 – 288 = 432 millin tonnes

In 1994, 700 – 340 = 360 million tonnes

In 1996, 660 – 450 = 210 million tonnes

□□□□

II Practice Exercise: II

Direction (Qs. 1 to 5): Study the diagram carefully and answer the questions given below it

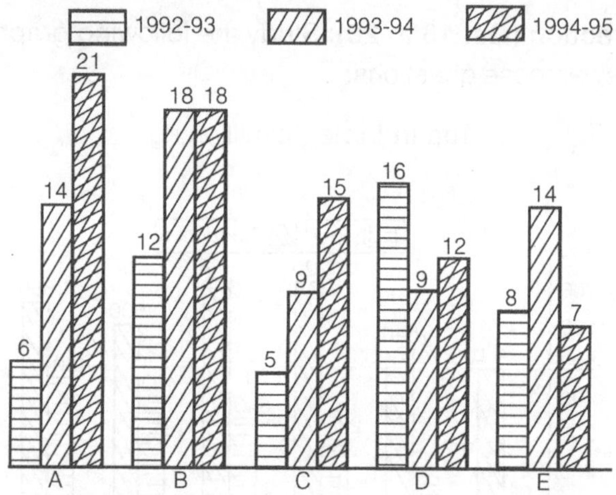

Cotton production of some states in,1000 tonnes

1. The production of state D in 1993-94 is how many times its production in 1994-95?
   (a) 1.33 (b) 0.75
   (c) 0.56 (d) 1.77

2. In which of the states there is a steady increase in the prodution of cotton during the given period?
   (a) A & B (b) A & C
   (c) B only (d) D & E

3. How many tonnes of cotton was produced by state E during the given period?
   (a) 2900 (b) 29000
   (c) 290000 (d) 29000000

4. How many states showing below average production in 1992-93 showed above average poduction in 1993-94?
   (a) 4 (b) 2
   (c) 3 (d) 1

5. Which of the following statement is in false?
   (a) States A and E showed the same production in 1993-94

(b) There was no improvement in the production of cotton in state B during 1994-95.
(c) State A has produced maximum cotton during the given period
(d) Production of States C and D together is equal to that of state B during 1993-94

Direction (Qs. 6 to 10): Study the diagram carefully and answer the question given below it

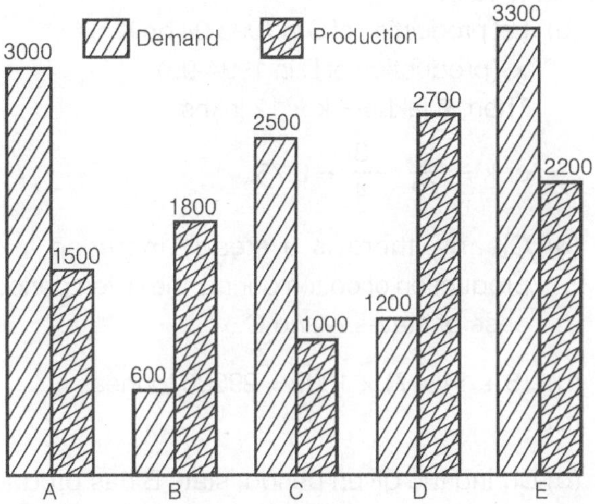

6. What is the ratio of companies having more demand than production to those having more production than demand?
(a) 2 : 3 (b) 4 : 1
(c) 2 : 2 (d) 3 : 2

7. What is the difference between average demand and average production of the five companies taken together?
(a) 1400 (b) 400
(c) 280 (d) 138

8. The production for company D is approximately how many times that of production of company A?
(a) 1.8 (b) 1.5
(c) 2.5 (d) 1.11

9. The demand for company B is approximately what percent of the demand for company C?
(a) 4 (b) 24
(c) 20 (d) 60

10. If company A desires to meet the demand by purchasing T.V. sets from a single company, which one of the following companies can meet the need adequately?
(a) B (b) C
(c) D (d) None of these

Direction (Qs. 11 to 15): Study the following bar charts (Figs) before answering the questions.

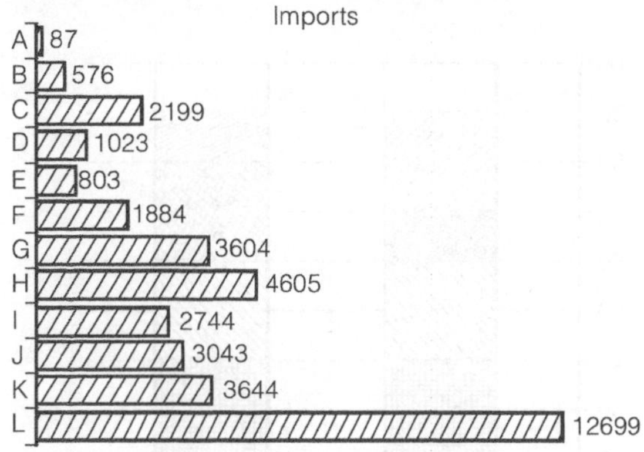

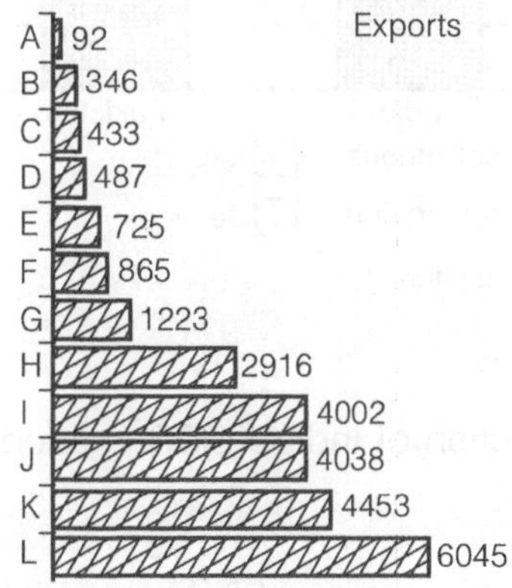

Fig. Foreign Trade by countries for year 1993-94

11. How many countries exhibited a trade surplus?
(a) 5 (b) 4
(c) 3 (d) 6

12. The higest trade deficit was shown by which country?
(a) C (b) G
(c) G (d) L

13. The ratio of Exports to Imports was highest for which country?
(a) A (b) I
(c) J (d) K

14. The total trade/surplus for all the countries put together was?
(a) 11286 surplus (b) 11286 deficit
(c) 10286 deficit (d) None of these

15. The ratio of the maximum exports to the minimum imports was closest to

(a) 64 (b) 69
(c) 74 (d) None of these

Direction (Qs. 16 to 20): Shows the compositions of the GDP of two countries (India and Pakistan).

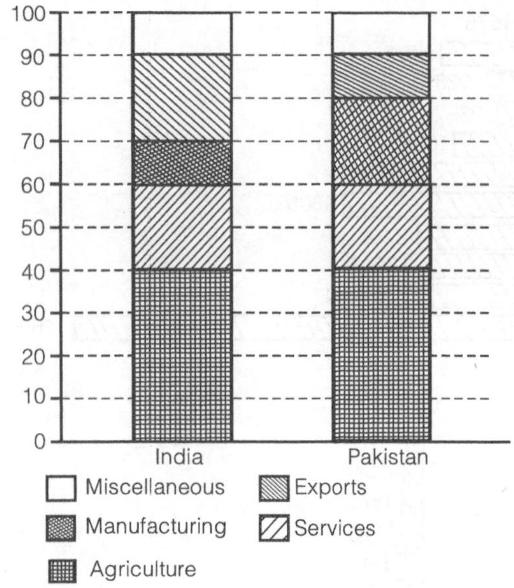

16. What fraction of India's GDP is accounted for by services?
(a) 6/33 (b) 1/5
(c) 2/3 (d) None of these

17. If the total GDP of Pakistan is Rs. 10,000 crore, then the GDP accounted by Manufacturing is
(a) Rs 200 crore (b) Rs 600 crore
(c) Rs 2,000 crore (d) Rs 6,000 crore

18. The total GDP of India is Rs 30,000 crores, then the GDP accounted for By Agriculture, Services and Miscellaneous is
(a) Rs. 18,500 crore (b) Rs 18,000 crore
(c) Rs. 21,000 crore (d) Rs. 15,000 crore

19. Which country accounts for higher earning out of Services and Miscelianeous together?
(a) India
(b) Pakistan
(c) Both spend equal amounts
(d) Cannot be determined

20. If the total GDP is the same for both the countries then what percentage is Pakistan's income through agriculture more than India's income through services?
(a) 100% (b) 200%
(c) 133.33% (d) None of these

Answers

1. (b)	2. (b)	3. (b)	4. (b)	5. (c)
6. (d)	7. (c)	8. (a)	9. (b)	10. (c)
11. (b)	12. (d)	13. (b)	14. (b)	15. (b)
16. (b)	17. (c)	18. (c)	19. (d)	20. (a)

Solutions

Answer 1 to 5:

1. (b) Let production of D in 1993-94 be K × (production of D in 1994-95).
Then, 9 lakhs = k × 12 lakhs
or $k = \frac{9}{12} = \frac{3}{4} = 0.75$.

2. (b) Clearly, there is a steady increase in the production of cotton during the given period in case of states A and C.

3. (b) (8 + 14 + 7) × 1000 = 29000 tonness

4. (b)

5. (c) During the given period, state B has produced 48000 tonnes while state A has produced only 41000 lakhs tonnes.
Hence, statement (c) is false

6. (d) The companies having more demand than production are A, C, E. Their number is 3. The companies having more production than demand are B and D. Their number is 2.

7. (c) Average demand
$= \left(\frac{3000 + 600 + 2500 + 1200 + 3300}{5}\right)$
= 2120.
Average production
$= \left(\frac{1500 + 1800 + 1000 + 2700 + 2200}{5}\right)$
= 1840.
∴ Difference in having average demand & production = (2120 – 1840) = 280.

8. (a) Let production of D
= k × production of A
Then, $2700 = k \times \frac{2700}{1500} = 1.8$.

9. (b) Let x% of demand for C = Demand for B.
Then, $\frac{x}{100} \times 2500 = 600$

or $x = \left(\frac{600 \times 100}{2500}\right) = 24\%$

10. (c) Since D produces highest number of sets and A desire to meet the demand by purchasing surplus sets from a single company, So, D can meet the demand of A.

11. (b) Out of a total of 12 countries, 8 showed a deficit while 4 showed a surplus.

12. (d) Visually it is clear that L has highest trade deficit.

13. (b) I has a ratio of 4002/2744 = 1.45, which is the highest.

14. (b) Sum of exports-sum of imports = deficit (11286)

15. (b) 6045/87

16. (b) Service accounts for 20%, i.e., 1/5th of the GDP of India

17. (c) 20% of 10000 = 2000

18. (c) (40 + 20 + 10) % of 30,000 = Rs. 21,0000 crore.

19. (d) Although the percentage on Service and Miscellaneous put together is equal for both the countries, but we cannot comment on this since we have no data about the respective GDPs.

20. (a) Since the GDP is same, the answer will be got by (40 – 20)/20 = 100%

■■■■

3.8 CHAPTER

Pie-Chart

In a Pie-chart

$$100\% \equiv 360^\circ$$
$$1\% = 3.6^\circ$$

or $$1\% = \left(\frac{18}{5}\right)$$

or vice versa $$1^\circ = \left(\frac{5}{18}\right)\%$$

Pie-Charts are also known as circuilar charts. It represent data in a very simple and lucid form such that interpretation become easier. Let's have a glimpse of Indian Economy through Pie-Charts.

Pie-Chart:I

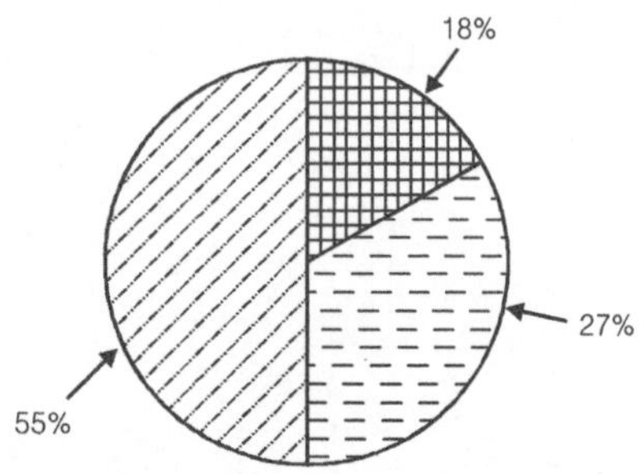

Sectoral Conbribution to GDP in Indian Economy (Financial Year 2008-09)

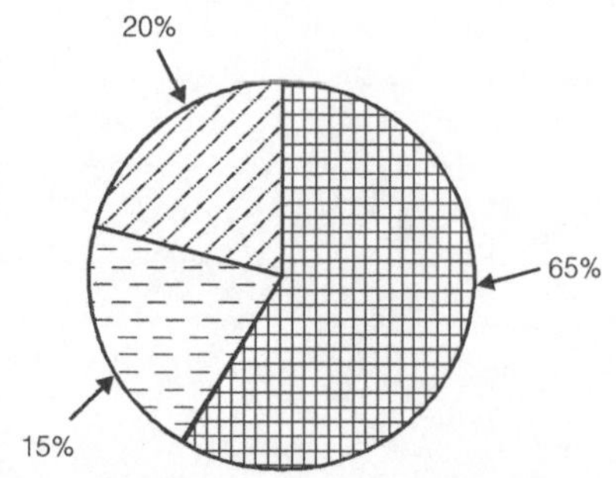

Work force employed in difference sectors in Indian Economy (Financial Year 2008-09)

where shaded regions for both previous pie chart are represented as below:

Pie-Chart:II

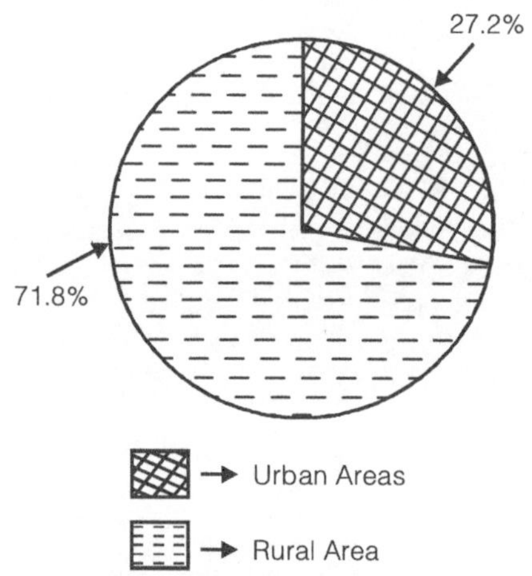

Demographic composition of Indian population living in Urban & Rural areas according to census 2001.

Pie-Chart:III

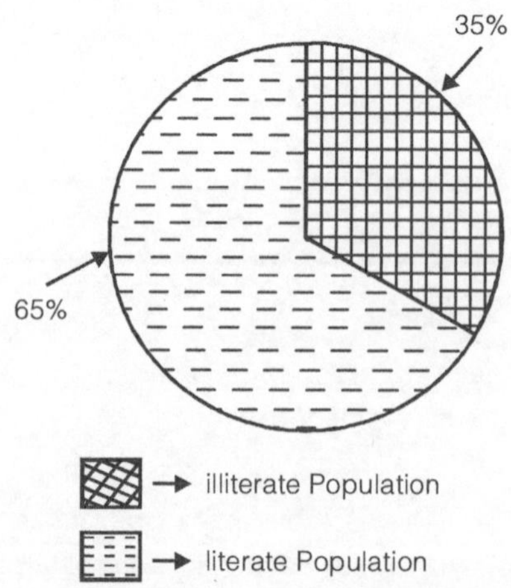

Literacy rates of Indian Population according to census 2001.

Pie-Chart:IV

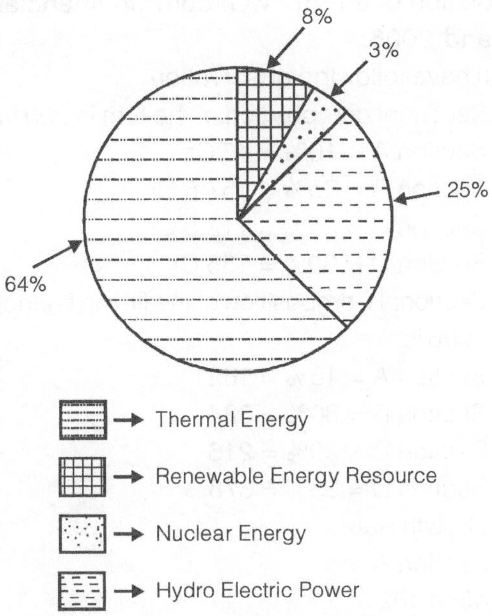

Source of Power in India (Installed Capacity)

Pie Chart-I

Shows that major contribution to GDP of Indian Economy is the share of services [Tertiary Sector] is 55% yet number of people employed in this sector is merely 20%. Share of agriculture is merely 18% yet approximately two third of population is still dependent on agriculture to support their livelihood.

Contribution of industries is 27% and approximately 15% people are employed in industries.

Pie Chart-II

It shows that more than 2/3 of our population are living in rural areas. Urban population is still less than 1/3. It need urgent attention to provide facilities of urban amenities in rural areas. Rural development should be given top most priority.

Pie Chart-III

Literacy rate is shown in pie chart III. It reveals that more than one-third of population do not know how to read and write. We need more resources and constitutional power to local bodies and state government so that they can increase their efforts to improve literacy rates.

Pie Chart-IV

It shows composition of different resources for power generation. We are still dependent much on Thermal Power to fulfil over energy demands. Nuclear energy and renewable resources of energy contributes merely 3% and 8% respectively.

Pie-Chart

There are two type of Pie Charts

1. Pie-Charts represented in Angular form
2. Pie-Charts represented in percent form

Conversion

$\because$ 100% = 360°

$\therefore$ $1\% = \frac{360}{100} = 3.6°$

Similarly

$\because$ 360° = 100%

$\therefore$ $1° = \frac{100}{360} = \left(\frac{10}{36}\right)^{\circ}$

Let us have an example on Angular form.

Example-I

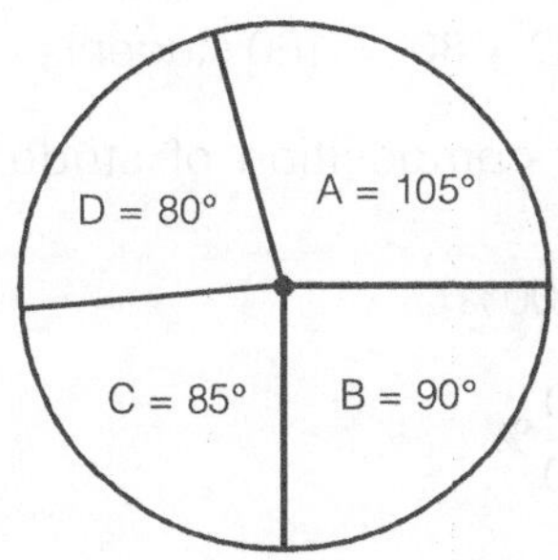

Fig. 5

Sectionwise composition of 720 students in a class in MADE EASY

From above pie chart we can answer following questions.

1. Students in section A is how many percent more than that in section B?

Sol. Section A = 105°

Section B = 90°

$\Rightarrow \frac{105-90}{90} \times 100$

$\Rightarrow \frac{15}{90} \times 100$

$\Rightarrow$ 16.66% Ans.

2. Students in section A is how many times than that of students in section D?

Sol. Students in A = 105°

Students in D = 80°

$\frac{105}{80} = \frac{21}{16} = 1.3125$ Ans.

3. Difference between studetns in section C and Section D is how many times the difference between students in Section A and B.

Sol. Difference between C & D $\Rightarrow 85° - 80° = 5°$
Difference between A & B

$\Rightarrow 105° - 90° = 15°$

$$\Rightarrow \frac{[C-D]}{[A-B]} = \frac{5}{15} = \frac{1}{3} \text{ times Ans.}$$

4. Individual number of students in different the sections.

Sol. 360° = 720 students

$$1° = \frac{720}{360} = 2 \text{ students}$$

Section A = 105° = 210 students
Section B = 90° = 180 students
Section C = 85° = 170 students
Section D = 80° = 160 students

5. Percent composition of students in various sections.

Sol. 360° = 100%

$$1° = \frac{100}{360}\%$$

$$\text{Section A} = 105° = \frac{105 \times 100}{360}\% = 29.167\%$$

$$\text{Section B} = 90° = \frac{90 \times 100}{360} = 25\%$$

$$\text{Section C} = 85° = \frac{85 \times 100}{360} = 23.611\%$$

$$\text{Section D} = 80° = \frac{80 \times 100}{360} = 22.22\%$$

Example-2

Rs. 450 Cr.

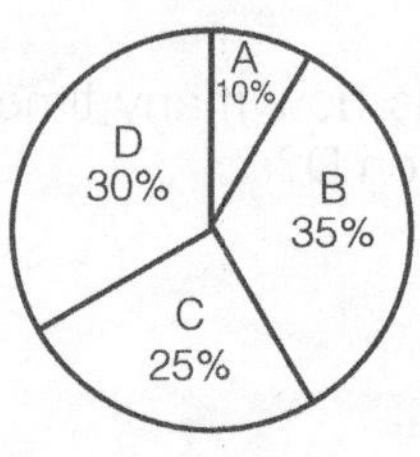

2001

Rs. 1080 Cr.

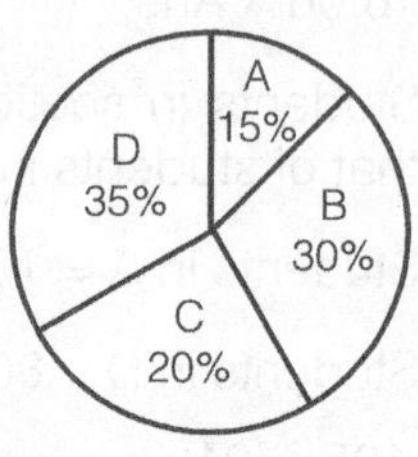

2006

Pie-Chart in above figures represent sectional composition of a firm "V.G. com" in financial years 2001 and 2006.

We can have following observation

1. Sectional composition of the firm in 2001 is
   Section A = 10% = 45 Cr.
   Section B = 35% = 157.5 Cr.
   Section C = 25% = 112.5 Cr.
   Section D = 30% = 135 Cr.
2. Sectional Composition of the firm in financial year 2006 is
   Section A = 15% = 162
   Section B = 30% = 324
   Section C = 20% = 216
   Section D = 35% = 378
3. Growth Rate
   Section A :
   45 to 162
   $$\Rightarrow \frac{162-45}{45} \times 100 = 260\%$$
   Section B :
   157.5 to 324
   $$\Rightarrow \frac{324-157.5}{157.5} \times 100 = 92\%$$
   Section C :
   112.5 to 216
   $$\Rightarrow \frac{216-112.5}{112.5} \times 100 = 105.7\%$$
   Section D : 135 to 378
   $$\Rightarrow \frac{378-135}{135} \times 100 = 180\%$$
   - Highest growth rate for the given duration 2001-2006 is observed by section A.
   - Minimum growth rate for the given duration 2001-2006 is observed 2001-2006 is observed by section C.
4. Growth rate for the duration 2001-2006 is
   $$\frac{1080-450}{405} \times 100 = 140\%$$
5. Average Annual Growth rate for duration 2001-2006 is
   $$\frac{1080-450}{405} \times \frac{100}{5} = 28\%$$

□□□□

Practice Exercise: I

Direction (Qs. 1 to 6): These questions are to be answered on the basis of the following pie chart, which gives marks scored by a student in different subjects-English, Hindi, Mathematics, Science and Social Science in an examination. Assuming that the total marks obtained for the examination are 540 answer the following questions:

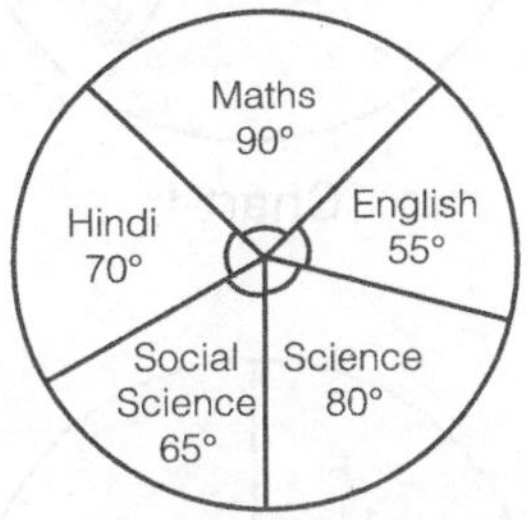

1. The marks scored by the student in Hindi and Mathematics exceed the marks scored in English and Social Science by
   (a) 60 (b) 75
   (c) 40 (d) 30

2. The subject in which the student scored approx 22.2% marks is
   (a) Hindi (b) Science
   (c) Social Science (d) English

3. The subject in which the student scored 105 marks is
   (a) Mathematics (b) Hindi
   (c) Science (d) Maths

4. The marks obtained in the three subjects : Englsih, Science and Social Science are what percentage of the total?
   (a) 45% (b) $44\frac{4}{9}\%$
   (c) 55% (d) $55\frac{5}{9}\%$

5. The marks obtained in Mathematics are what percentage of the total marks?
   (a) 20% (b) 30%
   (c) 35% (d) 25%

6. The difference of marks between English and Social Science is the same as between
   (a) Science and Hindi
   (b) Hindi and Social Science
   (c) English and Hindi
   (d) Social Science and Science

Direction (Qs. 7 to 11): Use the following chart to answer these questions on the age-wise distribution of personal income of a country in the year 2002.

Rs. 2499.4 Billions = 100%

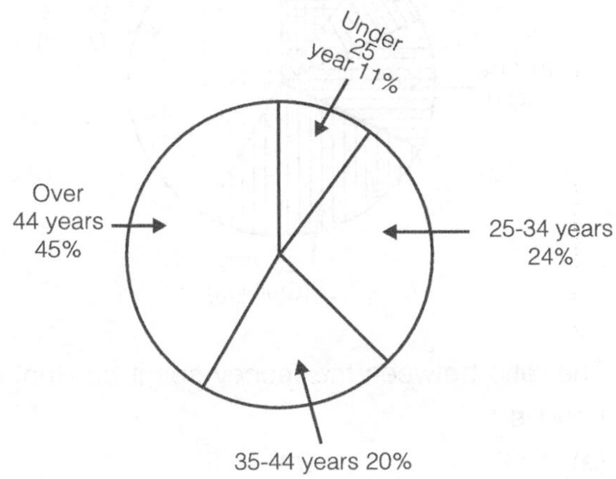

7. In 2002, what was the approximate personal income, in billions of rupees, of the age-group 35-44 years?
   (a) 500 (b) 600
   (c) 1125 (d) None of these

8. About how many degrees are there in the central angle devoted to the personal income of the age group 35-44
   (a) 20 (b) 40
   (c) 72 (d) None of these

9. What is the ratio of the personal income of the age group 25-34 to that of the age-group 35-44?
   (a) 9 : 4 (b) 5 : 6
   (c) 8 : 15 (d) None of these

10. If the total income in 2002 was 20 percent greater than it was in 2000, then what was the approximate total personal income in 2000 (in billions of rupees)?
    (a) 2200 (b) 2100
    (c) 2000 (d) None of these

11. If the total income in the year 2007 is projected to be one-and-a-half time of its current size in 2002, then what is the assumed annual compound rate of growth in income (in percent)?
    (a) 20 (b) 25
    (c) 15 (d) None of these

Direction (Qs. 12 to 16): These questions are based on the following pie chart showing percentage of money spend on household items by a representative family :

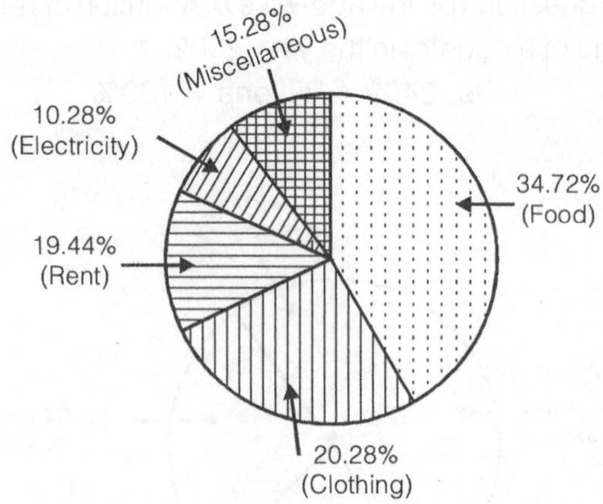

12. The ratio between the money spent on Rent and Food is
(a) 1 : 2 (b) 7 : 5
(c) 14 : 25 (d) None of these

13. If the income of a family is Rs. 12000 p.m. then the difference between the expenditure (in rupees) on Clothing and Electricity is nearly
(a) 8700 (b) 4400
(c) 3300 (d) None of these

14. If the income of a family is Rs. 20,000 p.m. then the ratio of the expenditure on Rent and Miscellaneous items is
(a) 3 : 2 (b) 7 : 5
(c) 7 : 6 (d) None of these

15. If the income of a family increase by Rs. 1000 p.m., then the, amount spent (in rupees) on Electricity should increase by nearly
(a) 100 (b) 200
(c) 300 (d) 400

16. If the income of a family is Rs. 8000 p.m., then the approximate difference of the expenditure on Electricity and Miscellaneous items in percent terms.
(a) 18 (b) 12
(c) 5 (d) None of these

Direction (Qs. 17 to 21): Study the following information to answer the given questions.

Percentage of total students in various courses (A,B,C,D,E,F) in the pie chart-I and percentage of girls in pie Chart-II

Total students : 1200
(800 girls + 400 boys)

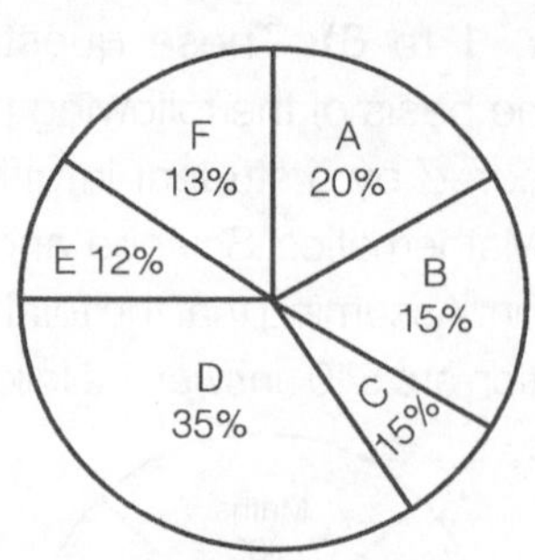

Chart-I

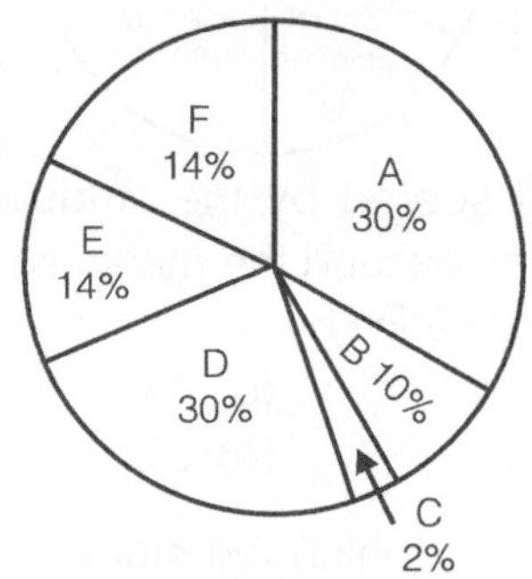

Chart-II

17. For course D, what is the ratio of boys and girls?
(a) 3 : 4 (b) 4 : 5
(c) 3 : 5 (d) 5 : 6

18. For which pair of course is the number of boys the same?
(a) E and F (b) A and D
(c) C and F (d) B and D

19. For course E, the number of girls is how much per cent more than the number of boys for course E?
(a) 250 (b) 350
(c) 150 (d) 80

20. For which course is the number of boys the minimum?
(a) E (b) F
(c) C (d) A

21. How many girls are there in course C?
(a) 44 (b) 16
(c) 40 (d) 160

Direction (Qs. 22 to 26): The pie chart represented below shows the spending by a family on various items during the year 1999. Study the graph and answer these questions.

Percent of money spent by a family on various items during 1999

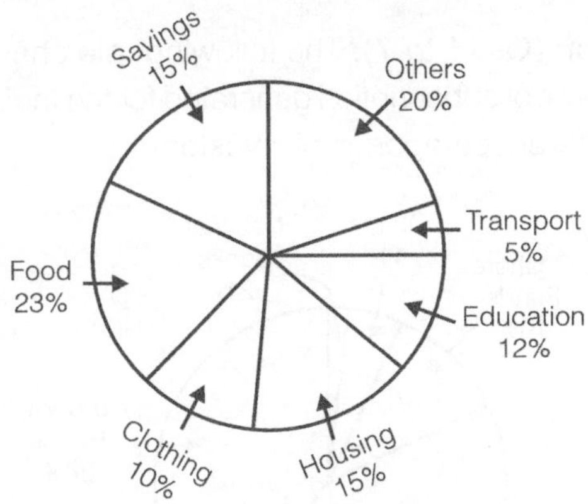

22. If the total amount spend during the year 1999 was Rs. 46000, the amount spent on food was
(a) Rs. 2000 (b) Rs, 10580
(c) Rs. 23000 (d) Rs. 2300

23. If the total amount spend was Rs 46000, how much money was spent on clothing and housing together?
(a) Rs. 11500 (b) Rs. 1150
(c) Rs. 10000 (d) Rs. 15000

24. The ratio of the total amount of money spent on housing to that spent on education was
(a) 5 : 2 (b) 2 : 5
(c) 4 : 5 (d) 5 : 4

25. Graph shows that the maximum amount was spent on
(a) Food (b) Housing
(c) Clothing (d) Others

26. If the total expenditure of the family for the year 1999 was Rs. 46000, the family saved during the year
(a) Rs. 1500 (b) Rs. 15000
(c) Rs. 6900 (d) Rs. 3067 approx

Answers

1. (a)	2. (b)	3. (b)	4. (d)	5. (d)
6. (a)	7. (a)	8. (c)	9. (d)	10. (b)
11. (d)	12. (c)	13. (d)	14. (d)	15. (a)
16. (c)	17. (a)	18. (c)	19. (a)	20. (d)
21. (b)	22. (b)	23. (a)	24. (d)	25. (a)
26. (c)				

Solutions

Answer 1 to 5:

1. (a) Marks scored in Hindi and Maths

$$= \frac{160}{360} \times 540 = 240$$

Marks scored in English and Social Science

$$= \frac{120}{360} \times 540 = 180.$$

2. (b) 100% = 360°

$$22.2\% = \frac{360}{100} \times 22.2 = 79.92° = 80°$$

Hence difference is
240 – 180 = 60.

3. (b) 540 = 360°

$$105 = \frac{360}{540} \times 105 = 70°.$$

4. (d) 360° = 100%

$$200° = \frac{100}{360} \times 200 = 55\frac{5}{9}\ \%.$$

5. (d)

6. (a)

7. (a) 20% of 2499.4 billions

$$= \frac{2499.4}{5} = 499.88 \simeq 500 \text{ billion.}$$

8. (c)

9. (d) Required ratio 24 : 20 = 6 : 5

10. (b) Total personal income in 2002 = 2499.4 billion
Total personal income in 2000 = x, say
∴ x + 20% of x = 2499.4

$$\Rightarrow x = \frac{5}{6} \times 2499.4 = 2082.83 = 2100.$$

11. (d) Total income project in the year 2007 = 150, when the income in the year 2002 was 100.
∴ Annual compound rate of growth is 8.5%

12. (c) Ratio between the money spent on rent and food

$$= \frac{19.44}{34.72} = \frac{1944}{3472} = \frac{14}{25}.$$

13. (d) Expenditure on clothing
= 20.28% of Rs. 12000 = Rs. 2433.60
Expenditure on electricity
=10.28% of 12000 = Rs. 1233.60
= Required difference = Rs. 1200.

Alternative

Difference of expenditure on clothing & Electricity is
20.28% – 10.28% = 10%
10% of 12000
= Rs. 1200 Ans.

14. (d) Required ratio = $\frac{19.44}{15.28} = 1.27,$
Whatever the family income may be

15. (a) Amount on electricity will increase by Rs. 102.80 = Rs. 100.

16. (c) Required difference will be 15.28% – 10.28% = 5%, irrespective of family income.

17. (a) Total no. of students for course D = 35% of 1200 = 420
No. of girl students for course D = 30% of 800 = 240
No. of boy students for course D = 420 – 240 = 180
Reqd. ratio = 180 : 240 = 3 : 4

18. (c) No. of boys for different courses are
A = 0; B = 100; C = 44; D = 180; E = 32; F = 44.
Hence C & F are same.

19. (a) Total students for E = 12% of 1200 = 144
No. of girls for course E = 14% of 800 = 112
So No. of boys for course E = 32
Reqd more % = $\frac{112-32}{32} \times 100 = 250\%$

20. (d) Using the information given in Q. No. (ii)

21. (b) No. of girls in courese C = 2% of 800 = 16.

22. (b) 23% of 46000 = Rs. 10580

23. (a) 25% of 46000 = Rs. 11500

24. (d) Required ratio = 15 : 12 = 5 : 4

25. (a) 23% are food.

26. (c) 15% of 46000 = Rs. 6900

II Practice Exercise: II

Direction (Qs. 1 to 7): The following pie chart shows the amount of subscription generated for the India Bonds from different categories of investors.

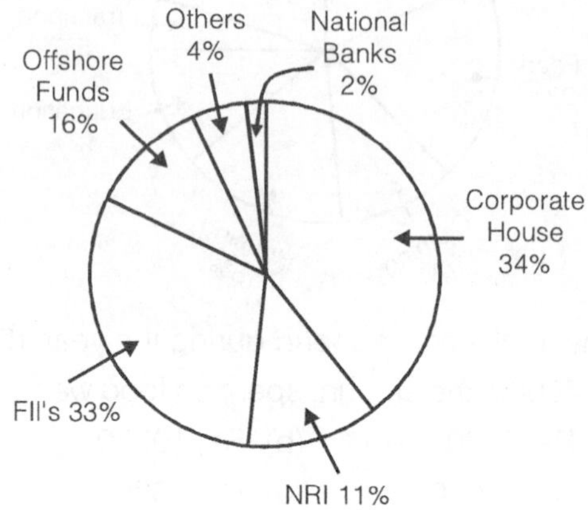

1. If the investments by NRI's are Rs. 4,000 crore, then the investment by corporate houses and FII's together is:
(a) 24,000 crore (b) 24,363 crore
(c) 25,423.4 crore (d) 25,643.3 crore

2. What percentage of the total investment is coming from either FIIs or NRIs?
(a) 33% (b) 11%
(c) 44% (d) 22%

3. If the total investment other than by FII and corporate house is Rs 335.000 crore, then the investment by NRIs and Offshore funds will be (approximately) :
(a) 274,100
(b) 285,600
(c) 293,000
(d) Can't be determined

4. What is the approximate ratio of investment flows into India Bonds from NRIs to corporate houses?
(a) 1 : 4
(b) 1 : 3
(c) 3 : 1
(d) Can't be determined

5. In the corporate sector, approximately how many degrees should be there in the central angle?
(a) 120 (b) 121
(c) 122 (d) 123

6. If the total investment flows from FII's were to be doubled in the next year and the investment flows

from all other sources had remained constant at their existing levels for this year, then what would be the proportion of FII investment in the total investment flows into India Bonds in the next year (Approximately)?
(a) 40%
(b) 50%
(c) 60%
(d) Can't be determined

7. If the inflow from the FIIs after the doubling (as given in Question 6) were approx US$ 500 million what would be the total investment into India Bonds next year (in US$ millions)
(a) 1000
(b) 1500
(c) 800
(d) Can't be determined

Direction (Qs. 8 to 11): (A) and (B) exhibit the out flow of the tourist traffic from India. The two charts show the tourist distribution by country (A) and the age wise (B) traffic of the tourist respectively. Study the charts carefully and answer the question which flow.

(A)

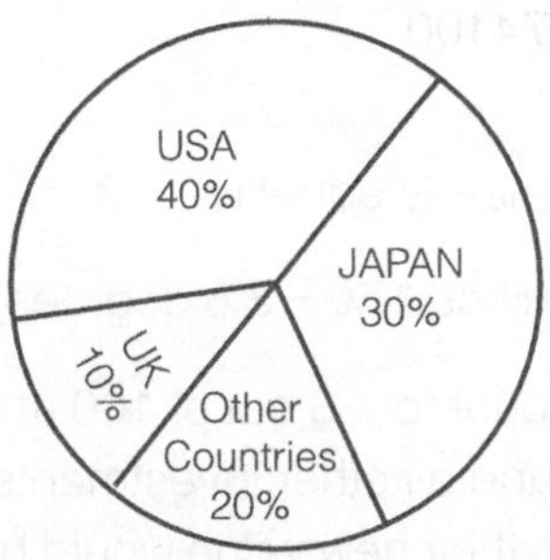

(B)

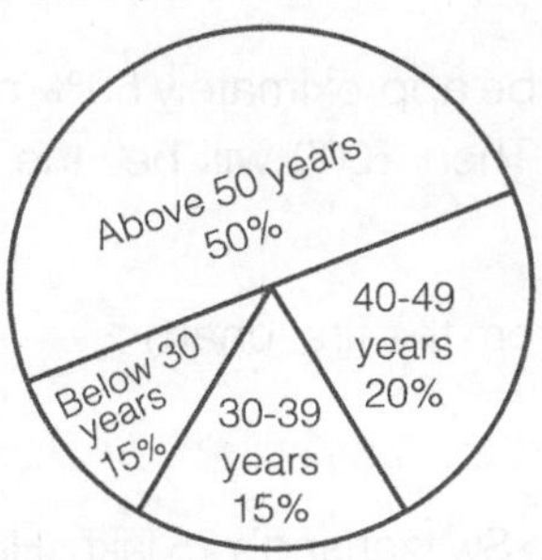

8. What percentage of the India tourist out flow to either USA or UK?
(a) 40% (b) 50%
(c) 60% (d) 70%

9. The ratio of the number of Indian tourists that went to USA to the number of Indian tourists who were below 30 years of age is:
(a) 2 : 1
(b) 8 : 3
(c) 3 : 8
(d) Can't be determined

10. If amongst other countries, Switzerland accounted for 25% of the Indian tourist traffic, and it is known from official Swiss records that a total of 25 lakh Indian tourist had gone to Switzerland during the year, then find the number of 30-39 years-old Indian tourists who went abroad in that year.
(a) 18.75 lakh (b) 25 lakh
(c) 50 lakh (d) 75 lakh

11. For Question 10, what was the volume of traffic of Indian tourists in the US?
(a) 150 lakh (b) 125 lakh
(c) 200 lakh (d) None of these

Direction (Qs. 12 to 16): Pie Chart given below shows the expenditure incurred in bringing out a book by a publisher

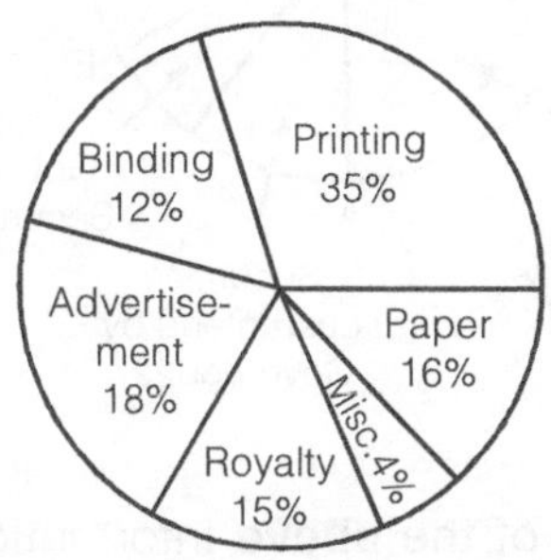

Study the graph carefully and answer the questions given below:

12. What is the central angle of the sector for the cost of the paper?
(a) 22.5° (b) 16°
(c) 54.8° (d) 57.6°

13. If the cost of printing is Rs. 17500, the royalty is:
(a) Rs. 8750 (b) Rs. 7500
(c) Rs. 3150 (d) Rs. 6300

14. If the miscellaneous charges are Rs. 9000, the advertisement charges are:
(a) Rs. 13500 (b) Rs. 20000
(c) Rs. 40500 (d) Rs. 1800

15. If 5500 copies are published and miscellaneous expenditure on them amount to Rs. 5544 and the publisher earns a profit of 25%, then marked price of each copy is

(a) Rs. 25.20 (b) Rs. 37.50
(c) Rs. 31.50 (d) Rs.30

16. Royalty on the book is less than the advertisement charges by:

(a) 3% (b) $16\frac{2}{3}\%$
(c) 20% (d) None of these

Direction (Qs. 17 to 21): The gross investment of Life Insurance Corporation of India (in crores of rupees) in different sectors are shown in the pie chart given below :

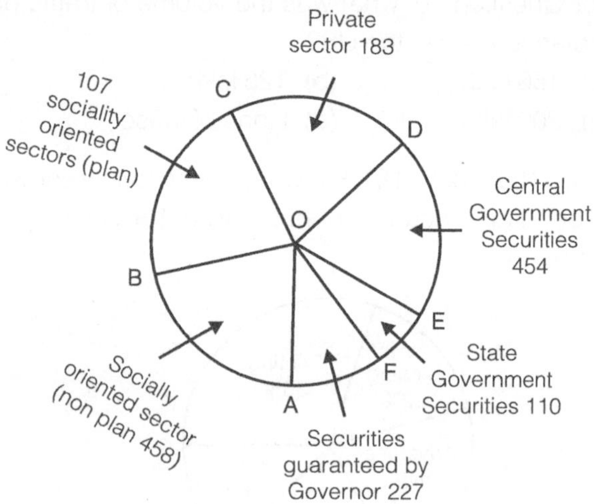

On the basic of the above information answer the following questions.

17. The percentage of gross investment in States Government Securities is nearly :
(a) 7.1% (b) 7.8%
(c) 8.6% (d) 9.2%

18. The magnitude of ∠AOC is nearly :
(a) 103° (b) 132°
(c) 126° (d) 115°

19. The investment in socially oriented sectors (plan and non plan) is ..... than the investment in Government securities (Central and State) by ...
(a) More, 4 crore (b) More, 1 crore
(c) More, 111 crore (d) Less, 106 crore

20. The investment in private sectors is nearly... percent higher than the investment in State Government Securities?
(a) 66 (b) 54
(c) 46 (d) 40

21. The ratio of the area of the circle above COF to the area of the circle below it is, nearly :
(a) 1 (b) 0.966
(c) 0.94 (d) 0.92

Answers

1. (b)	2. (c)	3. (a)	4. (b)	5. (c)
6. (b)	7. (a)	8. (b)	9. (b)	10. (d)
11. (c)	12. (d)	13. (b)	14. (c)	15. (c)
16. (b)	17. (a)	18. (b)	19. (b)	20. (a)
21. (c)				

Solutions

Answer 1 to 5 :

1. (b) (67/11) × 400 crore

2. (c) 33% + 11% = 44%

3. (a) Investment other than NRI and corporate houses is 33% = 335000. Also, invest-ment by offshore funds and NRIs is equal to 27%. Hence,

$$27 \times \frac{335,000}{33} = 274100$$

4. (b) 11 : 34 is approximately equal to 1 : 3

5. (c) 34 × 3.6 = 122.4 (since 1 % = 3.6 degrees).

6. (b) FII's currently account for 33 out of 100. If their value is doubled and all other investments are kept constant then their new value would be 66 out of 133 which is approximately equal to 50%

7. (a) 500 million would be approximately 50% of the total investment. Then 1000 will be the total investment.

8. (b) 40 + 10 = 50% (from the first chart)

9. (b) 40 : 15 = 8 : 3

10. (d) 5% corresponds to Switzerland's 25 lakh. Hence 15% will be 75 lakh

11. (c) US account for 40%, i.e., 8 times 5%. Since, Switzerland's 5% is 25 lakh, US will be 200 lakh

12. (d) Central angle for the cost of the paper

$$= \left(\frac{16}{100} \times 360\right)^\circ = 57.6^\circ.$$

13. (b) Let the royalty be Rs. x. Then
35 : 15 : : 17500 : x

$$x = \left(\frac{15 \times 17500}{35}\right) = \text{Rs. } 7500$$

14. (c) Let the advertisement charges be Rs. x.

The, 4 : 18 : : 9000 : x or $x = \left(\frac{18 \times 9000}{4}\right)$

= Rs. 40500.

15. (c) Let the total charges be Rs. x.
Then, 4 : 100 : : 5544 : x or x

$$= \left(\frac{18 \times 5444}{4}\right) = \text{Rs. } 138600.$$

∴ Total cost = Rs. 138600.
∴ Cost price of each copy

$$= \text{Rs. } \left(\frac{138600}{5500}\right) = \text{Rs. } 25.20.$$

∴ Marked price = 125% of Rs. 25.20

$$= \text{Rs. } \left(\frac{125}{100} \times 25.20\right) = \text{Rs. } 31.50$$

16. (b) Let royalty be Rs. 15. Then, advertistment charges = Rs. 18.

$$\therefore \text{Required \%} = \left(\frac{3}{18} \times 100\right)\% = \text{Rs. } 31.50$$

17. (a) Total investment = (458 + 107 + 183 + 454 + 110 + 227) crores

$$\therefore \text{Required percentage} = \left(\frac{110}{1539} \times 100\right)\%$$

= 7.1%

18. (b) $\angle AOC = \left[\frac{(458 + 107) \text{ crores}}{1539 \text{ crores}}\right]$

$$= \left(\frac{565}{1539} \times 360\right) = 135°.$$

19. (b) Investment in socially oriented sectors = (458 + 107) crores = 565 crores.
∴ Investment in socially oriented sectors is more than investment in Government securities by 1 crore.

20. (a) Investment in Private Sector = 183 crores.
Investment in State Government Securities = 110 crores.

$$\therefore \text{Required excess} = \left(\frac{73}{110} \times 100\right)\% = 66\%$$

21. (c) Required Ratio

$$= \frac{(183 + 454 + 110)}{107 + 458 + 227} = \frac{747}{792}$$

$$\Rightarrow \frac{83}{88} = 0.943$$

■■■■

CHAPTER 3.9

Miscellaneous Puzzles

Puzzles

A puzzle is a problem given that tests the ingenuity of the solver. In a basic puzzle, It solver should intended to piece together objects (puzzles pieces) in a logical way in order to come up with the desired shape, picture or solution.

Solutions to puzzles may require recognizing patterns and creating a particular order. People with a high reasoning aptitude may be better at solving these puzzles.

Types of Puzzles

The large number of puzzles that have been created can be divided into various categories for example logical puzzle using chess board.

Other categories include

- Puzzle or Numbers
- Mathematical
- Mathematical problem such as the missing square puzzle
- Picture puzzle
- Connect the dots
- Logical puzzle like Sudoko.
- Spot the difference etc.

Lets have a glimpse of different types of puzzles

Numbers

Example 1.

In a ten digit number first digit represents number of one present in the number. Second digit represents number of two present in the number. Similarly, third digit represents number of three present in the number and so on till ninth digit, which represents number of nine present in the number. Last digit represents number of zero present in the number. Find that number.

Sol.: Let the ten digit number be

a_1	a_2	a_3	a_4	a_5	a_6	a_7	a_8	a_9	a_0

The smallest possible ten digit number is 1,00,00,00,000

We will go on modifying this number, since last digit represent number of zeros it should be 9.

I[st] step : 1,00,00,00,009.

Since we have only 8 zeros last digit should be 8.

IInd step : 1,00,00,00,008

Since number of "8" present in the above case is one the next modification will be.

IIIrd step : 1,00,00,00,108.

Here number of zeros are only seven so

IVth step : 1,00,00,00,107.

Here, since the number of "7" present is one so next modification will be

V[th] step : 1,00,00,01,007.

Here, since number of "1" present is two so the next modification will be

VIth step : 2,00,00,01,007

Number of "2" present is be one so next modification will be

VIIth step : 2,10,00,01,007

Here, since number of zeros present is six so next modification will be

VIIIth step : 2,10,00,01,006.

Here, since number of "6" present is 1 so next modification will be

IXth step : 2,10,00,10,006.

This number 2,10,00,10,006 satisfy all the condition so this is the desired number.

Triangles

Example 2.

Find out the total number of triangles in the given figure.

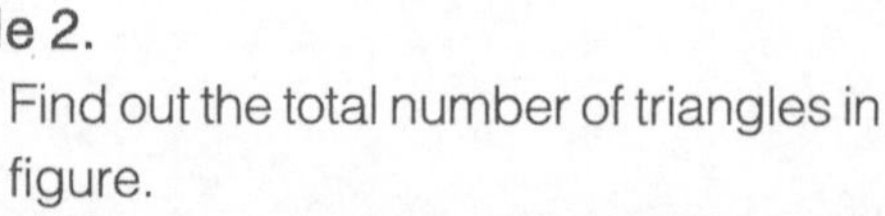

Fig. (1)

Sol. First of all we will count single triangle.

Number of single triangle = 12

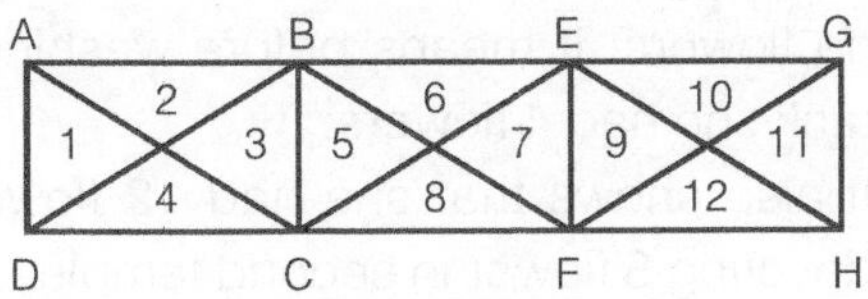

Fig. (2)

Now, we will count triangle made by joining two triangles for example in the below given fig. (3) we have 4 triangles made by joining two single triangles.

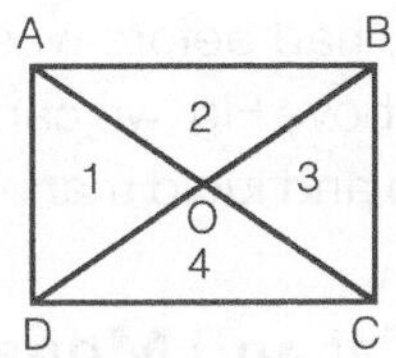

Fig. (3)

$\Delta ADB = \Delta 1 + \Delta 2$

$\Delta ADC = \Delta 1 + \Delta 2$

$\Delta BDC = \Delta 3 + \Delta 4$

$\Delta ABC = \Delta 2 + \Delta 3$

So in the given fig. (2) we will have 12 such "double triangles.

Now, we will count number of triangles made by joining four tirnalges.

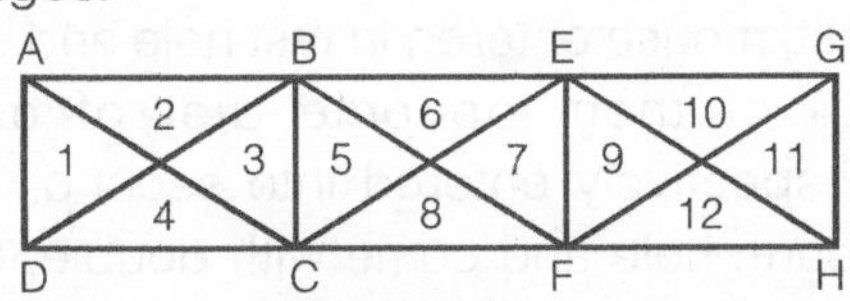

Fig. (4)

$\Delta ACE = \Delta 2 + \Delta 3 + \Delta 5 + \Delta 6$

$\Delta DBF = \Delta 4 + \Delta 3 + \Delta 5 + \Delta 8$

$\Delta CEH = \Delta 8 + \Delta 7 + \Delta 9 + \Delta 12$

$\Delta BFG = \Delta 6 + \Delta 7 + \Delta 9 + \Delta 10$

So we have

12 "Single triangle"

12 "Double triangle" and

4 "Four triangle"

⇒ 28 triangles Ans.

Example 3.

Find number of triangles present in given Fig. (5).

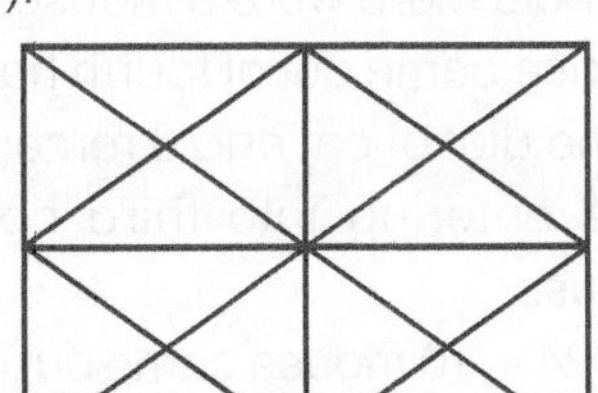

Fig. (5)

Sol. Number of "Single triangles" present in the Fig. (6) are 16.

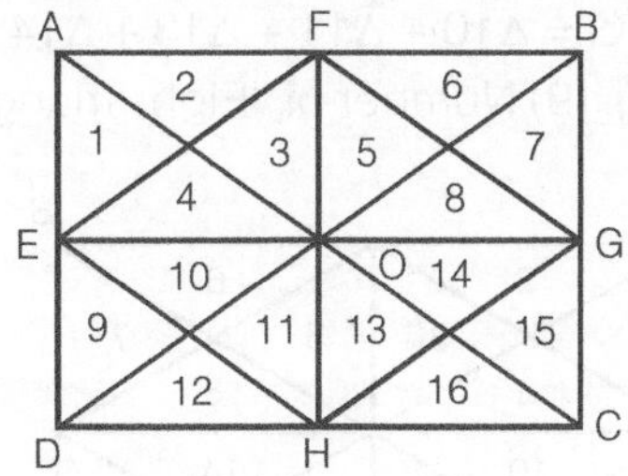

Fig. (6)

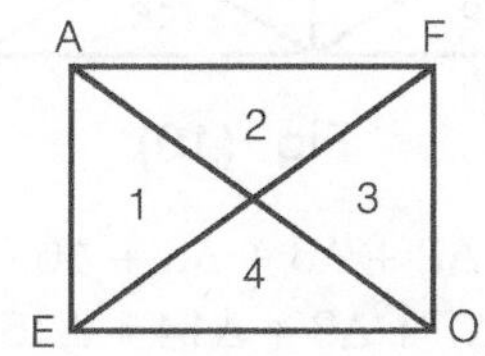

Fig. (7)

Now, we will find number of double triangles in Fig. (7) which is equal to 4 i.e.

$\Delta AFE = \Delta 1 + \Delta 2$

$\Delta AFO = \Delta 2 + \Delta 3$

$\Delta FOE = \Delta 3 + \Delta 4$

$\Delta AEO = \Delta 1 + \Delta 4$

So total "double triangle" in fig. (6) are 16.

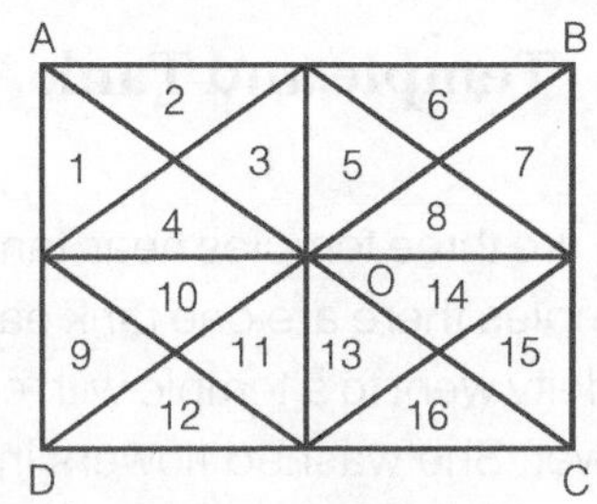

Fig. (8)

Number of "Four triangle" will be 4.

$\Delta AOD = \Delta 1 + \Delta 4 + \Delta 10 + \Delta 9$

$\Delta AOB = \Delta 2 + \Delta 3 + \Delta 5 + \Delta 6$

$\Delta BOC = \Delta 7 + \Delta 8 + \Delta 14 + \Delta 15$

$\Delta DOC = \Delta 12 + \Delta 11 + \Delta 13 + \Delta 16$

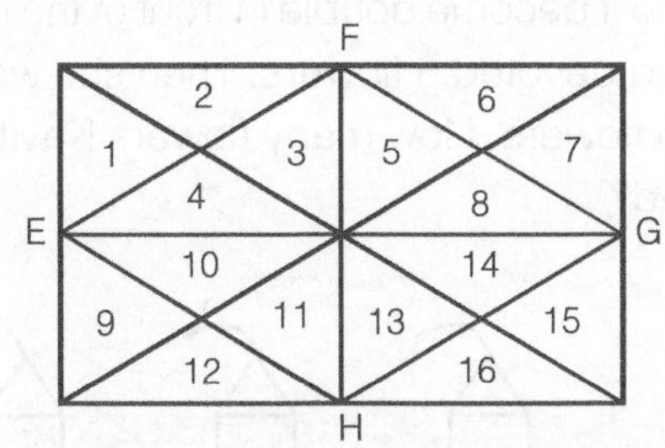

Fig. (9)

Number of "Four triangle" will be 4

$\Delta EFH = \Delta 3 + \Delta 4 + \Delta 10 + \Delta 11$

$\Delta EFG = \Delta 4 + \Delta 3 + \Delta 5 + \Delta 8$
$\Delta FGH = \Delta 5 + \Delta 8 + \Delta 14 + \Delta 13$
$\Delta EHG = \Delta 10 + \Delta 11 + \Delta 13 + \Delta 14$

In fig. (9) Number of "Eight Triangles" will be 4

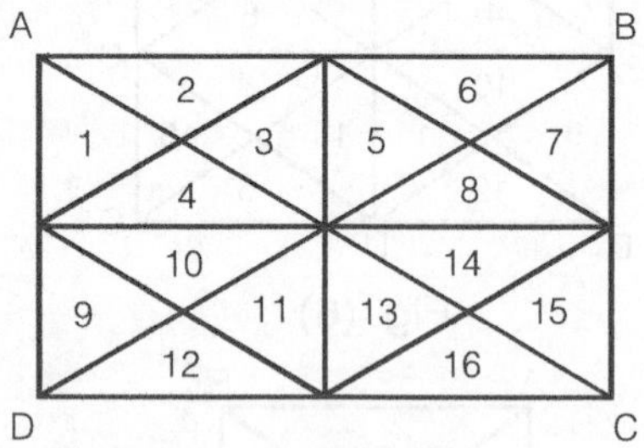

Fig. (10)

$\Delta ABC = \Delta 2 + \Delta 3 + \Delta 5 + \Delta 6 + \Delta 7 + \Delta 8 + \Delta 14 + \Delta 15$
$\Delta BCD = \Delta 7 + \Delta 8 + \Delta 14 + \Delta 15 + \Delta 16 + \Delta 13 + \Delta 11 + \Delta 12$
$\Delta ADC = \Delta 1 + \Delta 4 + \Delta 9 + \Delta 10 + \Delta 11 + \Delta 12 + \Delta 13 + \Delta 16$
$\Delta ADB = \Delta 1 + \Delta 2 + \Delta 3 + \Delta 4 + \Delta 5 + \Delta 6 + \Delta 9 + \Delta 10$

So, total number of triangles will be
16 + 16 + 4 + 4 +4 = 44 Ans.

Temple and Tank

Example 4.

There are three temples near Tanjaur. In front of all temples there are one tank each. Kavitha, a lady deity went to a temple with certain number of flower. She washed flowers in tanks in front of first temple and it became doubled. She devoted 8 flowers in that temple and moved to second temple. She again washed remaining flowers and again became doubled. She devoted 8 flowers in the second temple. She again washed remaining flowers in third tank and it become double in front of the last temple and devoted 8 flowers. Then she was left with no flowers. How many flowers Kavitha initially had?

Sol.

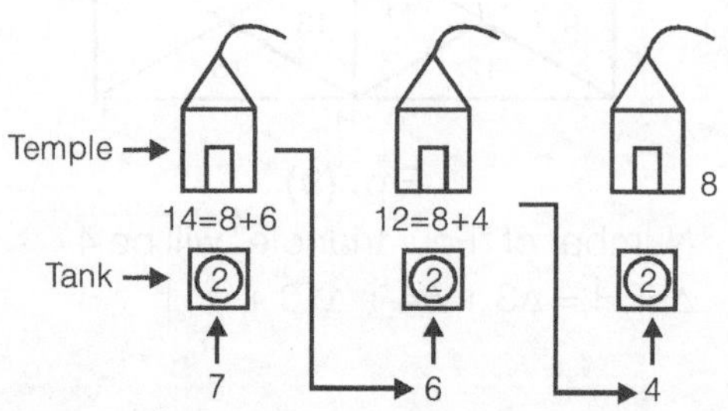

(1) We will start solving it from back side. After devoting 8 flowers in third temple Kavitha had no flowers. It means before washing in third tank she had 4 flowers.
(2) It also shows that she had 12 flower before devoting 8 flower in second temple.
(3) From this information we come to know that she had 6 flowers before washing it in second tank.
(4) It means she had (8 + 6) = 14 flowers before devoting to first temple.
(5) Now we can easily get the number of flower she initially had before washing in first tank.
(6) From the above Fig. we can easily find solution in a simple and lucid manner.

Cat and Mouse

Example 5.

There are four holes numbered 1, 2, 3 and 4. The unique properties of these holes is that number of mouse become double, triple and four times after entering into hole number 2, 3 and 4 respectively. While it remains same if they enter in first hole. One cat is running in search of few mouse. To save themselves from cat, mouse entered in first hole and came out. 24 of them become diet of cat. They respectively entered into second, third and fourth hole and come with double, triple and four times in number. After their exit from each hole i.e. second third and fourth, 24 mouse become diet of cat each time. At the end there is no mouse left. Find the initial number of mouse before entering into first hole.

Sol.:

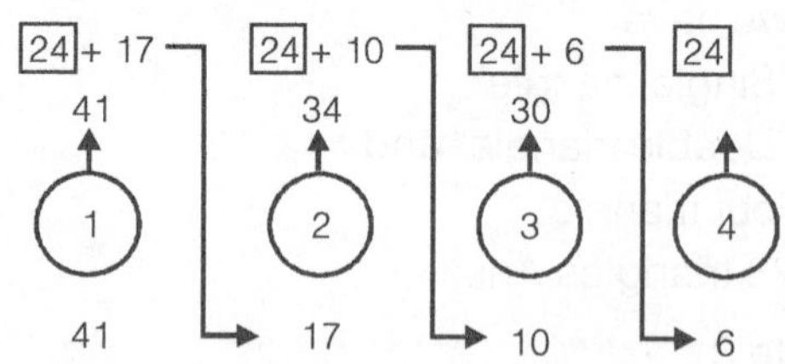

- From the fig it is clear that before entering into fourth hole there were 6 mouse.
- 30 mouse came out of fourth hole, because 24 become diet of cat and 6 remained.
- before entering into third hole there were 10 mouse.
- 34 i.e. 24 + 10 mouse came out of second hole, 24 become diet of cat and 10 remained.
- 17 mouse entered into second hole.

- 41 mouse came out of first hole.
- There were 41 mouse before entering into first hole.

Rectangles

Example 6.

Six parallel horizontal lines are intersecting perpendicularly to six verticle parallel lines. Find number of rectangles formed after such intersection.

Sol.:

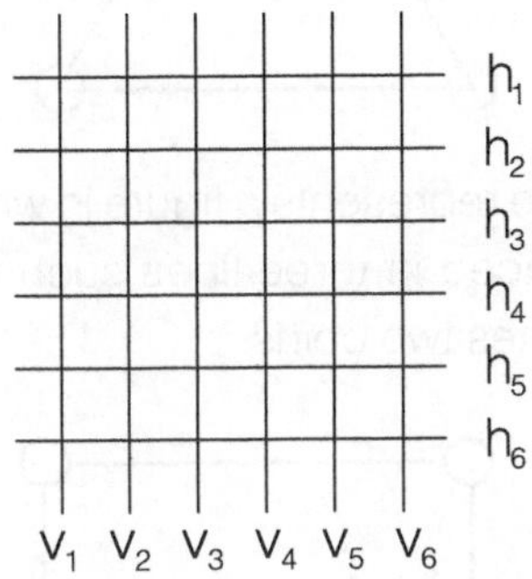

Here V_1 to V_6 represent six parallel vertical lines and h_1 to h_6 represent six parallel horizontal lines.

To form a rectangle we need to select any two horizontal parallel lines which intersect any two vertical parallel lines.

Thus

$^6C_2 \times {}^6C_2 = 15 \times 15 = 225$

225 rectangle will be formed.

Birthday Cake

Example 7.

In a Birthday party there is a cake, which is to be cut into different pieces. How many maximum pieces are possible if five vertical cuts are made given that no horizontal cut is allowed.

Sol.

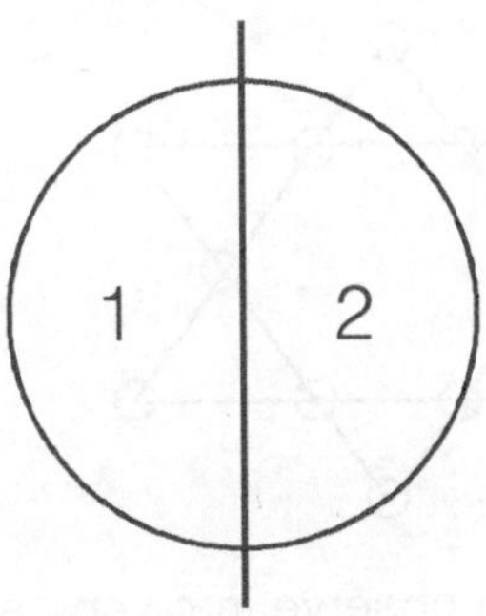

[Fig. First cut]

Total pieces = 1 + 1 = 2

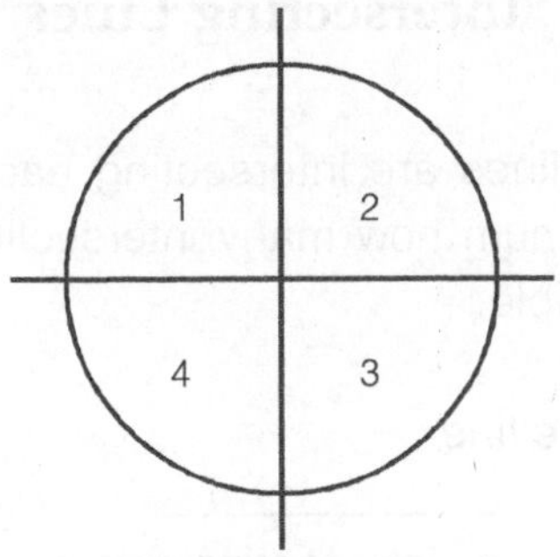

[Fig. Second cut]

Total pieces = 1 + 1 + 2 = 4

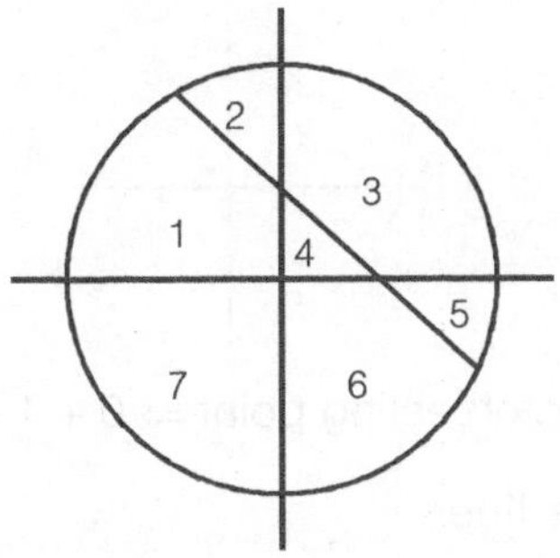

[Fig. Third cut]

Total pieces = 1 + 1 + 2 + 3 = 7

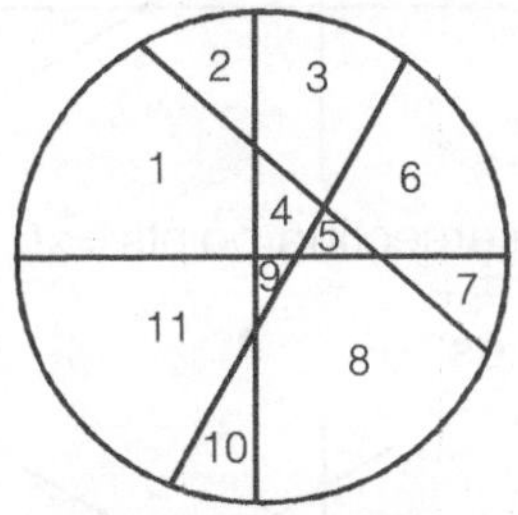

Fig. Fourth cut

Total pieces = 1 + 1 + 2 + 3 + 4 = 11

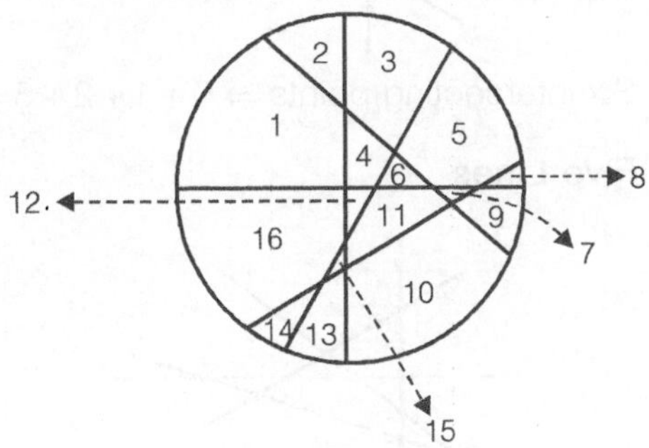

Fig. Fifth cut

Total pieces will be

= 1 + 1 + 2 + 3 + 4 + 5 = 16 Ans.

Similarly, for n cuts we get

1 + 1 + 2 + 3 + 4 ... n pieces.

$\Rightarrow$ 1 + Σn pieces.

- $1 + \frac{n(n+1)}{2}$ pieces is the generalize short-cut.

Intersecting Lines

Example 8.

Five lines are intersecting each other. Find maximum how many intersecting points are possible.

Sol.

1. Single line

no intersecting points

2. Two lines

One intersecting point $\Rightarrow 0 + 1 = 1$

3. Three lines

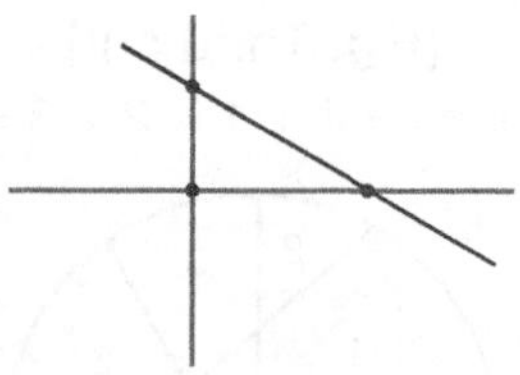

Three intersecting points $\Rightarrow 0 + 1 + 2 = 3$

4. Four lines

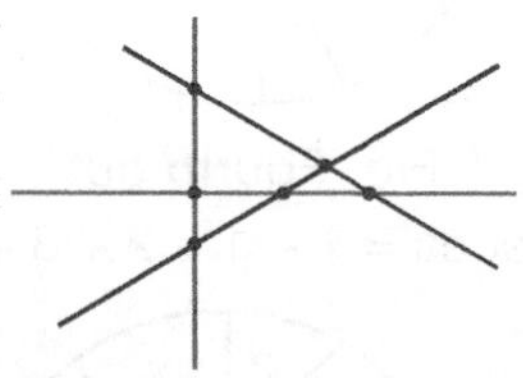

Six intersecting points $\Rightarrow 0 + 1 + 2 + 3 = 6$

5. Five Lines

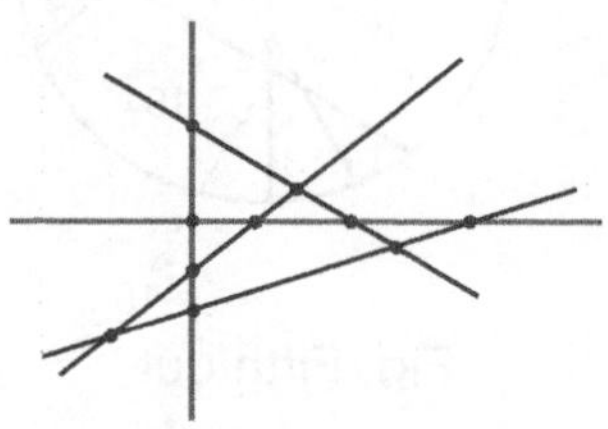

Ten intersecting points

$\Rightarrow 0 + 1 + 2 + 3 + 4 = 10$

Similar for n intersecting lines, maximum intersecting points will be

$0 + 1 + 2 + 3 + \ldots (n - 1) = \Sigma(n - 1)$

Coins

Example 9.

Ten coins are to be placed in five lines in such a manner that each line contains four coins.

Sol. 1. Triangles

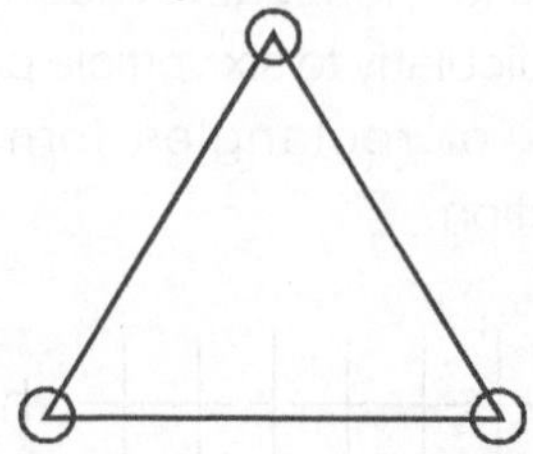

Triangle represents a figure in which three coins are placed in three lines such that each lines containes two coins.

2.

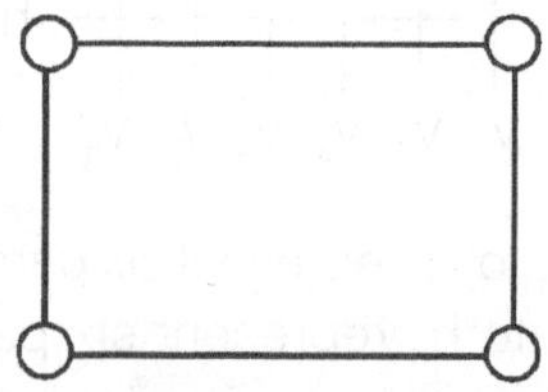

Square or rectangle represents a figure in which four coins are placed in four lines such that each lines contains two coins.

3.

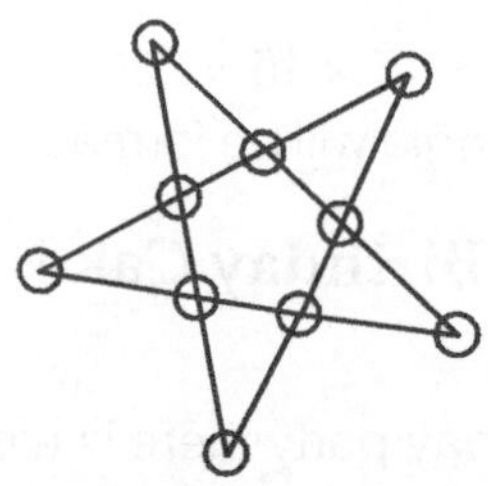

A Pentagonal star represents a figure in which ten coins are placed in four lines such that each lines contains four coins.

4.

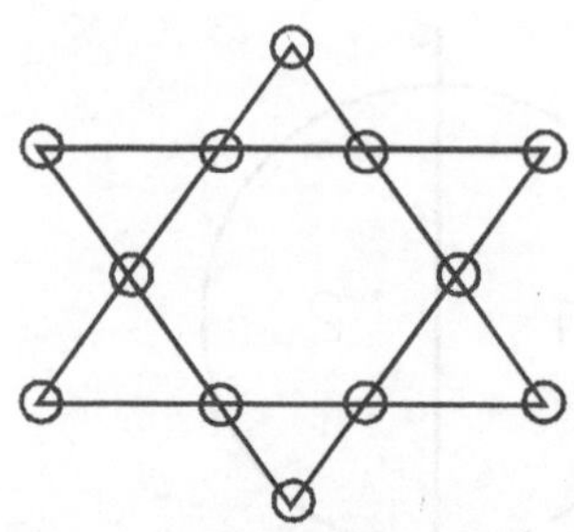

Similarly we can observe in given hexagonal star that twelve coins are present in six lines such that each lines contains four coins.

The Flag

Example 10.

Flag of a nation consists of six different vertical strips. How many such flags can be possible if we use six colours to fill the strips such that no two consequetive strips will have same colours?

Sol.

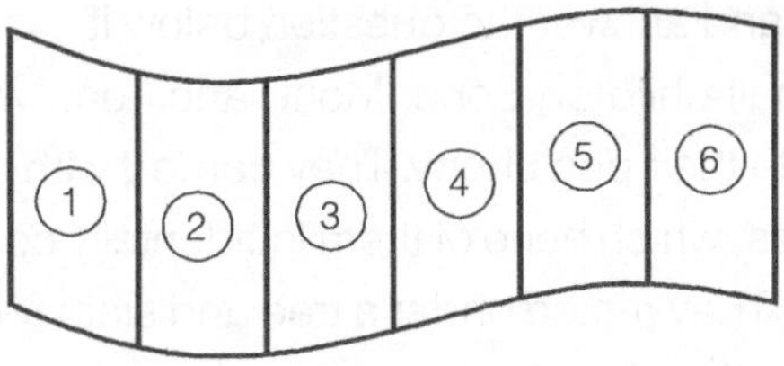

First strip can be filled with any six colours. Next strip can be filled with any five colours exept the colour filled in strip one. Successive strips will also be filled in the same manner i.e. 5 ways. Number of possible flags will be

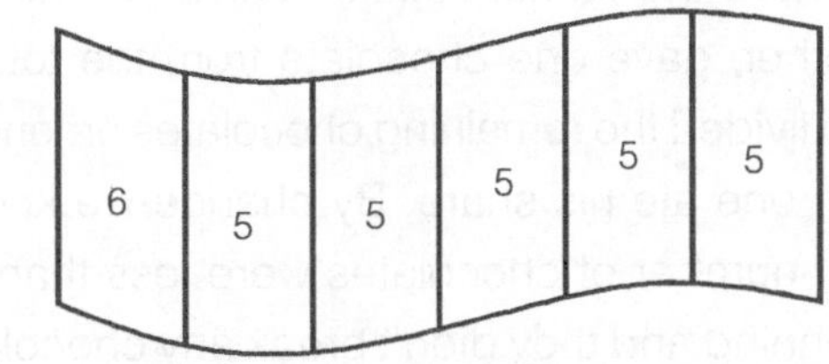

$6 \times 5 \times 5 \times 5 \times 5 \times 5 =$ **18750 Ans.**

□□□□

I Practice Exercise

Direction (Qs. 1 to 3): Read the following information carefully and answer the question below it.

Rahul brought some sweets on his 21st birthday. He offered one less than the half of total number sweets in the temple near the house. He also gave one sweet each to 3 beggars sitting on the stairs of 'temple'. On the way back to home, he stopped a big group of poor children and gave them half of what was left with him. After reaching home he shared the remaining two pieces of sweets with his younger brother.

1. How many sweets did he originally had?
   (a) 8 (b) 10
   (c) 12 (d) 14

2. How many sweets did he offered in the temple?
   (a) 3 (b) 4
   (c) 5 (d) 6

3. How many sweets did he gave to poor children?
   (a) 1 (b) 2
   (c) 30 (d) 42

Direction (Qs. 4 to 7): Read the following information carefully and answer the question below it.

Study the pyramid of the letters given below and answer the questions

B
M T R
G C S N P
Y Q H K E A I
J F U W X O V Z D

4. Which letter is missing in the pyramid?
   (a) F (b) I
   (c) L (d) P

5. If letters where to be studied vertically, which two letters happen to be neighbours that occur together in alphabetical order
   (a) WX (b) PR
   (c) UV (d) ST

6. Two letters in the above pyramid have exactly the same number of letters between them horizontally as they have in the alphabetical order. Which are they?
   (a) Y and Q (b) V and X
   (c) N and P (d) E and K

7. If all the horizontal lines were to be studied separately which neighbours in the alphabetical order are the farthest?
   (a) F and U (b) Q and Y
   (c) C and S (d) H and Q

Direction (Qs. 8 to 15): Read the following information carefully and answer the question below it.

The Selection Committee of a company laid down the following criteria of selection for the post of Manager-Accounts:

(i) The candidate must be a post-graduate in Commerce.

(ii) The candidate must have completed 21 years and should not be more than 35 years as on 31-07-1996.

(iii) If the candidate does not satisfy the criterion in (i) above but has completed his CA examination he will be referred to Director – Finance, who can allow the candidate to appear in the interview if otherwise eligible.

(iv) If the candidate fulfils all the criteria mentioned in (i) and (ii) above, he/she will be called for group discussion.

(v) The candidate must get 50% marks to qualify in the group discussion.

(vi) If the candidate qualifies in the group discussion he/she will be called for interview.

(vii) The candidate must get 30% marks in the interview (out of 50) to get finally selected.

Based on the above criteria, decide which of the following course of action should be taken in the case of candidate described in each of the following questions.

Mark answer as:

(1) If the candidate can be selected;

(2) If the candidate is to be referred to Director – Finance;

(3) If the candidate can be called for group discussion;

(4) If the candidate can be called for interview;

(5) If the candidate cannot be selected

8. Mr. Das is M.Com and his date of birth is 30-06-1961

9. Mrs. Krishnamurthy is a qualified CA and was 25 years old on 31-07-1996

10. Mr. Kant, a post-graduate with Commerce obtained 70% marks in group discussion and secured 20 marks in Interview. He was 35 years as on 31-07-1996

11. Mrs. Desai is a post-graduate in Economics and her date of birth is 3-04-1965

12. Mr. Patel, a post-graduate was permitted by Director Finance. He was born on 30-05-1963. He obtained 12 marks in interview

13. Mr. Mathur is Ph.D in Commerce and was 34 years old on 1-08-1995. He obtained 14 marks in interview.

14. Mr. Roy is a commerce graduate, has passed his CA examination and was born on 18-04-1974.

15. Mr. Subramaniam whose date of birth is 04-05-1965 is a post graduate with Commerce and has obtained 60% marks in the interview

Direction (Qs. 16 to 20): Read the following information carefully and answer the question below it.

Three small children Sonu, Monu and Tonu went on a picnic with their dog Jhony. They carried with them few chocolates, which none of them incidentally counted on their way. They rested under a tree and slept for a while. After some time Sonu woke up, gave one chocolate from the total to Jony and distributed the remaining into three equal parts, ate his share and slept. After some time, Monu woke up, gave one chocolate to Jony and distributed that remaining into three equal parts, ate his share and slept. After some time Tonu woke up and repeated the same. A little later all of them woke up together, gave one chocolate from the total to Jhony and divided the remaining chocolates among them and each one ate his share. By chance we know that the total number of chocolates were less than 150 in the beginning and they didn't break any chocolate.

16. How many chocolates were in the beginning?

(a) 66 (b) 84
(c) 118 (d) 79

17. What is the difference in the number of chocolates eaten by Monu and Tonu?

(a) 6 (b) 11
(c) 14 (d) 18

18. What is the difference in the number of chocolates eaten by Sonu and Tonu?

(a) 25 (b) 15
(c) 20 (d) 18

19. Monu and Tonu ate the chocolates in the ratio...

(a) 13 : 11 (b) 9 : 7
(c) 4 : 3 (d) 5 : 4

20. How many more chocolates did monu eat, than Jhony?

(a) 20 (b) 27
(c) 32 (d) 36

Direction (Qs. 21 to 25) : Read the following information carefully and answer the question below it.

Atul, Bhupinder, Charles and Deepak are four brothers playing a game where the loser doubles the money of each of the other players by giving them from his share at that point of time. They played four games and each brother lost one game in alphabetical order. At the end of the fourth game each brother had Rs. 64.

21. Who started with the lowest amount?
(a) Atul (b) Bhupinder
(c) Charles (d) Deepak

22. What was the amount left with Charles at the end of the second round?
(a) 72 (b) 144
(c) 132 (d) 136

23. How many rupees did Bhupinder start with?
(a) 64 (b) 136
(c) 68 (d) 72

24. At the end of the fourth round, who had the max. profit? (max. wining vis-a-vis opening amount)
(a) Atul (b) Bhupinder
(c) Charles (d) Deepak

25. Who had the minimum deviation from the opening amount?
(a) Atul (b) Bhupinder
(c) Charles (d) Deepak

Direction (Qs. 26 to 28): Read the following information carefully and answer the question below it.

Life was not easy before Aryabhatta. Zero was not invented and also people could not multiply numbers. Thus mathematics had only 9 digits (1 to 9) and after that came 11 and so on. Find the answer to following operations in mathematics used then.

26. What is 7 + 11 + 3 =?
(a) 20 (b) 21
(c) 22 (d) 23

27. What is 21 + 29 + 1 =?
(a) 50 (b) 51
(c) 52 (d) 54

28. What is 15 + 5 + 1 + 9 =?
(a) 30 (b) 31
(c) 32 (d) 33

Direction (Qs. 29 to 33): Read the following information carefully and answer the question below it.

Rahul was at a crossroads in his life. He went to an astrologer to learn what the future held for him. The astrologer explained that Rahul had many choices to make and he could only expalin what each choice lead to. On leaving the astrologer, Rahul could could take bus routes 213 or 231 to go to his next stop. The first led to a choice between two jobs and the second to a choice between two areas of business.

In all four cases, Rahul would go abroad. Rahul would marry a woman whose name begins with 'P' if he either took a job in production or had a business in readymade garments. Rahul would go to the US if he either had a job in marketing or a business in spare parts manufacture.

29. If Rahul has a wife called Pradnya and a production job, which bus did he take?
(a) 213 (b) 231
(c) (1) or (2) (d) cannot be determined

30. If Rahul took bus number 231 and married Rohini, what work does he do?
(a) marketing job
(b) production job
(c) garments business
(d) spare parts business

31. If Rahul took bus number 213 and is married to Rehana, which country does he live in?
(a) India (b) UK
(c) US (d) Canada

32. If Rahul has a spare parts business, which of the following could be his wife's name?
(a) Pradnya (b) Prachi
(c) Savita (d) Any of these

33. If Rahul works in Australia, which of the following is possible?
(a) Production job or spare parts business
(b) Production job, married to Poorva
(c) Garments business, married to Shruti
(d) Marketing job or garments business

Answers

1. (c)	2. (c)	3. (b)	4. (c)	5. (d)
6. (b)	7. (c)	8. (5)	9. (2)	10. (1)
11. (5)	12. (5)	13. (5)	14. (2)	15. (1)
16. (d)	17. (a)	18. (b)	19. (c)	20. (a)
21. (d)	22. (b)	23. (c)	24. (d)	25. (b)
26. (c)	27. (c)	27. (c)	29. (a)	30. (d)
31. (c)	32. (c)	33. (b)		

Solutions

Answer 1 to 3:

Let the number of sweets be x.

Number of sweets left with him after Temple.

$$= x - \left(\frac{x}{2} - 1\right) = \frac{x}{2} + 1$$

Number of sweets left with him after giving to the beggers $= \frac{x}{2} - 2$

Number of sweets left with him after giving to the poor children $= \frac{x}{4} - 1$

Number of sweets left with him after giving his brother $= \frac{x}{4} - 2 = 1$

This gives $x = 12$

1. (c)
2. (c)
3. (b)

Answer 4 to 8 :

4. (c) L is missing in the pyramid
5. (d) PR are not occurring together in alphabetical order. WX and UV are not vertical neighbours in a given pyramid. ST occur together in alphabetical order and also are vertical neighbours in a given pyramid.
6. (b) Y and Q; N and P; E and K are horizontal neigbours in the given pyramid, but name of these pairs does not occur together in alphabetical order. Both in the pyramid and in alphabetical order V and X are separated by exactly one letter.
7. (c) F and U are separated by 14 letters in an alphabetical order, while Q and Y are separated by 7 letters; C and S separated by 15 letters and H and Q are separated by 8 letters. Hence C and S are the farthest away from each other among given pairs.

Answer 8 to 15:

8. (5) Condition (ii) is not satisfied
9. (2) All the conditions of eligibility are satisfied
10. (1) All condition is are satisfied
11. (5) condition (ii) is not satisfied
12. (5) Condition satisfies all the five condition except conditon (vii)
13. (5) Condition satisfies all the five condition except conditon (vii)
14. (2) According to condition (iii) he should be refered to Director - Finance.
15. (1) He will be selected.

Answer 16 to 20:

Let the number of chocolates in the beginning be x.

So when Sonu woke up,

he left $\frac{2}{3}(x-1)$ for the others.

When Monu woke up he left.

$$\frac{2}{3}\left[\frac{2}{3}(x-1)-1\right] = \frac{4x-10}{9}$$

while Tonu left $= \frac{8x-38}{27}$

Finally they divided $\frac{8x-38}{27} - 1$ into three equal parts.

So, $\frac{8x-38-27}{27}$ or $\frac{8x-65}{27}$ was divided into three equal parts. This means $\frac{8x-65}{27}$ is divisible by 3 or $8x - 65$ is a multiple of 81. Let $8x - 65 = 81$ n where n is an integer

So 81n + 65 should be divisible by 8. It has to be an even number, so 81n should be odd number. Trying with 1, 3, 5 .... we get 7 and 15 as the two initial possible number but n = 15 will give us an initial starting number $x = 160$, which is not acceptable. So, n = 7 is the accepted value. So in the beginning there were 79 cholates with the given conditions. We can distribute them accordingly. The table below show number of chocolates eaten by each.

Rounds	Sonu	Monu	Tonu	Jhony
First	26	26	26	1
Second	17	17	17	1
Third	11	11	11	1
Last	7	7	7	1
Total = 79	33 ✓	24 ✓	18 ✓	4 ✓

Where numbers represent choclates distributed to each during four round. While squared number shown choclate eaten by an inidividual during that particular rounds and tick represent total choclates eaten by particular individual.

16. (d)

17. (a)

18. (b)

19. (c)

20. (a)

Answer 21 to 25:

The four brothers play in such a way that the loser doubles the money of the other players and they lose in an alphabetical order. Thus, working backwards all question can be answered.

	A	B	C	D
Initially	132	68	36	20
Round 1	8	136	72	40
Round 2	16	16	144	80
Round 3	32	32	32	160
Round 4	64	64	64	64

Refer to the table below the respective numbers represent the money after completion of the round mentioned.

21. (d)

22. (b)

23. (c)

24. (d)

25. (b)

Answer 26 to 28:

Replace 10 by 11, 20 by 22 and so on.

26. (c) 7 + 11 + 3 is actually 7 + 10 + 3 = 20 which is 22

27. (c) 21 + 29 + 1 is actually 19 + 27 + 1 = 47 which is 52

27. (c) 15 + 5 + 1 + 9 actually 14 + 5 + 1 + 9 = 29 which is 32.

Answer 29 to 33:

Draw a tree of possibilities for the predictions. You can read the answer from the tree

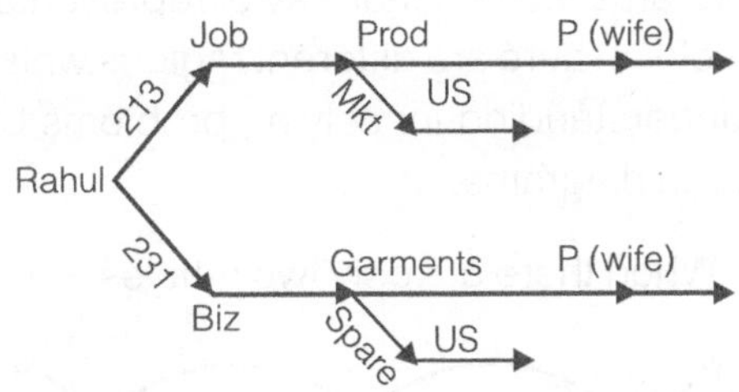

29. (a)

30. (d) Bus route 231, and not 'P' wife implies spare parts business.

31. (c) The bus number tell us that Rahul has a job, but if the wife is Rehana, it is a not a production job. Hence marketing job in US.

32. (c)

33. (b) As Rahul works in Australia he can do job in Production or Business in Garments and his wife's name will start P.

CHAPTER 3.10 Logical Venn Diagrams

Venn Diagrams

Venn Diagrams are pictorial way of representation of a set of articles. There are different regions which needs proper understanding for solving problems based on logical venn diagrams.

Case I : When there are only two articles

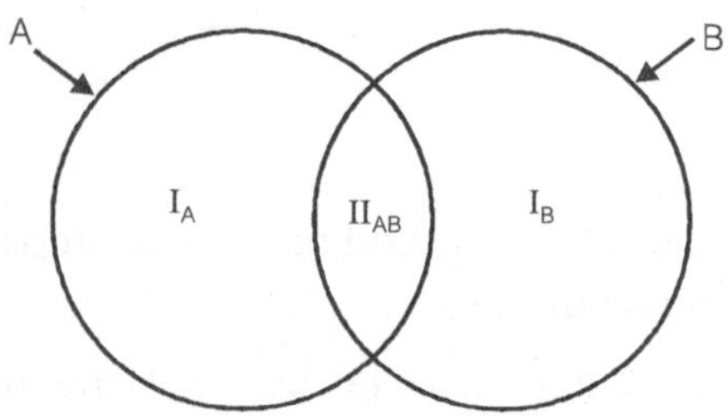

Here I_A represents A only. (i.e. not B)
I_B represents B only (i.e. not A)
II_{AB} represents A and B only i.e. both A and B.

Case II : When there are only three articles

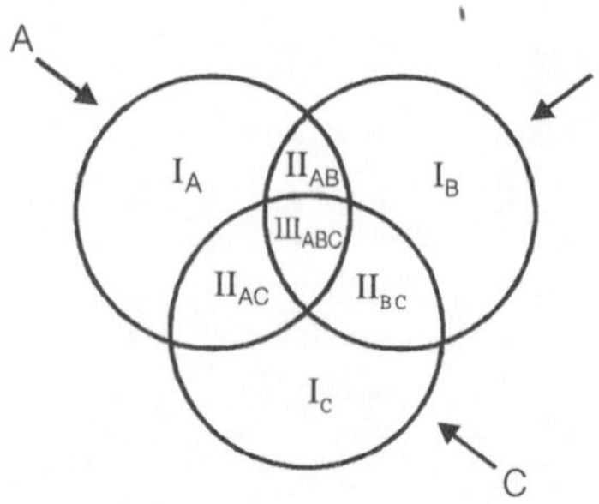

Here I_A represent A only (not B or C)
I_B represents B only (not A or C)
I_C represents C only (not A or B)
II_{AB} represents A and B only (not C)
II_{BC} represents B and C only (not A)
II_{AC} represents C and A only (not B) and
III_{ABC} represents A, B and C only i.e. region which is common to A, B and C.

Case III : When there are only four articles

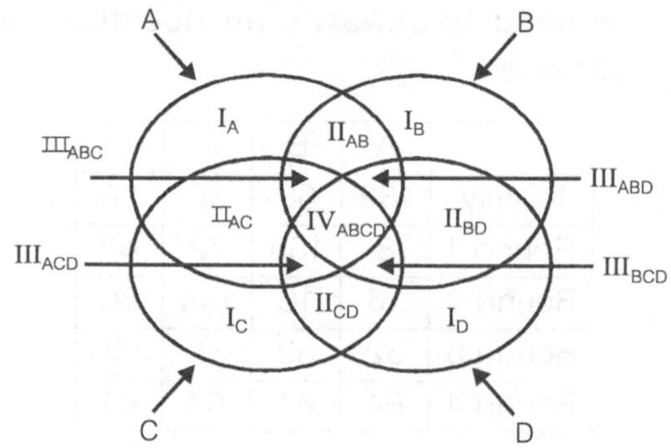

I_A represents A only (not B, C or D)
I_B represents B only (not A, C or D)
I_C represents C only (not A, B or D)
I_D represents D only (not A, B or C)
II_{AB} represents A and B only (not C or D)
II_{AC} represents A and C only (not B or D)
II_{BD} represents B and D only (not A or C)
II_{CD} represents C and D only (not A or B)
III_{ABC} represents A, B and C only (not D)
III_{BCD} represents B, C and D only (not A)
III_{ACD} represents A, C and D only (not B)
III_{ABD} represents A, B and D only (not C)
IV_{ABCD} represents A, B C and D all.
i.e. region which is common to all A, B, C and D.

Note: A, B, C and D can be represented in form of a triangle, circle, rectangle or square. The logic behind venn diagram will remain unchanged irrespective of the type of figure.

□□□□

I Practice Exercise

1. The triangle, square and circle shown below represent, the urban, hard working and educated people respectively. Which one of the areas marked I-VII is represented by the urban educated people who are not hard working?

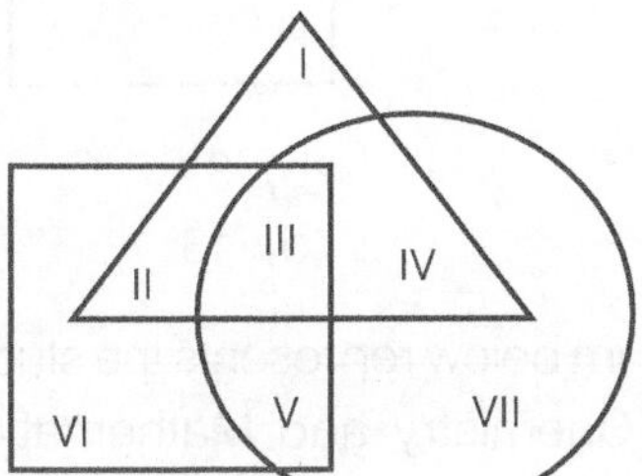

(a) II (b) I
(c) IV (d) III

Direction (Qs. 2 to 5): The figure given below consists of three interesting circles which represent sets of students who play Tennis, Badminton and Volleyball. Each region in the figure is represented by a small letter.

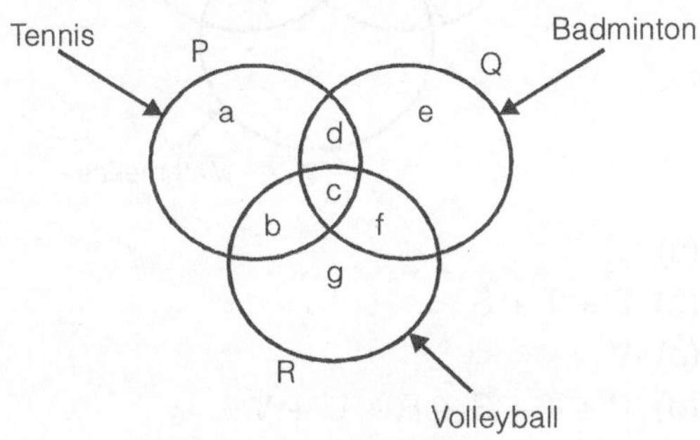

On the basis of the above figure, answer the questions given below.

2. Which letter represents the set of persons who play all the three games?
(a) b (b) c
(c) f (d) g

3. Which letter represents the set of persons who play Tennis and Volleyball but not Badminton?
(a) g (b) e
(c) c (d) b

4. Which letter represents the set of persons who play Tennis but neither Badminton nor Volleyball?
(a) a (b) b
(c) c (d) d

5. Which letter represents the set of persons who play Tennis and Badminton but not Volleyball?
(a) b (b) d
(c) c (d) f

6. In the given figure, the triangle represents girls, the square represents sports persons and the circle represents coaches. The portion in the figure which represents girls, who are sports persons but not coaches is the one labelled

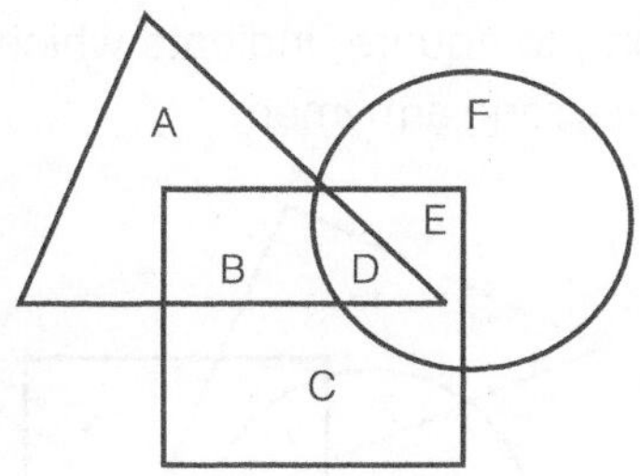

(a) A (b) B
(c) D (d) E

Direction (Qs. 7 to 11) : Below is given a figure with four intersecting circles, each representing a group of persons having the quality written against it. Study the figure carefully and answer the questions that follow.

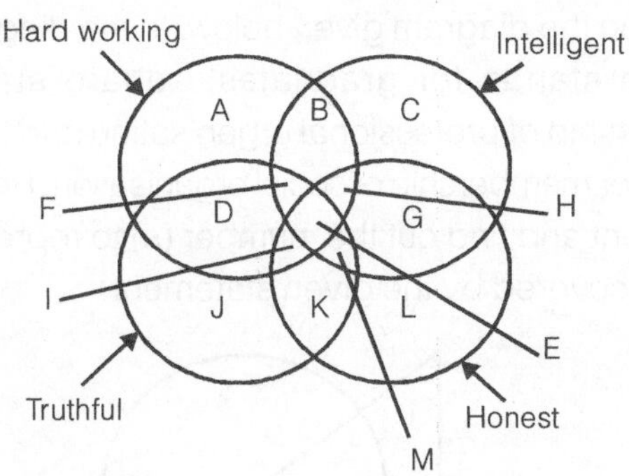

7. The region which represents the people who are intelligent, honest and truthful but not hard working is denoted by
(a) E (b) F
(c) M (d) I

8. The people possessing all the qualities are represented by
(a) I (b) H
(c) F (d) E

9. The region which represents people who are not honest but possess all other three qualities, is denoted by
(a) B (b) d
(c) F (d) I

10. People who are not hard working, intelligent and truthful are represented by
(a) G (b) H
(c) K (d) L

11. People who are not honest and truthful but are hard working and intelligent both, are represented by
(a) E (b) B
(c) M (d) I

12. If Tall is equivalent to circle, Armymen to triangle and Strong to square, indicate which number will represent strong armymen?

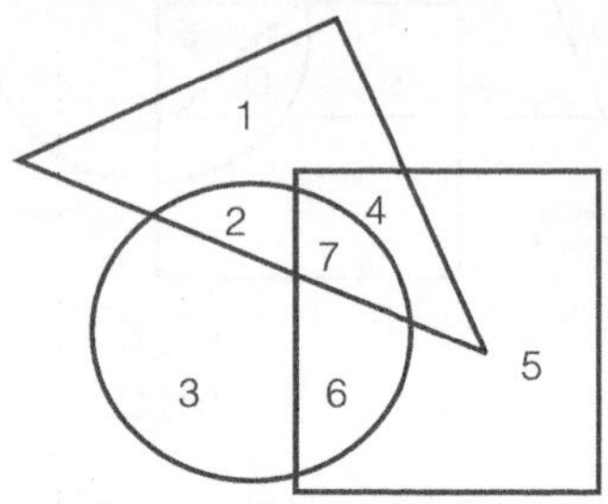

(a) 3 (b) 4
(c) 5 (d) 6

Direction (Qs. 13 to 15): The following questions are based on the diagram given below. In the diagram, the triangle stands for graduates, square stands for membership of professional organisation and the circle stands for membership of social organisation. Read each statement and find out the number (s) to represent the people covered by the given statement.

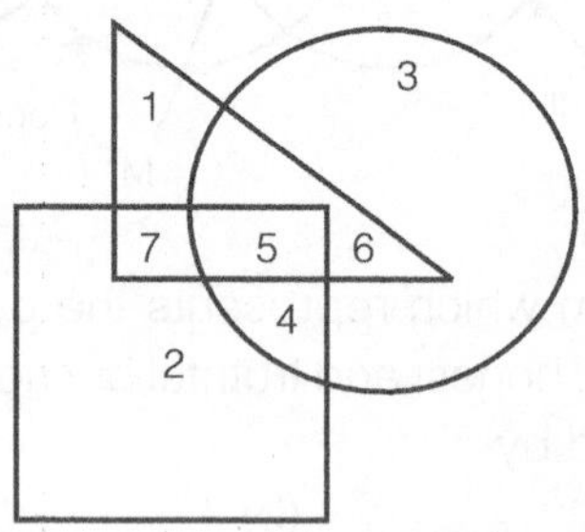

13. Number of graduates in social organisations
(a) 1 (b) 5
(c) 6 (d) 5 and 6

14. Number of a graduates in social organisations only
(a) 3 (b) 4
(c) 5 (d) 6

15. Number of graduates in professional organisation
(a) 5 and 7 (b) 5, 6 and 7
(c) 6 and 7 (d) 4, 5 and 6

16. In the following diagram, square represents women, triangle represents sub-inspectors of police and circle represents graduates. Which numbered area represents women graduate sub-inspector of police?

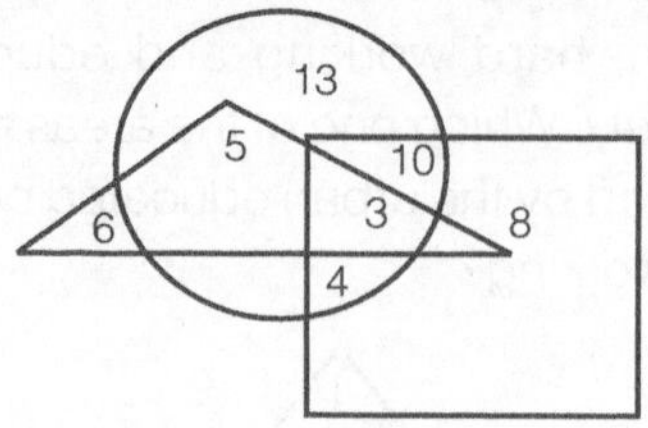

(a) 5 (b) 3
(c) 8 (d) 13

17. The diagram below represents the student who study Physics, Chemistry and Mathematics. Study the diagram and identify the region which represents the students who study Physics and Mathematics but not Chemistry.

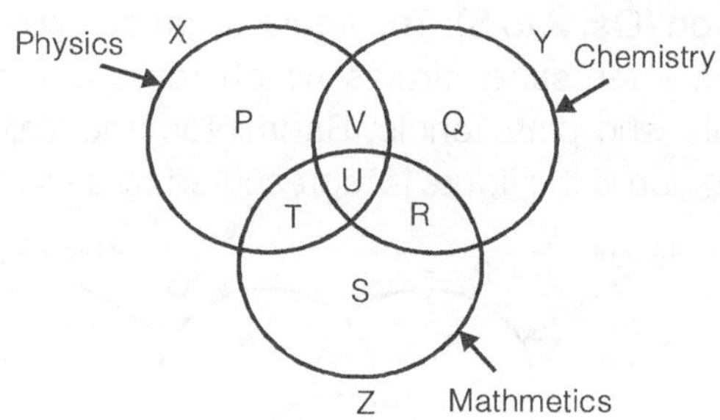

(a) T
(b) P + T + S
(c) V
(d) P + T + S + R + U + V

Direction (Qs. 18 to 25) : The following questions are based on the diagram given below:

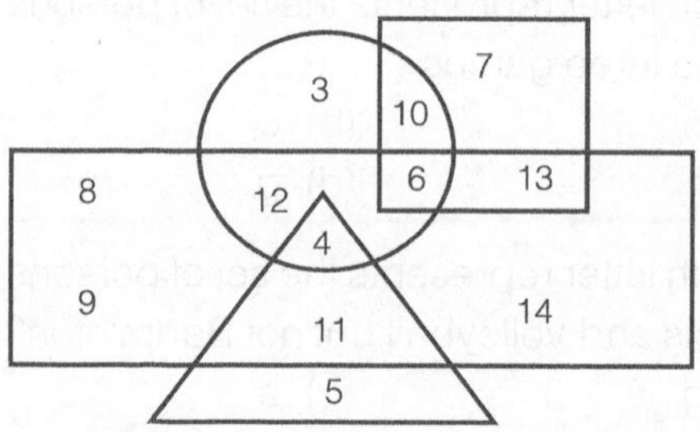

(1) Rectangle represents males
(2) Triangle represents educated
(3) Circle represents urban
(4) Square represents civil servants

18. Who among the following is an educated male who is not an urban resident?
(a) 4 (b) 5
(c) 9 (d) 11

19. Who among the following is neither a civil servant nor educated but is urban and not a male?
(a) 2 (b) 3
(c) 6 (d) 10

20. Who among the following is a female, urban resident and also a civil servant?
(a) 6 (b) 7
(c) 10 (d) 13

21. Who among the following is an educated male who hails from urban area?
(a) 4 (b) 2
(c) 11 (d) 5

22. Who among the following is uneducated and also an urban male?
(a) 2 (b) 3
(c) 11 (d) 12

23. Who among the following is only a civil servant but neither a male nor urban oriented and uneducated?
(a) 7 (b) 8
(c) 9 (d) 14

24. Who among the following is a male, urban oriented and also a civil servant but not educated?
(a) 13 (b) 12
(c) 6 (d) 10

25. Who among the following is a male civil servant, who is neither educated nor belongs to urban area?
(a) 7 (b) 13
(c) 4 (d) 1

Direction (Qs. 26 to 35): In the following figure, the circle stands for employed, the square stands for hard working, the triangle stands for rural and the rectangle stands for intelligent. Study the figure carefully and answer the questions that follow.

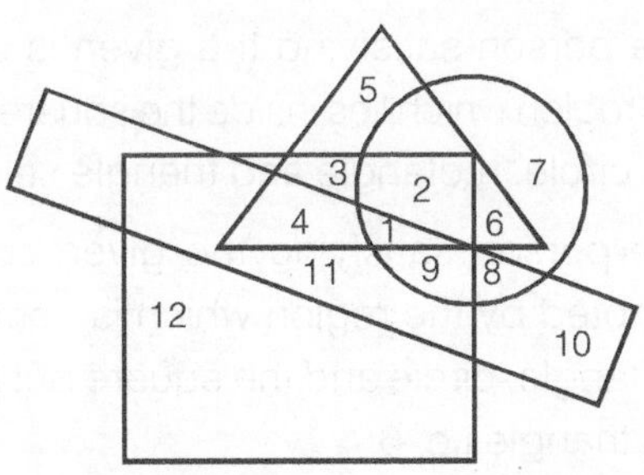

26. Non-rural, employed, hard working and intelligent people are indicated by region
(a) 8 (b) 9
(c) 10 (d) 11

27. Non-rural, employed people who are neither intelligent nor hard working are represented by region
(a) 12 (b) 11
(c) 10 (d) 7

28. Intelligent, employed and hard working non rural people are indicated by region
(a) 11 (b) 6
(c) 9 (d) 4

29. Hard working non-rural people who are neither employed nor intelligent are shown in region
(a) 8 (b) 7
(c) 6 (d) 12

30. Employed, hard working and intelligent rural people are indicated by region
(a) 1 (b) 2
(c) 3 (d) 4

31. Rural hard working people who are neither employed nor-intelligent are indicated by region
(a) 6 (b) 5
(c) 4 (d) 3

32. Rural employed people who are neither intelligent nor hard working are indicated by region
(a) 2 (b) 4
(c) 6 (d) 9

33. Rural people who are hard working and employed but not intelligent are indicated by region
(a) 1 (b) 2
(c) 3 (d) 4

34. Unemployed rural hard working and intelligent people are indicated by region
(a) 1 (b) 2
(c) 3 (d) 4

35. Rural employed people who are neither intelligent nor hard working are indicated by region
(a) 10 (b) 9
(c) 6 (d) 4

Answers

1. (c)	2. (b)	3. (d)	4. (a)	5. (b)
6. (b)	7. (c)	8. (d)	9. (c)	10. (d)
11. (b)	12. (b)	13. (d)	14. (d)	15. (a)
16. (b)	17. (a)	18. (d)	19. (b)	20. (c)
21. (a)	22. (d)	23. (a)	24. (c)	25. (b)
26. (b)	27. (d)	28. (c)	29. (d)	30. (a)
31. (d)	32. (c)	33. (b)	34. (d)	35. (c)

Solutions

1. (c) The required region is the one which is common to the triangle and the circle but is not a part of the square i.e. IV.

2. (b) The required region is the one common to all the three circles i.e. c.

3. (d) The required region is the one which is common to circles P and R but is not a part of circle Q i.e. b.

4. (a) The required region is the one which lies inside circle P but is not common to circle Q or circle R or both i.e. a.

5. (b) The required region is the one which is common to circles P and Q but lies outside circle R i.e. d.

6. (b) The required region is the region which is common to the triangle and square but lies outside the circle i.e., B.

7. (c) The required region is the one which is common to the circles 2, 3 and 4 but is not a part of circle 1 i.e. M.

8. (d) The required region is the one which is common to all the four circles i.e. E.

9. (c) The required region is the one which is common to the circles 1, 2 and 4, but lies outside circle 3 i.e., F.

10. (d) The required region is the one which does not lie inside circles 1, 2 and 4 i.e. L.

11. (b) The given conditions are satisfied by the persons denoted by the region which is common to circles 1 and 2 but is not a part of either circle 3 or circle 4 i.e. B.

12. (b) Strong armymen will be represented by the region which is common to the square and the triangle but lies outside the circle i.e. 4.

13. (d) The required region is the one common to the circle and triangle i.e regions 5 and 6

14. (d) The required region is the one common to the triangle and circle but lies outside the square i.e. 6.

15. (a) The required region is the one common to the triangle and square i.e. regions 5 and 7.

16. (b) The required region is the one common to the square, triangle and circle i.e. 3.

17. (a) The required region is the one which is common to circles X and Z but lies outside circles Y. i.e. T.

18. (d) The person satisfying the given conditions is represented by the region which is common to the triangle and the rectangle but lies outside the circle i.e. 11.

19. (b) The person satisfying the given conditions is represented by the region which lies inside the circle but outside the square, the rectangle and the triangle i.e. 3.

20. (c) The person satisfying the given conditions is represented by the region which lies outside the rectangle and is common to the circle and the square i.e. 10.

21. (a) The person satisfying the given conditions is represented by the region which is common to the triangle and the rectangle and also lies inside the circle i.e. 4.

22. (d) The person satisfying the given conditions is represented by the region which lies outside the triangle and is common to the circle and the rectangle i.e. 12.

Remember: The condition which is not mentioned shouldn't be considered or assumed. For instance, here, 6 also denotes the required region. But since it lies inside the square and there is no mention of 'civil servant', so it cannot be the answer.

23. (a) The person satisfying the given is denoted by the region which lies inside the square but outside the circle, rectangle and triangle i.e. 7.

24. (c) The person satisfying the given conditions is denoted by the region which is common to the rectangle, circle and the square but lies outside the triangle i.e. 6.

25.(b) The person satisfying the given conditions is represented by the region common to the rectangle and the square but lying outside the triangle and the circle i.e. 13.

26.(b) The required set of people is represented by the region which lies outside the triangle and is common to the circle, square and rectangle i.e. 9.

27.(d) The required set of people is represented by the region which lies outside the triangle, inside the circle but outside the rectangle and the square i.e. 7.

28.(c) The required set of people is represented by the region which is common to the rectangle, circle and square but lies outside the triangle i.e. 9.

29.(d) The required set of people of denoted by the region which lies inside the square but outside the triangle, circle and rectangle i.e. 12.

30.(a) The required set of people is denoted by the region common to the circle, square, rectangle and triangle i.e. 1.

31.(d) The required set of people is represented by the region which is common to the triangle and the square but lies outside the circle and rectangle i.e. 3.

32.(c) The required set of people is denoted by the region which is common to the triangle and the circle, but is not a part of either the rectangle or the square i.e. 6.

33.(b) The required set of people is represented by the region which is common to the triangle, square and circle but is not a part of the rectangle i.e. 2.

34.(d) The required set of people is represented by the region which lies outside the circle and is common to the triangle, square and rectangle i.e 4.

35.(c) The required set of people is denoted by the region which is common to the triangle and circle but is not a part of either the rectangle or the square i..e 6.

■■■■

3.11 CHAPTER

Analytical Reasoning

I Practice Exercise

Direction for Questions: (1 to 3)

(i) Six friends A, B, C, D, E and F are sitting along the sides of a hexagonal table for playing a game, though not necessarily in the same order.

(ii) F, who is sitting exactly opposite of A, is to the immediate right B.

(iii) D is between A and B and is exactly opposite to C.

1. A is sitting between which of the following pairs of persons?

(a) D and E (b) B and E
(c) E and C (d) None of these

2. Who is sitting opposite B?

(a) E (b) F
(c) A (d) C

3. Three of the following are alike in a certain way on the basis of sitting positions and so from a group. Which is the one that does not belong to the group?

(a) B, C (b) A, D
(c) B, F (d) E, A

Direction for Questions: (4 to 8)

A, B, C and D are four friends living together in a flat and they have an agreement that whatever edible comes they will share equally among themselves. One day A' s uncle came to him and gave a box of laddoos. Since no one was around, A divided the laddoos in four equal parts and ate his share after which he put the rest in the box. As he was closing the box, B walked in and took the box. He again divided remaining laddoos in four equal parts. A and B ate one part each and kept the remaining laddoos in the box. Suddenly C appeared and snatched the box. He again divided the laddoos in four equal parts, the three of them ate one part each and kept the remaining laddoos in the box. Later when D came, he again divided the laddoos in four equal parts and all four ate their respectively share. In total D ate 3 laddoos.

4. How many laddoos, in total did C eat?

(a) 12 (b) 15
(c) 39 (d) None of these

5. How many laddoos, in total did B eat?

(a) 24 (b) 15
(c) 39 (d) None of these

6. How many laddoos, in total did A eat?

(a) 56 (b) 68
(c) 71 (d) None of these

7. How many laddoos were given to A by his Uncle?

(a) 128 (b) 125
(c) 113 (d) None of these

8. How many laddoos, did A eat the first time?

(a) 32 (b) 24
(c) 15 (d) None of these

Direction for Questions: (9 to 13)

(i) A, B, C, D, E, F G and H are eight friends. Three of them play cricket and table tennis each and two of them play football. Each one of them has a different height

(ii) The tallest does not play football and the shortest does not play circket

(iii) F is taller than A and D but shorter than H and B. E who does not play cricket, is taller than B and is second to the tallest. G is shorter than D but taller than A.

(iv) H, who is fourth from the top, play table tennis with D.

(v) G does not play either cricket or football. B doest not play football.

9. Who is the tallest?
   (a) B (b) H
   (c) C (d) Data inadequate

10. Who is the shortest?
   (a) G (b) D
   (c) A (d) Data inadequate

11. Which of the following pairs of friends play football?
   (a) EF (b) EA
   (c) HF (d) Data inadequate

12. What is F's position from the top when they are arranged in descending order of their height?
   (a) Fifth (b) Fourth
   (c) Sixth (d) None of these

13. Which of the following group of friends play circket
   (a) CAE (b) CBF
   (c) CBA (d) None of these

Direction for Questions: (14 to 17)

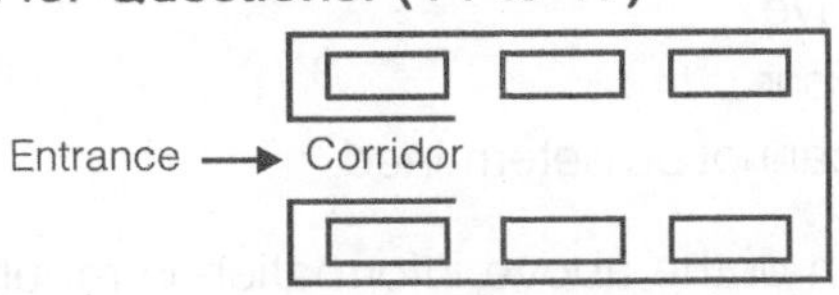

The plan above shows an office block for six officers– A, B, C, D, E and F. Both B and C occupy offices to the right of the corridor (as one enters the office block) and A occupies office to the left of the corridor. E and F occupy offices on opposite sides of the corridor but their offices do not face each other. The offices of C and D face each other. E does not have a corner office. F's office is further down the corridor than A's, but on the same side.

14. If E sits is his office and faces the corridor, whose office is to his left?
   (a) A (b) B
   (c) C (d) D

15. Whose office face A's office?
   (a) A (b) C
   (c) D (d) E

16. Who is/are F's neighbour (s)?
   (a) A only (b) A & B
   (c) C (d) B & C

17. D was heard telling someone to go further down the corridor to the last office on the right. To whose room was he trying to direct that person?
   (a) A (b) B
   (c) C (d) F

Direction (Qs.18 to 22)

A, B, C, D, E, F and G are travelling in three different vehicles. There are at least two passenger in each vehicle– Maruti, Santro, Opel– and only one of them is a male. There are two engineers, two doctors and three teachers among them.

(i) C is lady doctor and she does not travel with the pair of sisters, A and F
(ii) B, a male engineer, travels with only G, a teacher in Maruti
(iii) D is a male doctor
(iv) Two persons belonging to the same profession do not travel in the same vehicle.
(v) A is not an engineer and travels in Santro.

18. What is F's profession?
   (a) Engineer (b) Doctor
   (c) Teacher (d) Data inadequate

19. In which vehicle does C travel?
   (a) Maruti (b) Santro
   (c) Opel (d) Data inadequate

20. Which of the following represents the three teachers?
   (a) GEF (b) GEA
   (c) GBF (d) Data inadequate

21. How many lady members are there among them?
   (a) Three (b) Four
   (c) Three or four (d) Data inadequate

22. Which of the following is not correct
   (a) E-Male-Teacher
   (b) B-Male-Engineer
   (c) A-Female-Teacher
   (d) All are correct

Direction (Qs. 23 to 27)

Six friends A, B, C, D, E and F work in different companies namely– Pentasoft, Quark, Raymond's Sunmet, Trump & Gates and Udupi, and each wears company-sponsored different coloured shirts, viz. Blue, Green, Pink, Yellow, Purple and Red though not necessarily in the same order.

(i) The one wearing the Blue shirt works in Sunmet and the one wearing a Green shirt works in Pentasoft.
(ii) F does not work in Raymond's or Trump & Gates.
(iii) A wears Pink shirt and works in Quark
(iv) D does not work in Trump & Gates and Purple coloured shirt is not sponsored by Raymond's.
(v) E works in Udupi and neither D nor B works in Sunmet
(vi) Trump & Gates does not sponsor Purple or Yellow coloured shirts and C works in Pentasoft.

23. Which colour shirt is sponsored by Raymond's?
(a) Yellow
(b) Blue
(c) Pink
(d) Cannot be determined

24. Which pair is correctly matched?
(a) Red-Raymond's-A
(b) Red-Trump & Gates-B
(c) Green-Raymond's-C
(d) None of these

25. Which of the following is true?
(a) Udupi sponsors Green Shirt
(b) D is working in Trump & Gates
(c) E wears Red Shirt
(d) Red shirt is sponsored by Trump & Gates

26. What is the sequence of companies representing A, B, C, D, E & F?
(a) Quark, Pentasoft, Trump & Gates, Raymond's Udupi, Sunmet
(b) Quark, Trump & Gates, Pentasoft, Raymond's Udupi, Sunmet
(c) Quark, Pentasoft, Trump & Gates, Sunmet, Udupi, Raymond's
(d) None of these

27. If Raymond's and Sunmet decide to interchange the colours of sponsored shirts, then which two persons had to interchange their shirt?
(a) D & F (b) A & C
(c) D & E (d) B & D

Direction (Qs. 28 to 30)

Five courses – A, B, C, D and E each of one month duration are to be taught from January to May one after the other though not necessarily in the same order by lectures P, Q, R, S and T. P teaches course 'B' but not in the month of April or May. Q teaches course 'A' in the month of March. R teaches in the month of January but does not teach course 'C' or 'D'.

28. Which course is taught by S?
(a) C (b) E
(c) Either C or D (d) D

29. Which lecture's course immediately follows after course B?
(a) Q (b) P
(c) S (d) T

30. Which course is taught in the month of January?
(a) C (b) D
(c) E (d) Data inadequate

Direction (Qs. 31 to 34)

Rajeev planted some plants in his lawn but in a certain fixed pattern:
(i) In most of the rows there are neither Rose nor Marigold
(ii) There are two more row of Orchids than Tuilps and two more rows of Rose than Orchids.
(iii) There are four more rows of Rose than Tulips.
(iv) There aren't as many rows of Lilly as Fireball.
(v) There is one less Marigold row than Rose
(vi) There is just one row of Tulips
(vii) The maximum number of rows he planted is six.

31. How many rows of rose he planted?
(a) Two
(b) Five
(c) Four
(d) cannot be determined

32. Which of the above information is redundant and can be dispensed with?
(a) (i) (b) (iii)
(c) (i) and (iii) both (d) All are necessary

33. What is the sum of the rows of Orchids and Marigold he planted?
(a) Three
(b) Nine
(c) Seven
(d) Cannot be determined

34. How many rows of fireball did he plant
(a) Two (b) Six
(c) Two or Six (d) Data inadequate

Direction (Qs. 35 to 37)

(i) Five friends, Amar, Kapil, Sarvesh, Rohan and Nagesh wear trousers of different colours– red, yellow, blue, white and green (not necessarily in the same order)
(ii) Each one of them has different likings, viz, reading, playing travelling, singing and writting.
(iii) Kapil, who has liking for singing does not wear yellow trousers. Sarvesh wears red trouser and does not like reading or writing. Nagesh likes to play and does not wear blue or Yellow trousers. Amar has liking for writing and Rohan does not wear yellow or green trousers.

35. What is the colour of Kapil trousers?
(a) White (b) Blue
(c) Green (d) Data inadequate

36. What is the liking of Sarvesh?
(a) Writing (b) Travelling
(c) Reading (d) Data inadequate

37. Which of the following combintions of person–colour-liking is correct?
(a) Rohan–Blue–Reading
(b) Nagesh–White–Playing
(c) Amar–Yellow–Writing
(d) None of these

Direction (Qs. 38 to 43): Read the following information carefully and answer the questions given below it:
(i) There are six flat on a floor in two rows facing North and South are allotted to P, Q, R, S, T and U.
(ii) Q gets a North facing flat and is not next to S.
(iii) S and U get diagonally opposite flats
(iv) R, next to U, gets a South facing flat and T gets a North facing flat

38. Which of the following combinations get South facing flats?
(a) QTS (b) UTP
(c) URP (d) Data inadequate

39. Whose flat is between Q and S?
(a) T (b) U
(c) R (d) T

40. If the flats of T and P are interchanged, whose flat will be next to that of U?
(a) P (b) Q
(c) R (d) T

41. The flats of which of the other pairs than SU is diagonally opposite to each other?
(a) QP (b) QR
(c) PT (d) TS

42. To arrive at the answers to the above questions, which of the following statements can be dispensed with?
(a) None (b) (i) only
(c) (ii) only (d) (iii) only

43. Anju started walking positioning her back towards the sun. After sometime, she turned left, then turned right and then towards the left again. In which direction is she going now?
(a) North or South (b) East or West
(c) North or West (d) South or West

Direction (Qs. 44 to 48): The following questions are based on the diagram given below showing four persons stationed at the four corners of a square piece of plat as shown.

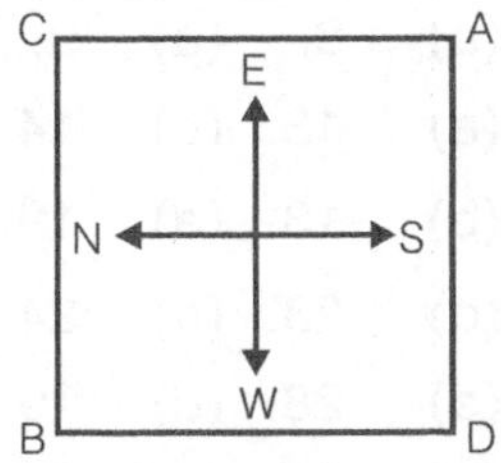

44. A starts crossing the field diagonally. After walking half the distance, he turns right, walks some distance and turns left. Which direction is A facing now?
(a) North-east (b) North-west
(c) North (d) South-east

45. From the original position given in the above figure, A and B move one arm length clockwise and then cross over to the corner diagonally opposite; C and D move one arm length anti-clockwise and cross over the corner diagonally opposite. The original configuration ADBC has now changed to
(a) CBDA (b) BDAC
(c) DACB (d) ABCD

46. From the original position, B and D move one and a half length of sides clockwise and anticlockwise respectively. Which one of the following statements is true?
(a) B and D are both at the midpoint between A and C
(b) D is at the midpoint between A and C, and B at the corner originally occupied by A.
(c) B is at the midpoint between A and C, and D at the corner originally occupied by A.
(d) B and D are both at the midpoint between A and D.

47. From the positions in original figure, C and A move diagonally to opposite corners and then one side each clockwise and anticlockwise respectively. B and D move two sides each clockwise and anticlockwise respectively. Where is A now?
(a) At the north-west corner
(b) At the north-east corner
(c) At the south-east corner
(d) At the south-west corner

48. After the movements given in Q.31 above, who is at the north-west corner?
(a) A (b) B
(c) C (d) None of these

Answers

1. (a)	2. (a)	3. (a)	4. (b)	5. (c)
6. (c)	7. (a)	8. (a)	9. (c)	10. (c)
11. (b)	12. (a)	13. (b)	14. (c)	15. (d)
16. (a)	17. (b)	18. (a)	19. (c)	20. (b)
21. (b)	22. (d)	23. (a)	24. (b)	25. (d)
26. (b)	27. (a)	28. (c)	29. (a)	30. (c)
31. (b)	32. (c)	33. (c)	34. (b)	35. (d)
36. (b)	37. (c)	38. (c)	39. (a)	40. (c)
41. (a)	42. (a)	43. (a)	44. (b)	45. (a)
46. (a)	47. (d)	48. (c)		

Solutions

Solution (1 to 3)

Construct the figure:

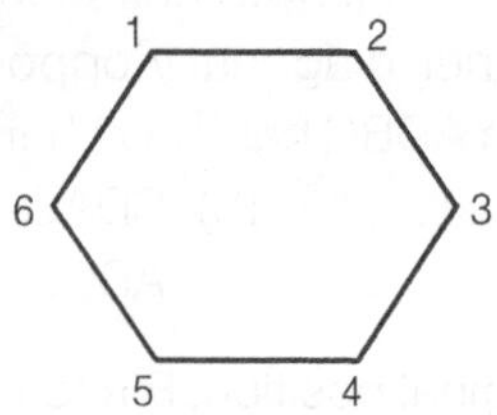

- From the second statement it is clear that F and A are sitting opposite to each other and F is to the immediate right of B.
  The sitting arrangement can be like this

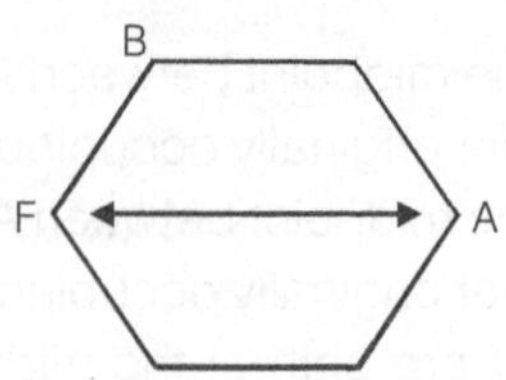

- From the third statement it is clear that D is sitting between A and B and is opposite to C. The sitting arrangement can be like this

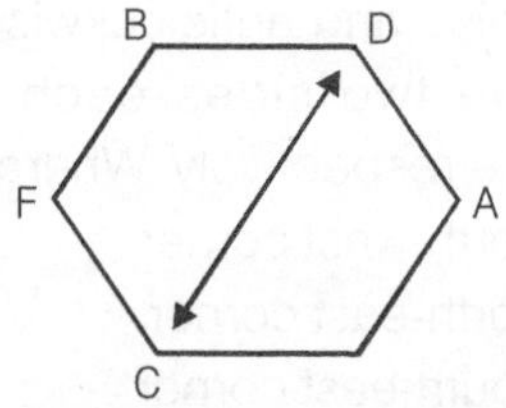

- There is only one vacant position for E, so the overall sitting arrangement will be like this

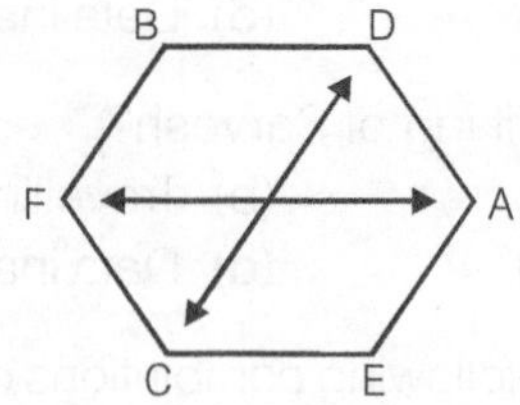

1. (a)
2. (a)
3. (a)

Solution (4 to 8)

We will start solving this problem from the last line. D ate 3 laddoos.

Before D ate the laddoos there should be 3 × 4 = 12 laddoos.

Before C ate the laddoos there should be 12 × 3 + 12 = 48 laddoos.

Before B ate the laddoos, there should be 24 × 2 + 48 = 96 laddoos. Before A ate laddos, these should be

$\frac{96 \times 4}{3}$ = 128 laddoos

At the Beginning 128 laddoos

	A	B	C	D	Remaining
A's share	32				96
B's share	24	24			48
C's share	12	12	12		12
D's share	3	3	3	3	0
	71	39	15	3	

Now we can answer all the question very easily.

4. (b)
5. (c)
6. (c)
7. (a)
8. (a)

Solution 9 to 13

We will read each statement one by one and start coding accordingly.

- Three of them play cricket and table tennis each and two of them play football.
  3 Cr 3 TT 2 Fb
  Each of them has a different height

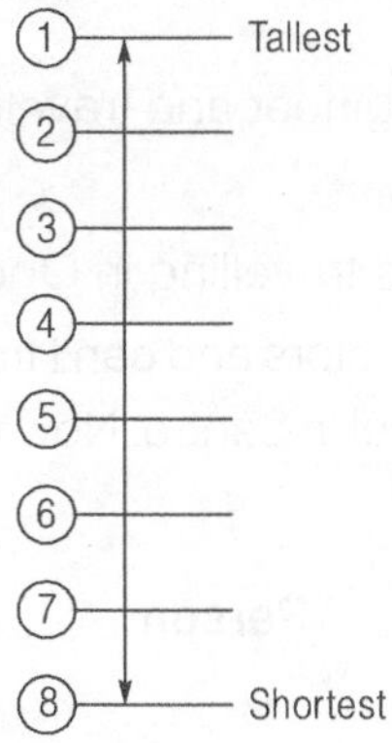

- The tallest does not play football and the shortest does not play cricket.

  (1) ~~Fb~~ (8) ~~Cr~~

- F is taller than A and D but shorter than H and B. The arrangement can be like this

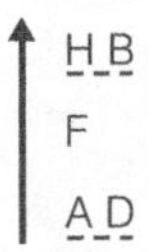

Note: Here the heights of H & B and A & D are not given.

E does not play cricket

E → ~~Cr~~

and is taller than B and also second to the tallest

So E's position is (2)

G is shorter than D but taller than A

So

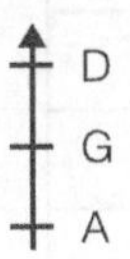

- H is fourth from the top so its position is

  H → (4)

  H plays TT with D so

  H → TT

  D → TT

- G does not play either cricket or football so G will play table tennis

  G → TT

  Now overall arrangements will be like this

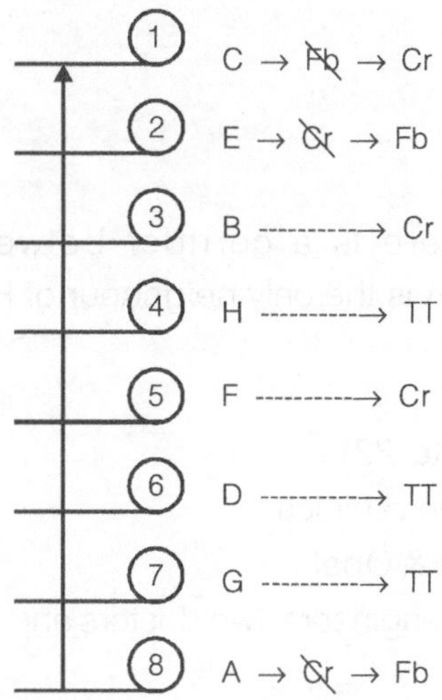

9. (c)

10. (c)

11. (b)

12. (a)

13. (b)

Solution (14 to 17)

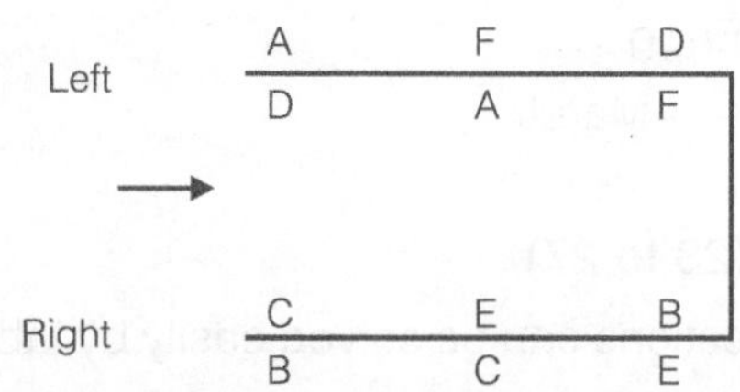

From the last sentence, we get that F is also on the left side, So E will be on the right side. D will be on the left side

E does not have a corner office so E must be in the middle.

E and F do not face each other and F's office is further down the corridor, so F occupies the last office on the left side.

Office of C and D face each other, so C will have first office on the right side and D will have the first office on the left side.

Final office arrangement will be like this

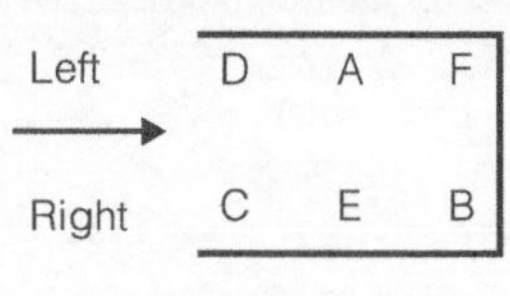

14. (c)

15. (d)

16. (a)

Since there is a corridor between office of F & B so A is the only neighbour of F.

17. (b)

Solution (18 to 22)

There are three vehicles

Maruti, Santro & Opel

There are two engineers, two doctors and three teachers among them

So 2 Engg. 2 Doc 3 Tea

- C is lady doctor and does not travel with the pair of sisters, A & F

  C → Doc

  C → ~~A~~ ~~F~~

- B is a male engineer and travels only with G a teacher in Maruti,

  so G must be female also

  G → Tea

  B → Engg.

  B G ----→ Maruti

- D is male doctor

  D → Doc

- A is not an engineer and travels in Santro so

  A F → Santro

  Obviously C is travelling in Opel.

  C and D are doctors and can't travel in same vehicle so D must travel in Santro. Now we will get following table.

Vehicle	Person	Profession
Maruti	B	Engg.
	G	Tea.
Santro	A	Tea.
	F	Engg.
	D	Doc.
Opel	C	Doc.
	E	Tea.

18. (a)

19. (c)

20. (b)

21. (b)

22. (d)

Solution (23 to 27)

These questions can be solved easily by tabular chart.

Companies \ Friends	A	B	C	D	E	F
Pentasoft						
Quark						
Raymonds						
Sunmet						
T & G						
Udupi						

Companies \ Colour	Blue	Green	Pink	Yellow	Purple	Red
Pentasoft						
Quark						
Raymonds						
Sunmet						
T & G						
Udupi						

We will start reading each conditions and mark a '✓' sign or '×' sign accordingly.

From the condition (i) we get following tables.

Companies \ Friends	A	B	C	D	E	F
Pentasoft						
Quark						
Raymonds						
Sunmet						
T & G						
Udupi						

Companies \ Colour	Blue	Green	Pink	Yellow	Purple	Red
Pentasoft	×	✓	×	×	×	×
Quark	×	×				
Raymonds	×	×				
Sunmet	✓	×	×	×	×	×
T & G	×	×				
Udupi	×	×				

Further applying condition (ii) we get following tables.

Companies \ Friends	A	B	C	D	E	F
Pentasoft						
Quark						
Raymonds						×
Sunmet						
T & G						×
Udupi						

Companies \ Colour	Blue	Green	Pink	Yellow	Purple	Red
Pentasoft	×	✓	×	×	×	×
Quark	×	×				
Raymonds	×	×				
Sunmet	✓	×	×	×	×	×
T & G	×	×				
Udupi	×	×				

Applying condition (iii) we get following tables.

Companies \ Friends	A	B	C	D	E	F
Pentasoft	×	×				
Quark	✓	×	×	×	×	×
Raymonds	×					×
Sunmet	×					
T & G	×					×
Udupi	×					

Companies \ Colour	Blue	Green	Pink	Yellow	Purple	Red
Pentasoft	×	✓	×	×	×	×
Quark	×	×	✓	×	×	×
Raymonds	×	×	×			
Sunmet	✓	×	×	×	×	×
T & G	×	×	×			
Udupi	×	×	×			

Further applying condition (iv) we get following tables.

Companies \ Friends	A	B	C	D	E	F
Pentasoft	×	×				
Quark	✓	×	×	×	×	×
Raymonds	×					×
Sunmet	×					
T & G	×			×		×
Udupi	×					

Companies \ Colour	Blue	Green	Pink	Yellow	Purple	Red
Pentasoft	×	✓	×	×	×	×
Quark	×	×	✓	×	×	×
Raymonds	×	×	×		×	
Sunmet	✓	×	×	×	×	×
T & G	×	×	×			
Udupi	×	×	×			

Further applying condition (v) we get following tables.

Companies \ Friends	A	B	C	D	E	F
Pentasoft	×					
Quark	✓	×	×	×	×	×
Raymonds	×				×	×
Sunmet	×	×		×	×	
T & G	×			×	×	×
Udupi	×	×	×	×	✓	×

Companies \ Colour	Blue	Green	Pink	Yellow	Purple	Red
Pentasoft	×	✓	×	×	×	×
Quark	×	×	✓	×	×	×
Raymonds	×	×	×		×	
Sunmet	✓	×	×	×	×	×
T & G	×	×	×			
Udupi	×	×	×			

Further applying (vi) we get following tables.

Companies \ Friends	A	B	C	D	E	F
Pentasoft	×	×	✓	×	×	×
Quark	✓	×	×	×	×	×
Raymonds	×		×		×	×
Sunmet	×	×	×	×	×	
T & G	×		×	×	×	×
Udupi	×	×	×	×	✓	×

Companies \ Colour	Blue	Green	Pink	Yellow	Purple	Red
Pentasoft	×	✓	×	×	×	×
Quark	×	×	✓	×	×	×
Raymonds	×	×	×		×	
Sunmet	✓	×	×	×	×	×
T & G	×	×	×	×	×	
Udupi	×	×	×			

Now we will mark '✓' on the only left space and mark '×' on the remaining spaces. Then the following table emerges.

Companies \ Friends	A	B	C	D	E	F
Pentasoft	×	×	✓	×	×	×
Quark	✓	×	×	×	×	×
Raymonds	×	×	×	✓	×	×
Sunmet	×	×	×	×	×	✓
T & G	×	✓	×	×	×	×
Udupi	×	×	×	×	✓	×

Companies \ Colour	Blue	Green	Pink	Yellow	Purple	Red
Pentasoft	×	✓	×	×	×	×
Quark	×	×	✓	×	×	×
Raymonds	×	×	×	✓	×	×
Sunmet	✓	×	×	×	×	×
T & G	×	×	×	×	×	✓
Udupi	×	×	×	×	✓	×

With the help of the above tables we can solve all the questions easily.

Note: After drawing table further modification can be done according to given conditions. There is no need to draw further tables. Tables drawn further is just to explain the steps in elaborate manner.

23. (a)

24. (b)

25. (d)

26. (b)

27. (a)

Solution (28 to 30)

We can be solved these questions with the help of tabular chart.

Courses \ Months	Jan	Feb	Mar	Apr	May
A					
B					
C					
D					
E					

Courses \ Lecturers	P	Q	R	S	T
A					
B					
C					
D					
E					

We will start reading each condition and mark '✓' sign or '×' sign accordingly.

- P teaches course B but not in the month of April or May

Courses \ Months	Jan	Feb	Mar	Apr	May
A					
B				×	×
C					
D					
E					

Courses \ Lecturers	P	Q	R	S	T
A	×				
B	✓	×	×	×	×
C	×				
D	×				
E	×				

- Q teaches course A in the month of March

Courses \ Months	Jan	Feb	Mar	Apr	May
A	×	×	✓	×	×
B			×	×	×
C			×		
D			×		
E			×		

Courses \ Lecturers	P	Q	R	S	T
A	×	✓	×	×	×
B	✓	×	×	×	×
C	×	×			
D	×	×			
E	×	×			

- R teaches in the month of January but does not teach course C or D.

Courses \ Months	Jan	Feb	Mar	Apr	May
A	×	×	✓	×	×
B	×	✓	×	×	×
C	×	×	×		
D	×	×	×		
E	✓	×	×	×	×

Courses \ Lecturers	P	Q	R	S	T
A	×	✓	×	×	×
B	✓	×	×	×	×
C	×	×	×		
D	×	×	×		
E	×	×	✓	×	×

Now, we can easily answer given questions with the help of above table.

Note: After drawing first table further modifications can be done according to conditions given. There is no need to draw further table. Further table drawn is just to explain the steps in elaborate manner.

Final table can be referred as.

28. (c)

29. (a)

30. (c)

Month	Subject	Teacher
January	E	R
February	B	P
March	A	Q
April	C/D	S/T
May	D/C	T/S

Solution (31 to 34)

$\rightarrow$ O = T + 2

R = O + 2

Also

R = M + 1

When T = 1, O = 3, R = 5, M = 4

Since from (iv) There are not as many rows of lily as fireball

$\therefore$ Lily < Fireball

So, Lily = 2, Fireball = 6

Now we can answer all the question easily.

31. (b)

32. (c)

33. (c)

34. (b)

Solution (35 to 37)

These questions can be solved easily by drawing tabular chart.

Friends \ Colour	Red	Yellow	Blue	White	Green
Amar					
Kapil					
Sarvesh					
Rohan					
Nagesh					

Friends \ Liking	Read	Play	Travel	Sing	Write
Amar					
Kapil					
Sarvesh					
Rohan					
Nagesh					

We will start reading each condition and mark '✓' sign or '×' sign accordingly.

- Kapil, who has liking for singing does not wear yellow trouser.

Friends \ Colour	Red	Yellow	Blue	White	Green
Amar					
Kapil		×			
Sarvesh					
Rohan					
Nagesh					

Friends \ Liking	Read	Play	Travel	Sing	Write
Amar				×	
Kapil	×	×	×	✓	×
Sarvesh				×	
Rohan				×	
Nagesh				×	

- Sarvesh wears read trouser and does not like reading or writing.

Friends \ Colour	Red	Yellow	Blue	White	Green
Amar	×				
Kapil	×	×			
Sarvesh	✓	×	×	×	×
Rohan	×				
Nagesh	×				

Friends \ Liking	Read	Play	Travel	Sing	Write
Amar				×	
Kapil	×	×	×	✓	×
Sarvesh	×			×	×
Rohan				×	
Nagesh				×	

- Nagesh like to play and does not wear blue or yellow trousers.

Friends \ Colour	Red	Yellow	Blue	White	Green
Amar	×				
Kapil	×	×			
Sarvesh	✓	×	×	×	×
Rohan	×				
Nagesh	×	×	×		

Friends \ Liking	Read	Play	Travel	Sing	Write
Amar		×		×	
Kapil	×	×	×	✓	×
Sarvesh	×	×		×	×
Rohan		×		×	
Nagesh	×	✓	×	×	×

- Amar has liking for weiting and Rohan does not wear yellow or green trousers.

Friends \ Colour	Red	Yellow	Blue	White	Green
Amar	×	✓			
Kapil	×	×			
Sarvesh	✓	×	×	×	×
Rohan	×	×			×
Nagesh	×	×	×		

Friends \ Liking	Read	Play	Travel	Sing	Write
Amar	×	×	×	×	✓
Kapil	×	×	×	✓	×
Sarvesh	×	×	✓	×	×
Rohan	✓	×	×	×	×
Nagesh	×	✓	×	×	×

Now, we can easily answer given questions with the help of above table.

Note: After drawing first table further modifications can be done according to conditions given there is no need to draw further table. Further table drawn is just to explain the steps in elaborate manner.

35. (d)

36. (b)

37. (c)

Answer 38 to 43

Q gets a North-facing flat and is not next to S means

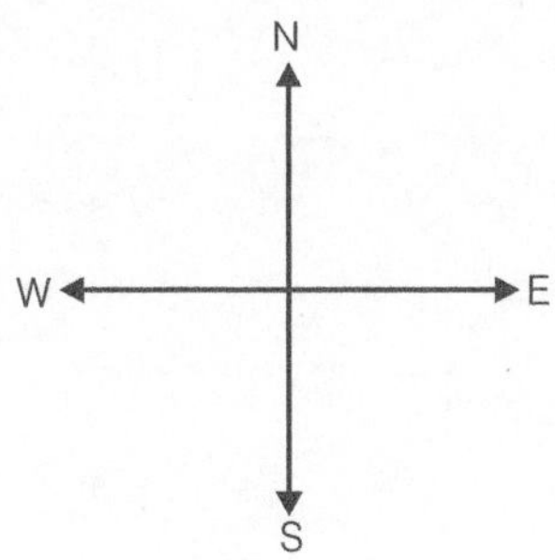

(South facing flats) ↓

Q		S

(North facing flats) ↑

S and U get diagonally opposite flats means

U		
Q		S

R, next to U, gets a South facing flat and T gets a north facing flat means

U	R	P
Q	T	S

So, the arrangement is:

South facing flats: U R P

North facing flats: Q T S

38. (c) The South facing flats are U, R, P

39. (a) T's flat is between Q and S

40. (c) The flat next to U's flat is of R, which remains unchanged if the flats of T and P are interchanged

41. (a) The diagonally opposite pairs are SU and QP

42. (a) Clearly, all the statements are necessary to answer the given questions

43. (a) Clearly, there are two possible movements of Anju as shown below:

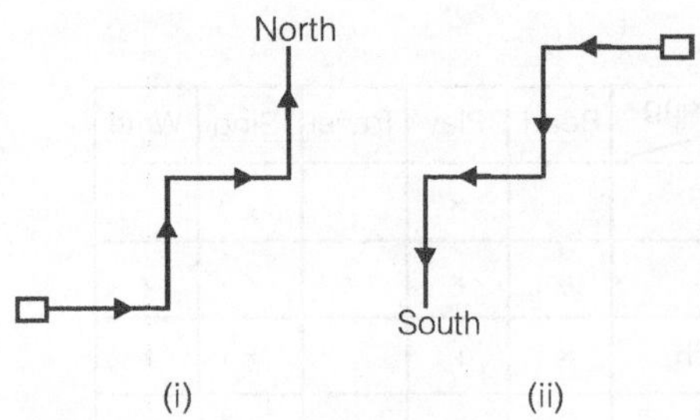

Thus, Anju is finally moving towards either North or South

44. (b) Clearly, the route of A is as shown Comparing it with the given diagram, the direction of A will be north-west

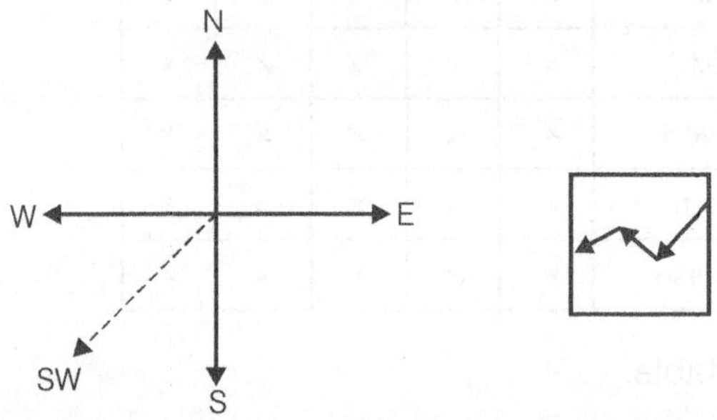

45. (a) Clearly, (i), (ii), (iii) and (iv) show the movements of A, B, C, and D respectively while the new arrangement so obtained is shown in (v). So, the congiruration changes to CBDA.

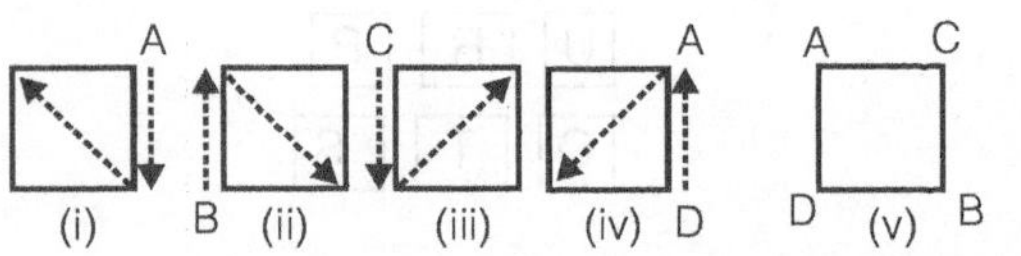

Final position

46. (a) The movements of B and D are clearly shown in the adjoining diagram
So, statements (a) is true.

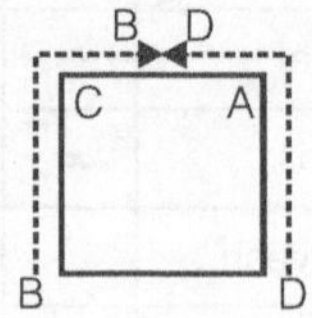

47. (d) The movements of A, C, B, and D are shown in figures (i), (ii), (iii) and (iv) respectively. The final configuration is shown in (v). Comparing (v) with the given diagram, A is in the south-west corner.

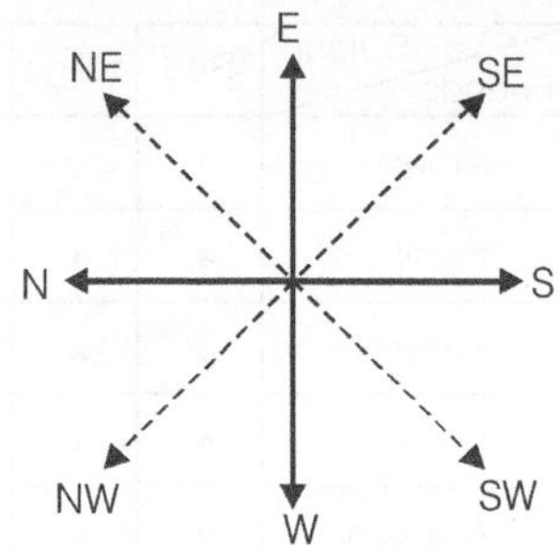

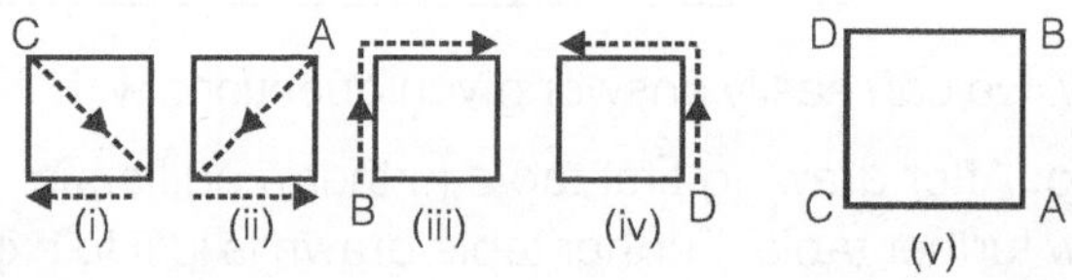

(Final Position)

48. (c) Clearly, C is at the north-west corner

CHAPTER 3.12

Figure Based Reasoning

These questions measure one's ability to infer a set of rules from a flowchart or similar diagram, and then to apply those rules to a new situation as illustrated in the solved examples in this chapter.

Directions (Q.1-Q.3): Figures given below are changing according to certain rule from left to right. Identity the missing figure of the series.

Q.1 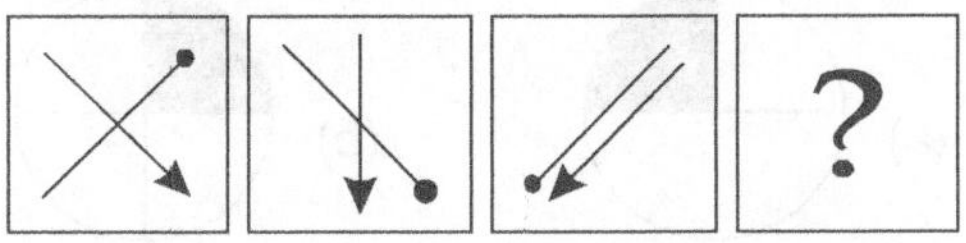

(a) 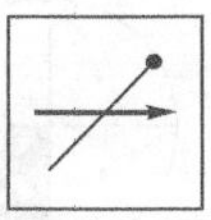 (b) 

(c) 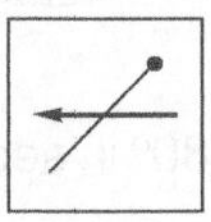 (d) 

Ans. (b)

Observe the pattern of both lines. Line with triangle at the end moves across half the side of square whereas line with circle at the end moves across full side of square.

Q.2 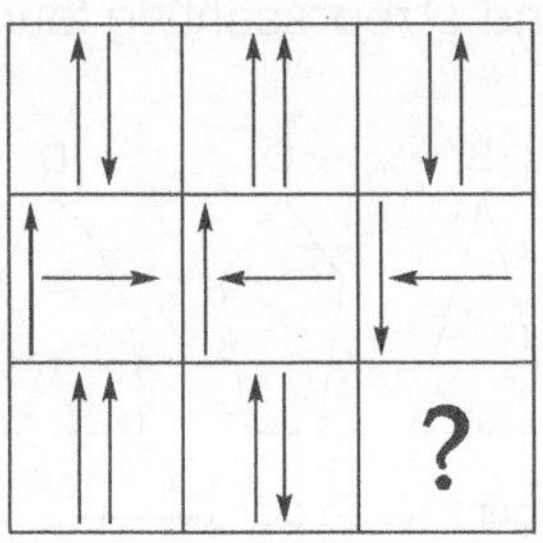

(a) 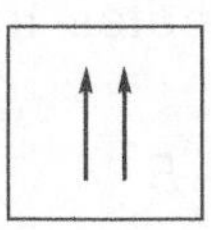 (b) 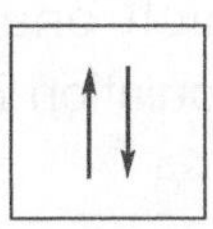

(c) 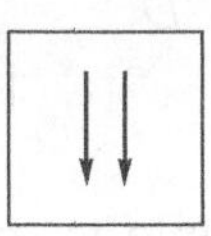 (d) 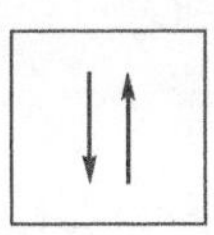

Ans. (c)

Observe the pattern of each arrow in a row. In the second column, first arrow remains in the same position while the second changes its orientation. In the third column first changes its orientation while second remains in same position.

Q.3 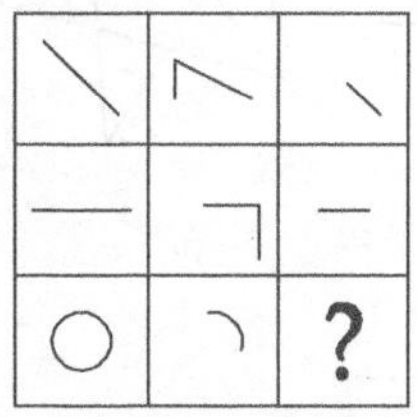

(a) 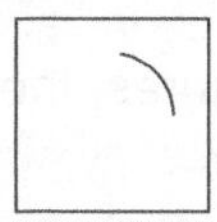 (b) 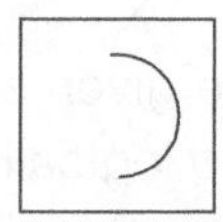

(c)  (d) 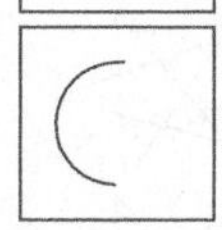

Ans. (a)

In every row, third figure in made out of the common parts of first and second figure.

Directions (Q.4-Q.5): Given are the scattered pieces of figure from the given options. Find out which figures can be made from these scattered figures.

Q.4 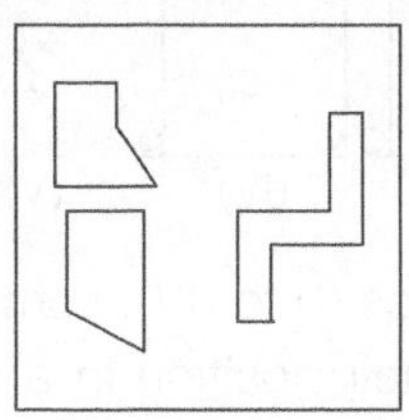

(a) 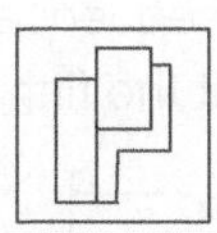 (b) 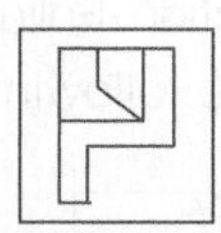

(c) 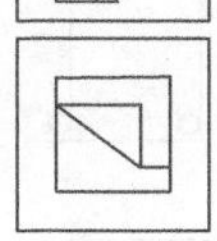 (d) 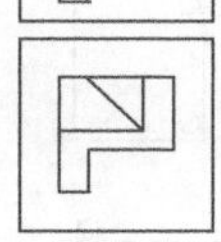

Ans. (b)

Q.5 

(a)  (b) 

(c)  (d) 

Ans. (a)

Q.6 Consider the figures given below:

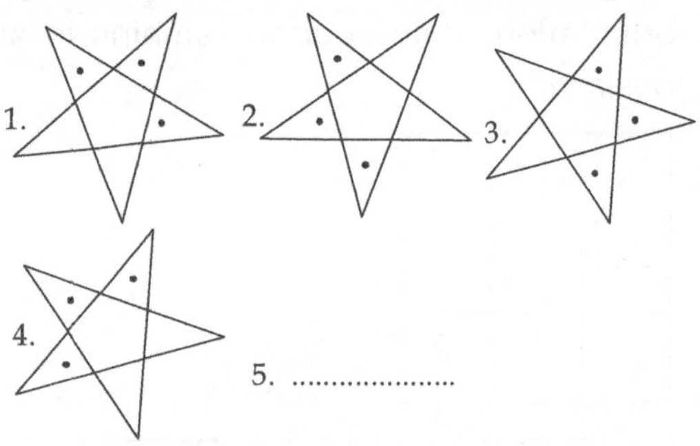

In the given sequence of figures, the figure that would logically appear at 5 is

(a) 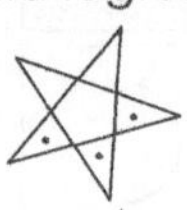 (b) 

(c)  (d) 

Ans. (a)

Circle are shifting in anticlockwise direction by two triangles.

Q.7 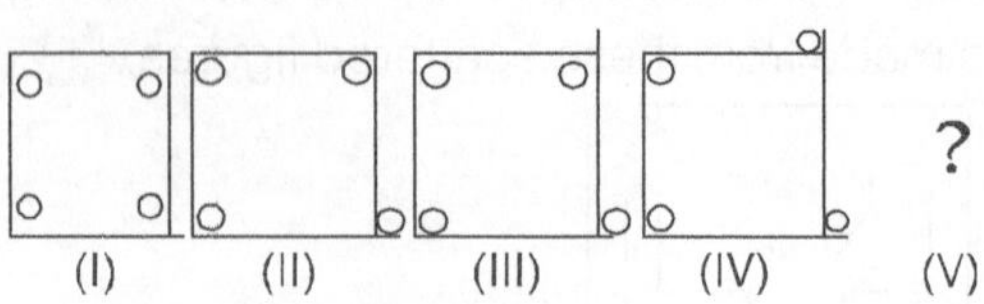

In the above set of figures (I) to (IV), some parts are shown to change their position in a regular direction. Following the same sequence, which one of the following will appear at the fifth stage?

(a) 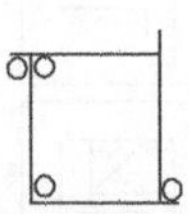 (b) 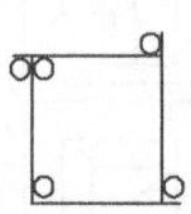

(c) 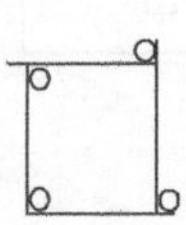 (d) 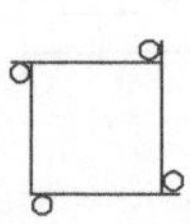

Ans. (c)

In alternating figure one side of square extends in anticlockwise direction and in the subsequent figure circle comes out.

Q.8 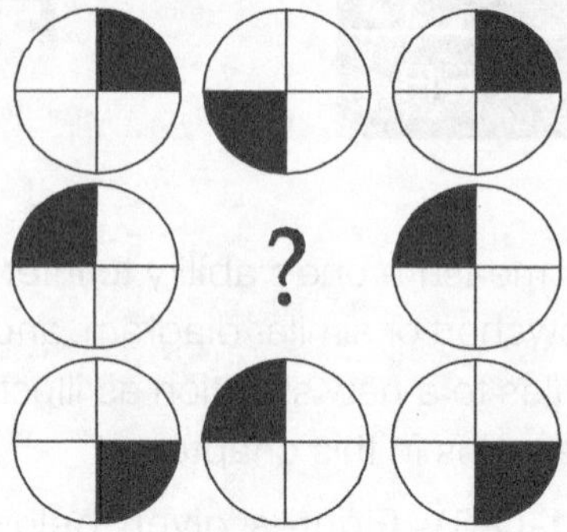

Which one of the figures shown below occupies the blank space (?) in the matrix given above?

(a) 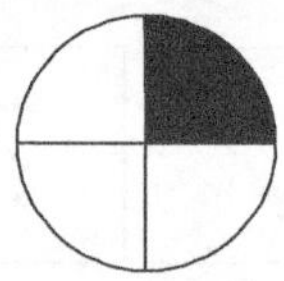 (b) 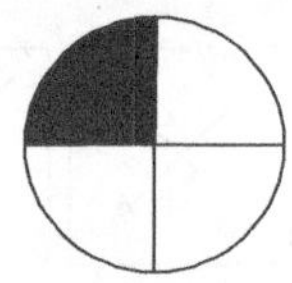

(c) 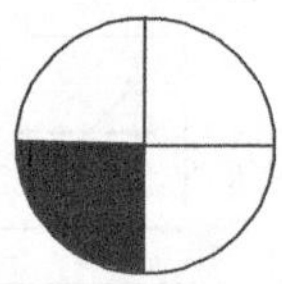 (d) 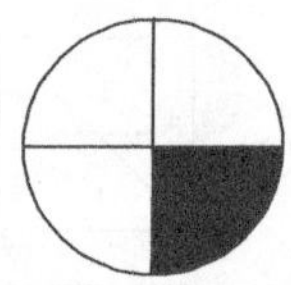

Ans. (d)

Dark park of circle rotates 180° in second figure of every row.

Previous Years Solved Questions

Q.1 Consider the four age pyramids given below namely *A*, *B*, *C* and *D* representing four different countries.

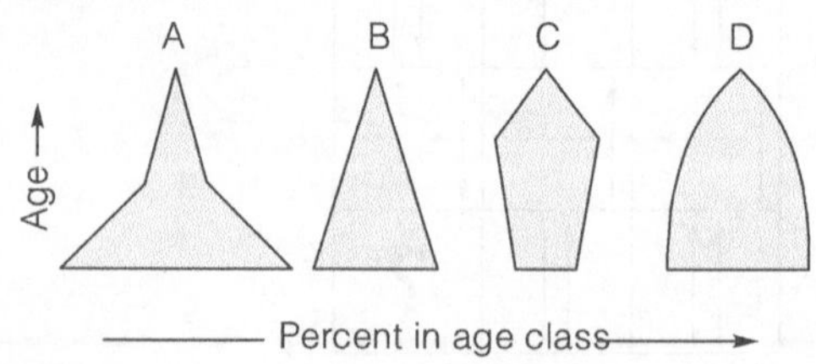

Which one of them indicates the declining population?

(a) *A* (b) *B*
(c) *C* (d) *D*

[UPSC 2011]

Ans. (c)

In *C* number of older people is more than the children born, therefore it will represent declining population.

Q.2 Consider the following figure and answer the item that follows:

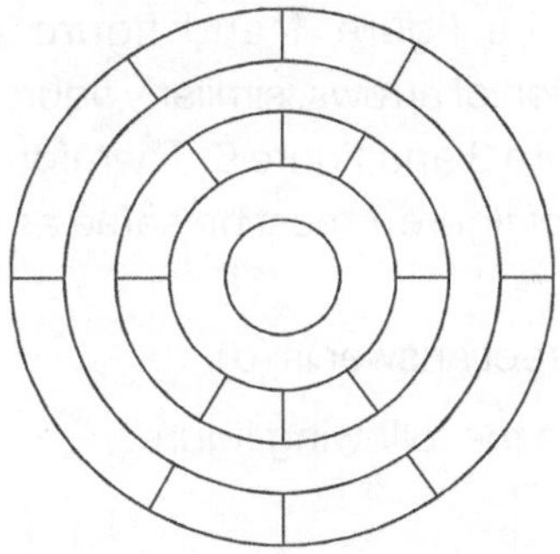

What is the minimum number of different colours required to paint the figure given above such that no two adjacent regions have the same colour ?

(a) 3 (b) 4

(c) 5 (d) 6

[UPSC 2011]

Ans. (a)

We can use three colors to paint the complete figure in the following way

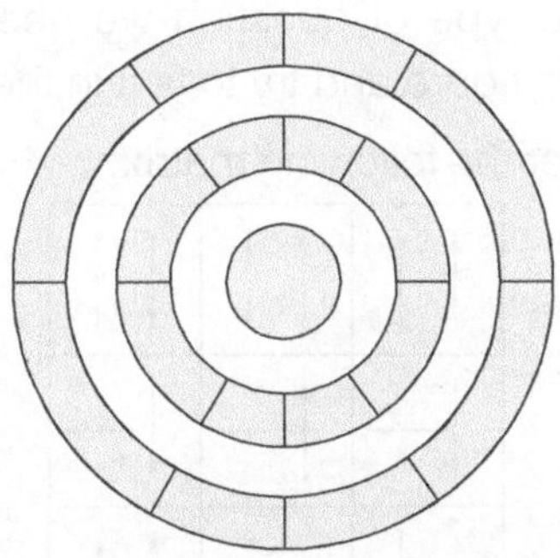

Q.3 Figures given below are changing with certain rules as we observe them from left to right:

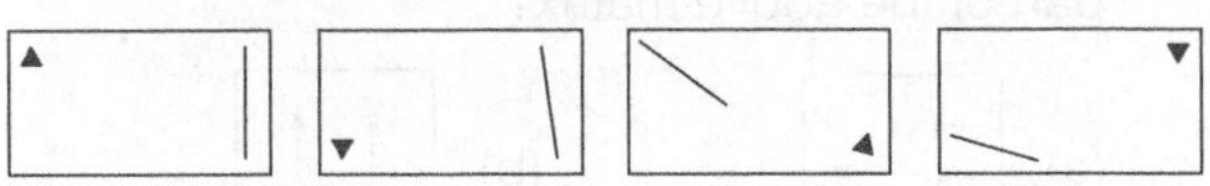

According to this rule which of the following would be the next figure if the changes were made in the same rule?

(a) 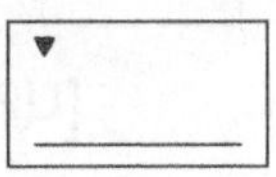 (b) 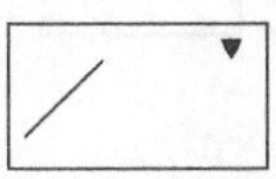

(c)  (d) 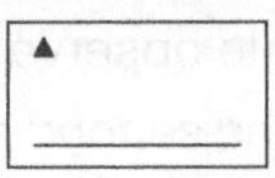

[UPSC 2012]

Ans. (d)

Follow the triangle. It is moving through the corners anticlockwise and turning upside down alternatively. The next figure must have triangle at the top left corner and must point towards up. Therefore (d) is the answer.

Q.4 Consider the following figure 1, 2, 3 and 4

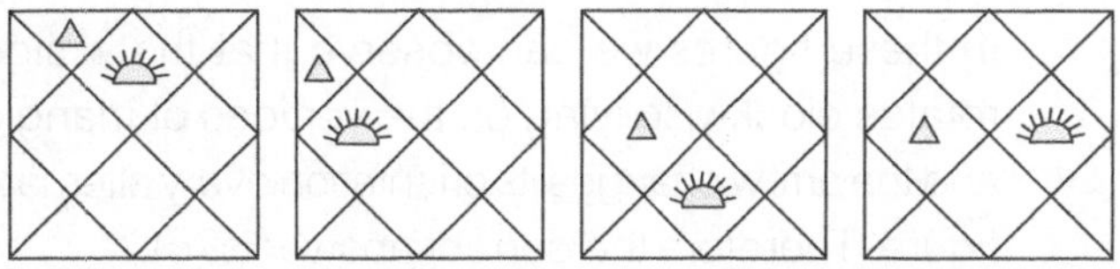

1 2 3 4

In the figure from 1 to 4 above, two symbols are shown to change their position in a regular direction. Following the same sequence, which one of the following will appear at the fifth stage?

(a)  (b) 

(c)  (d) 

[UPSC 2013]

Ans. (b)

In this sequence triangle moves anticlockwise and the other symbol moves anticlockwise but in the inner square only. So the obvious answer is option (b).

Directions (Q.5-Q.6): In each item, there are two sets of figures; first four figures named Problem figures and next four figures named Answer figures indicated as (a), (b), (c) and (d). The problem figures follow a particular sequence. In accordance with the same, which one of the four answer figures should appear as the fifth figure?

Q.5 Problem Figure:

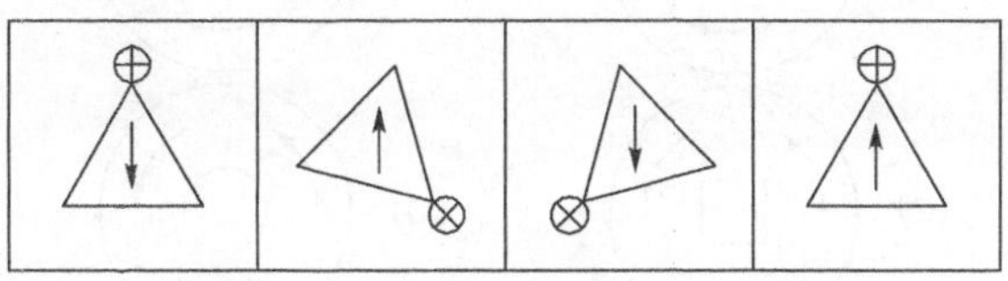

Answer Figure

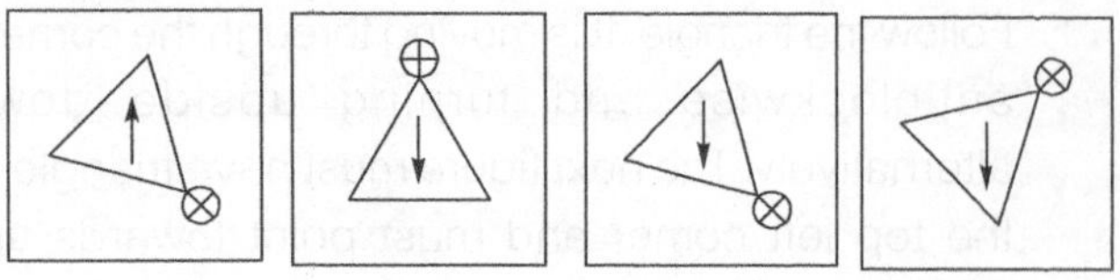

(a) (b) (c) (d)

[UPSC 2013]

Ans. (c)

In these figures we can observe that the + sign rotates clockwise lying on the vertices of triangle and the arrow changes its orientation every alternate figure. Therefore the correct answer is (c).

Q.6 Problem figure

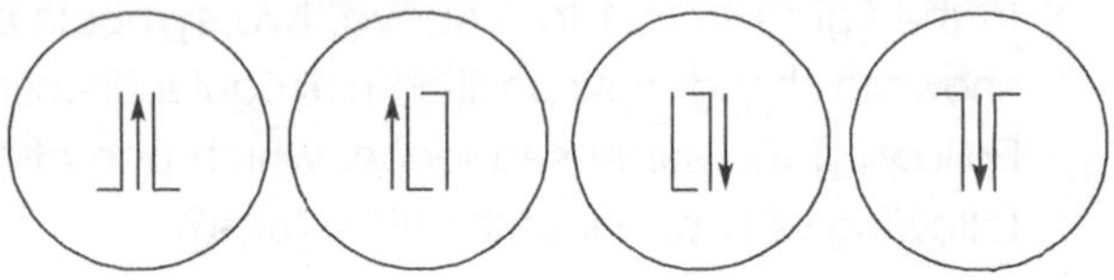

Answer Figure

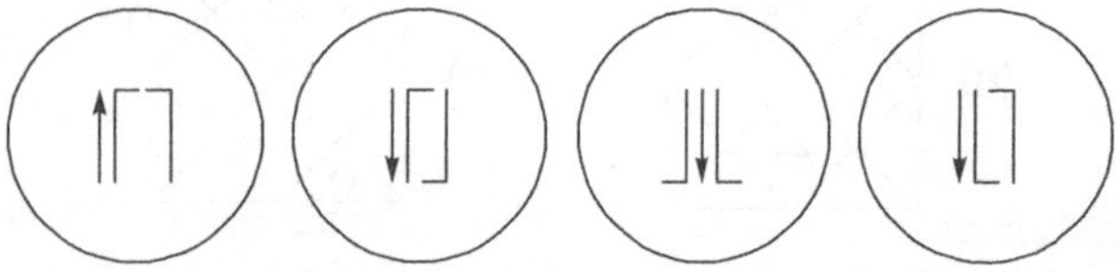

(a) (b) (c) (d)

[UPSC 2013]

Q.7

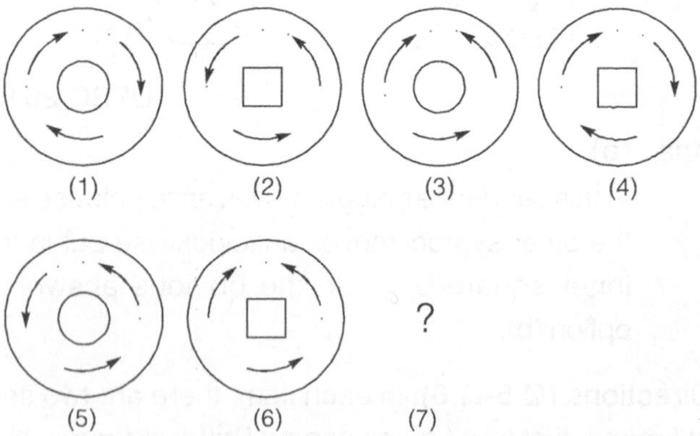

Which one of the following figures would logically come in the 7th position indicated above by a question mark?

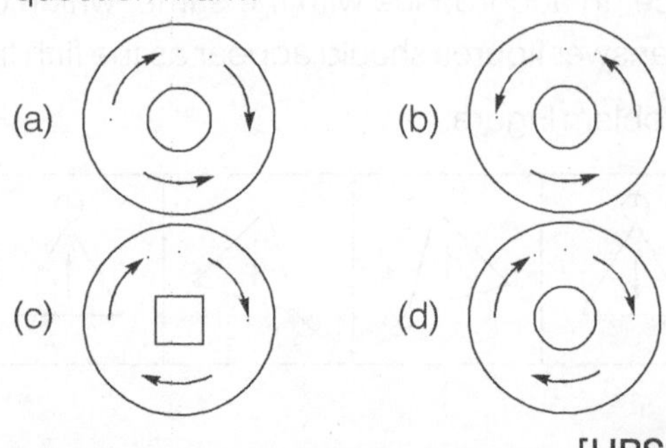

[UPSC 2013]

Ans. (d)

Circle and square comes in the alternate figures. The orientation of arrows repeats after every three figures i.e figure 1 and figure 4 have same orientation of arrows, similarly figure 2 and figure 5 and figure 3 and figure 6. Therefore orientation of arrows of figure 7 must be same as that of figure 4 and figure 1.

The correct answer is (d)

Q.8 Examine the following figure:

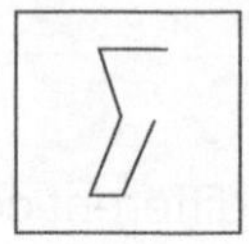

Which one of the following figures has the above figure embedded in it?

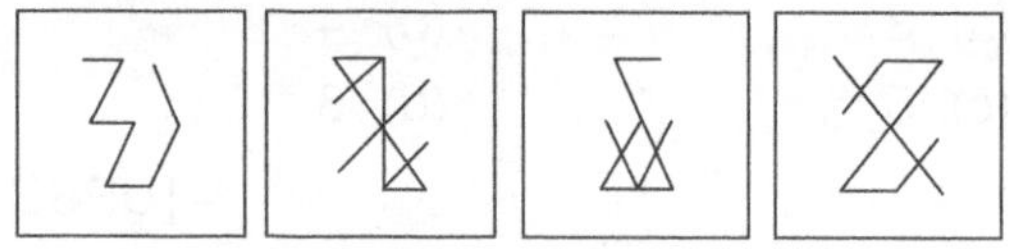

(a) (b) (c) (d)

[UPSC 2014]

Ans. (c)

In these type of question we need to take the question figure and try to fit it in the options.

Q.9 Consider the following matrix:

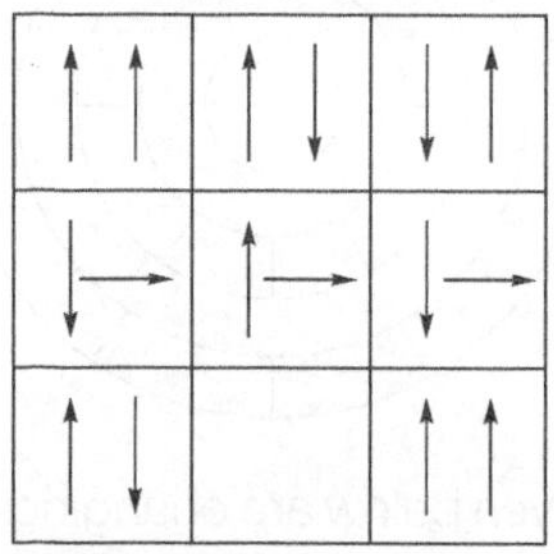

Which one of the following figures fits into the blank part of the above matrix?

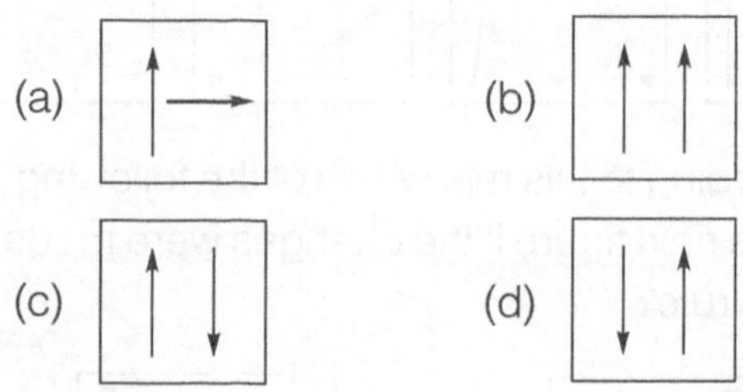

[UPSC 2014]

Ans. (b)

Try to observe the following in Row 1 and Row 2

In the second column, second arrow changes the direction while the first arrow remains in its original position. Now in the third column, both the arrows change their orientation vis a vis second column. Doing the same for the third row we will get (b) as the answer.

Q.10 Consider the following matrix with one empty block in the lower extreme corner:

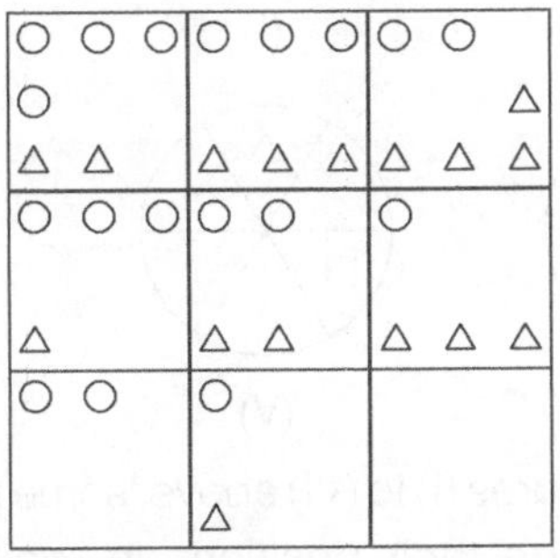

Which of the following figures could fit in the empty block and thus complete the matrix?

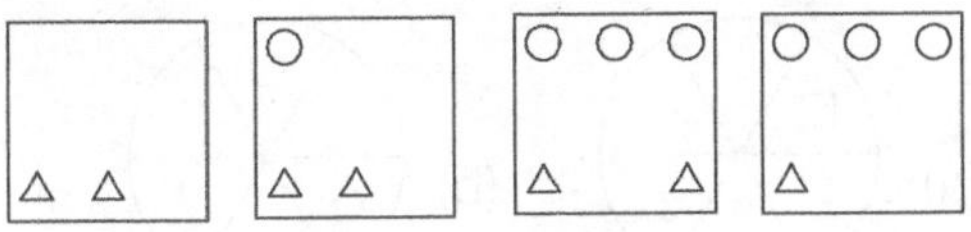

(a) (b) (c) (d)

[UPSC 2014]

Ans. (a)

It can be observed easily that the number of circles keeps on decreasing by one and the number of triangles increases by one in a particular row.

Therefore, the missing figure would have to have two triangles only.

Q.11 Consider the following figures:

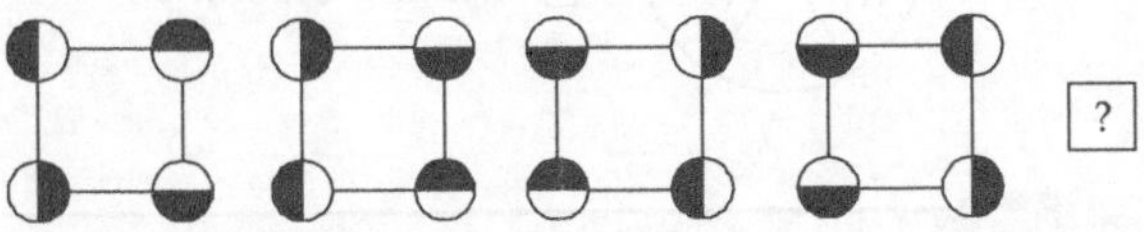

Change in positions of beads in the four figures above follows a sequence. Following the same sequence, which of the figures below should appear as the fifth figure above?

(a) 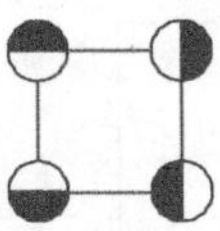 (b) 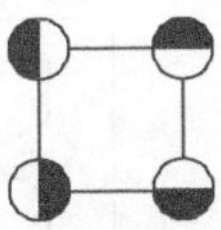

(c) 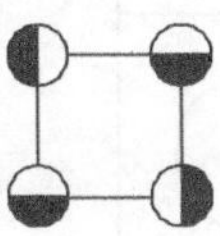 (d) 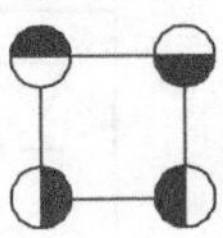

[UPSC 2014]

Ans. (b)

Q.12 Consider the figures given below:

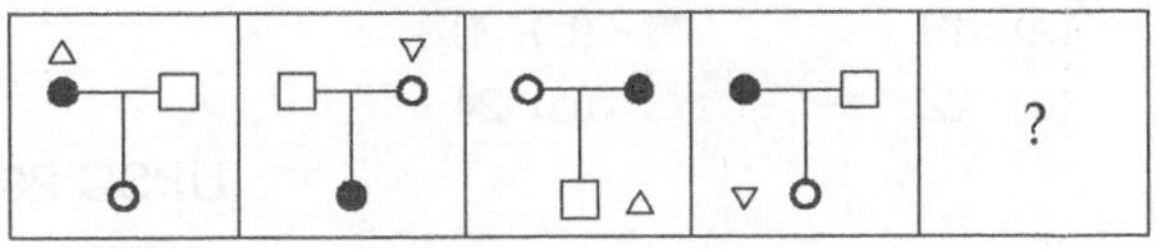

To fit the question mark, the correct answer is

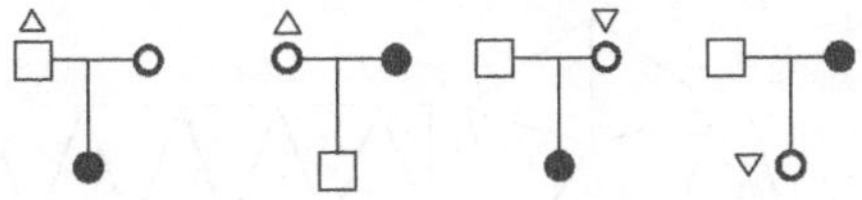

(a) (b) (c) (d)

[UPSC 2015]

Ans. (a)

Observe the pattern - Dark Circle moves anticlockwise. Triangle moves clockwise and changes orientation upside down. Square moves anti clockwise.

This gives us option (a) as correct answer.

Q.13 Four cardboard pieces of specific shapes are shown in the following figure:

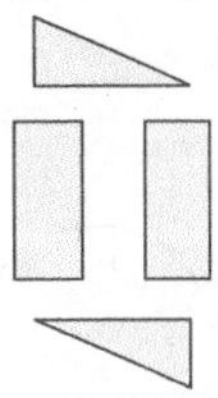

Which one of the following figures given can be formed by joining these pieces together?

(a) 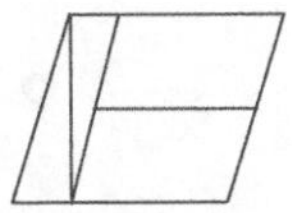 (b) 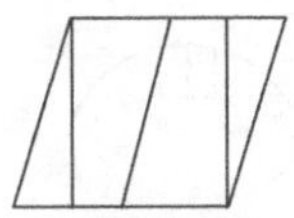

(c) 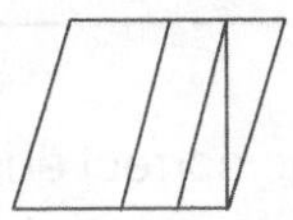 (d) 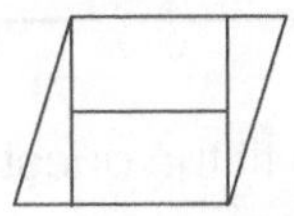

[UPSC 2015]

Ans. (d)

The answer is clearly option (d). In these types of questions we need to take the individual pieces and try to fit in the options.

Q.14 Consider the following three-dimensional figure:

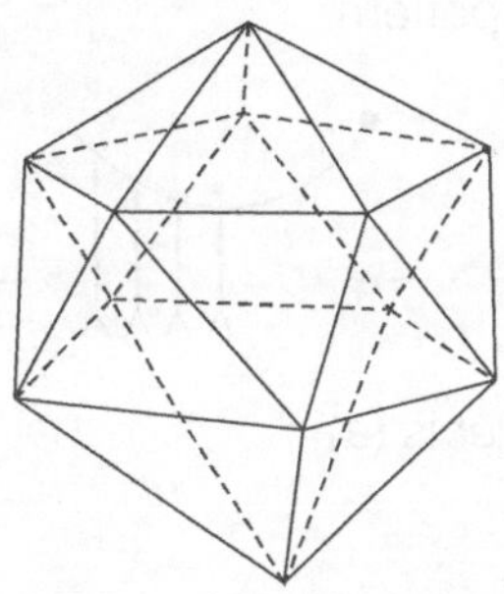

How many tringles does the above figure have?

(a) 18 (b) 20

(c) 22 (d) 24

[UPSC 2018]

Ans. (b)

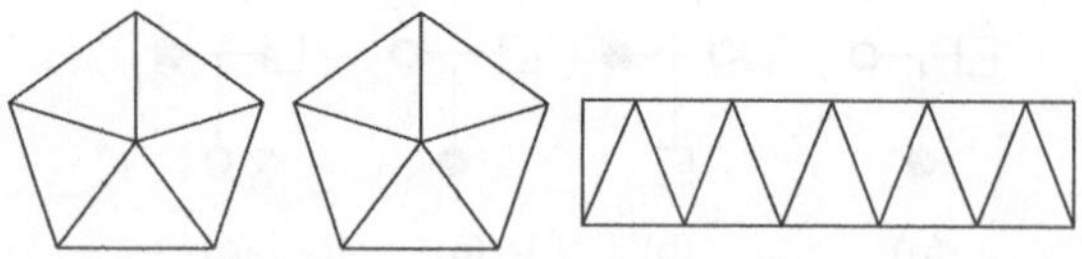

Five triangles are on top

Five triangles are in bottom

The middle layer is as shown with 10 triangles.

So, total = 5 + 5 + 10 = 20

Q.15 Consider the figures given below:

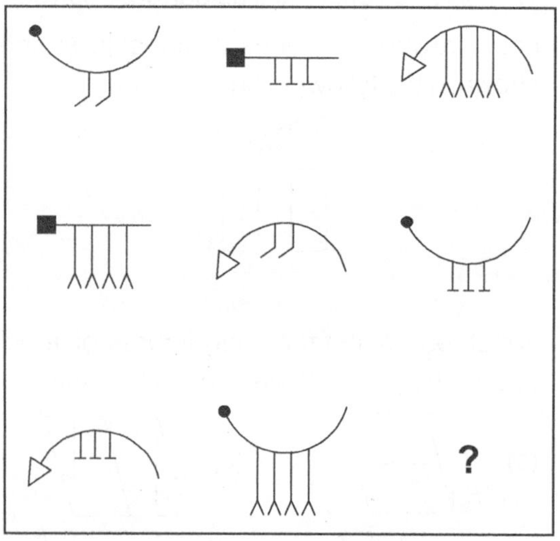

To fit the question mark, the correct answer is

(a) 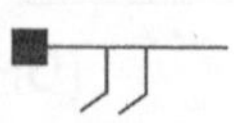 (b) 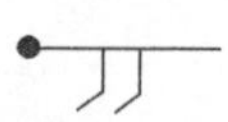

(c) 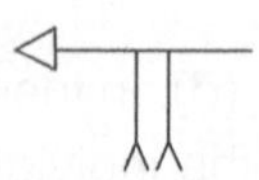 (d) 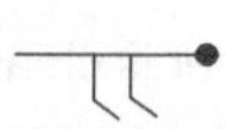

[UPSC 2018]

Ans. (a)

There is a pattern

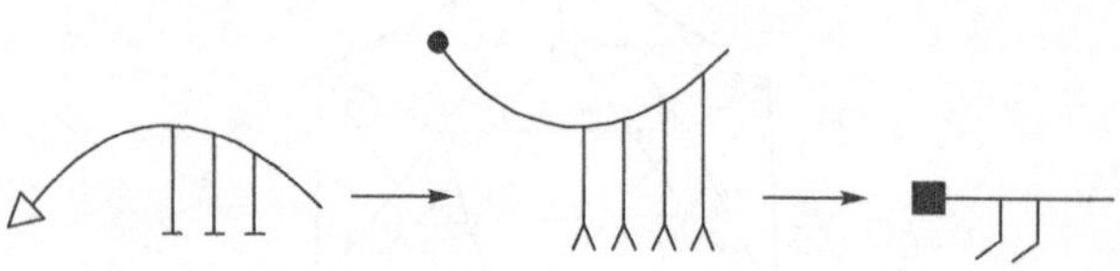

Best answer is (a).

Q.16 Consider the following figures :

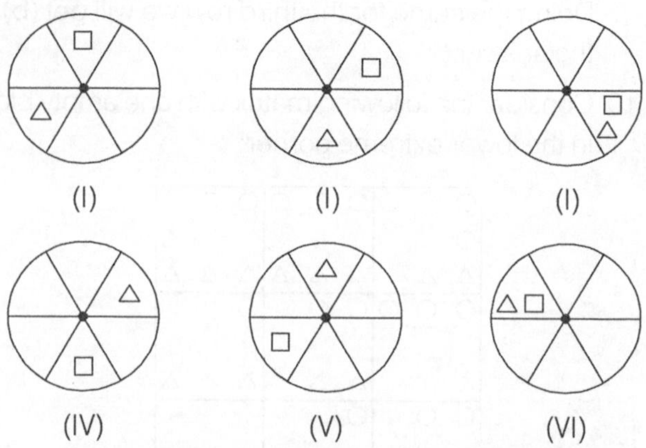

In the figures (I) to (VI) above, some parts are shown to change their positions in regular directions. Following the same sequence, which of the figures given below will appear at (VII) stage?

(a) 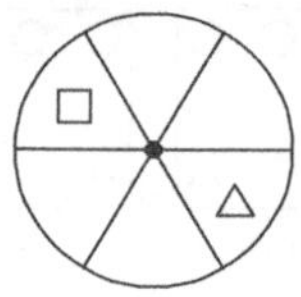 (b) 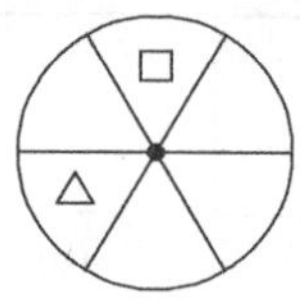

(c) 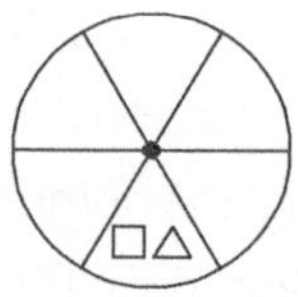 (d) 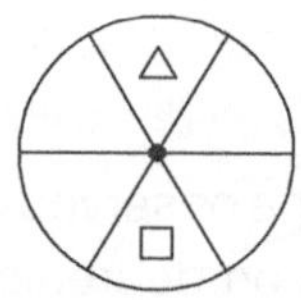

[UPSC 2018]

Ans. (b)

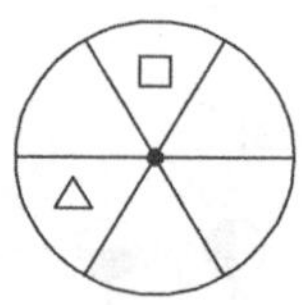

△ moving anti-clockwise

□ moving clockwise

PRACTICE SET : QUESTIONS

Q.1 Select a suitable figure from the four alternatives that would complete the figure matrix.

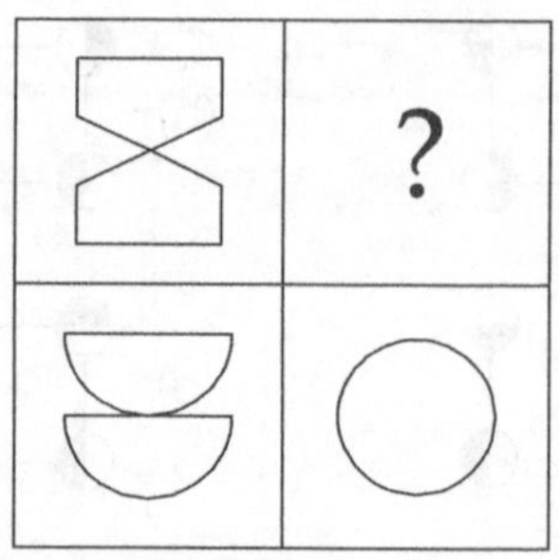

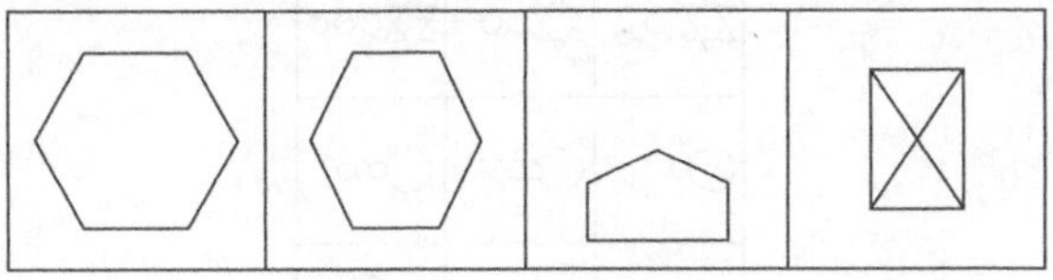

(1) (2) (3) (4)

(a) 1 (b) 2
(c) 3 (d) 4

Q.2 Select a suitable figure from the four alternatives that would complete the figure matrix.

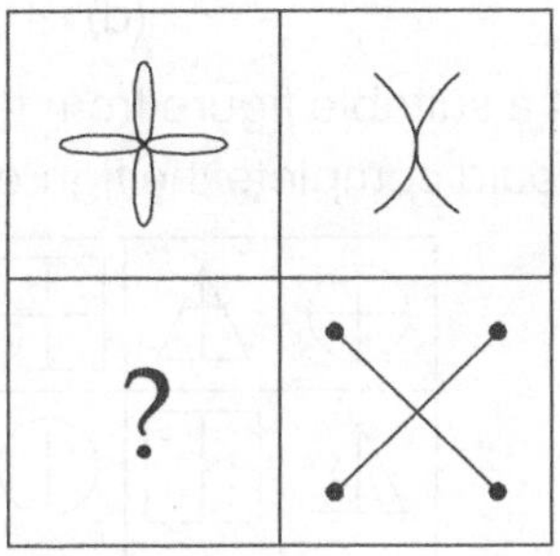

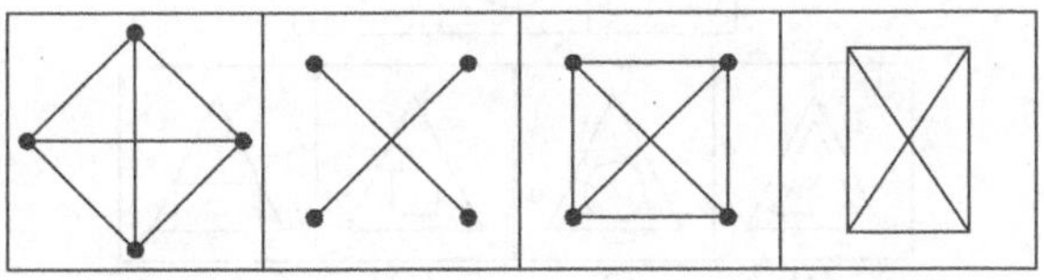

(1) (2) (3) (4)

(a) 1 (b) 2
(c) 3 (d) 4

Q.3 Select a suitable figure from the four alternatives that would complete the figure matrix.

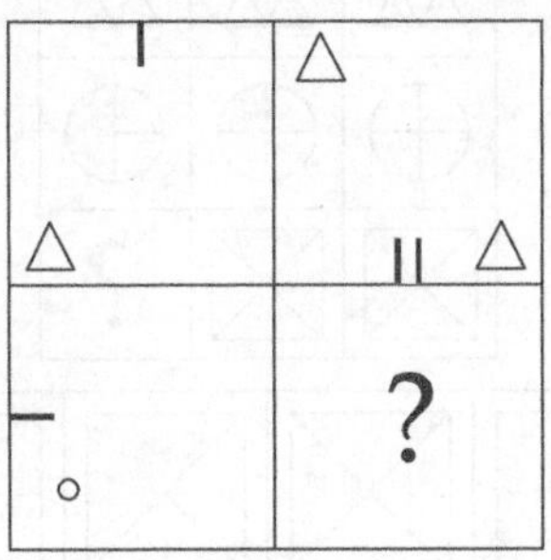

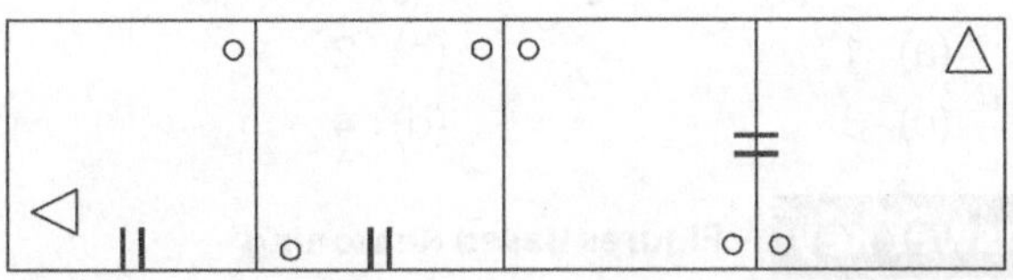

(1) (2) (3) (4)

(a) 1 (b) 2
(c) 3 (d) 4

Q.4 Select a suitable figure from the four alternatives that would complete the figure matrix.

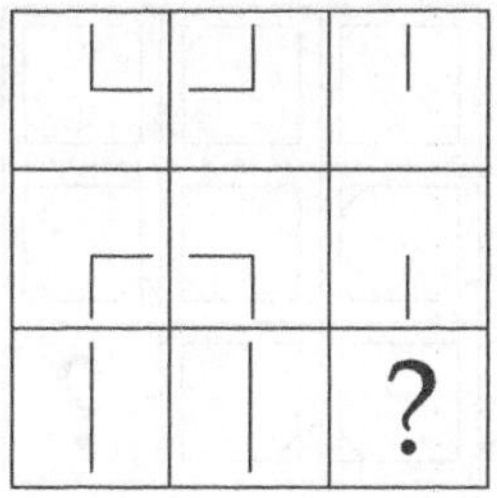

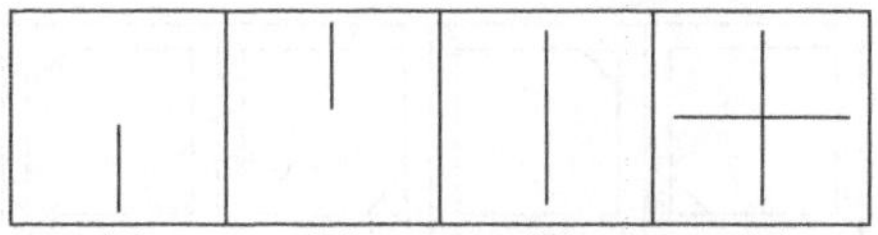

(1) (2) (3) (4)

(a) 1 (b) 2
(c) 3 (d) 4

Q.5 Select a suitable figure from the four alternatives that would complete the figure matrix.

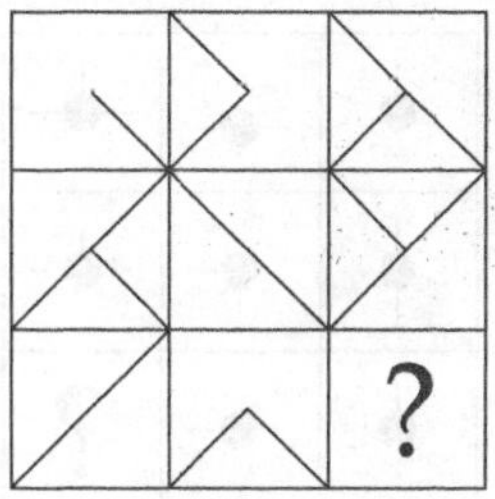

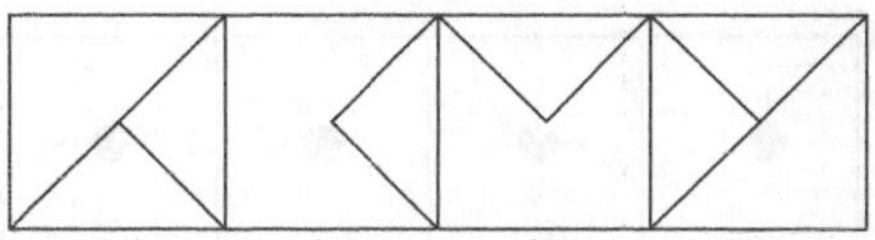

(1) (2) (3) (4)

(a) 1 (b) 2
(c) 3 (d) 4

Q.6 Select a suitable figure from the four alternatives that would complete the figure matrix.

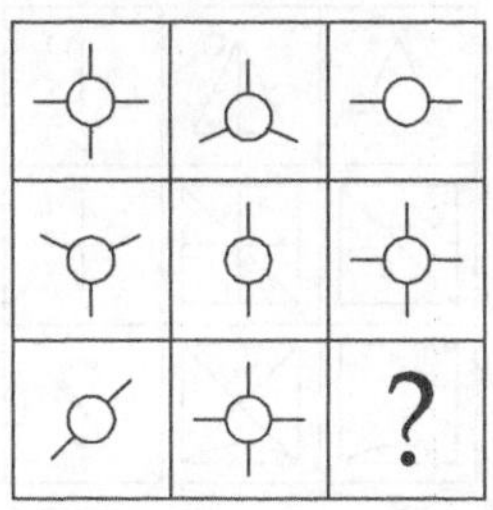

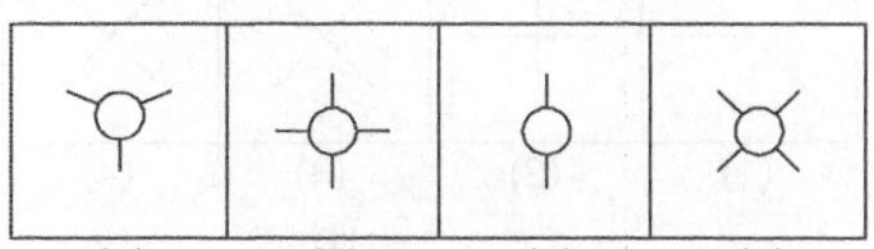

(1) (2) (3) (4)

(a) 1 (b) 2
(c) 3 (d) 4

Q.7 Select a suitable figure from the four alternatives that would complete the figure matrix.

(1) (2) (3) (4)

(a) 1 (b) 2
(c) 3 (d) 4

Q.8 Select a suitable figure from the four alternatives that would complete the figure matrix.

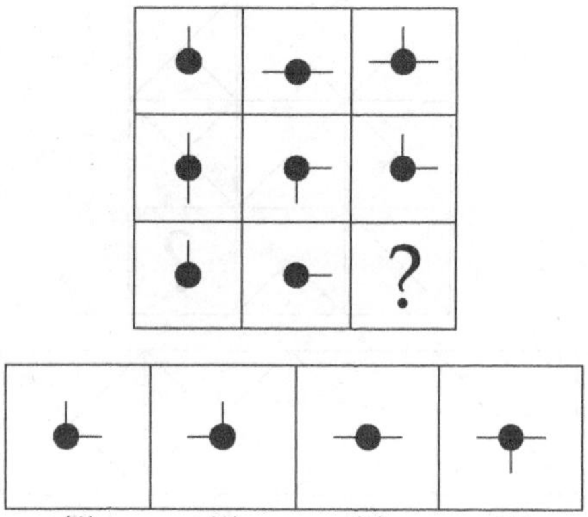

(1) (2) (3) (4)

(a) 1 (b) 2
(c) 3 (d) 4

Q.9 Select a suitable figure from the four alternatives that would complete the figure matrix.

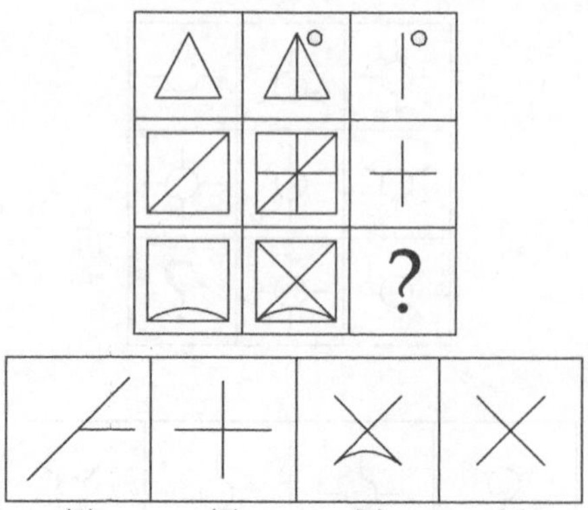

(1) (2) (3) (4)

(a) 1 (b) 2
(c) 3 (d) 4

Q.10 Select a suitable figure from the four alternatives that would complete the figure matrix.

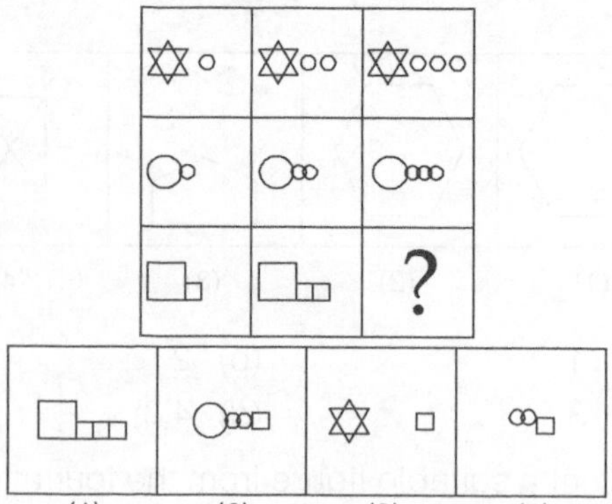

(1) (2) (3) (4)

(a) 1 (b) 2
(c) 3 (d) 4

Q.11 Select a suitable figure from the four alternatives that would complete the figure matrix.

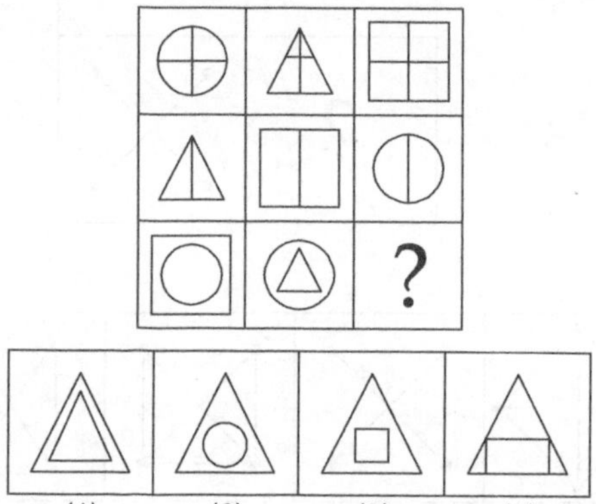

(1) (2) (3) (4)

(a) 1 (b) 2
(c) 3 (d) 4

Q.12 Select a suitable figure from the four alternatives that would complete the figure matrix.

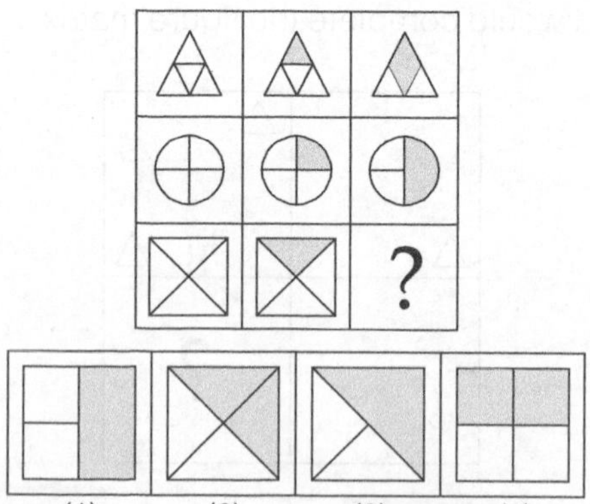

(1) (2) (3) (4)

(a) 1 (b) 2
(c) 3 (d) 4

Answer key: Figures Based Reasoning

1.	(b)	2.	(c)	3.	(c)	4.	(c)
5.	(b)	6.	(a)	7.	(b)	8.	(a)
9.	(d)	10.	(a)	11.	(c)	12.	(b)

■■■■

3.13 CHAPTER

Paper Cutting, Folding and Mirror Images

This type of questions is part of Non-verbal reasoning. In case of **mirror images**, two types of questions can be asked.

I. Finding mirror image of Sentences

II. Finding image of geometrical figures

In case of **paper folding and cutting**, a paper usually of square shape is cut and folded. We are expected to imagine the unfolded version of the paper.

Both kind of questions has been asked. Usually the frequency varies from 1-3. Such questions can be done by just looking at them carefully.

Basic Rule for finding Mirror Image

From the figure, it can be seen that **in a mirror image top and bottom of an image does not change** but LHS (Left Hand side) of a real image becomes RHS (Right Hand Side) in mirror image and RHS in real image become LHS in mirror image. So, it can be concluded that mirror image is lateral inversion of real image.

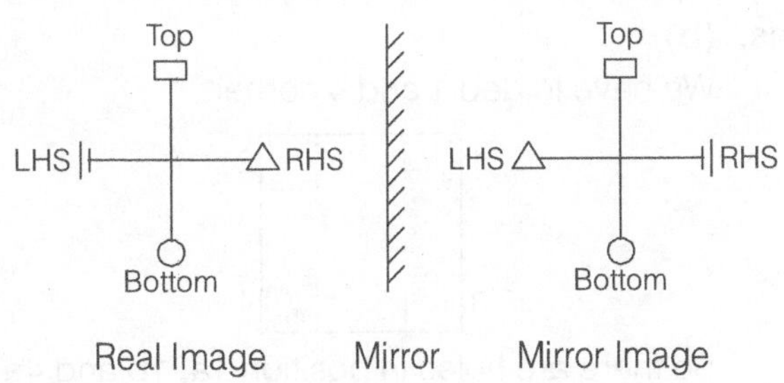

Mirror Image of Capital Alphabets

Letters	Mirror Images	Letters	Mirror Images	Letters	Mirror Images
A	A	J	Ⴑ	S	Ƨ
B	ᗺ	K	ꓘ	T	T
C	Ɔ	L	⅃	U	U
D	ᗡ	M	M	V	V
E	Ǝ	N	И	W	W
F	ꟻ	O	O	X	X
G	ᘐ	P	ꟼ	Y	Y
H	H	Q	Ϙ	Z	ᘔ
I	I	R	Я	-	-

Solved Examples

Q.1 Choose the option which closely relate to mirror image of given option.

SERVICE

(a) ƧEʁΛICE (b) ƧEЯΛIƆE

(c) ƧEЯΛICƎ (d) ƧEЯΛICE

Ans. (d)

In mirror image left and right portions get exchanged.

Q.2 Choose the mirror image of following figure.

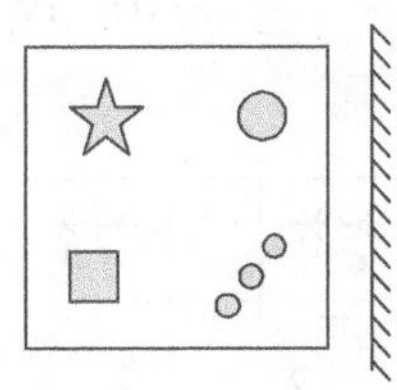

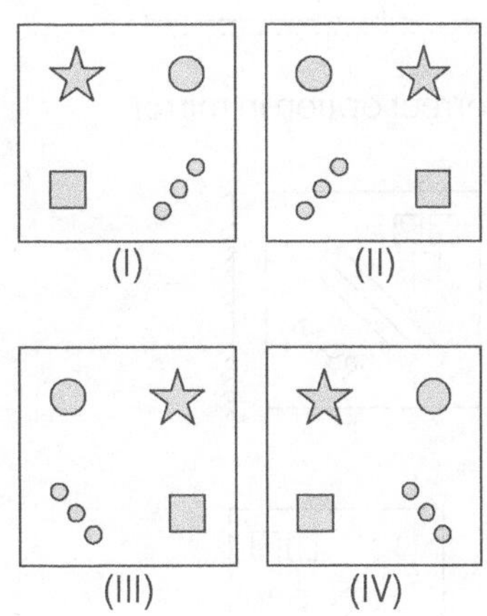

(a) I (b) II

(c) III (d) IV

Ans. (c)

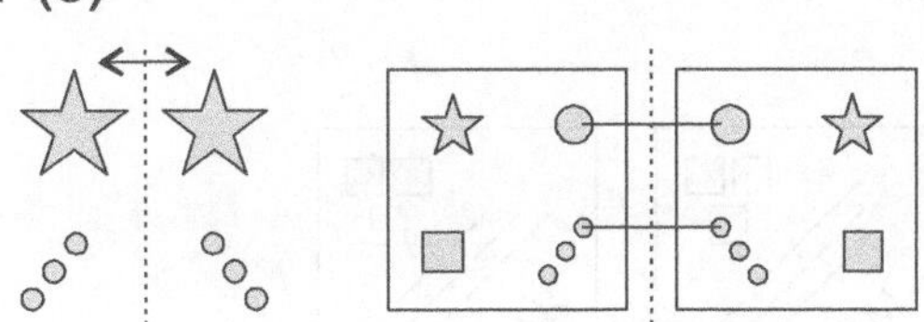

Q.3 Choose the correct option in mirror.

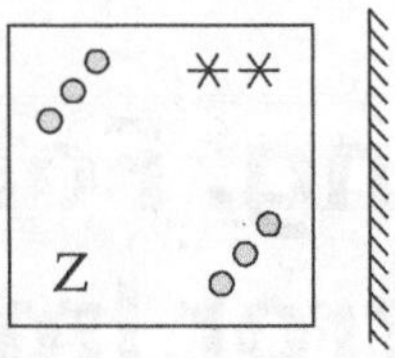

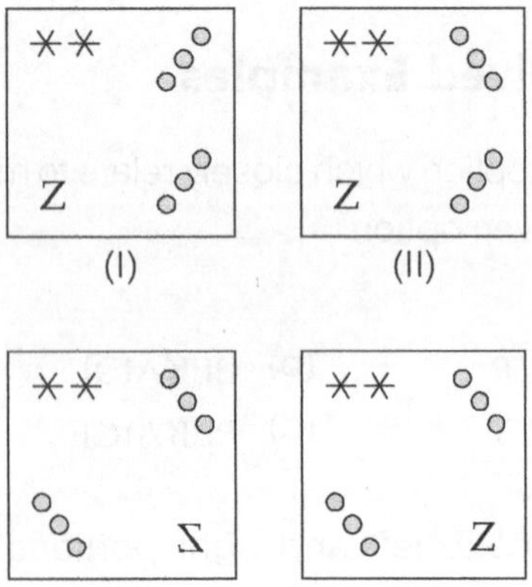

(I) (II) (III) (IV)

(a) I
b) II
(c) III
(d) IV

Ans. (c)

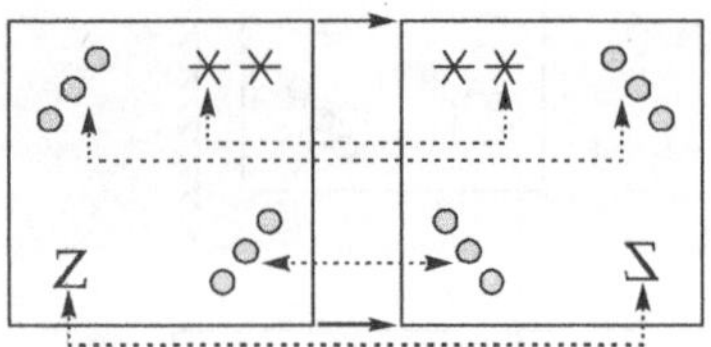

Q.4 Choose the correct option in mirror.

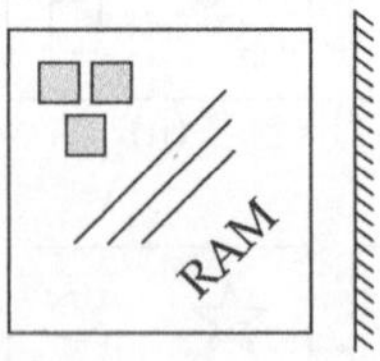

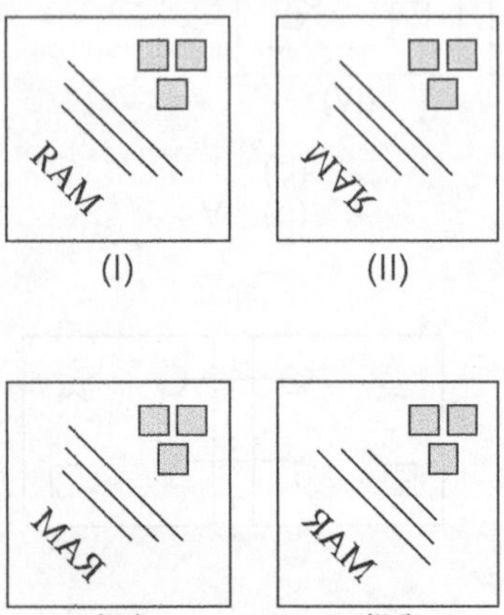

(I) (II) (III) (IV)

(a) I
(b) II
(c) III
(d) IV

Ans. (c)

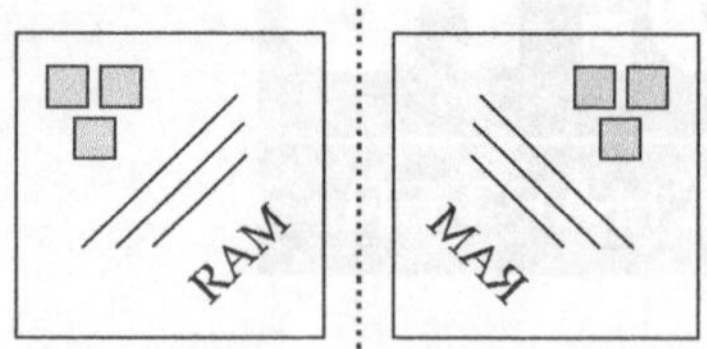

Q.5 Choose the figure which will appear as unfolded paper.

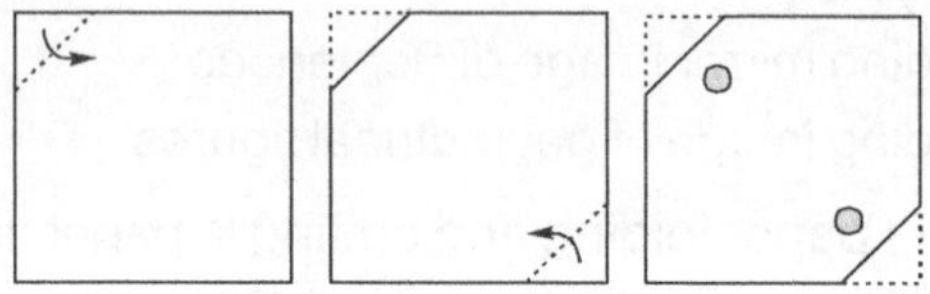

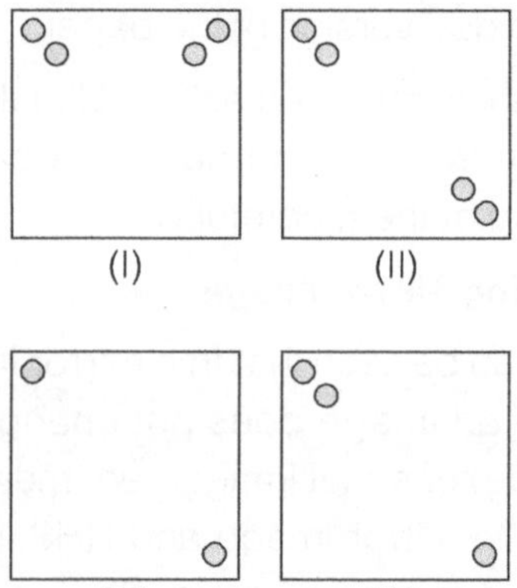

(I) (II) (III) (IV)

(a) I
(b) II
(c) III
(d) IV

Ans. (b)

We have folded 1 and 4 corner.

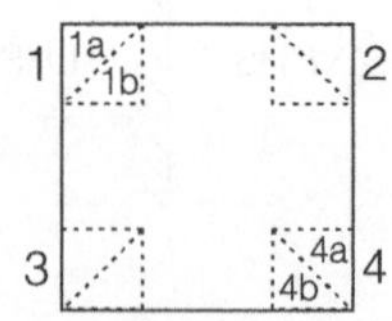

So there are holes in position 1a, 1b and 4a, 4b.

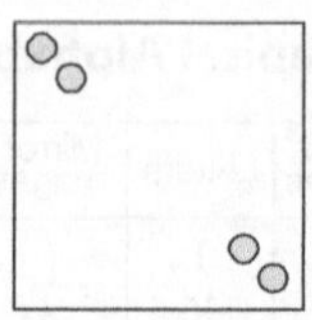

Q.6 Choose the correct option which will look as unfolded paper.

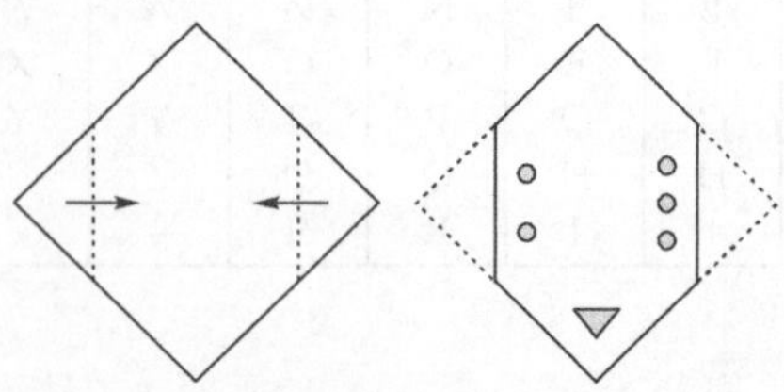

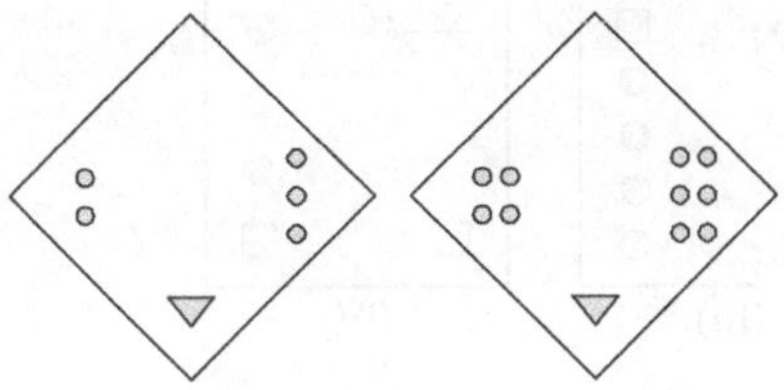

(I) (II)

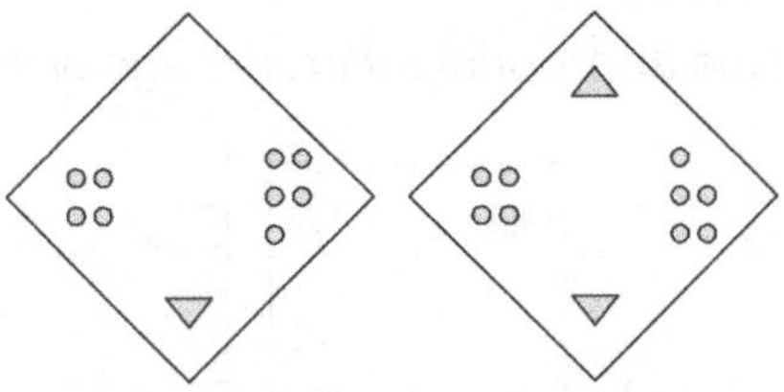

(III) (IV)

(a) I (b) II
(c) III (d) IV

Ans. **(b)**

Q.7 Which of the following most closely resembles to unfolded paper.

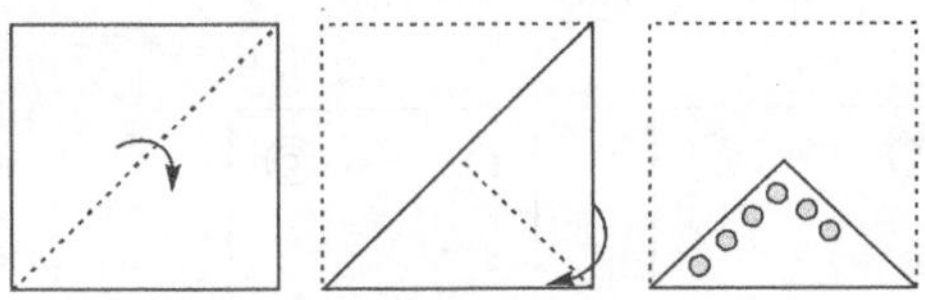

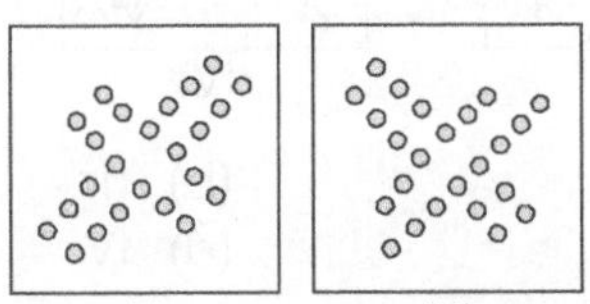

(I) (II)

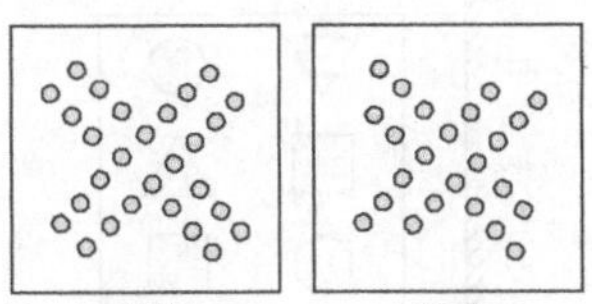

(III) (IV)

(a) I (b) II
(c) III (d) IV

Ans. **(c)**

Q.8 Which most closely resembles as mirror image of unfolded figure.

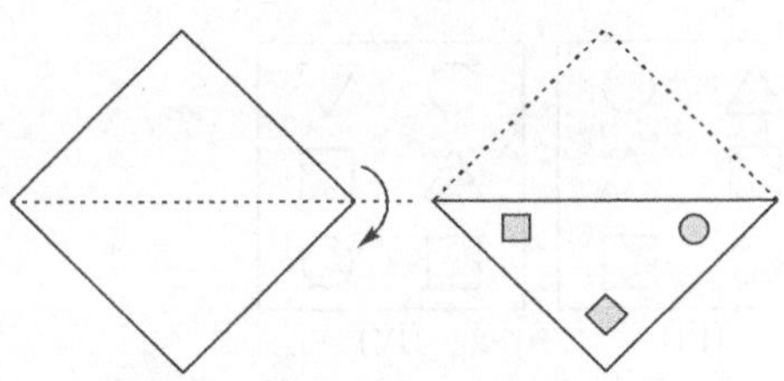

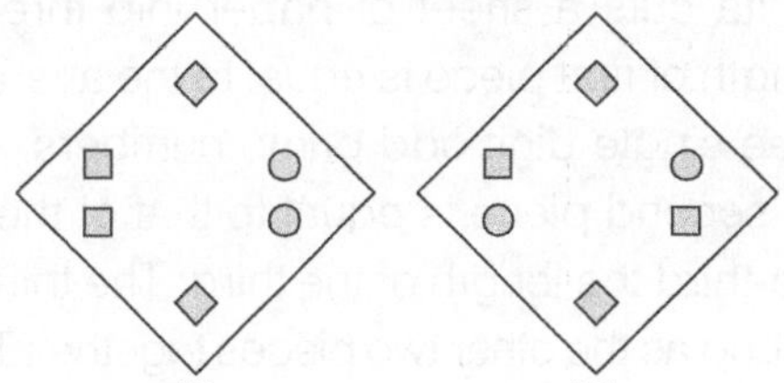

(I) (II)

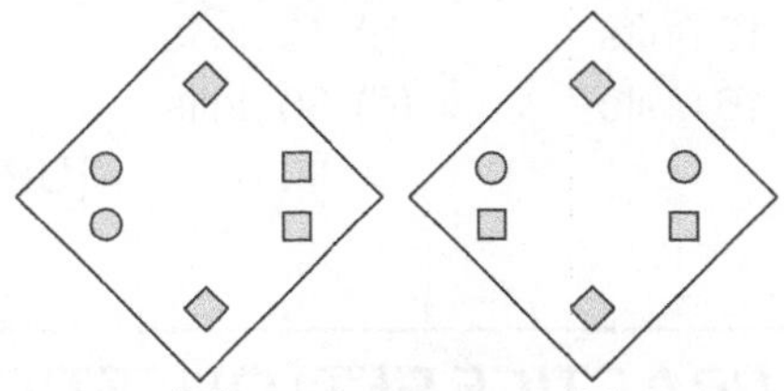

(III) (IV)

(a) I (b) II
(c) III (d) IV

Ans. **(a)**

Previous Years Solved Questions

Q.1 Consider the figure given below and answer the items that follows:

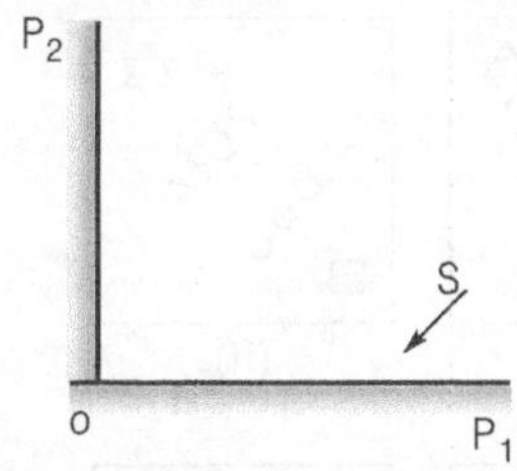

In the figure shown above, OP_1 and OP_2 are two plane mirrors kept perpendicular to each other. S is the direction of a beam of light falling on the mirror OP_1. The direction of the reflected beam of light from the mirror OP_2 will be:

(a) Perpendicular to the direction S.
(b) At 45° to the direction S.
(c) Opposite and parallel to the direction S.
(d) At 60° to the direction S. **[UPSC 2011]**

Ans. **(c)**

Suppose α is angle with the vertical axis.

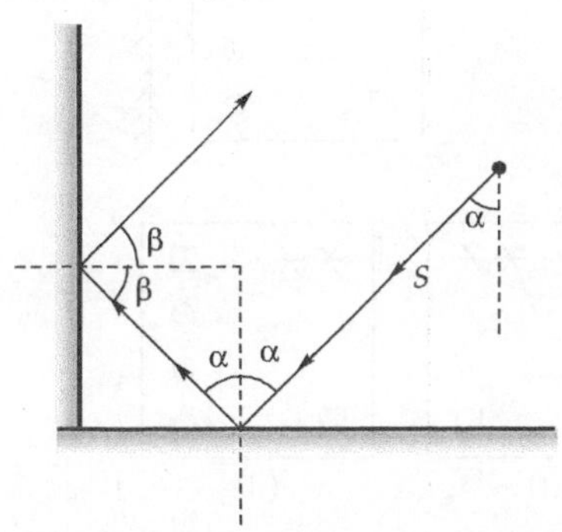

Q.2 Sunita cuts a sheet of paper into three pieces. Length of first piece is equal to the average of the three single digit odd prime numbers. Length of the second piece is equal to that of the first plus one-third the length of the third. The third piece is as long as the other two pieces together. The length of the original sheet of paper is

(a) 13 units (b) 15 units
(c) 16 units (d) 30 units

[UPSC 2019]

Ans. (d)

PRACTICE SET : QUESTIONS

Q.1 Find the appropriate mirror image of the figure.

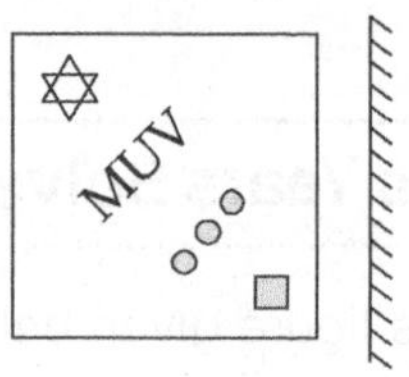

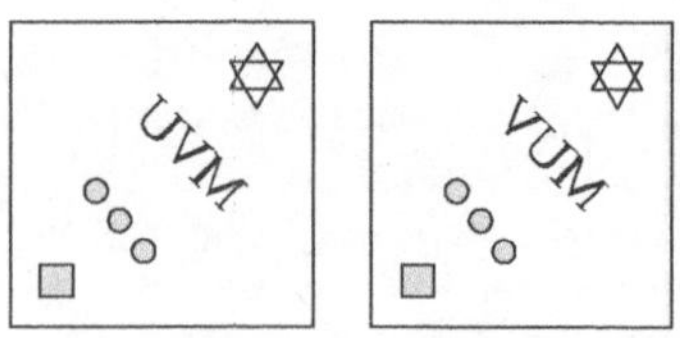

(I) (II)

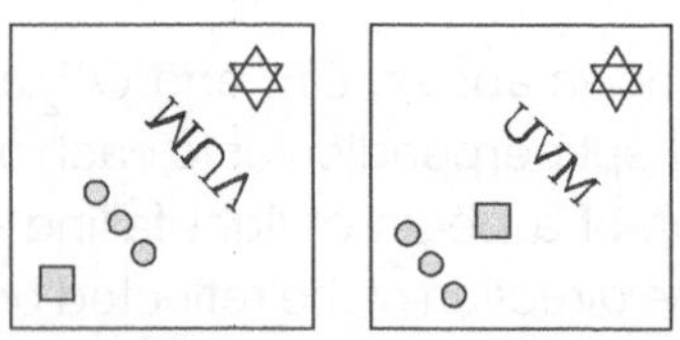

(III) (IV)

(a) I (b) II
(c) III (d) IV

Q.2 Find the appropriate mirror image of the figure.

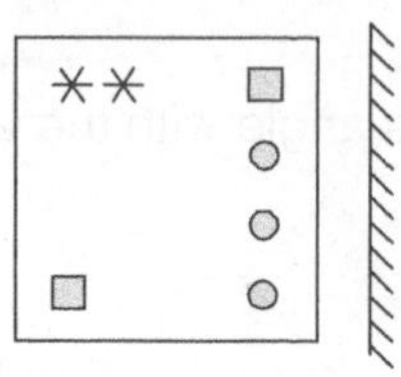

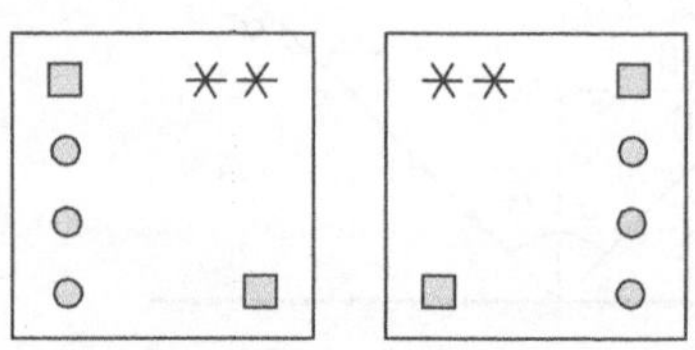

(I) (II)

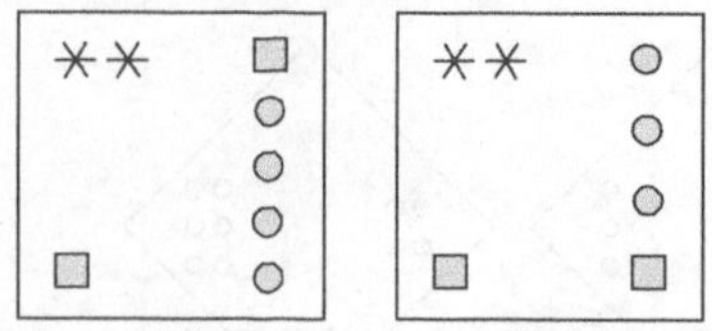

(III) (IV)

(a) I (b) II
(c) III (d) IV

Q.3 Find the appropriate mirror image of the figure.

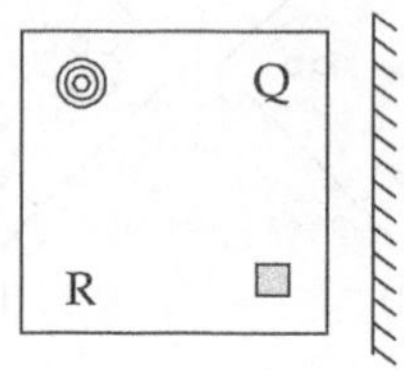

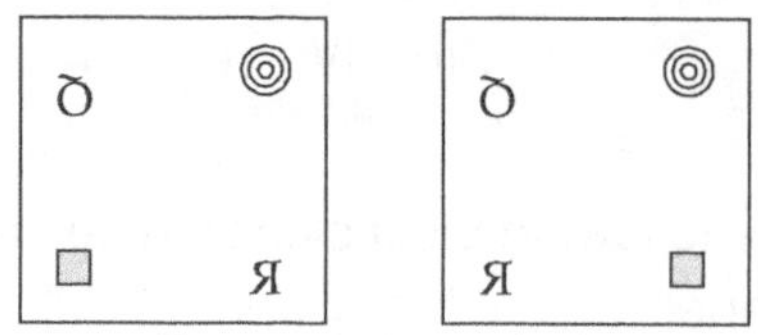

(I) (II)

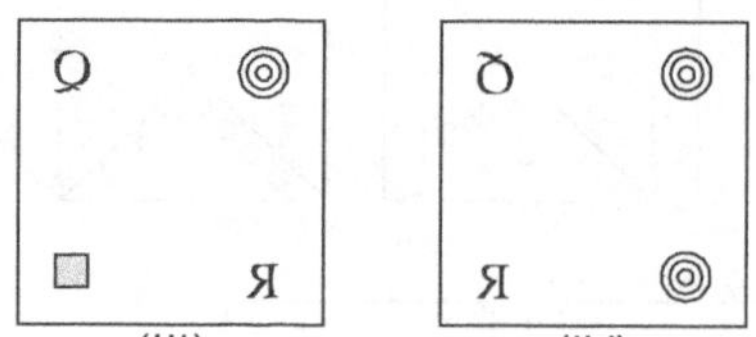

(III) (IV)

(a) I (b) II
(c) III (d) IV

Q.4 Find the appropriate mirror image of the figure.

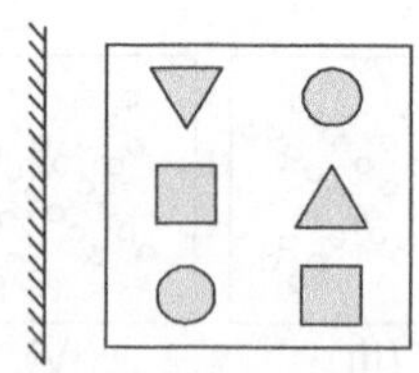

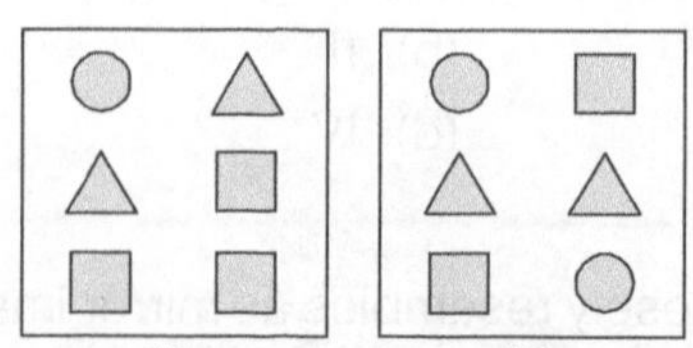

(I) (II)

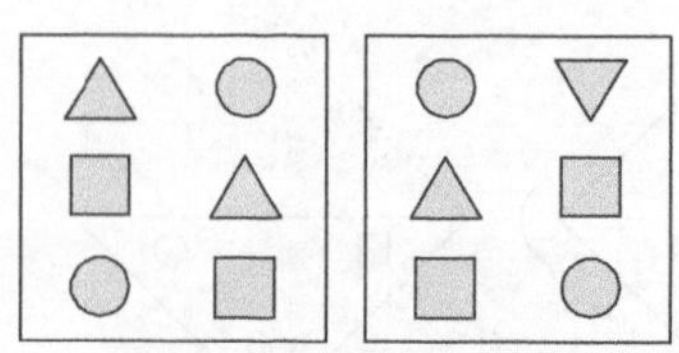

(III) (IV)

(a) I (b) II
(c) III (d) IV

Q.5 Find the appropriate mirror image of the figure.

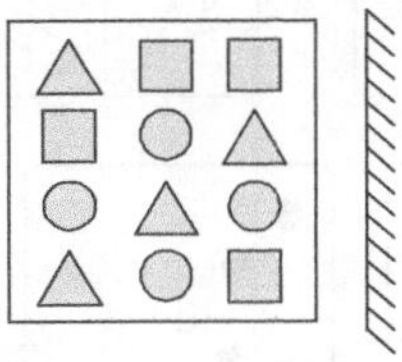

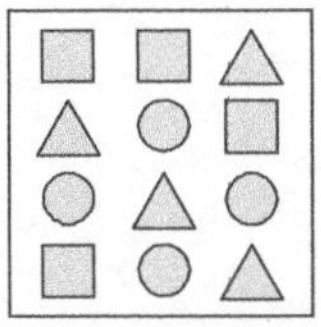

(I)

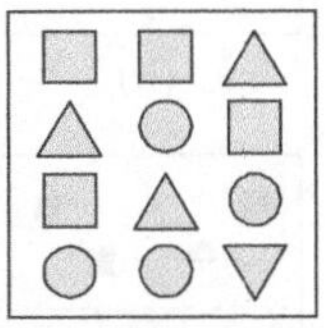

(II)

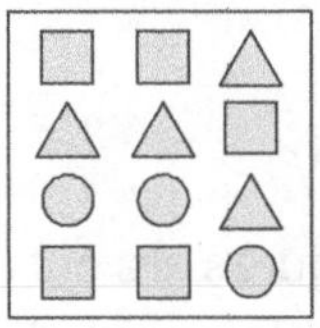

(III)

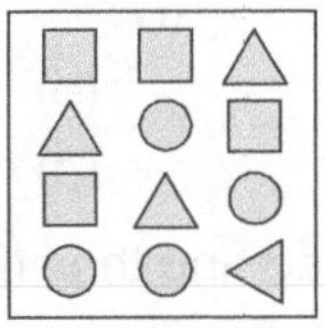

(IV)

(a) I (b) II
(c) III (d) None of these

Q.6 A paper is folded and then figures are cut (shown in dark). Which image resembles most closely to the unfolded paper.

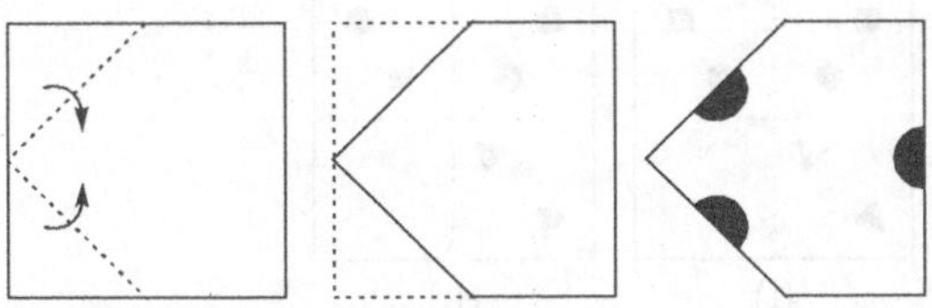

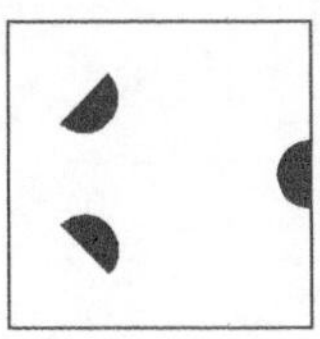

(I)

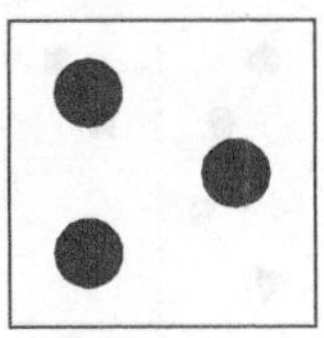

(II)

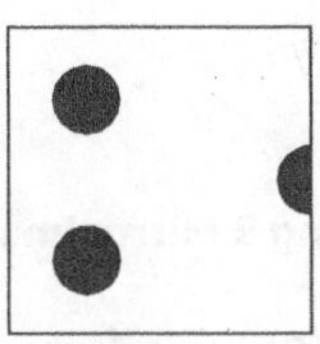

(III)

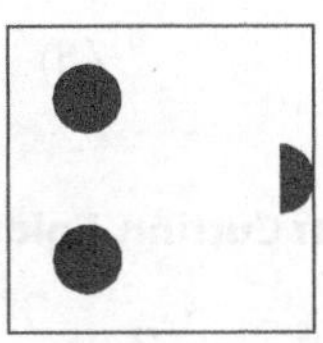

(IV)

(a) I (b) II
(c) III (d) IV

Q.7 A paper is folded and then figures are cut (shown in dark). Which image resembles most closely to the unfolded paper.

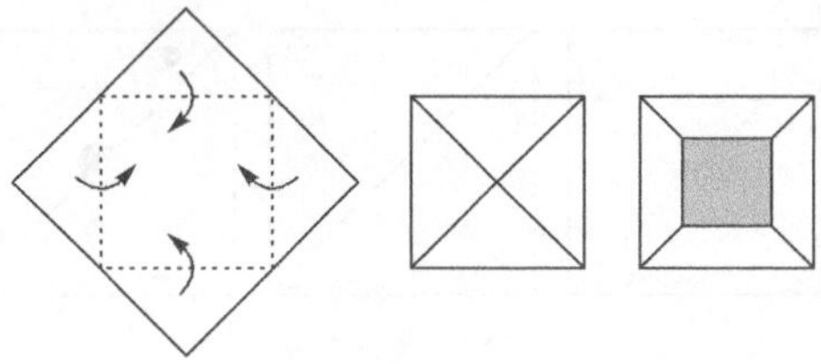

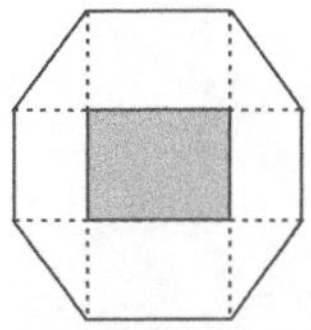

(I)

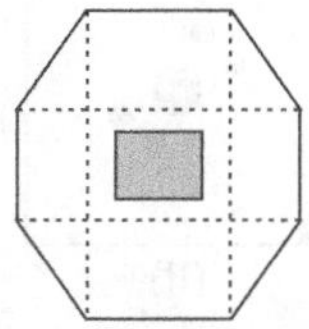

(II)

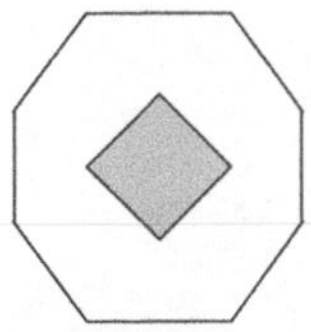

(III)

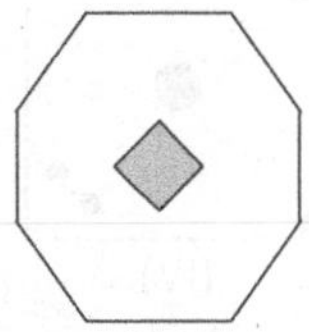

(IV)

(a) I (b) II
(c) III (d) IV

Q.8 A paper is folded and then figures are cut (shown in dark). Which image resembles most closely to the unfolded paper.

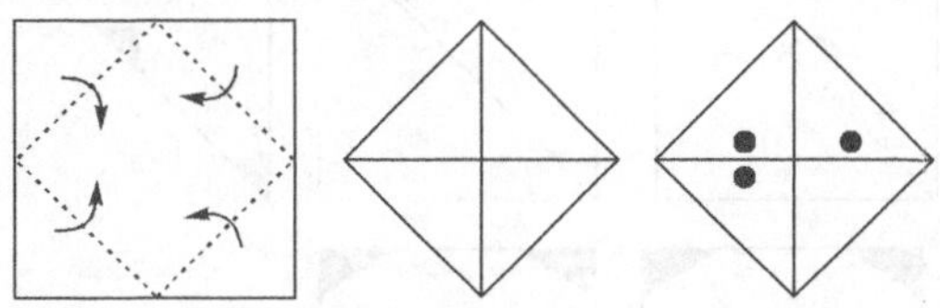

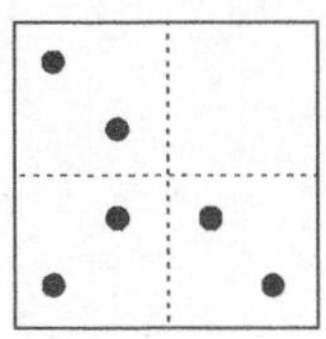

(I)

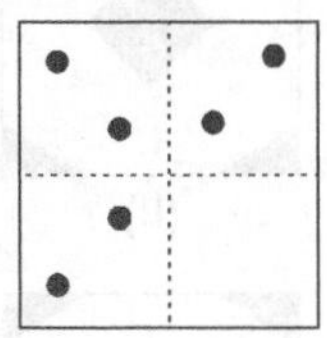

(II)

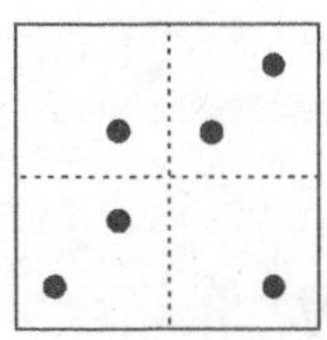

(III)

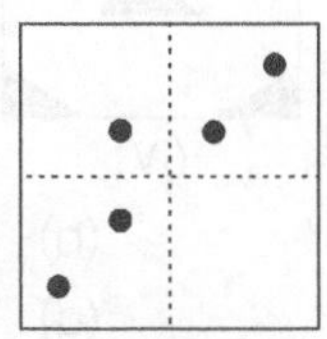

(IV)

(a) I (b) II
(c) III (d) IV

Q.9 A paper is folded and then figures are cut (shown in dark). Which image resembles most closely to the unfolded paper.

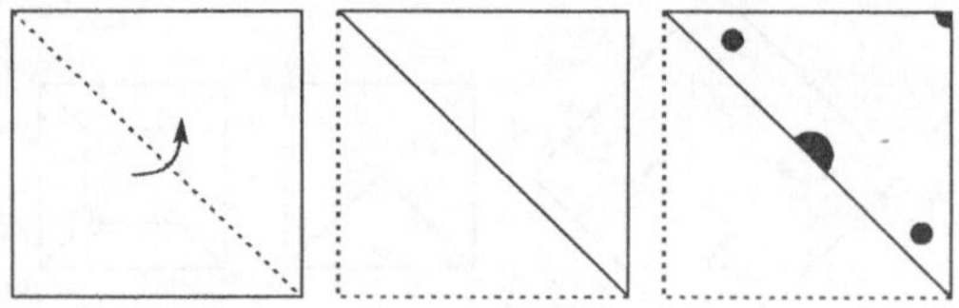

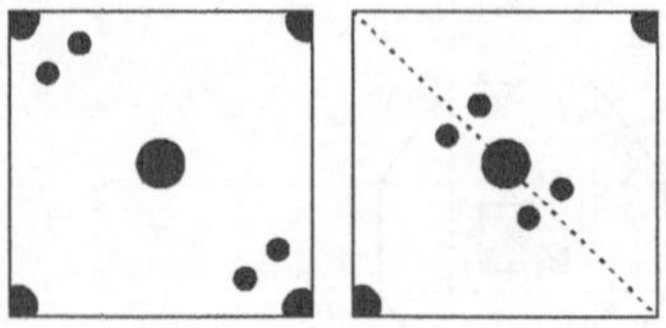

(I) (II)

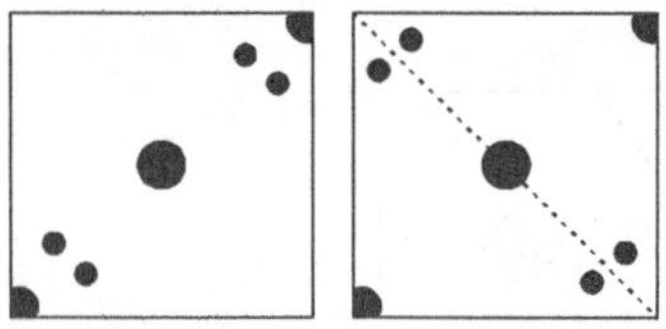

(III) (IV)

(a) I (b) II
(c) III (d) IV

Q.10 A paper is folded and then figures are cut (shown in dark). Which image resembles most closely to the unfolded paper.

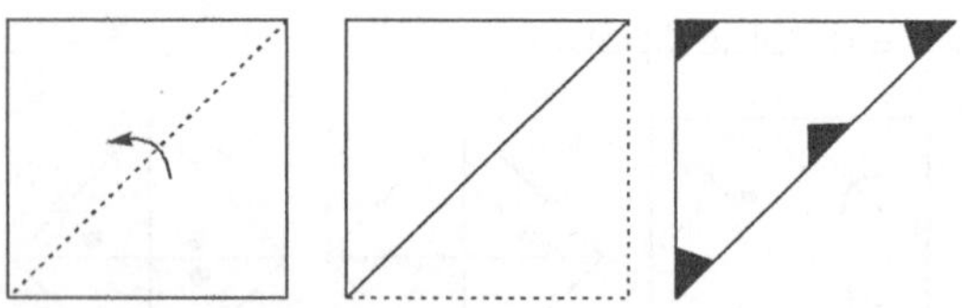

(I) (II)

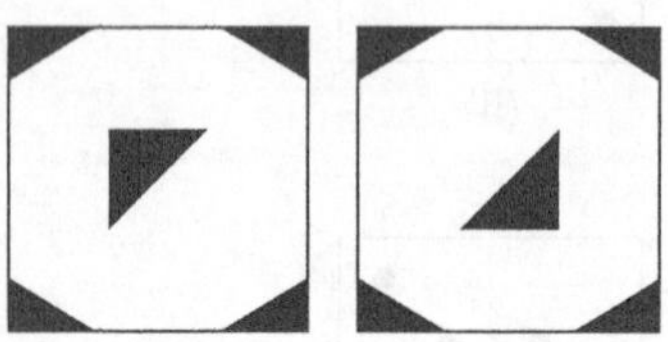

(III) (IV)

(a) I (b) II
(c) III (d) IV

Q.11 A paper is folded and then figures are cut (shown in dark). Which image resembles most closely to the unfolded paper.

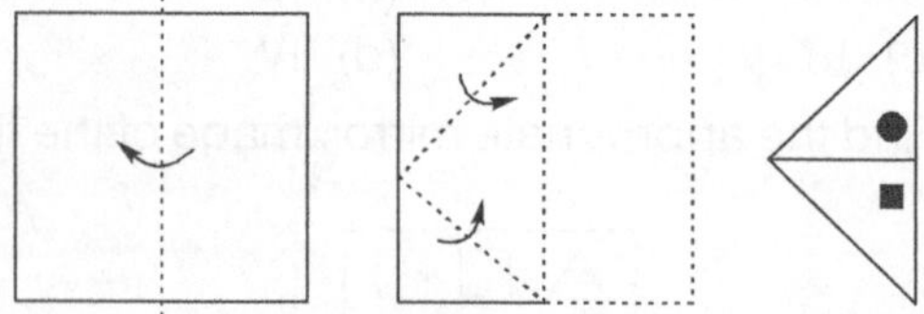

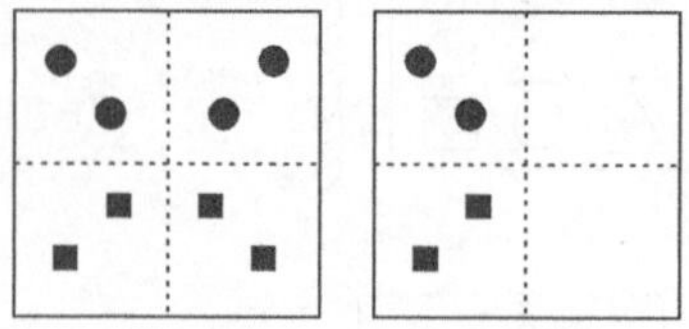

(I) (II)

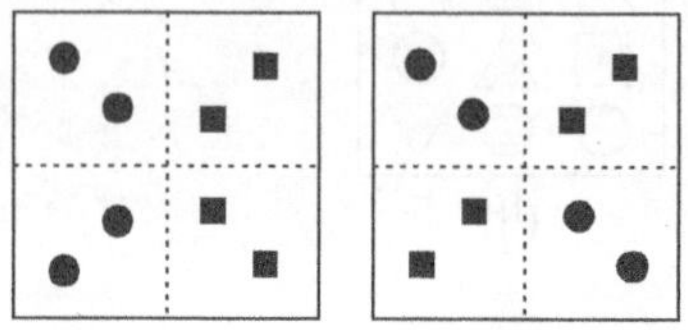

(III) (IV)

(a) I (b) II
(c) III (d) IV

Q.12 A paper is folded and then figures are cut (shown in dark). Which image resembles most closely to the unfolded paper.

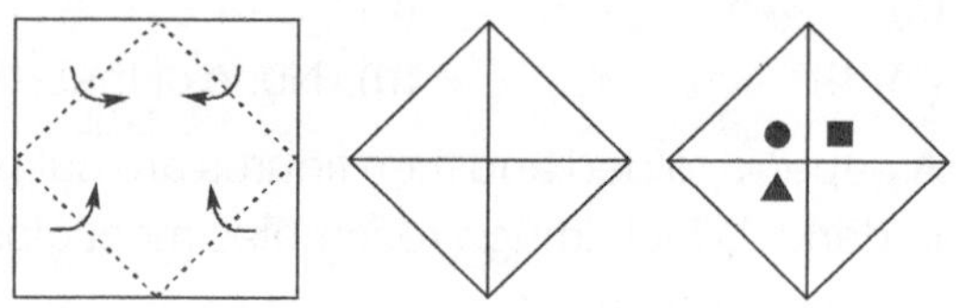

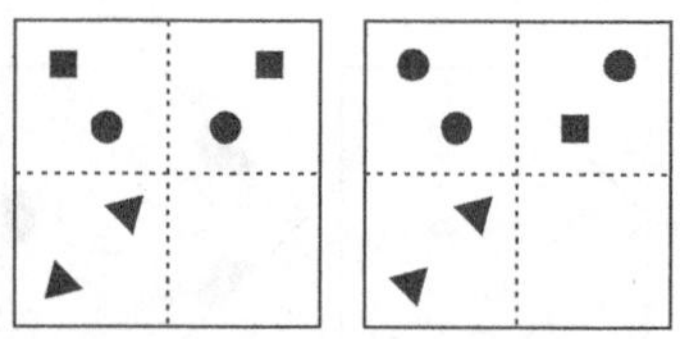

(I) (II)

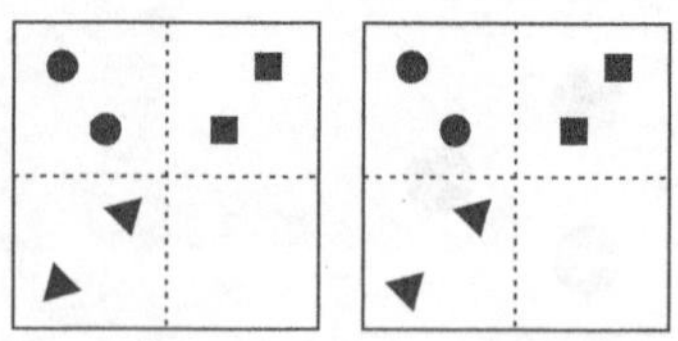

(III) (IV)

(a) I (b) II
(c) III (d) IV

Answer key: Paper Cutting, Folding & Mirror Images

1.	(b)	2.	(a)	3.	(c)	4.	(d)
5.	(a)	6.	(c)	7.	(b)	8.	(a)
9.	(d)	10.	(a)	11.	(a)	12.	(d)

D

Section

Previous GATE & ESE Solved Questions

Previous GATE Solved Questions
(General Aptitude)

1. 25 persons are in a room. 15 of them play hockey, 17 of them play football and 10 of them play both hockey and football. Then the number of persons playing neither hockey nor football is
   (a) 2 (b) 17
   (c) 13 (d) 3
   [2010, 1 Mark]

2. If 137 + 276 = 435 how much is 731 + 672?
   (a) 534 (b) 1403
   (c) 1623 (d) 1531
   [2010, 2 Marks]

3. 5 skilled workers can build a wall in 20 days; 8 semiskilled workers can build a wall in 25 days; 10 unskilled workers can build a wall in 30 days. If a team has 2 skilled, 6 semiskilled and 5 unskilled workers, how long will it take to build the wall?
   (a) 20 days (b) 18 days
   (c) 16 days (d) 15 days
   [2010, 2 Marks]

4. Given digits 2, 2, 3, 3, 3, 4, 4, 4, 4 how many distinct 4 digit numbers greater than 3000 can be formed?
   (a) 50 (b) 51
   (c) 52 (d) 54
   [2010, 2 Marks]

5. Hari (H), Gita (G), Irfan (I) and Saira (S) are siblings (i.e. brothers and sisters). All were born on 1st January. The age difference between any two successive siblings (that is born one after another) is less than 3 years. Given the following facts:
   1. Hari's age + Gita's age > Irfan's age + Saira's age.
   2. The age difference between Gita and Saira is 1 year. However, Gita is not the oldest and Saira is not the youngest.
   3. There are no twins.

   In what order were they born (oldest first)?
   (a) HSIG (b) SGHI
   (c) IGSH (d) IHSG
   [2010, 2 Marks]

6. If Log(P) = (1/2)Log(Q) = (1/3)Log(R), then which of the following options is TRUE?
   (a) $P^2 = Q^3R^2$ (b) $Q^2 = PR$
   (c) $Q^2 = R^3P$ (d) $R = P^2Q^2$
   [CE, ME, CS 2011, 1 Mark (Set-1)]

7. A container originally contains 10 litres of pure spirit. From this container 1 litre of spirit is replaced with 1 litre of water. Subsequently, 1 litre of the mixture is again replaced with 1 litre of water and this processes is repeated one more time. How much spirit is now left in the container?
   (a) 7.58 litres (b) 7.84 litres
   (c) 7 litres (d) 7.29 litres
   [CE, ME, CS 2011, 2 Marks (Set-1)]

8. The variable cost (V) of manufacturing a product varies according to the equation V = 4q, where q is the quantity produced. The fixed cost (F) of production of same product reduces with q according to the equation F = 100/q. How many units should be produced to minimize the total cost (V + F)?
   (a) 5 (b) 4
   (c) 7 (d) 6
   [CE, ME, CS 2011, 2 Marks (Set-1)]

9. P, Q, R and S are four types of dangerous microbes recently found in a human habitat. The area of each circle with its diameter printed in brackets represents the growth of a single microbe surviving human immunity system within 24 hours of entering the body. The danger to human beings varies proportionately with the toxicity, potency and growth attributed to a microbe shown in the figure below:

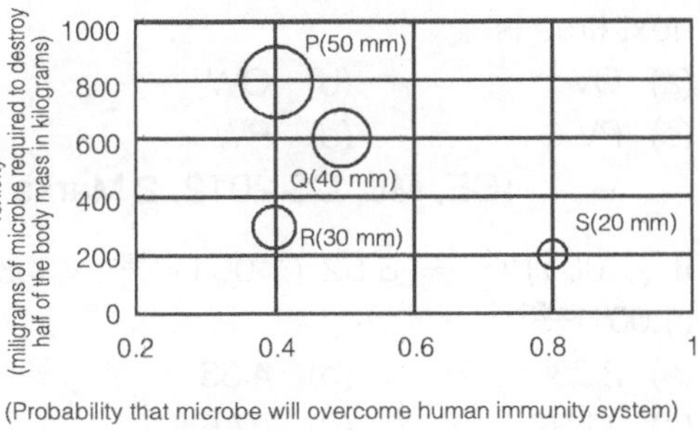

A pharmaceutical company is contemplating the development of a vaccine against the most dangerous microbe. Which microbe should the company target in its first attempt?

(a) P (b) Q

(c) R (d) S

[CE, ME, CS 2011, 2 Marks (Set-1)]

10. A transporter receives the same number of orders each day. Currently, he has some pending orders (backlog) to be shipped. If he uses 7 trucks, then at the end of the 4th day he can clear all the orders. Alternatively, if he uses only 3 trucks, then all the orders are cleared at the end of the 10th day. What is the minimum number of trucks required so that there will be no pending order at the end of the 5th day?

(a) 4 (b) 5

(c) 6 (d) 7

[CE, ME, CS 2011, 2 Marks (Set-1)]

11. There are two candidates P and Q in an election. During the campaign 40% of the voters promised to vote for P, and rest for Q. However, on the day of election 15% of the voters went back on their promise to vote for P and instead voted for Q. 25% of the voters went back on their promise to vote for Q and instead voted for P. Suppose, P lost by 2 votes, then what was the total number of voters?

(a) 100 (b) 110

(c) 90 (d) 95

[EE, EC 2011, 1 Mark (Set-2)]

12. The fuel consumed by a motorcycle during a journey while travelling at various speeds is indicated in the graph below

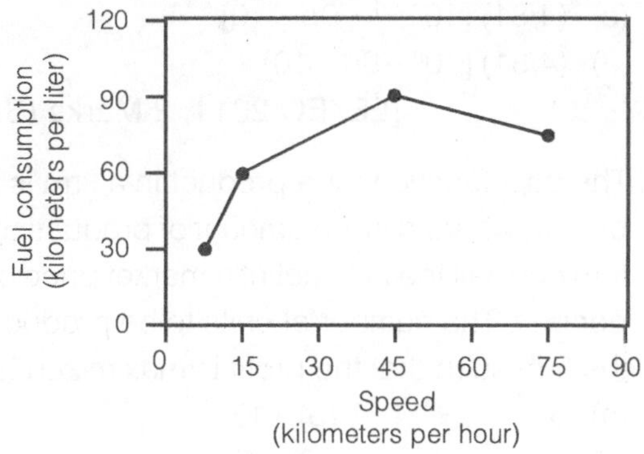

The distance covered during four laps of the journey are listed in the table below:

Lap	Distance (kilometers)	Average speed (kilometers per hour)
P	15	15
Q	75	45
R	40	75
S	10	10

From the given data, we can conclude that the fuel consumed per kilometre was least during the lap

(a) P (b) Q

(c) R (d) S

[EE, EC 2011, 2 Marks (Set-2)]

13. Three friends, R, S and T shared toffee from a bowl. R took 1/3rd of the toffees, but returned four to the bowl. S took 1/4th of what was left but returned three toffees to the bowl. T took half of the remainder but returned two back into the bowl. If the bowl had 17 toffees left, how many toffees were originally there in the bowl?

(a) 38 (b) 31

(c) 48 (d) 41

[EE, EC 2011, 2 Marks (Set-2)]

14. Given that $f(y) = |y|/y$, and q is any non-zero real number, the value of $|f(q) - f(-q)|$ is

(a) 0 (b) −1

(c) 1 (d) 2

[EE, EC 2011, 2 Marks (Set-2)]

15. The sum of n terms of the series 4 + 44 + 444 + ... is

(a) $(4/81)\,[10^{n+1} - 9n - 1]$

(b) $(4/81)\,[10^{n-1} - 9n - 1]$

(c) $(4/81)[10^{n+1} - 9n - 10]$
(d) $(4/81)[10^n - 9n - 10]$

[EE, EC 2011, 2 Marks (Set-2)]

16. The cost function for a product in a firm is given by $5q^2$, where q is the amount of production. The firm can sell the product at a market price of ₹ 50 per unit. The number of units to be produced by the firm such that the profit is maximized is

(a) 5 (b) 10
(c) 15 (d) 25

[CE, ME, CS 2012, 1 Mark (Set-1)]

17. A political party orders an arch for the entrance to the ground in which the annual conventions is being held. The profile of the arch follows the equation $y = 2x - 0.1x^2$ where y is the height of the arch in meters. The maximum possible height of the arch is

(a) 8 meters (b) 10 meters
(c) 12 meters (d) 14 meters

[CE, ME, CS 2012, 2 Marks (Set-1)]

18. An automobile plant contracted to buy shock absorbers from two supplies X and Y. X supplies 60% and Y supplies 40% of the shock absorbers. All shock absorbers are subjected to a quality test. The ones that pass the quality test are considered reliable. Of X's shock absorbers, 96% are reliable. OF Y's shock absorbers, 72% are reliable.

The probability that a randomly chosen shock absorber, which is found to be reliable, is made by Y is

(a) 0.288 (b) 0.334
(c) 0.667 (d) 0.720

[CE, ME, CS 2012, 2 Marks (Set-1)]

19. Which of the following assertions are CORRECT?

P: Adding 7 to each entry in a list adds 7 to the mean of the list

Q: Adding 7 to each entry in a list adds 7 to the standard deviation of the list

R: Doubling each entry in a list doubles the mean of the list

S: Doubling each entry in a list leaves the standard deviation of the list unchanged

(a) P, Q (b) Q, R
(c) P, R (d) R, S

[CE, ME, CS 2012, 2 Marks (Set-1)]

20. Given the sequence of terms, AD CG FK JP, the next time is

(a) OV (b) OW
(c) PV (d) PW

[CE, ME, CS 2012, 2 Marks (Set-1)]

21. If $(1.001)^{1259} = 3.52$ $(1.001)^{2062} = 7.85$, then $(1.001)^{3321}$

(a) 2.23 (b) 4.33
(c) 11.37 (d) 27.64

[EE, EC 2012, 1 Mark (Set-2)]

22. Raju has 14 currency notes in his pocket consisting of only ₹ 20 notes and ₹ 10 notes. The total money value of the notes is ₹ 230. The number of ₹ 10 notes that Raju has is

(a) 5 (b) 6
(c) 9 (d) 10

[EE, EC 2012, 2 Marks (Set-2)]

23. A and B friends. They decide to meet between 1 PM and 2 PM on a given day. There is a condition that whoever arrives first will not wait for the other for more than 15 minutes. The probability that they will meet on that day is

(a) 1/4 (b) 1/16
(c) 7/16 (d) 9/16

[EE, EC 2012, 2 Marks (Set-2)]

24. The data given in the following table summarizes the monthly budget of an average household.

Category	Amount
Food	4000
Clothing	1200
Rent	2000
Savings	1500
Others	1800

The approximate percentage of the monthly budget **NOT** spent on savings is

(a) 10% (b) 14%
(c) 81% (d) 86%

[EE, EC 2012, 2 Marks (Set-2)]

25. There are eight bags of rice looking alike, seven of which have equal and one is slightly heavier. The weighing balance is of unlimited capacity. Using this balance the minimum number of weighting required to identify the heavier bag is

(a) 2 (b) 3
(c) 4 (d) 8

[EE, EC 2012, 2 Marks (Set-2)]

26. A number much greater than 75 it is smaller than 117 is
(a) 91 (b) 93
(c) 89 (d) 96
[CE 2013, 1 Mark (Set-1)]

27. A firm is selling it product at ₹ 60/unit. The total cost of production is ₹ 100 and firm is earning total profit of ₹ 500, Later the total cost increased by 30%. By what percentage the price should be increased to maintain the same profit level.
(a) 5 (b) 15
(c) 10 (d) 30
[CE 2013, 2 Marks (Set-1)]

28. Following table provides figures (in rupees) on annual expenditure of a firm for two years 2010 and 2011

Category	2010	2011
Raw material	5200	6240
Power & fuel	7000	9450
Salary & wages	9000	12600
Plant & machinery	20000	25000
Advertising	15000	19500
Research & Development	22000	26400

In 2011, which of the following two categories have registered increase by same percentage?
(a) Raw material and salary and wages
(b) Salary and wages and advertising
(c) Power and fuel and advertising
(d) Raw material and research and development
[CE 2013, 2 Marks (Set-1)]

29. If $|4x - 7| = 5$ then the value of $2|x| - |-x|$ is
(a) $2, \frac{1}{3}$ (b) $\frac{1}{2}, 3$
(c) $\frac{2}{3}, \frac{1}{3}$ (d) $\frac{2}{9}, 3$
[CE 2013, 2 Marks (Set-1)]

30. x and y are two positive real numbers, such that equation
$2x + y \leq 6$; $x + 2y \leq 8$
For which values of (x, y), the function $f(x, y) = 3x + 6y$ will give maximum value
(a) 4/3, 10/3 (b) 8/3, 20/3
(c) 8/3, 10/3 (d) 4/3, 20/3
[CE 2013, 2 Marks (Set-1)]

31. What will be the maximum sum of 44,. 42, 40, ...?
(a) 502 (b) 504
(c) 506 (d) 500
[ME, PI & CS/IT 2013, 1 Mark (Set-2)]

32. Out of all the 2-digit integers between 1 and 100, a 2-digit number has to be selected at random. What is the probability that the selected number is not divisible by 7?
(a) 13/90 (b) 12/90
(c) 78/90 (d) 77/90
[ME, PI & CS/IT 2013, 2 Marks (Set-2)]

33. A tourist covers half of this journey by train at 60 km/h, half of the remainder by bus at 30 km/h and the rest by cycle at 10 km/h. The average speed of the tourist in km/h during his entire journey is
(a) 36 (b) 30
(c) 24 (d) 18
[ME, PI & CS/IT 2013, 2 Marks (Set-2)]

34. Find the sum of the expression

$$\frac{1}{\sqrt{1}+\sqrt{2}}+\frac{1}{\sqrt{2}+\sqrt{3}}+\frac{1}{\sqrt{3}+\sqrt{4}}+.....+\frac{1}{\sqrt{80}+\sqrt{81}}$$

(a) 7 (b) 8
(c) 9 (d) 10
[ME, PI & CS/IT 2013, 2 Marks (Set-2)]

35. The current erection cost of a structure is Rs. 13,200. If the labour wages per day increase by 1/5 of the current wages and the working hours decrease by 1/24 of the current period, then the new cost of erection in Rs., is
(a) 16, 500 (b) 15, 180
(c) 11, 000 (d) 10, 120
[ME, PI & CS/IT 2013, 2 Marks (Set-2)]

36. In the summer of 2012, in New Delhi, the mean temperature of Monday to Wednesday was 41°C and of Tuesday to Thursday was 43°C. If the temperature on Thursday was 15% higher than that of Monday, then the temperature in °C on Thursday was
(a) 40 (b) 43
(c) 46 (d) 49
[EE, EC, IN 2013, 1 Mark (Set-3)]

37. A car travels 8 km in the first quarter of an hour, 6 km in the second quarter and 16 km in the third quarter. The average speed of the car in km per hour over the entire journey is

(a) 30 (b) 36
(c) 40 (d) 24

[EE, EC, IN 2013, 2 Marks (Set-3)]

38. Find the sum to n terms of the series 10 + 84 + 734 +....

(a) $\frac{9(9^n+1)}{10}+1$ (b) $\frac{9(9^n-1)}{8}+1$

(c) $\frac{9(9^n-1)}{8}+n$ (d) $\frac{9(9^n-1)}{8}+n^2$

[EE, EC, IN 2013, 2 Marks (Set-3)]

39. The set of values of p for which the roots of the equation $3x^2 + 2x + p(p - 1) = 0$ are of opposite sign is

(a) $(-\infty, 0)$ (b) $(0, 1)$
(c) $(1, \infty)$ (d) $(0, \infty)$

[EE, EC, IN 2013, 2 Marks (Set-3)]

40. What is the change that a leap year, selected at random, will contain 53 Saturdays?

(a) $\frac{2}{7}$ (b) $\frac{3}{7}$

(c) $\frac{1}{7}$ (d) $\frac{5}{7}$

[EE, EC, IN 2013, 2 Marks (Set-3)]

41. The statistics of runs scored in a series by four batsmen are provided in the following table. Who is the most ***consistent*** batsman of these four?

Batsman	Average	Standard Deviation
K	31.2	5.21
L	46.0	6.35
M	54.4	6.22
N	17.9	5.90

(a) K (b) L
(c) M (d) N

[EC, ME 2014, 1 Mark (Set-1)]

42. What is the next number in the series?

12 35 81 173 357 ____.

[EC, ME 2014, 1 Mark (Set-1)]

43. Find the odd one from the following group:

WEKO IQWA FNTX NVBD

(a) WEKO (b) IQWA
(c) FNTX (d) NVBD

[EC, ME 2014, 2 Marks (Set-1)]

44. For submitting tax returns, all resident males with annual income below Rs 10 lakh should fill up Form P and all resident females with income below Rs 8 lakh should fill up Form Q. All people with incomes above Rs 10 lakh should fill up Form R, except non residents with income above Rs 15 lakhs, who should fill up Form S. All others should fill Form T. An example of a person who should fill Form T is

(a) a resident male with annual income Rs 9 lakh
(b) a resident female with annual income Rs 9 lakh
(c) a non-resident male with annual income Rs 16 lakh
(d) a non-resident female with annual income Rs 16 lakh

[EC, ME 2014, 2 Marks (Set-1)]

45. A train that is 280 metres long, travelling at a uniform speed, crosses a platform in 60 seconds and passes a man standing on the platform in 20 seconds. What is the length of the platform in metres?

[EC, ME 2014, 2 Marks (Set-1)]

46. The exports and imports (in crores of Rs.) of a country from 2000 to 2007 are given in the following bar chart. If the trade deficit is defined as excess of imports over exports, in which year is the trade deficit 1/5th of the exports?

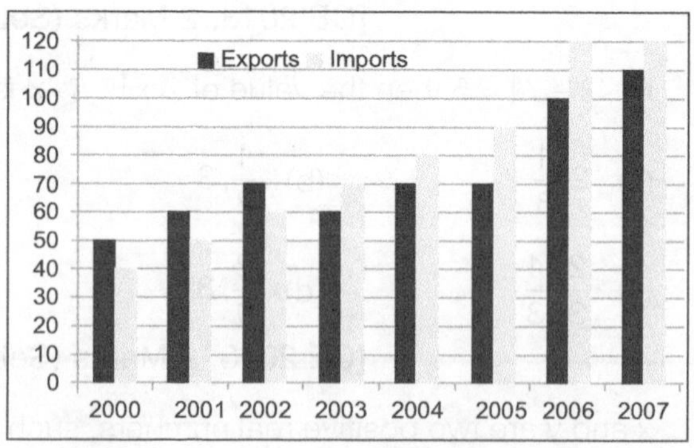

(a) 2005 (b) 2004
(c) 2007 (d) 2006

[EC, ME 2014, 2 Marks (Set-1)]

47. You are given three coins: one has heads on both faces, the second has tails on both faces, and the third has a head on one face and a tail on the other. You choose a coin at random and toss it, and it comes up heads. The probability that the other face is tails is

(a) 1/4 (b) 1/3
(c) 1/2 (d) 2/3

[EC, ME 2014, 2 Marks (Set-1)]

48. A regular die has six sides with numbers 1 to 6 marked on its sides. If a very large number of throws show the following frequencies of occurrence:

$1 \rightarrow 0.167$; $2 \rightarrow 0.167$; $3 \rightarrow 0.152$; $4 \rightarrow 0.166$; $5 \rightarrow 0.168$; $6 \rightarrow 0.180$. We call this die

(a) irregular (b) biased
(c) Gaussian (d) insufficient

[EC, ME 2014, 1 Mark (Set-2)]

49. Fill in the missing number in the series.

2 3 6 15 _?_ 157.5 630

[EC, ME 2014, 1 Mark (Set-2)]

50. Find the odd one in the following group

QWZB, BHKM, WCGJ, MSVX,

(a) QWZB (b) BHKM
(c) WCGJ (d) MSVX

[EC, ME 2014, 2 Marks (Set-2)]

51. The sum of eight consecutive odd numbers is 656. The average of four consecutive even numbers is 87. What is the sum of the smallest odd number and second largest even number?

[EC, ME 2014, 2 Marks (Set-2)]

52. The total exports and revenues from the exports of a country are given in the two charts shown below. The pie chart for exports shows the quantity of each item exported as a percentage of the total quantity of exports. The pie chart for the revenues shows the percentage of the total revenue generated through export of each item. The total quantity of exports of all the items is 500 thousand tonnes and the total revenues are 250 crore rupees. Which item among the following has generated the maximum revenue per kg?

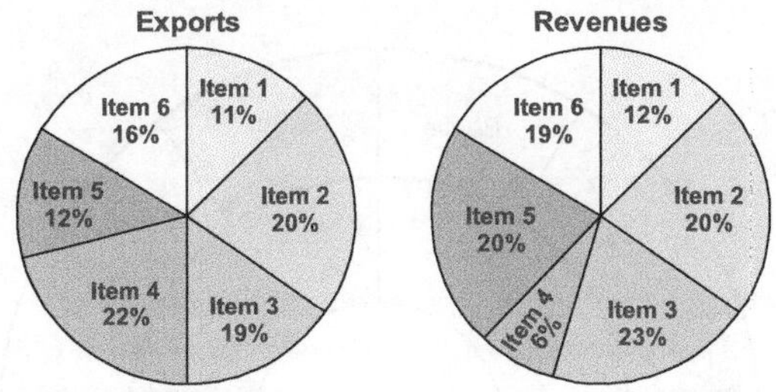

(a) Item 2 (b) Item 3
(c) Item 6 (d) Item 5

[EC, ME 2014, 2 Marks (Set-2)]

53. It takes 30 minutes to empty a half-full tank by draining it at a constant rate. It is decided to simultaneously pump water into the half-full tank while draining it. What is the rate at which water has to be pumped in so that it gets fully filled in 10 minutes?

(a) 4 times the draining rate
(b) 3 times the draining rate
(c) 2.5 times the draining rate
(d) 2 times the draining rate

[EC, ME 2014, 2 Marks (Set-2)]

54. The next term in the series 81, 54, 36, 24, ... is ________.

[EC, ME 2014, 1 Mark (Set-3)]

55. In which of the following option will the expression P < M be definitely true?

(a) $M < R > P > S$ (b) $M > S < P < F$
(c) $Q < M < F = P$ (d) $P = A < R < M$

[EC, ME 2014, 1 Mark (Set-3)]

56. Find the next term in the sequence:

7G, 11K, 13M, ____.

(a) 15Q (b) 17Q
(c) 15P (d) 17P

[EC, ME 2014, 2 Marks (Set-3)]

57. The multi-level hierarchical pie chart shows the population of animals in a reserve forest. The correct conclusions from this information are:

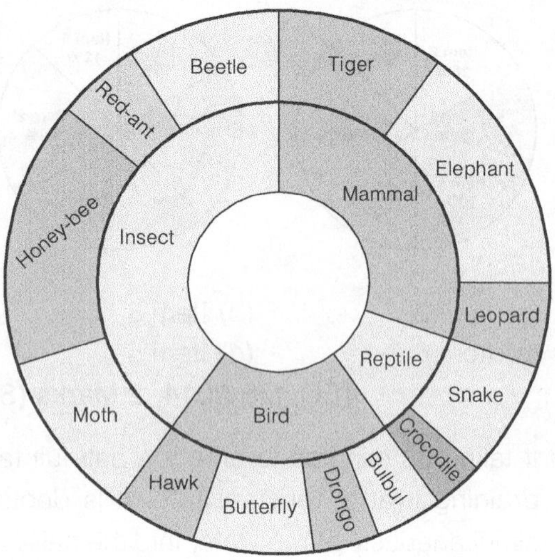

(i) Butterflies are birds

(ii) There are more tigers in this forest than red ants

(iii) All reptiles in this forest are either snakes or crocodiles

(iv) Elephants are the largest mammals in this forest

(a) (i) and (ii) only

(b) (i), (ii), (iii) and (iv)

(c) (i), (iii) and (iv) only

(d) (i), (ii) and (iii) only

[EC, ME 2014, 2 Marks (Set-3)]

58. A man can row at 8 km per hour in still water. If it takes him thrice as long to row upstream, as to row downstream, then find the stream velocity in km per hour.

[EC, ME 2014, 2 Marks (Set-3)]

59. A firm producing air purifiers sold 200 units in 2012. The following pie chart presents the share of raw material, labour, energy, plant & machinery, and transportation costs in the total manufacturing cost of the firm in 2012. The expenditure on labour in 2012 is Rs. 4,50,000. In 2013, the raw material expenses increased by 30% and all other expenses increased by 20%. If the company registered a profit of Rs. 10 lakhs in 2012, at what price (in Rs.) was each air purifier sold?

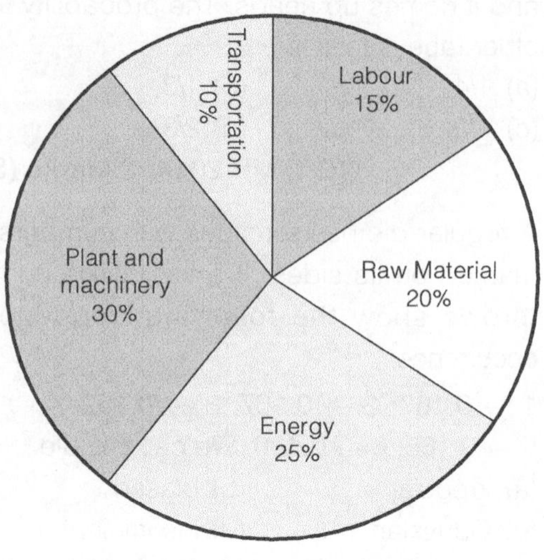

[EC, ME 2014, 2 Marks (Set-3)]

60. A batch of one hundred bulbs is inspected by testing four randomly chosen bulbs. The batch is rejected if even one of the bulbs is defective. A batch typically has five defective bulbs. The probability that the current batch is accepted is ______.

[EC, ME 2014, 2 Marks (Set-3)]

61. Let $f(x, y) = x^n y^m = P$. If x is doubled and y is halved, the new value of f is

(a) $2^{n-m} P$ (b) $2^{m-n} P$

(c) $2(n-m)P$ (d) $2(m-n)P$

[EC, ME 2014, 1 Mark (Set-4)]

62. In a sequence of 12 consecutive odd numbers, the sum of the first 5 numbers is 425. What is the sum of the last 5 numbers in the sequence?

[EC, ME 2014, 1 Mark (Set-4)]

63. Find the next term in the sequence:

13M, 17Q, 19S, ____.

(a) 21W (b) 21V

(c) 23W (d) 23V

[EC, ME 2014, 2 Marks (Set-4)]

64. If 'KCLFTSB' stands for 'best of luck' and 'SHSWDG' stands for 'good wishes', which of the following indicates 'ace the exam'?

(a) MCHTX (b) MXHTC

(c) XMHCT (d) XMHTC

[EC, ME 2014, 2 Marks (Set-4)]

65. Industrial consumption of power doubled from 2000-2001 to 2010-2011. Find the annual rate of increase in percent assuming it to be uniform over the years.
(a) 5.6 (b) 7.2
(c) 10.0 (d) 12.2
[EC, ME 2014, 2 Marks (Set-4)]

66. A firm producing air purifiers sold 200 units in 2012. The following pie chart presents the share of raw material, labour, energy, plant & machinery, and transportation costs in the total manufacturing cost of the firm in 2012. The expenditure on labour in 2012 is Rs. 4,50,000. In 2013, the raw material expenses increased by 30% and all other expenses increased by 20%. What is the percentage increase in total cost for the company in 2013?

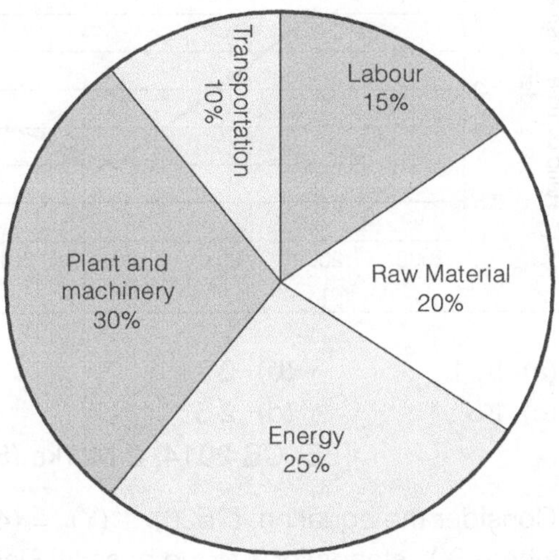

[EC, ME 2014, 2 Marks (Set-4)]

67. A five digit number is formed using the digits 1,3,5,7 and 9 without repeating any of them. What is the sum of all such possible five digit numbers?
(a) 6666660 (b) 6666600
(c) 6666666 (d) 6666606
[EC, ME 2014, 2 Marks (Set-4)]

68. If $\left(z+\frac{1}{z}\right)^2 = 98$, compute $\left(z^2+\frac{1}{z^2}\right)$.
[EE, CS 2014, 1 Mark (Set-1)]

69. The roots of $ax^2 + bx + c = 0$ are real and positive *a*, *b* and *c* are real. Then $ax^2 + b|x| + c = 0$ has
(a) no roots (b) 2 real roots
(c) 3 real roots (d) 4 real roots
[EE, CS 2014, 1 Mark (Set-1)]

70. Round-trip tickets to a tourist destination are eligible for a discount of 10% on the total fare. In addition, groups of 4 or more get a discount of 5% on the total fare. If the one way single person fare is Rs.100, a group of 5 tourists purchasing round trip tickets will be charged Rs. ______.
[EE, CS 2014, 2 Marks (Set-1)]

71. In a survey, 300 respondents were asked whether they own a vehicle or not. If yes, they were further asked to mention whether they own a car or scooter or both. Their responses are tabulated below. What percent of respondents do not own a scooter.

		Men	Women
Own vehicle	Car	40	34
	Scooter	30	20
	Both	60	46
Do not own vehicle		20	50

[EE, CS 2014, 2 Marks (Set-1)]

72. When a point inside of a tetrahedron (a solid with four triangular surfaces) is connected by straight lines to corners, how many (new) internal planes are created with these lines? ______
[EE, CS 2014, 2 Marks (Set-1)]

73. What is the average of all multiples of 10 from 2 to 198?
(a) 90 (b) 100
(c) 110 (d) 120
[EE, CS 2014, 1 Mark (Set-2)]

74. The value of $\sqrt{12+\sqrt{12+\sqrt{12+\ldots}}}$ is
(a) 3.464 (b) 3.932
(c) 4.000 (d) 4.444
[EE, CS 2014, 1 Mark (Set-2)]

75. If x is real and $|x^2 - 2x + 3| = 11$, then possible values of $|-x^3 + x^2 - x|$ include
(a) 2, 4 (b) 2, 14
(c) 4, 52 (d) 14, 52
[EE, CS 2014, 2 Marks (Set-2)]

76. The ratio of male to female students in a college for five years is plotted in the following line graph. If the number of female students doubled in 2009, by what percent did the number of male students increase in 2009?

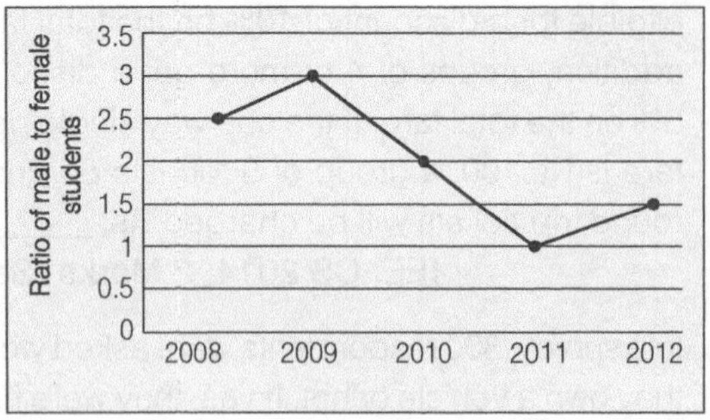

[EE, CS 2014, 2 Marks (Set-2)]

77. At what time between 6 a.m. and 7 a.m. will the minute hand and hour hand of a clock make an angle closest to 60°?

(a) 6:22 a.m. (b) 6:27 a.m.
(c) 6:38 a.m. (d) 6:45 a.m.

[EE, CS 2014, 2 Marks (Set-2)]

78. Which number does not belong in the series below?

2, 5, 10, 17, 26, 37, 50, 64

(a) 17 (b) 37
(c) 64 (d) 26

[EE, CS 2014, 1 Mark (Set-3)]

79. The table below has question wise data on the performance of students in an examination. The marks for each question are also listed. There is no negative or partial marking in the examination.

Q.No.	Marks	Answered Correctly	Answered Wrongly	Not Attemped
1	2	21	17	6
2	3	15	27	2
3	2	23	18	3

What is the average of the marks obtained by the class in the examination?

(a) 1.34 (b) 1.74
(c) 3.02 (d) 3.91

[EE, CS 2014, 1 Mark (Set-3)]

80. The Gross Domestic Product (GDP) in Rupees grew at 7% during 2012-2013. For international comparison, the GDP is compared in US Dollars (USD) after conversion based on the market exchange rate. During the period 2012-2013 the exchange rate for the USD increased from Rs. 50/ USD to Rs.60/ USD. India's GDP in USD during the period 2012-2013.

(a) Increased by 5%
(b) Decreased by 13%
(c) Decreased by 20%
(d) Decreased by 11%

[EE, CS 2014, 2 Marks (Set-3)]

81. The ratio of male to female students in a college for five years is plotted in the following line graph. If the number of female students in 2011 and 2012 is equal, what is the ratio of male students in 2012 to male students in 2011?

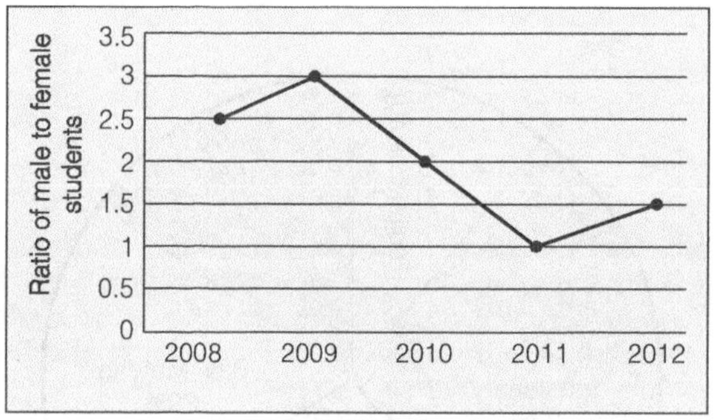

(a) 1 : 1 (b) 2 : 1
(c) 1.5 : 1 (d) 2.5 : 1

[EE, CS 2014, 2 Marks (Set-3)]

82. Consider the equation: $(7526)_8 - (Y)_8 = (4364)_8$, where $(X)_N$ stands for X to the base N. Find Y.

(a) 1634 (b) 1737
(c) 3142 (d) 3162

[EE, CS 2014, 2 Marks (Set-3)]

83. If $y = 5x^2 + 3$, then the tangent at $x = 0, y = 3$

(a) passes through $x = 0, y = 0$
(b) has a slope of +1
(c) is parallel to the x-axis
(d) has a slope of -1

[CE 2014, 1 Mark (Set-1)]

84. A foundry has a fixed daily cost of Rs 50,000 whenever it operates and a variable cost of Rs 800Q, where Q is the daily production in tonnes. What is the cost of production in Rs per tonne for a daily production of 100 tonnes?

[CE 2014, 1 Mark (Set-1)]

85. Find the odd one in the following group: ALRVX, EPVZB, ITZDF, OYEIK
(a) ALRVX (b) EPVZB
(c) ITZDF (d) OYEIK
[CE 2014, 2 Marks (Set-1)]

86. Anuj, Bhola, Chandan, Dilip, Eswar and Faisal live on different floors in a six-storeyed building (the ground floor is numbered 1, the floor above it 2, and so on). Anuj lives on an even-numbered floor. Bhola does not live on an odd numbered floor. Chandan does not live on any of the floors below Faisal's floor. Dilip does not live on floor number 2. Eswar does not live on a floor immediately above or immediately below Bhola. Faisal lives three floors above Dilip. Which of the following floor-person combination is correct?

	Anuj	Bhola	Chandan	Dilip	Eswar	Faisal
(a)	6	2	5	1	3	4
(b)	2	6	5	1	3	4
(c)	4	2	6	3	1	5
(d)	2	4	6	1	3	5

[CE 2014, 2 Marks (Set-1)]

87. The smallest angle of a triangle is equal to two thirds of the smallest angle of a quadrilateral. The ratio between the angles of the quadrilateral is 3:4:5:6. The largest angle of the triangle is twice its smallest angle. What is the sum, in degrees, of the second largest angle of the triangle and the largest angle of the quadrilateral?
[CE 2014, 2 Marks (Set-1)]

88. One percent of the people of country X are taller than 6 ft. Two percent of the people of country Y are taller than 6 ft. There are thrice as many people in country X as in country Y. Taking both countries together, what is the percentage of people taller than 6 ft?
(a) 3.0 (b) 2.5
(c) 1.5 (d) 1.25
[CE 2014, 2 Marks (Set-1)]

89. The monthly rainfall chart based on 50 years of rainfall in Agra is shown in the following figure. Which of the followings are true? (*k* percentile is the value such that k percent of the data fall below that value)

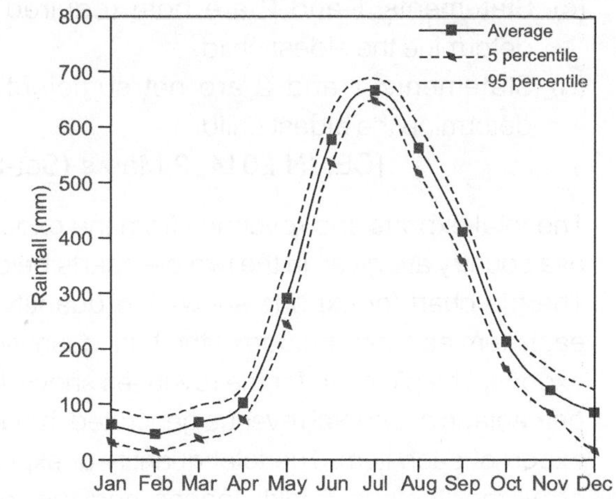

(i) On average, it rains more in July than in December
(ii) Every year, the amount of rainfall in August is more than that in January
(iii) July rainfall can be estimated with better confidence than February rainfall
(iv) In August, there is at least 500 mm of rainfall
(a) (i) and (ii) (b) (i) and (iii)
(c) (ii) and (iii) (d) (iii) and (iv)
[CE 2014, 2 Marks (Set-1)]

90. In any given year, the probability of an earthquake greater than Magnitude 6 occurring in the Garhwal Himalayas is 0.04. The average time between successive occurrences of such earthquakes is ____ years.
[CE, IN 2014, 1 Mark (Set-2)]

91. The population of a new city is 5 million and is growing at 20% annually. How many years would it take to double at this growth rate?
(a) 3-4 years (b) 4-5 years
(c) 5-6 years (d) 6-7 years
[CE, IN 2014, 1 Mark (Set-2)]

92. In a group of four children, Som is younger to Riaz. Shiv is elder to Ansu. Ansu is youngest in the group. Which of the following statements is/are required to find the eldest child in the group?

Statements

1. Shiv is younger to Riaz.
2. Shiv is elder to Som.

(a) Statement 1 by itself determines the eldest child.
(b) Statement 2 by itself determines the eldest child.

(c) Statements 1 and 2 are both required to determine the eldest child.
(d) Statements 1 and 2 are not sufficient to determine the eldest child.

[CE, IN 2014, 2 Marks (Set-2)]

93. The total exports and revenues from the exports of a country are given in the two pie charts below. The pie chart for exports shows the quantity of each item as a percentage of the total quantity of exports. The pie chart for the revenues shows the percentage of the total revenue generated through export of each item. The total quantity of exports of all the items is 5 lakh tonnes and the total revenues are 250 crore rupees. What is the ratio of the revenue generated through export of Item 1 per kilogram to the revenue generated through export of Item 4 per kilogram?

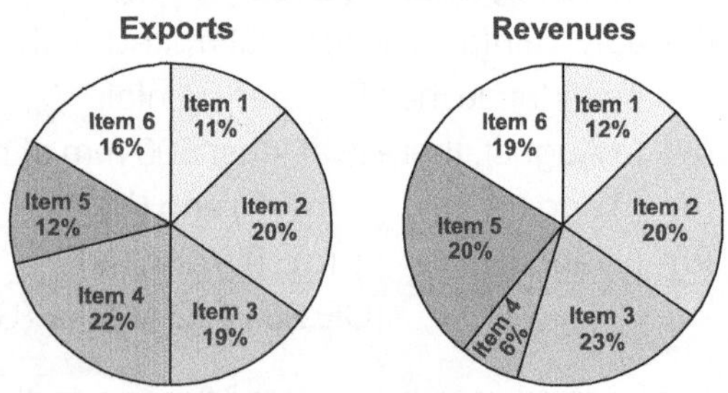

(a) 1 : 2
(b) 2 : 1
(c) 1 : 4
(d) 4 : 1

[CE, IN 2014, 2 Marks (Set-2)]

94. X is 1 km northeast of Y. Y is 1 km southeast of Z. W is 1 km west of Z. P is 1 km south of W. Q is 1 km east of P. What is the distance between X and Q in km?
(A) 1
(b) $\sqrt{2}$
(c) $\sqrt{3}$
(d) 2

[CE, IN 2014, 2 Marks (Set-2)]

95. 10% of the population in a town is HIV$^+$. A new diagnostic kit for HIV detection is available; this kit correctly identifies HIV$^+$ individuals 95% of the time, and HIV$^-$ individuals 89% of the time. A particular patient is tested using this kit and is found to be positive. The probability that the individual is actually positive is ______.

[CE, IN 2014, 2 Marks (Set-2)]

96. If ROAD is written as URDG, then SWAN should be written as:
(a) VXDQ
(b) VZDQ
(c) VZDP
(d) UXDQ

[CE (Set-1), CS (Set-3) 2015, 1 Mark]

97. A function $f(x)$ is linear and has a value of 29 at $x = -2$ and 39 at $x = 3$. Find its value at $x = 5$.
(a) 59
(b) 45
(c) 43
(d) 35

[CE (Set-1), CS (Set-3) 2015, 1 Mark]

98. The exports and imports (in crores of Rs.) of a country from the year 2000 to 2007 are given in the following bar chart. In which year is the combined percentage increase in imports and exports the highest?

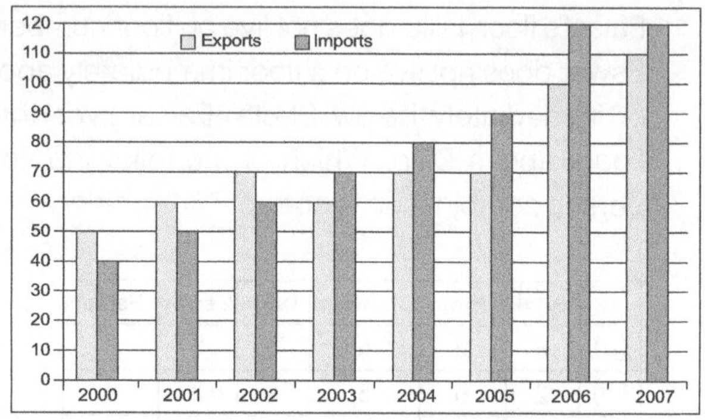

[CE (Set-1), CS (Set-3) 2015, 2 Marks]

99. Choose the most appropriate equation for the function drawn as a thick line, in the plot below:

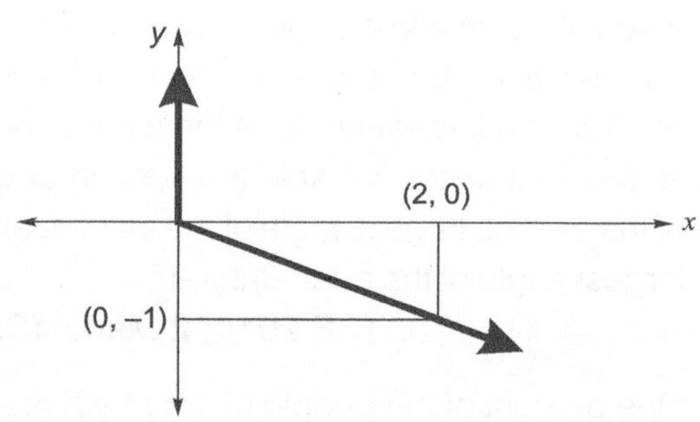

(a) $x = y - |y|$
(b) $x = -(y - |y|)$
(c) $x = y + |y|$
(d) $x = -(y + |y|)$

[CE (Set-1), CS (Set-3) 2015, 2 Marks]

100. The head of a newly formed government desires to appoint five of the six selected members *P*, *Q*, *R*, *S*, *T* and *U* to portfolios of Home, Power, Defense, Telecom, and Finance. *U* does not want any portfolio if *S* gets one of the five. *R* wants either Home or Finance or no portfolio. *Q* says that if *S* gets either Power or Telecom, then she must get the other one. *T* insists on a portfolio if *P* gets one.

Which is the valid distribution of portfolios?

(a) *P*-Home, *Q*-Power, *R*-Defence, *S*-Telecom, *T*-Finance

(b) *R*-Home, *S*-Power, *P*-Defence, *Q*-Telecom, *T*-Finance

(c) *P*-Home, *Q*-Power, *T*-Defence, *S*-Telecom, *U*-Finance

(d) *Q*-Home, *U*-Power, *T*-Defence, *R*-Telecom, *P*-Finance

[CE (Set-1), CS (Set-3) 2015, 2 Marks]

101. Four cards are randomly selected from a pack of 52 cards. If the first two cards are kings, what is the probability that the third card is a king?

(a) $\frac{4}{52}$ (b) $\frac{2}{50}$

(c) $\left(\frac{1}{52}\right)\times\left(\frac{1}{52}\right)$ (d) $\left(\frac{1}{52}\right)\times\left(\frac{1}{51}\right)\times\left(\frac{1}{50}\right)$

[CE 2015, 1 Mark (Set-2)]

102. Mr. Vivek walks 6 meters North-East, then turns and walks 6 meters South-East, both at 60 degrees to East. He further moves 2 meters South and 4 meters West. What is the straight distance in meters between the point he started from and the point he finally reached?

(a) $2\sqrt{2}$ (b) 2

(c) $\sqrt{2}$ (d) $\frac{1}{\sqrt{2}}$

[CE 2015, 1 Mark (Set-2)]

103. Read the following table giving sales data of five types of batteries for years 2006 to 2012

Year	Type I	Type II	Type III	Type IV	Type V
2006	75	144	114	102	108
2007	90	126	102	84	126
2008	96	114	75	105	135
2009	105	90	150	90	75
2010	90	75	135	75	90
2011	105	60	165	45	120
2012	115	85	160	100	145

Out of the following, which type of battery achieved highest growth between the years 2006 and 2012?

(a) Type V (b) Type III

(c) Type II (d) Type I

[CE 2015, 2 Marks (Set-2)]

104. How many four digit numbers can be formed with the 10 digits 0, 1, 2, ... 9 if no number can start with 0 and if repetitions are not allowed?

[CE 2015, 2 Marks (Set-2)]

105. The given question is followed by two statements; select the most appropriate option that solves the question

Capacity of a solution tank *A* is 70% of the capacity of tank *B*. How many gallons of solution are in tank *A* and tank *B*?

Statements:

I. Tank A is 80% full and tank B is 40% full.

II. Tank A if full contains 14,000 gallons of solution.

(a) Statement I alone is sufficient.

(b) Statement II alone is sufficient.

(c) Either statement I or II alone is sufficient.

(d) Both the statements I and II together are sufficient.

[CE 2015, 2 Marks (Set-2)]

106. There are 16 teachers who can teach Thermodynamics (TD), 11 who can teach Electrical Sciences (ES), and 5 who can teach both TD and Engineering Mechanics (EM). There are a total of 40 teachers. 6 cannot teach any of the three subjects, i.e. EM, ES or TD. 6 can teach only ES. 4 can teach all three subjects, i.e. EM, ES and TD. 4 can teach ES and TD. How many can teach both ES and EM but not TD?

(a) 1 (b) 2

(c) 3 (d) 4

[CE 2015, 2 Marks (Set-2)]

107. Consider a function $f(x) = 1 - |x|$ on $-1 \le x \le 1$. The value of *x* at which the function attains a maximum, and the maximum value of the function are:

(a) 0, – 1 (b) –1, 0

(c) 0, 1 (d) –1, 2

[CS (Set-1), EE (Set-2) 2015, 1 Mark]

108. Base on the given statements select the most appropriate option to solve the given question. What will be the total weight of 10 poles each of same weight?

Statements:

(I) One fourth of the weight of a pole is 5 kg.

(II) The total weight of these poles is 160 kg more than the total weight of two poles.

(a) Statement I alone is not sufficient
(b) Statement II alone is not sufficient
(c) Either I or II alone is sufficient
(d) Both statement I and II together are not sufficient

[CS (Set-1), EE (Set-2) 2015, 1 Mark]

109. If the list of letters, P, R, S, T, U is an arithmetic sequence, which of the following are also in arithmetic sequence?

(I) $2P, 2R, 2S, 2T, 2U$
(II) $P-3, R-3, S-3, T-3, U-3$
(II) P^2, R^2, S^2, T^2, U^2

(a) I only (b) I an II
(c) II and III (d) I and III

[CS (Set-1), EE (Set-2) 2015, 1 Mark]

110. If p, q, r, s are distinct integers such that

$f(p, q, r, s) = \max(p, q, r, s)$
$g(p, q, r, s) = \min(p, q, r, s)$
$h(p, q, r, s)$ = remainder of $(p \times q)/(r \times s)$ if $(p \times q) > (r \times s)$ or remainder of $(r \times s)/(p \times q)$ if $(r \times s) > (p \times q)$

Also a function $fgh(p, q, r, s) = f(p, q, r, s) \times g(p, q, r, s) \times h(p, q, r, s)$

Also the same operations are valid with two variable function of the form $f(p, q)$

What is the value of $fg(h(2, 5, 7, 3), 4, 6, 8)$?

[CS (Set-1), EE (Set-2) 2015, 2 Marks]

111. Four branches of a company are located at M, N, O, and P. M is north of N at a distance of 4 km; P is south of O at a distance of 2 km; N is southeast of O by 1 km. What is the distance between M and P in km?

(a) 5.34 (b) 6.74
(c) 28.5 (d) 45.49

[CS (Set-1), EE (Set-2) 2015, 2 Marks]

112. In a triangle PQR, PS is the angle bisector of $\angle QPR$ and $\angle QPS = 60°$. What is the length of PS?

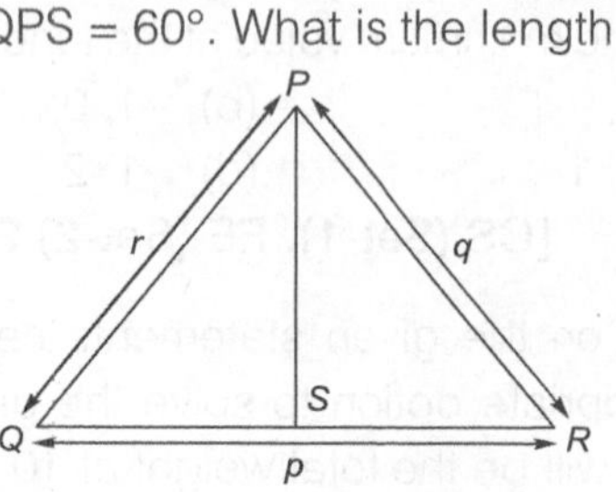

(a) $\dfrac{(q+r)}{qr}$ (b) $\dfrac{qr}{(q+r)}$

(c) $\sqrt{(q^2+r^2)}$ (d) $\dfrac{(q+r)^2}{qr}$

[CS (Set-1), EE (Set-2) 2015, 2 Marks]

113. Base on the given statements, select the most appropriate option to solve the given question. If two floors in a certain building are 9 feet apart, how many steps are there in a set of stairs that extends from the first floor to the second floor of the building?

Statements:

(I) Each step is 3/4 foot high.
(II) Each step is 1 foot wide.

(a) Statement-I alone is sufficient, but statement-II alone is not sufficient.
(b) Statement-II alone is sufficient, but statement I alone is not sufficient.
(c) Both statements together are sufficient, but neither statement alone is sufficient.
(d) Statement I and II together are not sufficient.

[CS (Set-2), EE (Set-1) 2015, 1 Mark]

114. Given Set A = {2, 3, 4, 5} Set B = {11, 12, 13, 14, 15}, two numbers are randomly selected, one from each set. What is the probability that the sum of the two numbers are equals 16?

(a) 0.20 (b) 0.25
(c) 0.30 (d) 0.33

[CS (Set-2), EE (Set-1) 2015, 1 Mark]

115. The pie chart below has the breakup of the number of students from different departments in an engineering college for the year 2012. The proportion of male to female students in each department is 5 : 4. There are 40 males in Electrical Engineering. What is the difference between the numbers of female students in the Civil department and the female students in the Mechanical department?

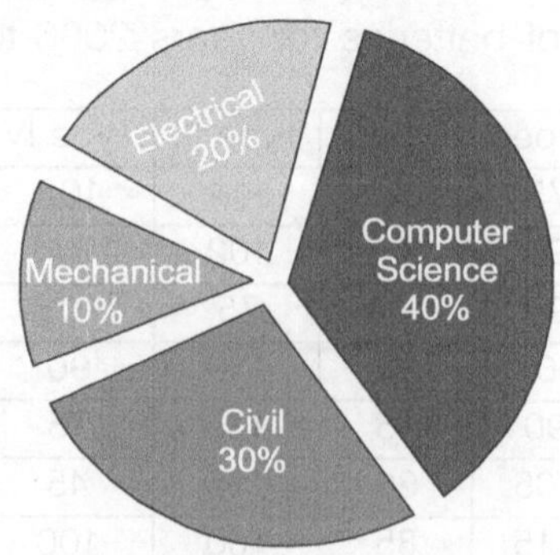

116. The probabilities that a student passes in Mathematics, Physics and Chemistry are m, p, and c respectively. Of these subjects, the student has 75% chance of passing in at least one, a 50% chance of passing in at least two and a 40%

chance of passing in exactly two. Following relations are drawn in *m*, *p*, *c*:

(I) $p + m + c = 27/20$

(II) $p + m + c = 13/20$

(III) $(p) \times (m) \times (c) = 1/10$

(a) Only relation I is true

(b) Only relation II is true

(c) Relations II and III are true

(d) Relations I and III are true

[CS (Set-2), EE (Set-1) 2015, 2 Marks]

117. The number of students in a class who have answered correctly, wrongly, or not attempted each question in an exam, are listed in the table below. The marks for each question are also listed. There is no negative or partial marking.

Q No.	Marks	Answered correctly	Answered Wrongly	Not Attempted
1	2	21	17	6
2	3	15	27	2
3	1	11	29	4
4	2	23	18	3
5	5	31	12	1

What is the average of the marks obtained by the class in the examination?

(a) 2.290 (b) 2.970

(c) 6.795 (d) 8.795

[CS (Set-2), EE (Set-1) 2015, 2 Marks]

118. If $\log_x (5/7) = -1/3$, then the value of x is

(a) 343/125 (b) 125/343

(c) –25/49 (d) –49/25

[EC 2015, 1 Mark (Set-1)]

119. Operates □, ◇ and → are defined by:

$$a \square b = \frac{a-b}{a+b};\ a \diamond b = \frac{a+b}{a-b};\ a \to b = ab.$$ Find

the value of $(66 \square 6) \to (66 \diamond 6)$

(a) –2 (b) –1

(c) 1 (d) 2

[EC 2015, 1 Mark (Set-1)]

120. A cube side 3 units is formed using a set of smaller cubes of side 1 unit. Find the proportion of the number of faces of the smaller cubes visible to those which are NOT visible.

(a) 1 : 4 (b) 1 : 3

(c) 1 : 2 (d) 2 : 3

[EC 2015, 2 Marks (Set-1)]

121. Fill in the missing value

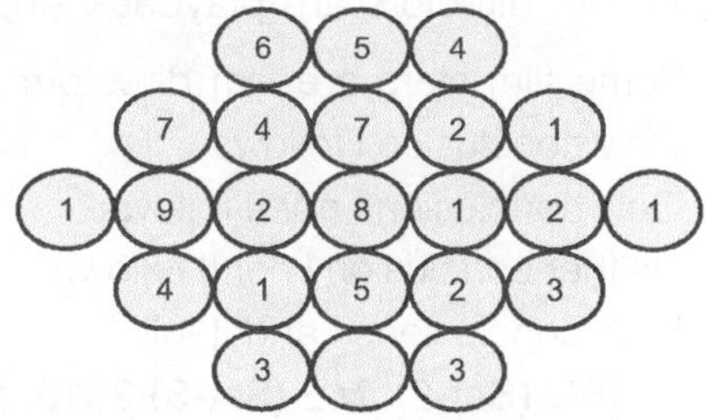

[EC 2015, 2 Marks (Set-1)]

122. An electric bus has onboard instruments that report the total electricity consumed since the start of the trip as well as the total distance covered. During a single day of operation, the bus travels on stretches *M*, *N*, *O* and *P*, in that order. The cumulative distance travelled and the corresponding electricity consumption are shown in the Table below:

Stretch	Cumulative distance (km)	Electricity used (kWh)
M	20	12
N	45	25
O	75	45
P	100	57

The stretch where the electricity consumption per km is minimum is

(a) *M* (b) *N*

(c) *O* (d) *P*

[EC (Set-2), ME (Set-3) 2015, 1 Mark]

123. Ram and Ramesh appeared in an interview for two vacancies in the same department. The probability of Ram's selection is 1/6 and that of Ramesh is 1/8. What is the probability that only one of them will be selected?

(a) 47/48 (b) 1/4

(c) 13/48 (d) 35/48

[EC (Set-2), ME (Set-3) 2015, 1 Mark]

124. Given below are two statements followed by two conclusions. Assuming these statements to be true, decide which one logically follows.

Statements:

I. All film star are playback singers.

II. All film directors are film stars.

Conclusions:

I. All film directors are playback singers.

II. Some film stars are film directors.

(a) Only conclusion I follows.
(b) Only conclusion I nor II follows.
(c) Neither conclusion I nor II follows.
(d) Both conclusions I and II follow.

[EC (Set-2), ME (Set-3) 2015, 2 Marks]

125. If $a^2 + b^2 + c = 1$, then $ab + bc + ac$ lies in the interval

(a) [1, 2/3] (b) [–1/2, 1]
(c) [–1, 1/2] (d) [2, –4]

[EC (Set-2), ME (Set-3) 2015, 2 Marks]

126. A tiger is 50 leaps of its own behind a deer. The tiger takes 5 leaps per minute to the deer's 4. If the tiger and the deer cover 8 metre and 5 metre per leap respectively, what distance in metres will the tiger have to run before it catches the deer?

[EC (Set-2), ME (Set-3) 2015, 2 Marks]

127. If $x > y > 1$, which of the following must be true?

(i) $ln\,x > ln\,y$ (ii) $e^x > e^y$
(iii) $y^x > x^y$ (iv) $\cos x > \cos y$
(a) (i) and (ii) (b) (i) and (iii)
(c) (iii) and (iv) (d) (ii) and (iv)

[EC (Set-3), ME (Set-2) 2015, 1 Mark]

128. Find the missing sequence in the letter series below:

A, CD, GHI?, UVWXY

(a) LMN (b) MNO
(c) MNOP (d) NOPQ

[EC (Set-3), ME (Set-2) 2015, 1 Mark]

129. Ms. X will be Bagdogra from 01/05/2014 to 20/05/2014 and from 22/05/2014 to 31/05/2014. On the morning of 21/05/2014, she will reach Kochi via Mumbai.

Which one of the statements below is logically valid and can be inferred from the above sentences?

(a) Ms. X will be in Kochi for one day, only in May.
(b) Ms. X will be in Kochi for only one day in May.
(c) Ms. X will be only in Kochi for one day in May.
(d) Only Ms. X will be in Kochi for one day in May.

[EC (Set-3), ME (Set-2) 2015, 2 Marks]

130. $\log \tan 1° + \log \tan 2° + ... + \log \tan 89°$ is ______.

(a) 1 (b) $\frac{1}{\sqrt{2}}$
(c) 0 (d) –1

[EC (Set-3), ME (Set-2) 2015, 2 Marks]

131. From a circular sheet of paper of radius 30 cm, a sector of 10% area is removed. If the remaining part is used to make a conical surface, then the ratio of the radius and height of the cone is ______.

[EC (Set-3), ME (Set-2) 2015, 2 Marks]

132. Five teams have to compete in a league, with every team playing every other team exactly once, before going to the next round. How many matches will have to held to complete the league round of matches?

(a) 20 (b) 10
(c) 8 (d) 5

[ME (Set-1), IN & PI 2015, 1 Mark]

133. Tanya is older than Eric.
Cliff is older than Tanya.
Eric is older than Cliff.

If the first two statements are true, then the third statement is

(a) True (b) False
(c) Uncertain (d) Data insufficient

[ME (Set-1), IN & PI 2015, 1 Mark]

134. In the given angle Q is a right angle, $PS : QS = 3 : 1$, $RT : QT = 5 : 2$ and $PU : UR = 1 : 1$. If area of triangle QTS is 20 cm², then the area of triangle PQR in cm² is ____.

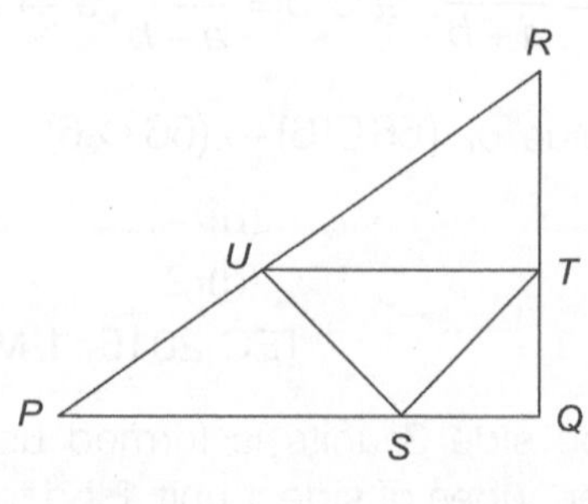

[ME (Set-1), IN & PI 2015, 2 Marks]

135. Given below are two statements followed by two conclusions. Assuming these statements to be true, decide which one logically follows:

Statements:
I. No manager is a leader.
II. All leaders are executives.

Conclusions:
I. No manager is an executive.
II. No executive is a manager
(a) Only conclusion I follows.
(b) Only conclusion II follows.
(c) Neither conclusion I nor II follows.
(d) Both conclusions I and II follow

[ME (Set-1), IN & PI 2015, 2 Marks]

136. A coin is tossed thrice. Let X be the even that head occurs in each of the first two tosses. Let Y be the event that a tail occurs on the third toss. Let Z be the event that two tails occur in three tosses. Based on the above information, which one of the following statements is TRUE?
(a) X and Y are not independent
(b) Y and Z are dependent
(c) Y and Z are independent
(d) X and Z are independent

[ME (Set-1), IN & PI 2015, 2 Marks]

137. Right triangle PQR is to be constructed in the xy-plane so that the right angle is at P and line PR is parallel to the x-axis. The x and y coordinates of P, Q, and R are to be integers that satisfy the inequalities:
$-4 \le x \le 5$ and $6 \le y \le 16$. How many different triangle could be constructed with these properties?
(a) 110 (b) 1100
(c) 9900 (d) 10000

[ME (Set-1), IN & PI 2015, 2 Marks]

138. In a huge pile of apples and oranges, both ripe and unripe mixed together, 15% are unripe fruits. Of the unripe fruits, 45% are apples. Of the ripe ones, 66% are oranges. If the pile contains a total of 5692000 fruits, how many of them are apples?
(a) 2029198 (b) 2467482
(c) 2789080 (d) 3577422

[EC, ME 2016, 1 Mark (Set-1)]

139. Michael lives 10 km away from where I live. Ahmed lives 5 km away and Susan lives 7 km away from where I live. Arun is farther away than Ahmed but closer than Susan from where I live. From the information provided here, what is one possible distance (in km) at which I live from Arun's place?
(a) 3.00 (b) 4.99
(c) 6.02 (d) 7.01

[EC, ME 2016, 1 Mark (Set-1)]

140. A person moving through a tuberculosis prone zone has a 50% probability of becoming infected. However, only 30% of infected people develop the disease. What percentage of people moving through a tuberculosis prone zone remains infected but does not show symptoms of disease?
(a) 15 (b) 33
(c) 35 (d) 37

[EC, ME 2016, 2 Marks (Set-1)]

141. Leela is older than her cousin Pavithra. Pavithra's brother Shiva is older than Leela. When Pavithra and Shiva are visiting Leela, all three like to play chess. Pavithra wins more often than Leela does. Which one of the following statement must be TRUE based on the above?
(a) When Shiva plays chess with Leela and Pavithra, he often loses.
(b) Leela is the oldest of the three
(c) Shiva is a better chess player than Pavithra
(d) Pavithra is the youngest of the three

[EC, ME 2016, 2 Marks (Set-1)]

142. If $q^{-a} = \frac{1}{r}$ and $r^{-b} = \frac{1}{s}$ and $s^{-c} = \frac{1}{q}$, the value of abc is______.
(a) $(rps)^{-1}$ (b) 0
(c) 1 (d) $r + q + s$

[EC, ME 2016, 2 Marks (Set-1)]

143. P, Q, R and S are working on a project. Q can finish the task in 25 days, working alone for 12 hours a day. R can finish the task in 50 days, working alone for 12 hours per day. Q worked 12 hours a day but took sick leave in the beginning for two days, R worked 18 hours a day on all days. What is the ratio of work done by Q and R after 7 days from the start of the project?
(a) 10 : 11 (b) 11 : 10
(c) 20 : 21 (d) 21 : 20

[EC, ME 2016, 2 Marks (Set-1)]

144. Given (9 inches)$^{1/2}$ = (0.25 yards)$^{1/2}$, which one of the following statement is TRUE?
(a) 3 inches = 0.5 yards
(b) 9 inches = 1.5 yards

(c) 9 inches = 0.25 yards
(d) 81 inches = 0.0625 yards
[EC (Set-2), ME (Set-3) 2016, 1 Mark]

145. *S*, *M*, *E* and *F* are working in shifts in a team to finish a project. *M* works with twice the efficiency of others but for half as many days as *E* worked. *S* and *M* have 6 hour shifts in a day, whereas *E* and *F* have 12 hours shifts. What is the ratio of contribution of *M* to contribution of *E* in the project?
(a) 1 : 1 (b) 1 : 2
(c) 1 : 4 (d) 2 : 1
[EC (Set-2), ME (Set-3) 2016, 1 Mark]

146. The Venn diagram shows the preference of the student population for leisure activities.

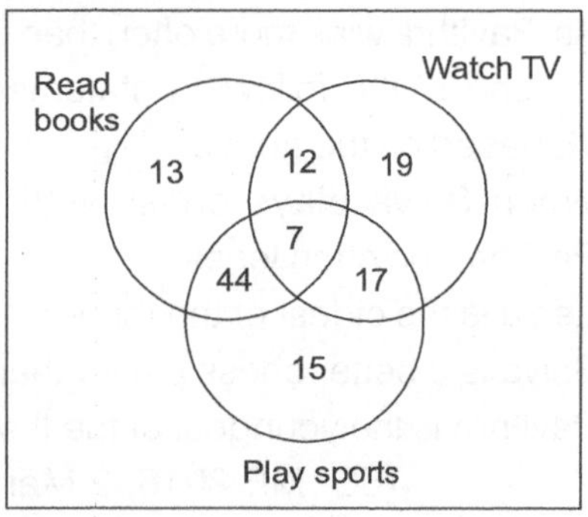

From the data given, the number of students who like to read books or play sports is_____.
(a) 44 (b) 51
(c) 79 (d) 108
[EC (Set-2), ME (Set-3) 2016, 2 Marks]

147. Two and a quarter hours back, when seen in a mirror, the reflection of a wall clock without number markings seemed to show 1 : 30. What is the actual current time shown by the clock?
(a) 8 : 15 (b) 11 : 15
(c) 12 : 15 (d) 12 : 45
[EC (Set-2), ME (Set-3) 2016, 2 Marks]

148. *M* and *N* start from the same location. *M* travels 10 km East and then 10 km North-East. *N* travels 5 km South and then 4 km South-East. What is the shortest distance (in km) between *M* and *N* at the end of their level.
(a) 18.60 (b) 22.50
(c) 20.61 (d) 25.00
[EC (Set-2), ME (Set-3) 2016, 2 Marks]

149. A wire of length 340 mm is to be cut into two parts. One of the parts is to be made into a square and the other into a rectangle where sides are in the ratio of 1 : 2. What is the length of the side of the square (in mm) such that the combined area of the square and the rectangle is a MINIMUM?
(a) 30 (b) 40
(c) 120 (d) 180
[EC (Set-2), ME (Set-3) 2016, 2 Marks]

150. The number that least fits this set: (324, 441, 97 and 64) is ________.
(a) 324 (b) 441
(c) 97 (d) 64
[EC (Set-3) & IN 2016, 1 Mark]

151. It takes 10s and 15s, respectively, for two trains travelling at different constant speeds to completely pass a telegraph post. The length of the first train is 120 m and that of the second train is 150 m. The magnitude of the difference in the speeds of the two trains (in m/s) is ______.
(a) 2.0 (b) 10.0
(c) 12.0 (d) 22.0
[EC (Set-3) & IN 2016, 1 Mark]

152. The velocity *V* of a vehicle along a straight line is measured in m/s and plotted as shown with respect to time in seconds. At the end of the 7 seconds, how much will the odometer reading increase by (in m)?

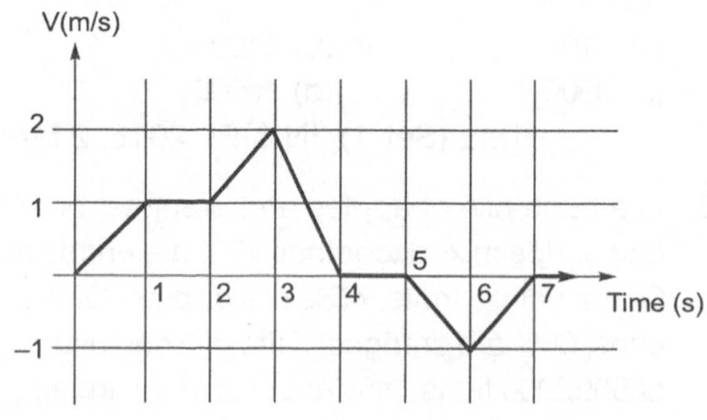

(a) 0 (b) 3
(c) 4 (d) 5
[EC (Set-3) & IN 2016, 2 Marks]

153. A flat is shared by four first year undergraduate students. They agreed to allow the oldest of them to enjoy some extra space in the flat. Manu is two months older than Sravan, who is three months younger than Trideep. Pavan is one month older than Sravan. Who should occupy the extra space in the flat?

(a) Manu (b) Sravan
(c) Trideep (d) Pavan

[EC (Set-3) & IN 2016, 2 Marks]

154. Find the area bounded by the lines $3x + 2y = 14$, $2x - 3y = 5$ in the first quadrant.

(a) 14.95 (b) 15.25
(c) 15.70 (d) 20.35

[EC (Set-3) & IN 2016, 2 Marks]

155. A straight line is fit to a data set ($\ln x$, y). This line intercepts the abscissa at $\ln x = 0.1$ and has a slope of -0.02. What is the value of y at $x = 5$ from the fit?

(a) -0.030 (b) -0.014
(c) 0.014 (d) 0.030

[EC (Set-3) & IN 2016, 2 Marks]

156. A cube is built using 64 cubic blocks of side one unit. After it is built, one cubic block is removed from every corner of the cube. The resulting surface area of the body (in square units) after the removal is _________.

(a) 56 (b) 64
(c) 72 (d) 96

[CE, CS 2016, 1 Mark (Set-1)]

157. A shaving set company sells 4 different types of razors, Elegance, Smooth, Soft and Executive. Elegance sells at Rs. 48, Smooth at Rs. 63, Soft at Rs. 78 and Executive at Rs. 173 per piece. The table below shows the numbers of each razor sold in each quarter of a year.

Quarter/ Product	Elegance	Smooth	Soft	Executive
Q1	27300	20009	17602	9999
Q2	25222	19392	18445	8942
Q3	28976	22429	19544	10234
Q4	21012	18229	16595	10109

Which product contributes the greatest fraction to the revenue of the company in that year?

(a) Elegance (b) Executive
(c) Smooth (d) Soft

[CE, CS 2016, 2 Marks (Set-1)]

158. Consider the following statements relating to the level of poker play of four players *P*, *Q*, *R* and S. I. *P* always beats *Q* II. *R* always beats *S* III. *S* loses to *P* only sometimes IV. R always loses to *Q*. Which of the following can be logically inferred from the above statements? (i) *P* is likely to beat all the three other players (ii) S is the absolute worst player in the set

(a) (i) only (b) (ii) only
(c) (i) and (ii) (d) neither (i) nor (ii)

[CE, CS 2016, 2 Marks (Set-1)]

159. If $f(x) = 2x^7 + 3x - 5$, which of the following is a factor $f(x)$?

(a) $(x^3 + 8)$ (b) $(x - 1)$
(c) $(2x - 5)$ (d) $(x + 1)$

[CE, CS 2016, 2 Marks (Set-1)]

160. In a process, the number of cycles to failure decreases exponentially with an increase in load. At a load of 80 units, it takes 100 cycles for failure. When the load is halved, it takes 10000 cycles for failure. The load for which the failure will happen in 5000 cycles is _______.

(a) 40.00 (b) 46.02
(c) 60.01 (d) 92.02

[CE, CS 2016, 2 Marks (Set-1)]

161. In a quadratic function, the value of the product of the roots (α, β) is 4. Find the value of

$$\frac{\alpha^n + \beta^n}{\alpha^{-n} + \beta^{-n}}$$

(a) n^4 (b) 4^n
(c) 2^{2n-1} (d) 4^{n-1}

[EE (Set-1), CS (Set-2) 2016, 1 Mark]

162. Pick the odd one from the following options.

(a) CADBE (b) JHKIL
(c) XVYWZ (d) ONPMQ

[EE (Set-1), CS (Set-2) 2016, 1 Mark]

163. Among 150 faculty members in an institute, 55 are connected with each other through Facebook® and 85 are connected through WhatsApp®. 30 faculty members do not have Facebook® or WhatsApp® accounts. The number of faculty members connected only through Facebook® accounts is __________.

(a) 35 (b) 45
(c) 65 (d) 90

[EE (Set-1), CS (Set-2) 2016, 2 Marks]

164. In a 2×4 rectangle grid shown below, each cell is a rectangle. How many rectangles can be observed in the grid?

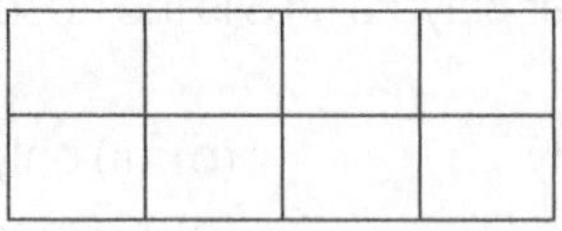

(a) 21 (b) 27
(c) 30 (d) 36

[EE (Set-1), CS (Set-2) 2016, 2 Marks]

165. Choose the correct expression for $f(x)$ given in the graph.

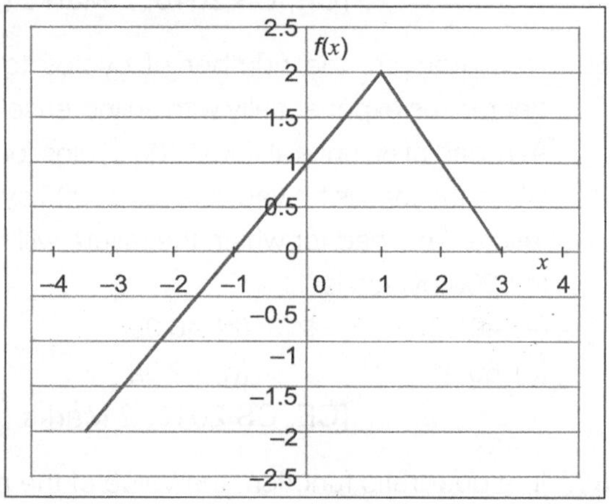

(a) $f(x) = 1 - |x - 1|$ (b) $f(x) = 1 + |x - 1|$
(c) $f(x) = 2 - |x - 1|$ (d) $f(x) = 2 + |x - 1|$

[EE (Set-1), CS (Set-2) 2016, 2 Marks]

166. (x% of y) + (y% of x) is equivalent to ____.
(a) 2% of xy (b) 2% of (xy/100)
(c) xy% of 100 (d) 100% of xy

[CE 2016, 1 Mark (Set-2)]

167. The sum of the digits of a two digit number is 12. If the new number formed by reversing the digits is greater than the original number by 54, find the original number.
(a) 39 (b) 57
(c) 66 (d) 93

[CE 2016, 1 Mark (Set-2)]

168. Two finance companies, P and Q declared fixed annual rates of interest on the amounts invested with them. The rates of interest offered by these companies may differ from year to year. Year-wise annual rates of interest offered by these companies are shown by the line graph provided below.

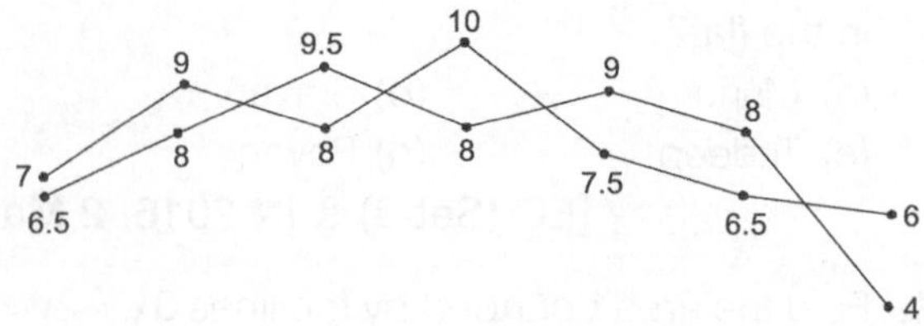

If the amounts invested in the companies, P and Q in 2006 are in the ratio 8 : 9, then the amounts received after one year as interest from companies P and Q would be the ratio
(a) 2 : 3 (b) 3 : 4
(c) 6 : 7 (d) 4 : 3

[CE 2016, 2 Marks (Set-2)]

169. Fact 1 : Humans are mammals.
Fact 2 : Some humans are engineers.
Fact 3 : Engineers build house.
If the above statements are facts, which of the following can be logically inferred?
I. All mammals build houses.
II. Engineers are mammals.
III. Some humans are not engineers.
(a) II only (b) III only
(c) I, II and III (d) I only

[CE 2016, 2 Marks (Set-2)]

170. A square pyramid has a base perimeter x and the slant height is half of the perimeter. What is the lateral surface area of the pyramid?
(a) x^2 (b) $0.75\,x^2$
(c) $0.50\,x^2$ (d) $0.25\,x^2$

[CE 2016, 2 Marks (Set-2)]

171. Ananth takes 6 hours and Bharath takes 4 hours to read a book. Both started reading copies of the book at the same time. After how many hours is the number of pages to be rad by Ananth, twice that to be read by Bharath? Assume Ananth and Bharath read all the pages with constant pace.
(a) 1 (b) 2
(c) 3 (d) 4

[CE 2016, 2 Marks (Set-2)]

172. Pick the odd one out in the following:
13, 23, 33, 43, 53
(a) 23 (b) 33
(c) 43 (d) 53

[EE 2016, 1 Mark (Set-2)]

173. If $|9y-6|=3$, then $y^2-\frac{4y}{3}$ is________.

(a) 0 (b) $+\frac{1}{3}$

(c) $-\frac{1}{3}$ (d) undefined

[EE 2016, 1 Mark (Set-2)]

174. Shaquille O' Neal is a 60% career free throw shooter, meaning that he successfully makes 60 free throws out of 100 attempts on average. What is the probability that he will successfully make exactly 6 free throws in 10 attempts?

(a) 0.2508 (b) 0.2816
(c) 0.2934 (d) 0.6000

[EE 2016, 2 Marks (Set-2)]

175. The numeral in the units position of $211^{870}+146^{127}\times 3^{424}$ is_____.

[EE 2016, 2 Marks (Set-2)]

176. The following graph represents the installed capacity for cement production (in tonnes) and the actual production (in tonnes) of nine cement plants of a cement company. Capacity utilization of a plant is defined as ratio of actual production of cement to installed capacity. A plant with installed capacity of at least 200 tonnes is called a large plant and a plant with lesser capacity is called a small plant. The difference between total production of large plants and small plants, in tonnes is ____.

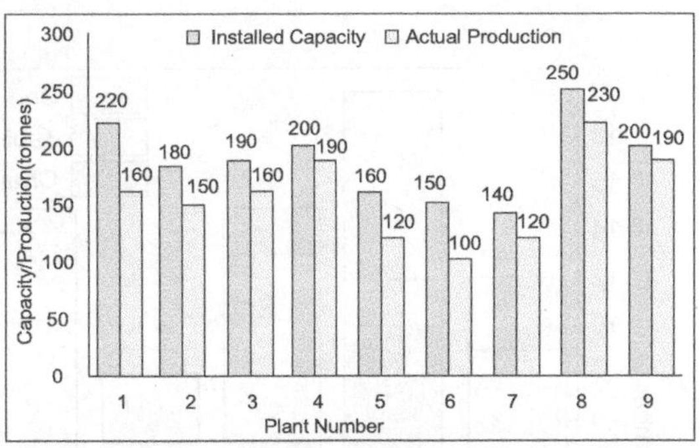

[EE 2016, 2 Marks (Set-2)]

177. **Fact:** If it rains, then the field is wet.
Read the following statements:

(i) It rains (ii) The field is not wet
(iii) The field is wet (iv) It did not rain

Which one of the option given below is **NOT** logically possible, based on the given fact?

(a) If (iii), then (iv). (b) If (i), then (iii).
(c) If (i), then (ii). (d) If (ii), then (iv).

[ME 2016, 1 Mark (Set-2)]

178. A window is made up of a square portion and an equilateral triangle portion above it. The base of the triangular portion coincides with the upper side of the square. If the perimeter of the window is 6 m, the area of the window in m^2 is__________.

(a) 1.43 (b) 2.06
(c) 2.68 (d) 2.88

[ME 2016, 1 Mark (Set-2)]

179. Students taking an exam are divided into two groups, *P* and *Q* such that each group has the same number of students. The performance of each of the students in a test was evaluated out of 200 marks. It was observed that the mean of group *P* was 105, while that of group *Q* was 85. The standard deviation of group *P* was 25, while that of group *Q* was 5. Assuming that the marks were distributed on a normal distribution, which of the following statement will have the highest probability of being TRUE?

(a) No student in group *Q* scored less marks than any student in group *P*.
(b) No student in group *P* scored less marks than any student in group *Q*.
(c) Most students of group *Q* scored marks in a narrower range than students in group *P*.
(d) The median of the marks of group *P* is 100.

[ME 2016, 2 Marks (Set-2)]

180. Find the missing sequence in the letter series.
B, FH, LNP, _ _ _ _.

(a) SUWY (b) TUVW
(c) TVXZ (d) TWXZ

[ME 2016, 2 Marks (Set-2)]

181. The binary operation □ is defined as $a \square b = ab + (a+b)$, where *a* and *b* are any two real numbers. The value of the identity element of this operation, defined as the number x such that $a \square x = a$, for any *a*, is________.

(a) 0 (b) 1
(c) 2 (d) 10

[ME 2016, 2 Marks (Set-2)]

182. Which of the following curve represents the function $y = \ln\left(\left|e^{[\sin(|x|)]}\right|\right)$ for $|x| < 2\pi$?
Here, x represents the abscissa and y represents the ordinate.

(a)
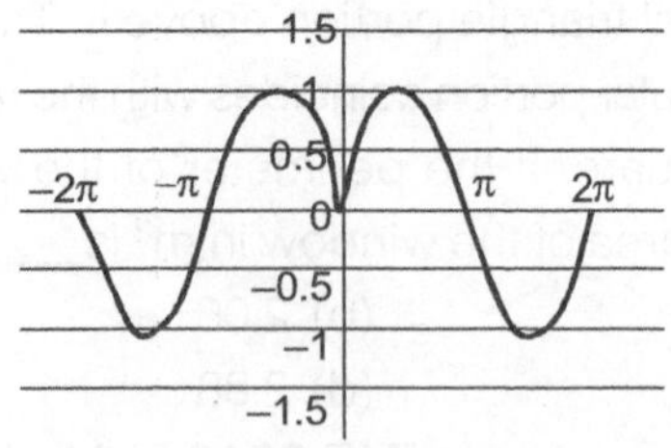

(b)
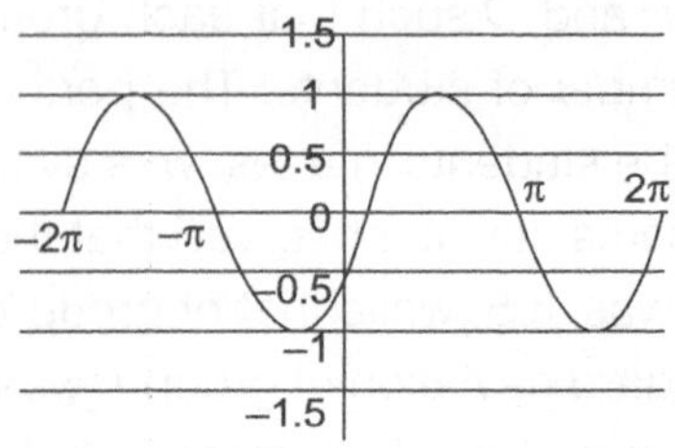

(c)
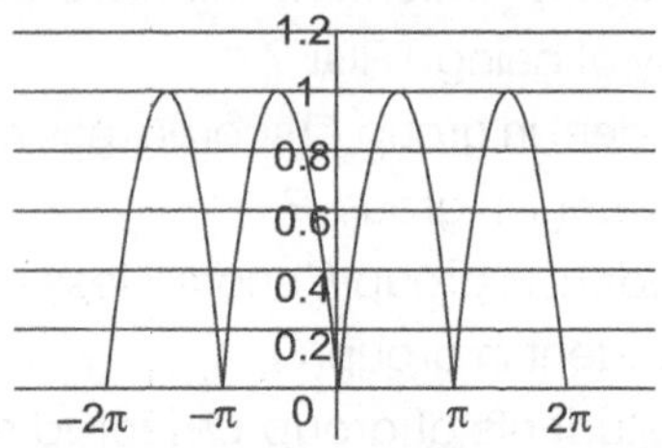

(d)
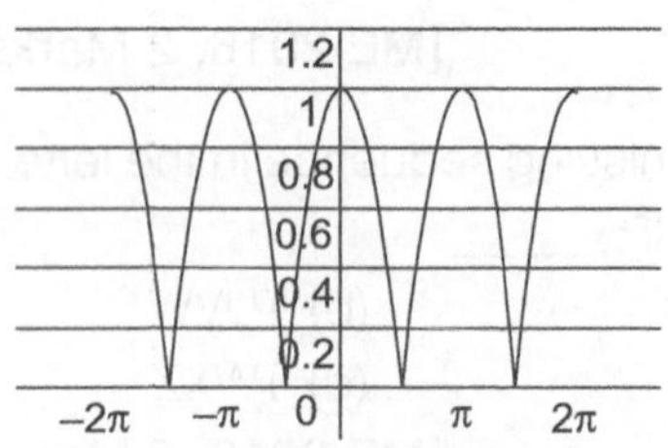

[ME 2016, 2 Marks (Set-2)]

183. The following sequence of numbers is arranged in increasing order: 1, x, x, x, y, y, 9, 16, 18. Given that the mean and median are equal, and are also equal to twice the mode, the value of y is
(a) 5 (b) 6
(c) 7 (d) 8
[CE 2017, 1 Mark (Set-1)]

184. Consider the following sentences:
All benches are beds. No bed is a bulb. Some bulbs are lamps.
Which of the following can be inferred?
(i) Some beds are lamps.
(ii) Some lamps are beds.
(a) Only (i) (b) Only (ii)
(c) Both (i) and (ii) (d) Neither (i) nor (ii)
[CE 2017, 1 Mark (Set-1)]

185. If the radius of a right circular cone is increased by 50%, its volume increases by
(a) 75% (b) 100%
(c) 125% (d) 237.5%
[CE 2017, 1 Mark (Set-1)]

186. Two machines M1 and M2 are able to execute any of four jobs, P, Q, R and S. The machines can perform one job on one object at a time, Jobs P, Q, R and S take 30 minutes, 20 minutes, 60 minutes and 15 minutes each respectively. There are 10 objects each requiring exactly 1 job. Job P is to be performed on 2 objects, Job Q on 3 objects, Job R on 1 object and Job S on 4 objects. What is the minimum time needed to complete all the jobs?
(a) 2 hours (b) 2.5 hours
(c) 3 hours (d) 3.5 hours
[CE 2017, 1 Mark (Set-1)]

187. The bar graph below shows the output of five carpenters over one month, each of whom made different items of furniture; chairs, tables and beds.

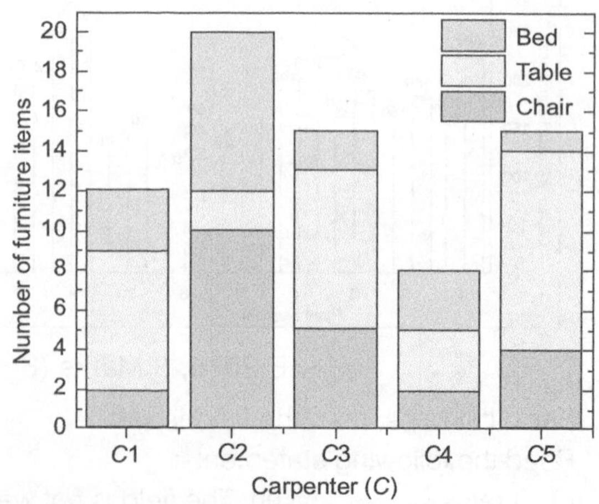

Consider the following statements:

(i) The number of beds made by carpenter C2 is exactly the same as the number of tables made by carpenter C3.

(ii) The total number of chairs made by all carpenters is less than the total number of tables.

Which one of the following is true?

(a) Only (i)
(b) Only (ii)
(c) Both (i) and (ii)
(d) Neither (i) nor (ii)

[CE 2017, 2 Marks (Set-1)]

188. Students applying for hostel rooms are alloted rooms in order of seniority. Students already staying in a room will move if they get a room in their preferred list. Preferences of lower ranked applicants are ignored during allocation.

Given the data below, which room will Ajit stay in?

Names	Student Seniority	Current room	Room preference list
Amar	1	P	R, S, Q
Akbar	2	None	R, S
Anthony	3	Q	P
Ajit	4	S	Q, P, R

(a) P
(b) Q
(c) R
(d) S

[CE 2017, 2 Marks (Set-1)]

189. The last digit of $(2171)^7 + (2172)^9 + (2173)^{11} + (2174)^{13}$ is

(a) 2
(b) 4
(c) 6
(d) 8

[CE 2017, 2 Marks (Set-1)]

190. Two dice are thrown simultaneously. The probability that the product of the numbers appearing on the top faces of the dice is a perfect square is

(a) $\frac{1}{9}$
(b) $\frac{2}{9}$
(c) $\frac{1}{3}$
(d) $\frac{4}{9}$

[CE, IN 2017, 1 Mark (Set-2)]

191. What is the value of x when

$$81 \times \left(\frac{16}{25}\right)^{x+2} \div \left(\frac{3}{5}\right)^{2x+4} = 144?$$

(a) 1
(b) –1
(c) –2
(d) Cannot be determined

[CE, IN 2017, 1 Mark (Set-2)]

192. The points in the graph below represent the halts of a lift for durations of 1 minute, over a period of 1 hour.

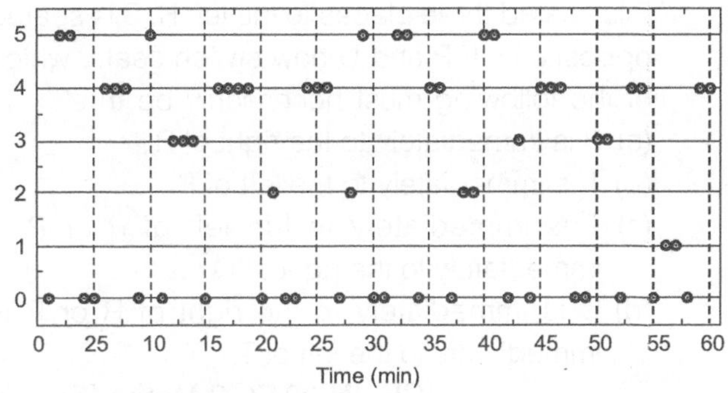

Which of the following statements are correct?

(i) The elevator never moves directly from any non-ground floor to another non-ground floor over the one hour period.

(ii) The elevator stays on the fourth floor for the longest duration over the one hour period.

(a) Only (i)
(b) Only (ii)
(c) Both (i) and (ii)
(d) Neither (i) nor (ii)

[CE, IN 2017, 2 Marks (Set-2)]

193. Budhan covers a distance of 19 km in 2 hours by cycling one fourth of the time and walking the rest. The next day he cycles (at the same speed as before) for half the time and walks the rest (at the same speed as before) and covers 26 km in 2 hours. The speed in km/h at which Budhan walks is

(a) 1
(b) 4
(c) 5
(d) 6

[CE, IN 2017, 2 Marks (Set-2)]

194. A map shows the elevations of Darjeeling. Gangtok, Kalimpong, Pelling and Siliguri. Kalimpong is at a lower elevation than Gangtok. Pelling is at a lower elevation than Gangtok. Pelling is at a higher elevation than Siliguri. Darjeeling is at a higher elevation than Gangtok. Which of the following statements can be inferred from the paragraph above?

(i) Pelling is at a higher elevation that Kalimpong.
(ii) Kalimpong is at a lower elevation than Darjeeling.
(iii) Kalimpong is at a higher elevation than Siliguri.
(iv) Siliguri is at a lower elevation than Gangtok.
(a) Only (ii) (b) Only (ii) and (iii)
(c) Only (ii) and (iv) (d) Only (iii) and (iv)

[CE, IN 2017, 2 Marks (Set-2)]

195. P, Q, R, S, T and U are seated around a circular table. R is seated two places to the right of Q, P is seated three places to the left R. S is seated opposite U. If P and U now switch seats, which of the following must necessarily be true?
(a) P is immediately to the right of R
(b) T is immediately to the left of P
(c) T is immediately to the left of P or P is immediately to the right of Q
(d) U is immediately to the right of R or P is immediately to the left of T

[CE, IN 2017, 2 Marks (Set-2)]

196. In the summer, water consumption is known to decrease overall by 25%. A Water Board official states that in the summer household consumption decreases by 20%, while other consumption increases by 70%.
Which of the following statement is correct?
(a) The ratio of household to other consumption is 8/17.
(b) The ratio of household to other consumption is 1/17.
(c) The ratio of household to other consumption is 17/8.
(d) There are errors in the official's statement.

[EC 2017, 1 Mark (Set-1)]

197. Some tables are shelves. Some shelves are chairs. All chairs are benches. Which of the following conclusions can be deduced from the preceding sentences?
(i) At least one bench is a table
(ii) At least one shelf is a bench
(iii) At least one chair is a table
(iv) All benches are chairs
(a) only (i) (b) only (ii)
(c) only (ii) and (iii) (d) only (iv)

[EC 2017, 1 Mark (Set-1)]

198. 40% of deaths on city roads may be attributed to drunken driving. The number of degrees needed to represent this as a slice of a pie chart is
(a) 120 (b) 144
(c) 160 (d) 212

[EC 2017, 1 Mark (Set-1)]

199. Trucks (10 m long) and cars (5 m long) go on a single lane bridge. There must be a gap of at least 20 m after each truck and a gap of at least 15 m after each car. Trucks and cars travel at a speed of 36 km/h. If cars and trucks go alternately. What is the maximum number of vehicles that can use the bridge in one hour?
(a) 1440 (b) 1200
(c) 720 (d) 600

[EC 2017, 2 Marks (Set-1)]

200. A contour line joins locations having the same height above the mean sea level. The following is a contour plot of a geographical region. Contour lines are shown at 25 m intervals in this plot.

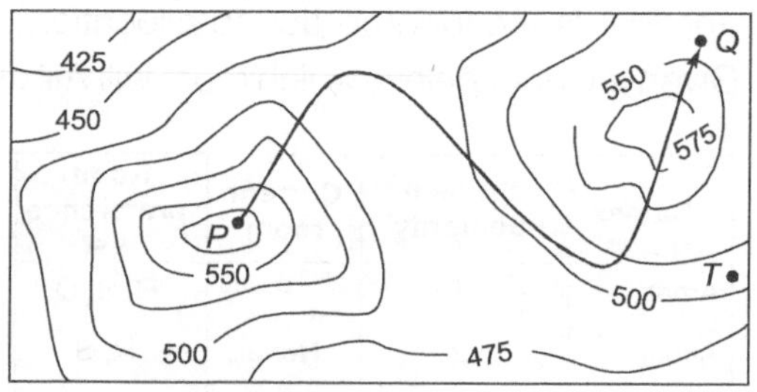

The path from *P* to *Q* is best described by
(a) Up-Down-Up-Down
(b) Down-Up-Down-Up
(c) Down-Up-Down
(d) Up-Down-Up

[EC 2017, 2 Marks (Set-1)]

201. There are 3 Indians and 3 Chinese in a group of 6 people. How many subgroups of this group can we choose so that every subgroup has at least one Indian?
(a) 56 (b) 52
(c) 48 (d) 44

[EC 2017, 2 Marks (Set-1)]

202. *S, T, U, V, W, X, Y* and *Z* are seated around a circular table. *T*'s neighbours are *Y* and *V*. *Z* is seated third to the left of *T* and second to the right of *S*. U's neighbours are *S* and *Y*; and *T* and *W* are not seated opposite each other. Who is third to the left of *V*?
(a) *X* (b) *W*
(c) *U* (d) *T*

[EC 2017, 2 Marks (Set-1)]

203. A rule states that in order to drink beer, one must be over 18 years old. In a bar there are 4 people. *P* is 16 years old, *Q* is 25 years old, *R* is drinking milkshake and *S* is drinking a beer. What must be checked to ensure that the rule is being followed?
(a) Only *P*'s drink
(b) Only *P*'s drink and *S*'s age
(c) Only *S*'s age
(d) Only *P*'s drink. *Q*'s drink and *S*'s age
[EC 2017, 1 Mark (Set-2)]

204. Fatima starts from point *P*, goes North for 3 km and then East for 4 km to reach point *Q*. She then turns to face point *P* and goes 15 km in that direction. She then goes North for 6 km. How far is she from point *P* and in which direction should she go to reach point *P*?
(a) 8 km, East (b) 12 km, North
(c) 6 km, East (d) 10 km, North
[EC 2017, 1 Mark (Set-2)]

205. 500 students are taking one or more courses out of Chemistry, Physics and Mathematics. Registration records indicate course enrolment as follows: Chemistry (329), Physics (186), Mathematics (295), Chemistry and Physics (83), Chemistry and Mathematics (217) and Physics and Mathematics (63). How many students are taking all 3 subjects?
(a) 37 (b) 43
(c) 47 (d) 53
[EC 2017, 1 Mark (Set-2)]

206. A contour line joins locations having the same height above the mean sea level. The following is a contour plot of a geographical region. Contour lines are shown at 25 m intervals in this plot.

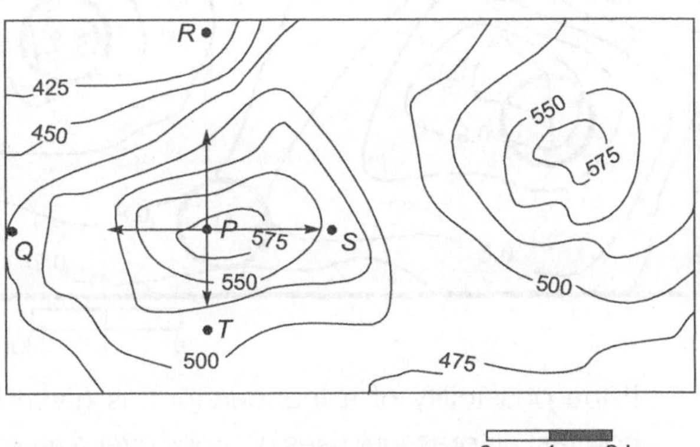

Which of the following is the steepest path leaving from *P*?
(a) *P* to *Q* (b) *P* to *R*
(c) *P* to *S* (d) *P* to *T*
[EC 2017, 2 Marks (Set-2)]

207. 1200 men and 500 women can build a bridge in 2 weeks, 900 men and 250 women will take 3 weeks to build the same bridge. How many men will be needed to build the bridge in one week?
(a) 3000 (b) 3300
(c) 3600 (d) 3900
[EC 2017, 2 Marks (Set-2)]

208. Each of *P*, *Q*, *R*, *S*, *W*, *X*, *Y* and *Z* has been married at most once. *X* and *Y* are married and have two children *P* and *Q*. *Z* is the grandfather of the daughter *S* of *P*. Further, *Z* and *W* are married and are parents of *R*. Which one of the following must necessarily be FALSE?
(a) *X* is the mother-in-law of *R*
(b) *P* and *R* are not married to each other
(c) *P* is a son of *X* and *Y*
(d) *Q* cannot be married to *R*
[EC 2017, 2 Marks (Set-2)]

209. The number of 3-digit numbers such that the digit 1 is never to the immediate right of 2 is
(a) 781 (b) 791
(c) 881 (d) 891
[EC 2017, 2 Marks (Set-2)]

210. The probability that a *k*-digit number does NOT contain the digits 0, 5 or 9 is
(a) 0.3^k (b) 0.6^k
(c) 0.7^k (d) 0.9^k
[EE, CS 2017, 1 Mark (Set-1)]

211. Find the smallest number *y* such that that y x 162 is a perfect cube.
(a) 24 (b) 27
(c) 32 (d) 36
[EE, CS 2017, 1 Mark (Set-1)]

212. Rahul, Murali, Srinivas and Arul are seated around a square table. Rahul is sitting to the left of Murali. Srinivas is sitting to the right of Arul. Which of the following pairs are seated opposite each other?
(a) Rahul and Murali
(b) Srinivas and Arul

(c) Srinivas and Murali
(d) Srinivas and Rahul

[EE, CS 2017, 1 Mark (Set-1)]

213. A contour line joins locations having the same height above the mean sea level. The following is a contour plot of a geographical region. Contour lines are shown at 25 m intervals in this plot. If in a flood, the water level rises to 525 m, which of the villages *P, Q, R, S, T* get submerged?

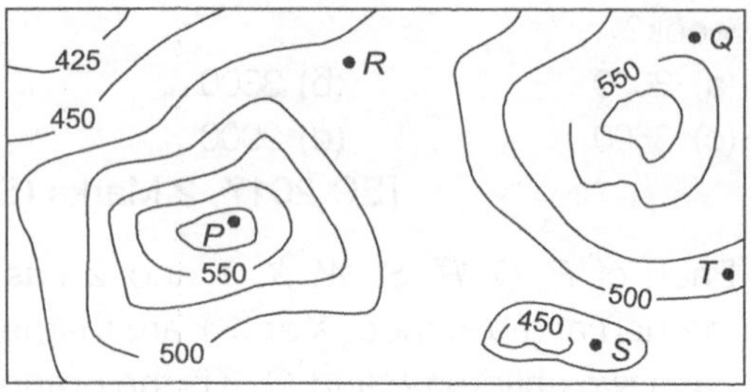

(a) *P, Q* (b) *P, Q, T*
(c) *R, S, T* (d) *Q, R, S*

[EE, CS 2017, 2 Marks (Set-1)]

214. The expression $\frac{(x+y)-|x-y|}{2}$ is equal to
(a) the maximum of x and y
(b) the minimum of x and y
(c) 1
(d) none of the above

[EE, CS 2017, 2 Marks (Set-1)]

215. Arun, Gulab, Neel and Shweta must choose one shirt each from a pile of four shirts coloured red, pink, blue and white respectively. Arun dislikes the colour red and Shweta dislikes the colour white. Gulab and Neel like all the colours. In how many different ways can they choose the shirts so that no one has a shirt with a colour he or she dislikes?
(a) 21 (b) 18
(c) 16 (d) 14

[EE, CS 2017, 2 Marks (Set-1)]

216. Six people are seated around a circular table. There are at least two men and two women. There are at least three right-handed persons. Every woman has a left-handed person to her immediate right. None of the women are right-handed. The number of women at the table is
(a) 2 (b) 3
(c) 4 (d) Cannot be determined

[EE, CS 2017, 2 Marks (Set-1)]

217. There are 3 red socks, 4 green socks and 3 blue socks. You choose 2 socks. The probability that they are of the same colour is
(a) 1/5 (b) 7/30
(c) 1/4 (d) 4/15

[EE, CS 2017, 1 Mark (Set-2)]

218. There are five buildings called V, W. X, Y and Z in a row (not necessarily in that order). V is to the West of W. Z is to the East of X and the West of V. W is to the West of Y. Which is the building in the middle?
(a) V (b) W
(c) X (d) Y

[EE, CS 2017, 1 Mark (Set-2)]

219. A test has twenty questions worth 100 marks in total. There are two types of questions. Multiple choice questions are worth 3 marks each and essay questions are worth 11 marks each. How many multiple choice questions do the exam have?
(a) 12 (b) 15
(c) 18 (d) 19

[EE, CS 2017, 1 Mark (Set-2)]

220. The number of roots of $e^x + 0.5x^2 - 2 = 0$ in the range [–5, 5] is
(a) 0 (b) 1
(c) 2 (d) 3

[EE, CS 2017, 2 Marks (Set-2)]

221. An air pressure contour line joins locations in a region having the same atmospheric pressure. The following is an air pressure contour plot of a geographical region. Contour lines are shown at 0.05 bar intervals in this plot.

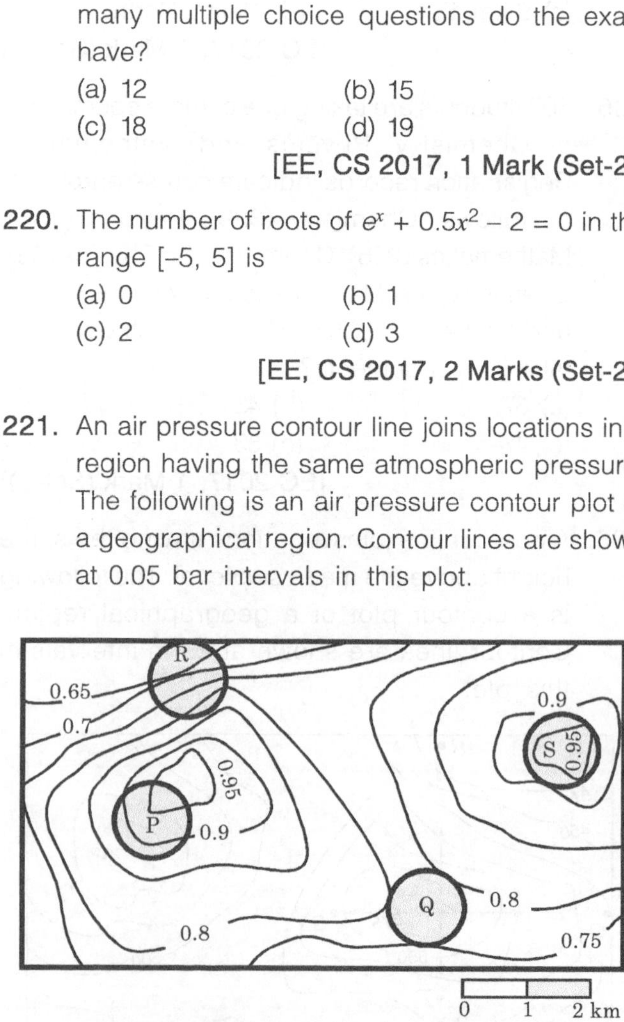

If the possibility of a thunderstorm is given by how fast air pressure rises or drops over a region, which of the following region is most likely to have a thunderstorm?

(a) P (b) Q
(c) R (d) S

[EE, CS 2017, 2 Marks (Set-2)]

222. X is a 30 digit number starting with the digit 4 followed by the digit 7. Then the number X^3 will have

(a) 90 digits (b) 91 digits
(c) 92 digits (d) 93 digits

[EE, CS 2017, 2 Marks (Set-2)]

223. There are three boxes. One contains apples, another contains oranges and the last one contains both apples and oranges. All three are known to be incorrectly labelled. If you are permitted to open just one box and then pull out and inspect only one fruit, which box would you open to determine the contents of all three boxes?

(a) The box labelled 'Apples'
(b) The box labelled 'Apples and Oranges'
(c) The box labelled 'Oranges'
(d) Cannot be determined

[EE, CS 2017, 2 Marks (Set-2)]

224. A right-angled cone (with base radius 5 cm and height 12 cm), as shown in the figure below, is rolled on the ground keeping the point P fixed until the point Q (at the base of the cone, as shown) touches the ground again.

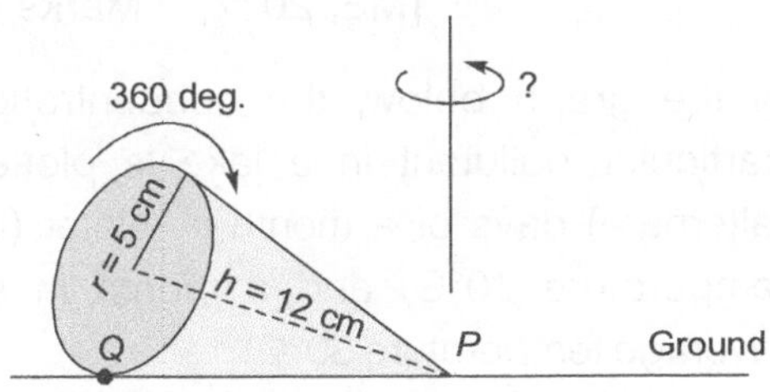

By what angle (in radians) about P does the cone travel?

(a) $\frac{5\pi}{12}$ (b) $\frac{5\pi}{24}$
(c) $\frac{24\pi}{5}$ (d) $\frac{10\pi}{13}$

[ME, 2017, 1 Mark (Set-1)]

225. P, Q, and R talk about S's car collection, P states that S has at least 3 cars. Q believes that S has less than 3 cars, R indicates that to his knowledge. S has at least one car. Only one of P, Q and R is right. The number of cars owned by S is

(a) 0 (b) 1
(c) 3 (d) Cannot be determined

[ME, 2017, 1 Mark (Set-1)]

226. In a company with 100 employees, 45 earn Rs. 20000 per month, 25 earn Rs. 30000, 20 earn Rs. 40000, 8 earn Rs. 60000, and 2 earn Rs. 150000. The median of the salaries is

(a) Rs. 20000 (b) Rs. 30000
(c) Rs. 32300 (d) Rs. 40000

[ME, 2017, 1 Mark (Set-1)]

227. What is the sum of the missing digits in the subtraction problem below?

$$\begin{array}{r} 5.......... \\ -48\_89 \\ \hline 1111 \end{array}$$

(a) 8 (b) 10
(c) 11 (d) Cannot be determined

[ME, 2017, 2 Marks (Set-1)]

228. Let S_1 be the plane figure consting of the points (x, y) given by the inequalities $|x-1| \le 2$ and $|y+2| \le 3$. Let S_2 be the plane figure given by the inequalities $x - y \ge -2$, $y \ge 1$, and $x \le 3$. Let S be the union of S_1 and S_2. The area of S is

(a) 26 (b) 28
(c) 32 (d) 34

[ME, 2017, 2 Marks (Set-1)]

229. Here, throughout the early 1820s, Stuart continued to fight his losing battle to allow his sepoys to year their caste-marks and their own choice of facial hair on parade, being again reprimanded by the commander-in-chief. His retort that A stronger instance than this of European prejudice with relation to this country has never come under my observations had no effect on his superiors. According to this paragraph, which of the statements below is most accurate?

(a) Stuart's commander-in-chief was moved by this demonstration of his prejudice.
(b) The Europeans were accommondating of the sepoys desire to were their caste-marks

(c) Stuart's losing battle refers to his inability to succeed in enabling sepoys to wear caste-marks.
(d) The commander-in-chief was exempt from the European prejudice that dictated how the sepoys were to dress

[ME, 2017, 2 Marks (Set-1)]

230. The growth of bacteria (lactobacillus) in milk leads to curd formation. A minimum bacterial population density of 0.8 (in suitable) is needed to form curd. In the graph below, the population density of lactobacillus in 1 litre of milk is plotted as a function of time, at two different temperature, 25°C and 37°C.

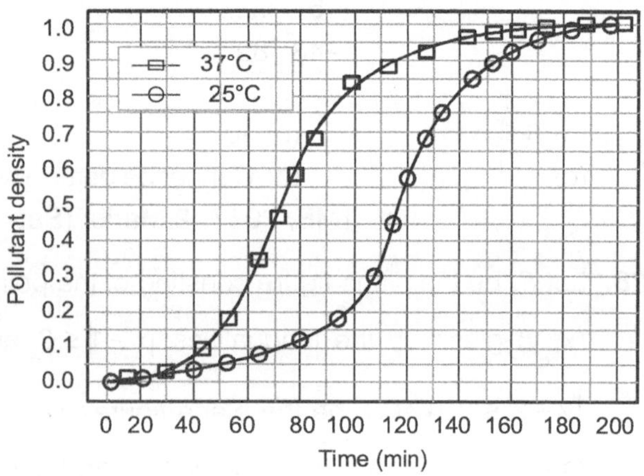

Consider the following statements based on the data shown above:
(i) The growth in bacterial population stops earlier at 37°C as compared to 25°C
(ii) The time taken for curd formation at 25°C is twice the time taken at 37°C

Which one of the following option is correct?
(a) only i (b) only ii
(c) both i and ii (d) neither i nor ii

[ME, 2017, 2 Marks (Set-1)]

231. If a and b are integers and $a - b$ is even, which of the following must always be even?
(a) ab (b) $a^2 + b^2 + 1$
(c) $a^2 + b + 1$ (d) $ab - b$

[ME, 2017, 1 Mark (Set-2)]

232. P looks at Q while Q looks at R, P is married, R is not. The number of pairs of people in which a married person is looking at an unmarried person is
(a) 0 (b) 1
(c) 2 (d) cannot be determined

[ME, 2017, 1 Mark (Set-2)]

233. A couple has 2 children. The probability that both children are boys if the older one is a boy is
(a) 1/4 (b) 1/3
(c) 1/2 (d) 1

[ME, 2017, 1 Mark (Set-2)]

234. All people in a certain island are either 'Knights' or 'Knaves' and each person knows every other person's identity. Knights NEVER lie, and knaves ALWAYS lie.
P says 'Both of us are knight'. Q says 'None of us are knaves'
Which one of the following can be logically inferred from the above?
(a) Both P and Q are knights
(b) P is a knight; Q is a knave
(c) Both P and Q are knaves
(d) The identities of P, Q cannot be determined

[ME, 2017, 2 Marks (Set-2)]

235. There are 4 women P, Q, R, S and 5 men V, W, X, Y, Z in a group. We are required to form pairs each consisting of one woman and one man. P is not to be paired with Z, and Y must necessarily be paired with someone. In how many ways can 4 such pairs be formed?
(a) 74 (b) 76
(c) 78 (d) 80

[ME, 2017, 2 Marks (Set-2)]

236. In the graph below, the concentration of a particular pollutant in a lake is plotted over (alternate) days of a month in winter (average temperature 10°C) and a month in summer (average temperature 30°C)

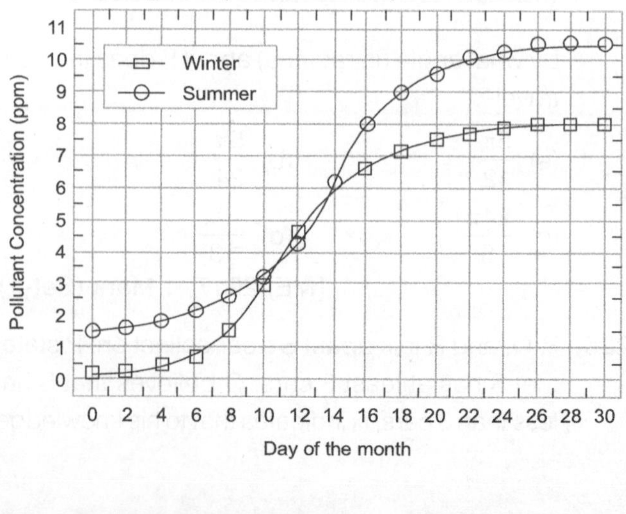

Consider the following statements based on the data shown above:

(i) Over the given month, the difference between the maximum and the minimum pollutant concentrations is the same in both winter and summer.

(ii) There are at least four days in the summer month such that the pollutant concentrations on those days are within 1 ppm of the pollutant concentrations on the corresponding days in the winter month.

Which one of the following option is correct?

(a) only i (b) only ii
(c) both i and ii (d) Neither i nor ii

[ME, 2017, 2 Marks (Set-2)]

237. X bullocks and Y tractors take 8 days to plough a field. If we halve the number of bullocks and double the number of tractors, it takes 5 days to plough the same field. How many days will it take X bullocks alone to plough the field?

(a) 30 (b) 35
(c) 40 (d) 45

[ME, 2017, 2 Marks (Set-2)]

238. If you are looking for a history of India, or for an account of the rise and fall of the British Raj, or for the reason of the cleaving of the subcontinent into two mutually antagonistic parts and the effects this mutilation will have in the respective sections, and ultimately on Asia, you will not find it in these pages; for though I have spent a lifetime in the country, I lived too near the seat of events, and was too intimately associated with the actors, to get the perspective needed for the impartial recording of these matters."

Which of the following is closest in meaning to 'cleaving'?

(a) deteriorating (b) arguing
(c) departing (d) splitting

[ME, 2017, 2 Marks (Set-2)]

239. The temperature T in a room varies as a function of the outside temperature T_0 and the number of persons in the room p, according to the relation $T = K(\theta p + T_0)$, where θ is K are constants. What would be the value of θ given the following data?

T_0	p	T
25	2	32.4
30	5	42.0

(a) 0.8 (b) 1.0
(c) 2.0 (d) 10.0

[CE, 2018, 1 Mark (Set-1)]

240. Tower A is 90 m tall and tower B is 140 m tall. They are 100 m apart. A horizontal skywalk connects the floors at 70 m in both the towers. If a taut rope connects the top of tower A to the bottom of tower B, at what distance (in meters) from tower A will the rope intersect the skywalk?

[CE, 2018, 1 Mark (Set-1)]

241. Hema's age is 5 years more than twice Hari's age. Suresh's age is 13 years less than 10 times Hari's age. If Suresh is 3 times as old as Hema. How old is Hema?

(a) 14 (b) 17
(c) 18 (d) 19

[CE, 2018, 1 Mark (Set-1)]

242. Consider a sequence of number a_1, a_2, a_3, ..., a_n where $a_n = \frac{1}{n} - \frac{1}{n+2}$, for each integer $n > 0$.

What is the sum of the first 50 terms?

(a) $\left(1+\frac{1}{2}\right)-\frac{1}{50}$

(b) $\left(1+\frac{1}{2}\right)+\frac{1}{50}$

(c) $\left(1+\frac{1}{2}\right)-\left(\frac{1}{51}+\frac{1}{52}\right)$

(d) $1-\left(\frac{1}{51}+\frac{1}{52}\right)$

[CE, 2018, 2 Marks (Set-1)]

243. A fruit seller sold a basket of fruits at 12.5% loss. Had he sold it for Rs. 108 more, he would have made a 10% gain. What is the loss in Rupees incurred by the fruit seller?

(a) 48 (b) 52
(c) 60 (d) 108

[CE, 2018, 2 Marks (Set-1)]

244. Each of the letters arranged as below represents a unique from 1 to 9. The letters are positioned in the figure such that (A × B × C), (B × G × E) and (D × E × F) are equal. Which integer among the following choices cannot be represented by the letters A, B, C, D, E, F or G?

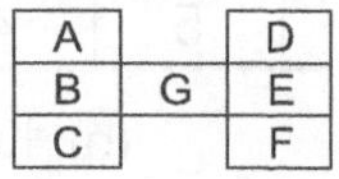

(a) 4 (b) 5
(c) 6 (d) 9

[CE, 2018, 2 Marks (Set-1)]

245. The price of a wire made of a superalloy material is proportional to the square of its length. The price of 10 m length of the wire is Rs. 1600. What would be the total price (in Rs.) of two wires of lengths 4 m and 6 m?

(a) 768 (b) 832
(c) 1440 (d) 1600

[CE, 2018, 2 Marks (Set-1)]

246. What of the following function(s) in an accurate description of the graph for the range(s) indicated?

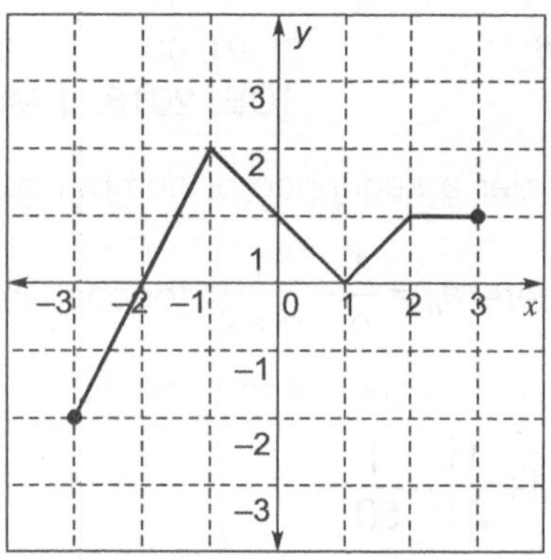

(a) $y = 2x + 4$ for $-3 \le x \le -1$
(b) $y = |x - 1|$ for $-1 \le x \le 2$
(c) $y = \left||x| - 1\right|$ for $-1 \le x \le 2$
(d) $y = 1$ for $2 \le x \le 3$

(a) (i), (ii) and (iii) only (b) (i), (ii) and (iv) only
(c) (i) and (iv) only (d) (ii) and (iv) only

[CE, 2018, 2 Marks (Set-1)]

247. A three-member committee has to be formed from a group of 9 people. How many such distinct committees can be formed?

(a) 27 (b) 72
(c) 81 (d) 84

[CE, 2018, 1 Mark (Set-2)]

248. For non-negative integers, *a*, *b*, *c*, what would be the value of $a + b + c$ if $\log a + \log b + \log c = 0$?

(a) 3 (b) 1
(c) 0 (d) –1

[CE, 2018, 1 Mark (Set-2)]

249. $\underbrace{a + a + a + \ldots + a}_{n \text{ times}} = a^2b$ and

$\underbrace{b + b + b + \ldots + b}_{m \text{ times}} = ab^2$, where *a*, *b*, *n* and *m* are natural numbers. What is the value of

$\left(\underbrace{m + m + m + \ldots + m}_{n \text{ times}}\right)\left(\underbrace{n + n + n + \ldots + n}_{m \text{ times}}\right)$?

(a) $2a^2b^2$ (b) a^4b^4
(c) $ab(a + b)$ (d) $a^2 + b^2$

[CE, 2018, 1 Mark (Set-2)]

250. In manufacturing industries, loss is usually taken to be proportional to the square of the deviation from a target. If the loss is Rs. 4900 for a deviation of 7 units, what would be the loss in Rupees for a deviation of 4 units from the target?

(a) 400 (b) 1200
(c) 1600 (d) 2800

[CE, 2018, 2 Marks (Set-2)]

251. Each of the letters in the figure below represents a unique integer from 1 to 9. The letters are positioned in the figure such that each of (*A*+*B*+*C*), (*C*+*D*+*E*), (*E*+*F*+*G*) and (*G*+*H*+*K*) is equal to 13. Which integer does *E* represent?

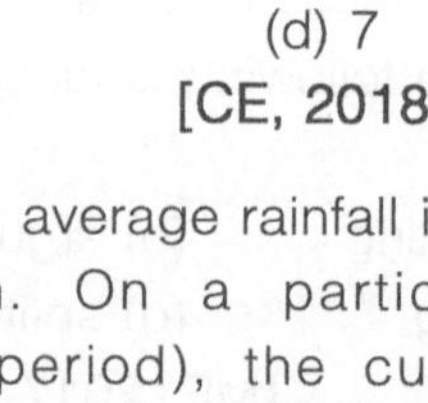

(a) 1 (b) 4
(c) 6 (d) 7

[CE, 2018, 2 Marks (Set-2)]

252. The annual average rainfall in a tropical city is 1000 mm. On a particular rainy day (24-hour period), the cumulative rainfall experienced by the city is shown in the graph. Over the 24-hour period. 50% of the rainfall falling on a rooftop, which had an obstruction-free area of 50 m^2, was harvested into a tank. What is the total volume of water collected in the tank in liters?

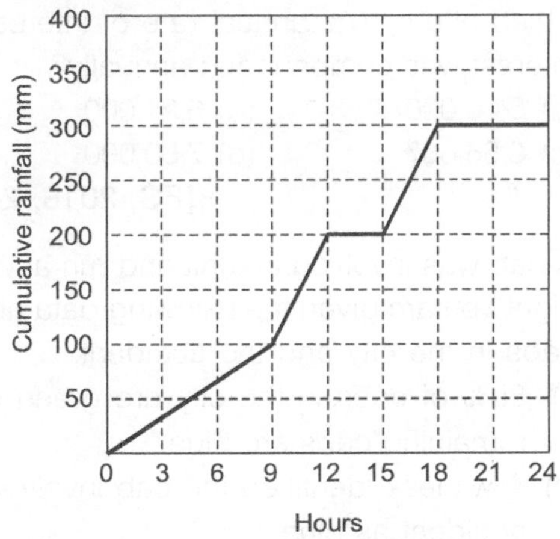

(a) 25,000 (b) 18,750
(c) 7,500 (d) 3,125

[CE, 2018, 2 Marks (Set-2)]

253. Given that $\frac{\log P}{y-z} = \frac{\log Q}{z-x} = \frac{\log R}{x-y} = 10$ for $x \neq y \neq z$, what is the value of the product PQR?

(a) 0 (b) 1
(c) xyz (d) 10^{xyz}

[CE, 2018, 2 Marks (Set-2)]

254. A faulty wall clock is known to gain 15 minutes every 24 hours. It is synchronized to the correct time at 9 AM on 11th July. What will be the correct time to the nearest minute when the clock shows 2 PM on 15th July of the same year?

(a) 12:45 PM (b) 12:58 PM
(c) 1:00 PM (d) 2:00 PM

[CE, 2018, 2 Marks (Set-2)]

255. What is the missing number in the following sequence?

2, 12, 60, 240, 720, 1440, ________, 0

(a) 2880 (b) 1440
(c) 720 (d) 0

[CS, 2018, 1 Mark]

256. What would be the smallest natural number which when divided either by 20 or by 42 or by 76 leaves a remainder of 7 in each case?

(a) 3047 (b) 6047
(c) 7987 (d) 63847

[CS, 2018, 1 Mark]

257. The area of a square is d. What is the area of the circle which has the diagonal of the square as its diameter?

(a) πd (b) πd^2
(c) $\frac{1}{4}\pi d^2$ (d) $\frac{1}{2}\pi d$

[CS, 2018, 1 Mark]

258. In $pqr \neq 0$ and $p^{-x} = \frac{1}{q}$, $q^{-y} = \frac{1}{r}$, $r^{-z} = \frac{1}{p}$, what is the value of the product xyz?

(a) –1 (b) $\frac{1}{pqr}$
(c) 1 (d) pqr

[CS, 2018, 2 Marks]

259. In appreciation of the social improvements completed in a town, a wealthy philanthropist decided to gift Rs 750 to each male senior citizen in the town and Rs 1000 to each female senior citizen. Altogether, there were 300 senior citizens eligible for this gift. However, only 8/9th of the eligible men and 2/3rd of the eligible women claimed the gift. How much money (in Rupees) did the philanthropist give away in total?

(a) 1,50,000 (b) 2,00,000
(c) 1,75,000 (d) 1,51,000

[CS, 2018, 2 Marks]

260. In the figure below, ∠DEC + ∠BFC is equal to ________.

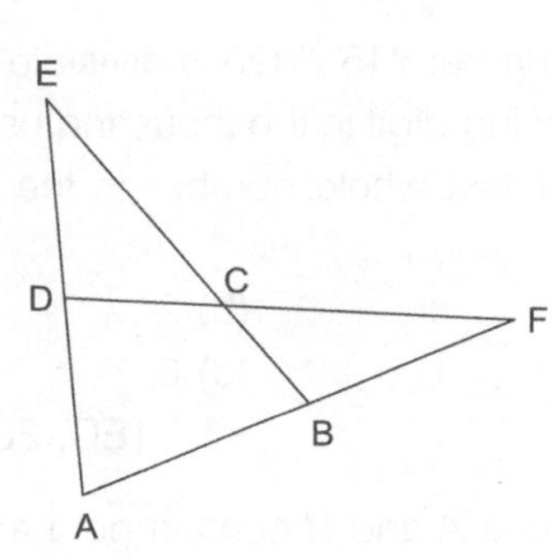

(a) ∠BCD – ∠BAD (b) ∠BAD + ∠BCF
(c) ∠BAD + ∠BCD (d) ∠CBA + ∠ADC

[CS, 2018, 2 Marks]

261. A six sided unbiased die with four green faces and two red faces is rolled seven times. Which of the following combinations is the most likely outcome of the experiment?

(a) Three green faces and four red faces.
(b) Four green faces and three red faces.

(c) Five green faces and two red faces.
(d) Six green faces and one red face.

[CS, 2018, 2 Marks]

262. In a party, 60% of the invited guests are male and 40% are female. If 80% of the invited guests attended the party and if all the invited female guests attended, what would be the ratio of males to females among the attendees in the party?
(a) 2 : 3 (b) 1 : 1
(c) 3 : 2 (d) 2 : 1

[CS, 2018, 2 Marks]

263. What is the value of $1+\frac{1}{4}+\frac{1}{16}+\frac{1}{64}+\frac{1}{256}+....?$
(a) 2 (b) $\frac{7}{4}$
(c) $\frac{3}{2}$ (d) $\frac{4}{3}$

[EC, 2018, 1 Mark]

264. A 1.5 m in tall person is standing at a distance of 3 m from a lamp post. The light from the lamp at the top of the post casts her shadow. The length of the shadow is twice her height. What is the height of the lamp post in meters?
(a) 1.5 (b) 3
(c) 4.5 (d) 6

[EC, 2018, 1 Mark]

265. If the number 715 ? 423 is divisible 3 (? denotes the missing digit in the thousandths place), then the smallest whole number in the place of ? is ______.
(a) 0 (b) 2
(c) 5 (d) 6

[EC, 2018, 1 Mark]

266. Two alloys *A* and *B* contain gold and copper in the ratios of 2 : 3 and 3 : 7 by mass, respectively. Equal masses of alloys *A* and *B* are melted to make an alloy *C*. The ratio of gold to copper in alloy *C* is ______.
(a) 5 : 10 (b) 7 : 13
(c) 6 : 11 (d) 9 : 13

[EC, 2018, 2 Marks]

267. Leila aspires to buy a car worth Rs.10,00,000 after 5 years. What is the minimum amount in Rupees that she should deposit now in a bank which offers 10% annual rate of interest, if the interest was compounded annually?
(a) 5,00,000 (b) 6,21,000
(c) 6,66,667 (d) 7,50,000

[EC, 2018, 2 Marks]

268. A cab was involved in a hit and run accident at night You are given the following data about the cabs in the city and the accident.
(i) 85% of cabs in the city are green and the remaining cabs are blue.
(ii) A witness identified the cab involved in the accident as blue.
(iii) It is known that a witness can correctly identify the cab colour only 80% of the time.

Which of the following options is closest to the probability that the accident was caused by a blue cab?
(a) 12% (b) 15%
(c) 41% (d) 80%

[EC, 2018, 2 Marks]

269. A coastal region with unparalleled beauty is home to many species of animals. It is dotted with coral reefs and unspoilt white sandy beaches. It has remained inaccessible to tourists due to poor connectivity and lack of accommodation. A company has spotted the opportunity and is planning to develop a luxury resort with helicopter service to the nearest major city airport. Environmentalists are upset that this would lead to the region becoming crowded and polluted like any other major beach resorts.

Which one of the following statements can be logically inferred from the information given in the above paragraph?
(a) The culture and tradition of the local people will be influenced by the tourists.
(b) The region will become crowded and polluted due to tourism.
(c) The coral reefs are on the decline and could soon vanish.
(d) Helicopter connectivity would lead to an increase in tourists coming to the region.

[EC, 2018, 2 Marks]

270. The Cricket Board has long recognized John's potential as a leader of the team. However, his on-field Temper has always been a matter of concern for them since his junior days. While this aggression has filled stadia with die-hard fans, it has taken a toll on his own batting. Until recently, it appeared that he found it difficult to convert his aggression into big scores. Over the past three seasons though, that picture of John has been replaced by a cerebral, calculative and successful batsman-captain. After many years, it appears that the team has finally found a complete captain.

Which of the following statements can be logically inferred from the above paragraph?

(i) Even as a junior cricketer, John was considered a good captain.

(ii) Finding a complete captain is a challenge.

(iii) Fans and the Cricket Board have differing views on what they want in a captain.

(iv) Over the past three seasons John has accumulated big scores.

(a) (i). (ii) and (iii) only

(b) (iii) and (iv) only

(c) (ii) and (iv) only

(d) (i), (ii), (iii) and (v)

[EC, 2018, 2 Marks]

271. For what values of k given below is $\frac{(k+2)^2}{k-3}$ an integer?

(a) 4, 8, 18 (b) 4, 10, 16

(c) 4, 8, 28 (d) 8, 26, 28

[EE, 2018, 1 Mark]

272. The three roots of the equation $f(x) = 0$ are $x = \{-2, 0, 3\}$. What are the three values of x for which $f(x - 3) = 0$?

(a) –5, –3, 0 (b) –2, 0, 3

(c) 0, 6, 8 (d) 1, 3, 6

[EE, 2018, 1 Mark]

273. Functions, $F(a, b)$ and $G(a, b)$ are defined as follows:

$F(a, b) = (a - b)^2$ and $G(a, b) = |a - b|$, where

$|x|$ represents the absolute value of x. What would be the value of $G(F(1, 3), G(1, 3))$?

(a) 2 (b) 4

(c) 6 (d) 36

[EE, 2018, 1 Mark]

274. In a certain code AMCF is written as EQGJ and NKUF is written as ROYJ. How will DHLP be written in the code?

(a) RSTN (b) TLPH

(c) HLPT (d) XSVR

[EE, 2018, 2 Marks]

275. An e-mail password must contain three characters. The password has to contain one numeral from 0 to 9, one upper case and one lower case character from the English alphabet. How many distinct passwords are possible?

(a) 6,760 (b) 13,520

(c) 40,560 (d) 1,05,456

[EE, 2018, 2 Marks]

276. A designer uses marbles of four different colours for his designs. The cost of each marble is the same, irrespective of the colour. The table below shows the percentage of marbles of each colour used in the current design. The cost of each marble increased by 25%. Therefore, the designer decided to reduce equal numbers of marbles of each colour to keep the total cost unchanged. What is the percentage of blue marbles in the new design?

Blue	Black	Red	Yellow
40%	25%	20%	15%

(a) 35.75 (b) 40.25

(c) 43.75 (d) 46.25

[EE, 2018, 2 Marks]

277. P, Q, R and S crossed a lake in a boat that can hold a maximum of two persons, with only one set of oars. The following additional facts are available

(i) The boat held two persons on each of the three forward trips across lake and one person on each of the two return trips.

(ii) P is unable to row when someone else is in the boat.

(iii) Q is unable to row with anyone else except R.

(iv) Each person rowed for at least one trip.

(v) Only one person can row during a trip.

Who rowed twice?

(a) P (b) Q

(c) R (d) S

[EE, 2018, 2 Marks]

278. A class of twelve children has two more boys than girls. A group of three children are randomly picked from this class to accompany the teacher on a field trip. What is the probability that the group accompanying the teacher contains more girls than boys?

(a) 0 (b) $\frac{325}{864}$

(c) $\frac{525}{864}$ (d) $\frac{5}{12}$

[EE, 2018, 2 Marks]

279. For $0 \le x \le 2\pi$, $\sin x$ and $\cos x$ are both decreasing functions in the interval ________.

(a) $\left(0, \frac{\pi}{2}\right)$ (b) $\left(\frac{\pi}{2}, \pi\right)$

(c) $\left(\pi, \frac{\pi}{2}\right)$ (d) $\left(\frac{3\pi}{2}, 2\pi\right)$

[IN, 2018, 1 Mark]

280. Arrange the following three-dimensional objects in the descending order of their volumes:
(i) A cuboid with dimensions 10 cm, 8 cm and 6 cm
(ii) A cube of side 8 cm
(iii) A cylinder with base radius 7 cm and height 7 cm
(iv) A sphere of radius 7 cm

(a) (i), (ii), (iii), (iv) (b) (ii), (i), (iv), (iii)
(c) (iii), (ii), (i), (iv) (d) (iv), (iii), (ii), (i)

[IN, 2018, 1 Mark]

281. The area of an equilateral triangle is $\sqrt{3}$. What is the perimeter of the triangle?

(a) 2 (b) 4
(c) 6 (d) 8

[IN, 2018, 1 Mark]

282. An automobile travels from city A to city B and returns to city A by the same route. The speed of the vehicle during the onward and return journeys were constant at 60 km/h and 90 km/h, respectively. What is the average speed in km/h for the entire journey?

(a) 72 (b) 73
(c) 74 (d) 75

[IN, 2018, 2 Marks]

283. In a detailed study of annual crow births in India, it was found that there was relatively no growth during the period 2002 to 2004 and a sudden spike from 2004 to 2005. In another unrelated study, it was found that the revenue from cracker sales in India which remained fairly flat from 2002 to 2004, saw a sudden spike in 2005 before declining again in 2006. The solid line in the graph below refers to annual sale of crackers and the dashed line refers to the annual crow births in India. Choose the most appropriate inference from the above data.

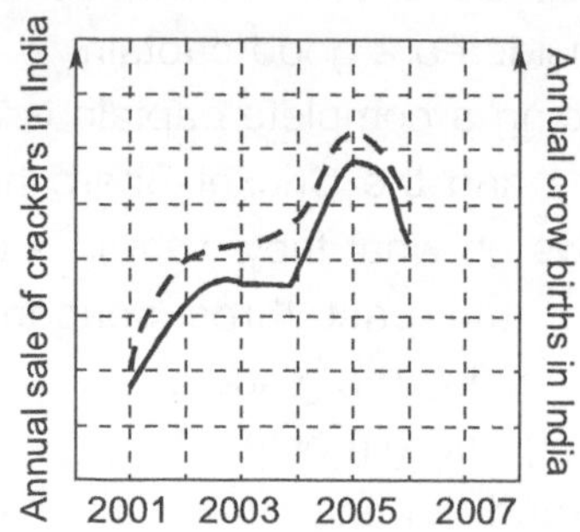

(a) There is a strong correlation between crow birth and cracker sales.
(b) Cracker usage increases crow birth rate.
(c) If cracker sale declines, crow birth will decline.
(d) Increased birth rate of crows will cause an increase in the sale of crackers.

[IN, 2018, 2 Marks]

284. If $x^2 + x - 1 = 0$ what is the value of $x^4 + \frac{1}{x^4}$?

(a) 1 (b) 5
(c) 7 (d) 9

[IN, 2018, 2 Marks]

285. To pass a test, a candidate needs or answer at least 2 out of 3 questions correctly. A total of 6,30,000 candidates appeared for the test. Question *A* was correctly answered by 3,30,000 candidates. Question *B* was answered correctly by 2,50,000 candidates. Question *C* was answered correctly by 2,60,000 candidates. Both questions *A* and *B* were answered correctly by 1,00,000 candidates. Both questions *B* and *C* were answered correctly by 90,000 candidates. Both questions *A* and *C* were answered correctly by 80,000 candidates. If the number of students answering all questions correctly is the same as

the number answering none, how many candidates failed to clear the test?

(a) 30,000 (b) 2,70,000
(c) 3,90,000 (d) 4,20,000

[IN, 2018, 2 Marks]

286. A set of 4 parallel lines intersect with another set of 5 parallel lines. How many parallelograms are formed?

(a) 20 (b) 48
(c) 60 (d) 72

[IN, 2018, 2 Marks]

287. A rectangle becomes a square when its length and breadth are reduced by 10 m and 5 m, respectively. during this process, the rectangle loses 650 m^2 of area. What is the area of the original rectangle in square meters?

(a) 1125 (b) 2250
(c) 2924 (d) 4500

[ME, 2018, 1 Mark (Set-1)]

288. A number consists of two digits. The sum of the digits is 9. If 45 is subtracted from the number, its digits are interchanged. What is the number?

(a) 63 (b) 72
(c) 81 (d) 90

[ME, 2018, 1 Mark (Set-1)]

289. Seven machines take 7 minutes to make 7 identical toys. At the same rate, how many minutes would it take for 100 machines to make 100 toys?

(a) 1 (b) 7
(c) 100 (d) 700

[ME, 2018, 1 Mark (Set-1)]

290. Given that a and b are integers and $a + a^2b^3$ is odd, which one of the following statements is correct?

(a) a and b are both odd
(b) a and b are both even
(c) a is even and b is odd
(d) a is odd and b is even

[ME, 2018, 2 Marks (Set-1)]

291. Consider the following three statements:

(i) Some roses are red
(ii) All red flowers fade quickly.
(iii) Some roses fade quickly

Which of the following statements can be logically inferred from the above statements?

(a) If (i) is true and (ii) is false, then (iii) is false.
(b) If (i) is true and (ii) is false, then (iii) is true.
(c) If (i) and (ii) are true, then (iii) is true.
(d) If (i) and (ii) are false, then (iii) is false.

[ME, 2018, 2 Marks (Set-1)]

292. Which of the following functions describe the graph shown in the below figure.

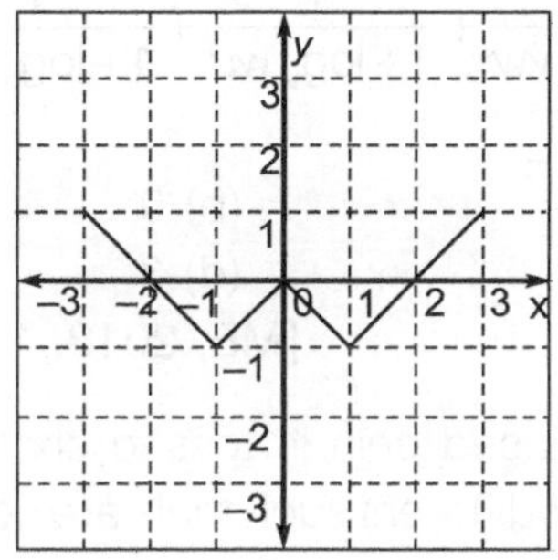

(a) $y = ||x| + 1| - 2$ (b) $y = ||x| - 1| - 1$
(c) $y = ||x| + 1| - 1$ (d) $y = ||x - 1| - 1|$

[ME, 2018, 2 Marks (Set-1)]

293. For integers a, b and c, what would be the minimum and maximum values respectively of $a + b + c$ if $\log |a| + \log |b| + \log |c| = 0$

(a) –3 and 3 (b) –1 and 1
(c) –1 and 3 (d) 1 and 3

[ME, 2018, 2 Marks (Set-1)]

294. From the time the front of a train enters a platform, it takes 25 seconds for the back of the train to leave the platform, while travelling at a constant speed of 54 km/h. At the same speed, it takes 14 seconds to pass a man running at 9 km/h in the same direction as the train. What is the length of the train and that of the platform in meters, respectively?

(a) 210 and 140 (b) 162.5 and 187.5
(c) 245 and 130 (d) 175 and 200

[ME, 2018, 2 Marks (Set-1)]

295. Find the missing group of letters in the following series:

BC, FGH, LMNO, _____

(a) UVWXY (b) TUVWX
(c) STUVW (d) RSTUV

[ME, 2018, 1 Mark (Set-2)]

296. The perimeters of a circle, a square and an equilateral triangle are equal. Which one of the following statements is true?
(a) The circle has the largest area
(b) The square has the largest area
(c) The equilateral triangle has the largest area
(d) All the three shapes have the same area
[ME, 2018, 1 Mark (Set-2)]

297. The value of the expression

$$\frac{1}{1+\log_u vw}+\frac{1}{1+\log_v wu}+\frac{1}{1+\log_w uv} \text{ is}$$

_______ .
(a) –1 (b) 0
(c) 1 (d) 3
[ME, 2018, 1 Mark (Set-2)]

298. An unbiased coin is tossed six times in a row and four different such trials are conducted. One trial implies six tosses of the coin. If *H* stands for head and *T* stands for tail, the following are the observations from the four trials:
(1) HTHTHT (2) TTHHHT (3) HTTHHT (4) HHHT __ __.
Which statement describing the last two coin tosses of the fourth trial has the highest probability of being correct?
(a) Two T will occur
(b) One H and one T will occur
(c) Two H will occur
(d) One H will be followed by one T
[ME, 2018, 2 Marks (Set-2)]

299. A contract is to be completed in 52 days and 125 identical robots were employed, each operational for 7 hours a day. After 39 days, five-seventh of the work was completed. How many additional robots would be required to complete the work on time, if each robot is now operational for 8 hours a day?
(a) 50 (b) 89
(c) 146 (d) 175
[ME, 2018, 2 Marks (Set-2)]

300. Forty students watched films, A, B and C over a week. Each student watched either only one film or all three. Thirteen students watched film A, sixteen students watched film B and nineteen students watched film C. How many students watched all three films?
(a) 0 (b) 2
(c) 4 (d) 8
[ME, 2018, 2 Marks (Set-2)]

301. A wire would enclose an area of 1936 m^2, if it is bent into a square. The wire is cut into two pieces. The longer piece is thrice as long as the shorter piece. The long and the short pieces are bent into a square and a circle, respectively. Which of the following choices is closest to the sum of the areas enclosed by the two pieces in square meters?
(a) 1096 (b) 1111
(c) 1243 (d) 2486
[ME, 2018, 2 Marks (Set-2)]

302. A house has a number which needs to be identified. The following three statements are given that can help in identifying the house number.
1. If the house number is a multiple of 3, then it is a number from 50 to 59.
2. If the house number is NOT a multiple of 4, then it is a number from 60 to 69.
3. If the house number is NOT a multiple of 6, then it is a number from 70 to 79.

Which is the house number?
(a) 54 (b) 65
(c) 66 (d) 76
[ME, 2018, 2 Marks (Set-2)]

303. If *E* = 10; *J* = 20; *O* = 30; and *T* = 40, what will be *P* + *E* + *S* + *T*?
(a) 51 (b) 120
(c) 82 (d) 164
[CE, 2019 : 1 Mark, Set-I]

304. On a horizontal ground, the base of a straight ladder is 6 m away from the base of a vertical pole. The ladder makes an angle of 45° to the horizontal. If the ladder is resting at a point located at one-fifth of the height of the pole from the bottom, the height of the pole is_______meters.
(a) 15 (b) 25
(c) 35 (d) 30
[CE, 2019 : 1 Mark, Set-I]

305. Suresh wanted to lay a new carpet in his new mansion with an area of 70 x 55 sq. mts. However an area of 550 sq. mts. had to be left out for flower pots. If the cost of carpet is Rs. 50 per sq. mts., how much money (in Rs.) will be spent by Suresh

for the carpet now?
(a) Rs. 1,65, 000 (b) Rs. 2,75, 000
(c) Rs. 1,92, 500 (d) Rs. 1,27, 500
[CE, 2019 : 1 Mark, Set-II]

306. A retaining wall with measurements 30 m x 12 m x 6 m was constructed with bricks of dimensions 8 cm x 6 cm x 6 cm If 60% of the wall consists of bricks, the number of bricks used for the construction is_______lakhs.
(a) 30 (b) 40
(c) 75 (d) 45
[CE, 2019 : 1 Mark, Set-II]

307. A square has sides 5 cm smaller than the sides of a second square. The area of the larger square is four times the area of the smaller square. The side of the larger square is_______cm.
(a) 15.10 (b) 8.50
(c) 18.50 (d) 10.00
[CE, 2019 : 2 Marks, Set-I]

308. P, Q, R, S and T are related and belong to the same family. P is the brother of S. Q is the wife of P. R and T are the children of the siblings P and S respectively. Which one of the following statements is necessarily FALSE?
(a) S is the sister-in-law of Q
(b) S is the aunt of R
(c) S is the brother of P
(d) S is the aunt of T
[CE, 2019 : 2 Marks, Set-I]

309. In a sports academy of 300 people, 105 play only cricket, 70 play only hockey, 50 play only football, 25 play both cricket and hockey, 15 play both hockey and football and 30 play both cricket and football. The rest of them play all three sports. What is the percentage of people who play at least two sports?
(a) 25.00 (b) 28.00
(c) 23.30 (d) 50.00
[CE, 2019 : 2 Marks, Set-I]

310. Mohan, the manager, wants his four workers to work in pairs. No pair should work for more than 5 hours. Ram and John have worked together for 5 hours. Krishna and Amir have worked as a team for 2 hours. Krishna does not want to work with Ram. Whom should Mohan allot to work with John, if he wants all the workers to continue working?
(a) Amir (b) Krishna
(c) Ram (d) None of the three
[CE, 2019 : 2 Marks, Set-II]

311. Population of state X increased by x% and the population of state Y increased by y% from 2001 to 2011. Assume that x is greater than y. Let P be the ratio of the population of state X to state Y in a given year. The percentage increase in P from 2001 to 2011 is_________.
(a) $\frac{x}{y}$ (b) $\frac{100(x-y)}{100+y}$
(c) $x-y$ (d) $\frac{100(x-y)}{100+x}$
[CE, 2019 : 2 Marks, Set-II]

312. An oil tank can be filled by pipe X in 5 hours and pipe Y in 4 hours, each pump working on its own. When the oil tank is full and the drainage hole is open, the oil is drained in 20 hours. If initially the tank was empty and someone started the two pumps together but left the drainage hole open, how many hours will it take for the tank to be filled? (Assume that the rate of drainage is independent of the Head)
(a) 4.00 (b) 1.50
(c) 2.00 (d) 2.50
[CE, 2019 : 2 Marks, Set-II]

313. Two cars start at the same time from the same location and go in the same direction. The speed of the first car is 50 km/h and the speed of the second car is 60 km/h. The number of hours it takes for the distance between the two cars to be 20 km is _______.
(a) 1 (b) 2
(c) 3 (d) 6
[CS, 2019 : 1 Mark]

314. Ten friends planned to share equally the cost of buying a gift for their teacher. When two of them decided not to contribute, each of the other friends had to pay Rs. 150 more. The cost of the gift was Rs. _______.
(a) 12000 (b) 6000
(c) 3000 (d) 666
[CS, 2019 : 1 Mark]

315. In a college, there are three student clubs. Sixty students are only in the Drama club. 80 students are only in the Dance club, 30 students are only in the Maths club. 40 students are in both Drama and Dance clubs. 12 students are in both Dance and Maths clubs, 7 students are in both Drama and Maths clubs, and 2 students are in all the clubs. If 75% of the students in the college are not in any of these clubs, then the total number of students in the college is ______.

(a) 225 (b) 1000
(c) 975 (d) 900

[CS, 2019 : 2 Marks]

316. The police arrested four criminals – P, Q, R and S. The criminals knew each other. They made the following statements:

P says "Q committed the crime."
Q says "S committed the crime."
R says "I did not do it."
S says "What Q said about me is false."

Assume only one of the arrested four committed the crime and only one of the statements made above is true. Who committed the crime?

(a) R (b) P
(c) S (d) Q

[CS, 2019 : 2 Marks]

317. In the given diagram, teachers are represented in the triangle, researchers in the circle and administrators in the rectangle. Out of the total number of the people, the percentage of administrators shall be in the range of _______.

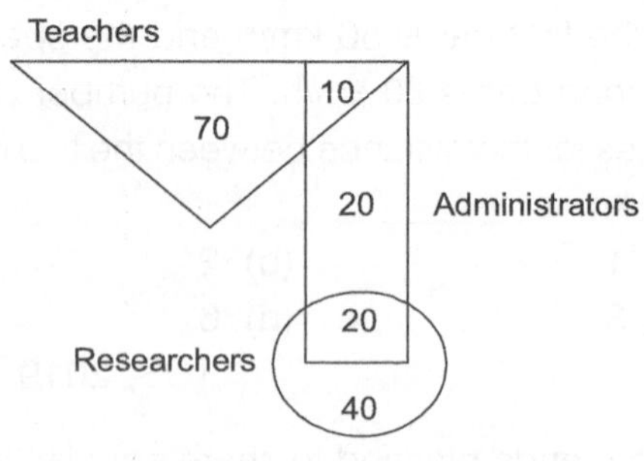

(a) 46 to 60 (b) 0 to 15
(c) 31 to 45 (d) 16 to 30

[CS, 2019 : 2 Marks]

318. Three of the five students allocated to a hostel put in special requests to the warden. Given the floor plan of the vacant rooms, select the allocation plan that will accommodate all their requests.

Request by X : Due to pollen allergy. I want to avoid a wing next to the garden.

Request by Y : 1 want to live as far from the washrooms as possible, since I am very sensitive to smell.

Request by Z : I believe in Vaastu and so want to stay in the South-West wing.

The shaded rooms are already occupied. WR is washroom.

(a)

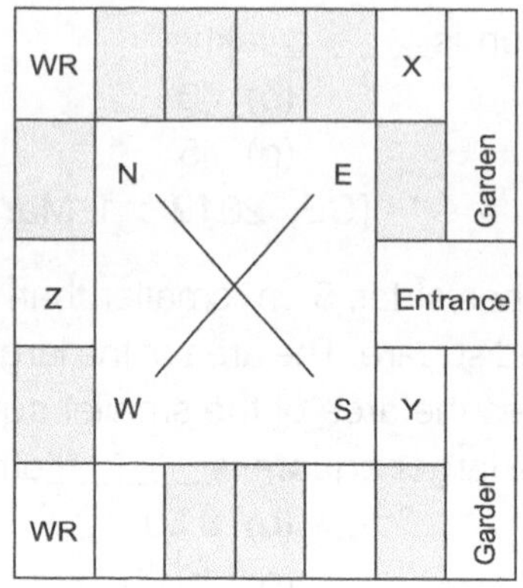

(b)

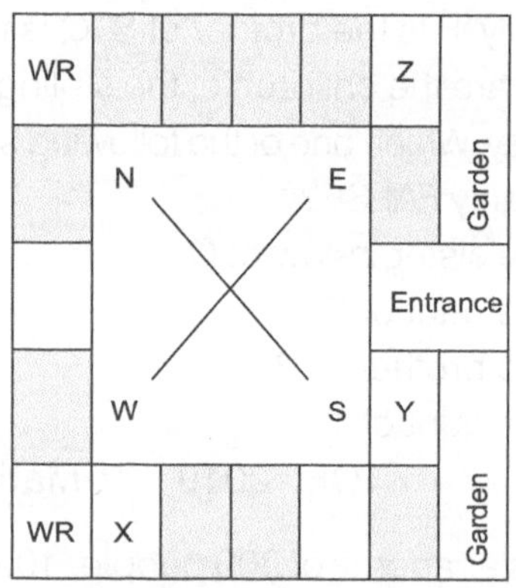

(c)

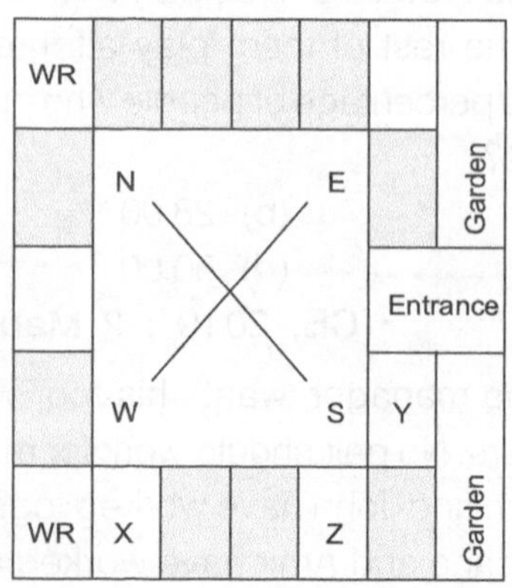

(d) 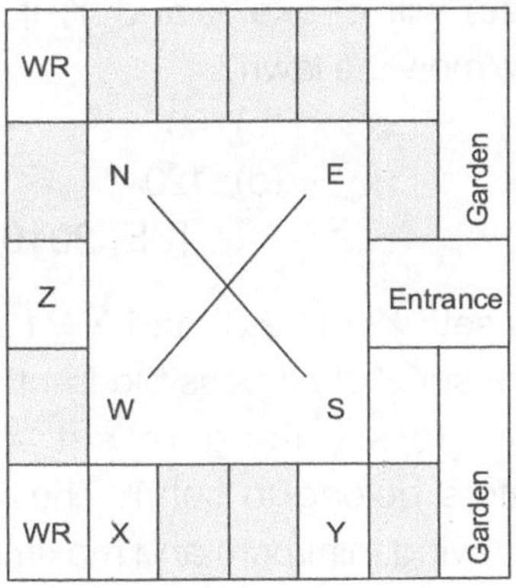

[CS, 2019 : 2 Marks]

319. It would take one machine 4 hours to complete a production order and another machine 2 hours to complete the same order. If both machines work simultaneously at their respective constant rates, the time taken to complete the same order is ____ hours.

(a) 7/3 (b) 3/4
(c) 2/3 (d) 4/3

[EC, 2019 : 1 Mark]

320. Five different books (P, Q, R, S, T) are to be arranged on a shelf. The books R and S are to be arranged first and second, respectively from the right side of the shelf. The number of different orders in which P, Q and T may be arranged is _____.

(a) 12 (b) 2
(c) 120 (d) 6

[EC, 2019 : 1 Mark]

321. The bar graph in Panel (a) shows the proportion of male and female illiterates in 2001 and 2011. The proportions of males and females in 2001 and 2011 are given in Panel (b) and (c), respectively. The total population did not change during this period. The percentage increase in the total number of literates from 2001 to 2011 is ____.

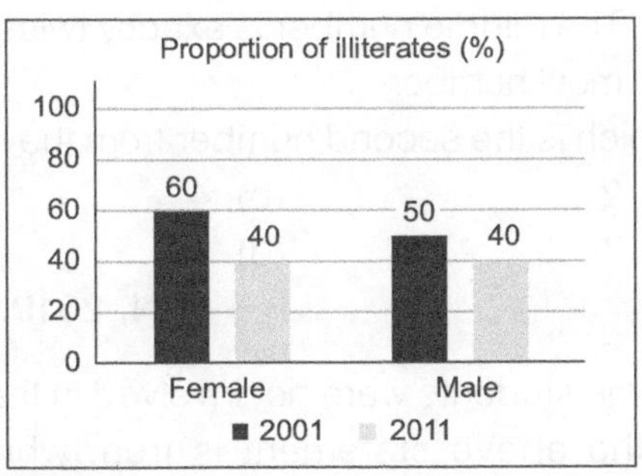

Panel (a)

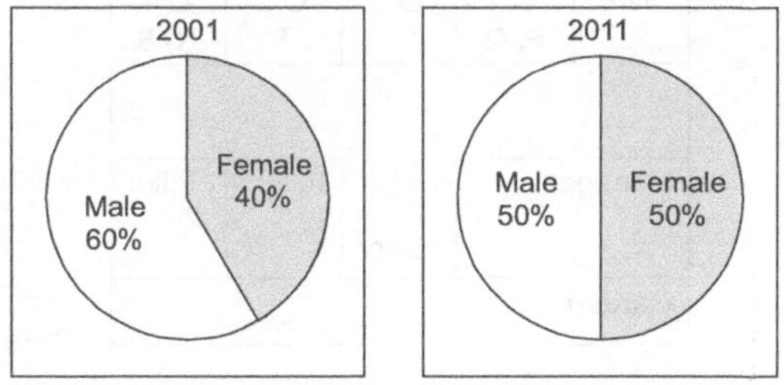

Panel (b) Panel (c)

(a) 35.43 (b) 34.43
(c) 30.43 (d) 33.43

[EC, 2019 : 2 Marks]

322. Two design consultants, P and Q, started working from 8 AM For a client. The client budgeted a total of USD 3000 for the consultants. P stopped working when the hour hand moved by 210 degrees on the clock. Q stopped working when the hour hand moved by 240 degrees. P took two lea breaks of 15 minutes each during her shift, but took no lunch break. Q took only one lunch break for 20 minutes, but no tea breaks. The market rate for consultants is USD 200 per hour and breaks are not paid. After paying the consultants, the client shall have USD ________ remaining in the budget.

(a) 433.33 (b) 000.00
(c) 300.00 (d) 166.67

[EC, 2019 : 2 Marks]

323. Five people P, Q, R, S and T work in a bank. P and Q don't like each other but have to share an office till T gets a promotion and moves to the big office next to the garden. R, who is currently sharing an office with T wants to move to the adjacent office with S, the handsome new intern. Given the floor plan, what is the current location of Q, R and T?

(O = Office, WR = Washroom)

(a) 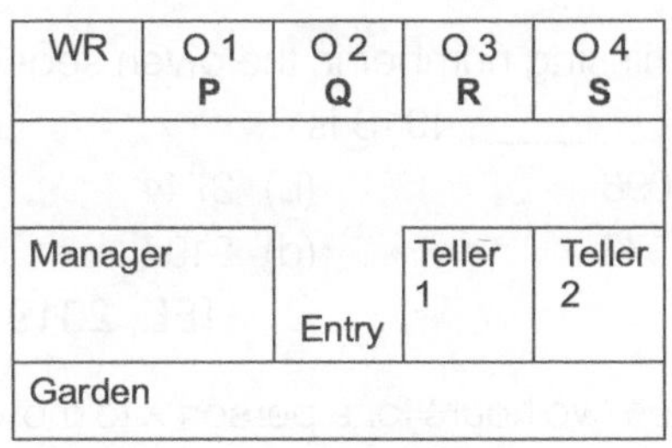

(b)

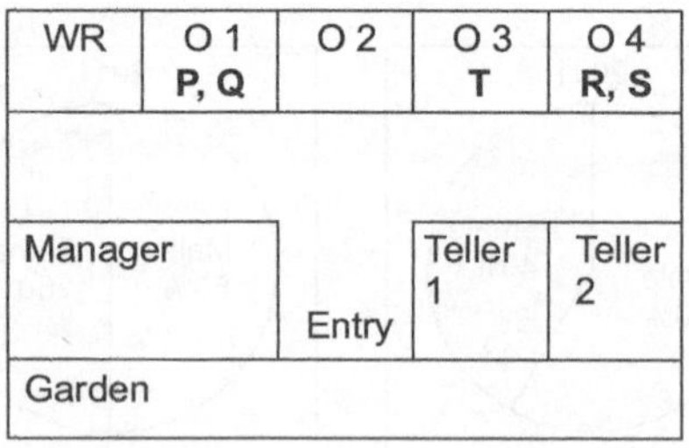

(c)

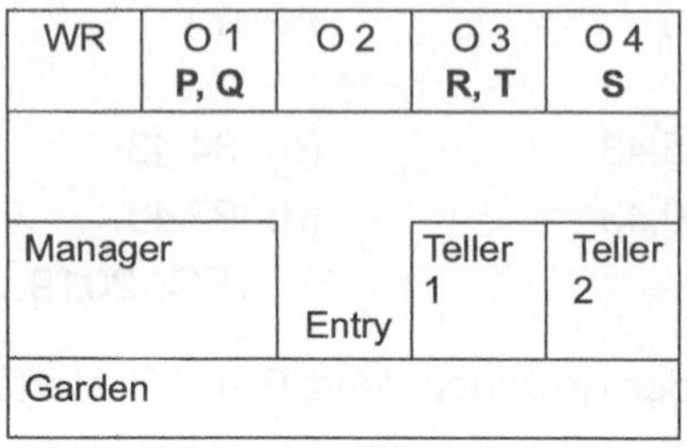

(d)

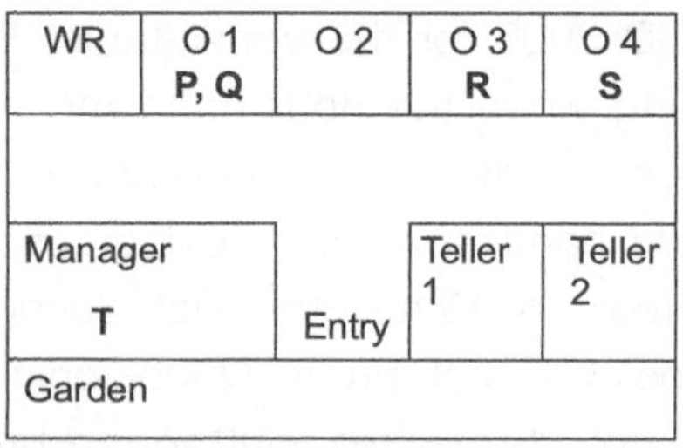

[EC, 2019 : 2 Marks]

324. Four people are standing in a line facing you. They are Rahul, Mathew, Seema and Lohit. One is an engineer, one is a doctor, one a teacher and another a dancer. You are told that:

1. Mathew is not standing nest to Seema.
2. There are two people standing between Lohit and the engineer.
3. Rahul is not a doctor.
4. The teacher and the dancer are standing next to each other.
5. Seema is turning to her right to speak to the doctor standing next to her.

Who among them is an engineer?

(a) Lohit (b) Rahul
(c) Seema (d) Mathew

[EC, 2019 : 2 Marks]

325. The missing number in the given sequence 343, 1331, _______, 4913 is

(a) 4096 (b) 2744
(c) 3375 (d) 2197

[EE, 2019 : 1 Mark]

326. It takes two hours for a person X to mow the lawn. Y can mow the same lawn in four hours. How long (in minutes) will it take X and Y, if they work together to mow the lawn?

(a) 90 (b) 60
(c) 80 (d) 120

[EE, 2019 : 1 Mark]

327. Given two sets $X = \{1,2,3\}$ and $Y = \{2, 3, 4\}$, we construct a set Z of all possible fractions where the numerators belong to set X and the denominators belong to set Y. The product of elements having minimum and maximum values in the set Z is _____.

(a) $\frac{1}{6}$ (b) $\frac{1}{12}$
(c) $\frac{3}{8}$ (d) $\frac{1}{8}$

[EE, 2019 : 1 Mark]

328. How many integers are there between 100 and 1000 all of whose digits are even?

(a) 90 (b) 60
(c) 100 (d) 80

[EE, 2019 : 2 Marks]

329. The ratio of the number of boys and girls who participated in an examination is 4 : 3. The total percentage of candidates who passed the examination is 80 and the percentage of girls who passed is 90. The percentage of boys who passed is _______.

(a) 72.50 (b) 90.00
(c) 80.50 (d) 55.50

[EE, 2019 : 2 Marks]

330. Five numbers 10, 7, 5,4 and 2 are to be arranged in a sequence from left to right following the directions given below:

1. No two odd or even numbers are next to each other.
2. The second number from the left is exactly half of the left-most number.
3. The middle number is exactly twice the right-most number.

Which is the second number from the right?

(a) 2 (b) 4
(c) 7 (d) 10

[IN, 2019 : 1 Mark]

331. Some students were not involved in the strike.

If the above statement is true, which of the following conclusions is/are logically necessary?

1. Some who were involved in the strike were students.
2. No student was involved in the strike.
3. At least one student was involved in the strike.
4. Some who were not involved in the strike were students.

(a) 1 and 2 (b) only 3
(c) only 4 (d) 2 and 3

[IN, 2019 : 1 Mark]

332. The radius as well as the height of a circular cone increases by 10%. The percentage increase in its volume is_______.

(a) 17.1 (b) 21.0
(c) 33.1 (d) 72.8

[IN, 2019 : 1 Mark]

333. In a country of 1400 million population 70% own mobile phones. Among the mobile phone owners, only 294 million access the Internet. Among these Internet users, only half buy goods from e-commerce portals. What is the percentage of these buyers in the country?

(a) 10.50 (b) 14.70
(c) 15.00 (d) 50.00

[IN, 2019 : 2 Mark]

334. Two trains started at 7 AM from the same point. The first train travelled north at a speed of 80 km/h and the second train travelled south at a speed of 100 km/h. The time at which they were 540 km apart is _______AM.

(a) 9 (b) 10
(c) 11 (d) 11.30

[IN, 2019 : 2 Mark]

335. A worker noticed that the hour hand on the factory clock had moved 225 degrees during her stay at the factory. For how long does she stay in the factory?

(a) 8.5 hours
(b) 7.5 hours
(c) 3.75 hours
(d) 4 hours and 15 minutes

[ME, 2019 : 1 Mark, Set-1]

336. The sum and product of two integers are 26 and 165 respectively. The difference between these two integers is _______.

(a) 2 (b) 6
(c) 3 (d) 4

[ME, 2019 : 1 Mark, Set-1]

337. A firm hires employees at five different skill levels P, Q, R, S, T. The shares of employment at these skill levels of total employment in 2010 is given in the pie chart as shown. There were a total of 600 employees in 2010 and the total employment increased by 15% from 2010 to 2016. The total employment at skill levels P, Q, and R remained unchanged during this period. If the employment at skill level S increased by 40% from 2010 to 2016, how many employees were there at skill level T in 2016?

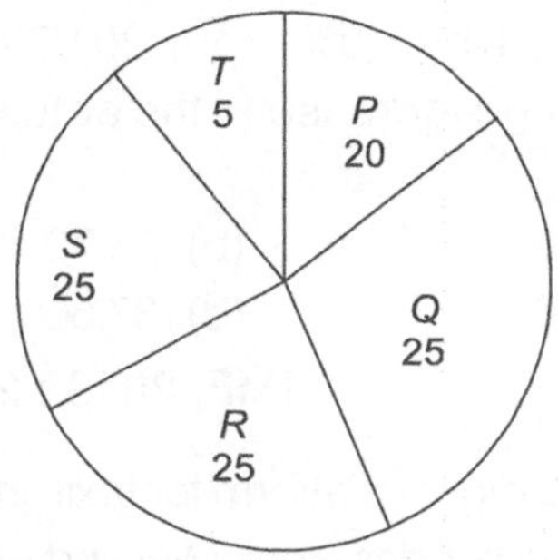

Percentage share of skills in 2010

(a) 35 (b) 30
(c) 72 (d) 60

[ME, 2019 : 2 Mark, Set-1]

338. M and N had four children P, Q, R and S. Of them, only P and R were married. They had children X and Y respectively. If Y is a legitimate child of W, which one of the following statements is necessarily FALSE?

(a) W is the wife of R
(b) R is the father of Y
(c) M is the grandmother of Y
(d) W is the wife of P

[ME, 2019 : 2 Mark, Set-1]

339. A person divided an amount of Rs. 100,000 into two parts and invested in two different schemes. In one he got 10% profit and in the other he got 12%. If the profit percentages are interchanged with these investments he would have got Rs. 120 less. Find the ratio between his investments in the two schemes.

(a) 9 : 16 (b) 37 : 63
(c) 47 : 53 (d) 11 : 14

[ME, 2019 : 2 Mark, Set-1]

340. If IMHO = JNIP; IDK = JEL; and SO = TP, then IDC = _______.

(a) JDE (b) JED
(c) JDC (d) JCD

[ME, 2019 : 1 Mark, Set-2]

341. The product of three integers X, Y and Z is 192. *Z* equal to 4 and P is equal to the average of X and Y. What is the minimum possible value of P?
(a) 8 (b) 7
(c) 9.5 (d) 6

[ME, 2019 : 1 Mark, Set-2]

342. Fiscal deficit was 4% of the GDP in 2015, and that increased to 5% in 2016. If the GDP increased by 10% from 2015 to 2016, then percentage increase in the actual fiscal deficit is ______
(a) 25.00 (b) 35.70
(c) 10.00 (d) 37.50

[ME, 2019 : 2 Mark, Set-2]

343. Mola is a digital platform for taxis in a city. It offers three type of rides-Pool, Mini and Prime. The table below presents the number of rides for the past four months. The platform earns one US dollar per ride. What is the percentage share of revenue contributed by prime to the total revenues of Mola, for the entire duration?

Type	Month			
	January	February	March	April
Pool	170	320	215	190
Mini	110	220	180	70
Prime	75	180	120	90

(a) 38.74 (b) 23.97
(c) 25.86 (d) 16.24

[ME, 2019 : 2 Mark, Set-2]

344. Two pipes P and Q can fill a tank in 6 hours and 9 hours respectively, while a third pipe R can empty the tank in 12 hours. Initially, P and R are open for 4 hours. Then P is closed and Q is opened. After 6 more hours R is closed. The total time taken to fill the tank (in hours) is _____
(a) 16.50 (b) 15.50
(c) 14.50 (d) 13.50

[ME, 2019 : 2 Mark, Set-2]

345. The total expenditure of a family, on different activities in a month, is shown in the pie-chart. the extra money spent on education as compared to transport (in percent) is ________.

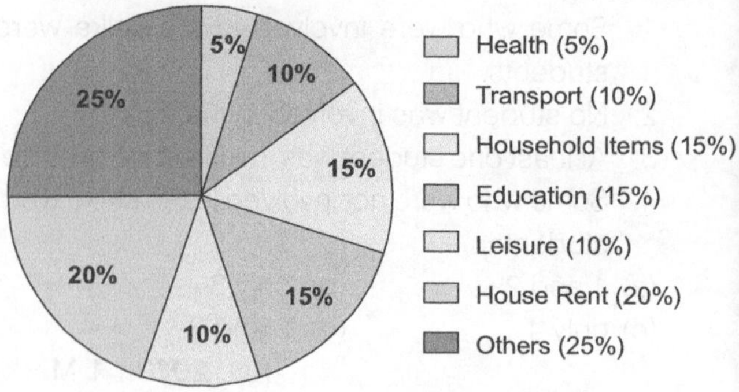

(a) 33.3 (b) 50
(c) 100 (d) 5

[CE, 2020 : 2 Marks, Set-1]

346. The unit's place in 26591749^{110016} is ______.
(a) 6 (b) 1
(c) 3 (d) 9

[CE, 2020 : 2 Marks, Set-1]

347. The sum of two positive numbers is 100. After subtracting 5 from each number, the product of the resulting numbers is 0. One of the original numbers is ______.
(a) 95 (b) 90
(c) 85 (d) 80

[CE, 2020 : 1 Mark, Set-1]

348. Five friends P, Q, R, S and T went camping. At night, they had to sleep in a row inside the tent. P, Q and T refused to sleep next to R since he snored loudly. P and S wanted to avoid Q as he usually hugged people in sleep.
Assuming everyone was satisfied with the sleeping arrangements, what is the order in which they slept?
(a) RSPTQ (b) QRSPT
(c) QTSPR (d) SPRTQ

[CE, 2020 : 2 Marks, Set-1]

349. Insert seven numbers between 2 and 34, such that the resulting sequence including 2 and 34 is an arithmetic progression. The sum of these inserted seven numbers is ________.
(a) 124 (b) 120
(c) 126 (d) 130

[CE, 2020 : 2 Marks, Set-1]

350. If 0, 1, 2, ..., 7, 8, 9 are coded as O, P, Q, ..., V, W, X, then 45 will be coded as ______.
(a) ST (b) SS
(c) SU (d) TS

[CE, 2020 : 1 Mark, Set-1]

351. The monthly distribution of 9 Watt LED bulbs sold by two firms X and Y from January to June 2018 is shown in the pie-chart and the corresponding table. If the total number of LED bulbs sold by two firms during April-June 2018 is 50000, then the number of LED bulbs sold by the firm Y during April-June 2018 is ________.

Months	Ratio of LED bulbs sold by two firms X : Y
January	7 : 8
February	2 : 3
March	2 : 1
April	3 : 2
May	1 : 4
June	9 : 11

January	15%
February	20%
March	30%
April	15%
May	10%
June	10%

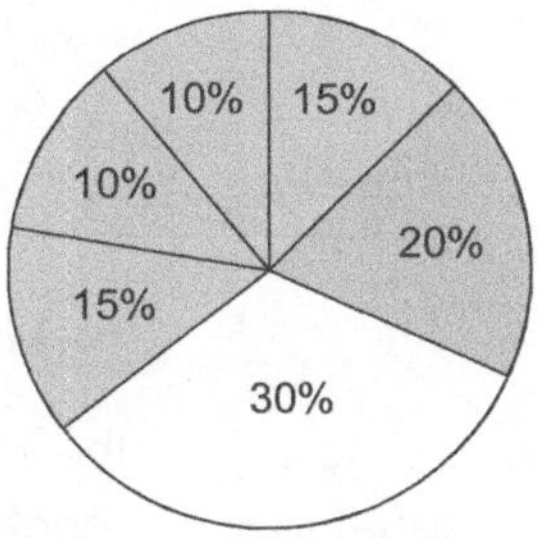

(a) 11250 (b) 9750
(c) 8250 (d) 8750

[CE, 2020 : 2 Marks, Set-2]

352. After the inauguration of the new building, the head of department (HOD) collated faculty preferences for office space. P wanted a room adjacent to the lab. Q wanted to be close to the lift. R wanted a view of the playground and S wanted a corner office.

Assuming that everyone was satisfied, which among the following shows a possible allocation?

(a)

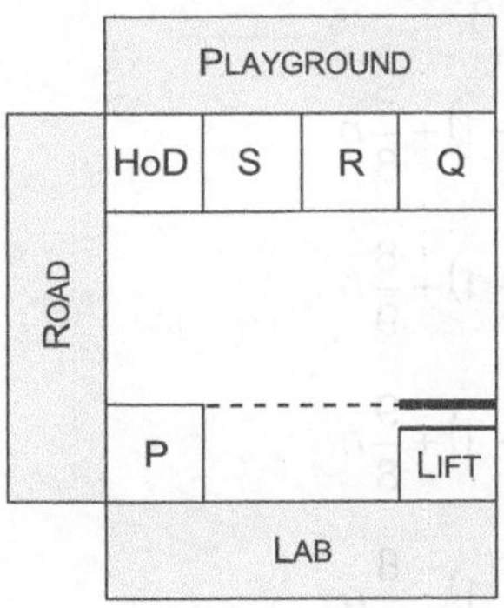

(b)

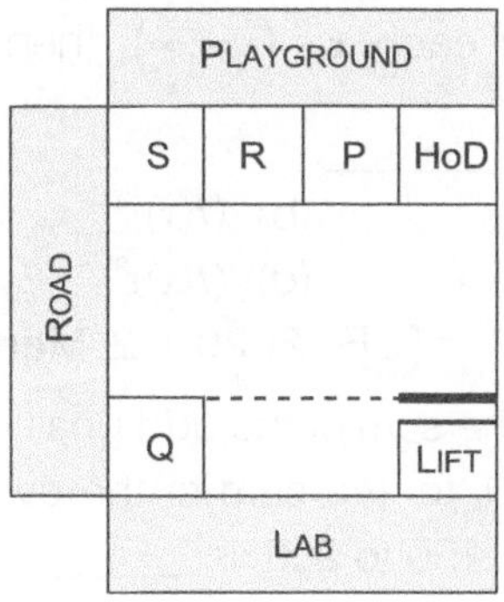

(c)

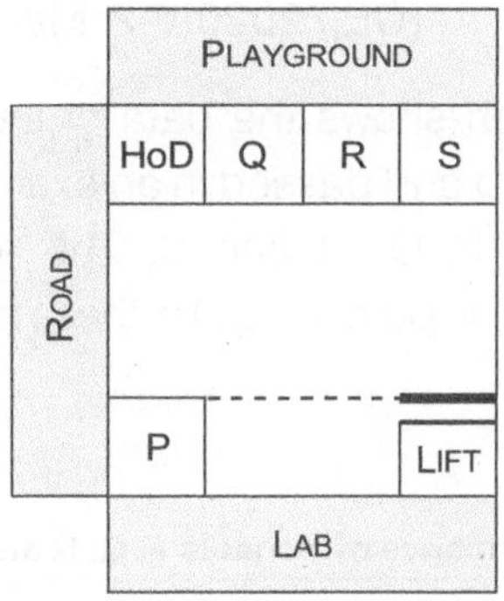

(d)

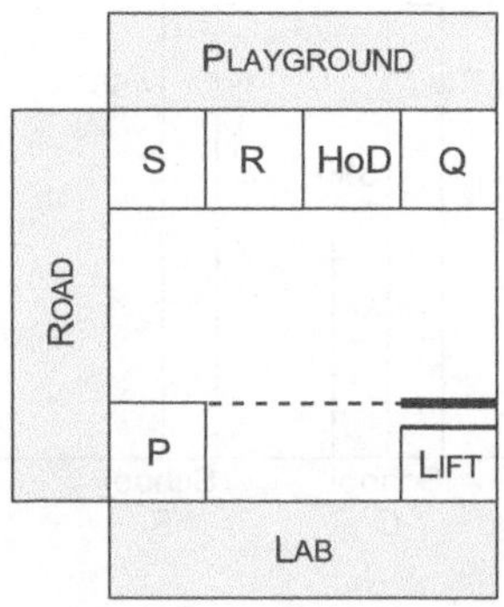

[CE, 2020 : 1 Mark, Set-2]

353. In a school of 1000 students, 300 students play chess and 600 students play football. If 50 students play both chess and football, the number of students who play neither is ________.

(a) 150 (b) 50
(c) 100 (d) 200

[CE, 2020 : 1 Mark, Set-2]

354. For the year 2019, which of the previous year's calendar can be used?

(a) 2011 (b) 2013
(c) 2012 (d) 2014

[CE, 2020 : 2 Marks, Set-2]

355. If $f(x) = x^2$ for each $x \in (-\infty, \infty)$, then $\dfrac{f(f(f(x)))}{f(x)}$ is equal to ________.

(a) $f(x)$ (b) $(f(x))^4$

(c) $(f(x))^2$ (d) $(f(x))^3$

[CE, 2020 : 2 Marks, Set-2]

356. The ratio of 'the sum of the odd positive integers from 1 to 100' to 'the sum of the even positive integers from 150 to 200' is ________.

(a) 50 : 91 (b) 1 : 1

(c) 1 : 2 (d) 45 : 95

[CE, 2020 : 2 Marks, Set-2]

357. The bar graph shows the data of the students who appeared and passed in an examination for four schools P, Q, R and S. The average of success rates (in percentage) of these four schools is ________.

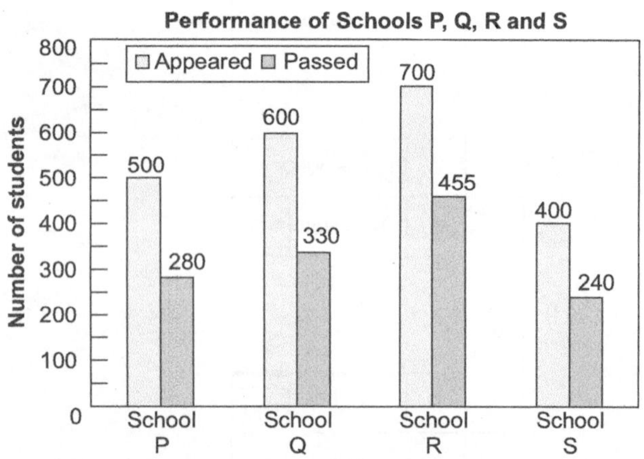

(a) 58.8 (b) 59.0

(c) 58.5 (d) 59.3

[ME, 2020 : 2 Marks, Set-1]

358. Select the graph that schematically represents BOTH $y = x^m$ and $y = x^{1/m}$ properly in the interval $0 \le x \le 1$. For integer values of m, where $m > 1$.

(a)

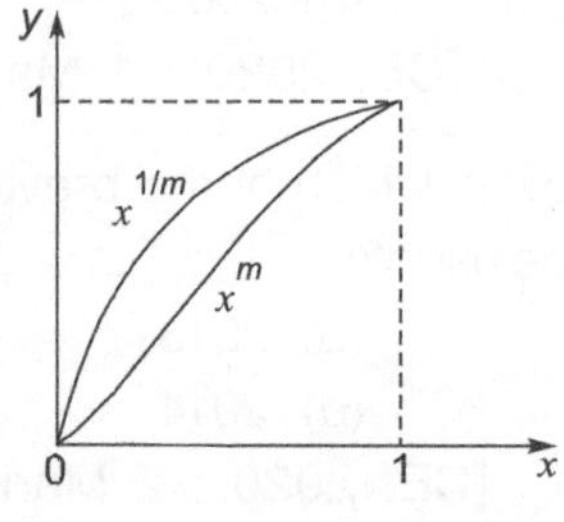

(b)

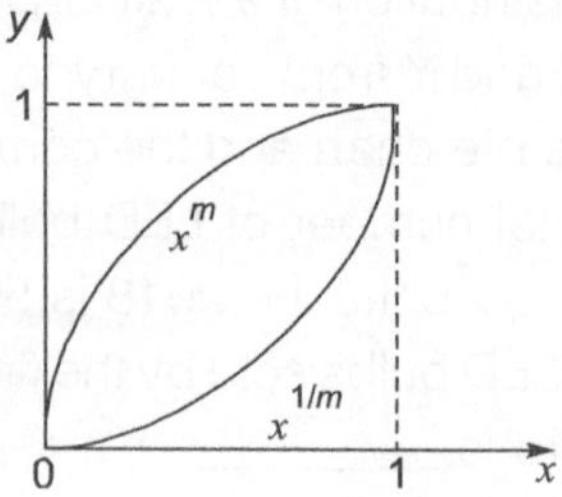

(c)

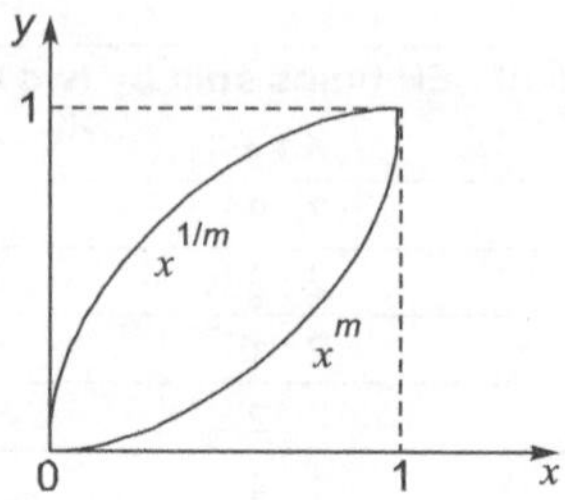

(d)

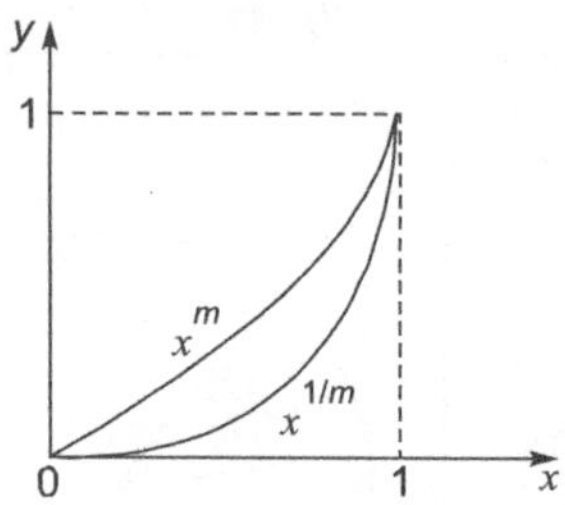

[ME, 2020 : 2 Marks, Set-1]

359. Define $[x]$ as the greatest integer less than or equal to x, for each $x \in (-\infty, \infty)$. If $y = [x]$, then area under y for $x \in [1, 4]$ is ______.

(a) 3 (b) 1

(c) 6 (d) 4

[ME, 2020 : 2 Marks, Set-1]

360. P, Q, R and S are to be uniquely coded using α and β. If P is coded as $\alpha\alpha$ and Q as $\alpha\beta$, then R and S, respectively, can be coded as ________.

(a) $\beta\alpha$ and $\beta\beta$ (b) $\alpha\beta$ and $\beta\beta$

(c) $\beta\alpha$ and $\alpha\beta$ (d) $\beta\beta$ and $\alpha\alpha$

[ME, 2020 : 2 Marks, Set-1]

361. The sum of the first n terms in the sequence 8, 88, 888, 8888, ... is ______.

(a) $\dfrac{81}{80}(10^n - 1) + \dfrac{9}{8}n$

(b) $\dfrac{80}{81}(10^n - 1) + \dfrac{8}{9}n$

(c) $\dfrac{81}{80}(10^n - 1) + \dfrac{9}{8}n$

(d) $\dfrac{80}{81}(10^n - 1) - \dfrac{8}{9}n$

[ME, 2020 : 2 Marks, Set-1]

362. It was estimated that 52 men can complete a strip in a newly constructed highway connecting cities *P* and *Q* in 10 days. Due to an emergency, 12 men were sent to another project. How many number of days, more than the original estimate, will be required to complete the strip?

(a) 3 days (b) 5 days
(c) 10 days (d) 13 days

[ME, 2020 : 2 Marks, Set-2]

363. An engineer measures THREE quantities *X*, *Y* and *Z* in an experiment. She finds that they follow a relationship that is represented in the figure below: (the product of *X* and *Y* linearly varies with *Z*)

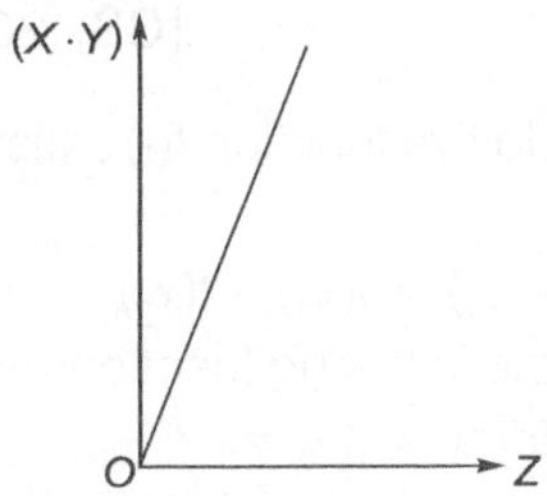

Then, which of the following statements is FALSE?

(a) For fixed *Y*; *X* is proportional to *Z*
(b) *XY*/*Z* is constant
(c) For fixed *X*; *Z* is proportional to *Y*
(d) For fixed *Z*; *X* is proportional to *Y*

[ME, 2020 : 2 Marks, Set-2]

364. There are five levels {*P*, *Q*, *R*, *S*, *T*} in a linear supply chain before a product reaches customers, as shown in the figure.

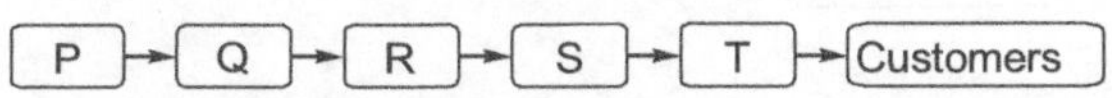

At each of the five levels, the price of the product is increased by 25%. If the product is produced at level *P* at the cost of Rs. 120 per unit, what is the paid (in rupees) by the customers?

(a) 292.96 (b) 187.50
(c) 234.38 (d) 366.21

[ME, 2020 : 2 Marks, Set-2]

365. The two pie-charts given below show the data of total students and only girls registered in different streams in a university. If the total number of students registered in the university is 5000, and the total number of the registered girls is 1500; then, the ratio of boys enrolled in Arts to the girls enrolled in Management is ________.

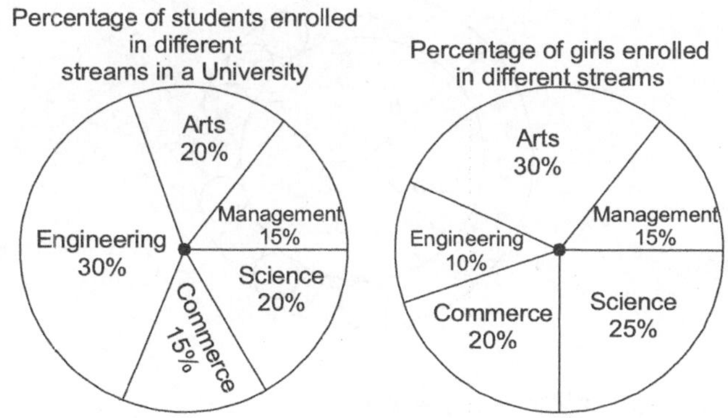

(a) 9 : 22 (b) 11 : 9
(c) 2 : 1 (d) 22 : 9

[ME, 2020 : 2 Marks, Set-2]

366. Find the missing element in the following figure:

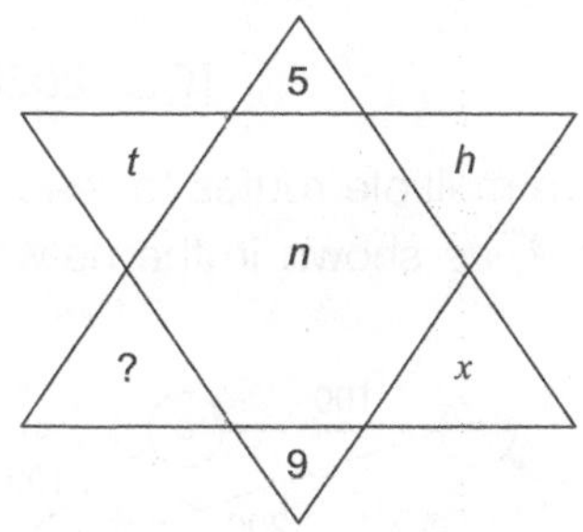

(a) *d* (b) *w*
(c) *e* (d) *y*

[ME, 2020 : 2 Marks, Set-2]

367. If P = 3, R = 27, T = 243, then Q + S = ________.

(a) 110 (b) 80
(c) 90 (d) 40

[CS, 2020 : 2 Marks]

368. Two straight lines are drawn perpendicular to each other in X-Y plane. If α and β are the acute angles the straight lines make with the X-axis, then $\alpha + \beta$ is ________.

(a) 180° (b) 120°
(c) 90° (d) 60°

[CS, 2020 : 2 Marks]

369. The figure below shows an annular ring with outer and inner radii as *b* and *a*, respectively. The annular space has been painted in the form of blue colour circles touching the outer and inner periphery of annular space. If maximum *n* number of circles can be painted, then the unpainted area available in annular space is ________.

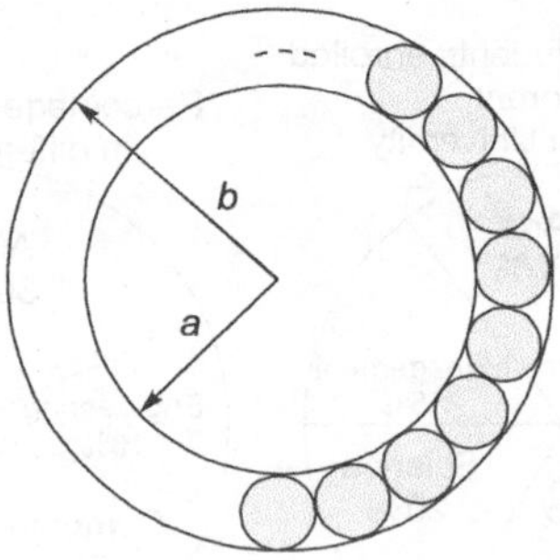

(a) $\pi[(b^2 - a^2) + n(b - a)^2]$

(b) $\pi\left[(b^2 - a^2) + \frac{\pi}{4}(b - a)^2\right]$

(c) $\pi[(b^2 - a^2) - n(b - a)^2]$

(d) $\pi\left[(b^2 - a^2) - n\frac{\pi}{4}(b - a)^2\right]$

[CS, 2020 : 2 Marks]

370. There are multiple routes to reach from node 1 to node 2, as shown in the network.

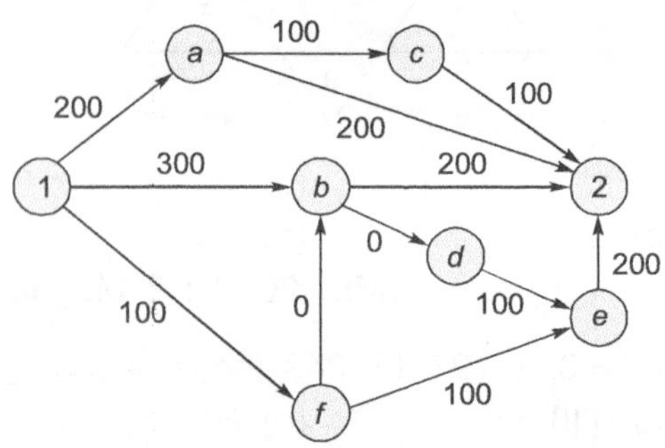

The cost of travel on an edge between two nodes is given in rupees. Nodes 'a' 'b', 'c', 'd', 'e' and 'f' are toll booths. The toll price at toll booths marked 'a' and 'c' is Rs. 200. and is Rs. 100 for the other toll booths. Which is the cheapest route from node 1 to node 2?

(a) 1-a-C-2 (b) 1-b-2
(c) 1-f-b-2 (d) 1-f-e-2

[CS, 2020 : 1 Mark]

371. The total revenue of a company during 2014-2018 is shown in the bar graph. If the total expenditure of the company in each year is 500 million rupees, then the aggregate profit loss (in percentage) on the total expenditure of the company during 2014-2018 is ______.

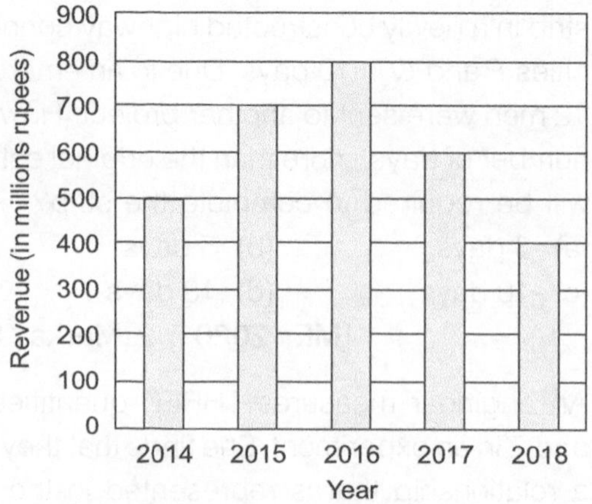

(a) 16.67% profit (b) 20% loss
(c) 16.67% loss (d) 20% profit

[CS, 2020 : 2 Marks]

372. A superadditive function $f(\cdot)$ satisfies the following property

$$f(x_1 + x_2) \geq f(x_1) + f(x_2)$$

Which of the following functions is a superadditive function for $x > 1$?

(a) $\sqrt{x}$ (b) e^{-x}
(c) e^x (d) $1/x$

[EC, 2020 : 1 Marks]

373. The following figure shows the data of students enrolled in 5 years (2014 to 2018) for two schools P and Q. During this period, the ratio of the average number of the students enrolled in school P to the average of the difference of the number of students enrolled in schools P and Q is ______.

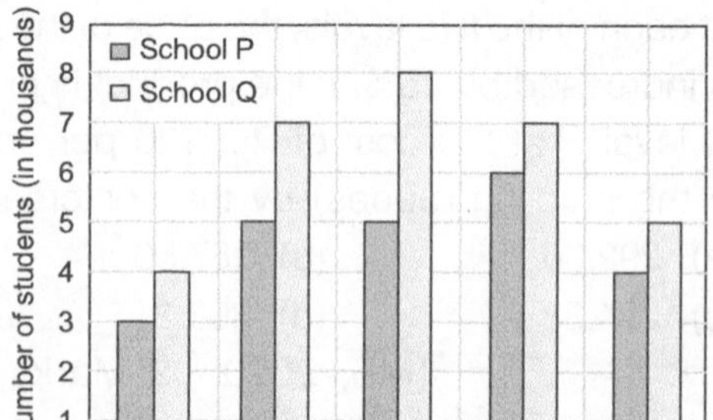

(a) 8 : 23 (b) 23 : 8
(c) 23 : 31 (d) 31 : 23

[EC, 2020 : 2 Marks]

374. a, b, c are real numbers. The quadratic equation $ax^2 - bx + c = 0$ has equal roots, which is β, then

(a) $\beta^2 = ac$ (b) $\beta^3 = bc/(2a^2)$
(c) $\beta = b/a$ (d) $b^2 \neq 4ac$

[EC, 2020 : 2 Marks]

375. It is quarter past three in your watch. The angle between the hour hand and the minute hand is ________.

(a) 22.5° (b) 7.5°
(c) 0° (d) 15°

[EC, 2020 : 2 Marks]

376. A circle with centre O is shown in the figure. A rectangle PQRS of maximum possible area is inscribed in the circle. If the radius of the circle is a, then the area of the shaded portion is ________.

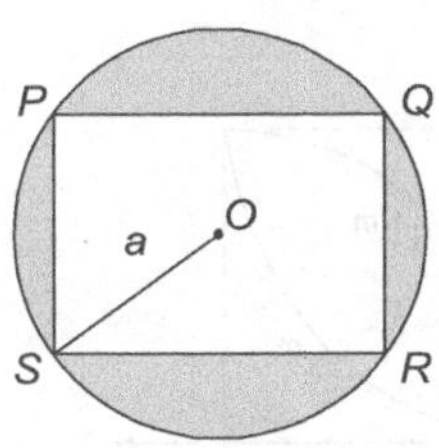

(a) $\pi a^2 - 2a^2$ (b) $\pi a^2 - 3a^2$
(c) $\pi a^2 - \sqrt{2}a^2$ (d) $\pi a^2 - a^2$

[EC, 2020 : 2 Marks]

377. Select the next element of the series: Z, WV, RQP, ______.

(a) JIHG (b) KJIH
(c) NMLK (d) LKJI

[EE, 2020 : 2 Marks]

378. If P, Q, R, S are four individuals. How many teams of size exceeding one can be formed, with Q as a member?

(a) 8 (b) 6
(c) 5 (d) 7

[EE, 2020 : 1 Mark]

379. In four-digit integer numbers from 1001 to 9999, the digit group "37" (in the same sequence) appears ______ times.

(a) 299 (b) 270
(c) 280 (d) 279

[EE, 2020 : 2 Marks]

380. Non-performing Assets (NPAs) of a bank in India is defined as an asset, which remains unpaid by a borrower for a certain period of time in terms of interest, principal, or both. Reserve Bank of India (RBI) has changed the definition of NPA thrice during 1993-2004. in terms of the holding period of loans. The holding period was reduced by one quarter each time. In 1993, the holding period was four quarters (360 days). Based on the above paragraph, the holding period of loans in 2004 after the third revision was ______ days.

(a) 45 (b) 135
(c) 90 (d) 180

[EE, 2020 : 2 Marks]

381. Given a semicircle with O as the centre; as shown in the figure, the ratio $\frac{\overline{AC} + \overline{CB}}{\overline{AB}}$ is ________.

Where $\overline{AC}, \overline{CB}$ and $\overline{AB}$ are chords.

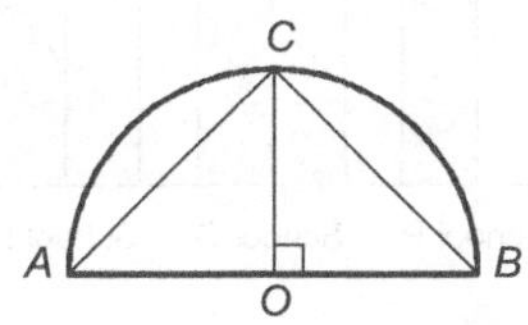

(a) $\sqrt{2}$ (b) $\sqrt{3}$
(c) 2 (d) 3

[EE, 2020 : 2 Marks]

382. The revenue and expenditure of four different companies P, Q, R and S in 2015 are shown in the figure. If the revenue of company Q in 2015 was 20% more than that in 2014, and company Q had earned a profit of 10% on expenditure in 2014, then its expenditure (in million rupees) in 2014 was _______.

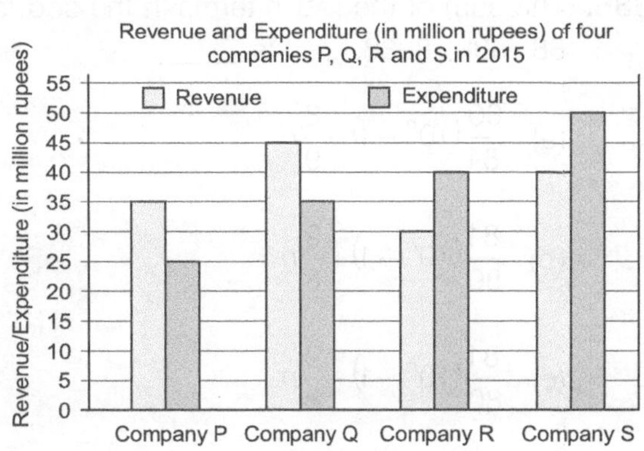

(a) 34.1 (b) 32.7
(c) 33.7 (d) 35.1

[EE, 2020 : 2 Marks]

383. *P*, *Q*, *R* and *S* are to be uniquely coded using α and β. If *P* is coded as $\alpha\alpha$ and *Q* as $\alpha\beta$, then *R* and *S*, respectively, can be coded as _____.

(a) $\beta\alpha$ and $\alpha\beta$ (b) $\alpha\beta$ and $\beta\beta$
(c) $\beta\alpha$ and $\beta\beta$ (d) $\beta\beta$ and $\alpha\alpha$

[IN, 2020 : 2 Marks]

384. The bar graph shows the data of students who appeared and passed in an examination for four schools P, Q, R and S. The average of success rates (in percentage) of these four schools is ____.

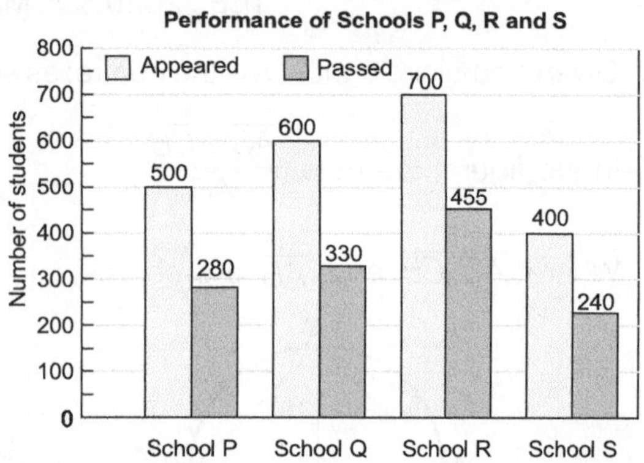

(a) 58.8% (b) 58.5%
(c) 59.0% (d) 59.3%

[IN, 2020 : 2 Marks]

385. Define $[x]$ as the greatest integer less than or equal to x, for each $x \in (-\infty, \infty)$. If $y = [x]$ then area under *y* for $x \in [1, 4]$ is _____.

(a) 4 (b) 1
(c) 6 (d) 3

[IN, 2020 : 1 Mark]

386. The sum of the first *n* terms in the sequence 8, 88, 888, 8888, ... is _____.

(a) $\frac{80}{81}(10^n - 1) - \frac{8}{9}n$

(b) $\frac{81}{80}(10^n - 1) + \frac{9}{8}n$

(c) $\frac{81}{80}(10^n - 1) - \frac{9}{8}n$

(d) $\frac{80}{81}(10^n - 1) + \frac{8}{9}n$

[IN, 2020 : 2 Marks]

387. Select the graph that schematically represents BOTH $y = x^m$ and $y = x^{1/m}$ properly in the interval $0 \le x \le 1$, for integer values of *m*, where $m > 1$.

(a)
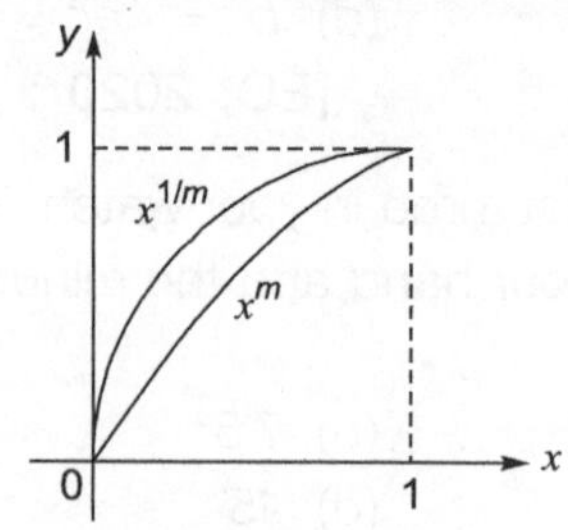

(b)
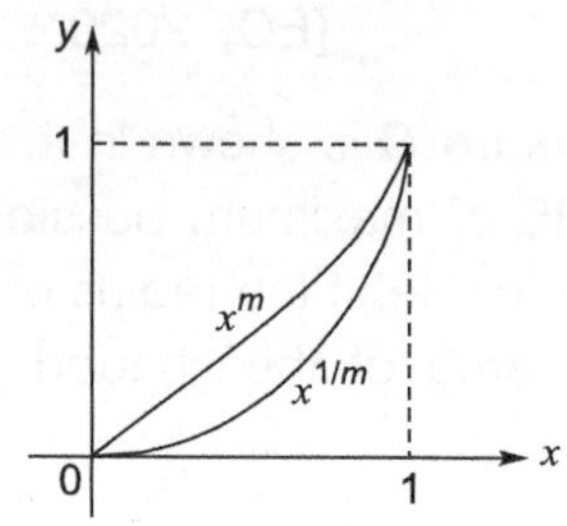

(c)
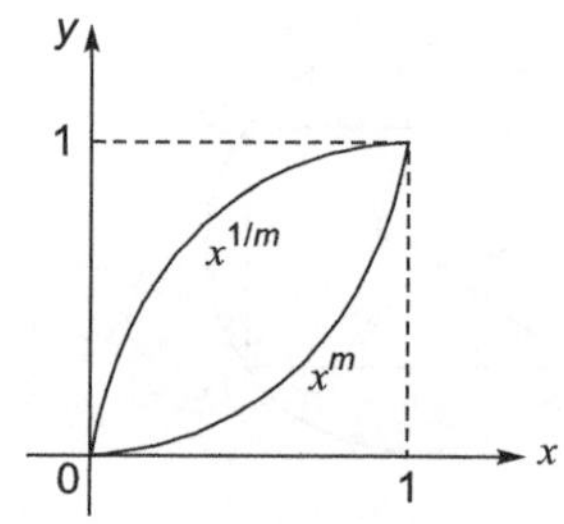

(d)
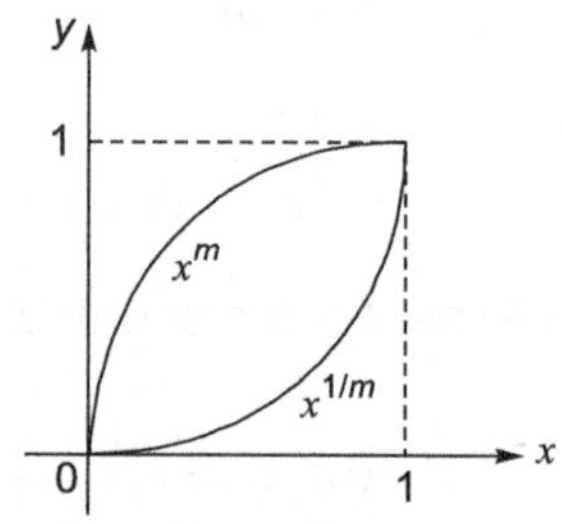

[IN, 2020 : 2 Marks]

388. It was estimated that 52 men can complete a strip in a newly constructed highway connecting cities P and Q in 10 days. Due to an emergency, 12 men were sent to another project. How many number of days, more than the original estimate, will be required to complete the strip?

(a) 5 days (b) 10 days
(c) 13 days (d) 3 days

[PI, 2020 : 2 Marks]

389. An engineer measures THREE quantities *X*, *Y* and *Z* in an experiment. She finds that they follow a relationship that is represented in the figure below : (the product of *X* and *Y* linearly varies with *Z*)

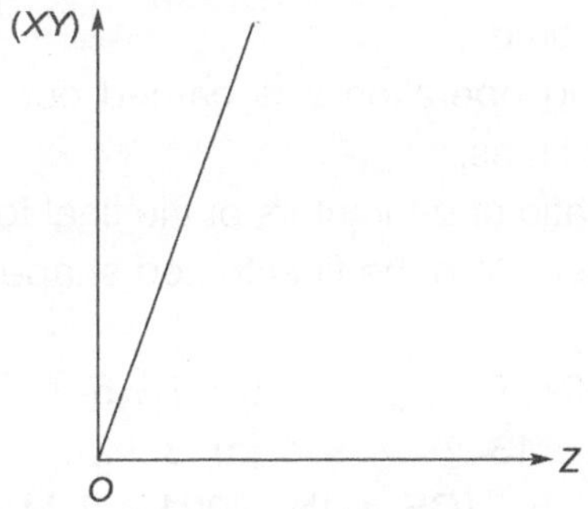

Then, which of the following statements is FALSE?

(a) For fixed Z; X is proportional to Y

(b) $\frac{XY}{Z}$ is constant

(c) For fixed X; Z is proportional to Y

(d) For fixed Y; X is proportional to Z

[PI, 2020 : 2 Marks]

390. Find the missing element in the following figure:

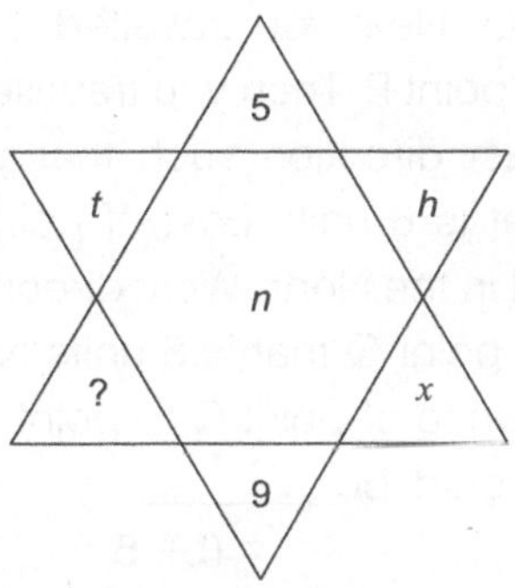

(a) w (b) e
(c) y (d) d

[PI, 2020 : 2 Marks]

391. There are five levels {P, Q, R, S, T} in a linear supply chain before a product reaches customers, as shown in the figure :

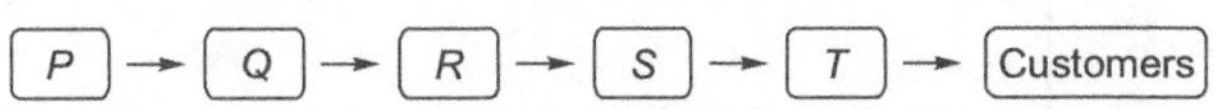

At each of the five levels, the price of the product is increased by 25%. If the product is produced at level P at the cost of Rs. 120 per unit, what is the price paid (in rupees) by the customers?

(a) 366.21 (b) 292.96
(c) 234.38 (d) 187.50

[PI, 2020 : 2 Marks]

392. The two pie-charts given below show the data of total students and only girls registered in different streams in a university. If the total number of students registered in the university is 5000, and the total number of the registered girls is 1500; then the ratio of boys enrolled in Arts to the girls enrolled in Management is ______

Percentage of students enrolled in different streams in a University

Percentage of girls enrolled in different streams

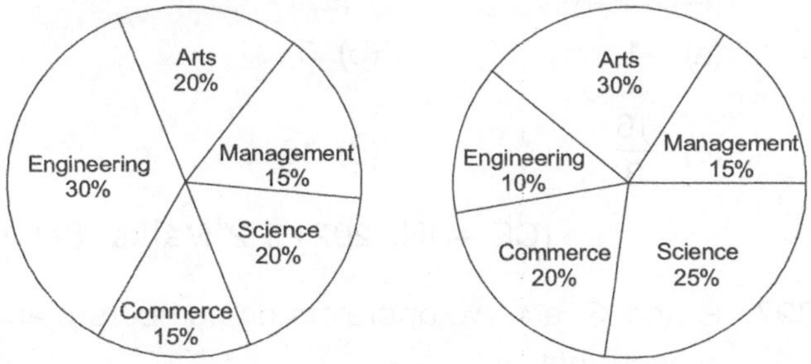

(a) 9 : 22 (b) 11 : 9
(c) 22 : 9 (d) 2 : 1

[PI, 2020 : 2 Marks]

393. The mirror image of the about text about the x-axis is

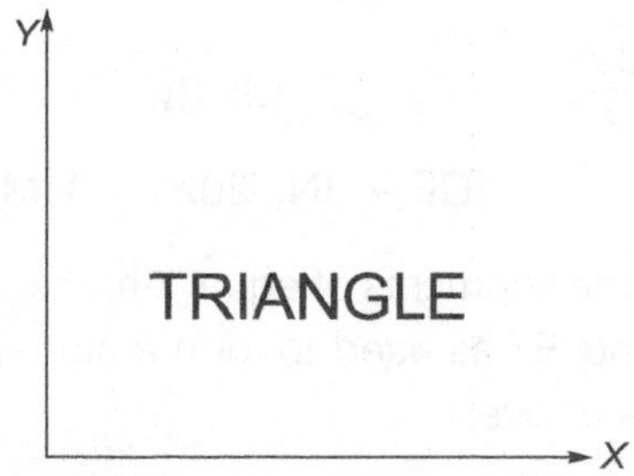

(a)  (b) 
(c)  (d) 

[CE + IN, 2021 : 1 Mark, Set-1]

394. In a company, 35% of the employees drink coffee, 40% of the employees drink tea and 10% of the employees drink both tea and coffee. What percentage of employees drink neither tea nor coffee?

(a) 15 (b) 25
(c) 35 (d) 40

[CE + IN, 2021 : 1 Mark, Set-1]

395. Four persons P, Q, R and S are to be seated in a row. All facing the same direction, but not necessarily in the same order. P and R cannot sit adjacent to each other. S should be seated to the right of Q. The number of distinct seating arrangements possible is:

(a) 4 (b) 2
(c) 8 (d) 6

[CE + IN, 2021 : 1 Mark, Set-1]

396. A function λ, is defined by

$$\lambda(p,q)=\begin{cases}(p-q)^2, & \text{if } p\geq q\\ p+q, & \text{if } p<q\end{cases}$$

The value of expression $\dfrac{\lambda[-(-3+2),(-2,+3)]}{[-(-2+1)]}$ is:

(a) –1 (b) 0

(c) $\dfrac{16}{3}$ (d) 16

[CE + IN, 2021 : 2 Marks, Set-1]

397. $\oplus$ and $\odot$ are two operators on numbers p and q such that

$$p\oplus Q=\frac{p^2+q^2}{pq} \text{ and } p\odot q=\frac{p^2}{q};$$

If $x \oplus y = 2\odot 2$, then x =

(a) y (b) $\dfrac{y}{2}$

(c) $\dfrac{3y}{2}$ (d) $2y$

[CE + IN, 2021 : 1 Mark, Set-1]

398. Five line segments of equal lengths, *PR*, *PS*, *QS*, *QT* and *RT* as used to form a star shown in the figure above.

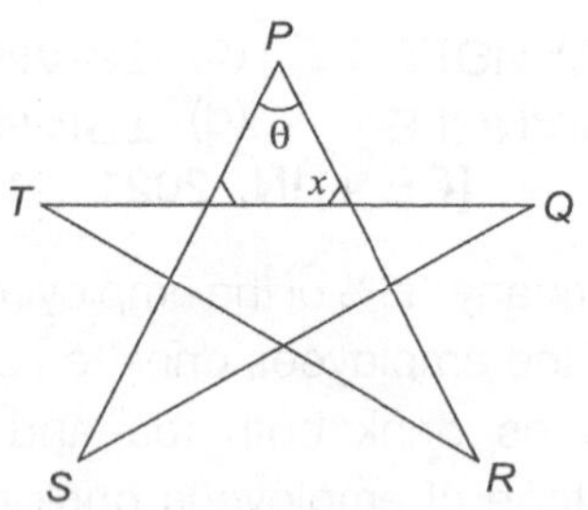

The value of θ, in degrees, is______.

(a) 72 (b) 108

(c) 36 (d) 45

[CE + IN, 2021 : 2 Marks, Set-1]

399. Consider two rectangular sheets, Sheet *M* and Sheet *N* of dimensions 6 cm × 4 cm each.

Folding operation 1: The sheet is folded into half by joining the short edges of the current shape,

Folding operation 2: The sheet is folded into half by joining the long edges of the current shape.

Folding operation 1 is carried out on Sheet *M* three times,

Folding operation 2 is carried out on Sheet *N* three times,

The ratio of perimeters of the final folded shape of Sheet *N* to the final folded shape of Sheet *M* is______.

(a) 13 : 7 (b) 7 : 5

(c) 5 : 13 (d) 3 : 2

[CE + IN, 2021 : 2 Marks, Set-1]

400. $\oplus$ and $\odot$ are two operators on numbers p and q such that

$$p\odot q=p-q, \text{ and } p\oplus q=p\times q$$

Then, $(9\odot(6\oplus 7))\odot(7\oplus(6\odot 5))=$

(a) –33 (b) 40

(c) –40 (d) –26

[CE, 2021 : 1 Mark, Set-2]

401. On a planar field, you travelled 3 units East from a point O. Next you travelled 4 units South to arrive at point P. Then you travelled from P in the North-East direction such that you arrive at a point that is 6 units East of point O. Next, you travelled in the North-West direction, so that you arrive at point Q that is 8 units North of point P. The distance of point Q to point O, in the same units, should be ______.

(a) 6 (b) 5

(c) 4 (d) 3

[CE, 2021 : 2 Marks, Set-2]

402.

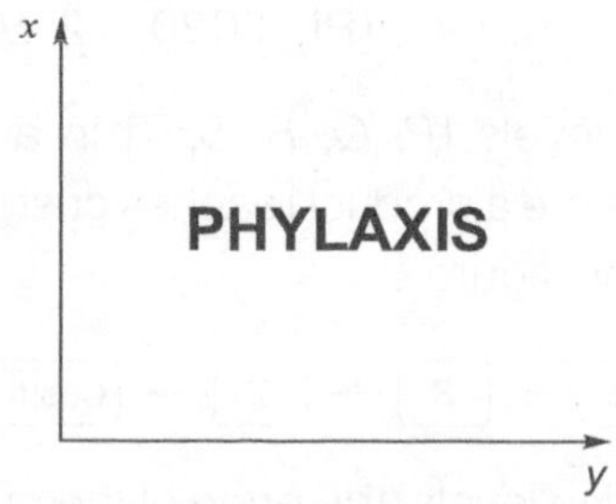

The mirror image of the above text about the *x*-axis is

(a) 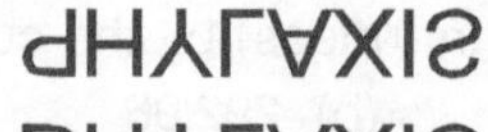

(b) 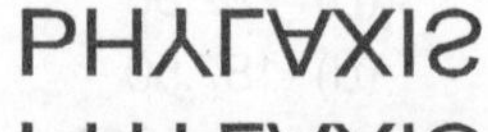

(c) 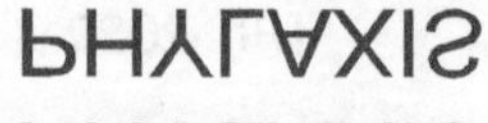

(d) 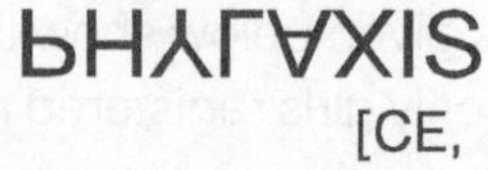

[CE, 2021 : 1 Mark, Set-2]

403. In an equilateral triangle PQR, side PQ is divided into four equal parts, side QR is divided into six equal parts and side PR is divided into eight equal parts. The length of each subdivided part in cm is an integer.
The minimum area of the triangle PQR possible, in cm^2, is

(a) 24 (b) $144\sqrt{3}$
(c) $48\sqrt{3}$ (d) 18

[CE, 2021 : 2 Marks, Set-2]

404. 1. Some football players play cricket.
2. All cricket players play hockey.

Among the options given below, the statement that logically follows from the two statements 1 and 2 above, is:

(a) All hockey players play football.
(b) Some football players play hockey.
(c) All football players play hockey.
(d) No football player plays hockey.

[CE, 2021 : 2 Marks, Set-2]

405.

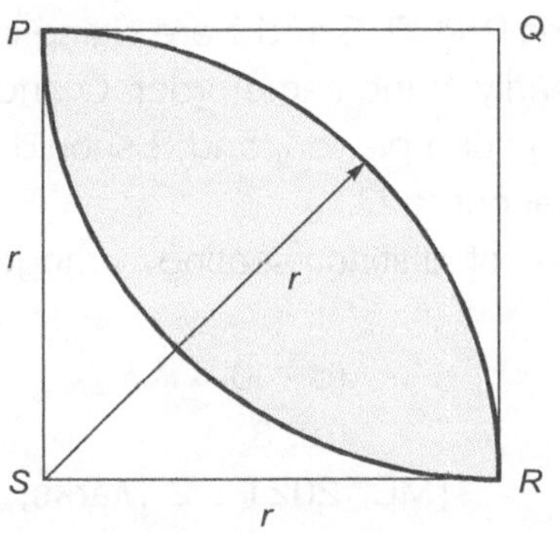

In the figure shown above, PQRS is a square. The shaded portion is formed by the intersection of sectors of circles with radius equal to the side of the square and centers at S and Q.
The probability that any point picked randomly within the square falls in the shaded area is ______.

(a) $\frac{\pi}{2}-1$ (b) $4-\frac{\pi}{2}$
(c) $\frac{\pi}{4}$ (d) $\frac{1}{2}$

[CE, 2021 : 2 Marks, Set-2]

406. Four persons P, Q, R and S are to be seated in a row. R should not be seated at the second position from the left end of the row. The number of distinct seating arrangements possible is:

(a) 24 (b) 9
(c) 6 (d) 18

[CE, 2021 : 1 Mark, Set-2]

407. Two identical cube shaped dice each with faces numbered 1 to 6 are rolled simultaneously. The probability that an even number is rolled out on each dice is:

(a) $\frac{1}{8}$ (b) $\frac{1}{36}$
(c) $\frac{1}{4}$ (d) $\frac{1}{12}$

[CE, 2021 : 1 Mark, Set-2]

408.

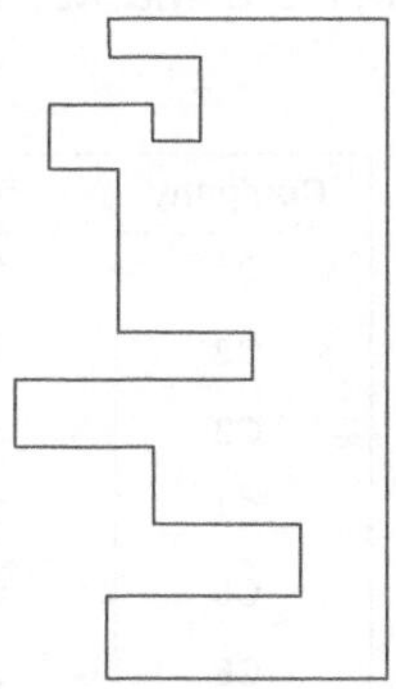

A jigsaw puzzle has 2 pieces. One of the pieces is shown above. Which one of the given options for the missing piece when assembled will form a rectangle? The piece can be moved, rotated or flipped to assemble with the above piece.

(a)

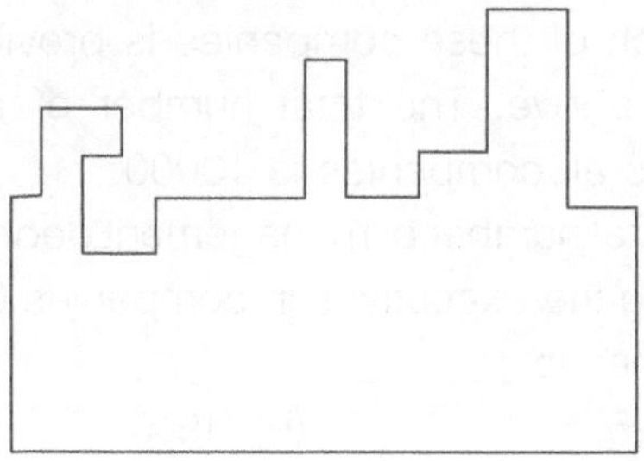

(b)

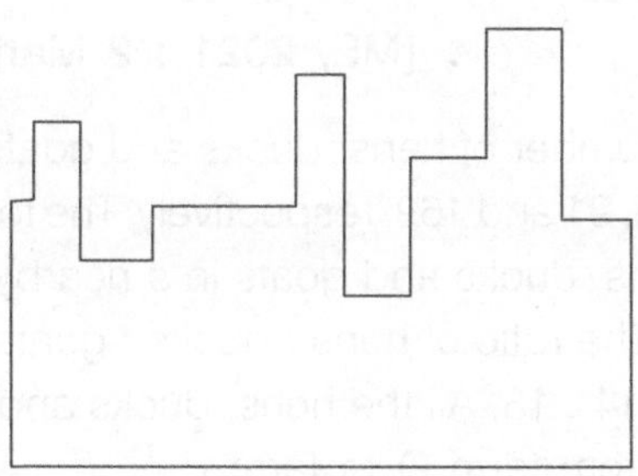

(c) 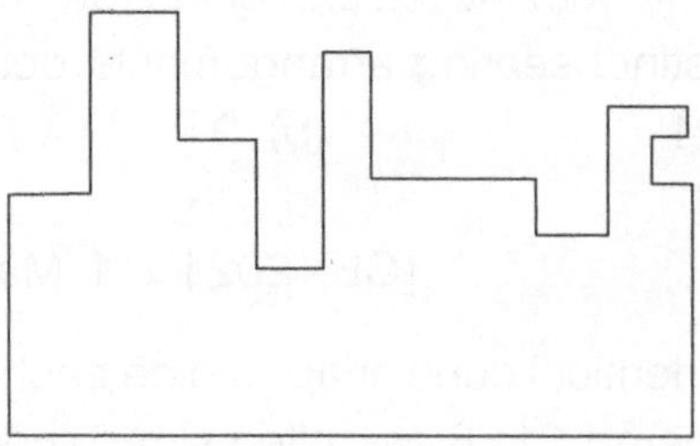

(d) 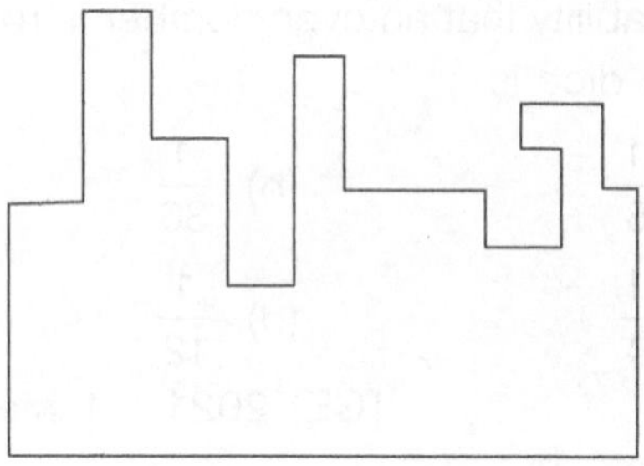

[ME, 2021 : 2 Marks, Set-1]

409. 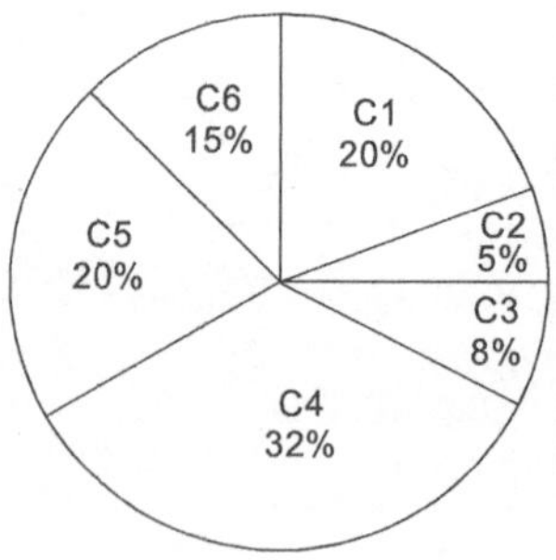

Company	Ratio
C1	3 : 2
C2	1 : 4
C3	5 : 3
C4	2 : 3
C5	9 : 1
C6	3 : 4

The distribution of employees at the rank of executives, across different companies C1, C2, ..., C6 is presented in the chart given above. The ratio of executives with a management degree to those without a management degree in each of these companies is provided in the table above. The total number of executives across all companies is 10000.

The total number of management degree holders among the executives in companies C2 and C5 together is ______.

(a) 225 (b) 1900
(c) 2500 (d) 600

[ME, 2021 : 2 Marks, Set-1]

410. The number of hens, ducks and goats in farm P are 65, 91 and 169, respectively. The total number of hens, ducks and goats in a nearby farm Q is 416. The ratio of hens : ducks : goats in farm Q is 5 : 14 : 13. All the hens, ducks and goats are sent from farm Q to farm P.

The new ratio of hens : ducks : goats in farm P is ______.

(a) 5 : 7 : 13 (b) 21 : 10 : 26
(c) 5 : 14 : 13 (d) 10 : 21 : 26

[ME, 2021 : 2 Marks, Set-1]

411. 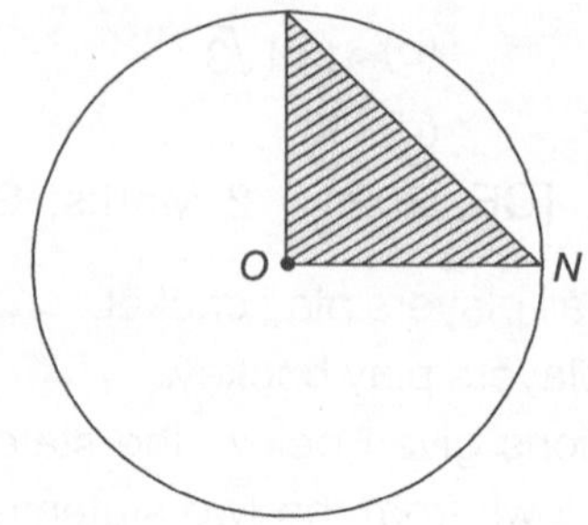

In the above figure, O is the center of the circle and, M and N lie on the circle.

The area of the right triangle MON is 50 cm^2.

What is the area of the circle in cm^2?

(a) 2π (b) 100π
(c) 50π (d) 75π

[ME, 2021 : 1 Mark, Set-1]

412. Five persons P, Q, R, S and T are sitting in a row not necessarily in the same order. Q and R are separated by one person, and S should not be seated adjacent to Q.

The number of distinct seating arrangements possible is

(a) 4 (b) 10
(c) 16 (d) 8

[ME, 2021 : 2 Marks, Set-1]

413. If $\begin{cases} "\oplus" \text{ means} "-", \\ "\otimes" \text{ means} "\div", \\ "\Delta" \text{ means} "+", \\ "\nabla" \text{ means} "\times", \end{cases}$

then, the value of expression $\Delta 2 \oplus 3 \Delta ((4 \otimes 2) \nabla 4) =$

[ME, 2021 : 1 Mark, Set-1]

414. Consider a square sheet of side 1 unit. The sheet is first folded along the main diagonal. This is followed by a fold along its line of symmetry. The resulting folded shape is again folded along its line of symmetry. The area of each face of the final folded shape, in square units, equal to ______.

(a) $\frac{1}{8}$ (b) $\frac{1}{32}$

(c) $\frac{1}{4}$ (d) $\frac{1}{16}$

[ME, 2021 : 2 Marks, Set-2]

415. Five persons P, Q, R, S and T are to be seated in a row, all facing the same direction, but not necessarily in the same order. P and T cannot be seated at either end of the row. P should not be seated adjacent to S. R is to be seated at the second position from the left end of the row. The number of distinct seating arrangements possible is:

(a) 2 (b) 3

(c) 4 (d) 5

[ME, 2021 : 1 Mark, Set-2]

416. The front door of Mr. X's house faces East. Mr. X leaves the house, walking 50 m straight from the back door that is situated directly opposite to the front door. He then turns to his right, walks for another 50 m and stops. The direction of the point Mr. X is now located at with respect to the starting point is ______.

(a) South-East (b) North-West

(c) North-East (d) West

[ME, 2021 : 2 Marks, Set-2]

417. A box contains 15 blue balls and 45 black balls. If 2 balls are selected randomly, without replacement, the probability of an outcome in which the first selected is a blue ball and the second selected is a black ball, is _____.

(a) $\frac{45}{236}$ (b) $\frac{1}{4}$

(c) $\frac{3}{16}$ (d) $\frac{3}{4}$

[ME, 2021 : 2 Marks, Set-2]

418. A digital watch X beeps every 30 seconds while watch Y beeps every 32 seconds. They beeped together at 10 AM.
The immediate next time that they will beep together is _____.

(a) 11.00 AM (b) 10.08 AM

(c) 10.42 AM (d) 10.00 PM

[ME, 2021 : 1 Mark, Set-2]

419. The ratio of the area of the inscribed circle to the area of circumscribed circle of an equilateral triangle is ______.

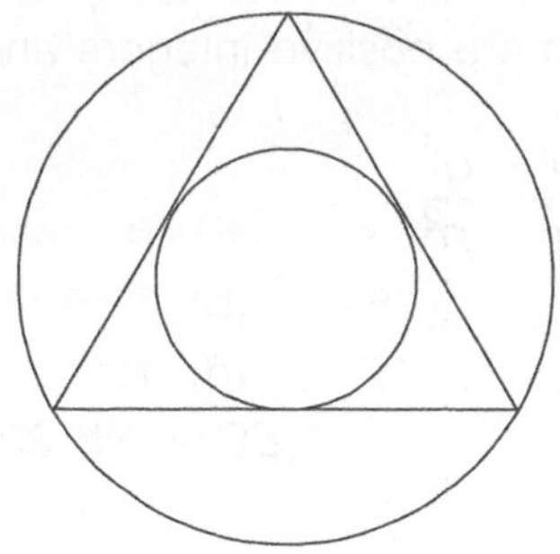

(a) $\frac{1}{4}$ (b) $\frac{1}{2}$

(c) $\frac{1}{6}$ (d) $\frac{1}{8}$

[ME, 2021 : 2 Marks, Set-2]

420. Given below are two statements 1 and 2, and two conclusions I and II.

Statement 1: All entrepreneurs are wealthy.

Statement 2: All wealthy are risk seekers.

Conclusion I: All risk seekers are wealthy.

Conclusion II: Only some entrepreneurs are risk seekers.

Based on the above statements and conclusions, which one of the following options is CORRECT?

(a) Only conclusion I is correct

(b) Only conclusion II is correct

(c) Neither conclusion I nor II is correct

(d) Both conclusions I and II are correct

[ME, 2021 : 2 Marks, Set-2]

421. If $\oplus \div \odot = 2$; $\oplus \div \Delta = 3$; $\odot + \Delta = 5$; $\Delta \times \otimes = 10$,

Then, the value of $(\otimes - \oplus)^2$, is:

(a) 4 (b) 16

(c) 1 (d) 0

[ME, 2021 : 1 Mark, Set-2]

422.

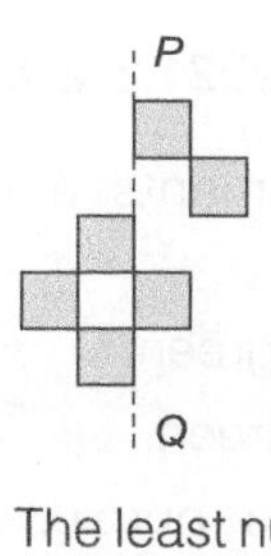

The least number of squares that must be added so that the line *P-Q* becomes the line of symmetry is _____.

(a) 3 (b) 7
(c) 4 (d) 6

[EC + PI, 2021 : 1 Mark]

423. p and q are positive integers and $\frac{p}{q}+\frac{q}{p}=3,$

then, $\frac{p^2}{q^2}+\frac{q^2}{p^2}=$

(a) 9 (b) 7
(c) 3 (d) 11

[EC + PI, 2021 : 1 Mark]

424.

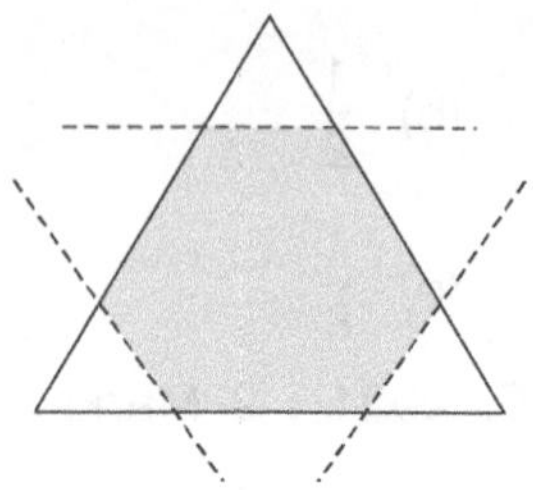

Corners are cut from an equilateral triangle to produce a regular convex hexagon as shown in the figure above.

The ratio of the area of the regular convex hexagon to the area of the original equilateral triangle is

(a) 3 : 4 (b) 4 : 5
(c) 2 : 3 (d) 5 : 6

[EC + PI, 2021 : 2 Marks]

425. Consider a square sheet of side 1 unit. In the first step, it is cut along the main diagonal to get two triangles. In the next step, one of the cut triangles is revolved about its short edge to form a solid cone. The volume of the resulting cone, in cubic units, is ____.

(a) $\frac{3\pi}{2}$ (b) 3π

(c) $\frac{\pi}{3}$ (d) $\frac{2\pi}{3}$

[EC + PI, 2021 : 2 Marks]

426. Given below are two statements and two conclusions.

Statement 1 : All purple are green.

Statement 1 : All black are green.

Conclusion I : Some black are purple.

Conclusion II : No black is purple.

Based on the above statements and conclusions, which one of the following options is logically CORRECT?

(a) Only conclusion II is correct.
(b) Only conclusion I is correct.
(c) Either conclusion I or II is correct.
(d) Both conclusion I and II are correct.

[EC + PI, 2021 : 2 Marks]

427.

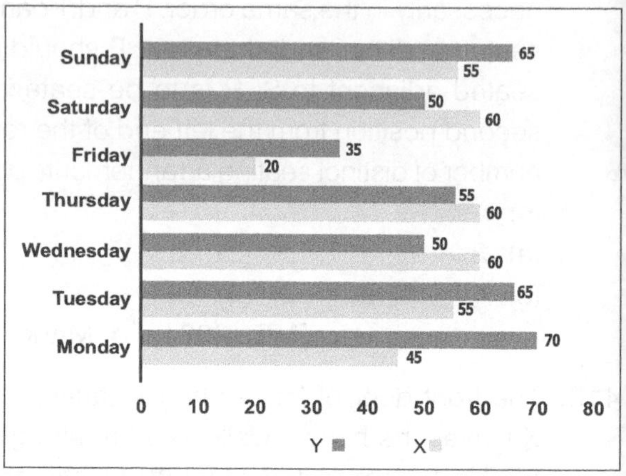

The number of minutes spent by two students, X and Y, exercising every day in a given week are shown in the bar chart above.

The number of days in the given week in which one of the students spent a minimum of 10% more than the other student, on a given day, is

(a) 7 (b) 6
(c) 5 (d) 4

[EC + PI, 2021 : 2 Marks]

428. The current population of a city is 11,02,500. If it has been increasing at the rate of 5% per annum, what was its population 2 years ago?

(a) 9,95,006 (b) 10,00,000
(c) 12,51,506 (d) 9,92,500

[EC + PI, 2021 : 1 Mark]

429. Seven cars P, Q, R, S, T, U and V are parked in a row not necessarily in that order. The cars T and U should be parked next to each other. The cars S and V also should be parked next to each other, whereas P and Q cannot be parked next to each other. Q and S must be parked next to each other. R is parked to the immediate right of V. T is parked to the left of U.

Based on the above statements, the only INCORRECT option given below is:

(a) There are two cars parked in between Q and V.
(b) Q and R are not parked together.
(c) V is the only car parked in between S and R
(d) Car P is parked at the extreme end.

[EE, 2021 : 2 Marks]

430. 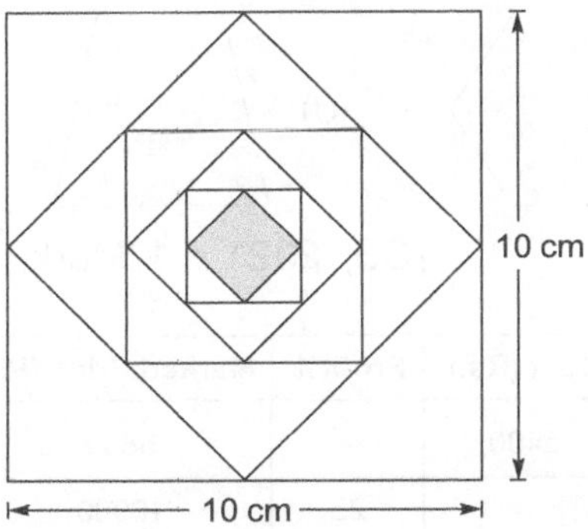

In the figure shown above, each inside square is formed by joining the midpoints of the sides of the next larger square. The area of the smallest square (Shaded) as shown, in cm^2 is

(a) 3.125 (b) 12.50
(c) 1.5625 (d) 6.25

[EE, 2021 : 2 Marks]

431. Which one of the following numbers is exactly divisible by $(11^{13} + 1)$?

(a) $11^{33} + 1$ (b) $11^{52} - 1$
(c) $11^{26} + 1$ (d) $11^{39} + 1$

[EE, 2021 : 1 Mark]

432. Let X be a continuous random variable denoting the temperature measured. The range of temperature is (0, 100) degree Celsius and let the probability density function of X be $f(x) = 0.01$ for $0 \leq X \leq 100$. The mean of X is _______ .

(a) 5.0 (b) 50.0
(c) 25.0 (d) 2.5

[EE, 2021 : 2 Marks]

433. For a regular polygon having 10 sides, the interior angle between the sides of the polygon, in degrees, is :

(a) 396 (b) 144
(c) 324 (d) 216

[EE, 2021 : 1 Mark]

434. 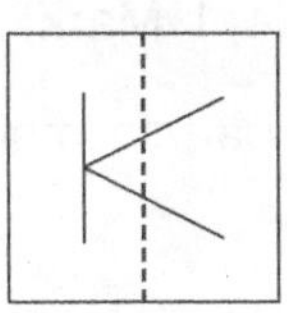

A transparent square sheet above is folded along the dotted line. The folded sheet will look like _______ .

(a) 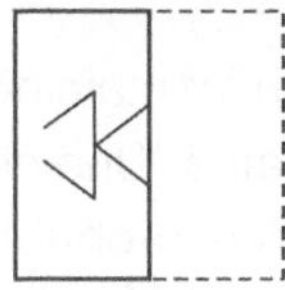 (b) 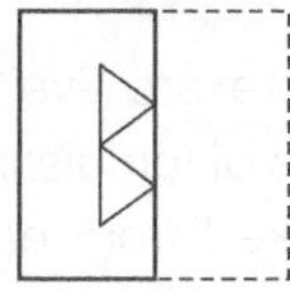

(c) 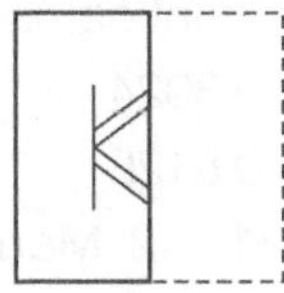 (d) 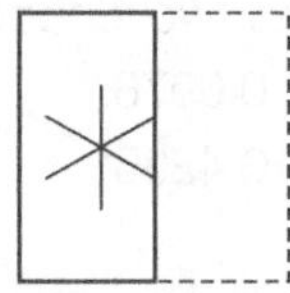

[EE, 2021 : 1 Mark]

435. 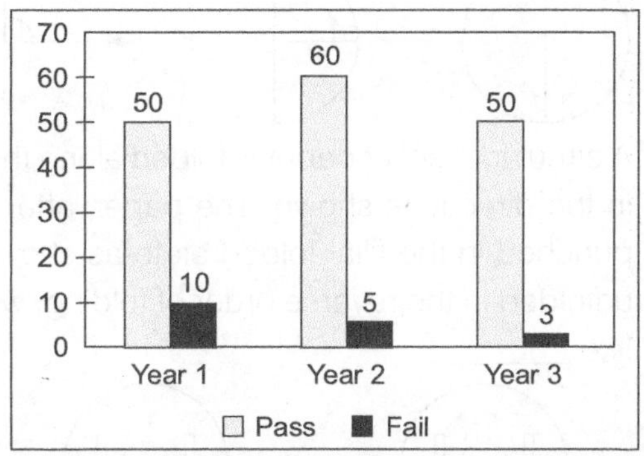

The number of students passing or failing in an exam for a particular subject is presented in the bar chart above. Students who pass the exam cannot appear for the exam again. Students who fail the exam in the first attempt must appear for the exam in the following year. Students always pass the exam in their second attempt.

The number of students who took the exam for the first time in the year 2 and the year 3 respectively, are _______ .

(a) 65 and 53 (b) 55 and 53
(c) 60 and 50 (d) 55 and 48

[EE, 2021 : 2 Marks]

436. Given below are two statements I and II and two conclusions I and II:

Statement:

I. All bacteria are microorganisms.
II. All pathogens are microorganisms.

Conclusions:

I. Some pathogens are bacteria.
II. All pathogens are not bacteria.

Based on the above statements and conclusions, which one of the following options is logically CORRECT?

(a) Only conclusion I is correct
(b) Either conclusion I or II is correct.

(c) Only conclusion II is correct
(d) Neither conclusion I nor II is correct.

[CS, 2021 : 2 Marks, Set-1]

437. There are five bags each containing identical sets of ten distinct chocolates. One chocolate is picked from each bag. The probability that at least two chocolates are identical is ________.

(a) 0.6976 (b) 0.3024
(c) 0.4235 (d) 0.8125

[CS, 2021 : 2 Marks, Set-1]

438. 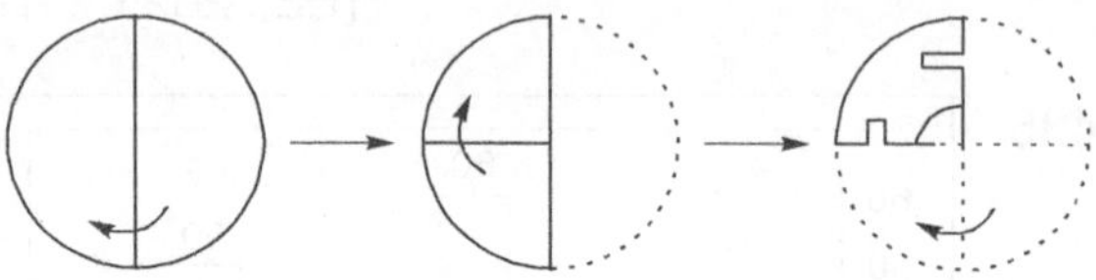

A circular sheet of paper is folded along the lines in the directions shown. The paper, after being punched in the filial folded state as shown and unfolded in the reverse order of folding, will look like ________.

(a) 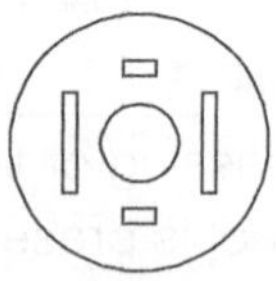 (b) 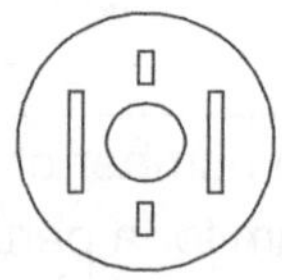

(c) 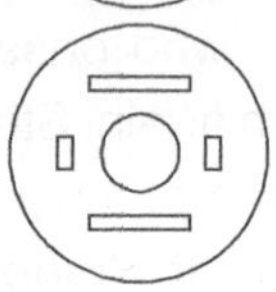 (d) 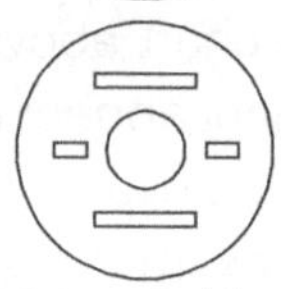

[CS, 2021 : 1 Mark, Set-1]

439. We have 2 rectangular sheets of paper, M and N, of dimensions 6 cm × 1 cm each. Sheet M is rolled to form an open cylinder by bringing the short edges of the sheet together. Sheet N is cut into equal square patches and assembled to form the largest possible closed cube. Assuming the ends of the cylinder are closed, the ratio of the volume of the cylinder to that of the cube is ________.

(a) $\frac{9}{\pi}$ (b) 3π
(c) $\frac{\pi}{2}$ (d) $\frac{3}{\pi}$

[CS, 2021 : 2 Marks, Set-1]

440. A polygon is convex if, for every pair of points, P and Q belonging to the polygon, the line segment PQ lies completely inside or on the polygon.

Which one of the following is NOT a convex polygon?

(a) 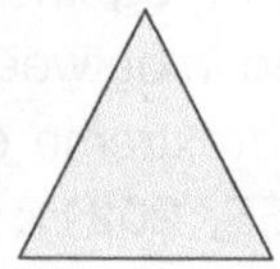 (b) 

(c) 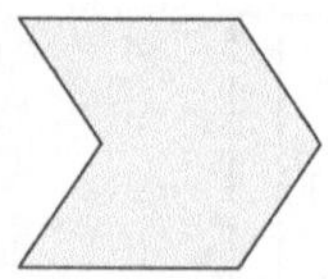 (d) 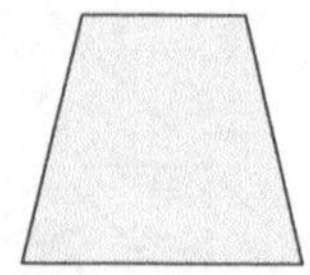

[CS, 2021 : 1 Mark, Set-1]

441.

Items	Cost (Rs.)	Profit%	Marked Price (Rs.)
P	5400	–	5860
Q	–	25	10000

Details of prices of two items P and Q are presented in the above table. The ratio of cost of item P to cost of item Q is 3 : 4. Discount is calculated as the difference between the marked price and the selling price. The profit percentage is calculated as the ratio of the difference between selling price and cost, to the cost

$$\left(\text{Profit\%} = \frac{\text{Selling price} - \text{Cost}}{\text{Cost}} \times 100\right).$$

The discount on item Q, as a percentage of its marked price, is ________.

(a) 25 (b) 10
(c) 5 (d) 12.5

[CS, 2021 : 2 Marks, Set-1]

442. The ratio of boys to girls in a class is 7 to 3. Among the options below, an acceptable value for the total number of students in the class is:

(a) 37 (b) 50
(c) 21 (d) 73

[CS, 2021 : 1 Mark, Set-1]

443. If $\left(x-\frac{1}{2}\right)^2 - \left(x-\frac{3}{2}\right)^2 = x+2$, then the value of x is :

(a) 2 (b) 8
(c) 4 (d) 6

[CS, 2021 : 1 Mark, Set-2]

444. What will be the remaining part to make it a rectangle?

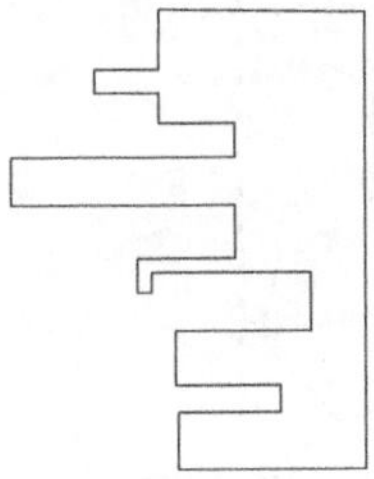

(a) 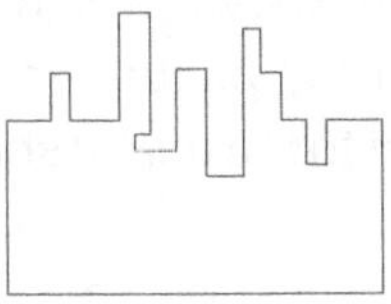 (b) 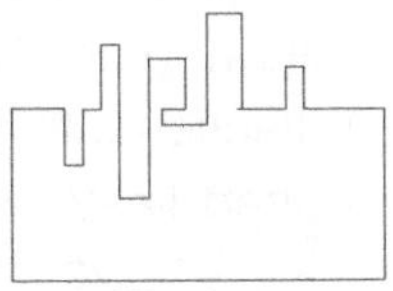

(c) 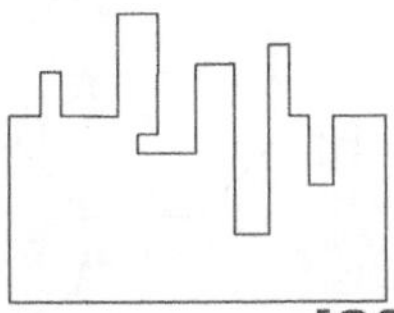 (d) 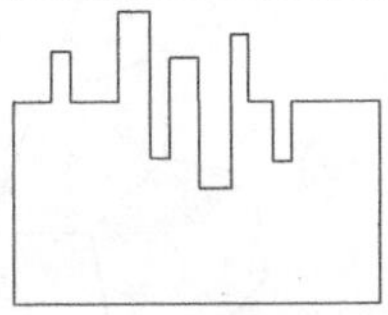

[CS, 2021 : 2 Marks, Set-2]

445. If θ is the angle, in degrees, between the longest diagonal of the cube and any one of the edges of the cube, then cos θ = ____.

(a) $\frac{1}{\sqrt{3}}$ (b) $\frac{1}{2}$

(c) $\frac{1}{\sqrt{2}}$ (d) $\frac{\sqrt{3}}{2}$

[CS, 2021 : 1 Mark, Set-2]

446. Six students P, Q, R, S, T and U, with distinct heights, compare their heights and make the following observations :

Observation I : S is taller than R.

Observation II : Q is the shortest of all.

Observation III : U is taller than only one student.

Observation IV : T is taller than S but is not the tallest.

The number of students that are taller than R is the same as the number of students shorter than ___.

(a) R (b) P

(c) S (d) T

[CS, 2021 : 2 Marks, Set-2]

447. The number of student in three classes is in the ratio 3 : 13 : 6. If 18 students are added to each class, the ratio changes to 15 : 35 : 21.

The total number of students in all the three classes in the beginning was:

(a) 88 (b) 110

(c) 66 (d) 22

[CS, 2021 : 2 Marks, Set-2]

448. The number of units of a product sold in three different years and the respective net profits are presented in the figure below. The cost/unit in year 3 was Re. 1, which was half the cost/unit in year 2. The cost/unit in year 3 was one-third of the cost/unit in year 1. Taxes were paid on the selling price at 10%, 13% and 15% respectively for the three years. Net profit is calculated as the difference between the selling price and the sum of cost and taxes paid in that year.

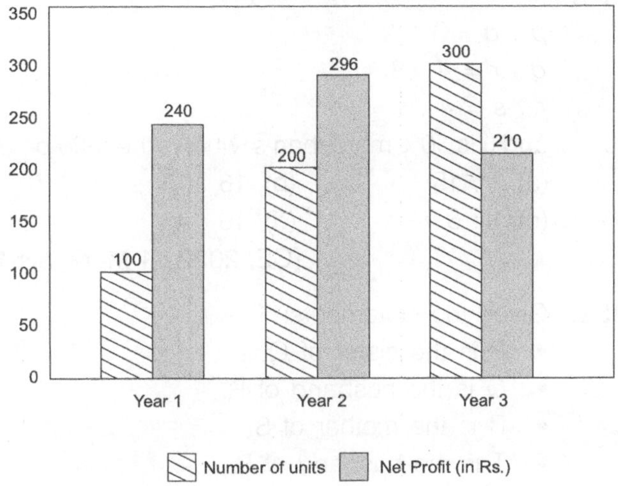

The ratio of the selling price in Year 2 to the selling price in Year 3 is ________.

(a) 3 : 4 (b) 1 : 2

(c) 1 : 1 (d) 4 : 3

[CS, 2021 : 2 Marks, Set-2]

449. A transparent square sheet shown above is folded along the dotted line. The folded sheet will look like _____.

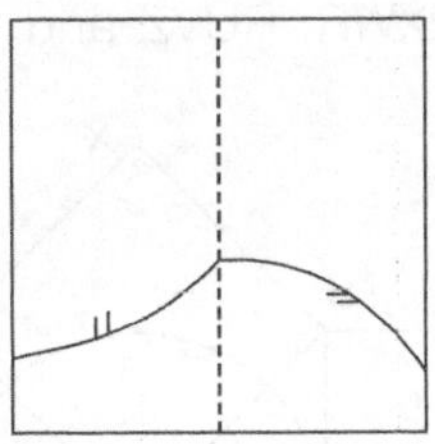

(a) 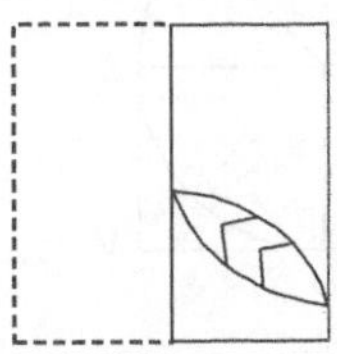 (b) 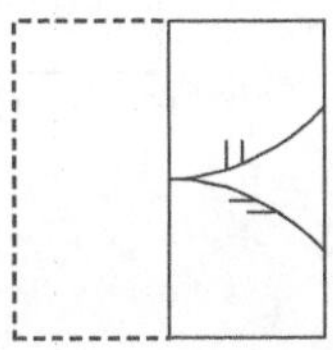

(c) 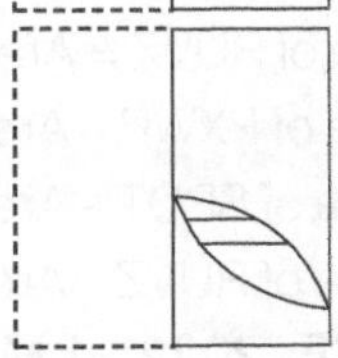 (d) 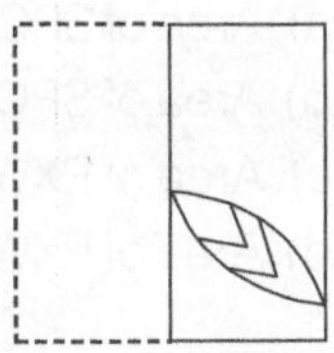

[CS, 2021 : 1 Mark, Set-2]

450. Two straight lines pass through the origin $(x_0, y_0) = (0, 0)$. One of them passes through the point $(x_1, y_1) = (1, 3)$ and the other passes through the point $(x_2, y_2) = (1, 2)$.
What is the area enclosed between the straight lines in the interval [0, 1] on the x-axis?
(a) 0.5 (b) 1.5
(c) 1.5 (d) 2.0
[CE, 2022 : 1 Mark, Set-I]

451. If
$p : q = 1 : 2$
$q : r = 4 : 3$
$r : s = 4 : 5$
and u is 50% more than s, what is the ratio $p : u$?
(a) 2 : 15 (b) 16 : 15
(c) 1 : 5 (d) 16 : 45
[CE, 2022 : 1 Mark, Set-I]

452. Given the statements:
- P is the sister of Q.
- Q is the husband of R.
- R is the mother of S.
- T is the husband of P.

Based on the above information, T is ______ of S.
(a) the grandfather (b) an uncle
(c) the father (d) a brother
[2022: 1 Mark, Set-I]

453. In the following diagram, the point R is the center of the circle. The lines PQ and ZV are tangential to the circle. The relation among the areas of the squares, PXWR, RUVZ and SPQT is

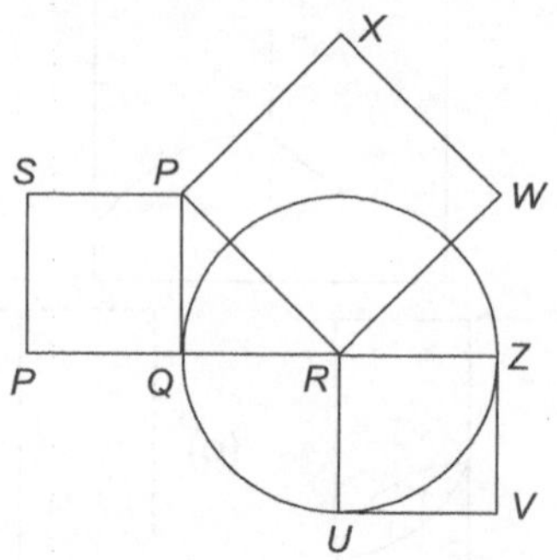

(a) Area of SPQT = Area of RUVZ = Area of PXWR
(b) Area of SPQT = Area of PXWR – Area of RUVZ
(c) Area of PXWR = Area of SPQT – Area of RUVZ
(d) Area of PXWR = Area of RUVZ – Area of SPQT
[CE, 2022 : 1 Mark, Set-I]

454.

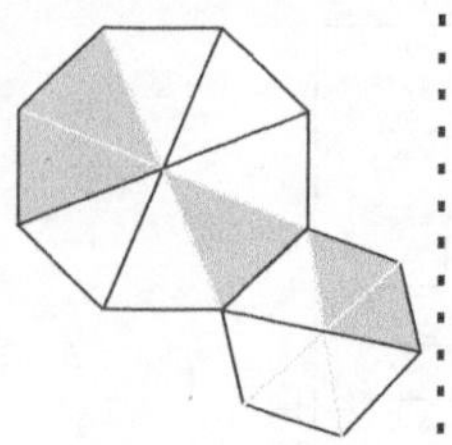

For the picture shown above, which one of the following is the correct picture representing reflection with respect to the mirror shown as the dotted line?

(a) 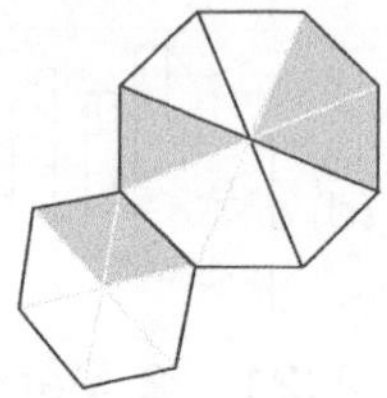 (b) 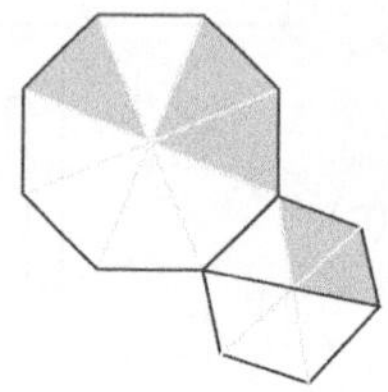

(c) 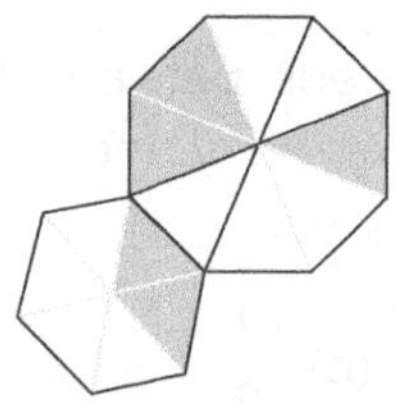 (d) 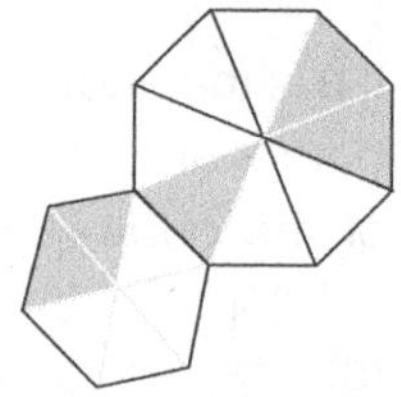

[CE, 2022 : 1 Mark, Set-II]

455. An ant walks in a straight line on a plane leaving behind a trace of its movement. The initial position of the ant is at point P facing East.
The ant first turns 72° anticlockwise at P, and then does the following two steps in sequence exactly FIVE times before halting.

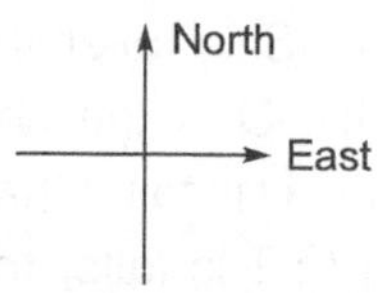

1. moves forward for 10 cm.
2. turns 144° clockwise

The pattern made by the trace left behind by the ant is

(a) 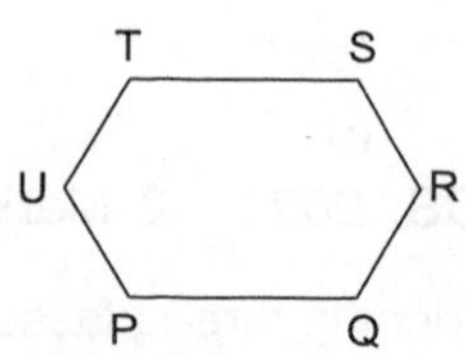

PQ = QR = RS = ST = TP = 10 cm

(b) 

SQ = QT = TR = RP = PS = 10 cm

(c) 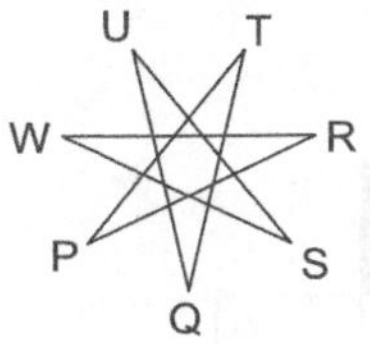

SW = WR = RP = PT = TQ = QU = US = 10 cm

(d) 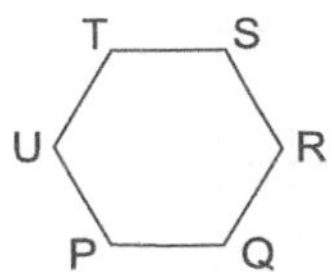

PQ = QR = RS = ST = TU = UP = 10 cm

[CE, 2022 : 1 Mark, Set-II]

456. In a partnership business the monthly investment by three friends for the first six months is m the ratio 3 :4 : 5. After six months, they had to increase their monthly investments by 10%, 15% and 20%, respectively, of then initial monthly investment. The new investment ratio was kept constant for the next six months.

What is the ratio of their shares in the total profit (m the same order) at the end of the year such that the share is proportional to their individual total investment over the year?

(a) 63 : 86 : 110 (b) 33 : 46 : 60
(c) 22 : 23 : 24 (d) 22 : 33 : 50

[CE, 2022 : 1 Mark, Set-II]

457. A survey of 450 students about their subjects of interest resulted in the following outcome.

- 150 students are interested m Mathematics.
- 200 students are interested in Physics.
- 175 students are interested in Chemistry.
- 50 students are interested in Mathematics and Physics.
- 60 students are interested in Physics and Chemistry
- 40 students are interested in Mathematics and Chemistry.
- 30 students are interested in Mathematics, Physics and Chemistry.
- Remaining students are interested in Humanities.

Based on the above information, the number of students interested in Humanities is

(a) 45 (b) 10
(c) 40 (d) 30

[CE, 2022 : 1 Mark, Set-II]

458. P invested ₹ 5000 per month for 6 months of a year and Q invested ₹ x per month for 8 months of the year in a partnership business. The profit is shared in proportion to the total investment made in that year.

If at the end of that investment year, Q receives $\frac{4}{9}$ of the total profit, what is the value of (in ₹)?

(a) 2500 (b) 3000
(c) 4687 (d) 8437

[CE, 2022 : 2 Marks, Set-I]

459.

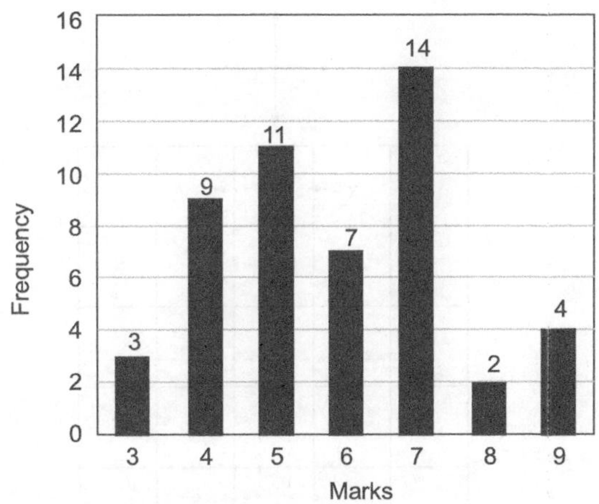

The above frequency chart shows the frequency distribution of marks obtained by a set of students in an exam.

From the data presented above, which one of the following is CORRECT?

(a) mean > mode > median
(b) mode > median > mean
(c) mode > mean > median
(d) median > mode > mean

[CE, 2022 : 2 Marks, Set-I]

460. In the square grid shown on the left, a person standing at P2 position is required to move to P5 position.

The only movement allowed for a step involves, "two moves along one direction followed by one move in a perpendicular direction". The permissible directions for movement are shown as dotted arrows in the right.

For example, a person at a given position Y can move only to the positions marked X on the right.

Without occupying any of the shaded squares at the end of each step, the minimum number of steps required to go from P2 to P5 is

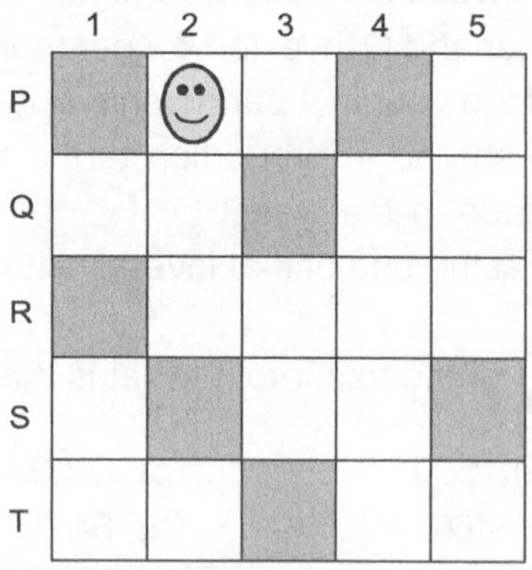

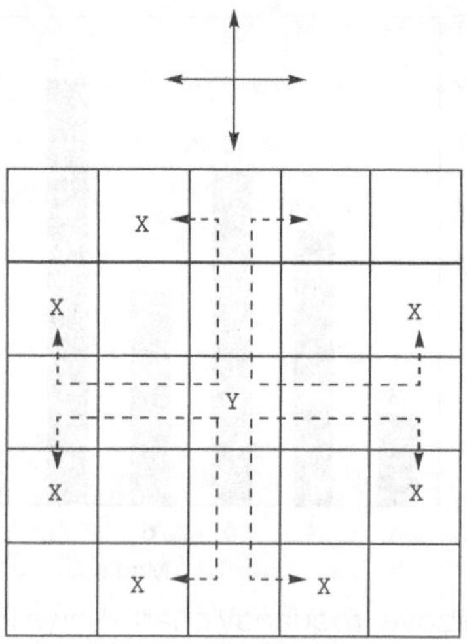

(a) 4 (b) 5
(c) 6 (d) 7

[CE, 2022 : 2 Marks, Set-I]

461.

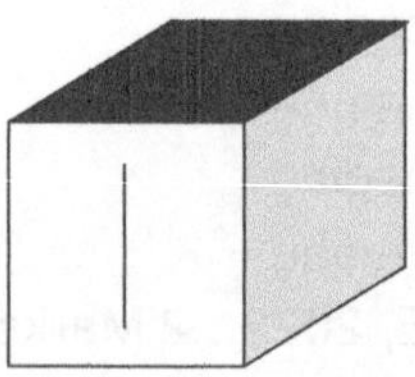

Consider a cube made by folding a single sheet of paper of appropriate shape.

The interior faces of the cube are all blank. However, the exterior faces that are not visible in the above view may not be blank.

Which one of the following represents a possible unfolding of the cube?

(a)

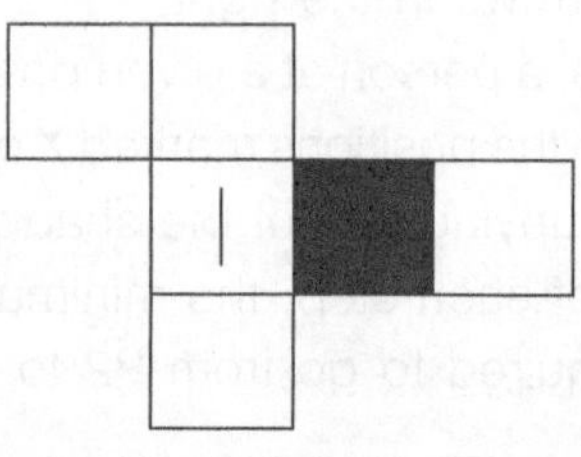

(b)

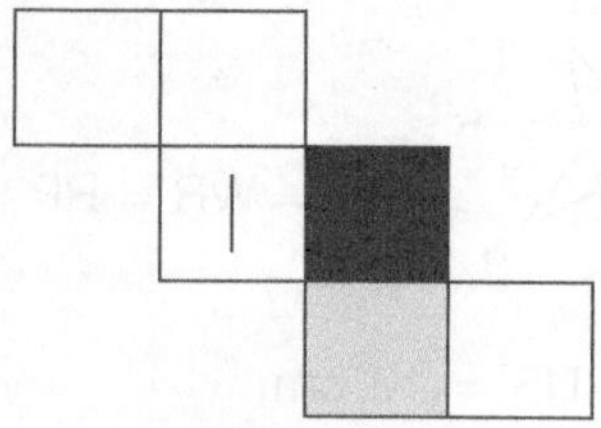

(c)

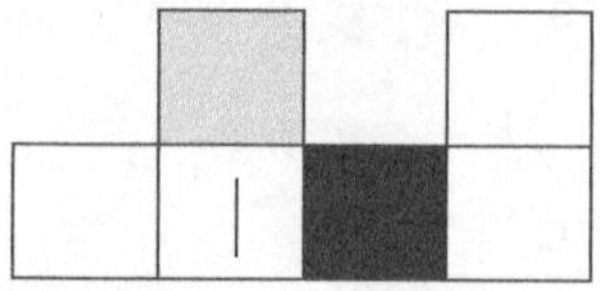

(d)

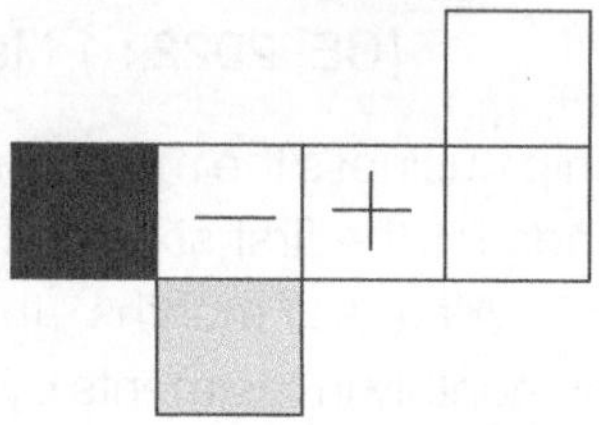

[CE, 2022 : 2 Marks, Set-I]

462. $x : y : z = \frac{1}{2} : \frac{1}{3} : \frac{1}{4}$

What is the value of $\frac{x+z-y}{y}$ = ?

(a) 1.25 (b) 0.75
(c) 3.25 (d) 2.25

[CE, 2022 : 2 Marks, Set-II]

463. Both the numerator and the denominator of $\frac{3}{4}$ are increased by a positive $\frac{15}{17}$ integer, x, and those of — are decreased by the same integer. This operation results m the same value for both the fractions. What is the value of x?

(a) 3 (b) 4
(c) 1 (d) 2

[CE, 2022 : 2 Marks, Set-II]

464. Consider the following equations of straight lines:

Line L1 : $2x - 3y = 5$
Line L2 : $3x + 2y = 8$
Line L3 : $4x - 6y = 5$
Line L4 : $6x - 9y = 6$

Which one of the following is the correct statement?

(a) L1 is parallel to L2 and L1 is perpendicular to L3
(b) L2 is parallel to L4 and L2 is perpendicular to L1

(c) L4 is perpendicular to L2 and L4 is parallel to L3

(d) L3 is perpendicular to L4 and L3 is parallel to L2

[CE, 2022 : 2 Marks, Set-II]

465. Given below are two statements and four conclusions drawn based on the statements.

Statement 1: Some soaps are clean.

Statement 2: All clean objects are wet.

Conclusion I: Some clean objects are soaps.

Conclusion II: No clean object is a soap.

Conclusion III: Some wet objects are soaps.

Conclusion IV: All wet objects are soaps.

Which one of the following options can be logically inferred?

(a) Either conclusion I or conclusion II is correct

(b) Either conclusion III or conclusion IV is correct

(c) Only conclusion I and conclusion III are correct

(d) Only conclusion I is correct

[CE, 2022 : 2 Marks, Set-II]

466. A function $y(x)$ is defined in the interval [0, 1] on the x-axis as

$$y(x)=\begin{cases} 2 \text{ if } 0 \le x < \dfrac{1}{3} \\ 3 \text{ if } \dfrac{1}{3} \le x < \dfrac{3}{4} \\ 1 \text{ if } \dfrac{3}{4} \le x < 1 \end{cases}$$

Which one of the following is the area under the curve for the interval [0, 1] on the x-axis?

(a) $\frac{5}{6}$ (b) $\frac{6}{5}$

(c) $\frac{13}{6}$ (d) $\frac{6}{13}$

[CS, 2022 : 1 Mark]

467. Let r be a root of the equation $x^2 + 2x + 6 = 0$. Then the value of the expression $(r + 2)(r + 3)(r + 4)(r + 5)$ is

(a) 51 (b) –51

(c) 126 (d) –126

[CS, 2022 : 1 Mark]

468. Given below are four statements:

Statement 1: All students are inquisitive.

Statement 2: Some students are inquisitive.

Statement 3: No student is inquisitive.

Statement 4: Some students are not inquisitive.

From the given four statements, find the two statements that **CANNOT BE TRUE** simultaneously, assuming that there is at least one student in the class.

(a) Statement 1 and Statement 3

(b) Statement 1 and Statement 2

(c) Statement 2 and Statement 4

(d) Statement 3 and Statement 4

[CS, 2022 : 1 Mark]

469. A palindrome is a word that reads the same forwards and backwards. In a game of words, a player has the following two plates painted with letters.

| A | | D |

From the additional plates given in the options, which one of the combinations of additional plates would allow the player to construct a five-letter palindrome. The player should use all the five plates exactly once. The plates can be rotated in their plane.

(a) | D | | ᗡ | | J |

(b) | R | | ∀ | | ᴚ |

(c) | Z | | Ǝ | | D |

(d) | I | | ˥ | | Y |

[CS, 2022 : 1 Mark]

470. In a recently conducted national entrance test, boys constituted 65% of those who appeared for the test. Girls constituted the remaining candidates and they accounted for 60% of the qualified candidates.

Which one of the following is the correct logical inference based on the information provided in the above passage?

(a) Equal number of boys and girls qualified

(b) Equal number of boys and girls appeared for the test

(c) The number of boys who appeared for the test is less than the number of girls who appeared

(d) The number of boys who qualified the test is less than the number of girls who qualified

[CS, 2022 : 2 Marks]

471. A box contains five balls of same size and shape. Three of them are green coloured balls and two of them are orange coloured balls. Balls are drawn from the box one at a time. If a green ball is drawn, it is not replaced. If an orange ball is drawn, it is replaced with another orange ball. First ball is drawn. What is the probability of getting an orange ball in the next draw?

(a) $\frac{1}{2}$ (b) $\frac{8}{25}$

(c) $\frac{19}{50}$ (d) $\frac{23}{50}$

[CS, 2022 : 2 Marks]

472. The corners and mid-points of the sides of a triangle are named using the distinct letters P, Q, R, S, T and U, but not necessarily in the same order. Consider the following statements:

- The line joining P and R is parallel to the line joining Q and S.
- P is placed on the side opposite to the corner T.
- S and U cannot be placed on the same side.

Which one of the following statements is correct based on the above information?

(a) P cannot be placed at a corner
(b) S cannot be placed at a corner
(c) U cannot be placed at a mid-point
(d) R cannot be placed at a corner

[CS, 2022 : 2 Marks]

473. A plot of land must be divided between four families. They want their individual plots to be similar in shape, not necessarily equal in area. The land has equally spaced poles, marked as dots in the below figure. Two ropes, R1 and R2, are already present and cannot be moved. What is the least number of additional straight ropes needed to create the desired plots? A single rope can pass through three poles that are aligned in a straight line.

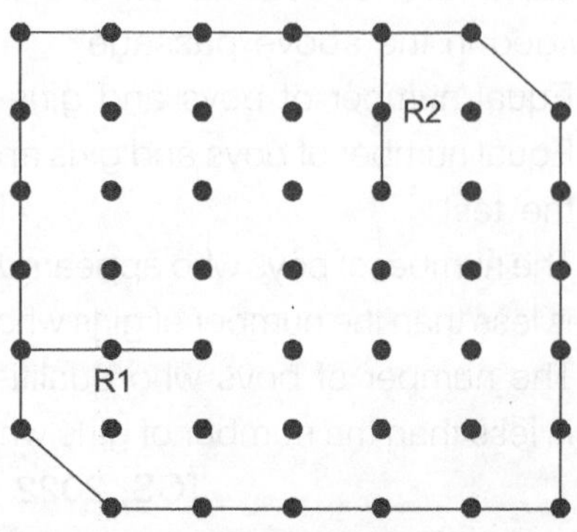

(a) 2 (b) 4
(c) 5 (d) 3

[CS, 2022 : 2 Marks]

474. A sum of money is to be distributed among *P*, *Q*, *R*, and *S* in the proportion 5 : 2 : 4 : 3, respectively.

If *R* gets Rs. 1000 more than S, what is the share of Q (in Rs.)?

(a) 500 (b) 1000
(c) 1500 (d) 2000

[EC, 2022 : 1 Mark]

475. A trapezium has vertices marked as *P*, *Q*, *R* and *S* (in that order anticlockwise). The side *PQ* is parallel to side *SR*.

Further, it is given that, *PQ* = 11 cm, *QR* = 4 cm, *RS* = 6 cm and *SP* = 3 cm.

What is the shortest distance between PQ and SR (in cm)?

(a) 1.80 (b) 2.40
(c) 4.20 (d) 5.76

[EC, 2022 : 1 Mark]

476. The figure shows a grid formed by a collection of unit squares. The unshaded unit square in the grid represents a hole.

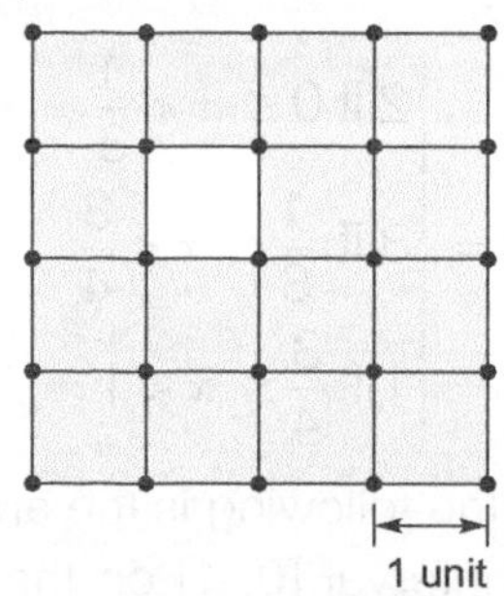

What is the maximum number of squares without a "hole in the interior" that can be formed within the 4 × 4 grid using the unit squares as building blocks?

(a) 15 (b) 20
(c) 21 (d) 26

[EC, 2022 : 1 Mark]

477. Consider the following inequalities.

(i) $2x - 1 > 7$
(ii) $2x - 9 < 1$

Which one of the following expressions below satisfies the above two inequalities?

(a) $x \leq -4$ (b) $-4 < x \leq 4$
(c) $4 < x < 5$ (d) $x \geq 5$

[EC, 2022 : 2 Marks]

478. Four points $P(0, 1)$, $Q(0, -3)$, $R(-2, -1)$, and $S(2, -1)$ represent the vertices of a quadrilateral.

What is the area enclosed by the quadrilateral?

(a) 4 (b) $4\sqrt{2}$

(c) 8 (d) $8\sqrt{2}$

[EC, 2022 : 2 Marks]

479. Consider the following square with the four corners and the center marked as P, Q, R, S and T respectively.

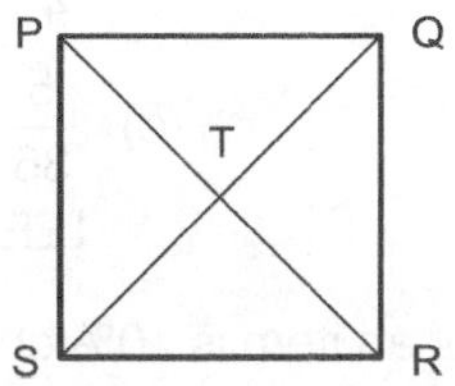

Let X, Y and Z represent the following operations:

X: rotation of the square by 180 degree with respect to the S-Q axis.

Y: rotation of the square by 180 degree with respect to the P-R axis.

Z: rotation of the square by 90 degree clockwise with respect to the axis perpendicular, going into the screen and passing through the point T.

Consider the following three distinct sequences of operation (which are applied in the left to right order).

(1) XYZZ

(2) XY

(3) ZZZZ

Which one of the following statements is correct as per the information provided above?

(a) The sequence of operations (1) and (2) are equivalent

(b) The sequence of operations (1) and (3) are equivalent

(c) The sequence of operations (2) and (3) are equivalent

(d) The sequence of operations (1), (2) and (3) are equivalent

[EC, 2022 : 2 Marks]

480. An art gallery engages a security guard to ensure that the items displayed are protected. The diagram below represents the plan of the gallery where the boundary walls are opaque. The location the security guard posted is identified such that all the inner space (shaded region in the plan) of the gallery is within the line of sight of the security guard. If the security guard does not move around the posted location and has a 360° view, which one of the following correctly represents the set of ALL possible locations among the locations P, Q, R and S, where the security guard can be posted to watch over the entire inner space of the gallery.

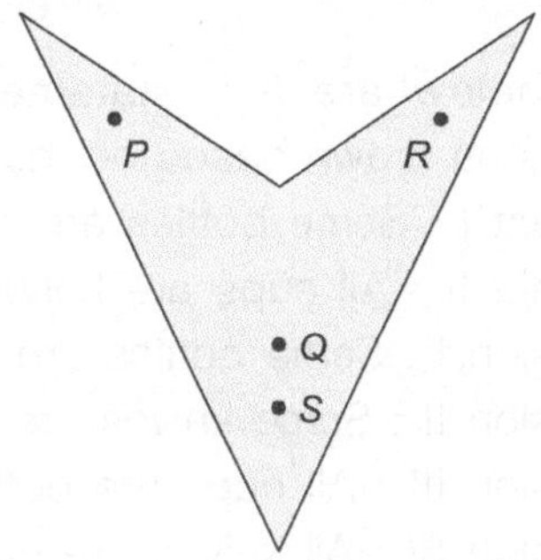

(a) *P* and *Q* (b) *Q*

(c) *Q* and *S* (d) *R* and *S*

[EC, 2022 : 1 Mark]

481. In a class of five students P, Q, R, S and T, only one student is known to have copied in the exam. The disciplinary committee has investigated the situation and recorded the statements from the students as given below.

Statement of P: R has copied in the exam.

Statement of Q: S has copied in the exam.

Statement of R: P did not copy in the exam.

Statement of S: Only one of us is telling the truth.

Statement of T: R is telling the truth.

The investigating team had authentic information that S never lies.

Based on the information given above, the person who has copied in the exam is

(a) R (b) P

(c) Q (d) T

[EC, 2022 : 2 Marks]

482. In a 500 m race, P and Q have speeds in the ratio of 3 : 4. Q starts the race when P has already covered 140 m.

What is the distance between P and Q (in m) when P wins the race?

(a) 20 (b) 40

(c) 60 (d) 140

[EE, 2022 : 1 Mark]

483. Three bells P, Q and R are rung periodically in a school. P is rung every 20 minutes; Q is rung every 30 minutes and R is rung every 50 minutes. If all the three bells are rung at 12:00 PM, when will the three bells ring together again the next time?

(a) 5:00 PM (b) 5:30 PM
(c) 6:00 PM (d) 6:30 PM

[EE, 2022 : 1 Mark]

484. Given below are two statements and four conclusions drawn based on the statements.

Statement I : Some bottles are cups.
Statement II : All cups are knives.
Conclusion I : Some bottles are knives.
Conclusion II : Some knives are cups.
Conclusion III : All cups are bottles.
Conclusion IV : All knives are cups.

Which one of the following options can be logically inferred?

(a) Only conclusion I and conclusion II are correct.
(b) Only conclusion II and conclusion III are correct.
(c) Only conclusion II and conclusion IV are correct.
(d) Only conclusion III and conclusion IV are correct.

[EE, 2022 : 1 Mark]

485. The figure below shows the front and rear view of a disc, which is shaded with identical patterns. The disc is flipped once with respect to any one of the fixed axes 1–1, 2–2 or 3–3 chosen uniformly at random.

What is the probability that the disc DOES NOT retain the same front and rear views after the flipping operation?

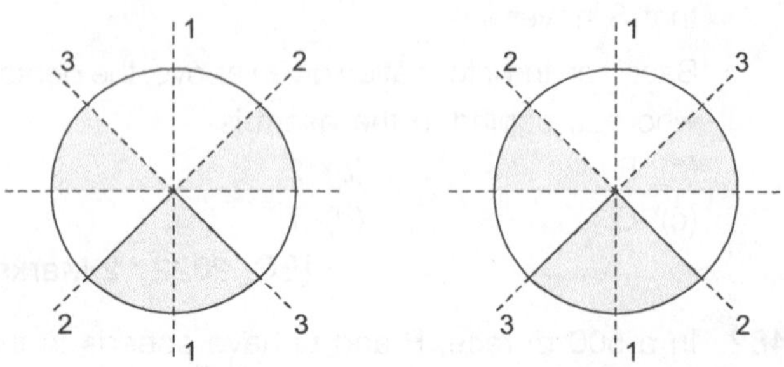

Front View Rear View

(a) 0 (b) $\frac{1}{3}$
(c) $\frac{2}{3}$ (d) 1

[EE, 2022 : 1 Mark]

486. There are two identical dice with a single letter on each of the faces. The following six letters : Q, R, S, T, U and V, one on each of the faces. Any of the six outcomes are equally likely.

The two dice are thrown once independently at random.

What is the probability that the outcomes on the dice were composed only of any combination of the following possible outcomes : Q, U and V?

(a) $\frac{1}{4}$ (b) $\frac{3}{4}$
(c) $\frac{1}{6}$ (d) $\frac{5}{36}$

[EE, 2022 : 2 Marks]

487. The price of an item is 10% cheaper in an online store S compared to the price at another online store M. Store S charges Rs. 150 for delivery. There are no delivery charges for orders from the store M. A person bought the item from the store S and saved Rs. 100.

What is the price of the item at the online store S (in Rs.) if there are no other charges than what is described above?

(a) 2500 (b) 2250
(c) 1750 (d) 1500

[EE, 2022 : 2 Marks]

488. The letters P, Q, R, S, T and U are to be placed one per vertex on a regular convex hexagon, but not necessarily in the same order.

Consider the following statements :

- The line segment joint R and S is longer than the line segment joining P and Q.
- The line segment joining R and S is perpendicular to the line segment joining P and Q.
- The line segment joining R and U is parallel to the line segment joining T and Q.

Based on the above statements, which one of the following options is CORRECT?

(a) The line segment joining R and T is parallel to the line segment joining Q and S.
(b) The line segment joining T and Q is parallel to the line joining P and U.
(c) The line segment joining R and P is perpendicular to the line segment joining U and Q.
(d) The line segment joining Q and S is perpendicular to the line segment joining R and P.

[EE, 2022 : 2 Marks]

489. An ant is at the bottom-left corner of a grid (point P) as shown below in the figure. It aims to move to the top-right corner of the grid. The ant moves only along the lines marked in the grid such that the current distance to the top-right corner strictly decreases.
Which one of the following is a part of a possible trajectory of the ant during the movement?

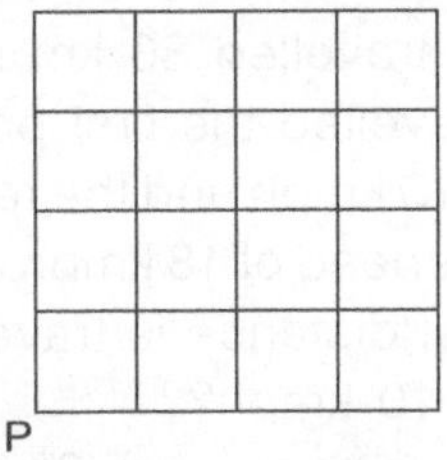

(a) 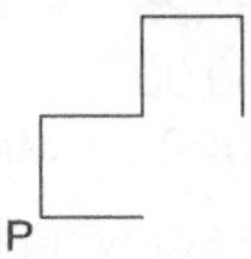
(b) 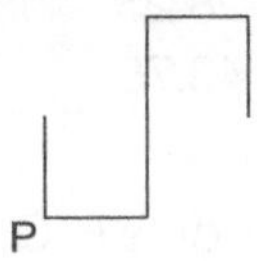
(c) 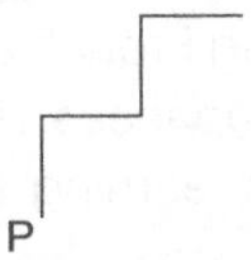
(d) 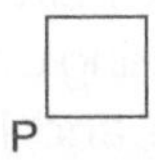

[EE, 2022 : 2 Marks]

490. A sphere of radius r cm is packed in a box of cubical shape.

What should be the minimum volume (in cm^3) of the box that can enclose the sphere?

(a) $\frac{r^3}{8}$ (b) r^3
(c) $2r^3$ (d) $8r^3$

[IN + PI + CH, 2022 : 1 Mark]

491. Pipes P and Q can fill a storage tank in full with water in 10 and 6 minutes, respectively. Pipe R draws the water out from the storage tank at a rate of 34 litres per minute. P, Q and R operate at a constant rate.
If it takes one hour to completely empty a full storage tank with all the pipes operating simultaneously, what is the capacity of the storage tank (in litres)?

(a) 26.8 (b) 60.0
(c) 120.0 (d) 127.5

[IN + PI + CH, 2022 : 1 Mark]

492. Six persons P, Q, R, S, T and U are sitting around a circular table facing the center not necessarily in the same order. Consider the following statements:

- P sits next to S and T.
- Q sits diametrically opposite to P.
- The shortest distance between S and R is equal to the shortest distance between T and U.

Based on the above statements, Q is a neighbor of

(a) U and S (b) R and T
(c) R and U (d) P and S

[IN + PI + CH, 2022 : 1 Mark]

493. A building has several rooms and doors as shown in the top view of the building given below. The doors are closed initially.
What is the minimum number of doors that need to be opened in order to go from the point P to the point Q?

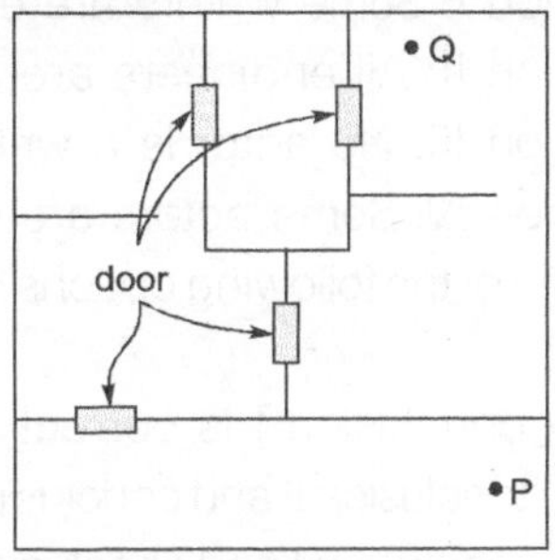

(a) 4 (b) 3
(c) 2 (d) 1

[IN + PI + CH, 2022 : 1 Mark]

494. A game consists of spinning an arrow around a stationary disk as shown below. When the arrow comes to rest, there are eight equally likely outcomes. It could come to rest in any one of the sectors numbered 1, 2, 3, 4, 5, 6, 7 or 8 as shown. Two such disks are used in a game where their arrows are independently spun.
What is the probability that the sum of the numbers on the resulting sectors upon spinning the two disks is equal to 8 after the arrows come to rest?

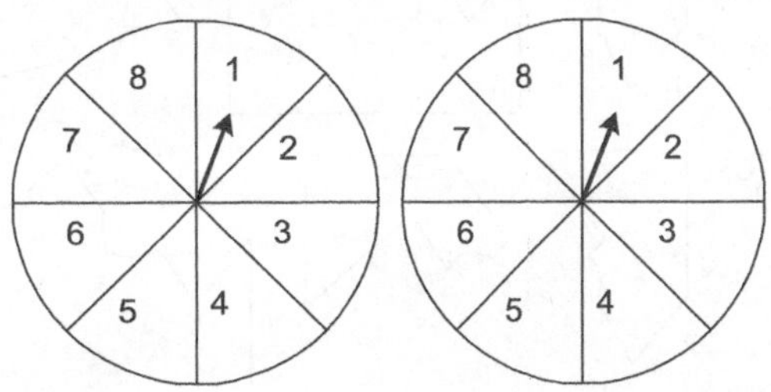

(a) $\frac{1}{16}$ (b) $\frac{5}{64}$
(c) $\frac{3}{32}$ (d) $\frac{7}{64}$

[IN + PI + CH, 2022 : 2 Marks]

495. Consider the following inequalities.
(i) $3p - q < 4$
(ii) $3q - p < 12$
Which one of the following expressions below satisfies the above two inequalities?
(a) $p + q < 8$ (b) $p + q = 8$
(c) $8 \le p + q < 16$ (d) $p + q \ge 16$
[IN + PI + CH, 2022 : 2 Marks]

496. Given below are three statements and four conclusions drawn based on the statements.
Statement 1: Some engineers are writers.
Statement 2: No writer is an actor.
Statement 3: All actors are engineers.
Conclusion I: Some writers are engineers.
Conclusion II: All engineers are actors.
Conclusion III: No actor is a writer.
Conclusion IV: Some actors are writers.
Which one of the following options can be logically inferred?
(a) Only conclusion I is correct
(b) Only conclusion II and conclusion III are correct
(c) Only conclusion I and conclusion III are correct
(d) Either conclusion III or conclusion IV is correct
[IN + PI + CH, 2022 : 2 Marks]

497. Which one of the following sets of pieces can be assembled to form a square with a single round hole near the center? Pieces cannot overlap.

(a) 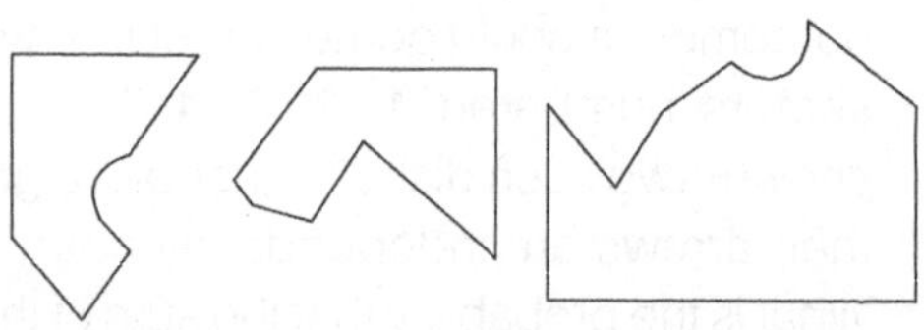

(b) 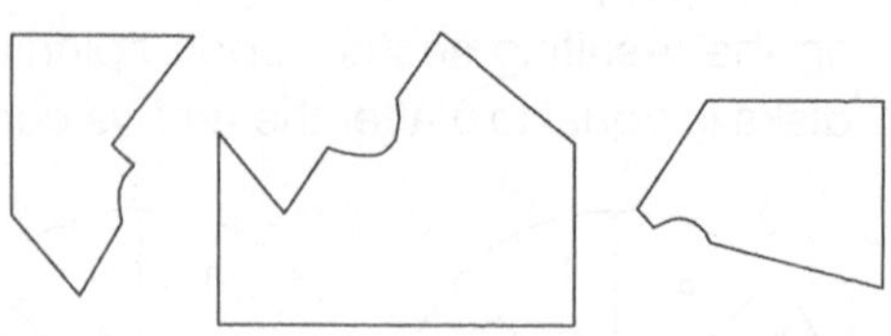

(c) 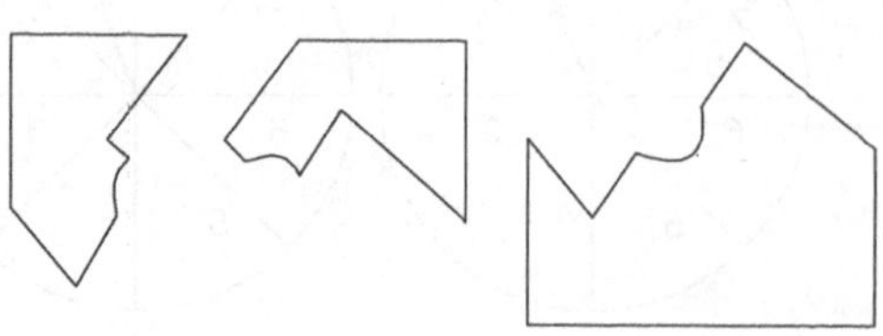

(d) 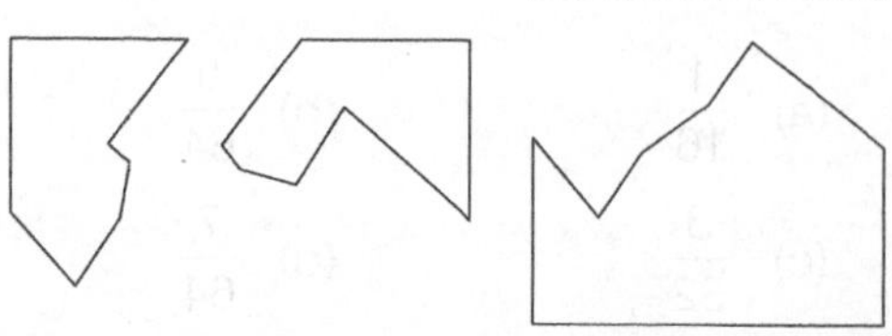

[IN + PI + CH, 2022 : 2 Marks]

498. The average of the monthly salaries of M, N and S is ₹4000. The average of the monthly salaries of N, S and P is ₹5000. The monthly salary of P is ₹6000.
What is the monthly salary of M as a percentage of the monthly salary of P?
(a) 50% (b) 75%
(c) 100% (d) 125%
[ME, 2022 : 1 Mark, Set-1]

499. A person travelled 80 km in 6 hours. If the person travelled the first part with a uniform speed of 10 kmph and the remaining part with a uniform speed of 18 kmph. What percentage of the total distance is travelled at a uniform speed of 10 kmph?
(a) 28.25 (b) 37.25
(c) 43.75 (d) 50.00
[ME, 2022 : 1 Mark, Set-1]

500. Four girls P, Q, R and S are studying languages in a University. P is learning French and Dutch. Q is learning Chinese and Japanese. R is learning Spanish and French. S is learning Dutch and Japanese.
Given that: French is easier than Dutch; Chinese is harder than Japanese; Dutch is easier than Japanese, and Spanish is easier than French.
Based on the above information, which girl is learning the most difficult pair of languages?
(a) P (b) Q
(c) R (d) S
[ME, 2022 : 1 Mark, Set-1]

501.

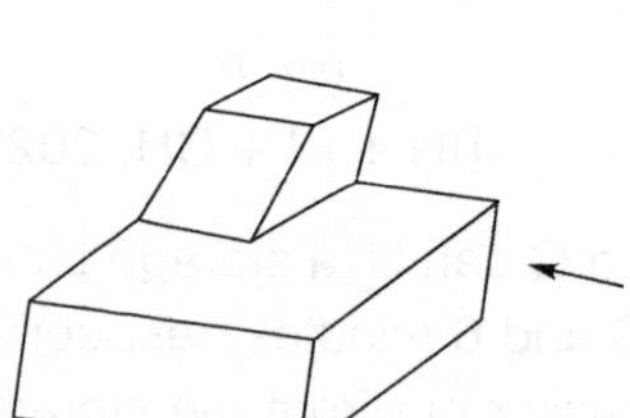

A block with a trapezoidal cross-section is placed over a block with rectangular cross section as shown above. Which one of the following is the correct drawing of the view of the 3D object as viewed in the direction indicated by an arrow in the above figure?

(a) 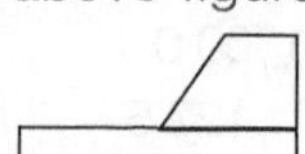 (b) 

(c) 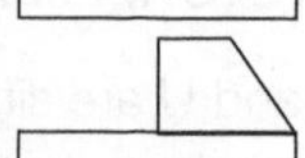 (d) 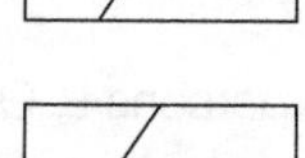

[ME, 2022 : 1 Mark, Set-1]

502. A rhombus is formed by joining the midpoints of the sides of a unit square. What is the diameter of the largest circle that can be inscribed within the rhombus?

(a) $\frac{1}{\sqrt{2}}$ (b) $\frac{1}{2\sqrt{2}}$

(c) $\sqrt{2}$ (d) $2\sqrt{2}$

[ME, 2022 : 2 Marks, Set-1]

503. An equilateral triangle, a square and a circle have equal areas. What is the ratio of the perimeters of the equilateral triangle to square to circle?

(a) $3\sqrt{3}:2:\pi$ (b) $\sqrt{(3\sqrt{3})}:2:\sqrt{\pi}$

(c) $\sqrt{(3\sqrt{3})}:4:2\sqrt{\pi}$ (d) $\sqrt{(3\sqrt{3})}:2:2\sqrt{\pi}$

[ME, 2022 : 2 Marks, Set-1]

504. Given below are three conclusions drawn based on the following three statements.

Statement 1: All teachers are professors.

Statement 2: No professor is a male.

Statement 3: Some males are engineers.

Conclusion I: No engineer is a professor.

Conclusion II: Some engineers are professors.

Conclusion III: No male is a teacher.

Which one of the following options can be logically inferred?

(a) Only conclusion III is correct

(b) Only conclusion I and conclusion II are correct

(c) Only conclusion II and conclusion III are correct

(d) Only conclusion I and conclusion III are correct

[ME, 2022 : 2 Marks, Set-1]

505. In a 12-hour clock that runs correctly, how many times do the second, minute, and hour hands of the clock coincide, in a 12-hour duration from 3 PM in a day to 3 AM the next day?

(a) 11 (b) 12

(c) 144 (d) 2

[ME, 2022 : 2 Marks, Set-1]

506. Which one of the following is a representation (not to scale and in bold) of all values of x satisfying the inequality $2 - 5x \leq -\left(\frac{6x-5}{3}\right)$ on the real number line?

(a) 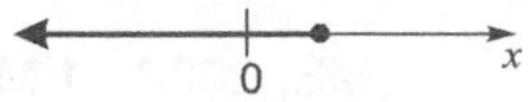

(b) 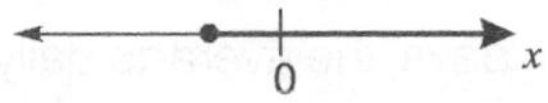

(c) 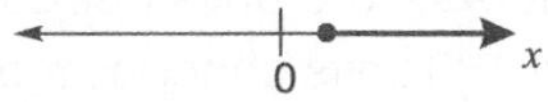

(d) 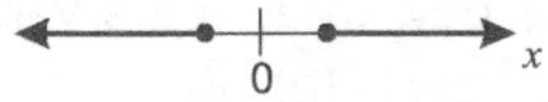

[ME, 2022 : 1 Mark, Set-2]

507. If $f(x) = 2\ln\left(\sqrt{e^x}\right)$, what is the area bounded by $f(x)$ for the interval [0, 2] on the x-axis?

(a) $\frac{1}{2}$ (b) 1

(c) 2 (d) 4

[ME, 2022 : 1 Mark, Set-2]

508. A person was born on the fifth Monday of February in a particular year. Which one of the following statements is correct based on the above information?

(a) The 2nd February of that year is a Tuesday.

(b) There will be five Sundays in the month of February in that year

(c) The 1st February of that year is Sunday.

(d) All Mondays of February in that year have even dates.

[ME, 2022 : 1 Mark, Set-2]

509. Which one of the groups given below can be assembled to get the shape that is shown above using each piece only once without overlapping with each other? (rotation and translation operations may be used).

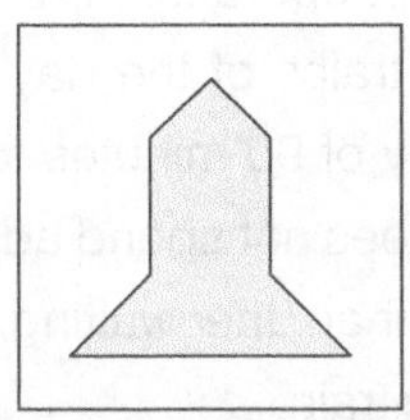

(a) 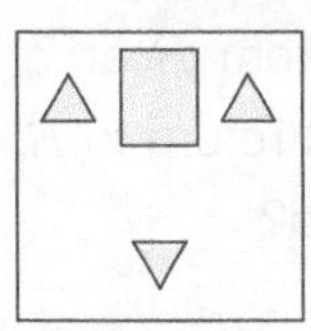

(b) 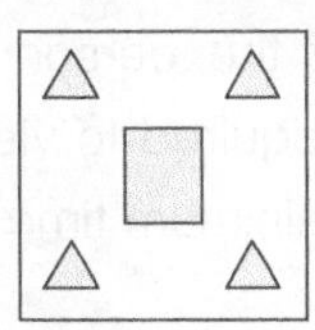

(c) 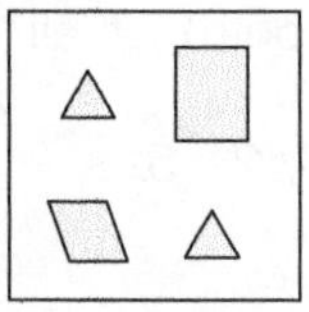 (d) 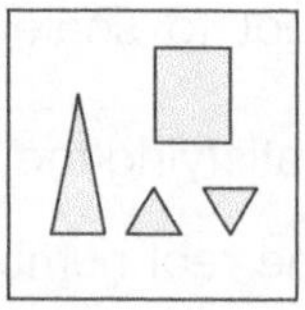

[ME, 2022 : 1 Mark, Set-2]

510. For the past m days, the average daily production at a company was 100 units per day. If today's production of 180 units changes the average to 110 units per day, what is the value of m?

(a) 18 (b) 10

(c) 7 (d) 5

[ME, 2022 : 2 Marks, Set-2]

511. Consider the following functions for non-zero positive integers, p and q.

$$f(p,q) = \underbrace{p \times p \times p \times \ldots\ldots \times p}_{q \text{ terms}} = p^q;\ \ f(p,1) = p$$

$$g(p,q) = p^{p^{p^{p^{p^{\cdot^{\cdot^{\cdot}}}}}}} \text{(upto to } q \text{ terms)} \quad ;g(p,1) = p$$

Which one of the following options is correct based on the above?

(a) $f(2, 2) = g(2, 2)$

(b) $f(g(2, 2), 2) < f(2, g(2, 2))$

(c) $g(2, 1) \neq f(2, 1)$

(d) $f(3, 2) > g(3, 2)$

[ME, 2022 : 2 Marks, Set-2]

512. Four cities P, Q, R and S are connected through one-way routes as shown in the figure. The travel time between any two connected cities is one hour. The boxes beside each city name describe the starting time of first train of the day and their frequency of operation. For example, from city P, the first trains of the day start at 8 AM with a frequency of 90 minutes to each of R and S. A person does not spend additional time at any city other than the waiting time for the next connecting train.

If the person starts from R at 7 AM and is required to visit S and return to R, what is the minimum time required?

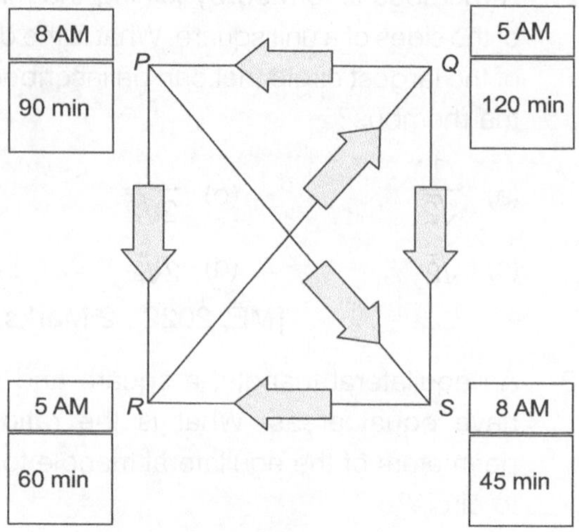

(a) 6 hours 30 minutes

(b) 3 hours 45 minutes

(c) 4 hours 30 minutes

(d) 5 hours 15 minutes

[ME, 2022 : 2 Marks, Set-2]

513. Equal sized circular regions are shaded in a square sheet of paper of 1 cm side length. Two cases, case M and case N, are considered as shown in the figures below. In the case M, four circles are shaded in the square sheet and in the case N, nine circles are shaded in the square sheet as shown.

What is the ratio of the areas of unshaded regions of case M to that of case N?

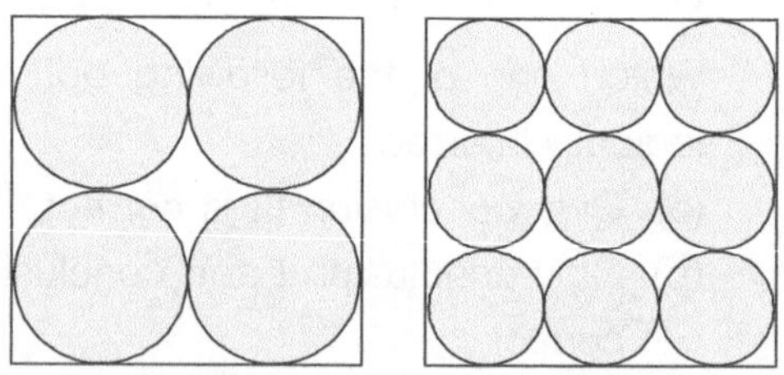

Case M **Case N**

(a) 2 : 3 (b) 1 : 1

(c) 3 : 2 (d) 2 : 1

[ME, 2022 : 2 Marks, Set-2]

514. Given a fair six-faced dice where the faces are labelled '1', '2', '3', '4', '5' and '6', what is the probability of getting a '1' on the first roll of the dice and a '4' on the second roll?

(a) $\frac{1}{36}$ (b) $\frac{1}{6}$

(c) $\frac{5}{6}$ (d) $\frac{1}{3}$

[EE, 2023 : 1 Mark]

515. How many triangles are present in the given figure?

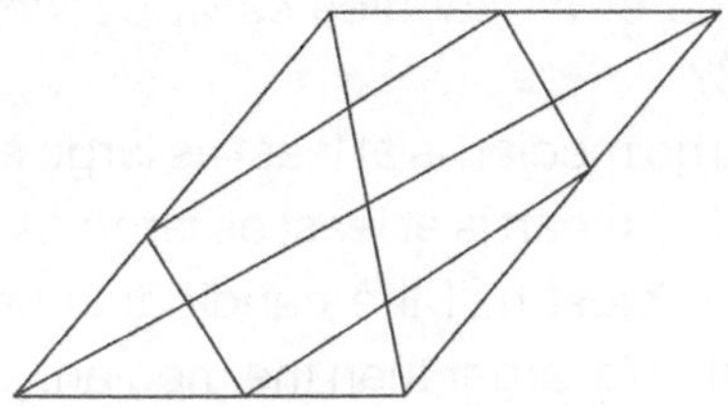

(a) 12 (b) 16
(c) 20 (d) 24

[EE, 2023 : 1 Mark]

516. Which one of the following options represents the given graph?

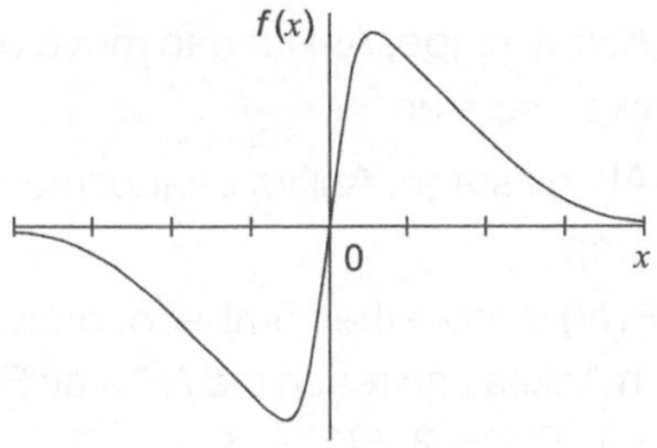

(a) $f(x) = x^2 2^{-|x|}$ (b) $f(x) = x 2^{-|x|}$
(c) $f(x) = |x| 2^{-x}$ (d) $f(x) = x 2^{-x}$

[EE, 2023 : 2 Marks]

517. The digit in the unit's place of the product $3^{999} \times 7^{1000}$ is _____.

(a) 7 (b) 1
(c) 3 (d) 9

[EE, 2023 : 2 Marks]

518. A square with sides of length 6 cm is given. The boundary of the shaded region is defined by two semi-circles whose diameters are the sides of the square, as shown.
The area of the shaded region is _____ cm^2.

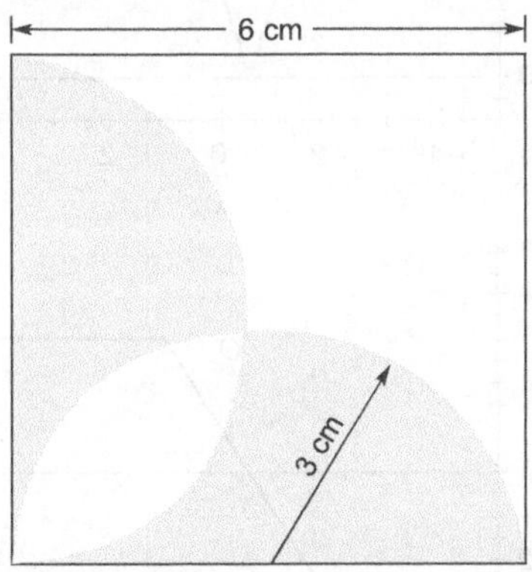

(a) 6π (b) 18
(c) 20 (d) 9π

[EE, 2023 : 2 Marks]

519. A certain country has 504 universities and 25951 colleges. These are categorised into Grades I, II, and III as shown in the given pie charts.
What is the percentage, correct to one decimal place, of higher education institutions (colleges and universities) that fall into Grade III?

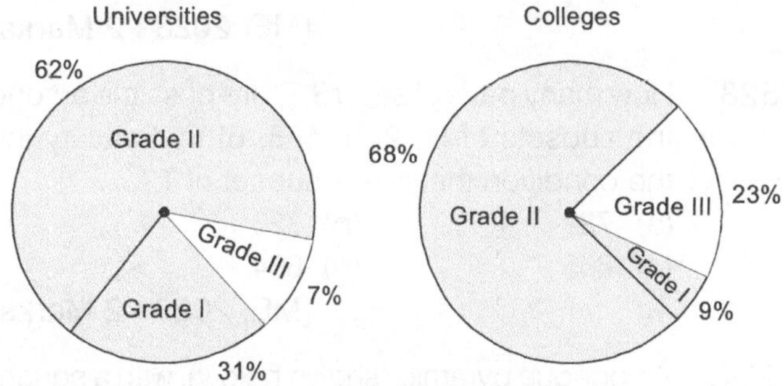

(a) 22.7 (b) 23.7
(c) 15.0 (d) 66.8

[ME, 2023 : 1 Mark]

520. The minute hand and second-hand of a clock cross each other _____ times between 09:15:00 AM and 09:45:00 AM on a day.

(a) 30 (b) 15
(c) 29 (d) 31

[ME, 2023 : 1 Mark]

521. The symbols ○, ∗, △, and □ are to be filled, one in each box, as shown below.
The rules for filling in the four symbols are as follows:

1. Every row and every column must contain each of the four symbols.
2. Every 2 × 2 square delineated by bold lines must contain each of the four symbols.

Which symbol will occupy the box marked with '?' in the partially filled figure?

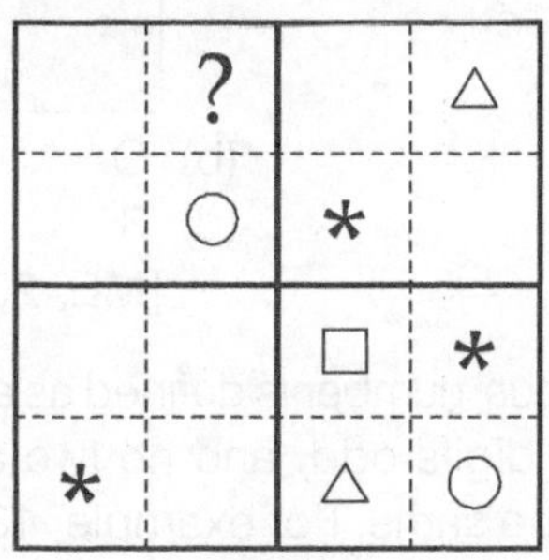

(a) ○ (b) ∗
(c) △ (d) □

[ME, 2023 : 1 Mark]

522. Consider the following inequalities:
$p^2 - 4q < 4$
$3p + 2q < 6$
where p and q are positive integers.
The value of $(p + q)$ is

(a) 2 (b) 1
(c) 3 (d) 4

[ME, 2023 : 2 Marks]

523. How many pairs of sets (S,T) are possible among the subsets of {1, 2, 3, 4, 5, 6} that satisfy the the condition that S is a subset of T?

(a) 729 (b) 728
(c) 665 (d) 664

[ME, 2023 : 2 Marks]

524. An opaque pyramid (shown below), with a square base and isosceles faces, is suspended in the path of a parallel beam of light, such that its shadow is cast on a screen oriented perpendicular to the direction of the light beam. The pyramid can be reoriented in any direction within the light beam. Under these conditions, which one of the shadows P, Q, R and S is NOT possible?

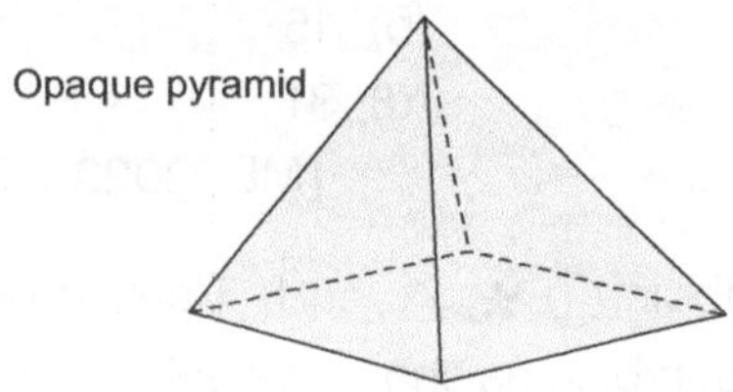

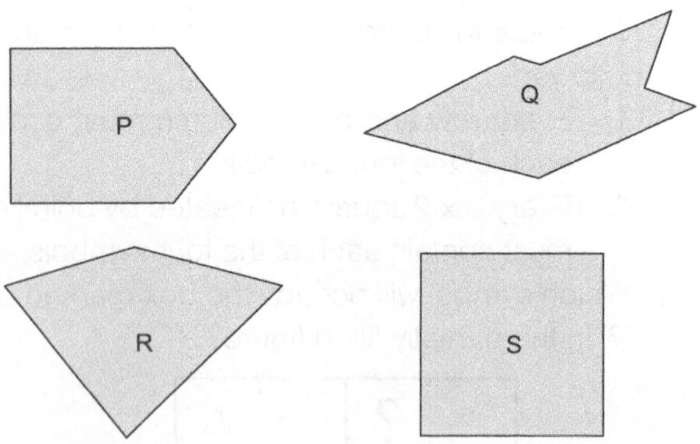

(a) P (b) Q
(c) R (d) S

[ME, 2023 : 2 Marks]

525. A 'frabjous' number is defined as a 3 digit number with all digits odd, and no two adjacent digits being the same. For example, 137 is a frabjous number, while 133 is not. How many such frabjous numbers exist?

(a) 125 (b) 720
(c) 60 (d) 80

[IN, 2023 : 1 Mark]

526. Which one among the following statements must be TRUE about the mean and the median of the scores of all candidates appearing for GATE 2023?

(a) The median is at least as large as the mean.
(b) The mean is at least as large as the median.
(c) At most half the candidates have a score that is larger than the median.
(d) At most half the candidates have a score that is larger than the mean.

[IN, 2023 : 1 Mark]

527. Ankita has to climb 5 stairs starting at the ground, while respecting the following rules:

1. At any stage, Ankita can move either one or two stairs up.
2. At any stage, Ankita cannot move to a lower step.

Let $F(N)$ denote the number of possible ways in which Ankita can reach the Nth stair For example, $F(1)=1$, $F(2)= 2$, $F(3) = 3$.
The value of $F(5)$ is ______.

(a) 8 (b) 7
(c) 6 (d) 5

[IN, 2023 : 2 Marks]

528. Which one of the given figures P, Q, R and S represents the graph of the following function?

$$f(x) = \left| |x + 2| - |x - 1| \right|$$

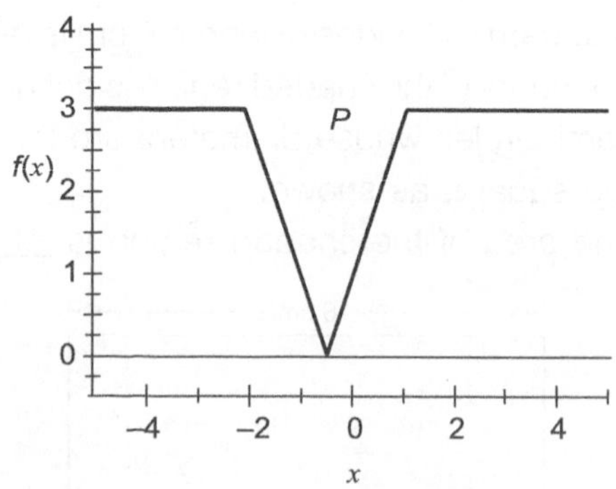

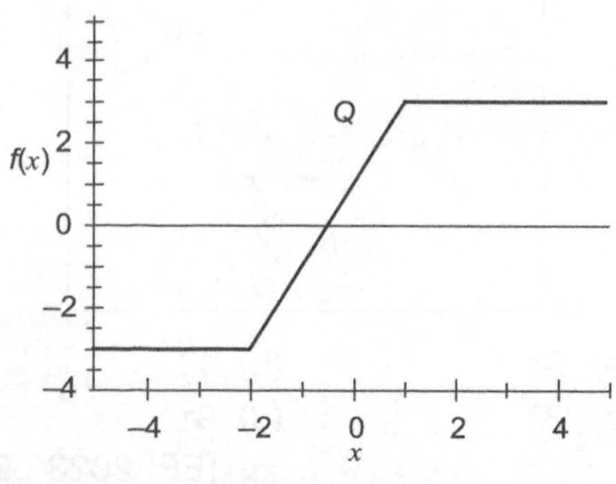

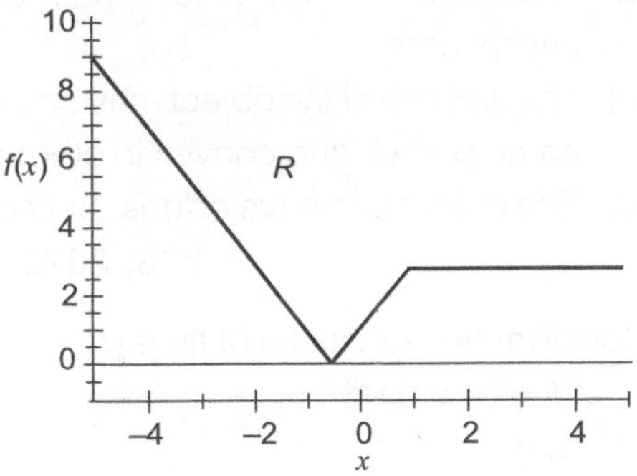

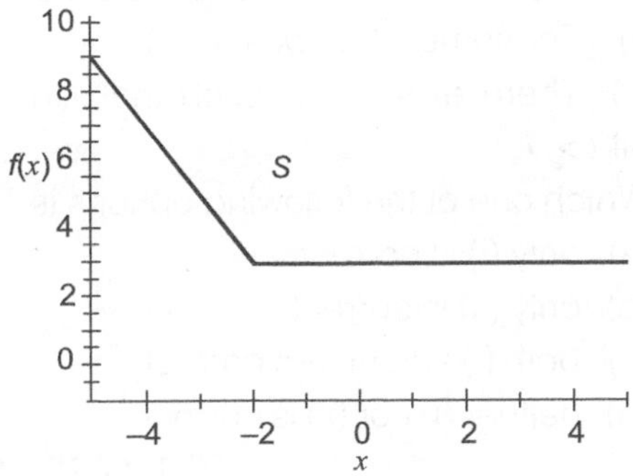

(a) P (b) Q
(c) R (d) S

[IN, 2023 : 2 Marks]

529. In the given diagram, ovals are marked at different heights (h) of a hill. Which one of the following options P, Q, R, and S depicts the top view of the hill?

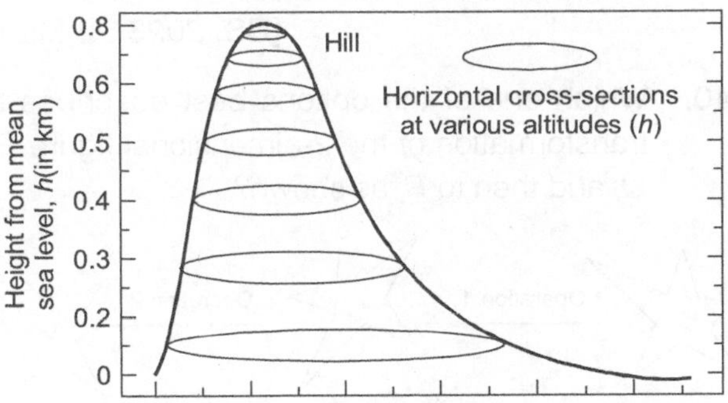

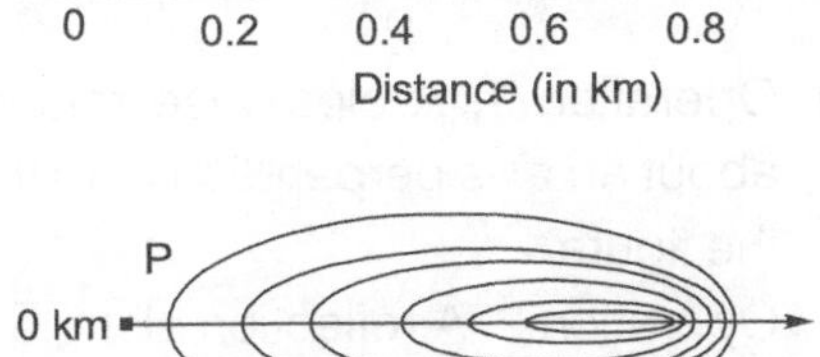

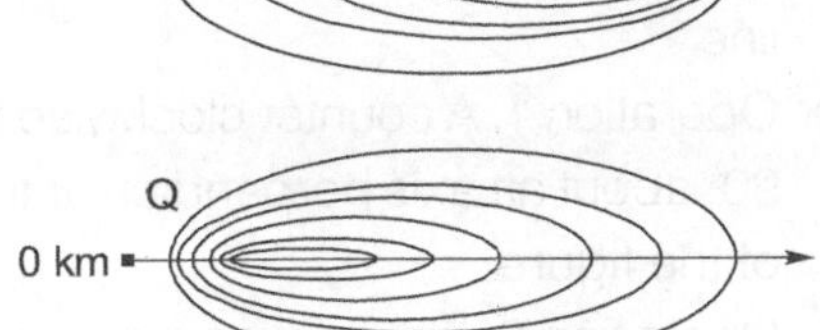

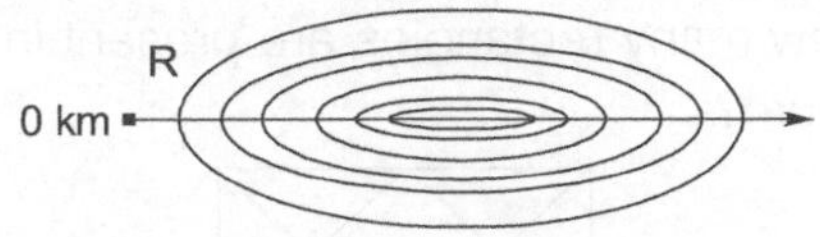

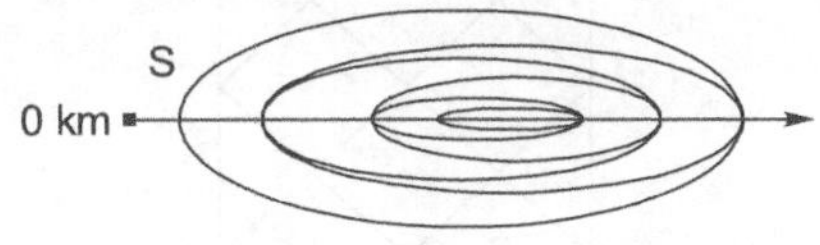

(a) P (b) Q
(c) R (d) S

[IN, 2023 : 1 Mark]

530. An opaque cylinder (shown below) is suspended in the path of a parallel beam of light, such that its shadow is cast on a screen oriented perpendicular to the direction of the light beam. The cylinder can be reoriented in any direction within the light beam. Under these conditions, which one of the shadows *P*, *Q*, *R*, and *S* is NOT possible?

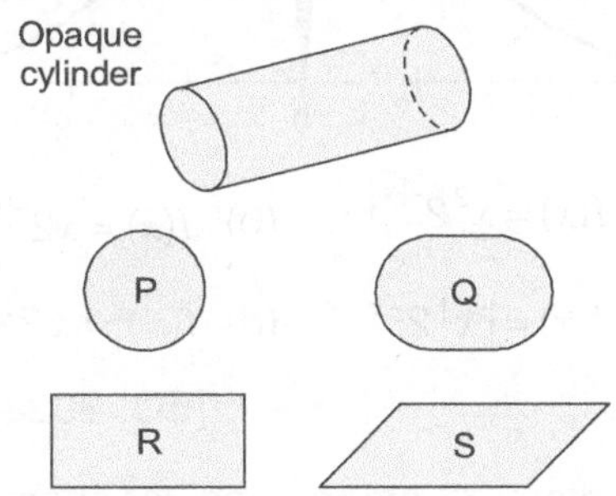

(a) P (b) Q
(c) R (d) S

[IN, 2023 : 2 Marks]

531. What is the smallest number with distinct digits whose digits add up to 45?

(a) 123555789 (b) 123457869
(c) 123456789 (d) 99999

[EC, 2023 : 1 Mark]

532. In a class of 100 students,

(i) there are 30 students who neither like romantic movies nor comedy movies,
(ii) the number of students who like romantic movies is twice the number of students who like comedy movies, and
(iii) the number of students who like both romantic movies and comedy movies is 20.

How many students in the class like romantic movies?

(a) 40 (b) 20
(c) 60 (d) 30

[EC, 2023 : 1 Mark]

533. How many rectangles are present in the given figure?

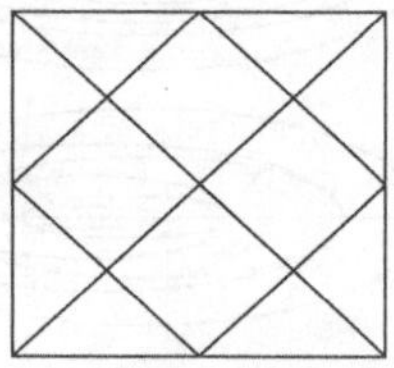

(a) 8 (b) 9
(c) 10 (d) 12

[EC, 2023 : 1 Mark]

534. Which one of the following options represents the given graph?

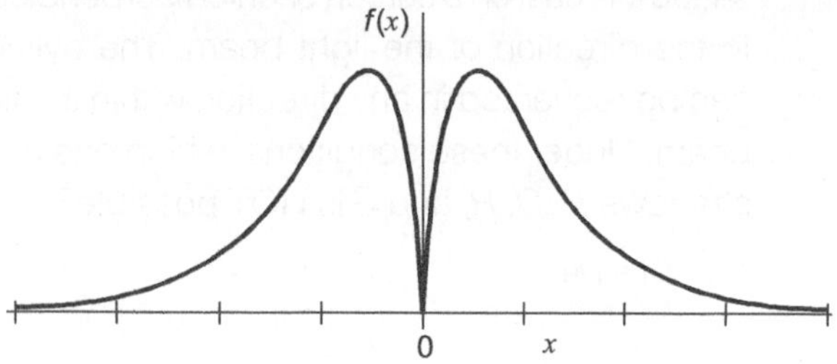

(a) $f(x) = x^2 2^{-|x|}$ (b) $f(x) = x2^{-|x|}$

(c) $f(x) = |x| 2^{-x}$ (d) $f(x) = x2^{-x}$

[EC, 2023 : 2 Marks]

535. 100 cm × 32 cm rectangular sheet is folded 5 times. Each time the sheet is folded, the long edge aligns with its opposite side. Eventually, the folded sheet is a rectangle of dimensions 100 cm × 1 cm.

The total number of creases visible when the sheet is unfolded is ______.

(a) 32 (b) 5
(c) 31 (d) 63

[EC, 2023 : 2 Marks]

536. A series of natural numbers $F_1, F_2, F_3, F_4, F_5, F_6, F_7, \ldots$ obeys $F_{n+1} = F_n + F_{n+1}$ for all integers $n \geq 2$.

If $F_6 = 37$, and $F_7 = 60$, then what is F_1?

(a) 4 (b) 5
(c) 8 (d) 9

[CS, 2023 : 1 Mark]

537. Looking at the surface of a smooth 3-dimensional object from the outside, which one of the following options is TRUE?

(a) The surface of the object must be concave everywhere.
(b) The surface of the object must be convex everywhere.
(c) The surface of the object may be concave in some places and convex in other places.
(d) The object can have edges, but no corners.

[CS, 2023 : 1 Mark]

538. Consider two functions of time (t),

$$f(t) = 0.01t^2$$
$$g(t) = 4t$$

where $0 < t < \infty$.

Now consider the following two statements:

(i) For some $t > 0$, $g(t) > f(t)$
(ii) There exists a T, such that $f(t) > g(t)$ for all $t > T$.

Which one of the following options is TRUE?

(a) only (i) is correct
(b) only (ii) is correct
(c) both (i) and (ii) are correct
(d) neither (i) nor (ii) is correct

[CS, 2023 : 2 Marks]

539. $f(x)$ and $g(y)$ are functions of x and y, respectively, and $f(x) = g(y)$ for all real values of x and y Which one of the following options is necessarily TRUE for all x and y?

(a) $f(x) = 0$ and $g(y) = 0$
(b) $f(x) = g(y) =$ constant
(c) $f(x) \neq$ constant and $g(y) \neq$ constant
(d) $f(x) + g(y) = f(x) - g(y)$

[CS, 2023 : 2 Marks]

540. Which one of the options best describes the transformation of the 2-dimensional figure P to Q, and then to R, as shown?

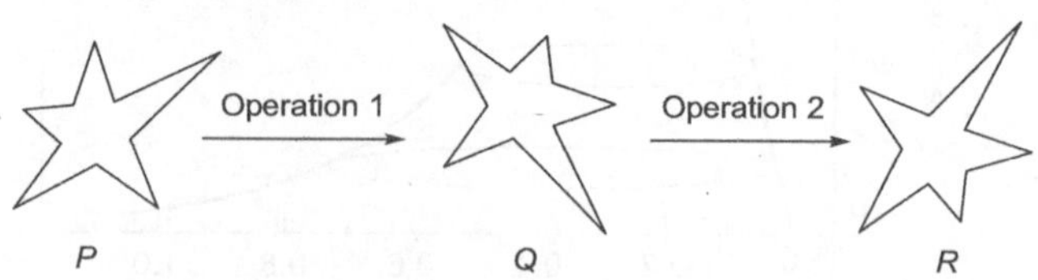

(a) Operation 1: A clockwise rotation by 90° about an axis perpendicular to the plane of the figure
Operation 2: A reflection along a horizontal line
(b) Operation 1: A counter clockwise rotation by 90° about an axis perpendicular to the plane of the figure
Operation 2: A reflection along a horizontal line

(c) Operation 1: A clockwise rotation by 90° about an axis perpendicular to the plane of the figure
Operation 2: A reflection along a vertical line

(d) Operation 1: A counter clockwise rotation by 180° about an axis perpendicular to the plane of the figure
Operation 2: A reflection along a vertical line

[CS, 2023 : 2 Marks]

541. In the given figure, *PQRS* is a parallelogram with *PS* = 7 cm, *PT* = 4 cm and *PV* = 5 cm. What is the length of *RS* in cm? (The diagram is representative.)

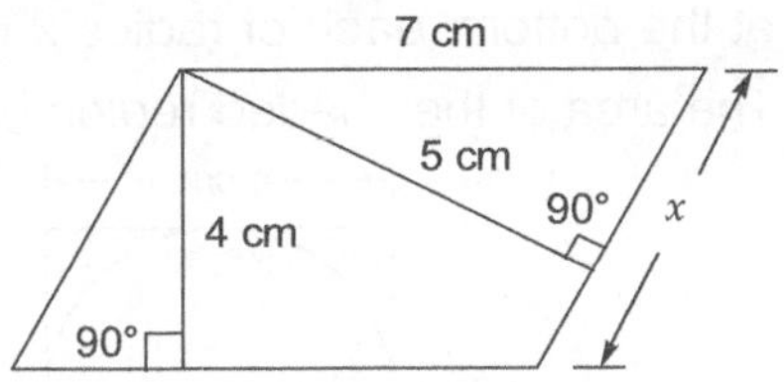

(a) $\frac{20}{7}$ (b) $\frac{28}{5}$
(c) $\frac{9}{2}$ (d) $\frac{35}{4}$

[CH + PI 2023 : 1 Mark]

542. A line of symmetry is defined as a line that divides a figure into two parts in a way such that each part is a mirror image of the other part about that line.
The given figure consists of 16 unit squares arranged as shown. In addition to the three black squares, what is the minimum number of squares that must be coloured black, such that both PQ and MN form lines of symmetry? (The figure is representative)

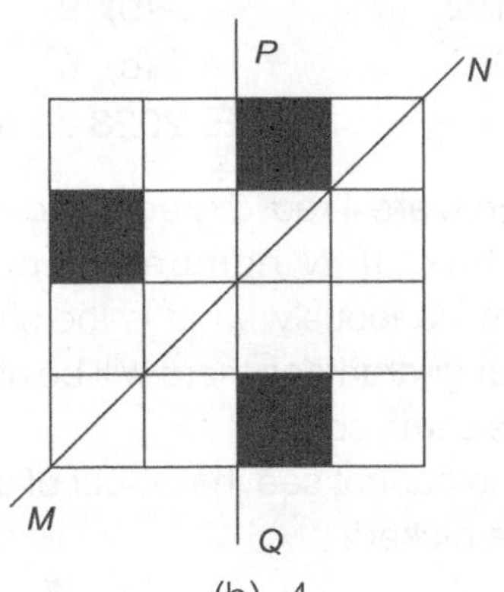

(a) 3 (b) 4
(c) 5 (d) 6

[CH + PI 2023 : 1 Mark]

543. To construct a wall, sand and cement are mixed in the ratio of 3 : 1. The cost of sand and that of cement are in the ratio of 1 : 2.
If the total cost of sand and cement to construct the wall is 1000 rupees, then what is the cost (in rupees) of cement used?

(a) 400 (b) 600
(c) 800 (d) 200

[CH + PI 2023 : 2 Marks]

544. The coefficient of x^4 in the polynomial $(x - 1)^3 (x - 2)^3$ is equal to ______.

(a) 33 (b) –3
(c) 30 (d) 21

[CH + PI 2023 : 2 Marks]

545. Which one of the following shapes can be used to tile (completely cover by repeating) a flat plane, extending to infinity in all directions, without leaving any empty spaces in between them? The copies of the shape used to tile are identical and are not allowed to overlap.

(a) circle (b) regular octagon
(c) regular pentagon (d) rhombus

[CH + PI 2023 : 2 Marks]

546. In the given figure, PQRSTV is a regular hexagon with each side of length 5 cm. A circle is drawn with its centre at V such that it passes through P. What is the area (in cm^2) of the shaded region? (The diagram is representative)

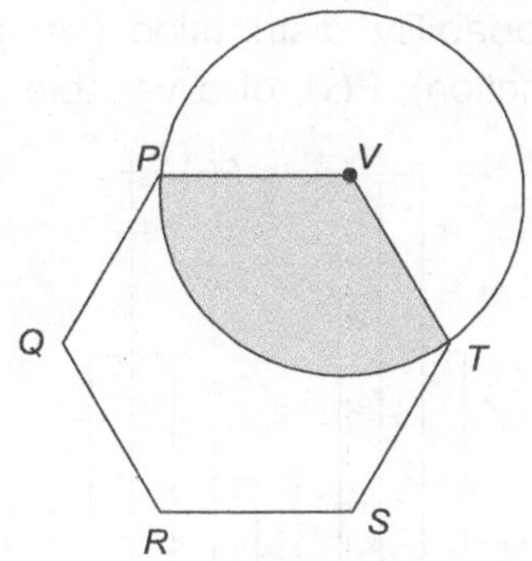

(a) $\frac{25\pi}{3}$ (b) $\frac{20\pi}{3}$
(c) 6π (d) 7π

[CE, 2023 : 1 Mark, Forenoon]

547. A line of symmetry is defined as a line that divides a figure into two parts in a way such that each part is a mirror image of the other part about that line.

The figure below consists of 20 unit squares arranged as shown. In addition to the given black squares, upto 5 more may be coloured black. Which one among the following options depicts the minimum number of boxes that must be coloured black to achieve two lines of symmetry? (The figure is representative)

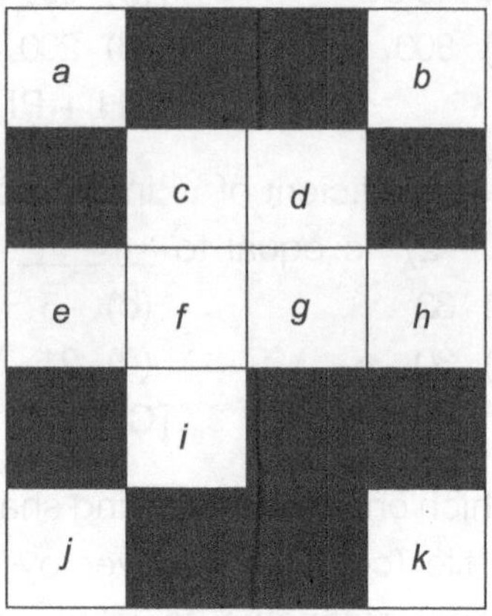

(a) d
(b) c, d, i
(c) c, i
(d) c, d, i, f, g

[CE, 2023 : 1 Mark, Forenoon]

548. Based only on the truth of the statement 'Some humans are intelligent', which one of the following options can be logically inferred with certainty?
(a) No human is intelligent.
(b) All humans are intelligent.
(c) Some non-humans are intelligent.
(d) Some intelligent beings are humans.

[CE, 2023 : 2 Marks, Forenoon]

549. Which one of the options can be inferred about the mean, median, and mode for the given probability distribution (i.e. probability mass function), P(x), of a variable x?

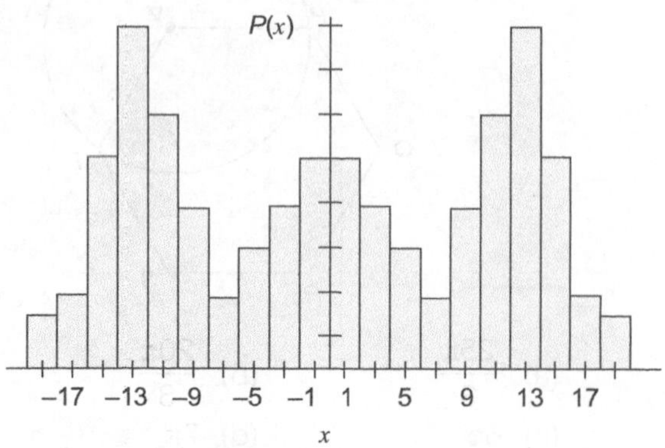

(a) mean = median ≠ mode
(b) mean = median = mode
(c) mean ≠ median = mode
(d) mean ≠ mode = median

[CE, 2023 : 2 Marks, Forenoon]

550. Let $a = 30!$, $b = 50!$ and $c = 100!$. Consider the following numbers:

$$\log_a c, \log_c a, \log_b a, \log_a b$$

Which one of the following inequalities is CORRECT?
(a) $\log_c a < \log_b a < \log_a c < \log_a c$
(b) $\log_c a < \log_a b < \log_b a < \log_b c$
(c) $\log_c a < \log_b a < \log_a c < \log_a b$
(d) $\log_b a < \log_c a < \log_a b < \log_a c$

[CE, 2023 : 2 Marks, Forenoon]

551. A square of side length 4 cm is given. The boundary of the shaded region is defined by one semi-circle on the top and two circular arcs at the bottom, each of radius 2 cm, as shown. The area of the shaded region is ______ cm^2.

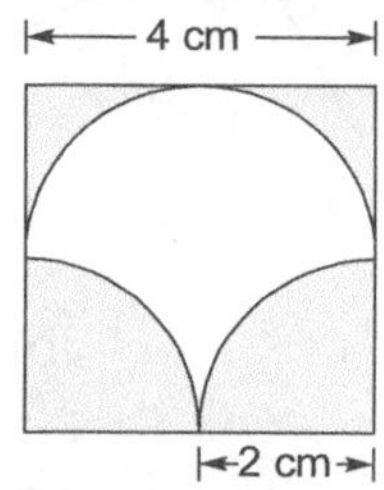

(a) 8
(b) 4
(c) 12
(d) 10

[CE, 2023 : 2 Marks, Forenoon]

552. In how many ways can cells in a 3 × 3 grid be shaded, such that each row and each column have exactly one shaded cell? An example of one valid shading is shown.

(a) 2
(b) 9
(d) 3
(d) 6

[CE, 2023 : 1 Mark, Afternoon]

553. There are 4 red, 5 green, and 6 blue balls inside a box. If N number of balls are picked simultaneously, what is the smallest value of N that guarantees there will be at least two balls of the same colour?
One cannot see the colour of the balls until they are picked.
(a) 4
(b) 15
(c) 5
(d) 2

[CE, 2023 : 1 Mark, Afternoon]

554. Consider a circle with its centre at the origin (O), as shown. Two operations are allowed on the circle.
Operation 1: Scale independently along the x and y axes.
Operation 2: Rotation in any direction about the origin.
Which figure among the options can be achieved through a combination of these two operations on the given circle?

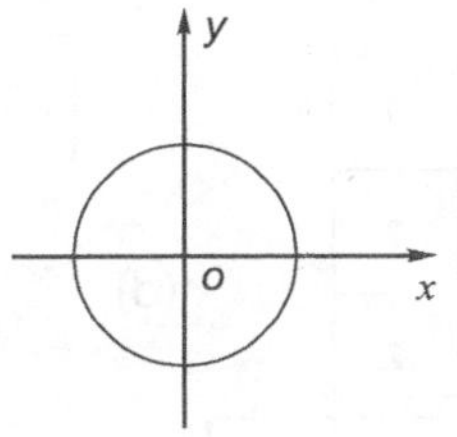

(a) 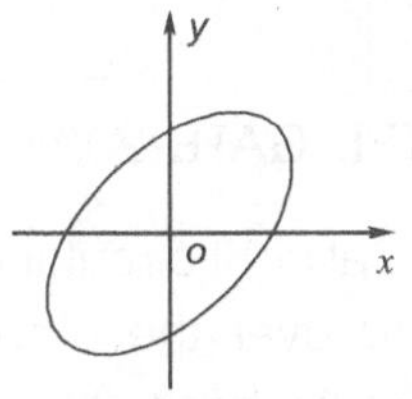
(b) 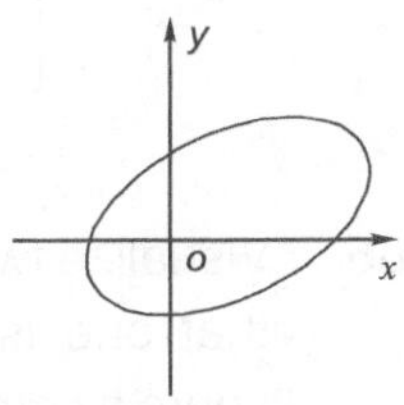
(c) 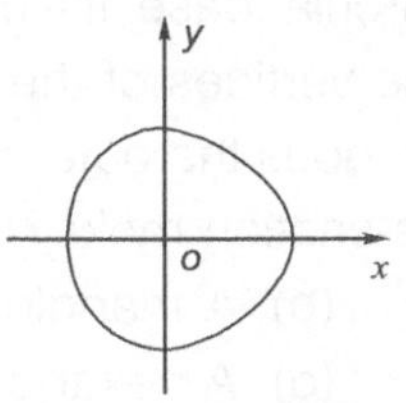
(d) 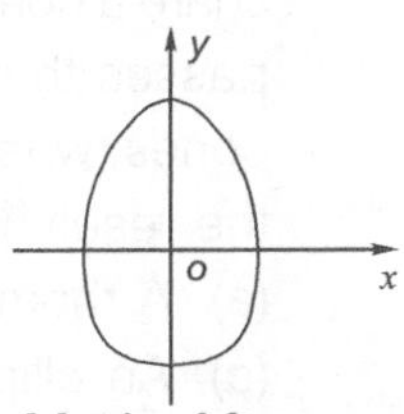

[CE, 2023 : 1 Mark, Afternoon]

555. Three husband-wife pairs are to be seated at a circular table that has six identical chairs. Seating arrangements are defined only by the relative position of the people. How many seating arrangements are possible such that every husband sits next to his wife?
(a) 16 (b) 4
(c) 120 (d) 720

[CE, 2023 : 2 Marks, Afternoon]

556. If x satisfies the equation $4^{8^x} = 256$, then x is equal to ______.
(a) $\frac{1}{2}$ (b) $\log_{16} 8$
(c) $\frac{2}{3}$ (d) $\log_4 8$

[CE, 2023 : 2 Marks, Afternoon]

557. Consider a spherical globe rotating about an axis passing through its poles. There are three points P, Q and R situated respectively on the equator, the north pole, and midway between the equator and the north pole in the northern hemisphere. Let P, Q and R move with speeds v_P, v_Q, and v_R, respectively.
Which one of the following options is CORRECT?
(a) $v_P < v_R < v_Q$ (b) $v_P < v_Q < v_R$
(c) $v_P > v_R > v_Q$ (d) $v_P = v_R \neq v_Q$

[CE, 2023 : 2 Marks, Afternoon]

558 The 15 parts of the given figure are to be painted such that no two adjacent parts with shared boundaries (excluding corners) have the same color. The minimum number of colors required is

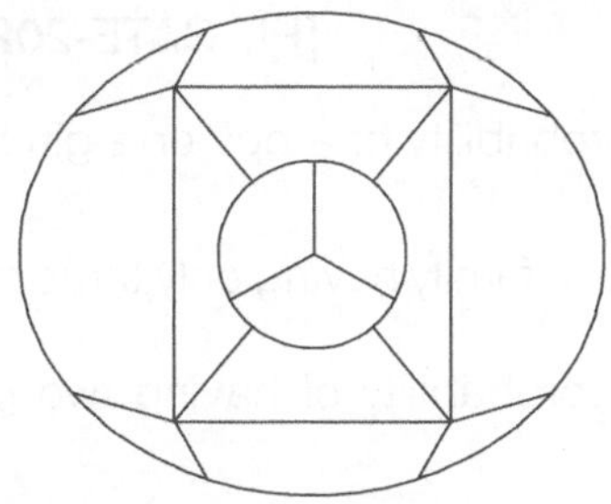

(a) 4 (b) 3
(c) 5 (d) 6

[PI, GATE-2024 : 1 Mark]

559. How many 4-digit positive integers divisible by 3 can be formed using only the digits {1, 3, 4, 6, 7}, such that no digit appears more than once in a number?
(a) 24 (b) 48
(c) 72 (d) 12

[PI, GATE-2024 : 1 Mark]

560. The sum of the following infinite series is

$$2+\frac{1}{2}+\frac{1}{3}+\frac{1}{4}+\frac{1}{8}+\frac{1}{9}+\frac{1}{16}+\frac{1}{27}+....$$

(a) $\frac{11}{3}$ (b) $\frac{7}{2}$
(c) $\frac{13}{4}$ (d) $\frac{9}{2}$

[PI, GATE-2024 : 1 Mark]

561. In an election, the share of valid votes received by the four candidates A, B, C, and D is represented by the pie chart shown. The total number of votes cast in the election were 1,15,000, out of which 5,000 were invalid.

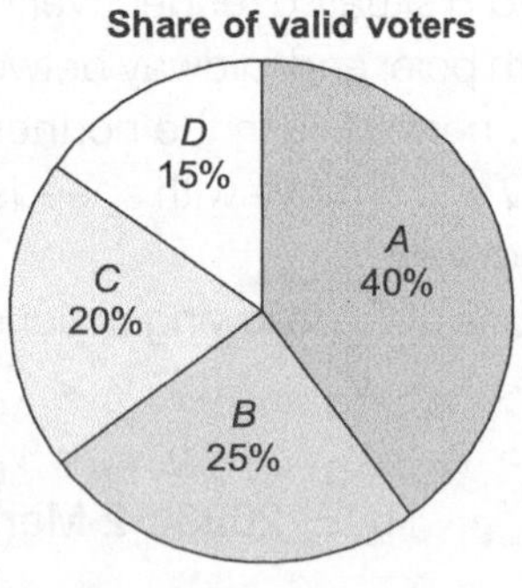

Based on the data provided, the total number of valid votes received by the candidates B and C is

(a) 45000 (b) 49500
(c) 51750 (d) 54000

[PI, GATE-2024 : 1 Mark]

562. The probability of a boy or a girl being born is $\frac{1}{2}$. For a family having only three children, what is the probability of having two girls and one boy?

(a) $\frac{3}{8}$ (b) $\frac{1}{8}$
(c) $\frac{1}{4}$ (d) $\frac{1}{2}$

[PI, GATE-2024 : 2 Marks]

563. Person 1 and Person 2 invest in three mutual funds A, B, and C. The amounts they invest in each of these mutual funds are given in the table.

	Mutual fund A	Mutual fund B	Mutual fund C
Person 1	₹ 10000	₹ 20000	₹ 20000
Person 2	₹ 20000	₹ 15000	₹ 15000

At the end of one year, the total amount that Person 1 gets is ₹500 more than Person 2. The annual rate of return for the mutual funds B and C is 15% each. What is the annual rate of return for the mutual fund A?

(a) 7.5% (b) 10%
(c) 15% (d) 20%

[PI, GATE-2024 : 2 Marks]

564. Three different views of a dice are shown in the figure below.

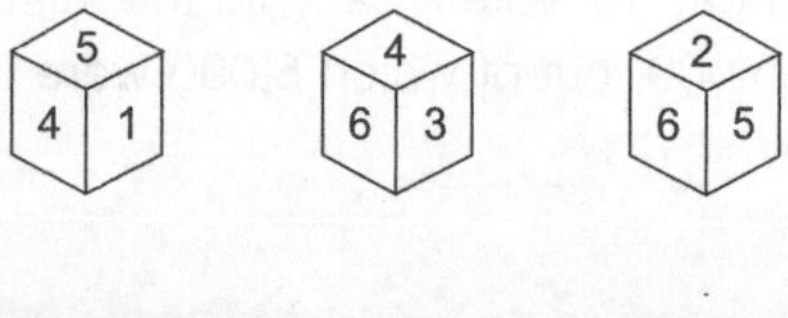

The piece of paper that can be folded to make this dice is

(a)

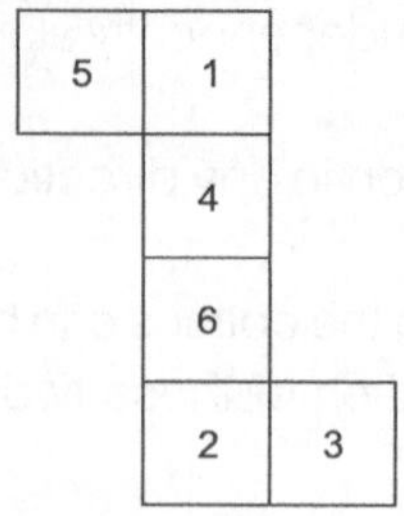

(b)

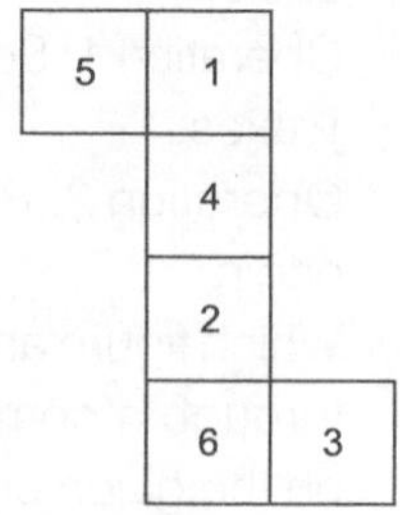

(c)

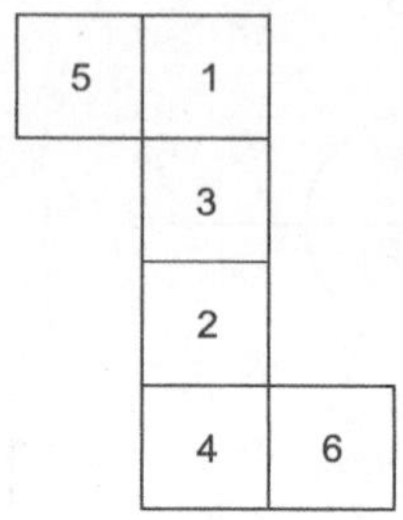

(d)

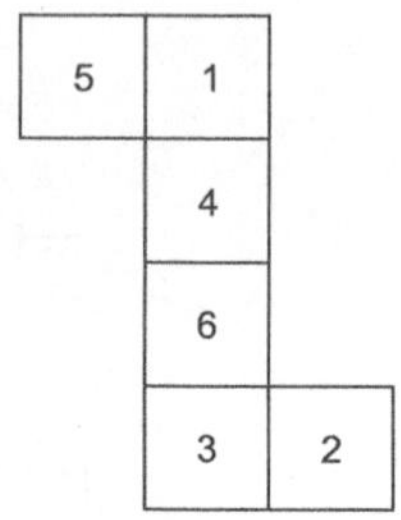

[PI, GATE-2024 : 2 Marks]

565. Visualize two identical right circular cones such that one is inverted over the other and they share a common circular base. If a cutting plane passes through the vertices of the assembled cones, what shape does the outer boundary of the resulting cross-section make?

(a) A rhombus (b) A triangle
(c) An ellipse (d) A hexagon

[PI, GATE-2024 : 2 Marks]

566. The real variable x, y, z, and the real constants p, q, r satisfy

$$\frac{x}{pq-r^2} = \frac{y}{qr-p^2} = \frac{z}{rp-q^2}$$

Given the denominator are non-zero, the value of $px + qy + rz$.

(a) pqr (b) 1
(c) $p^2 + q^2 + r^2$ (d) 0

[ME, GATE-2024 : 1 Mark]

567. Four equilateral triangles are used to form a regular closed three-dimensional object by joining along the edges. The angle between any two faces is

(a) 45° (b) 30°
(c) 60° (d) 90°

[ME, GATE-2024 : 1 Mark]

568. How many combinations of non-null sets A, B, C are possible from the subsets of {2, 3, 5} satisfying the condition: (i) A is a subset of B, and (ii) B is a subset of C?

(a) 18 (b) 27
(c) 28 (d) 19

[ME, GATE-2024 : 1 Mark]

569. The bar chart gives the batting averages of VK and RS for 11 calender years from 2012 to 2022. Considering that 2015 and 2019 are world cup years, which one of the following options is true?

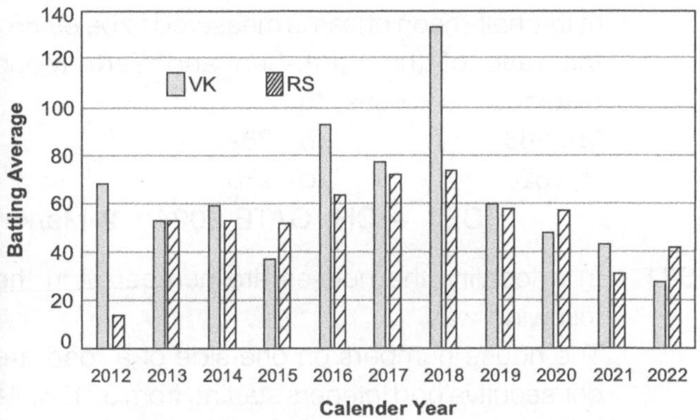

(a) VK has a higher yearly batting average than that of RS in every world cup year.
(b) VK's yearly batting average is consistently higher than that of RS between the two world cup years.
(c) RS's yearly batting average is consistently higher than that of VK in the last three years.
(d) RS has a higher yearly batting average than that of VK in every world cup year.

[ME, GATE-2024 : 1 Mark]

570. Take two long dice (rectangular parallelepiped), each having four rectangular faces labelled as 2, 3, 5, and 7. If thrown, the long dice cannot land on the square faces and has 1/4 probability of landing on any of the four rectangular faces. The lebel on the top face of the dice is the score of the throw.

If thrown together, what is the probability of getting the sum of the two long dice scores greater than 11?

(a) $\frac{3}{16}$ (b) $\frac{1}{16}$
(c) $\frac{1}{8}$ (d) $\frac{3}{8}$

[ME, GATE-2024 : 2 Marks]

571. Find the odd one out in the set: {19, 37, 21, 17, 23, 29, 31, 11}

(a) 23 (b) 21
(c) 29 (d) 37

[ME, GATE-2024 : 2 Marks]

572. A planar rectangular paper has two V-shaped pieces attached as shown below:

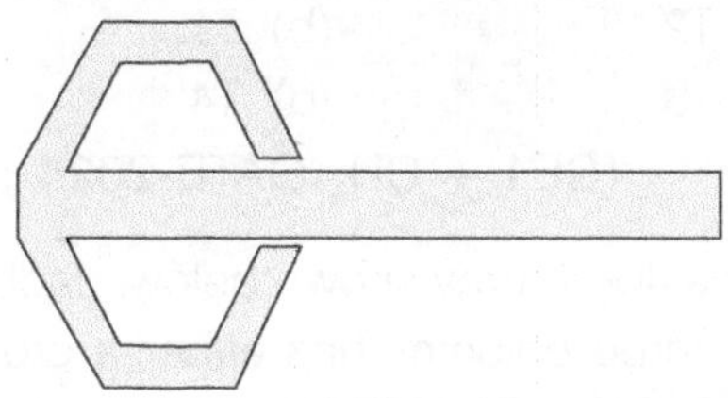

This piece of paper is folded to make the following closed three-dimensions object.

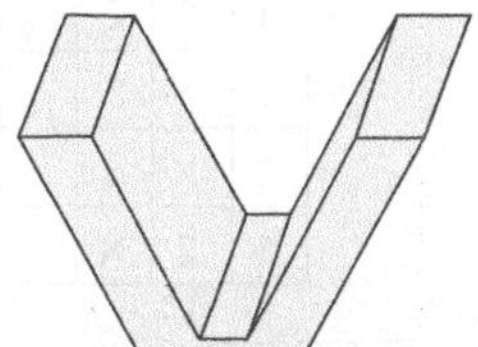

The number of folds required to form the above object is

(a) 8 (b) 7
(c) 9 (d) 11

[ME, GATE-2024 : 2 Marks]

573. In the following series, identify the number that needs to be changed to form the Fibonacci series.
1, 1, 2, 3, 6, 8, 13, 21, ...

(a) 13 (b) 8
(c) 21 (d) 6

[ME, GATE-2024 : 2 Marks]

574. The chart given below compares the Installed Capacity (MW) of four power generation technologies, T1, T2, T3 and T4 and their Electricity Generation (MWh) in a time of 1000 hours (h).

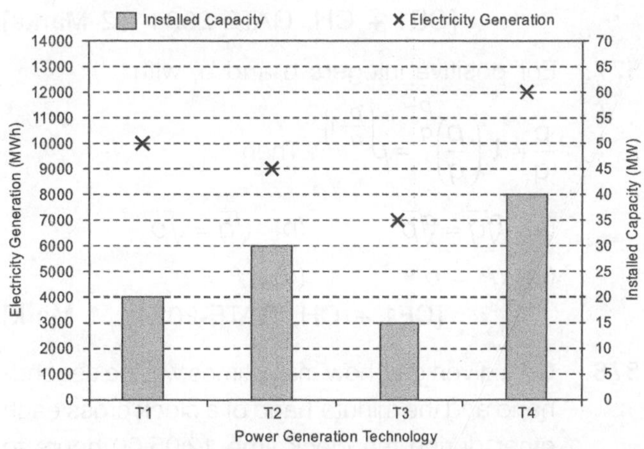

The capacity factor of a power generation technology is:

$$\text{Capacity Factor} = \frac{\text{Electricity Generation (MWh)}}{\text{Installed Capacity (MW)} \times 1000\text{(h)}}$$

Which one of the given technologies has the highest Capacity Factor?
(a) T2 (b) T1
(c) T3 (d) T4
[CE1 + CH, GATE-2024 : 2 Marks]

575. In the 4 × 4 array shown below, each cell of the first three columns has either a cross (X) or a number, as per the given rule.

1	1	2	
2	X	3	
2	X	4	
1	2	X	

Rule: The number in a cell represents the count of crosses around its immediate neighboring cells (left, right, top, bottom, diagonals).
As per this rule, the maximum number of crosses possible in the empty column is
(a) 3 (b) 0
(c) 1 (d) 2
[CE1 + CH, GATE-2024 : 2 Marks]

576. Three distinct sets of indistinguishable twins are to be seated at a circular table that has 8 identical chairs.. Unique seating arrangements are defined by the relative positions of the people.

How many unique seating arrangements are possible such that each person is sitting next to their twin?
(a) 12 (b) 10
(c) 14 (d) 28
[CE1 + CH, GATE-2024 : 2 Marks]

577. For positive integers p and q, with

$\frac{p}{q} \neq 1, \left(\frac{p}{q}\right)^{\frac{p}{q}} = p^{\left(\frac{p}{q}-1\right)}$. Then,

(a) $\sqrt[p]{q} = \sqrt[q]{p}$ (b) $\sqrt{q} = \sqrt{p}$
(c) $q^p = p^{2q}$ (d) $q^p = p^q$
[CE1 + CH, GATE-2024 : 1 Mark]

578. On a given day, how many times will the second-hand and the minute-hand of a clock cross each other during the clock time 12:05:00 hours to 12:55:00 hours?
(a) 51 (b) 49
(c) 50 (d) 55
[CE1 + CH, GATE-2024 : 1 Mark]

579. Which one of the given options is a possible value of x in the following sequence?
3, 7, 15, x, 63, 127, 255
(a) 35 (b) 31
(c) 45 (d) 40
[CE1 + CH, GATE-2024 : 1 Mark]

580. During a half-moon phase, the Earth-Moon-Sun form a right triangle. If the Moon-Earth-Sun angle at this half-moon phase is measured to be 89.85", the ratio of the Earth-Sun and Earth-Moon distances is closest to
(a) 283 (b) 382
(c) 328 (d) 238
[CE1 + CH, GATE-2024 : 2 Marks]

581. In a locality, the houses are numbered in the following way:
The house-numbers on one side of a road are consecutive odd integers starting from 301, while the house-numbers on the other side of the road are consecutive even numbers starting from 302. The total number of houses is the same on both sides of the road.
If the difference of the sum of the house-numbers between the two sides of the road is 27, then the number of houses on each side of the rood is
(a) 52 (b) 54
(c) 26 (d) 27
[CE1 + CH, GATE-2024 : 1 Mark]

582. **Statements:**
1. All heroes are winners.
2. All winners are lucky people.

Inferences:
I. All lucky people are heroes.
II. Some lucky people are heroes.
III. Some winners are heroes.
Which of the above inferences can be logically deduced from statements 1 and 2?
(a) Only I and II (b) Only II and III
(c) Only I and III (d) Only III
[IN + CE2, GATE-2024 : 1 Mark]

583. A student was supposed to **multiply** a positive real number *p* with another positive real number *q*. Instead, the student **divided** *p* by *q*. If the percentage error in the student's answer is 80%, the value of *q* is
(a) 5 (b) $\sqrt{2}$
(c) 2 (d) $\sqrt{5}$
[IN + CE2, GATE-2024 : 1 Mark]

584. If the sum of the first 20 consecutive positive odd numbers is divided by 20^2, the result is

(a) 1 (b) 20
(c) 2 (d) 1/2

[IN + CE2, GATE-2024 : 1 Mark]

585. The ratio of the number of girls to boys in class VIII is the same as the ratio of the number of boys to girls in class IX. The total number of students (boys and girls) in classes VIII and IX is 450 and 360, respectively. If the number of girls in classes VIII and IX is the same, then the number of girls in each class is

(a) 150 (b) 200
(c) 250 (d) 175

[IN + CE2, GATE-2024 : 1 Mark]

586. Seven identical cylindrical chalk-sticks are fitted tightly in a cylindrical container. The figure below shows the arrangement of the chalk-sticks inside the cylinder.

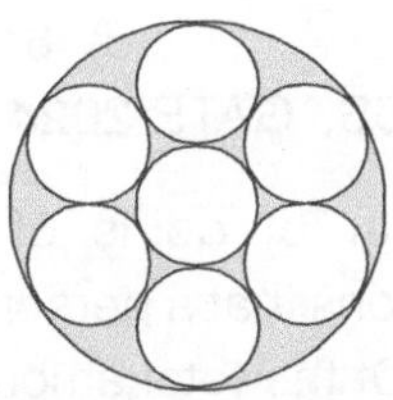

The length of the container is equal to the length of the chalk-sticks. The ratio of the occupied space to the empty space of the container is

(a) $\frac{5}{2}$ (b) $\frac{7}{2}$
(c) $\frac{9}{2}$ (d) 3

[IN + CE2, GATE-2024 : 2 Marks]

587. The plot below shows the relationship between the mortality risk of cardiovascular disease and the number of steps a person walks per day. Based on the data, which one of the following options is true?

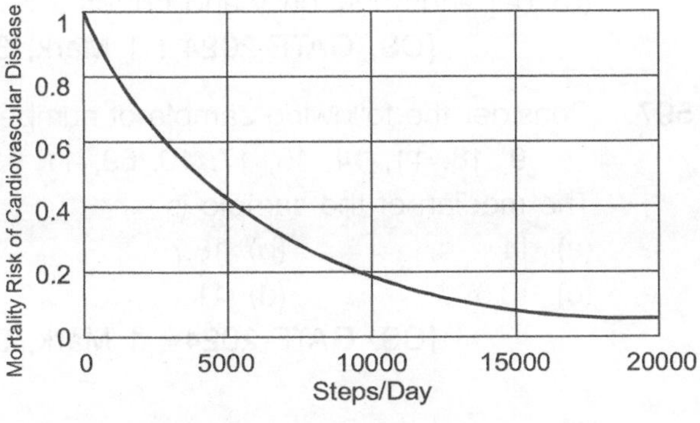

(a) The risk reduction on increasing the steps/day from 0 to 10000 is less than the risk reduction on increasing the steps/day from 10000 to 20000.

(b) The risk reduction on increasing the steps/day from 0 to 5000 is less than the risk reduction on increasing the steps/day from 15000 to 20000.

(c) For any 5000 increment in steps/day the largest risk reduction occurs on going from 0 to 5000.

(d) For any 5000 increment in steps/day the largest risk reduction occurs on going from 15000 to 20000.

[IN + CE2, GATE-2024 : 2 Marks]

588. Five cubes of identical size and another smaller cube are assembled as shown in Figure A. If viewed from direction X, the planar image of the assembly appears as Figure B.

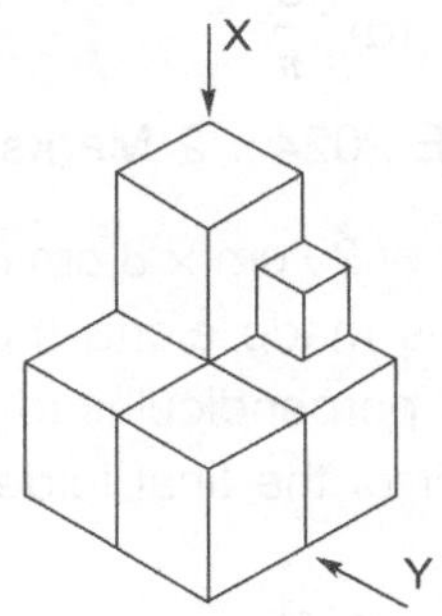

Figure A

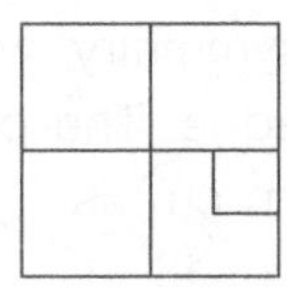

Figure B

If viewed from direction Y, the planar image of the assembly (Figure A) will appear as

(a)
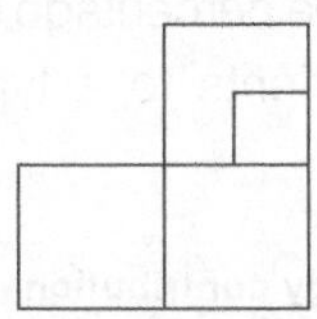

(b)
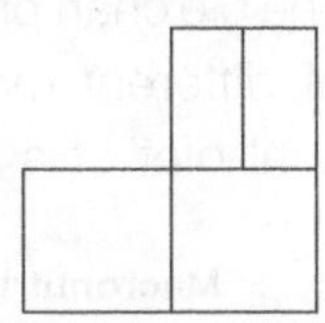

(c)
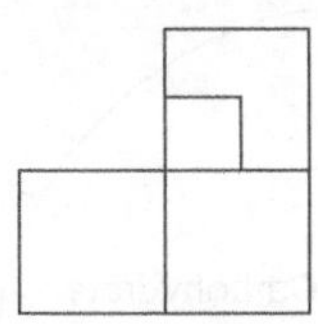

(d)
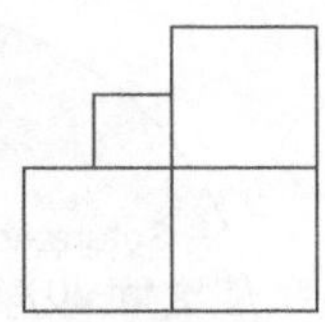

[IN + CE2, GATE-2024 : 2 Marks]

589. Visualize a cube that is held with one of the four body diagonals aligned to the vertical axis. Rotate the cube about this axis such that its view remains unchanged. The magnitude of the minimum angle of rotation is

(a) 120° (b) 60°
(c) 90° (d) 180°

[IN + CE2, GATE-2024 : 2 Marks]

590. For positive non-zero real variables p and q, if $\log(p^2 + q^2) = \log p + \log q + 2\log 3$,

then, the value of $\frac{p^4 + q^4}{p^2q^2}$

(a) 9 (b) 79
(c) 81 (d) 83

[CS, GATE-2024 : 1 Mark, Set-1]

591. A rectangular paper sheet of dimensions 54 cm × 4 cm is taken. The two longer edges of the sheet are joined together to create a cylindrical tube. A cube whose surface area is equal to the area of the sheet is also taken.

Then, the ratio of the volume of the cylindrical tube to the volume of the cube is

(a) $\frac{2}{\pi}$ (b) $\frac{1}{\pi}$
(c) $\frac{4}{\pi}$ (d) $\frac{3}{\pi}$

[CS, GATE-2024 : 2 Marks, Set-1]

592. A rectangular paper of 20 cm × 8 cm is folded 3 times. Each fold is made along the line of symmetry, which is perpendicular to its long edge. The perimeter of the final folded sheet (in cm) is

(a) 18 (b) 20
(c) 24 (d) 21

[CS, GATE-2024 : 2 Marks, Set-1]

593. The pie chart present the percentage contribution of different macronutrients to a typical 2,000 kcal diet of a person.

Macronutrient energy contribution

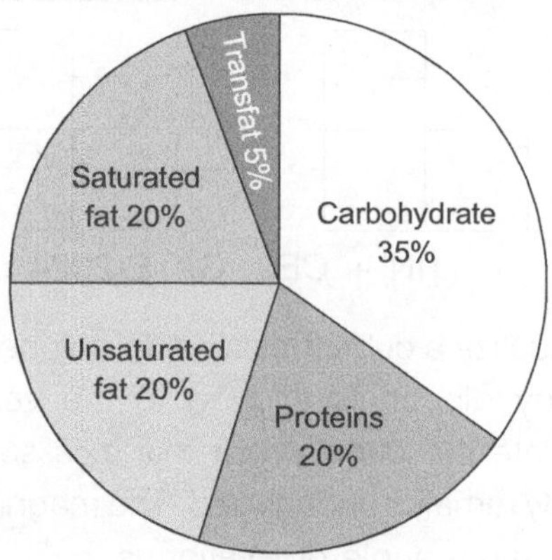

The typical energy density (kcal/g) of these macronutrients is given in the table.

Macronutrient	Energy density (kcal/g)
Carbohydrates	4
Proteins	4
Unsaturared fat	9
Saturared fat	9
Trans fat	9

The total fat (all three types), in grams, this person consumes is

(a) 3,600 (b) 100
(c) 44.4 (d) 77.8

[CS, GATE-2024 : 2 Marks, Set-1]

594. The least number of squares to be added in the figure to make AB a line of symmetry is

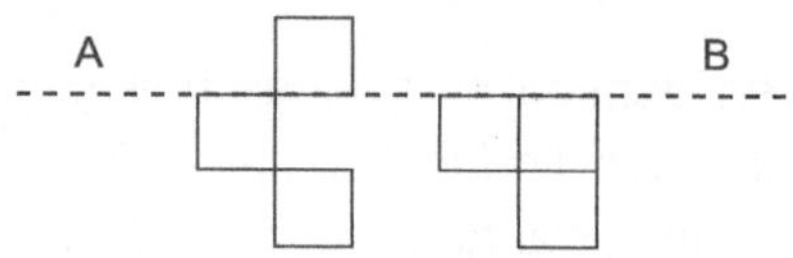

(a) 4 (b) 5
(c) 7 (d) 6

[CS, GATE-2024 : 2 Marks, Set-1]

595. The number of coins of ₹1, ₹5, and ₹10 denominations that a person has are in the ratio 5 : 3 : 13. Of the total amount, the percentage of money in ₹5 coins is

(a) $14\frac{2}{7}\%$ (b) 10%
(c) 21% (d) 30%

[CS, GATE-2024 : 1 Mark, Set-1]

596. If two distinct non-zero real variables x and y are such that $(x + y)$ is proportional to $(x - y)$ then the value of $\frac{x}{y}$

(a) is a constant
(b) depends on xy
(c) depends only on x and not on y
(d) depends only on y and not on x

[CS, GATE-2024 : 1 Mark, Set-1]

597. Consider the following sample of numbers:

9, 18, 11, 14, 15, 17, 10, 69, 11, 13

The median of the sample is

(a) 14 (b) 18.7
(c) 13.5 (d) 11

[CS, GATE-2024 : 1 Mark, Set-1]

598. In the 4 × 4 array shown below, each cell of the first three rows has either a cross (X) or a number.

1	X	4	3
X	5	5	4
3	X	6	X

The number in a cell represents the count of the immediate neighboring cells (left, right, top, bottom, diagonals) NOT having a cross (X). Given that the last row has no crosses (X), the sum of the four numbers to be filled in the last row is

(a) 9 (b) 11
(c) 12 (d) 10

[CS, GATE-2024 : 2 Marks, Set-2]

599. In the sequence 6, 9, 14, x, 30, 41, a possible value of x is

(a) 18 (b) 21
(c) 25 (d) 20

[CS, GATE-2024 : 1 Mark, Set-2]

600. Two wizards try to create a spell using all the four elements, *water*, *air*, *fire*, and *earth*. For this, they decide to mix all these elements in all possible orders. They also decide to work independently. After trying all possible combination of elements, they conclude that the spell does not work. How many attempts does each wizard make before coming to this conclusion, independently?

(a) 12 (b) 24
(c) 16 (d) 48

[CS, GATE-2024 : 1 Mark, Set-2]

601. A person sold two different items at the same price. He made 10% profit in one item, and 10% loss in the other item. In selling these two items, the person made a total of

(a) 1% profit (b) 1% loss
(c) 2% profit (d) 2% loss

[CS, GATE-2024 : 2 Marks, Set-2]

602. A cube is to be cut into 8 pieces of equal size and shape. Here, each cut should be straight and it should not stop till it reaches the other end of the cube.
The minimum number of such cuts required is

(a) 8 (b) 3
(c) 7 (d) 4

[CS, GATE-2024 : 2 Marks, Set-2]

603. Sequence the following sentences in a coherent passage.

P : This fortuitous geological event generated a colossal amount of energy and heat that resulted in the rocks rising to an average height of 4 km across the contact zone.

Q : Thus, the geophysicists tend to think of the Himalayas as an active geological event rather than as a static geological feature.

R : The natural process of the cooling of this massive edifice absorbed large quantities of atmospheric carbon dioxide, altering the earth's atmosphere and making it better suited for life.

S : Many millennia ago, a breakaway chunk of bedrock from the Antarctic Plate collided with the massive Eurasian Plate.

(a) SRPQ (b) SPRQ
(c) QPSR (d) QSPR

[CS, GATE-2024 : 2 Marks, Set-2]

604. In an engineering college of 10,000 students, 1,500 like neither their core branches nor other branches. The number of students who like their core branches is $\frac{1}{4}^{th}$ of the number of students who like other branches. The number of students who like both their core and other branches is 500.
The number of students who like their core branches is

(a) 1800 (b) 1500
(c) 3500 (d) 1600

[CS, GATE-2024 : 1 Mark, Set-2]

605. The pie charts depict the shares of various power generation technologies in the total electricity generation of a country for the years 2007 and 2023.

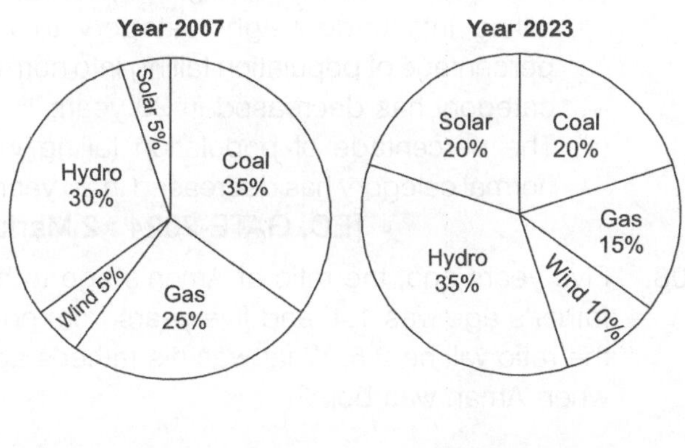

The renewable sources of electricity generation consist of Hydro, Solar and Wind. Assuming that the total electricity generated remains the same from 2007 to 2023, what is the percentage increase in the share of the renewable sources of electricity generation over this period?
(a) 77.5% (b) 50%
(c) 62.5% (d) 25%
[CS, GATE-2024 : 2 Marks, Set-2]

606. For positive non-zero real variables x and y, if $\ln\left(\frac{x+y}{2}\right)=\frac{1}{2}[\ln(x)+\ln(y)]$ then, the value of $\frac{x}{y}+\frac{y}{x}$ is
(a) 4 (b) 1/2
(c) 2 (d) 1
[CS, GATE-2024 : 1 Mark, Set-2]

607. The bar chart shows the data for the percentage of population falling into different categories based on Body Mass Index (BMI) in 2003 and 2023.

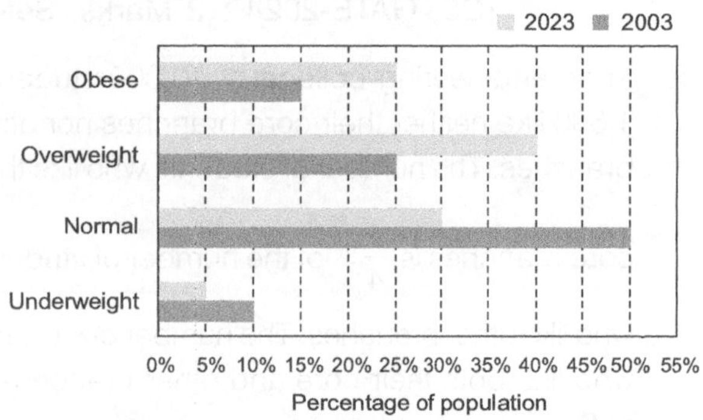

Based on the data provided, which one of the following options is INCORRECT?
(a) The ratio of the percentage of population falling into overweight category to the percentage of population falling into normal category has increased in 20 years.
(b) The ratio of the percentage of population falling into obese category to the percentage of population falling into normal category has decreased in 20 years.
(c) The ratio of the percentage of population falling into underweight category to the percentage of population falling into normal category has decreased in 20 years.
(d) The percentage of population falling into normal category has decreased in 20 years.
[EC, GATE-2024 : 2 Marks]

608. Five years ago, the ratio of Aman's age to his father's age was 1:4, and five years from now, the ratio will be 2:5. What was his father's age when Aman was born?
(a) 32 years (b) 35 years
(c) 28 years (d) 30 years
[EC, GATE-2024 : 1 Mark]

609. The greatest prime factor of $(3^{199} - 3^{196})$ is
(a) 11 (b) 13
(c) 3 (d) 17
[EC, GATE-2024 : 1 Mark]

610. Four identical cylindrical chalk-sticks, each of radius $r = 0.5$ cm and length $l = 10$ cm. are bound tightly together using a duct tape as shown in the following figure

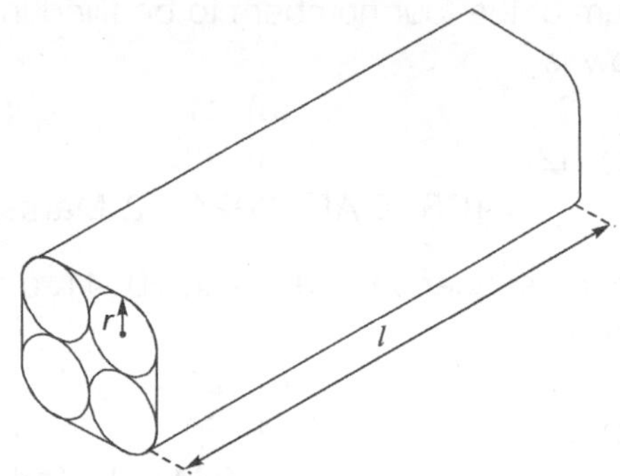

The width of the duct tape is equal to the length of the chalk-stick. The area (in cm²) of the duct tape required to wrap the bundle of chalk-sticks once, is
(a) $10(8 + \pi)$ (b) $20(4 + \pi)$
(c) $10(4 + \pi)$ (d) $20(8 + \pi)$
[EC, GATE-2024 : 2 Marks]

611. Two identical sheets A and B, of dimensions 24 cm × 16 cm, can be folded into half using two distinct operations. FO1 or FO2.
In FO1, the axis of folding remains parallel to the initial long edge, and in FO2, the axis of folding remains parallel to the initial short edge.
If sheet A is folded twice using FO1, and sheet B is folded twice using FO2, the ratio of the perimeters of the final shapes of A and B is
(a) 11 : 18 (b) 18 : 11
(c) 11 : 14 (d) 14 : 11
[EC, GATE-2024 : 2 Marks]

612. For a real number $x > 1$,
$$\frac{1}{\log_2 x}+\frac{1}{\log_3 x}+\frac{1}{\log_4 x}=1$$
The value of x is
(a) 24 (b) 12
(c) 4 (d) 36
[EC, GATE-2024 : 1 Mark]

613. Examples of mirror and water reflections are shown in the figures below:

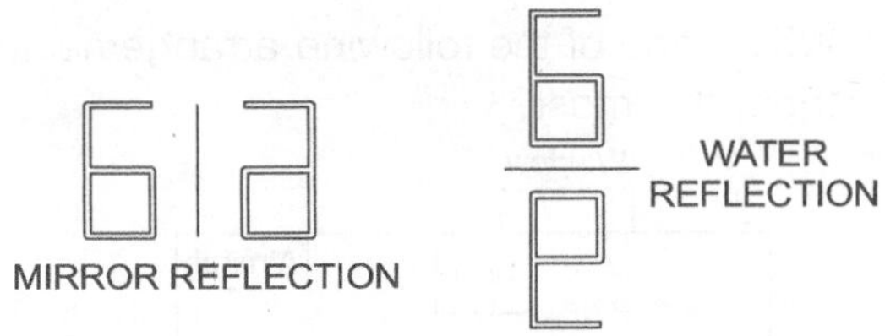

An object appears as the following image after first reflecting in a minor and then reflecting on water.

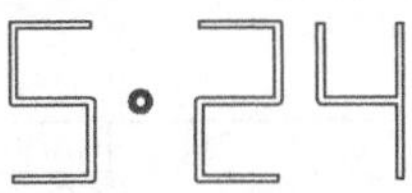

The original object is

(a) 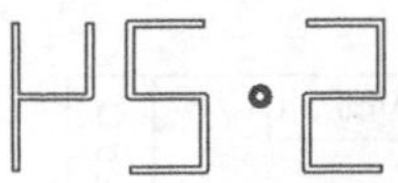 (b) 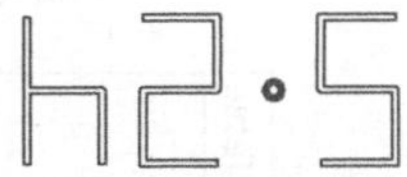

(c) 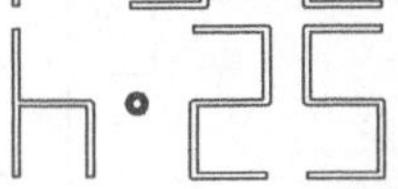 (d) 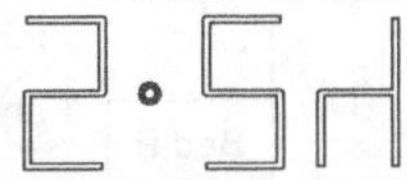

[EC, GATE-2024 : 2 Marks]

614. P, Q, R, S, and T have launched a new startup. Two of them are siblings. The office of the startup has just three rooms. All of them agree that the siblings should not share the same room.

If S and Q are single children, and the room allocations shown below are acceptable to all,

PR	TS	Q

PQ	RT	S

then, which one of the given options is the siblings?

(a) T and R (b) T and Q

(c) P and S (d) P and T

[EC, GATE-2024 : 1 Mark]

615. The asembly shown below has three teethed circular objects (Pinions) and two teethed flat objects (Racks), which are perfectly mating with each other. Pinions can only rotate clockwise or anti-clockwise staying at its own center. Racks can translate towards the left (←) and the right (→) direction.

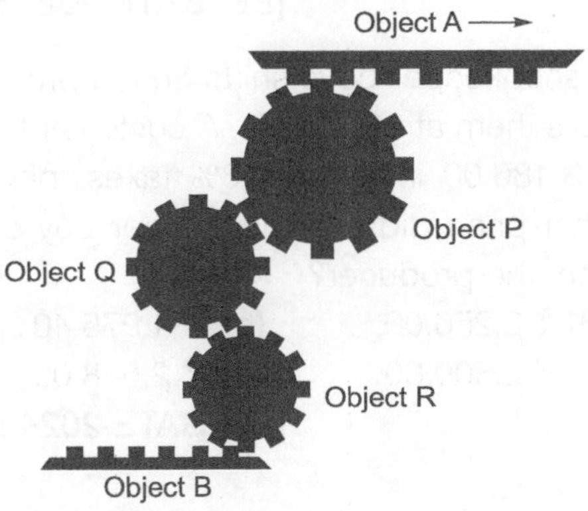

If the object A (Rack) is translating towards the right (→) direction, the correct statement among the following is

(a) Object B translates towards the right direction.

(b) Object Q rotates in the clockwise direction.

(c) Object B translates towards the left direction.

(d) Object R rotates in the anti-clockwise direction.

[EE, GATE-2024 : 2 Marks]

616. If, for non-zero real variables x, y and real parameter $a > 1$,

$$x : y = (a + 1) : (a - 1),$$

then, the ratio $(x^2 - y^2) : (x^2 + y^2)$ is

(a) $a : (a^2 - 1)$ (b) $a : (a^2 + 1)$

(c) $2a : (a^2 + 1)$ (d) $2a : (a^2 - 1)$

[EE, GATE-2024 : 1 Mark]

617. The decimal number system uses the characters 0, 1, 2, ..., 8, 9, and the octal number system uses the characters 0, 1, 2, ..., 6, 7.

For example, the decimal number 12 ($= 1 \times 10^1 + 2 \times 10^0$) is expressed as 14 ($= 1 \times 8^1 + 4 \times 8^0$) in the octal number system.

The decimal number 108 in the octal number system is

(a) 150 (b) 168

(c) 108 (d) 154

[EE, GATE-2024 : 1 Mark]

618. In the following figure,

CD = 5 cm, BE = 10 cm, AE = 12 cm

$\angle DAB = \angle DCB$ and $\angle DAE = \angle DBC = 90°$

Points AFCD create a rhombus.

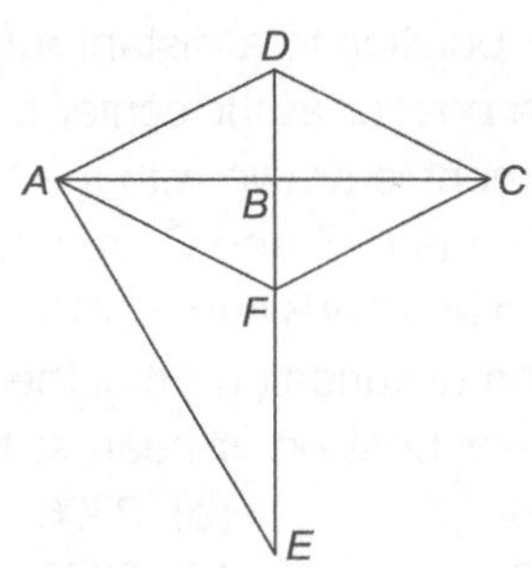

The length of BF (in cm) is

(a) 2 (b) 6

(c) 4 (d) 3

[EE, GATE-2024 : 2 Marks]

619. The chart below shows the data of the number of cars bought by Millennials and Gen X people in a country from the year 2010 to 2020 as well as the yearly fuel consumption of the country (in Million liters).

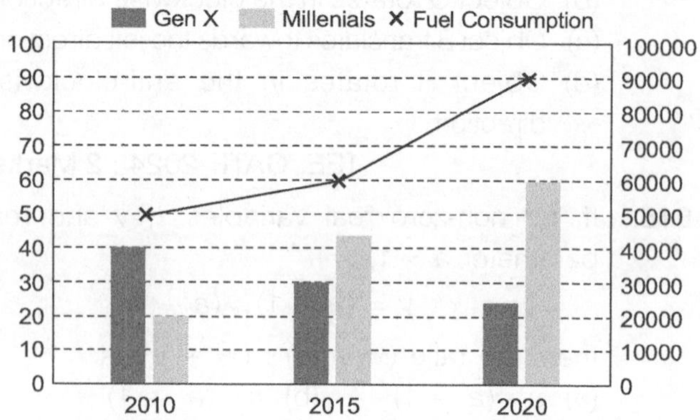

Considering the data presented in the chart, which one of the following options is true?

(a) The increase in the number of Millennial car buyers from 2015 to 2020 is less than the decrease in the number of Gen X car buyers from 2010 to 2015.

(b) The decrease in the number of Gen X car buyers from 2015 to 2020 is more than the increase in the number of Millennial car buyers from 2010 to 2015.

(c) The increase in the number of Millennial car buyers from 2010 to 2015 is more than the decrease in the number of Gen X car buyers from 2010 to 2015.

(d) The percentage increase in fuel consumption from 2010 to 2015 is more than the percentage increase in fuel consumption from 2015 to 2020.

[EE, GATE-2024 : 2 Marks]

620. A surveyor has to measure the horizontal distance from her position to a distant reference point C. Using her position as the center, a 200 m horizontal line segment is drawn with the two endpoints A and B. Points A, B and C are not collinear. Each of the angles ∠CAB and ∠CBA are measured as 87.8°. The distance (in m) of the reference point C from her position is nearest to

(a) 2606 (b) 2306
(c) 2603 (d) 2063

[EE, GATE-2024 : 2 Marks]

621. P and Q have been allotted a hostel room with two beds, a study table, and an almirah. P is an avid bird-watcher and wants to sit at the table and watch birds outside the window. Q does not mind that as long as his bed is close to the ceiling fan.

Which one of the following arrangements suits them the most?

(a)

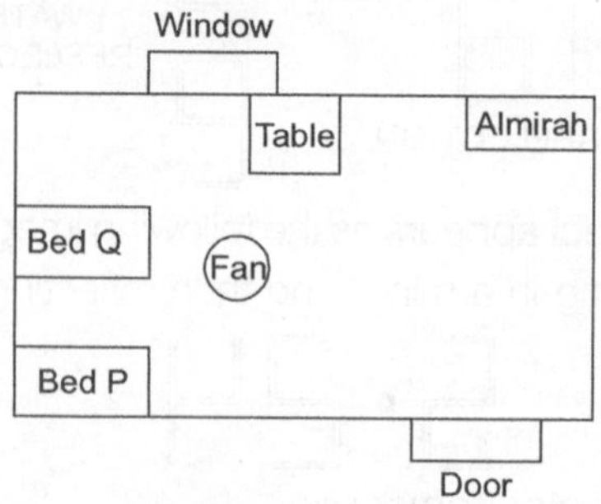

(b)

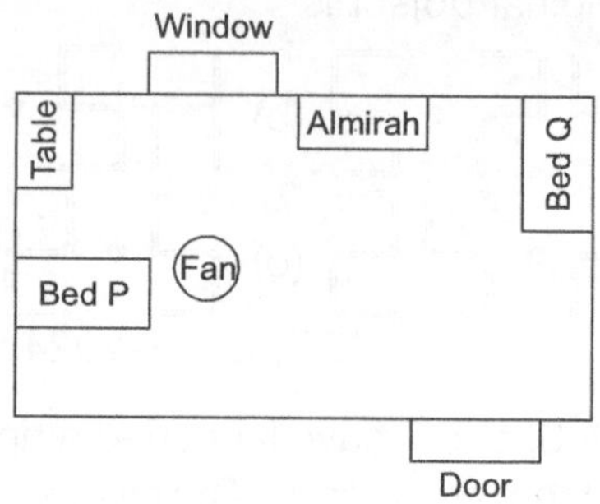

(c)

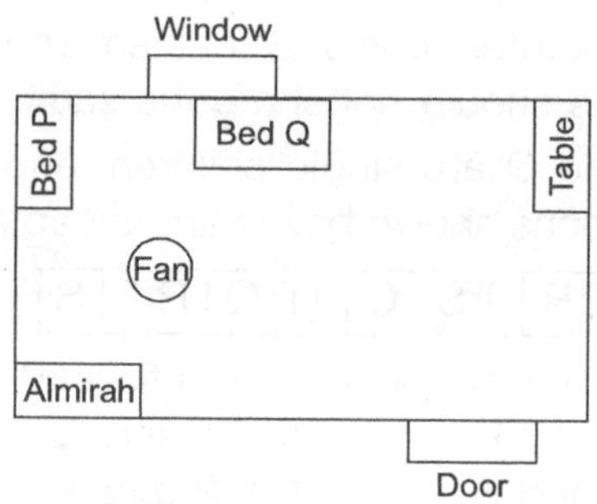

(d)

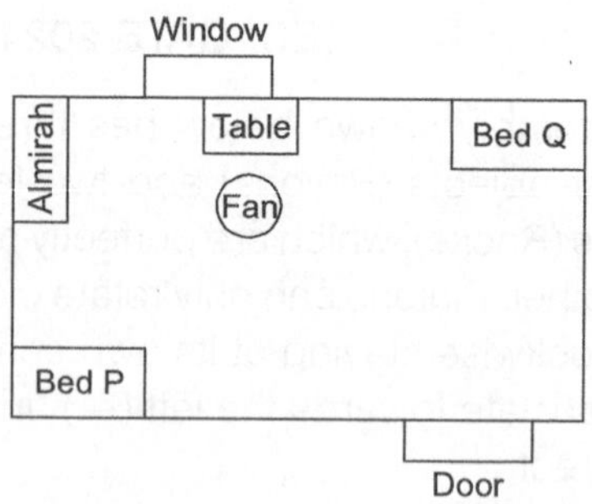

[EE, GATE-2024 : 1 Mark]

622. A shopkeeper buys shirts from a producer and sells them at 20% profit. A customer has to pay ₹ 3,186.00 including 18% taxes, per shirt. At what price did the shopkeeper buy each shirt from the producer?

(a) ₹ 2,250.00 (b) ₹ 1,975.40
(c) ₹ 2,500.00 (d) ₹ 2,548.00

[EE, GATE-2024 : 1 Mark]

■■■■

ANSWER KEY

1. (d)
2. (c)
3. (d)
4. (b)
5. (b)
6. (b)
7. (d)
8. (a)
9. (d)
10. (c)
11. (a)
12. (b)
13. (c)
14. (d)
15. (c)
16. (a)
17. (b)
18. (b)
19. (c)
20. (a)
21. (d)
22. (a)
23. (c)
24. (d)
25. (a)
26. (d)
27. (a)
28. (d)
29. (b)
30. (a)
31. (c)
32. (d)
33. (c)
34. (b)
35. (b)
36. (c)
37. (c)
38. (d)
39. (b)
40. (a)
41. (a)
42. (16)
43. (d)
44. (b)
45. (560)
46. (d)
47. (b)
48. (b)
49. (45)
50. (c)
51. (163)
52. (d)
53. (a)
54. (16)
55. (d)
56. (b)
57. (d)
58. (4)
59. (20000)
60. (0.81)
61. (a)
62. (495)
63. (c)
64. (b)
65. (b)
66. (22)
67. (b)
68. (96)
69. (d)
70. (850)
71. (48)
72. (6)
73. (b)
74. (c)
75. (d)
76. (140)
77. (a)
78. (c)
79. (c)
80. (d)
81. (c)
82. (c)
83. (c)
84. (1300)
85. (d)
86. (b)
87. (180)
88. (d)
89. (b)
90. (25)
91. (a)
92. (a)
93. (d)
94. (c)
95. (0.4896)
96. (b)
97. (c)
98. (4.54)
99. (b)
100. (b)
101. (b)
102. (a)
103. (d)
104. (4536)
105. (d)
106. (a)
107. (c)
108. (c)
109. (b)
110. (8)
111. (a)
112. (b)
113. (a)
114. (a)
115. (32)
116. (c)
117. (c)
118. (a)
119. (c)
120. (c)
121. (3)
122. (d)
123. (b)
124. (d)
125. (b)
126. (800)
127. (a)
128. (c)
129. (b)
130. (c)
131. (2.064)
132. (b)
133. (b)
134. (280)
135. (c)
136. (b)
137. (c)
138. (a)
139. (c)
140. (c)
141. (d)
142. (c)
143. (c)
144. (c)
145. (b)
146. (d)
147. (d)
148. (c)
149. (b)
150. (c)
151. (a)
152. (d)
153. (c)
154. (b)
155. (a)
156. (d)
157. (b)
158. (d)
159. (b)
160. (b)
161. (a)
162. (d)
163. (a)
164. (c)
165. (c)
166. (a)
167. (a)
168. (d)
169. (b)
170. (d)
171. (c)
172. (b)
173. (c)
174. (a)
175. (7)
176. (120)
177. (c)
178. (b)
179. (c)
180. (c)
181. (a)
182. (c)
183. (d)
184. (d)
185. (c)
186. (a)
187. (c)
188. (b)
189. (b)
190. (b)
191. (b)
192. (d)
193. (d)
194. (c)
195. (c)
196. (d)
197. (b)
198. (b)
199. (a)
200. (c)
201. (a)
202. (a)
203. (b)
204. (a)
205. (d)
206. (b)
207. (c)
208. (b)
209. (c)
210. (c)
211. (d)
212. (c)
213. (c)
214. (b)
215. (d)
216. (a)
217. (d)
218. (a)
219. (b)
220. (c)
221. (c)
222. (a)

223. (b)
224. (d)
225. (a)
226. (b)
227. (d)
228. (c)
229. (c)
230. (a)
231. (d)
232. (b)
233. (c)
234. (d)
235. (c)
236. (b)
237. (a)
238. (d)
239. (b)
240. (22.22)
241. (d)
242. (c)
243. (c)
244. (b)
245. (b)
246. (b)
247. (d)
248. (a)
249. (b)
250. (c)
251. (b)
252. (c)
253. (b)
254. (b)
255. (b)
256. (c)
257. (d)
258. (c)
259. (b)
260. (a)
261. (c)
262. (b)
263. (d)
264. (b)
265. (b)
266. (b)
267. (b)
268. (c)
269. (b)
270. (c)
271. (c)
272. (d)
273. (a)
274. (c)
275. (c)
276. (c)
277. (c)
278. (b)
279. (b)
280. (d)
281. (c)
282. (a)
283. (a)
284. (c)
285. (d)
286. (c)
287. (b)
288. (b)
289. (b)
290. (d)
291. (c)
292. (b)
293. (a)
294. (d)
295. (b)
296. (a)
297. (c)
298. (b)
299. (*)
300. (c)
301. (c)
302. (d)
303. (b)
304. (d)
305. (a)
306. (d)
307. (d)
308. (d)
309. (a)
310. (b)
311. (b)
312. (d)
313. (b)
314. (b)
315. (d)
316. (a)
317. (c)
318. (c)
319. (d)
320. (d)
321. (c)
322. (d)
323. (c)
324. (d)
325. (d)
326. (c)
327. (c)
328. (c)
329. (a)
330. (c)
331. (c)
332. (c)
333. (a)
334. (b)
335. (b)
336. (d)
337. (d)
338. (d)
339. (c)
340. (b)
341. (b)
342. (d)
343. (b)
344. (c)
345. (b)
346. (b)
347. (a)
348. (a)
349. (c)
350. (a)
351. (*)
352. (d)
353. (a)
354. (b)
355. (d)
356. (a)
357. (b)
358. (c)
359. (c)
360. (a)
361. (d)
362. (a)
363. (d)
364. (d)
365. (d)
366. (a)
367. (c)
368. (c)
369. (d)
370. (c)
371. (d)
372. (c)
373. (b)
374. (b)
375. (b)
376. (a)
377. (b)
378. (d)
379. (c)
380. (c)
381. (a)
382. (a)
383. (c)
384. (c)
385. (c)
386. (a)
387. (c)
388. (d)
389. (a)
390. (d)
391. (a)
392. (c)
393. (b)
394. (c)
395. (d)
396. (b)
397. (a)
398. (c)
399. (a)
400. (c)
401. (b)
402. (c)
403. (b)
404. (b)
405. (a)
406. (d)
407. (c)
408. (d)
409. (b)
410. (d)
411. (b)
412. (c)
413. (7)
414. (a)
415. (b)
416. (b)
417. (a)
418. (b)
419. (a)
420. (c)
421. (c)
422. (d)
423. (b)
424. (c)
425. (c)
426. (c)
427. (b)
428. (b)
429. (a)
430. (a)
431. (b)
432. (b)
433. (b)
434. (b)
435. (d)
436. (d)
437. (a)
438. (c)
439. (a)
440. (c)
441. (b)
442. (b)
443. (c)
444. (c)
445. (a)
446. (c)
447. (a)
448. (d)
449. (a)
450. (a)
451. (d)
452. (b)
453. (b)
454. (d)
455. (b)
456. (a)

457. (a)
458. (b)
459. (b)
460. (b)
461. (d)
462. (a)
463. (a)
464. (c)
465. (c)
466. (c)
467. (d)
468. (a)
469. (b)
470. (d)
471. (d)
472. (b)
473. (d)
474. (d)
475. (b)
476. (b)
477. (c)
478. (c)
479. (b)
480. (c)
481. (b)
482. (a)
483. (a)
484. (a)
485. (c)
486. (a)
487. (b)
488. (a)
489. (c)
490. (d)
491. (c)
492. (c)
493. (c)
494. (d)
495. (a)
496. (c)
497. (c)
498. (a)
499. (c)
500. (b)
501. (a)
502. (a)
503. (b)
504. (a)
505. (a)
506. (c)
507. (c)
508. (a)
509. (b)
510. (c)
511. (a)
512. (a)
513. (b)
514. (a)
515. (c)
516. (b)
517. (a)
518. (b)
519. (a)
520. (a)
521. (b)
522. (a)
523. (a)
524. (b)
525. (d)
526. (c)
527. (a)
528. (a)
529. (b)
530. (d)
531. (c)
532. (c)
533. (c)
534. (a)
535. (c)
536. (a)
537. (c)
538. (c)
539. (b)
540. (a)
541. (b)
542. (c)
543. (a)
544. (a)
545. (d)
546. (a)
547. (b)
548. (d)
549. (a)
550. (a)
551. (a)
552. (d)
553. (a)
554. (a)
555. (a)
556. (c)
557. (c)
558. (a)
559. (b)
560. (b)
561. (b)
562. (a)
563. (b)
564. (a)
565. (a)
566. (d)
567. (*)
568. (*)
569. (b)
570. (a)
571. (b)
572. (c)
573. (d)
574. (b)
575. (d)
576. (a)
577. (d)
578. (c)
579. (b)
580. (b)
581. (d)
582. (b)
583. (d)
584. (a)
585. (b)
586. (b)
587. (c)
588. (a)
589. (a)
590. (b)
591. (b)
592. (a)
593. (b)
594. (d)
595. (b)
596. (a)
597. (c)
598. (b)
599. (b)
600. (b)
601. (b)
602. (b)
603. (b)
604. (a)
605. (c)
606. (c)
607. (b)
608. (d)
609. (b)
610. (c)
611. (d)
612. (a)
613. (b)
614. (d)
615. (c)
616. (c)
617. (d)
618. (*)
619. (c)
620. (c)
621. (a)
622. (a)

EXPLANATIONS

1. **(d)**
Using the set theory formula
n(A) : Number of people who play hockey = 15
n(B) : Number of people who play football = 17
n $(A \cap B)$: Persons who play both hockey and football = 10
n $(A \cup B)$: Persons who play either hockey or football or both
Using the formula

$$n(A \cup B) = n(A) + n(B) - n(A \cap B)$$
$$n(A \cup B) = 15 + 17 - 10 = 22$$

Thus people who play neither hockey nor football = 25 − 22 = 3

Alternative Method

Refer to Venn diagram given below:

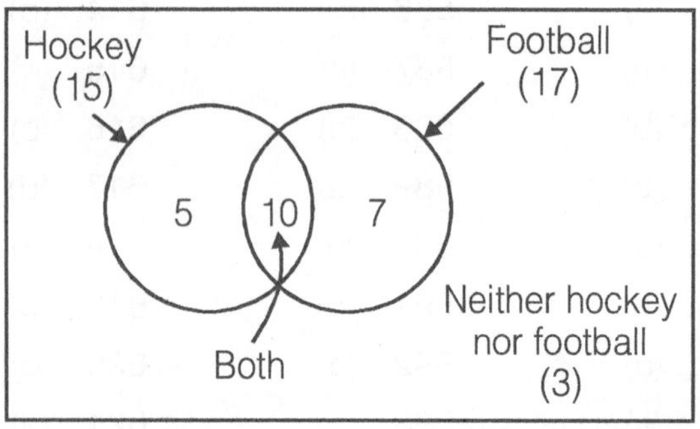

Number of people playing nither of the two games is equal to 3.

2. **(c)**
137 + 276 = 435
This an addition on base 8.
Hence, 731 + 672(8) = 1623
Alternative Method
7 and 6 added is becoming five means the given two numbers are added on base 8.

$$\begin{array}{r} (137)_8 \\ +(276)_8 \\ \hline (435)_8 \end{array}$$

Hence we have to add the another two given set of numbers also on base 8.

$$\begin{array}{r} (731)_8 \\ +(672)_8 \\ \hline (1623)_8 \end{array}$$

Hence the overall problem was based on identyfying base, which was 8, and adding number on base 8.

3. **(d)**

Per day work or rate of 5 skilled workers = $\frac{1}{20}$

⇒ Per day work or rate of one skill worker

$= \frac{1}{5 \times 20} = \frac{1}{100}$

Similarly Per day work or rate of 8 semiskilled workers = $\frac{1}{25}$

⇒ Per day work or rate of one semi-skill worker

$= \frac{1}{8 \times 25} = \frac{1}{200}$

And per day work or rate of 10 unskilled workers

$= \frac{1}{30}$

⇒ Per day work or rate of one semi-skill worker

$= \frac{1}{10 \times 30} = \frac{1}{300}$

Thus total per day work of 2 skilled, 6 semiskilled and 5 unskilled workers

$$= \frac{2}{100} + \frac{6}{200} + \frac{5}{300} = \frac{12+18+10}{600}$$

$$= \frac{40}{600} = \frac{1}{15}$$

Thus time to complete the work is 15 days.

Alternative Method
Let one day work of skilled semi-skilled and unskilled worker be a, b, c units respectively.
5a × 20 = 8b + 25 = 10c × 30 = Total unit of work

$$100a = 200b = 300c$$
$$a = 2b = 3c$$

$$\Rightarrow \quad b = \frac{a}{2} \text{ and } c = \frac{a}{3}$$

Given that 2 skilled, 6 semi-skilled and 5 unskilled workers are working. Let they finish the work in 'x' days.

$(2a + 6b + 5c)x = 5a + 20$

= Total units of work

$$\left(2a + 3a + \frac{5}{3}a\right)x = 5a \times 20$$

$$\frac{20a}{3}x = 5a \times 20$$

$$x = 15 \text{ days}$$

4. (b)

We have to make 4 digit numbers, so the number should be start with 3 or 4, two cases possible;

Case (1) thousands digit is 3

Now other three digits may be any of 2, 2, 3, 3, 4, 4, 4, 4.

(a) Using 2, 2, 3
$\Rightarrow$ 223, 232, 322 -------

$\left(\frac{3!}{2!} = 3 \text{ numbers are possible}\right)$

(b) Using 2, 2, 4 $\Rightarrow$ 224, 242, 422

$\left(\frac{3!}{2!} = 3 \text{ numbers are possible}\right)$

(c) Using 2, 3, 3 $\Rightarrow$ 233, 323, 332

$\left(\frac{3!}{2!} = 3 \text{ numbers are possible}\right)$

(d) Using 2, 3, 4 $\Rightarrow$ 234, 243, 324, 342, 423, 432

(3! = 6 numbers are possible)

(e) Using 2, 4, 4 $\Rightarrow$ 244, 424, 442

$\left(\frac{3!}{2!} = 3 \text{ numbers are possible}\right)$

(f) Using 3, 3, 4 $\Rightarrow$ 334, 343, 433

$\left(\frac{3!}{2!} = 3 \text{ numbers are possible}\right)$

(g) Using 3, 4, 4 $\Rightarrow$ 344, 434, 443

$\left(\frac{3!}{2!} = 3 \text{ numbers are possible}\right)$

(h) Using 4, 4, 4 $\Rightarrow$ 444

$\left(\frac{3!}{3!} = 1 \text{ numbers are possible}\right)$

Total 4 digit numbers in case 1 = 3 + 3 + 3 + 6 + 3 + 3 + 3 + 1 = 25

Case (2) thousands digit is 4 ; Now other three digits may be any of 2, 2, 3, 3, 3, 4, 4, 4.

(a) Using 2, 2, 3 $\Rightarrow$ 223, 232, 322

$\left(\frac{3!}{2!} = 3 \text{ numbers are possible}\right)$

(b) Using 2, 2, 4 $\Rightarrow$ 224, 242, 422

$\left(\frac{3!}{2!} = 3 \text{ numbers are possible}\right)$

(c) Using 2, 3, 3 $\Rightarrow$ 233, 323, 332

$\left(\frac{3!}{2!} = 3 \text{ numbers are possible}\right)$

(d) Using 2, 3, 4 $\Rightarrow$ 234, 243, 324, 342, 423, 432

$\left(\frac{3!}{2!} = 3 \text{ numbers are possible}\right)$

(e) Using 2, 4, 4 $\Rightarrow$ 244, 424, 442

$\left(\frac{3!}{2!} = 3 \text{ numbers are possible}\right)$

(f) Using 3, 3, 3 $\Rightarrow$ 333

$\left(\frac{3!}{3!} = 1 \text{ number is possible}\right)$

(g) Using 3, 3, 4 $\Rightarrow$ 334, 343, 433

$\left(\frac{3!}{2!} = 3 \text{ numbers are possible}\right)$

(h) Using 3, 4, 4 $\Rightarrow$ 344, 434, 443

$\left(\frac{3!}{2!} = 3 \text{ numbers are possible}\right)$

(i) Using 4, 4, 4 $\Rightarrow$ 444

$\left(\frac{3!}{3!} = 1 \text{ number is possible}\right)$

Total 4 digit numbers in case 2 = 3 + 3 + 3 + 6 + 3 + 3
= 1 + 3 + 1 = 26

Thus total 4 digits numbers using case (1) and case (2)
= 25 + 26 = 51

*** Alternative Method / Shortcut method**

As the number is greater than 3000. So thousand's place can be tiehr 3 or 4.

Let's consider the following two cases

Case (I) When thousand's place is 3.

<u>3</u> <u>a</u> <u>b</u> <u>c</u>

If there is no restricton on number of two's, three's and four's. Then each of a, b, c can be filled with 2 or 3 or 4 each in 3 ways.
So $3 \times 3 \times 3 = 27$ numbers are there. Out of which 3222, 3333 are invalid as 2 can be used twice & three thrice only so number of such valid numbers begining with 3 are $27 - 2 = 25$. ...(i)

Case (II) When thousand's place is 4

4 a b c

Without restriction on number of 2's, 3's and 4's a, b, c (as explained in case I) can be filled in 27 ways.
Out of these 27 numbers, 4 2 2 2 is only invalid as two have to be used twice only.
So valid numbers are $27 - 1 = 26$. ...(ii)
Total numbers from Case (I) & Case (II) 25 + 26 = 51.

5. (b)

Suppose: Hari's age : H, Gita's age : G, Saira's age : S, Irfan's age : I

- $H + G > I + S$
- Using Statement (2) both $G - S = 1$ or $S - G = 1$; G can't be oldest and S can't be youngest.
- There are no twins thus using statement (2) either GS or SG possible.

(A) HSIG: not possible as there is I between S and G which is not possible using statement (3)
(B) SGHI: SG order is possible, $S > G > H > I$ and $G + H > S + I$ (possible) Because if { $S = G + 1$; and $G = H+1$ and $H = I + 2$ then $G + (I+2) > (G+1) + I$ }
(C) IGSH: according to this $I > G$ and $S > H$ thus adding these both inequalities we get $I + S > G + H$ which is opposite of statement (2) thus not possible.
(D) IHSG: according to this $I > H$ and $S > G$ thus adding both inequalities $I + S > H + G$ which is opposite of statement (2). Thus not possible.

6. (b)

$$\log(P) = \frac{1}{2}\log(Q) = \frac{1}{3}\log(R)$$

$$\Rightarrow \log(P) = \log(Q)^{1/2} = \log(R)^{1/3} = K$$

$$\Rightarrow P = (Q)^{1/2} = (R)^{1/3} = K$$

$$\Rightarrow P = K, Q = K^2, R = K^3 \quad ...(i)$$

Now, only option (B) $Q^2 = PR$ satisfies

$Q^2 = (K^2)^2 = K^4$ From ...(i)

$P.R = K * K^3 = K^4$ From...(i)

Here, $Q^2 = PR$ holds true

7. (d)

Shortcut Method

Every time if we take 1 litre of mixture out and replace with water, content of pure spirit will keep on reducing by 10%.
So, final quantity of spirit after 3 such operations are

$$10 \times 0.9 \times 0.9 \times 0.9 = 7.29 \text{ litres}$$

Alternative Solution

$$\frac{\text{Quantity of spirit left after } n^{th} \text{ operation}}{\text{Intial quantity of spirit}}$$

$$= \left(\frac{a-b}{a}\right)^n = \left(1-\frac{b}{a}\right)^n$$

where 'a' is initial quantity of pure spirit and 'b' is quantity taken out and replaced every time.
Hence, quantity of spirit left after 3rd operation

$$= \text{initial quantity} \times \left(1-\frac{1}{10}\right)^3$$

$$= 10 \times 0.9 \times 0.9 \times 0.9$$

$$= 7.29 \text{ litres}$$

8. (a)

(T.C.) Total cost = V + f

$$\boxed{T.C = 4q + \frac{100}{q}}$$

As we have to minimize total cost.
Using options

(a) $q = 5$, T.C. $= 4 \times 5 + \frac{100}{5} = 40$

(b) $q = 4$, T.C. $= 4 \times 4 + \frac{100}{4} = 41$

(c) $q = 7$, T.C. $= 4 \times 7 + \frac{100}{7} = 42.285$

(d) $q = 6$, T.C. $= 4 \times 6 + \frac{100}{6} = 40.\overline{66}$

Hence, T.C. is minimum at q = 5, **(a) ans.**
* always put options to get answer fast.

Alternative Solution

or $\quad T.C. = 4q + \dfrac{100}{q}$

for minimum $\dfrac{d}{dq}\left(4q + \dfrac{100}{q}\right) = 0$

$\Rightarrow \quad q^2 = 25$ or $q = 5$

$$\frac{d}{dq}\left(4q + \frac{100}{q}\right)_{q=5} > 0$$

Hence, TC (Total cost) is minimum at q = 5 **(a) ans.**

9. **(d)**
As we can understand that the danger of a microbe to human being will be directly proportional to potency and growth. At the same time, it will be inversely proportional to toxicity defined (more dangerous will a microbe be if lesser of its milligram is required). So level of dangerous

$\alpha \dfrac{P\uparrow \times G\uparrow}{T\downarrow}$ where P, G and T are the potency, growth and toxicity as defined in question.

So $D_i = \dfrac{KPG}{T}$ ...(i)

Where K is constant of proportionality.
So level of dangerous of S will be maximum given

by $D_S = \dfrac{0.8 \times \pi(10mm)^2}{200}$.

Similar calculation for D_P, D_Q, D_R can be done based on (i) to find out that D_S is maximum and so most dangerous among them.

10. **(c)**
Let 'y' we the backlog with transporter and 'x' be the number of orders each day.
So, as per conditions given in question

$4x + y = 28$ ...(i)

$10x + y = 30$ ...(ii)

Solving (i) and (ii),

$$x = \frac{1}{3} \text{ and } y = \frac{80}{3}$$

Since, we need to find out number of trucks so that no pending order will be there at the end of 5th day.

$5x + y = n \times 5$

So we need to find 'n', where 'n' is the number of trucks required.

$$n = \frac{5x+y}{5} = \frac{5\times\frac{1}{3}+\frac{80}{3}}{5} = \frac{\frac{85}{3}}{5}$$

Hence, $5.\overline{66}$ truck will be required. As number of trucks have to be natural number. Hence, 6 trucks will be required.

11. **(a)**
Let there be overall 'x' candidates. Initially those who decided to vote for P and Q are. $0.4x$ and $0.6x$ respectively. On the day of election 15% of $0.4x = 0.06x$ went from P's side to Q's side and 25% of $0.6x = 0.15x$ went from Q's side to P's side.
Now, after transfer 'p' has $0.4x - 0.06x + 0.15x = 0.49x$
and after transfer 'q' has $0.6x - 0.15x + 0.06x = 0.51x$
Given, in the question that P lost by 2 votes

$$Q - P = 2 \text{ votes}$$

$$0.51x - 0.49x = 0.02x = 2 \text{ votes}$$

Hence, $(x = 100)$
Total number of votes are 100.

Alternative Solution
Using options:
Lets take option (a) 100 as total number of voters.
Initially P and Q has 40 and 60 voters on their sides respectively.
15% went from P's side to Q and 25% went from Q's side to P.

P	Q
(40)	(60)
(–) 15% of 40	– 25% of (60)
+ 25% of 60	+ 15% of (40)
40 – 6 + 15	60 – 15 + 6
= 49	= 51

Hence Q and P has 51 and 49 respectively difference in number of votes = 2. Hence answer (a).

12. **(b)**
Fuel consumed per km will we least, when mileage (kilometres per litres) mentioned on 'y' axis of graph will be maximum irrespective of number of kilometres travelled.
From the graph ('y' axis) we can observe that mileage (kilometres per litres) is maximum when vehicle is driven at 45 kilometres per hour. Hence

the stretch which 'Q' covered at 45 kmph, mileage was highest and fuel consumption per kilometre was lowest.

13. (c)

Shortcut Method

The question can be done orally as R took 1/3rd of tofees initially. Hence total number of tofees have to be multiple of 3. There is only one option (c) as 48 which is multiple of 3.

Alternative Solution

Using options:

If we use options. For e.g. option (c), if total number of tofees are 48. R took 1/3rd if 48 and returned 4. So R has 16 – 4 = 12 tofees. Out of left over 36 tofees S took 1/4th of 36 and returned 3. So, S has 9 – 3 = 6 and remaining 30 will be left in bowl. Now T took 1/2 of 30 and returned two back. So, the bowl has 17 left.

*Always solve such questions by options.

14. (d)

$$f(y) = \frac{|y|}{y}$$

As we know,

$$|y| = \begin{rcases} y & y \geq 0 \\ -y & y < 0 \end{rcases} \quad ...(i)$$

So from (i),

$$f(y) = \frac{y}{y} = 1 \quad \text{If } y \geq 0$$

$$f(y) = \frac{-y}{y} = -1 \quad \text{If } y < 0$$

Hence, $|f(q) - f(-q)| = |1 - (-1)| = 2$

Alternative Solution

Otherwise if $f(y) = \frac{|y|}{y}$

$$f(1) = \frac{|1|}{1} = 1$$

$$f(-1) = \frac{|-1|}{-1} = -1$$

For calculating $|f(q) - f(-q)|$

Let, $q = 1$, $|f(q) - f(-q)|$

$$|f(1) - f(-1)| = 1 - (-1) = 2$$

15. (c)

4 + 44 + 444...............

If we see first term of series = 4

Hence, sum upto 1st term is also = 4

Put, n = 1 (first term) only option (c) satisfies.

$$\frac{4}{81}\left[10^{n+1} - 9n - 10\right] = \frac{4}{81}\left[10^{1+1} - 9 \times 1 - 10\right]$$

$$= \frac{4}{81} \times 81 = 4$$

Hence, option (c) is correct answer. Always solve these questions by putting values.

Alternative Solution

4 + 44 + 444......................

$$\frac{4}{9}[9 + 99 + 999.............]$$

$$= \frac{4}{9}\left[(10-1) + (10^2 - 1) + (10^3 - 1).......\right]$$

$$= \frac{4}{9}\left[(10 + 10^2 + 10^3 .... 10^n) - (1 + 1 + 1.....\right]$$

$$= \frac{4}{81}\left[10^{n+1} - 9n - 10\right]$$

16. (a)

Total Cost $C = 5q^2$

Total Sales Revenue,

$S = 50q$

Profit (P)=(S – C) = $50q - 5q^2$

As the profit have to be maximized

$P = 50q - 5q^2$

Using options

(a) q = 5

$P = 50 \times 5 - 5 \times 5^2 = 125$

(b) q = 10

$P = 50 \times 10 - 5 \times 10^2 = 0$

(c) q = 15

$P = 50 \times 15 - 5 \times 15^2 =$ –ive

(d) q = 25

$P = 50 \times 25 - 5 \times 25^2 =$ –ive

Hence the maximum profit will happen at q = 5

Alternative solution:

$$P = 50q - 5q^2$$

For max/min $\frac{dp}{dq} = 0,$

$$\Rightarrow \quad 50 - 10q = 0$$

$$q = 10$$

$$\left.\frac{d^2p}{dq^2}\right]_{q=5} = -10 \text{ which is negative}$$

Hence maximum profit will happen at

$$q = 5$$

17. (b)

$$y = 2x - 0.1x^2$$

For y (height) to be maximum $\frac{dy}{dx} = 0$

$$\frac{dy}{dx} = 2x - 2x = 0$$

$$\Rightarrow \quad x = 10$$

$$\left.\frac{d^2y}{dx^2}\right]_{x=10} = -0.2 \text{ (–ive)}, \quad \frac{d^2y}{dx^2} < 0$$

Hence maximum height will be at x = 10

$$y = 2 \times 10 - 0.1 \times (10)^2 = 10$$

Maximum height = 10 meters.

18. (b)

Let total 100 shock absorbes are supplied

so X supplies = 60 (40% of 100)

Y supplies = 40 (60% of 100)

Reliable supply by X = 96% of 60 = 57.6

Reliable supply by Y = 72% of 40 = 28.8

So the required probability that random chosen reliable shock absorber is made by Y is

Required probability

$$= \frac{28.8}{57.6 + 28.8} = 0.333$$

$$\approx 0.334$$

19. (c)

Adding each and every observation by constant increases the arithmetic mean by same constant. Also multiplying each and every observation by constant mutiplies the arithmetic mean by same constant. While the standard deviation will not show similar effect.

Alternative Solution:

$$\frac{x_1 + x_2 + x_3 \cdots x_N}{N} = \bar{x}$$

$$\frac{(x_1 + 7) + (x_2 + 7) + \cdots + (x_n + 7)}{N} = \bar{x} + 7$$

and $$\frac{2(x_1 + x_2 \cdots x_N)}{N} = 2\bar{x}$$

Hence only P, R are correct.

20. (a)

AD, CG, FK, JP are given terms. The first letter in each term A, C, F, J are respectively 1, 3, 6, 10 letter of alphabetical sequence.

A = 1; 1st letter of alphabetical sequence

C = 1 + 2 = 3 ; 3rd letter of alphabetical sequence

F = 1 + 2 + 3 = 6 ; 6th letter of alphabetical sequence

G = 1 + 2 + 3 + 4 = 10; 10th letter of alphabetical sequence

The next term should begin with

1 + 2 + 3 + 4 + 5 = 15th letter of alphabetical sequence. So next term should begin with O.

Also in the series AD, CG, FK, JP, two letters in each term are separated by 2, 3, 4, 5 letters in between them respectively. Hence next term should be OV as there should be a gap of 6 alphabets between O and V.

Hence answer is OV.

21. (d)

$(1.001)^{1259} = 3.52$ and $(1.001)^{2062} = 7.85$

$$(1.001)^{1259} \times (1.001) = (1.001)^{3321}$$

$$\downarrow \qquad\qquad \downarrow$$

$$3.52 \times 7.85 = (1.001)^{3321} \quad (\text{as } a^m \times a^n = a^{m+n})$$

Hence $(1.001)^{3321} = 27.632 \approx 27.64$

22. (a)

Let the number of ₹ 20 notes be 'a' and ₹ 10 notes be 'b' respectively.

So, a + b = 14

20 a + 10 b = 230

on solving a = 9, b = 5

Hence number of ₹10 notes are 5.

Alternative Solution:

If we use option (a)

Number of ten ₹ notes are 5.

₹ 10 × 5 = ₹ 50

The remaining nine notes have to be ₹ 20 notes

₹ 20 × 9 = 180

and hence 50 + 180 = 230 (total money value)

Hence option (a) number of ten ₹ notes are 5.

23. (c)

Probability that A & B will meet will be given by the graphical representation where shaded region represents favorable area

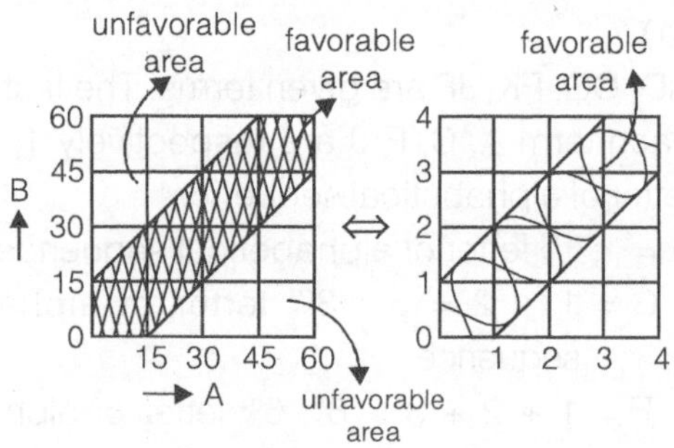

Required probability

$$= \frac{\text{fabourable area}}{\text{total area}} = \frac{7}{16}$$

24. (d)

Category	Amount
Food	4000
Clothing	1200
Rent	2000
Savings	1500
Others	1800
Total	10500

So total expenses are 10,500.
Expenses excluding savings are 9,000.
Hence required percentage

$$= \frac{9{,}000}{10{,}500} \times 100 = 85.714\% \approx 86\%$$

25. (a)

Divide 8 bags into three parts.
3, 3 and 2 respectively. Now two cases arise.

Case (i): Firstly, if we compare the 3, 3 bags on pans of a balance. We can identify on which side is the lighter bag placed. And then we will need only one more weighing for identifying faulty bag. So only two weighings are required.

Case (ii): Secondly, if the 3, 3 compared bags are equal, then to be sure faulty bag is one among the remaing two bags. So one more weighing is required to find out faulty bag from these two bags. Hence overall two weighings are required. Thus in either cases only two weighings are required.

Alternative Solution:

In case of a weighing balance (i.e. beam balance) the following model can be observed

1 – 3 → 1 weghing required
4 – 9 → 2 weghing required
10 – 27 → 3 weghing required
28 – 81 → 4 weghing required
and the process will go so on.

As there are eight objects. Minimum number of weighings required will be only 2.

28. (d)

% of raw material are salary and wages

$$= \frac{6240 - 5200}{5200} \times 100 = 20\%$$

% of raw material and research and development

$$= \frac{26400 - 22000}{22000} \times 100 = 20\%$$

29. (b)

$$|4x - 7| = 5$$
$$4x - 7 = 5$$
and $$4x - 7 = -5$$
$$\Rightarrow \quad x = 3$$
and $$x = 0.5$$
$$\therefore \quad 2|x| - |-x| = 2 \times 3 - 3 = 3$$
and $$2|x| - |-x| = 2 \times \frac{1}{2} - \frac{1}{2} = \frac{1}{2}$$
$$\Rightarrow \quad = \frac{1}{2}, 3$$

30. (a)

$$2x + y \le 6$$
$$x + 2y \le 8$$
$$x \le 1.33$$
$$y \le 3.33$$

∴ option (b) and (c) is rejected because $x \not> 1.33$.
Option (d) is rejected $x \not> 10/3$.

31. (c)

For sum to be maximum, we will consider only positive terms

$$44 + 42 + \ldots\ldots\ldots\ldots 2$$
or $$2 + 4 + 6 \ldots\ldots\ldots\ldots 42 + 44$$
$$2[1 + 2 + 3 \ldots\ldots\ldots\ldots 22]$$
$$= \frac{2[22 \times 23]}{2} = 506$$

Alternative:

Sum of even Natural Numbers

$$2 + 4 + 6 \ldots 44 = n(n + 1)$$
$$22 \times 23 = 506$$

32. (d)

There are Ninety (90) two digited numbers between 1 and 100 (typically 10 to 99). Among these numbers which are divisible by 7 are (14, 21, ...98) only 13 numbers.

$$\left[\because \left(\frac{100}{7}\right) - 1 = 13\right]$$

So, the required probability that the number is not divisible by

$$1 - \frac{13}{90} = \frac{77}{90}$$

33. (c)

$$\text{Average speed} = \frac{\text{Total distance}}{\text{Total time}}$$

$$= \frac{D}{\frac{D/2}{60} + \frac{D/4}{30} + \frac{D/4}{10}} = 24 \text{ kmph}$$

34. (b)

The question can be written as..

$$\frac{1}{\sqrt{2}+\sqrt{1}} + \frac{1}{\sqrt{3}+\sqrt{2}} + \frac{1}{\sqrt{4}+\sqrt{3}} \cdots \frac{1}{\sqrt{8}+\sqrt{80}}$$

Rationalize each term by multiplying it with its congugate in both numerator and denominator.

$$\left[\frac{1}{\sqrt{2}+\sqrt{1}} \times \frac{\sqrt{2}-\sqrt{1}}{\sqrt{2}-\sqrt{1}}\right] \times \left[\frac{1}{\sqrt{3}+\sqrt{2}} \times \frac{\sqrt{3}-\sqrt{2}}{\sqrt{3}-\sqrt{2}}\right] \times \ldots\ldots.$$

$$\ldots\ldots.. \left[\frac{1}{\sqrt{8}\times\sqrt{80}} \times \frac{\sqrt{81}-\sqrt{80}}{\sqrt{81}-\sqrt{80}}\right]$$

Which can be resolved as

$$\left[\sqrt{2}-\sqrt{1}\right] + \left[\sqrt{3}-\sqrt{2}\right] \ldots\ldots\ldots \left[\sqrt{81}-\sqrt{80}\right]$$

(as denominator in each term will become 1)

Now cancelling like terms we will be left with

$$\sqrt{81} - \sqrt{1} = (9 - 1) = 8$$

35. (b)

$$\text{Now cost} = 13200\left[1 + \frac{1}{5}\right]\left[1 - \frac{1}{24}\right]$$

$$= 13200\left[\frac{6}{5}\right] \times \left[\frac{23}{24}\right]$$

$$= 15,180$$

36. (c)

$$\frac{M + Tue + W}{3} = 41$$

$\Rightarrow$ $M + Tue + W = 123$ ...(i)

$$\frac{Tue + W + Th}{3} = 43$$

$\Rightarrow$ $Tue + W + Th = 129$ ...(ii)

From eq. (ii) – eq. (i)

$Th - M = 6$

Also $Th = 1.15\,M$

$1.15\,M - M = 6$

$\Rightarrow$ $15\,M = 6$

$\Rightarrow$ $M = 40$

$Th = 1.15\,m$

$\Rightarrow$ $Th = 1.15 \times 40$

$\Rightarrow$ $Th = 46$

37. (c)

$$\text{Average speed} = \frac{\text{Total distance}}{\text{Total time}}$$

$$\frac{\text{Total distance}}{\text{Total time}} = \frac{(8+6+16)\text{km}}{\left(\frac{1}{4}+\frac{1}{4}+\frac{1}{4}\right)\text{hr}}$$

$$\frac{30\text{km}}{\frac{3}{4}\text{hr}} = 40 \text{ km/hr}$$

38. (d)

$S = 10 + 84 + 73 + \ldots$

$S = (9 + 1) + (9^2 + 3) + (9^3 + 5)\ldots$

$S = \{9 + 9^2 + 9^3 + \ldots 9^n\} + \{1 + 3 + 5 \ldots (2n - 1)\}$

$$S = \frac{9(9^n - 1)}{(9-1)} + n^2$$

$$\left(\begin{array}{l}\text{As } a + ar + ar^2 \ldots\ldots\ldots ar^{n-1} \\ = \dfrac{a(r^n - 1)}{(r-1)} \quad \text{for} \quad r > 1\end{array}\right)$$

$$\left(\begin{array}{l}\text{and } 1 + 3 + 5 \ldots\ldots\ldots\ldots (2n-1) = n^2 \\ \text{Sum of odd natural numbers } = n^2\end{array}\right)$$

$$\text{So } S = \frac{9(9^n - 1)}{8} + n^2$$

Alternative Method:

Put n = 2 in all options and check wheather sum = 10 + 84 = 94 is in which option.

39. (b)

$3x^2 + 2x + p(p - 1) = 0$

If roots are of opposite signs product of the roots will be negative.

$$\frac{p(p-1)}{3} < 0$$

$$p(p - 1) < 0$$

$$0 < p < 1$$

$$P \in (0, 1)$$

40. (a)
Every leap year will have 366 days.

1 leap year = 366 days
= 52 × 7 + 2 days

Every leap year will have 52 complete weeks. For 53rd Saturday there are two favourable cases from 7 possible cases

Fri	Sat
Sat	Sun
Sun	Mon
Mon	Tue
Tue	Wed
Wed	Thu
Thu	Fri

(Fri Sat, Sat Sun) → 2 favourable cases

Required probability = $\frac{2}{7}$

42. Solution :
12, 35, 81, 173, 357, _____
Difference between consecutive terms of the series are : 13, 46, 92, 184, _____.

$$35-12 = 23 = 23\times1$$
$$81-35 = 46 = 23\times2$$
$$173-81 = 92 = 23\times4$$
$$357-173 = 184 = 23\times8$$

So the next difference have to be 23 × 8 = 368, therefore next term have to 357 + 368

$$= 725$$
$$= 23\times16$$
$$725-357 = 368 = 23\times16$$

43. (d)
Difference between 1st and 4th, 2nd and 3rd letters in each series respectively is same except NVBD.

44. (b)
A resident female with annual income Rs. 9 Lakh should satisfy eligibility for filling form T. So answer is option B is answer. Option A, C , D corresponds to filling form P, S, D respectively.

45. Solution :
Let the length of platform is x and velocity of train is V.
Distance = Vel. × time

$$(280+x) = V\times60 \quad \text{... (i)}$$
$$280 = V\times20 \quad \text{... (ii)}$$

From equation (i) and (ii), we get

$$x = 560 \text{ m}$$

46. (d)
Deficit should be equal to (1/5)th of exports which is possible in 2006

$$\text{Deficit} = \frac{\text{Imports} - \text{Exports}}{\text{Exports}}$$
$$= \frac{120-100}{100} = \frac{20}{100} = \frac{1}{5}$$

47. (b)
As one of the face is already shown as head os the possibility of second coin is not there (because second coin has both tail). Now sample space will contain now events (H, H, T). So the probability for another face as tail will be 1/3.

48. (b)
As the numbers appearing on the faces are not equally likely because their probability of occurrence over a large number of throws are different (as given) so we can term it to be biased die.

49. Solution :
2, 3, 6, 15, _____, 157.5, 630

$$2\times1.5 = 3$$
$$3\times2 = 6$$
$$6\times2.5 = 15$$
$$15\times3 = 45$$
$$45\times3.5 = 157.5$$
$$157.5\times4 = 630$$

Hence the answer is 45.
45 is missing term of the series.

50. (c)
Adding 6, 3 and 2 to the consecutive letters gives the only (c) does not follow.

51. Solution :
Let eight consecutive odd numbers be
a, a + 2 .... a + 14
and four consecutive over numbers be
b, b + 2, b + 14, b + 16
Given (a) + (a + 2), ... + (a + 14) = 656

$$8a + 56 = 656$$
$$a = 75$$

Also given
(b) + (b + 2)+ (b + 4) + (b + 6) = 87 × 4
= 348

$$4b + 12 = 348$$
$$b = 84$$

Sum of smallest odd no (a) and second largest even no ($b + 4$) will be

$$= (a) + (b + 4)$$
$$= 75 + (84 + 4) = 163$$

52. (d)

For maximum revenue per kg. Let us solve through options

A. Item 2 = $\frac{20\% \text{ of } 250}{20\% \text{ of } 500} = 0.5$

B. Item 3 = $\frac{23\% \text{ of } 250}{19\% \text{ of } 500} = 0.6$

C. Item 6 = $\frac{19\% \text{ of } 250}{16\% \text{ of } 500} = 0.59$

D. Item 5 = $\frac{20\% \text{ of } 250}{12\% \text{ of } 500} = 0.83$

Clearly item 4 is largest.

53. (a)

Drainage pipe drains half of the tank in 30 minutes or full tank is 68 minutes. Now the tank have to be filled in 10 minutes (when both drainage and filling pipes are in action)

Drainage pipe would drained 1/6th of capacity of tank in 10 minutes. So filling pipe have to fill 1/2 of capacity of 1/6 of capacity more i.e. 2/3rd of the tank in 10 minutes. or complete tank is 15 minutes. That's why the filling tank should have rate 4 times the draining rate.

Alternatively :

Drainage pipe can drain complete tank in 60 min. Let the filling pipe can fill the tank in 'x' min. Now both are operational and tank is half filled. Also remaining half needs to get filled in 10 minute.

$$10\left[\frac{1}{x} - \frac{1}{60}\right] = \frac{1}{2}$$
$$x = 15 \text{ minutes}$$

So filling pipe needs to have rate be times drainage pipe's rate.

54. Solution :

81, 54, 36, 24, ____

Alternate terms are perfect squares 9^2, 6^2, 4^2 and in between terms are products 9×6, 6×4m ____ and so on missing terms is

$$4^2 = 16$$

55. (d)

$$P = A < R < M$$

P is surely less than M only in option (d) else in any other option it cannot be conveyed surely.

56. (b)

7 G, 11 K, 13 M, ____

Next term will be Q which is 17 letter of English alphabet.

Among option only (b) (i.e. 17 Q) is correct.

57. (d)

On seeing the given figure in question i, ii and iii can be concluded.

58. Solution :

Let the velocity of stream is 'V' time taken during downstream is 't' and distance is 'x'.

$$t = \frac{x}{8+V} \quad \text{...(i)}$$

During upstream

$$3t = \frac{x}{8-V} \quad \text{...(ii)}$$

Dividing equation (ii) by (i), we get

$$V = 4 \text{ m/s}$$

59. Solution :

Since,

$$\because \quad 15\% = 4,50,000$$
$$\therefore \quad 100\% = \frac{4,50,000 \times 100}{15}$$
$$= 30,00,000$$

Selling price of purifier

$$= \frac{30,00,000 + 10,00,000}{200}$$
$$= 20,000$$

60. Solution :

Probability of bulb being defective = $\frac{5}{100}$

$$= 0.05$$

Probability of bulb being non defective

$$= 1 - 0.05$$
$$= 0.95$$

Probability that the batch is accepted

= None of the four bulb being defective

$$= (0.95)^4 = 0.81450625$$
$$\approx 0.81$$

61. (a)

$$f(x, y) = (2x)^n \times \left(\frac{y}{n}\right)^m = 2^{n-m} \times P$$

62. Solution :

Let the 12 consecutive odd numbers be

a, a + 2, a + 4, a + 6 ..... a + 22

Sum of first five (a) + (a + 2) + (a + 4) + (a + 6) + (a + 8) = 425

5a + 20 = 425 ⇒ a = 81

Sum of last five (a + 14) + (a + 16) + (a+18) + (a + 20) + (a + 22)

5a + 90 = 5 × 81 + 90 = 495

63. (c)

13 M, 17 Q, 19 S

among options only W is correct as W is 23rd letter in English language.

64. (b)

BEST OF LUCK

B–ST –F L–CK

GOOD WISHES

G—D W–SH–S

ACE THE EXAM

–C– TH_ –X–M

So answer is MXHTC (vowel is missing).

65. (b)

The value doubled over 10 years

$$A = P\left(1+\frac{R}{100}\right)^n$$

$$2P = P\left(1+\frac{R}{100}\right)^{10}$$

$$R \approx 7.2$$

Alternative :

Simple growth rate for a quantity to double in 10 years will be 10%. For compounded growth answer have to be less than 10. Among options 7.2 (Ans. (b)) is correct as ((a) (5.6)) will be very much less than 10.

((c) (10)), ((d) (12) cannot be answer as compared growth rate have to be less than 10.

66. (22)

In 2012, total cost

$$= \frac{4,50,000 \times 100}{15}$$

$$= 30,00,000$$

Cost of raw material in 2012

$$= 30,00,000 \times \frac{20}{100}$$

$$= 6,00,000$$

Cost of raw material in 2013

$$= 600000 + 600000 \times \frac{30}{100}$$

$$= 7,80,000$$

other expenditure in 2012

$$= 3000000 - 600000$$

$$= 2400000$$

Increased cost of other expenditure in 2013

$$= 2400000 + 2400000 \times \frac{20}{100}$$

$$= 2880000$$

Total cost in 2013

$$= 2880000 + 780000$$

$$= 3660000$$

% increase in total cost in 2013

$$= \frac{3660000 - 3000000}{3000000} \times 100$$

$$= 22\%$$

67. (b)

Overall there will be 5! = 120 numbers. Digit I will appear at unity, tens, hundred, thousand and ten thousand place 4! times each. So sum received due to all 1's will be 4! × (1 + 10 + 100 + 1000 + 10000) = 4! × 1111

Similarly sum received due to all 3's

4! × (3 + 30 + 300 + 3000 + 30000) = 4! × 3333

Similarly for digits 3, 4 and 5

So, overall sum received due to all digits

= 4! × 1111 [1 + 3 + 5 + 7 + 9]

= 24 × 1111 [25] = 66666600

68. (96)

$$\left(z+\frac{1}{z}\right)^2 = 98$$

So, $$z^2 + \frac{1}{z^2} = 96$$

69. (d)

Given $ax^2 + bx + c = 0$

roots are real and positive

For e.g.

$x^2 - 5x + 6 = 0$ roots are 2, 3

Then roots of $ax^2 + b|x| + c = 0$

then roots of $x^2 - 5|x| + 6 = 0$

$\therefore |x|^2 - 5|x| + 6 = 0$

$$\left.\begin{array}{l}\text{if } x > 0 \\ |x| = x\end{array}\right\} \quad x^2 - 5x + 6 = 0$$

$(x - 2)(x - 3) = 0$ $\quad$ x = 2, 3 are roots

$$\left.\begin{array}{l}\text{If } x < 0 \\ |x| = -x\end{array}\right\} \quad x^2 + 5x + 6 = 0$$

$(x + 2)(x + 3) = 0$ $\quad$ x = –2, –3 are roots

So if $x^2 - 5x + 6 = 0$ has two real roots

then $|x|^2 - 5|x| + 6 = 0$ has four real roots

$ax^2 + bx + c = 0$ has real roots

then $ax^2 + b|x| + c = 0$ has 4 real roots.

70. (850)

One way fare of single person = 100

Undiscounted round trip fare of singe person = 200

Undiscounted round trip fare of five persons = 1000

Round trip will result in a discount of 10%. Also additional discount of 5% on total fare for 4 or more person.

So a term of 5 person will be eligible for overall discount of (10% + 5%) = 15% discount

Discounted round trip for 5 persons

$1000 \times 0.85 = 850$ Rs.

71. (48)

Following table can be redrawn:

		Men	Women	Total
Own vehicle	Car	40	34	74
	Scooter	30	20	50
	Both	60	46	106
Do not own vehicle		20	50	70

Now following venn-diagram can be drawn

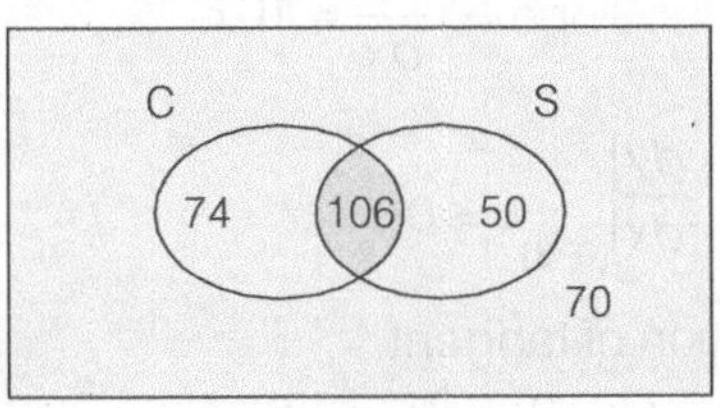

So % of people not owning a scooter

$$\left[\left(\frac{74+70}{300}\right) \times 100\right] = \frac{144}{300} \times 100 = 48\%$$

73. (b)

There are 19 multiples of 10 from 2 to 198 typically 10, 20, 30 ··· 190.

Average of these number will be

$$= \frac{10 + 20 + 30 + \cdots + 190}{19}$$

$$= 10\frac{[1 + 2 + 3 + \cdots + 19]}{19}$$

$$= 10\frac{[19 \times 20]}{19 \times 2} = 100$$

74. (c)

Let, $\quad y = \sqrt{12 + \sqrt{12 + \sqrt{12 + \cdots}}}$

So, $\quad y = \sqrt{12 + y}$

Squaring $\quad y^2 = 12 + y$

$y^2 - y - 12 = 0$

$\Rightarrow \quad (y - 4)(y + 3) = 0$

$y = 4$

Option (c) $\rightarrow$ 4.000

75. (d)

Given x is real

$|x^2 - 2x + 3| = 11$

$x^2 - 2x - 8 = 11$

$x^2 - 2x - 8 = 0$

$(x - 4)(x + 2) = 0$

$x = 4, -2$

or $x^2 - 2x + 3 = -11$

$x^2 - 2x + 14 = 0$

Will not give real roots as $D = b^2 - 4ac < 0$

Now $|-x^3 + x^2 - x|$ can attach

If x = 4,

$|-x^3 + x^2 - x| = |-64 + 16 - 4| = 52$

If x = –2,

$|-x^3 + x^2 - x| = |8 + 4 + 2| = 14$

76. (140)

Given ratio of male to female is 2.5 in year 2008.

So let us assume male as 500 and female as 200.

So ratio $\frac{M}{F}$ in 2008 = 2.5

Now female doubles up in 2009

Number of female in 2009 = 2 × 200 = 400

$\frac{M}{F}$ ratio in 2009 is 3

$$\frac{M}{400} = 3$$

$\Rightarrow$ M = 1200

So number of males in 2009 is 1200.

% increase in number of males from 2008 to 2009

% increase = $\frac{1200-500}{500} \times 100 = 140\%$

77. (a)

Any angle and corresponding time can be driven by

$$\left[5x - \left(\frac{D^0}{6}\right)\right] \times \frac{12}{11}$$

Between 6 and 7 pm angle have to be 60°

Put x = 6 and D = 60°

$$\left[5 \times 6 - \frac{60°}{6}\right] \times \frac{12}{11} = 20 \times \frac{12}{11} = 21\frac{9}{11}$$

So at $4:21\frac{9}{11}$ angle between minutes and hour hand will be 60°.

In given question angle have to close to 60° nearest option is 6 : 22.

78. (c)

2, 5, 10, 17, 26, 37, 50, 64.

$2 = 1^2 + 1$, $5 = 2^2 + 1$, $10 = 3^2 + 1$,

$17 = 4^2 + 1$, $26 = 5^2 + 1$, $37 = 6^2 + 1$,

$50 = 7^2 + 1$, $\boldsymbol{64 = 8^2}$

So 64 is the incorrect term.

79. (c)

Q.No.	Marks	Answered Correctly	Answered Wrongly	Not Attemped	Total Students
1	2	21	17	6	44
2	3	15	27	2	44
3	2	23	18	3	44

Average = $\frac{\text{Total marks scored}}{\text{Total number of students}}$

$= \frac{21 \times 2 + 15 \times 3 + 23 \times 2}{44} = 3.022727$

$\simeq 3.02$

80. (d)

Let GDP before year 2012-2013 be 100 Rs.

During 2012-2013 GDP increased by 7%.

So, GDP in year 2013-2013 = 107 Rs.

Now exchange rate for USD change from Rs. 50/USD to Rs. 60/USD.

i.e. value of Rupee reduced to $\frac{5}{6}$ of original value.

So, new GDP (in USD) = $107 \times \frac{5}{6} = 89.16$

Effective change is from 100 to 89.16

= 10.83% decrease $\simeq$ 11% decrease

81. (c)

Ratio of male to female 2001 is given.

Let there be 100 males and 100 female in 2011.

Now number of female in 2012 is same as number of females in 2012 as given in question.

So female in 2012 is also 100.

Ratio of male of female in 2012 is 1.5

$$\frac{M}{F} = 1.5$$

$$\Rightarrow \frac{M}{F} = 1.5$$

$\Rightarrow$ M = 150

So there are 150 males n 2012.

Ratio of male in 2012 to male in 2011

$$= \frac{150}{100} = \frac{1.5}{1}$$

82. (c)

$(7526) - (Y)_8 = (4364)_8$

So, $Y = (7526)_8 - (4364)_8$

$= (7526 - 4364)_8$

$= (31420)_8$

83. (c)

$$\text{Slope} = \frac{dy}{dx} = 10x$$

$$\left.\frac{dy}{dx}\right|_{(0,\,3)} = 0$$

Equation of tangent

$(y - 3) = 0\,(x - 0)$

$y = 3$

Line parallel to x-axis.

84. **Solution**

Total cost of production

= 50000 + 800 × 100

= ₹ 130000

Cost of production per tonne = $\frac{130000}{100}$ = ₹ 1300

88. (d)

Let the population of county Y is P

∴ Population of country X is 3P

% of people taken than 6 ft

$$= \frac{\frac{3P \times 1}{100} + \frac{P \times 2}{100}}{3P + P} = 1.25$$

90. **Solution**

Average time interval = $T = \frac{1}{P}$ = 25 years

91. (a)

$$P_{new} = P_{old}\left(1 + \frac{r}{100}\right)^n$$

$$\Rightarrow \quad 10 = 5\left(1 + \frac{20}{100}\right)^n$$

$$\Rightarrow \quad 2 = (1.2)^n$$

$$\Rightarrow \quad \frac{\log 2}{\log 1.2} = n$$

$$\Rightarrow \quad n = 3.8 \text{ year}$$

93. (d)

Total quantity of export = 5 lakh tonnes

Quantity of export of item-1

= 0.11 × 5 = 0.55 lakh tonnes

Quantity of export of item-4 = 0.22 × 5

= 1.1 lakh tonnes

Total revenues generated from item-1

= 250 × 0.12 = 30 crore

Total revenues generated from item-4

= 0.06 × 250 = 15 crore

Revenue generated through export of item-1 per

$$\text{kg} = \frac{30 \times 1000000}{0.55 \times 1000000}$$

Rs = 545.45/kg

Revenue generated through export of item-4 per

$$\text{kg} = \frac{15 \times 10^6}{101 \times 10^5}$$

Rs = 136.36/kg

$$\frac{\text{R.G. item 1}}{\text{R.G. item 4}} = \frac{4}{1}$$

94. (c)

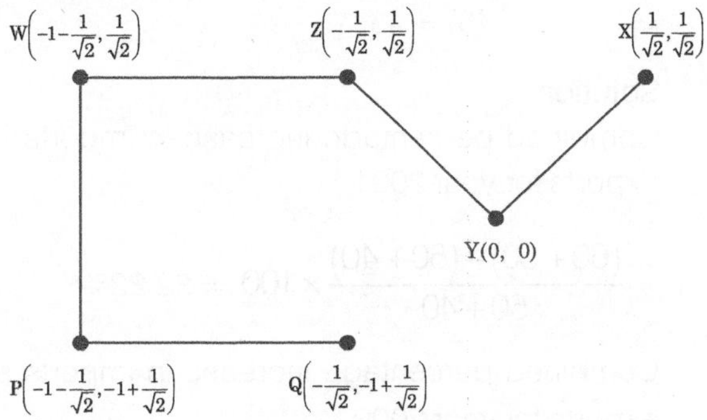

$$XQ = \sqrt{\left(\frac{1}{\sqrt{2}} + \frac{1}{\sqrt{2}}\right)^2 + (-1)^2}$$

$$= \sqrt{3}$$

95. **Solution**

10% HIV$^+$ → 95% correct, 5% incorrect

90% HIV$^-$ → 89% correct, 11% incorrect

N(declared HIV$^+$) tree diagram is as shown above = 0.1 × 0.95 + 0.9 × 0.11

= 0.194

N(actual HIV$^+$) = 0.1 × 0.95

= 0.095

$$P(\text{actual HIV}^+) = \frac{0.095}{0.194} = 0.4896$$

96. (b)

R O A D
+3↓ +3↓ +3↓ +3↓
U R D G

Add 3 to respective letters

So,

S W A N
+3↓ +3↓ +3↓ +3↓
V Z D Q

97. (c)

Let $f(x) = ax + b$ (linear function)

at $x = -2$, $f(x) = 29$

$\Rightarrow \quad -2a + b = 29 \quad$ ...(i)

$x = 3$, $f(x) = 39$

$\Rightarrow \quad 3a + b = 39 \quad$ ...(ii)($a = 2$, $b = 33$)

Giving $f(x) = 2x + 33$

$\Rightarrow \quad f(5) = 43$

98. Solution

Combined percentage increase in imports and exports for year 2001

$$= \frac{(60+50)-(50+40)}{50+40} \times 100 = 22.22\%$$

Combined percentage increase in imports and exports for year 2002

$$= \frac{(70+60)-(60+50)}{60+50} \times 100 = 18.18\%$$

Combined percentage increase in imports and exports for year 2003

$$= \frac{(70+60)-(60+70)}{70+60} \times 100 = 0$$

Combined percentage increase in imports and exports for year 2004

$$= \frac{(70+80)-(60+70)}{60+70} \times 100 = 15.38\%$$

Combined percentage increase in imports and exports for year 2005

$$= \frac{(70+90)-(70-80)}{70+80} \times 100 = 6.67\%$$

Combined percentage increase in imports and exports for year 2006

$$= \frac{(100+120)-(70+90)}{70+90} \times 100 = 37.5\%$$

Combined percentage increase in imports and exports for year 2007

$$= \left| \frac{(100+120)-(110+120)}{100+120} \times 100 \right| = 4.54\%$$

$\therefore$ year 2006 has shown in highest percentage combined growth in imports and exports.

99. (b)

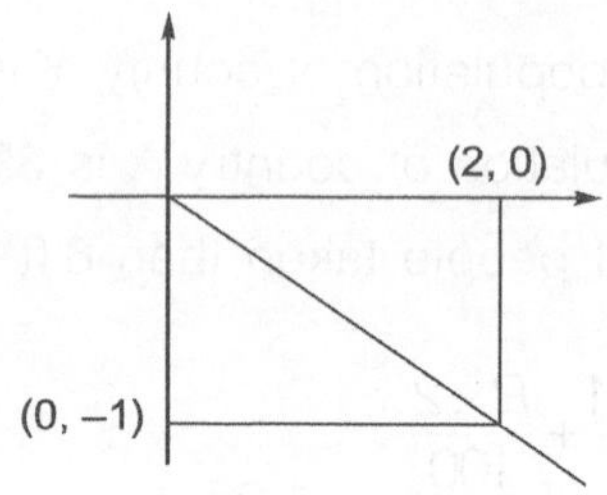

Option (c) is not possible.

$\because \quad x = 0$ for +ve value of y in graph.

Option (a), through valid for +ve values of y, is not correct for –ve values of 'y'.

Option (d) is also not true since x is –ve for +ve values of y which is contradictory to the given graph.

100. (b)

Option (a) is not correct as R is given defence. As given in question R wants only Finance or Home.

Option (d) is not corect as R is alloted Telecom (same reason as above).

Option (c) is not possible because S & U cannot be together as per question.

So only option (b) is possible.

101. (b)

Before drawing third card, only two kings remain among the remaining 50 cards.

$$\text{Probability} = \frac{^{2}C_1}{^{50}C_1} = \frac{2}{50}$$

102. (a)

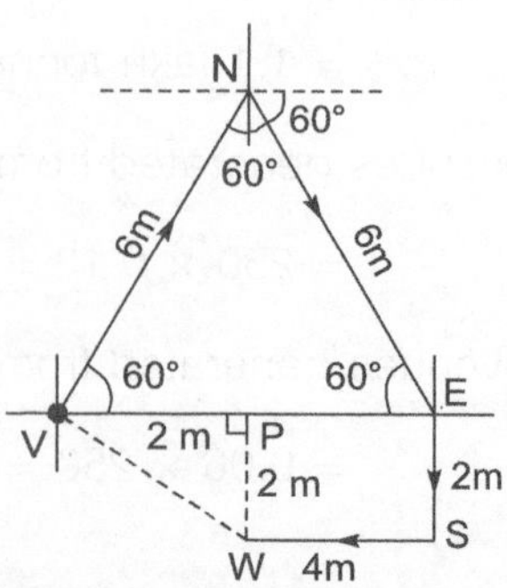

ΔVNE is equilateral Δ

So $VE = 6$

Also $VP = VE - PE = 6 - 4 = 2$

$VP = 2, PW = 2$

So, $VW = \sqrt{VP^2 + PW^2}$

$= 2\sqrt{2}$

103. (d)

Type I:

$$= \left(\frac{115-75}{75}\right) \times 100 = 53.3\%$$

Type II:

$$= \left(\frac{85-144}{144}\right) \times 100 = -40.97\%$$

Type III:

$$= \left(\frac{160-114}{114}\right) \times 100 = 40.3\%$$

Type IV:

$$= \left(\frac{145-108}{108}\right) \times 100 = 14.25\%$$

Type V:

$$= \left(\frac{100-102}{102}\right) \times 100 = 1.9\%$$

So maximum for Type I.

104. Solution

A B C D, left most digit cannot be zero.

$\therefore$ A can have 9 district values B, C, D can have 9, 8, 7 values.

$\therefore \; 9 \times 9 \times 8 \times 7 = 4536.$

105. (d)

Both statements will be required as capacity is required and what portion of capacity is filled is also required.

From (ii) we get full capacity of tank A = 14000

So capacity of B is 20000 $\left(= \frac{14000}{7}\right)$

But for getting quality of solution we also need how much it is full.

Statement I gives A is 80% full and B is 40% full.

i.e., Vol. in A occupied = 80% of 14000

= 11,200

Vol. in B occupied = 40% of 20000 = 8000

Hence both statements will be required for finding out gallons of solution in A and B respectively.

106. (a)

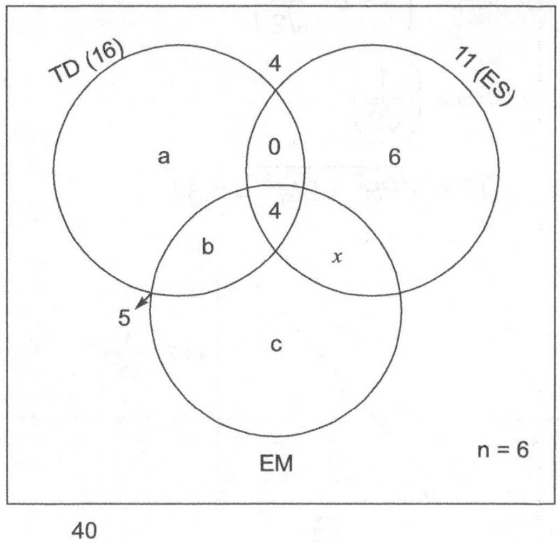

See set ES

$0 + 4 + x + 6 = 11$

$x = 1$

No. of teachers ES & EM but not TD is $x = 1$

107. (c)

$f(x) = 1 - |x|$, use options

Put $x = 0$, $f(x) = 1$ is maximum value

109. (b)

If subtract or multiply every term of an A.P. by a constant resulting sequence will be also A.P. as common difference will still be constant.

110. (8)

$fg(h(2, 5, 7, 3), 4, 6, 8)$

$$fg\left[\left(\text{Re}\,\frac{7 \times 3}{2 \times 5}\right), 4, 6, 8\right] = fg[1, 4, 6, 8]$$

$= f[1, 4, 6, 8] \times g[1, 4, 6, 8] = 8 \times 1 = 8$

111. (a)

Refer to the figure below.

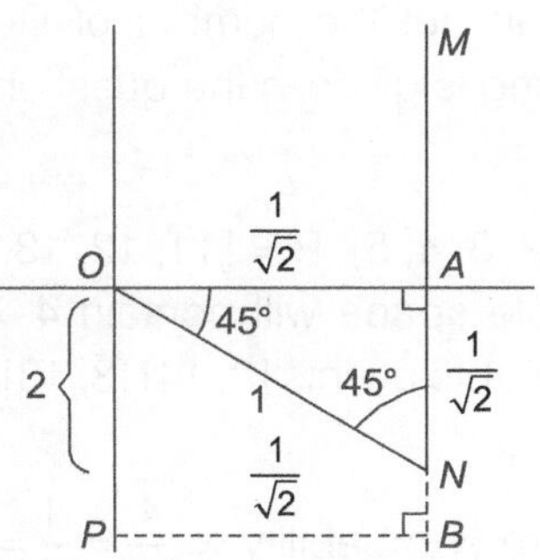

As $ON = 1$, $OA = AN = \frac{1}{\sqrt{2}} = PB$

So, $MB = \left(4 + 2 - \frac{1}{\sqrt{2}}\right)$

and $PB = \left(\frac{1}{\sqrt{2}}\right)$

$$PM = \sqrt{PB^2 + BM^2} = 5.34$$

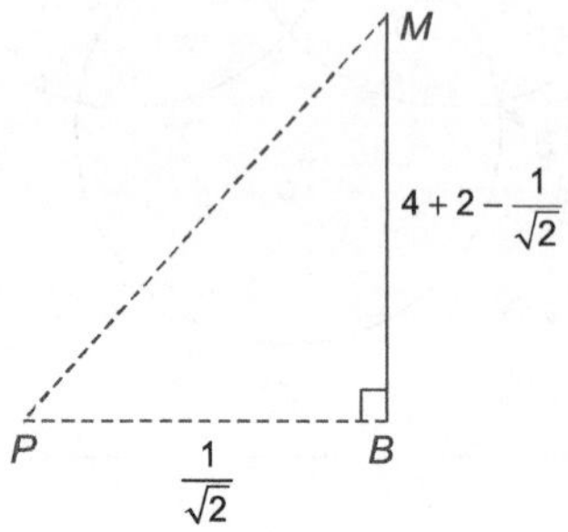

112. (b)

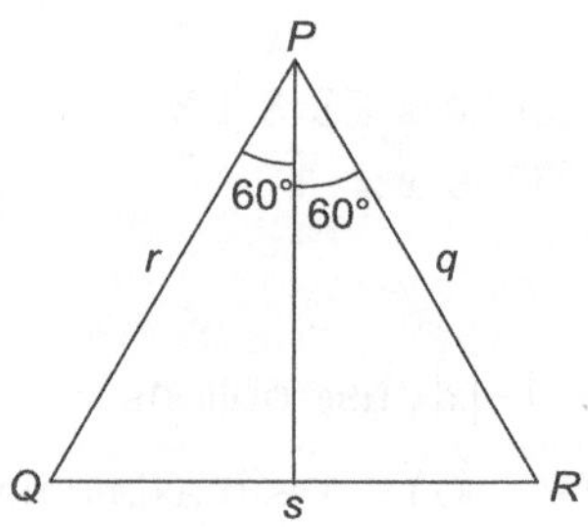

In $\Delta PQR \Rightarrow Ar\Delta PQR = Ar(\Delta QPS) + Ar(\Delta PSR)$

$$\frac{1}{2}r \times q \times \sin 120° = \frac{1}{2}r \times PS \times \sin 60° + \frac{1}{2}q \times PS \sin 60°$$

$$\frac{1}{2}r \times q \times \sin(180 - 60°) = \frac{1}{2}PS \sin 60°(r + q)$$

$$\frac{1}{2}r \times q \times \sin 60° = \frac{1}{2}PS \sin 60°(r + q)$$

$$PS = \left[\frac{r \times q}{r + q}\right]$$

113. (a)
From statement A, if height of each step is known, we can get the number of steps as height of building is given in the question.

114. (a)
$A = \{2, 3, 4, 5\}$, $B = [11, 12, 13, 14, 15]$
Sample space will contain $4 \times 5 = 20$ events favourable events [2, 14] [3, 13] [4, 12] [5, 11] for sum 16

required probability $= \frac{4}{20} = \frac{1}{5} = 0.2$

115. (32)
Let the total number of students in electrical be T_e. Male students in electrical is 40.

So, $\frac{5}{9}T_e = 40 \Rightarrow T_e = 72$

Also, let total number of students (overall) = T.
Electrical is 20% of T
$\Rightarrow$ 20% of $T = 72 \Rightarrow T = 360$
Now number of female students in Civil

$\Rightarrow$ 30% of 360 $\times \frac{4}{9} = 48$

and number of female students in Mechanical

$\Rightarrow$ 10% of 360 $\times \frac{4}{9} = 16$

Difference = 48 − 16 = 32

116. (c)

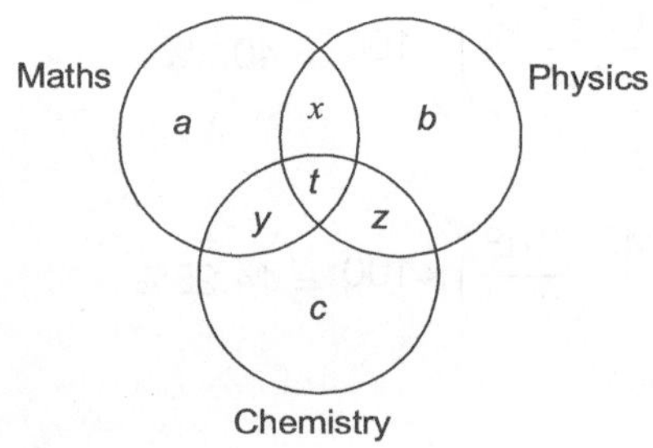

According to question
75% = $a + b + c + (x + y + z) + t$
Also, $x + y + z + t = 50\%$
$x + y + z = 40\%$
so, $t = 10\%$
so, (*iii*) is correct.
and $a + b + c = 25\%$
Also, $m + c + ch = (a + x + y + t) + (b + x + z + t) + (c + y + z + t)$
$m + c + ch = (a + b + c) + 2(x + y + z) + 3t$ 25

$+ 2(40) + 3 \times 10 = 135\% = \frac{27}{30}$

117. (c)
Marks are alloted only for correct answer

$$\text{Average marks} = \frac{\text{Total marks}}{\text{No. of students}}$$

$$= \frac{21 \times 2 + 15 \times 3 + 11 \times 1 + 23 \times 2 + 31 \times 5}{21 + 17 + 6}$$

$$= \frac{299}{44} = 6.795$$

118. (a)

If $x^a = m$

$$a \log x = \log m \text{ then } a = \frac{\log m}{\log x}$$

$$a = \log_x^m$$

$$\log_x\left(\frac{5}{7}\right) = -\frac{1}{3}$$

$$\Rightarrow \quad (x)^{-1/3} = \frac{5}{7}$$

$$\Rightarrow \quad x = \left(\frac{7}{5}\right)^3$$

$$\Rightarrow \quad x = \frac{343}{125}$$

119. (c)

(66 □ 6) → (66 ◇ 6)

$$\left(\frac{66-6}{66+6}\right) \times \left(\frac{66+6}{66-6}\right) = 1$$

120. (c)

Number of faces visible = 9 × 6 = 54

(9 on each face × 6 faces)

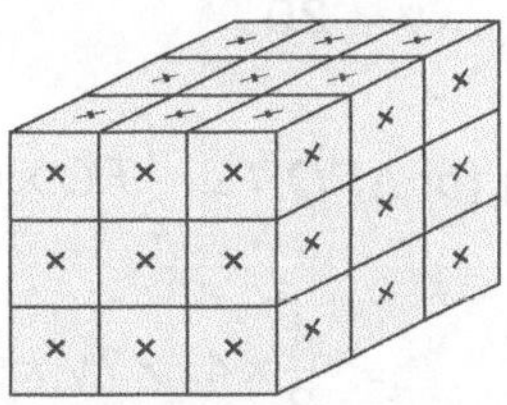

$$\text{Total faces} = 27 \times 6$$

$$\text{Not visible} = \text{Total face} - \text{visible faces}$$

$$= 27 \times 6 - 9 \times 6 = 108$$

$$\text{Required ratio} = \frac{\text{Visible}}{\text{Not visible}} = \frac{54}{108} = \frac{1}{2}$$

121. (3)

Look at first row middle term is $\left(\frac{\text{sum}}{2}\right)$ of terms an either side it $\frac{6+4}{x} = 5.$

Second row middle term is $\left(\frac{\text{sum}}{2}\right)$ of terms an either side of it $\frac{7+4+2+1}{2} = 7$ is equal to middle term in 2nd row.

Similarly 3rd row middle term

$$\frac{1+9+2+1+2+1}{2} = 8$$

So last row middle term = $\frac{3+3}{2} = 3.$

122. (d)

Following chart can be drawn from commulative chart:

Stretch	Distance	Electricity used	
M	20	12	12/20 = 0.6
N	45	25	13/25 = 0.52
O	75	45	20/30 = 0.66
P	100	57	12/25 = 0.48

Hence minimum for *P*.

123. (b)

Required probability

$$= \frac{1}{6} \times \left(\frac{7}{8}\right) + \frac{1}{8} \times \left(\frac{5}{6}\right) = \frac{12}{48} = \frac{1}{4}$$

124. (d)

Both the conclusions I and II follow.

125. (b)

We know,

$(a + b + c)^2 = a^2 + b^2 + c^2 + 2[ab + bc + ca]$

$(a + b + c)^2 - 1 = 2[ab + bc + ca]$

[as $a^2 + b^2 + c^2 = 1$]

For $ab + bc + ca$ to be min. $(a + b + c) = 0$

Min. $(ab + bc + ca) = -\frac{1}{2}$

[an putting $a + b + c = 0$]

Only option (b) has min. value as $-\frac{1}{2}$.

126. (800)

Tiger is 50 leap of its own behind deer. Given tiger's one leap is 8 m. So initial separation of Tiger and deer = 50 ×8 = 400 m/min

Tiger goes 5 leap in a minute which is equal to 5 ×8 = 40 m/min.

Deer goes 4 leap in a minute which is equal to 4 ×5 = 20 m/min.

Now relative distance of 400 m have to be covered with (40 – 20) = 20 m/min.

$$\text{Required time} = \frac{400\,\text{m}}{(40-20)\,\text{m/min}}$$

$$\frac{\text{(Distance)}}{\text{Relative speed}} = 20 \text{ min.}$$

So distance travelled by Tiger

$$\frac{40 \text{ m}}{\text{min}} \times 20 \text{ min} = 800 \text{ m/min.}$$

127. (a)

If $1 < y < x$ so $\log x > \log y$ a as well $e^x > e^y$ other two will not always be true.

128. (c)

A, CD, GHI ?, UVWXY

Series is as follows: Write first letter A, leave aside nearby one *B*, write next two *CD*. Leave aside net two *EF*. Write next three *GHI* leave aside next *JKL* then we have to write *MNOP*.

130. (c)

$\log \tan 1° + \log \tan 2° + ... + \log \tan 89°$

$= \log(\tan 1° \times \tan 2° \times \cdots \tan 89°)$

As $[\log m + \log n = \log mn]$

$= \log(\tan 1° \times \tan 2° \cdots \tan(90° - 2°)) \tan(90° - 1°)$

$= \log(\tan 1° \times \tan 2° \times \cdots \tan 45° \times \cot 44 \cdots \cot 2° \cot 1°)$

$= \log(\tan 45°)$

$= \log(1) = 0$

As $\tan(90° - \theta) = \cot\theta$ and

$\tan 1° \times \cot 1° = \tan 2° \times \cot 2° = \cdots 1$

131. (2.064)

The shade of part of circular sheet is folded to form a cone (Figure below).

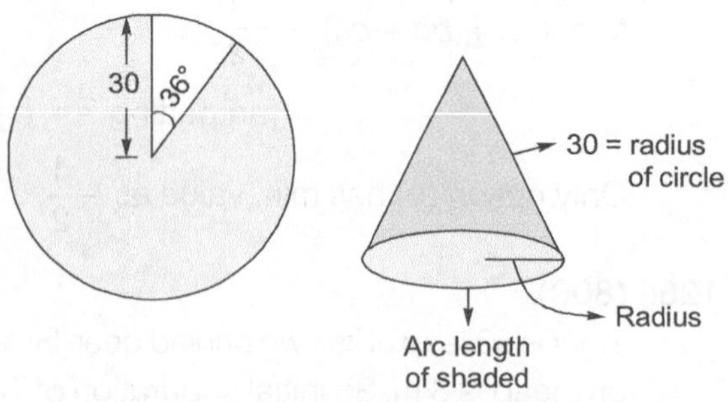

$$2\pi R' = \frac{9}{10} 2\pi \times 30 \quad R' = 27$$

$$h = \sqrt{30^2 - 27^2} = \sqrt{900 - 729}$$

$$= \sqrt{171}$$

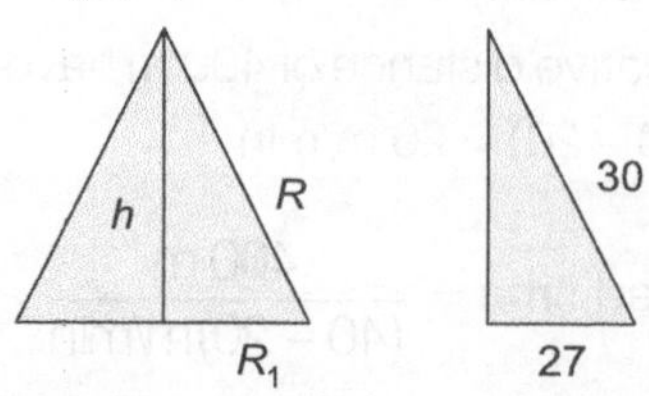

Required ratio $\frac{R'}{h}$

$$= \frac{27}{\sqrt{171}} = 2.064$$

132. (b)

Total number of matches $^5C_2 = 10$.

133. (b)

T > E, C > T, ⇒ C > T > E. So Eric is actually order older than Cliff so third statement is false.

134. (280)

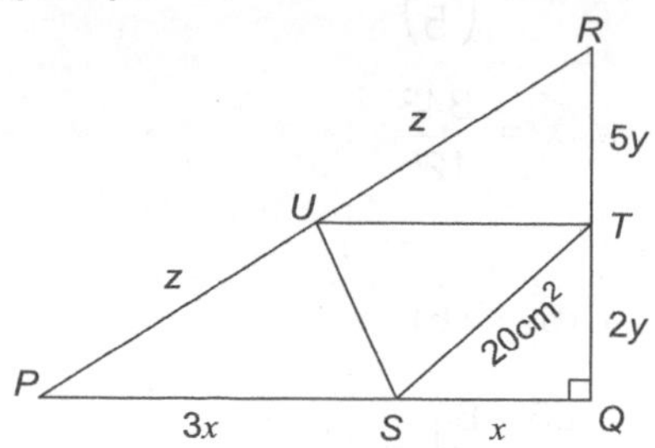

$$\text{Area } \Delta QTS = \frac{1}{2} SQ + QT \sin 90°$$

$$20 = \frac{1}{2} \times x \times 2y \times 1$$

$$\Rightarrow \quad xy = 20$$

$$\text{Now Area for } \Delta PQR = \frac{1}{2} PQ \times QR \times \sin 90°$$

$$= \frac{1}{2} \times 4x \times 7y \times \sin 90°$$

$$= 14\,xy = 280 \text{ cm}^2$$

135. (c)

S1: 

S2: 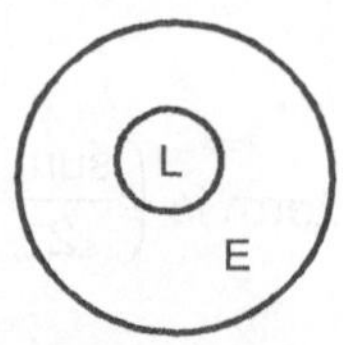

Therefore concluding diagram can be

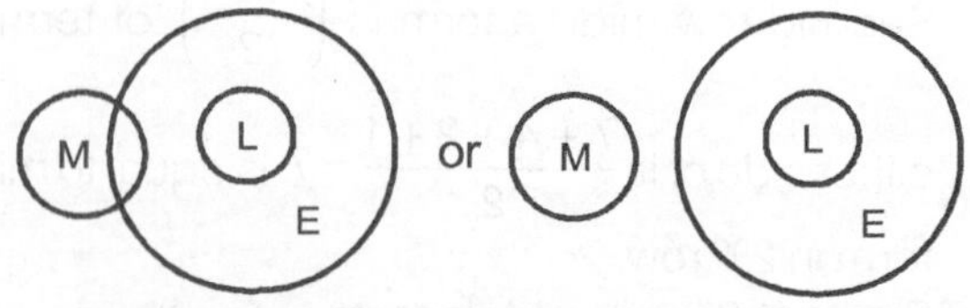

Fig. (a) Fig. (b)

1. Some manager can be executive.
2. Some executive can be manage.

137. (c)

PR is parallel to x axis
$-4 \le x \le 5$
$6 \le y \le 16$

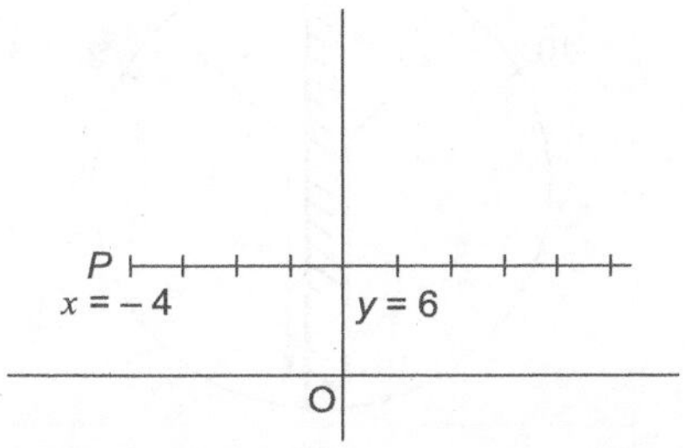

P at $x = -4$
Δ's = with base 1 unit y can take values (7, 8, ..., 16) : 10 values
Δ's = with base 2 unit y varies (7.... 16) : 10 values
Δ's = with base 9 unit y varies (7.... 16) : 10 values
Similarly P at $x = -3$ we will get 80 and 50 on
So [90 + 80 + ... 10] = 9[1 + 2 .... 10]

$$= 10\sum_{n=1}^{9} n$$

Similarly changing ordinate of P we will respectively get

$$[81 + 725 \text{ .... } 9]\ \Delta\text{'s} = 9\ [1 + \text{....}9] = 9\sum_{n=1}^{9} n$$

and further (64 + 56 .... 8) Δ's = $8\sum_{n=1}^{8} n$ and so on

If we sum all of the above

$$10\sum_{n=1}^{9} n + 9\sum_{n=1}^{9} n + 8\sum_{n=1}^{8} n.\Sigma n.......$$

$$9\sum_{n=1}^{9} n[1+2+3.....10] = \sum_{n=1}^{9} n \times \sum_{n=1}^{10} n$$

But there are overall four orientation for rt.

____ as shown 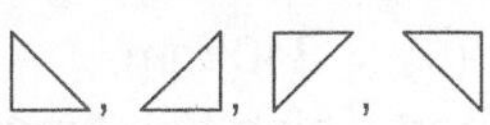, so actual

answer will be $4 \times \sum_{n=1}^{9} n \times \sum_{n=1}^{10} n$

$$4 \times \left(\frac{9 \times 10}{2}\right) \times \left(\frac{10 \times 11}{2}\right) = 9900.$$

Alternative Method :
$-4 \le x \le 5,\ 6 \le y \le 16$
PQ ll y axis and PR ll x axis $\angle QPR$ is 90°
As all co-ordinates of P, Q, R, are integers so, we can choose (x, y) integer co-ordinates for vertex P in 10 × 11 = 110 ways.
Now with respect to fixed y co-ordinate for R we can choose its x-coordinate in only 9 ways as one point is already occupied (two points of cannot concide) overall 9 ways of doing it.
Now choosing y-coordinate for Q because x-coordinate of Q is already fixed (it has same x-coordinate as R) it can take 10 different values.
SO, overall 110 × 9 × 10 = 9900 ways 9900 such Δ's will be possible.

138. (a)

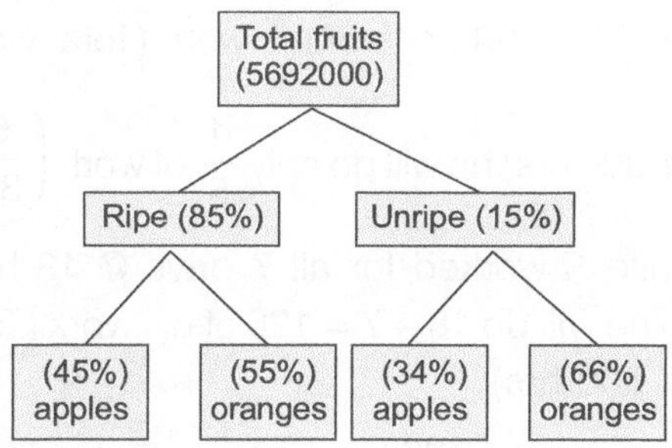

Total number of apples = Ripe apples + Unripe apples
= (0.85 × 0.45 + 0.15 × 0.34) × 569200
= **2029198**

139. (c)
Following line with respective distances can be drawn

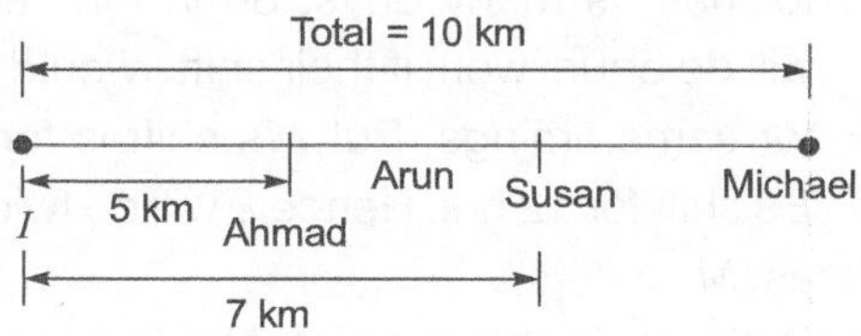

Arun can reside anywhere between Ahmed and Susan *i.e.* between 5 km and 7 km from I.
Hence ans (c) **6.02**
5 < 6.02 < 7

140. (c)
The required probability

$$0.5 \times 0.7 = 0.35 \approx \frac{35}{100} = 35\%$$

141. (d)

$L > P$ (Leela is older thean Pavithra)

$S > L$ (Shiv is older than Leela)

So Pavithra is youngest

142. (c)

$q^a = r,\ r^b = s,\ s^c = q$

$a \log q = \log r,$

$b \log r = \log s,$

$c \log s = \log q$

$$\text{So,} a \times b \times c = \frac{\log r}{\log q} \times \frac{\log s}{\log r} \times \frac{\log q}{\log s} = 1$$

143. (c)

Q can do work in $25 \times 12 = 300$ hrs

R can do work in $50 \times 12 = 600$ hrs

So we can say Q is twice efficient as R

Now Q worked only for 5 days @ 12 hrs/day. So he will do 60 units of his work (Total work for Q *i.e.* 300 hrs) he will do only $\frac{1}{5}$ of work $\left(\frac{60}{300} = \frac{1}{5}\right)$

While R worked for all 7 days @ 18 hrs/day

So he will do $18 \times 7 = 126$ of his work (Total work for 600 hrs)

He will do $\left(\frac{126}{600}\right) = (0.21)$ of his work

So required ratio $\frac{1}{5} : \frac{126}{600}$

120 : 126

20 : 21

145. (b)

M works with twice efficiency as E but worked for half as many days. So in this repsect they will do equal work if their shifts would have been for same timings. But M's shift is for hrs, while E's shift for 12 hrs. Hence E will do twice the work as M.

Ratio of contribution of $M : E$ in work, 1 : 2

146. (d)

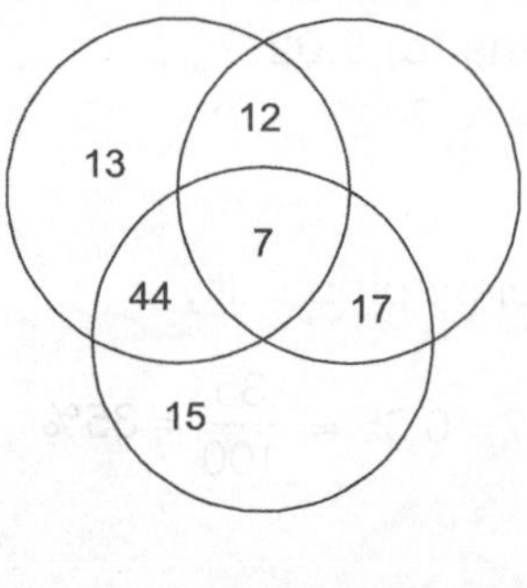

The number of student who like to read books or play sports have been shown

$= 13 + 12 + 44 + 7 + 15 + 1$

$= \mathbf{108}$

147. (d)

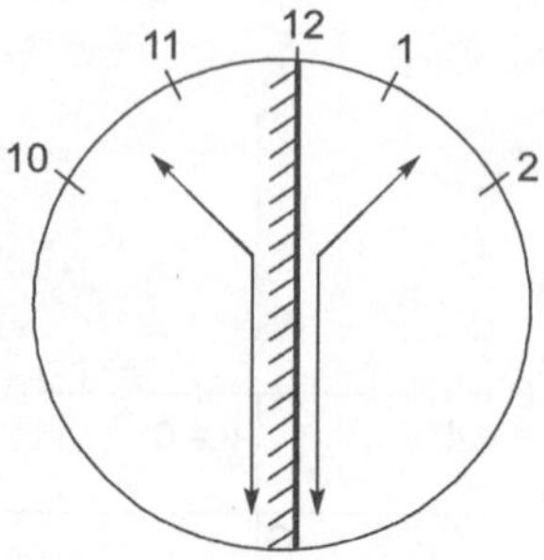

Mirror image of 1 : 30 is 10 : 30

10 : 30 was the time two and quarter hour back so time now will be 12 : 45

148. (c)

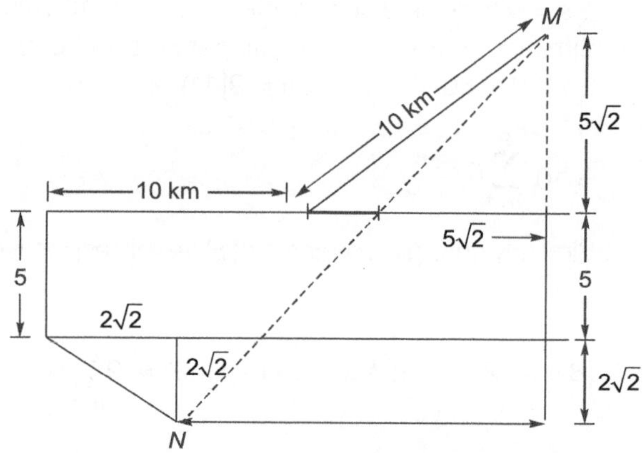

See the adjoining figure for solution

$$MM' = 5\sqrt{2} + 5 + 2\sqrt{2} = 5 + 7\sqrt{2}$$

$$NM' = 10 + 5\sqrt{2} - 2\sqrt{2} = 10 + 3\sqrt{2}$$

$$MN = \sqrt{(MM')^2 + (NM')^2}$$

$$= \sqrt{(5 + 7\sqrt{2}) + (10 + 3\sqrt{2})^2} \approx \mathbf{20.61}$$

149. (b)

Let the two parts length is x and y

so $x + y = 340$ mm ...(1)

The part which has length x is used in making a square so each side will be length $x/4$, and the part which has length y is used in making a rectangle with sides in ratio 1 : 1 so sides have length ($y/6$: $y/3$)

Area of square is $x^2/16$

Area of rectangle is $y^2/16$

So combined area is $\left(\frac{x^2}{16}+\frac{y^2}{18}\right)$ ...(2)

Since the combined area should be minimum so using equation 1 in 2, we get

$$\frac{d}{dx}\left(\frac{x^2}{16}+\frac{(340-x)^2}{18}\right) = 0$$

$$\text{so} \left[\frac{x}{8}+\frac{x}{9}-\frac{680}{18}\right] = 0$$

or $x = 160$ mm

and length of side of square is

$$\frac{x}{4} = 40 \text{ mm}$$

150. (c)

97 is not a perfect square.

152. (d)

The odometer reading increases from starting point to end point

Magnitude of area of the given diagram = Odometer reading

Magnitude of area of the velocity and time graph per second

1st sec $\Rightarrow$ triangle $= \frac{1}{2}\times 1\times 1 = \frac{1}{2}$

2nd sec $\Rightarrow$ square $= 1 \times 1 = 1$

3rd sec $\Rightarrow$

square + triangle $= 1+1+\frac{1}{2}\times 1\times 1 = 1\frac{1}{2}$

4th sec $\Rightarrow$ triangle $= \frac{1}{2}\times 1\times 2 = 1$

5th sec $\Rightarrow$ straight line = 0

6th sec $\Rightarrow$ triangle $= \frac{1}{2}\times 1\times 1 = \frac{1}{2}$

7th sec $\Rightarrow$ triangle $= \frac{1}{2}\times 1\times 1 = \frac{1}{2}$

Total Odometer reading at 7 seconds

$$= \frac{1}{2}+1+1\frac{1}{2}+1+0+\frac{1}{2}+\frac{1}{2} = 5$$

153. (c)

Manu's age = Sravan's age + 2 months

Manu's age = Trideep's age – 3 months

Pavan's age = Sravan's age + 1 month

From this Trideep's age > Manu > Pavan > Sravan

∴ Trideep can occupy the extra spane in the flat.

154. (b)

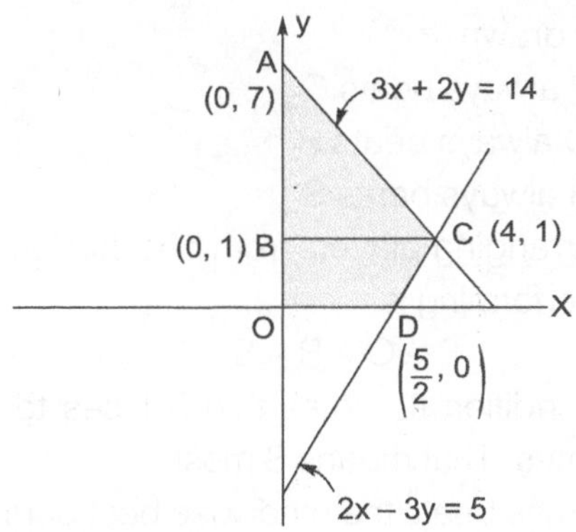

Required area = Area of ΔABC + Area of trapezoid BCDO

$$= \frac{1}{2}\times 4\times 6+\frac{1}{2}\times(4+2.5)\times 1$$

$$= 12 + 3.25 = 15.25 \text{ sq. units}$$

155. (a)

Straight line equation $y = mx + c$

m = slope = –0.02 $(\ln x, y)$

If $\ln x = X$, then set (x, y) → 0.1, 0

$y = mX + C$

$o = -0.02 \times 0.1 + C$

∴ $C = 0.002$

$y = mX + C$

$y = -0.02 \times \log x + C$

at $x = 5$

$y = -0.02 \times \log 5 + 0.002$

$= -0.030$

156. (d)

Original surface area = 6 $(4)^2$ = 96

If corner cubes are removed, three exposed surfaces are removed which will create 3 new surfaces in original large cube. So surface area will remain unchanged, i.e. 96.

157. (b)

Elegance

(27300 + 25222 + 28976 + 21012) × 48 = A

Executive

(999 + 8942 + 10234 + 10234 + 10109) × 173 = B

Smooth

(20009 + 9392 + 22429 + 18229) × 63 = C

Soft

(17602 +18445 + 19544 + 16595) × 78 = D

Which is highest for B (executive).

158. (d)

From conditions I, II and IV, following inference can be drawn

P always beats Q

Q always beats R

and R always beats S

Thus, arranging players from best performing to worst performing is

P – Q – R – S

Now, condition III states that S loses to P only sometimes. That means S most of the times beats the otherwise best performing player P.

Thus, statement (i) P is likely to beat all the other three players is FALSE, As P is likely to lose from S.

Statement (ii) S is absolutely the worst player in the set is also FALSE as S beats P most of the times.

159. (b)

$$f(x) = 2x^7 + 3x - 5$$

Use option (b), if $(x - 1)$ will be a factor then on putting $x - 1 = 0$ i.e. $(x = 1)$ in $f(x)$.

$$f(1) = 2(1)^7 + 3 - 5 = 5 - 5 = 0$$

161. (b)

$$\frac{\alpha^n + \beta^n}{\alpha^{-n} + \beta^{-n}} = \frac{\alpha^n + \beta^n}{\frac{1}{\alpha^n} + \frac{1}{\beta^n}} = \frac{\alpha^n + \beta^n}{\left(\frac{\beta^n + \alpha^n}{\alpha^n \beta^n}\right)} = (\alpha\beta)^n$$

given, $\alpha \times \beta = 4$

Hence, $(\alpha\beta)^n = 4^n$

162. (d)

$$\underbrace{C\ A\ D\ B\ E}_{-2,\ +3,\ -2,\ 3}$$

Same pattern follows in all options A, B, C only (d) doesn't follow this pattern , Hence odd number out (D).

163. (a)

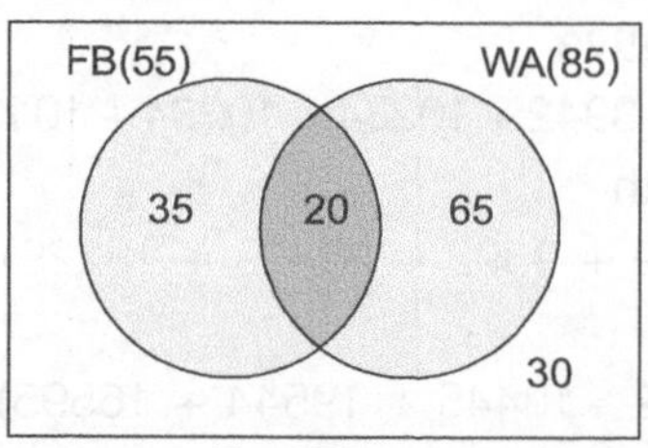

T = 150

Following venn diagram can be drawn as

$$n(A \cup B) = n(A) + n(B) - n(A \cap B)$$

$$120 = 55 + 85 - n(A \cap B)$$

$$n(A \cap B) = 20$$

So only facebook 35(A)

164. (c)

Number of rectangules will be

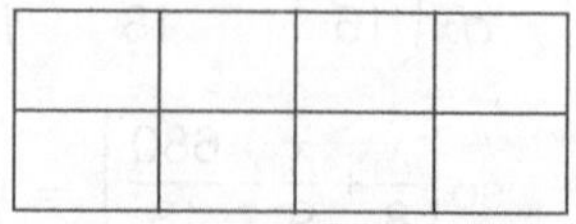

$${}^5C_2 \times {}^3C_2$$

$$10 \times 3 = 30$$

As whenever any two horizontal and any two vertical lines are choosen their intersection will produce a rectangle.

172. (b)

13, 23, 43, 53 are all prime numbers.

only 33 is composite 33, (11 × 3)

odd one out is 33.

173. (c)

$$|9y - 6| = 3$$

⇒ Either $(9y - 6 = 3)$ or $(9y - 6 = -3)$

$$\Rightarrow [y = 1] \text{ or } \left[y = \frac{1}{3}\right] \left[y^2 - \frac{4y}{3}\right]$$

put $y = \frac{1}{3}$, $\left(\frac{1}{3}\right)^2 - \frac{4}{3} \times \frac{1}{3} = -\frac{1}{3}$ or $y^2 - \frac{4y}{3}$

put $y = 1$, $1^2 - \frac{4}{3} = -\frac{1}{3}$

174. (a)

Probability of free throw $= \frac{60}{100} = 0.6$

Probability of NOT free throw = 1 – 0.6 = 0.4

So required probability of exactly 6 throws in 10 attempts will be given by

$${}^{10}C_6 (0.6)^6 \times (0.4)^4 = 0.2508$$

175. Solution:

Unit digit of $211^{870} + 146^{127} \times 3^{424}$ is $1 + 6 \times 1 = 7$

178. (b)

Following figure can be drawn:

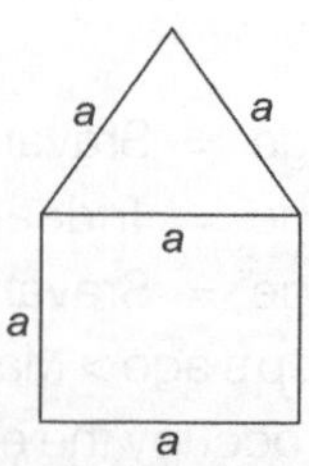

Perimeter of window,

$5a = 6$ m

$a = 1.2$ m

Complete area for window = Area of square + Area of equi Δ

$$= (1.2)^2 + \frac{\sqrt{3}}{4}(1.2)^2 \approx \mathbf{2.06\ m^2}$$

179. (c)

Lesser standard deviation for group Q will mean that the observation sin group Q are closer and hence narrower is the range of marks scored in group Q as compared to group P. Other options (a), (b), (d) remains unconfined.

180. (c)

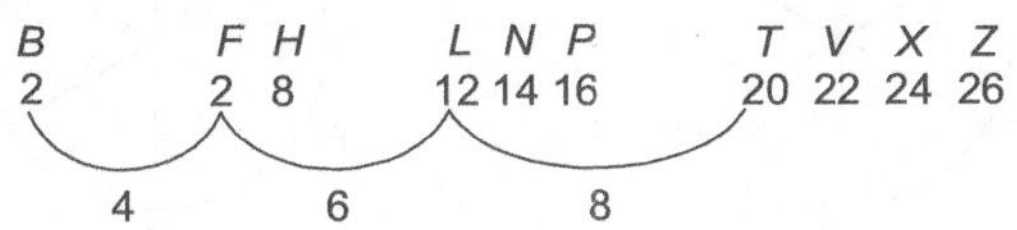

So the required alphabets have to start with 20 is T and there after common difference in term will be 2 each hence answer is $TVXZ$

181. (a)

Use options only $(A) \rightarrow 0$, satisfies

$a \square b = ab - (a + b)$

$a \square b = a \times 0 + (a + 0) = a$

Given $a \square x = a$ which satisfied for $(x = 0)$

182. (c)

$$y = \ln\left|e^{[\sin|x|]}\right|$$

$$= |\sin|x||$$

$$y = \begin{cases} -\sin x & x < 0 \\ \sin x & x > 0 \end{cases}$$

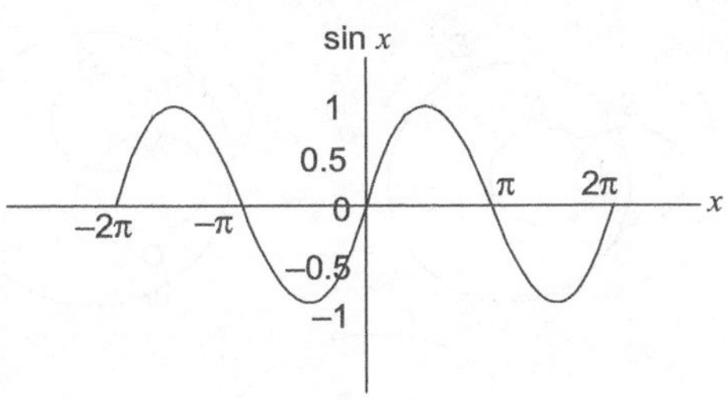

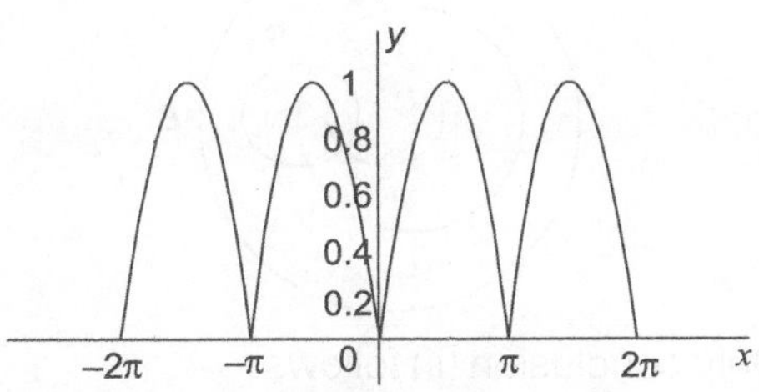

183. (d)

1, x, x, x, y, y, 9, 16, 18

$$\text{Mean} = \frac{1 + 3x + 2y + 9 + 16 + 18}{9} = \frac{3x + 2y + 44}{9} \quad ...(i)$$

Median ⇒ As number of terms are odd i.e. 9, so median will be middle i.e. y.

Mode ⇒ Number showing maximum frequency of repetition. So mode is x.

Given ⇒ Mean = Median = 2 Mode

$$\Rightarrow \quad y = 2x \quad ...(ii)$$

From equation (i) and (ii), we get

$$x = 4, \ y = 8$$

Value of y is 8

184. (d)

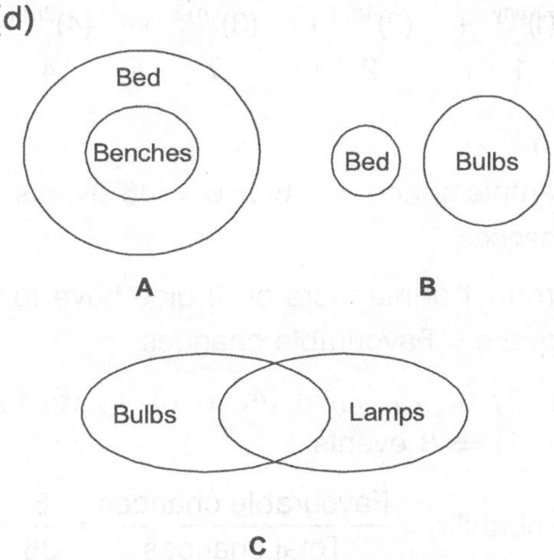

Since there is no direct relation given between **lamps** and **beds**. So, neither will be correct.

185. (c)

$$V = \frac{1}{3}\pi(R)^2 H$$

$$V' = \frac{1}{3}\pi(1.5R)^2 H$$

$$= \frac{1}{3}\pi\, 2.25\, R^2\, H = 2.25\ V$$

Hence increase 125% as

$$\left[\frac{2.25V - V}{V}\right] \times 100 = 125\%$$

186. (a)

P 30 min × 2 = 60 min

Q 20 min × 3 = 60 min

R 60 min × 1 = 60 min

S 15 min × 4 = 60 min

M_1 [P] [Q] = 2 hrs

M_2 [R] [S] = 2 hrs

187. (c)

Clearly we can see from graph both (i) and (ii) are correct.

188. (b)

Amar ⟶ R
Akbar ⟶ S
Antony ⟶ P   As per their preferences given.
Ajit ⟶ Q

189. (b)

Unit digit

$(2171)^7 + (2172)^9 + (2173)^{11} + (2174)^{13}$

$(1)^{any} + (2)^{4n+1} + (3)^{4n+3} + (4)^{2n+1}$

$1 + 2 + 7 + 4 = 14$

190. (b)

Sample space = 6 × 6 = 36 events = Total chances

Product of numbers on 2 dice have to perfect square = Favourable chances

(1, 1) (2, 2) (3, 3) (4, 4) (5, 5) (6, 6) (1, 4) (4, 1) ⇒ 8 events

$$\text{Probability} = \frac{\text{Favourable chances}}{\text{Total chances}} = \frac{8}{36} = \frac{2}{9}$$

191. (b)

$$\text{Put } (x = -1)\frac{81\left[\frac{16}{25}\right]^1}{\left[\frac{3}{5}\right]^2} = \frac{\frac{3^4 \times 2^4}{5^2}}{\frac{3^2}{5^2}} = 3^2 \times 2^2 = 144$$

On putting $(x = -1)$

⇒ LHS = RHS

Hence option (b) is correct.

192. (d)

(i) is incorrect as it has move directly.

(ii) is incorrect as it stayed for maximum duration on ground floor.

193. (d)

Let cycling speed = C

and walking speed = W

$$C\left(\frac{1}{2}\right) + W\left(\frac{3}{2}\right) = 19 \quad \text{...(i)}$$

$$C + W = 26 \quad \text{...(ii)}$$

On Solving eq. (i) and (ii) We get,

W = 6 km/hr

194. (c)

Following information is given

$K < G$, $P < G$, $S < P$ and $G < D$

From these it can be inferred

$S < P < G < D$ and also $(K < G)$

So, $K < D$ as well but no comparison between K & S or K & P is given

So (ii) and (iv) follows.

195. (c)

Following two possibilities can be drawn

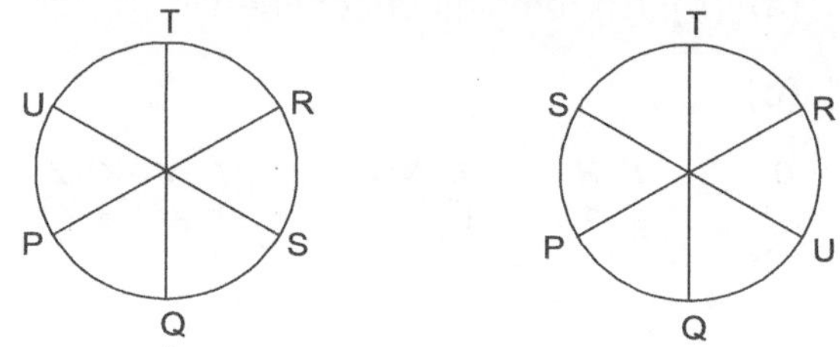

Given in question P & U interchange then new diagram can be drawn.

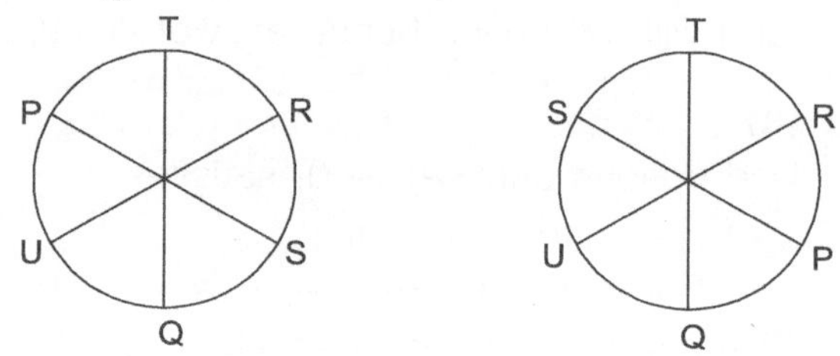

Now verify (c) only follows.

196. (d)

Let H is household consumption and P be other consumption.

$H \times 0.8 + P \times 1.7 = (H + P) \times 0.75$

(According to given condition)

∴ from here ratio is negative.

197. (b)

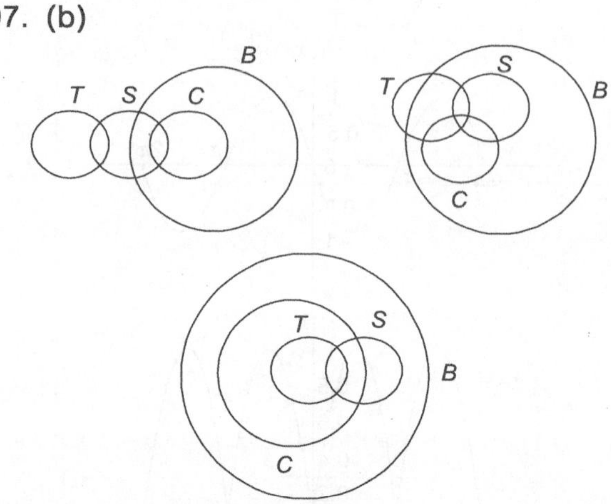

Only conclusion (ii) follows.

198. (b)

40% of $360° = \dfrac{40}{100} \times 360° = 144°$

So, the angle subtended on pie chart will be $= 144°$.

199. (a)

Length of Truck + gap required = 10 + 20 = 30 m

Length of Car + gap required = 5 + 15 = 20 m

Alternative pairs of Truck and Car needs

30 + 20 = 50 m

Let 'n' be the number of repetition of (Truck + Car) in 1 hour (3600 secs)

Given speed = 36 km/hr = 10 m/sec

$$\frac{50\,\text{m} \times n}{3600\,\text{secs}} = 36\ \text{km/hr}$$

$$\frac{50n}{3600}\ \text{m/sec} = 10\ \text{m/sec}$$

$$n = \frac{36000}{50} = 720\ \text{(Trucks + Cars)}$$

So, 720 (Trucks + Cars) passes = 720 × 2 = 1440 vehicles

200. (c)

Contour lines can be observed to cross region with height

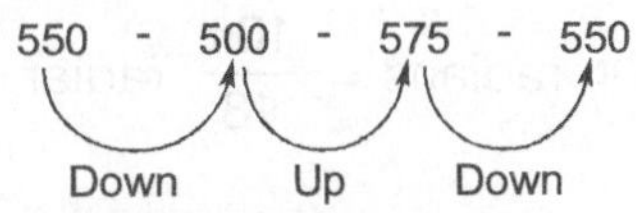

Down-Up-Down satisfies.

201. (a)

Subgroups containing only Indians

$= {}^3C_1 + {}^3C_2 + {}^3C_3 = 3 + 3 + 1 = 7$

Subgroups containing one Indian and rest Chinese

$= {}^3C_1\left[{}^3C_1 + {}^3C_2 + {}^3C_3\right]$

$= 3[3 + 3 + 1] = 21$

Subgroups containing two Indian and remaining

Chinese $= {}^3C_2\left[{}^3C_1 + {}^3C_2 + {}^3C_3\right] = 21$

Subgroups containing three Indian and remaining

Chinese $= {}^3C_3\left[{}^3C_1 + {}^3C_2 + {}^3C_3\right] = 7$

Total number of subgroups

= 7 + 21 + 21 + 7 = 56.

202. (a)

Following circular seating arrangement can be drawn

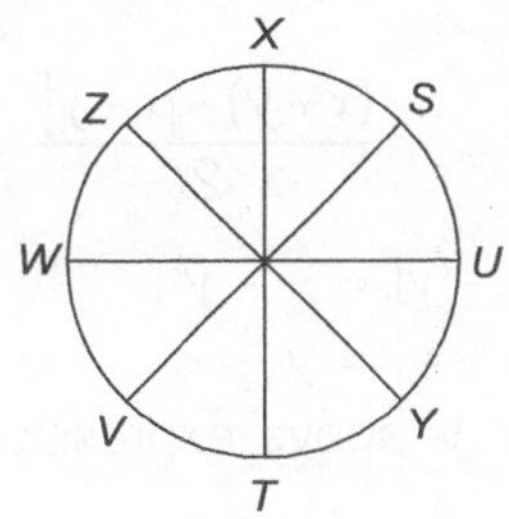

Only one such arrangement can be drawn.

The person on third to the left of V is X.

210. (c)

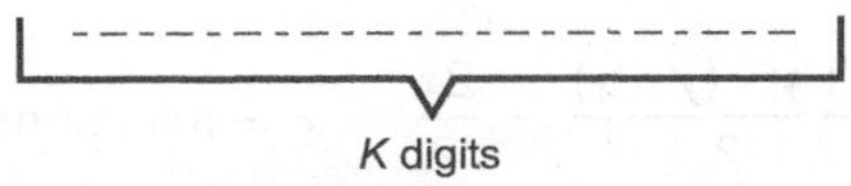

Each digit can be filled in 7 ways as 0, 5 and 9 is not allowed, so each of these places can be filled by 1, 2, 3, 4, 6, 7, 8 so required

probability is $\left(\dfrac{7}{10}\right)^K$ or 0.7^K

211. (d)

$y \times 162$ = Perfect cube

Option,

$y = 24 \Rightarrow 2^3 \times 3\ (2 \times 81) \neq$ Not perfect cube

$y = 27 \Rightarrow 3^3 \times (2 \times 3^4) \neq$ Not perfect cube

$y = 32 \Rightarrow 2^5 \times 2 \times 3^4 \neq$ Not perfect cube

$y = 36 \Rightarrow 2^2 \times 3^2 \times 2 \times 3^4 = 2^3 \times 3^6 = (2 \times 3^2)^3$ is a perfect cube

Hence the answer is, $y = 36$

212. (c)

Following seating arrangement can be drawn

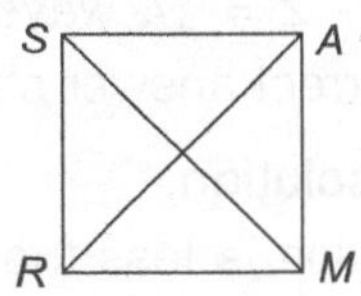

Therefore correct option is (c).

213. (c)

P will not get submerged as it is @ 550. So P is not present in correct options. Hence (i) and (ii) options are incorrect. Now compare Q and T. As T is between 500 and 525. T will get submerged. Hence ans (c) as among option C and D. T is present in only option (c).

214. (b)

$$\frac{(x+y)-|x-y|}{2} \qquad ...(i)$$

If $x > y |x-y| = x-y$;
If $x < y |x-y| = y-x$
Now if $x > y$, above expression (i) becomes

$$\frac{(x+y)-(x-y)}{2} = \frac{2y}{2}$$

$= y =$ minimum of (x, y) as $x > y$

Now if $x < y$;

$$\frac{x+y-(y-x)}{2} = \frac{2x}{2} = x = \text{minimum of } (x, y)$$

as $x < y$
Therefore correct answer is option (b).

Alternative solution:
Use easy values,

$x = 1$ and $y = -2$

Now, $\frac{(1-2)-|1-(-2)|}{2} = -2$

or $x = 2$ and $y = -1$

$$\frac{(2-1)-|2-(-1)|}{2} = \frac{1-3}{2} = -1$$

which is minimum of (x, y).
Therefore, correct answer is option (b).

215. (d)

As there are 4 people A, G, N, S and four colours so without any restriction total ways have to be $4 \times 4 = 16$.
Now, Arun → dislikes Red and Shweta dislikes white so $16 - 2 = 14$ ways.
Therefore correct answer should be option (d).

Alternative solution:
Only one option is less than 16.
Therefore correct answer should be option (d).

216. (a)

Out of six people. 3 place definitely occupied by right handed people as at least 2 women are there so these two will sit adjacently. Now as only on seat is left it will be occupied by a left handed man because on right side of this seat is sitting an right handed man.

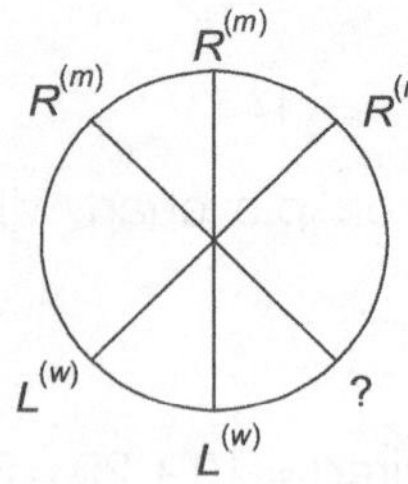

? have to be a left handed man.

Here, $R^{(m)}$ indicates right handed man and $L^{(w)}$ indicates left handed women.
Therefore answer should be only 2 women.

224. (d)

Distance travelled by cone in one rotation

$= 2\pi r = 2\pi\,(5) = 10\pi \qquad ...\text{(i)}$

⇒ point Q touches ground again after going (10π) distance.
2π order to make complete revolution now slant height will be acting as a radius

$R = \sqrt{5^2 + 12^2} = 13$ (Radius of larger complete revolution)

$2\pi R = 2\pi \times 13 = (26\pi) \qquad ...\text{(ii)}$

Now this one rotation angle inscribed in radians will be (i)/(ii)

$$\frac{10\pi}{20\pi} \times 2\pi \text{ radians} = \frac{10\pi}{13} \text{ radians}$$

225. (a)

P says → S has atleast 3 Cars
Q says → S has less than 3 Cars
R says → S has atleast 1 Car
Solve by options

A (0)		B (1)		C (3)	
	P ✗		P ✗		P ✓
	Q ✓		Q ✓		Q ✗
	R ✗		R ✓		R ✓

as given in question only has to be right so answer is 0 (car).

226. (b)

45p → 20K
25p → 30K
20p → 40K
8p → 60K
2p → 150K

Median will be arithmatic mean 50th and 51st term, so both are in 30K bracket.

So average or arithmatic mean of,

$$\frac{30000+30000}{2} = \text{Rs. } 30000$$

227. (d)

The following Two possibiles exists.

$$\begin{array}{r} 5\,0\,1\,0\,0 \\ -\,4\,8\,9\,8\,9 \\ \hline 0\,1\,1\,1\,1 \end{array} \quad \text{and} \quad \begin{array}{r} 5\,0\,0\,0\,0 \\ -\,4\,8\,8\,8\,9 \\ \hline 0\,1\,1\,1\,1 \end{array}$$

Sum = 1 + 9 = 10 Sum = 0 + 8 = 8

228. (c)

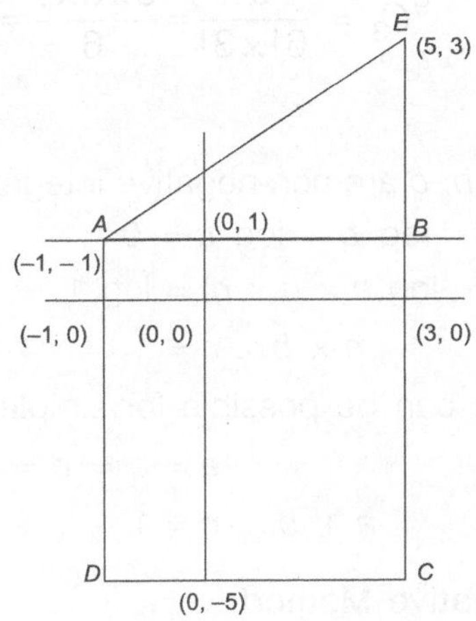

$$|x-1| \le 2 \qquad |y+2| \le 3 \qquad x - y \ge -2$$

$$x \le 3 \qquad y \le 1$$

$$x \ge -1 \qquad -5 \le y$$

Total area= Area of rectangle ABCD

+ Area of ΔABE

$$= 4\times 6+\frac{1}{2}(4\times 4) = 32$$

230. (a)

We can see from the graph that only (i) follows.

232. (b)

P → Married

R → Unmarried

Q → Status Not given

P → Looks @ Q

Q → Looks @ R

Now two possibility arise,

1. Let Q be married

$P^{\text{married}} \rightarrow Q^{\text{married}}$

$Q^{\text{married}} \rightarrow R^{\text{unmarried}}$

only one pair in which married person is looking @ unmarried.

2. Let Q be unmarried

$P^{\text{married}} \rightarrow Q^{\text{unmarried}}$

$Q^{\text{unmarried}} \rightarrow R^{\text{unmarried}}$

Then also only one pair exists.

So, only one pair exists in which married person is looking @ unmarried.

237. (a)

$x \rightarrow$ Bullocks

y → Tractors

$$(x + y)8 = \left(\frac{x}{2}+2y\right)5$$

$$8x + 8y = \frac{5x}{2}+10y$$

$$\Rightarrow \quad 5.5x = 2y \text{ or } y = 2.75x$$

Now, x bullocks are alone working or 'd' days.

$$x \times d = (x + y)8$$

$$x \times d = (x + 2.75x)8$$

$$x \times d = 3.75x \times 8$$

$$d = 30 \text{ days}$$

238. (d)

Cleave – means to divide/split

239. (b)

$$32.4 = K(2\theta + 25) \quad ...(i)$$

$$42 = K(5\theta + 30) \quad ...(ii)$$

$$\theta = 1$$

240. (22.22)

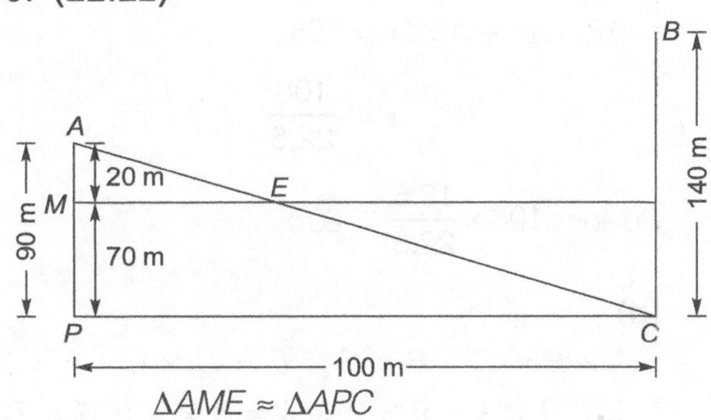

$$\Delta AME \approx \Delta APC$$

$$\frac{AM}{AP} = \frac{ME}{PC}$$

$$\Rightarrow \quad \frac{20}{90} = \frac{ME}{100}$$

$$\Rightarrow \quad ME = 22.22$$

241. (d)

Using options (a) and (c) cannot be answer.

As Hema's age = $2n$ +5, where n is a natural number.

$\left.\begin{array}{l}(2n + 5 \neq 14)\\(2n + 5 \neq 18)\end{array}\right\}$ for any natural number n

(b) and (d) remains

Let us take Hema's age as option (d) which is 19.

So, Hari's age = 7 and Suresh age = 57

Verifies all condition "answer (d)"

Alternative Method

Hema = 2 Hari + 5

Suresh = 10 Hari − 13 = 3 Hema

Solving equations.

Hari = 7

Hema = 19

Suresh = 57

242. (c)

Sum of series will be

$$\left(1-\frac{1}{3}\right)+\left(\frac{1}{2}-\frac{1}{4}\right)+\left(\frac{1}{3}-\frac{1}{5}\right)\ldots\ldots.$$

$$\left(\frac{1}{48}-\frac{1}{50}\right)+\left(\frac{1}{49}-\frac{1}{51}\right)+\left(\frac{1}{50}-\frac{1}{52}\right)$$

All like terms will cancel out and we will be left with

$$\left(1+\frac{1}{2}\right)-\left(\frac{1}{51}+\frac{1}{52}\right)$$

243. (c)

$$12.5\%x + 10\%x = 108$$

$$x = \frac{108}{22.5}$$

So loss $108 \times \dfrac{12.5}{22.5} = 60$

244. (b)

$$A \times B \times C = B \times G \times E = D \times E \times F = 72$$

$$8 \times 9 \times 1 = 9 \times 2 \times 4 = 3 \times 4 \times 6 = 72$$

Any of A, B, C, D, E, F, G cannot be 5.

245. (b)

$$C \propto W^2$$

$$C = kW^2$$

$$\Rightarrow \quad C = k(10)^2 = 100k = 1600 \Rightarrow k = 16$$

$$C_1 = k(4)^2 = 16k$$

$$C_2 = k(6)^2 = 36k$$

Now total cost = $52k = 52 \times 16 = 832$

246. (b)

Put value and verify

(i) $y = 2x + 4$ is true in $-3 \le x \le -1$

On putting $x = -3$, $y = -2$ and $x = -2$, $y = 0$ and $x = -1$, $y = 2$

(ii) $y = |x - 1|$ is also true ($x = -1$, $y = 2$), ($x = 0$, $y = 1$) and ($x = 1$, $y = 0$)

(iv) $y = 1$ in $(2 \le x \le 3)$ always true

(i), (ii) and (iv) are true.

247. (d)

$${}^9C_3 = \frac{9!}{6! \times 3!} = \frac{9 \times 8 \times 7}{6} = 84$$

248. (a)

As a, b, c are non-negative integers and given

$\log a + \log b + \log c = 0$

$$\log(a \times b \times c) = \log 1$$

$$\Rightarrow \quad a \times b \times c = 1$$

Which can be possible for simple values.

$$a = b = c = 1$$

Hence $\quad a + b + c = 1 + 1 + 1 = 3$

Alternative Method:

Given,

$\log a + \log b + \log c = 0$

As we know $\log 1 = 0$, so each one of them can be zero if $a = b = c = 1$

$\log 1 + \log 1 + \log 1 = 0$

By putting $a = b = 1$ equation satisfies

$$a + b + c = 1 + 1 + 1 = 3$$

249. (b)

$$\underbrace{a + a + a + \ldots + a}_{n\text{ times}} = na = a^2b$$

$$\Rightarrow \quad n = ab \quad \ldots(i)$$

$$\underbrace{b + b + b + \ldots + b}_{m\text{ times}} = mb = b^2a$$

$$\Rightarrow \quad m = ab \quad \ldots(ii)$$

So, $\underbrace{[m + m + \ldots\ldots + m]}_{n\text{ times}} \times \underbrace{[n + n + \ldots\ldots + n]}_{m\text{ times}}$

i.e., $\quad mn \times mn = (mn)^2$

from (i) and (ii) $\quad mn = a^2b^2$

So, result, $\quad (mn)^2 = (a^2b^2) = a^4b^4$

250. (c)

Loss = kd^2 For duration of 7 units

$4900 = k(7)^2 \Rightarrow k = 100$

Loss $= kd^2$ For duration of 4 units

$= k(4)^2 \Rightarrow 16k = 1600$

251. (b)

$A + B + C = 13$ ...(i)

$C + D + E = 13$ ...(ii)

$E + F + G = 13$ ...(iii)

$G + H + K = 13$ ...(iv)

Adding [(i) + (ii) + (iii) + (iv)]

$A + B + C + D + E + F + G + H + K + (C + E + G) = 13 \times 4 = 52$ ...(v)

Also A, B, C, D, E, F, G, H & K represents natural numbers from (1 to 9)

There sum will be given by $\frac{n(n+1)}{2} = 45$

Substituting (iv) $C + E + G = 7$ ...(vi)

Only possibly for sum 7 will be (1, 2, 4)

Now, $C + E$ cannot be (1 and 2)

As eq. (ii) is $C + D + E = 13$

Now, D will become equal to 10 (which is not possible because digits 1 to 9 given)

C & E can be either $\left.\begin{matrix}(1,4)\\(2,4)\end{matrix}\right]$ ...(vii)

If $C = H$ from eq. (vi) $C + E + G = 7$

Now, $E + G = 3$

(Not possible in eq. (iii) $E + F + G = 10$, $F = 10$ which is not possible)

So from eq. (vii) only possibility remains is $E = H$.

252. (c)

Cumulative rainfall = 300 mm

50% of rainfall $= 300 \times \frac{50}{100} = 150$ mm

Area $= 50\ m^2$

$\Rightarrow$ Volume stored in tank $= 150 \times 10^{-3} \times 50\ m^3$

or $= 7500\ l$

253. (b)

$\log P = 10(y - z)$

$\log Q = 10(z - x)$

$\log R = 10(x - y)$

$\log P + \log Q + \log R = 0$

$\log(PQR) = \log 1$

$PQR = 1$

254. (b)

9 Am of 11 July of 2 PM on 15th July = 101 hours

$\left(24 + \frac{15}{60}\right)$ hours of incorrect clock = 24 hours of correct clock

$\left(24 + \frac{15}{60}\right)$ hours of IC

= 24 hours of correct clock

1 hour of IC $= \frac{96}{97}$ hours of correct clock

101 hour of IC $= \frac{96}{97} \times 101$ hours of correct clock

= 99.958 hours of correct clock

$= 99$ hours $+ 0.95876 \times 60$ minutes of correct clock

= 99 hours + 57.525 minutes

= 99 hours and approx 58 minutes

So, correct time will be

2 PM, 11th july + (99 hours and 58 minutes)

= 12 : 58 PM on 15th July

255. (b)

2 → 12 → 60 → 240 → 720 → 1440 → 1440 → 0

(2 × 6) (12 × 5) (60 × 4) (240 × 3) (720 × 2) (1440 × 1) (1440 × 0)

So, 1440 must be answer.

256. (c)

Number is divided by either by 20 or 42 or by 76

K × LCM(20, 42, 76) + constant difference

= 7980 K + 7 (K is natural number)

Least number will be 7980 + 7 = 9987.

257. (d)

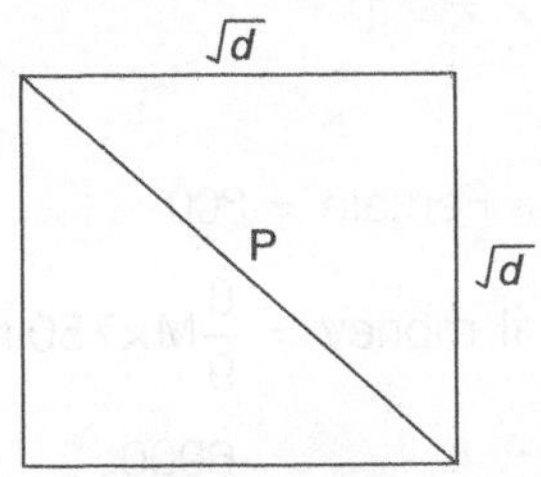

Area of square = d

Side one side of square $= \sqrt{d}$

Diagonal of square $= \sqrt{d+d} = \sqrt{2d}$

Area of circle $= \pi r^2$

$= \pi\left(\frac{\sqrt{2d}}{2}\right)^2$

$$= \pi\frac{d}{2}$$

$$= \frac{1}{2}\pi d$$

258. (c)

$$p^{-x} = \frac{1}{q},\ q^{-y} = \frac{1}{r},\ r^{-z} = \frac{1}{p}$$

Put $x = 1,\ y = 1,\ z = 1$

$$p^{-1} = \frac{1}{q},\ q^{-1} = \frac{1}{r},\ r^{-1} = \frac{1}{p}$$

i.e. $\frac{1}{p} = \frac{1}{q},\ \frac{1}{q} = \frac{1}{r},\ \frac{1}{r} = \frac{1}{p}$

Which is true i.e.

$$\frac{1}{p} = \frac{1}{q} = \frac{1}{r}$$

So, $xyz = 1$

Alternative Solution:

$$p^x = q$$
$$q^y = r$$
$$r^z = p$$
$$x \log p = \log q$$
$$y \log q = \log r$$
$$z \log r = \log p$$

$$x = \frac{\log q}{yp}$$

$$y = \frac{yr}{\log q}$$

$$z = \frac{\log p}{yr}$$

$$x \times y \times z = 1$$

259. (b)

Male + Female = 300 ...(i)

$$\text{Total money} = \frac{8}{9}M\times750+\frac{2}{3}F\times1000$$

$$= \frac{6000}{9}M + \frac{6000}{9}F$$

$$= \frac{6000}{9}(M+F)$$

From equation (i)

$$\text{Total money} = \frac{6000}{9}(300) = 2,00,000$$

260. (a)

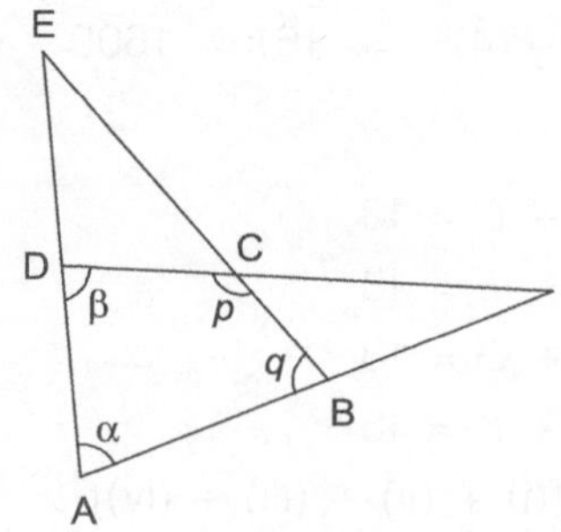

$\angle E + \angle F = ?$

$\alpha + q + E = 180$ ...(i)

$\alpha + \beta + F = 180$ ...(ii)

$\alpha + \beta + p + q = 360$ ...(iii)

Equation (i) + (ii) = (iii)

$$\alpha + q + E + \alpha + \beta + F = \alpha + \beta + p + q$$

$$E + F = p - \alpha$$

261. (c)

Four green, two red face

$$P(G) = \frac{4}{6} = \frac{2}{3}$$

$$q(R) = \frac{1}{3}$$

$$n = 7$$

Option (1), $P(G = 3) = {}^7C_3\left(\frac{2}{3}\right)^3\left(\frac{1}{3}\right)^4$

$$= \frac{35\times2^3}{(3)^7} = \frac{35\times2^3}{(3)^7}$$

(2), $P(G = 4) = {}^7C_4\times\left(\frac{2}{3}\right)^4\times\left(\frac{1}{3}\right)^3$

$$= \frac{35\times2^4}{(3)^7} = \frac{35\times2^4}{(3)^7}$$

(3), $P(G = 5) = {}^7C_5\times\left(\frac{2}{3}\right)^5\times\left(\frac{1}{3}\right)^2$

$$= \frac{21\times2^5}{(3)^7} = \frac{42\times2^4}{(3)^7}$$

(4), $P(G = 6) = {}^7C_6\times\left(\frac{2}{3}\right)^6\times\left(\frac{1}{3}\right)$

$$= \frac{7\times2^6}{(3)^7} = \frac{28\times2^4}{(3)^7}$$

Option 3 is maximum value.

So, five green faces and two red faces.

262. (b)

Let total number of people are 100.

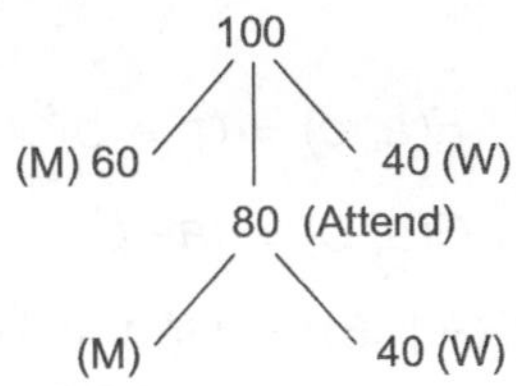

So, M must be 80 – 40 = 40

Ratio of male to female

40 : 40

1 : 1

263. (d)

This is a infinite Geometric progression with first term (a) = 1

and common ratio (r) = $\frac{1}{4}$

The formula for finding the sum of infinite G.P is given by $\frac{a}{1-r}$

$$\frac{1}{1-\frac{1}{4}} = \frac{1}{3/4} = \frac{4}{3}$$

264. (b)

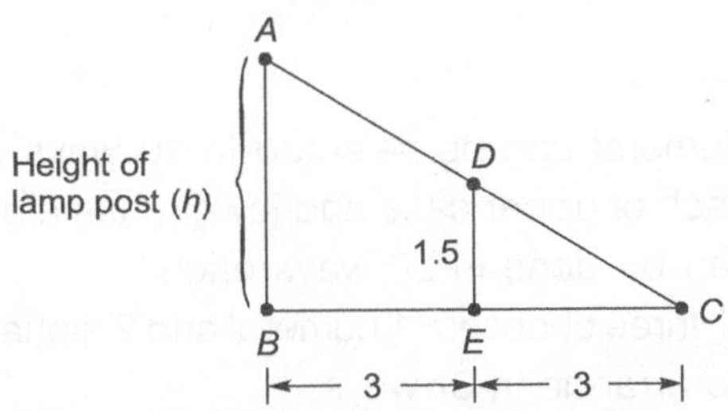

ABC and DEC are similar triangles,

$$\frac{AB}{DE} = \frac{BC}{EC}$$

$$\frac{h}{1.5} = \frac{6}{3}$$

$$\frac{h}{1.5} = 2$$

h = 3 meter

265. (b)

715 ? 423

We know that the divisibility rule for 3 is sum of all digits should be divisible by 3.

7 + 1 + 5 + 4 + 2 + 3 = 22

So, the next number after sum 22 which are divisible by 3 are 24. 27, 30 etc.

So, 22 + 2 = 24

22 + 5 = 27

But according to minimum condition 2 is right answer.

266. (b)

Alloy A contains Gold and Copper.

Let $2x : 3x$

Which is same as $4x : 6x$

Alloy B contains Copper $3x : 7x$

As masses of Alloy A is equal to Alloy of mass B.

i.e.

		Gold		Copper	
i.e.	Alloy 1	$4x$	:	$6x$	$\Rightarrow 10x$ (mass)
	Alloy 2	$3x$	:	$7x$	$\Rightarrow 10x$ (mass)

∴ Ratio of Gold to Copper is,

$$\frac{\text{Gold from alloy 1} + \text{Gold of alloy 2}}{\text{Copper from alloy 1} + \text{Copper of alloy 2}}$$

$$= \frac{4x+3x}{6x+7x} = \frac{7x}{13x} = \frac{7}{13}$$

267. (b)

$$A = P\left(1+\frac{r}{100}\right)^n$$

$$1000000 = P\left(1+\frac{10}{100}\right)^5$$

$$1000000 = P\left(\frac{11}{10}\right)^5$$

$$1000000 = P\left(\frac{11^5}{10^5}\right)$$

$$P = 1000000 \times \frac{10^5}{11^5} = \frac{10^{11}}{11^5}$$

$$= 620921.323 \approx 621000$$

268. (c)

Let us suppose total number of cabs are 100.

85 are green and rest 15 are blue.

Witness is correct 80% of times in identifying.

Total number of blue cabs identified correctly.

(80% of 15) $\Rightarrow \frac{80}{100} \times 15 = 12$

Witness is incorrect 20% of times

That incorrectness should happen for green

(20% of 85) $\Rightarrow \frac{20}{100} \times 85 = 17$

Witness identifies 12 + 17 = 29 Cabs in total.

$$\text{Probability} = \frac{\text{Required}}{\text{Total}} = \frac{12}{12+17} = \frac{12}{29}$$
$$= 41.3\% \approx 41\%$$

269. (b)

(a) is beyond the scope of given information option (c) can also be discarded on the same grounds.The argument deals with the coastal region becoming crowded and polluted because of the upcoming luxury resort. Option (b) precisely underlines the theme of the para.

270. (c)

Statement (i) is not true as nowhere it is mentioned that John was a captain in junior team. The introductory line emphasizes on the board recognizing John's potential (Latent quality/ possibility) as leader of the team.

Statement (iii) also manipulates the facts mentioned in the argument.

The 3rd statement of the argument while this aggression has filled stadia with die-hard fans does not indicate fans expectations from John as a caption.

Statement (ii) The concluding statement of the para suggests that finding a completer captain is a tough task as it took John many years to become a successful and calculative batsman - captain.

Statement (iv) can be explicitly concluded from the last 4 lines of the para.

271. (c)

$$K = 4 \Rightarrow \frac{(4+2)^2}{4-3} = 36$$
$$K = 8 \Rightarrow \frac{(8+2)^2}{8-3} = \frac{100}{5} = 20$$
$$K = 28 \Rightarrow \frac{(28+2)^2}{28-3} = \frac{900}{25} = 36$$

} Integers

So, option (c) ⇒ 4, 8, 28

272. (d)

$$f(x) = 0$$
$$x = \{-2, 0, 3\}$$
$$f(-2) = 0,\ f(0) = 0,\ f(3) = 0$$
$$f(x - 3) = 0$$

1, 3, 6

$$f(1 - 3) = f(-2) = 0$$
$$f(3 - 3) = f(0) = 0$$
$$f(6 - 3) = f(3) = 0$$

273. (a)

$$F(a, b) = (a - b)^2$$
$$G(a, b) = |a - b|$$
$$G(F(1, 3)\ G(1, 3)) = G((1 - 3)^2, |1 - 3|)$$
$$= G(4, 2)$$
$$= |4 - 2| = 2$$

274. (c)

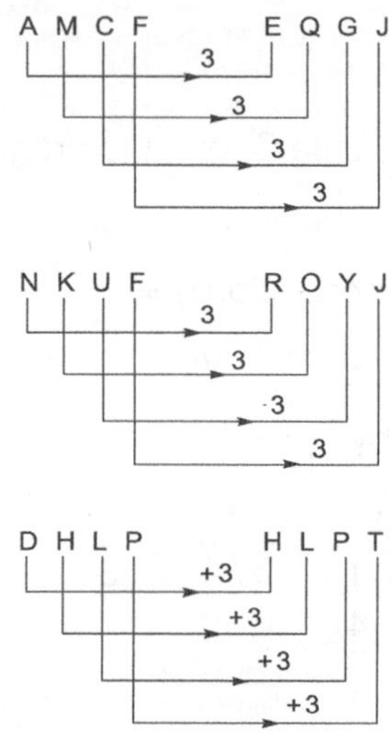

275. (c)

Numeral can be selected in 10 ways (0 - 9).

Each of upper case and lower case alphabhats can be done in 26 ways each.

All three choosen (1 numeral and 2 alphabat) can be arrange in 3! ways.

So, total number of ways will be 10 × 26 × 26 × 3! = 40560 ways

276. (c)

Assume number of marbles = 100

Cost of marbles increased = 25

New cost = Rs. 125

Number of marbles in Rs.100

$$= \frac{100}{125} \times 100 = 80 \text{ marbles}$$

Now, $(40 - x) + (25 - x) + (20 - x) + (15 - x) = 80$

$100 - 4x = 80$

$x = 5$

% blue marbles in new design

$$= \frac{(40-5)}{80} \times 100 = 35 \times \frac{5}{4} = 43.75\%$$

277. (c)

(i) Q and R moves first.
In forward trip Q rowed.
In return trip R rowed.

(ii) P and R moves in second trip
R rowed in forward trip
P rowed in return trip

(iii) P and S moves in last trip
S rowed in forward trip.
R rowed twice.

278. (b)

$$B + G = 12$$
$$B = G + 2$$
$$\Rightarrow \quad B = 7$$
$$G = 5$$

7 Boys and 5 Girls are there is 12 students. Among 3 students selected boys have to be more than girls.

So only two cases arise.

G	B
2	1
3	0

As case (i)

$$GGG \rightarrow \frac{5}{12}\times\frac{5}{12}\times\frac{5}{12}$$

$$\text{case (ii)}\left.\begin{matrix} G & G & B \\ G & B & G \\ B & G & G \end{matrix}\right\} \rightarrow 3\times\frac{5}{12}\times\frac{5}{12}\times\frac{7}{12}$$

$$\text{Required prob.} = \left(\frac{5}{12}\right)^3 + 3\left(\frac{5}{12}\right)^2\times\left(\frac{7}{12}\right)$$

$$= \frac{650}{1728} = \frac{325}{864}$$

$$= 0.3761574 \approx 0.376$$

Alternative Method :

There are $7B$ and $5G$

Through as question is stating 3 students are taken at random.

This can be a possible way to approach it.

G	B
2	1
3	0

$$= \frac{{}^5C_2\times{}^7C_1 + {}^5C_3\times{}^7C_0}{{}^{12}C_3}$$

$$= \frac{10\times7+10\times1}{\frac{12\times10\times11}{6}} = \frac{80}{220} = \frac{4}{11}$$

$$\approx 0.3636$$

279. (b)

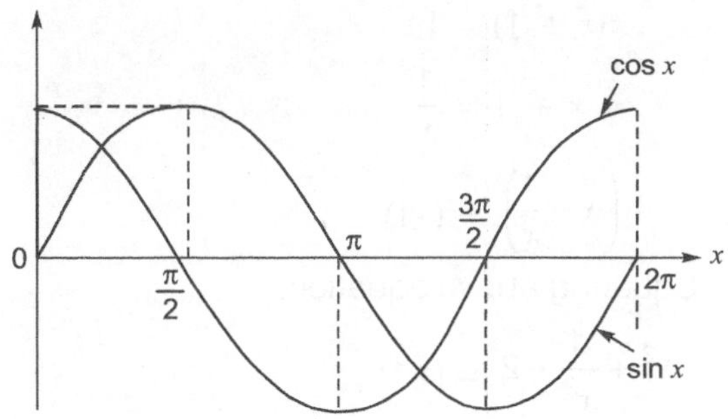

From the curve it is clear that sin$x$ and cos$x$ both are decreasing in the interval $\left(\frac{\pi}{2}, \pi\right)$.

280. (d)

(i) Cuboid volume = $8 \times 10 \times 6 = 480$ cm^3

(ii) Cube volume = $8 \times 8 \times 8 = 512$ cm^3

(iii) Cylinder volume = $\pi r^2 h$

$$= \frac{22}{7}\times7\times7\times7 = 1078 \text{ cm}^3$$

(iv) Sphere volume

$$= \frac{4}{3}\pi r^3 = \frac{4}{3}\pi\times(7)^3 = 1436.75 \text{ cm}^3$$

Hence, the descending order of their volumes is (iv), (iii), (ii), (i).

281. (c)

Area of equilateral triangle = $\sqrt{3}$

$$\frac{\sqrt{3}}{4}a^2 = \sqrt{3}$$

$$a^2 = 4$$
$$a = 2$$

Perimeter = $3a = 3 \times 2 = 6$

282. (a)

Total distance = x

(Onward journey) S_1 = 60 km/h

(Return journey) S_2 = 90 km/h

$$\text{Average speed} = \frac{\text{Total distance}}{\text{Total time}}$$

$$= \frac{x+x}{\frac{x}{60}+\frac{x}{90}} = \frac{2x}{x\left[\frac{3+2}{90\times2}\right]}$$

$$= \frac{2\times90\times2}{5} = 72 \text{ km/h}$$

284. (c)

$$x^2 + x = 1$$
$$x(x + 1) = 1$$
$$x + 1 = \frac{1}{x}$$
$$\left(x - \frac{1}{x}\right) = (-1)$$

Squaring above equation,

$$x^2 + \frac{1}{x^2} - 2 = (+1)$$
$$x^2 + \frac{1}{x^2} = 3$$

Squaring above equation,

$$x^4 + \frac{1}{x^4} + 2 = 9$$
$$x^4 + \frac{1}{x^4} = 7$$

285. (d)

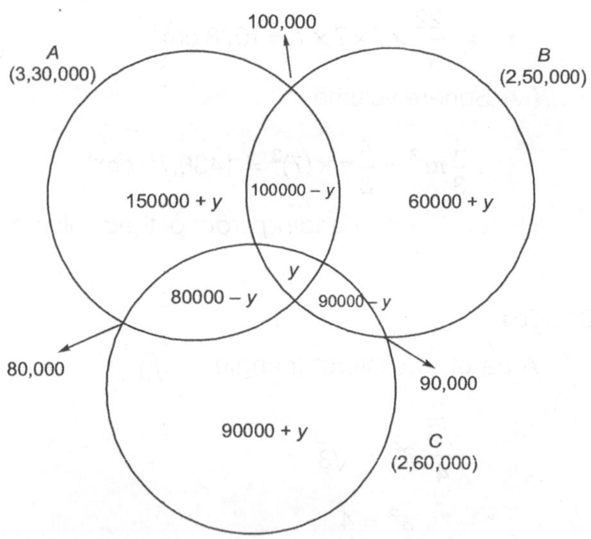

6,30,000 = 2y + 1,50,000 + 100000 + 80000 + 60000 + 90000 + 90000

6,30,000 – 5,70,000 = 2y

$y = 30000$

Students who failed to clear the test = 150000 + 60000 + 90000 + 4y

= 300000 + 4 × 30000

Students who failed to clear the test = 420000

286. (c)

Number of parallelogram

$$= {}^4C_2 \times {}^5C_2$$
$$= 6 \times 10 = 60$$

287. (b)

Rectangle initial length is l and breath is b. If l is reduced by 10 m and breath reduced by 5 m then it becomes square.

So, $l - 10 = b - 5$

$l - b = 5$

Given, initial, $l \times b = A$ ...(i)

2nd condition,

$$(l - 10)(b - 5) = A - 650$$
$$lb - 10b - 5l + 50 = A - 650$$
$$A - 10b - 5l = A - 700$$
$$10b + 5l = 700$$
$$10b + 5(b + 5) = 700$$
$$15b + 25 = 700$$
$$15b = 700 - 25$$
$$15b = 675 \Rightarrow b = 45 \text{ m}$$
$$l = 45 + 5 = 50 \text{ m}$$

Area of original rectangle = 45 × 50 = 2250 m^2

288. (b)

Let two digit no is xy

$x + y = 9$ ...(i)

$10x + y - 45 = 10y + x$

or $x - 5 = y$

$x - y = 5$ ...(ii)

Adding (i) and (ii)

$x = 7$

Subtracting (i) and (ii)

$y = 2$

Therefore the number is 72.

289. (b)

7 machine → 7 toys → 7 minutes

1 machine → 1 toy → 7 minutes

Because one machine takes 7 minute for making 1 toy.

So, 100 machines will take 7 minute for making 100 toys.

290. (d)

Given, a and b are integer

$a + a^2b^3$ is odd

$a(1 + ab^3)$ is odd

Multiplication of odd and odd number is odd.

So, a is odd and $1 + ab^3$ is odd.

$1 + ab^3$ is odd, so ab^3 will be even.

Because a is odd so for ab^3 to be even b must be even.

So, a is odd and b is even.

291. (c)

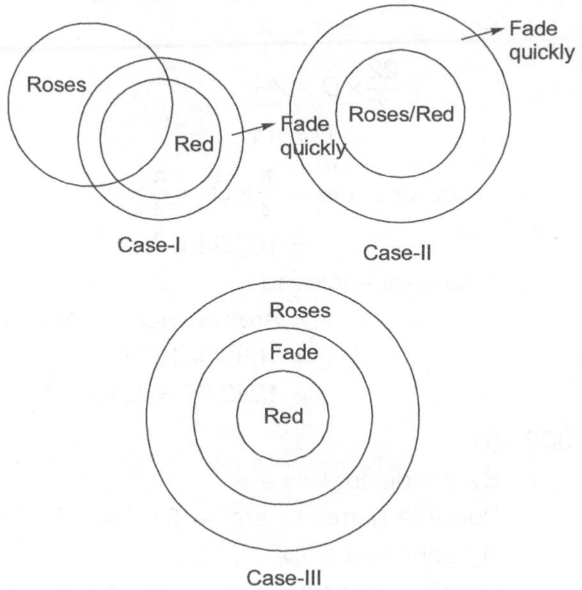

292. (b)

x	0	± 1	± 2
y	0	–1	0

Solving through options,

(a) $y = ||x|+1|-2$

Putting, $x = 1$

$y = 2 - 2 = 0$ Not satisfied

(b) $y = ||x|-1|-1$

Putting, $x = 1$

$y = |1 - 1| - 1$

$y = -1$ Satisfied

(c) $y = ||x|+1|-1$

Putting, $x = 1$

$y = |1 + 1| - 1$

$y = -1$ Not satisfied

(d) $y = ||x-1|-1|$

Putting, $x = 1$

$y = ||1 - 1| - 1$

$y = |0 - 1|$

$y = 1$ Not satisfied

293. (a)

$\log|a| + \log|b| + \log|c| = 0$

It is possible only,

if $|a|, |b|$ and $|c|$ all are equal to 1.

So, a, b, c may be ±1, ±1, ±1 respectively.]

For minimum value all three will be negative.

So, minimum value = –3

For maximum value all three will be positive,

So, maximum value = +3.

294. (d)

Train speed = 54 km/h

Time = 25 sec for travelling length of train and length of platform

Man speed = 9 km/h

Speed of train to man = 45 km/h

Time = 14 sec

So,length of train= time × speed

$$= 14 \times 45 \times \frac{5}{18}$$

Length of train = 35 × 5 m = 175 m

Length of platform + length of train = speed × time

$$= 54 \times \frac{5}{18} \times 25 = 15 \times 25 = 375 \text{ m}$$

Length of platform = 375 – 175 = 200 m

295. (b)

BC, FGH, LMNO, TUVWX

296. (a)

$$P_C = P_S = P_T$$
$$\pi D = 4a = 3s$$

now, Area of circle $= \frac{\pi}{4}D^2$ $[\pi D = 4a]$

Area of square $= a^2$

$$= \frac{\pi^2 D^2}{16} = \frac{\pi^2}{16}D^2$$

Area of triangle $= \frac{\sqrt{3}}{4}s^2 = \frac{\sqrt{3}}{4}\left(\frac{\pi D}{3}\right)^2$ $\{\because \pi D = 3s\}$

$$= \frac{(\sqrt{3})\pi^2 D^2}{36}$$

$$\because \quad \frac{\pi}{4} > \frac{\pi^2}{16} > \frac{\sqrt{3}\pi^2}{36}$$

So, circle has the largest area.

297. (c)

$$\frac{1}{1+\log_u vw} + \frac{1}{1+\log_v wu} + \frac{1}{1+\log_w uv}$$

$$= \frac{1}{\log_u u + \log_u vw} + \frac{1}{\log_v v + \log_v wu} + \frac{1}{\log_w w + \log_w uv}$$

$$= \frac{1}{\log_u uvw} + \frac{1}{\log_v uvw} + \frac{1}{\log_w uvw}$$

$$= \log_{uvw} u + \log_{uvw} v + \log_{uvw} w = \log_{uvw} uvw = 1$$

298. (b)

For remaining two trials highest probability is, that

H H

H T

T H

T T

Out of 4 possible cases one head and T will be have the highest probability of occurance.

299. (*)

$$125 \times 7 \times 39 = \frac{5}{7}W$$

$$W = 47775 \text{ Robot hrs.}$$

$$\text{Left work} = 47775 - 125 \times 7 \times 39$$

$$= 13650 \text{ Robot hrs.}$$

$$13 \times x \times 8 = 13650$$

$$x = 131.25$$

Addition robot required

$$= 131.25 - 125 = 6.25 \approx 7$$

300. (c)

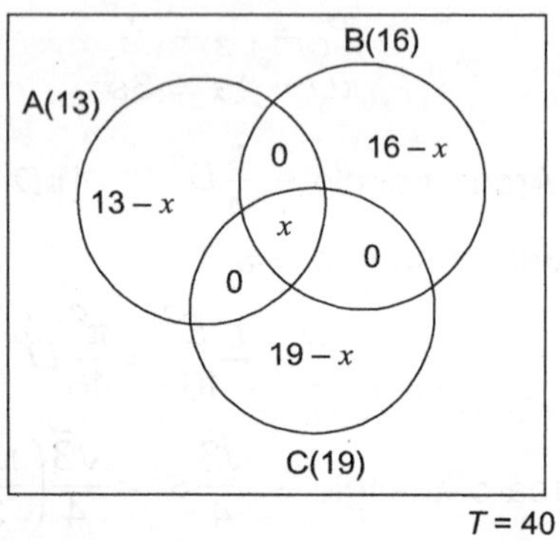

Total student = 40

$13 - x + 16 - x + 19 - x + x = 40$

Students watches all three movies,

$$x = 4$$

301. (c)

$$\text{Area} = 1936 \text{ m}^2$$

$$a^2 = 1936 \text{ m}^2$$

$$a = 44 \text{ m}$$

$$\text{Length of wire} = 4a$$

$$= 4 \times 44 = 176 \text{ m}$$

Now, part-1 length

$$= 3 \times 44 = 132 \text{ m}$$

$$\text{part-2 length} = 1 \times 44 = 44 \text{ m}$$

Long wire is bent in square,

So, $4a = 132$

$$a = 33 \text{ m}$$

Area of square = $33^2 = 1089$ m^2

Now, small wire is bent in circle,

So, $\pi D = 44$

$$\frac{22}{7} \times D = 44$$

$$D = 14 \text{ m}$$

$$\text{Area of circle} = \frac{\pi}{4} \times D^2 = \frac{\pi}{4} \times 14^2$$

$$= 153.94 \text{ m}^2$$

Total area enclosed

$$= \text{Area of square} + \text{Area of circle}$$

$$= 1089 + 153.94$$

$$= 1242.97 \approx 1243 \text{ m}^2$$

302. (d)

By condition 1 ⇒

Possible numbers are → 51, 54, 57

By condition 2 ⇒

Possible numbers are → 61, 62, 63, 65, 66, 67, 69

By condition 3 ⇒

Possible numbers are → 70, 71, 73, 74, 75, 76, 77, 79

⇒ 66 is multiple of 3, it does not belong to 50 to 59. So it is not the answer.

⇒ 54 is multiple of 3 and 6. But it is not the multiple of 4. So according to second condition it cannot be the answer.

⇒ Because 65 is not the multiple of 6. So condition 3 is not satisfied. So it cannot be the answer.

⇒ For 76. All three conditions are satisfied. So it is the answer.

303. (b)

According to given coding

$P = 32,\ E = 10,\ S = 38,\ T = 40$

$P + E + S + T = 32 + 10 + 38 + 40 = 120$

304. (d)

Here tan 45° = 1

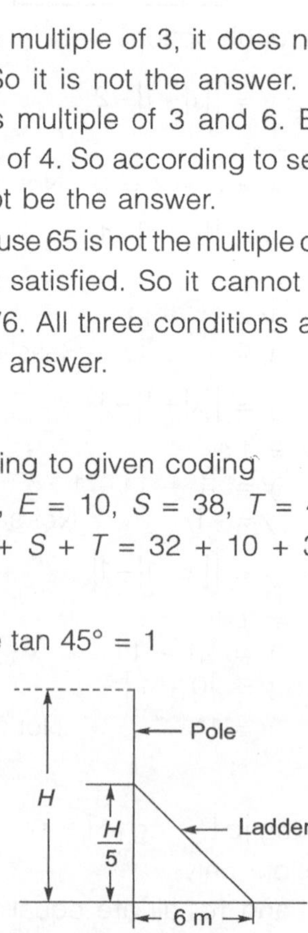

$$\frac{H}{5} = 6; \quad H = 30 \text{ m}$$

Height of wall = 30 m

305. (a)

Cost of carpet = [70 × 55 − 550] × 50 = Rs. 1,65,000

306. (d)

No. of bricks = x

$\Rightarrow 30 \times 12 \times 6 \times 10^6 \times 0.6 = 8 \times 6 \times 6 \times x$

$x = 4.5 \times 10^6 = 45 \times 10^5$ = 45 lakhs bricks

307. (d)

Let side of larger square = x

Side of smaller square = $(x - 5)$

Area of larger square = x^2

Area of smaller square = $(x - 5)^2$

Given, Area of larger square

= 4 times of area of smaller square

$$x^2 = 4(x-5)^2$$

$$\therefore \quad x = 10$$

Side of larger square = 10 cm

308. (d)

P is brother of S

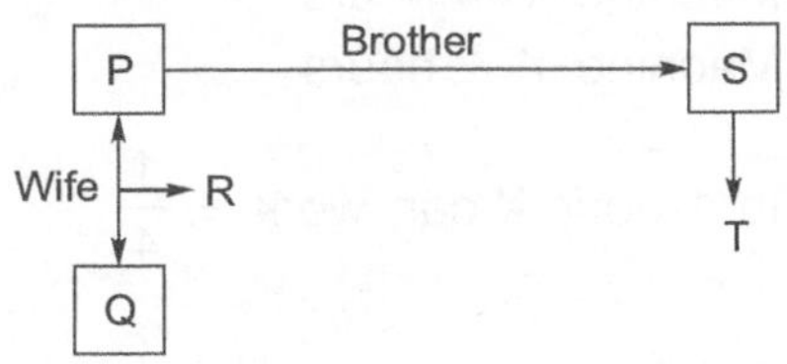

P is male,

S is male or Female

S is sister-in-law of Q may be possible

S is brother of P may be possible

S is aunt of R may be possible

S is the aunt of T is impossible in this case because given T is child of S.

309. (a)

All the three sports = 300 − 295

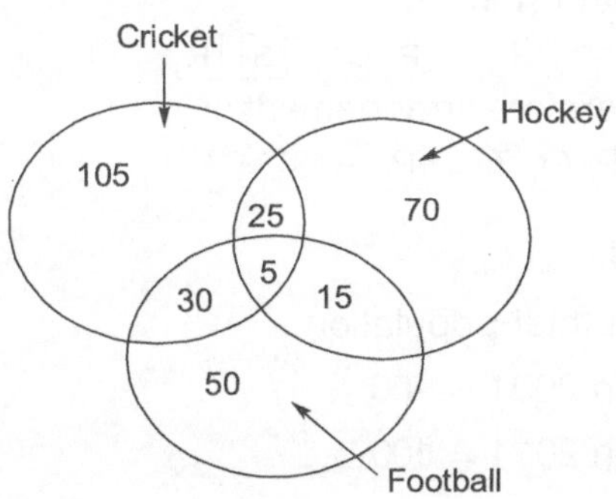

= 5

No. of players playing at least 2 sports

= 25 + 5 + 30 + 15 = 75

Percentage of players playing at least 2 sports

$$= \frac{75}{300} \times 100 = 25$$

310. (b)

Ram and John = 5 hours

Krishna and Amir = 2 hours

Krishna does not work with Ram.

Now pair will be

John and Krishna

Ram and Amir

Given all the workers to continue working.

John and Krishna will be one pair.

311. (b)

Let a, b be initial population

Given $\frac{a}{b} = p$ (ratio earlier)

$$\frac{a\left(1+\frac{x}{100}\right)}{b\left(1+\frac{y}{100}\right)} = p' \text{ (new ratio)}$$

So, required % change,

$$\frac{p'-p}{p} \times 100 = \left[\frac{\frac{a}{b}\left(\frac{100+x}{100+y}\right) - \frac{a}{b}}{\frac{a}{b}}\right] \times 100$$

$$= \frac{100(x-y)}{100+y}$$

312. (d)

Letinlet pipes, A=4 hours

B = 5 hours

outlet pipe, C = 20 hours

In 1 hour $(A + B + C)$ can fill

$$= \frac{1}{4} + \frac{1}{5} - \frac{1}{20} = \frac{8}{20} = \frac{2}{5}$$

$(A + B + C)$ together

$$= \frac{5}{2} = 2.5 \text{ hours}$$

313. (b)

Speed of car A = 50 km/hr

Speed of car B = 60 km/hr

Since, both cars A and B are moving in same direction, the relative speed = 60 − 50 = 10 km/hr

Distance required between them = 20 km

$$\therefore \quad \text{Time} = \frac{\text{Distance}}{\text{Speed}} = \frac{20}{10} = 2 \text{ hrs}$$

314. (b)

Let share of each student = x

Total cost of gift = $10 \times x$

$x = 8(x + 150)$

$x = 600$

Total cost = $10 \times 600 = 6000$

315. (d)

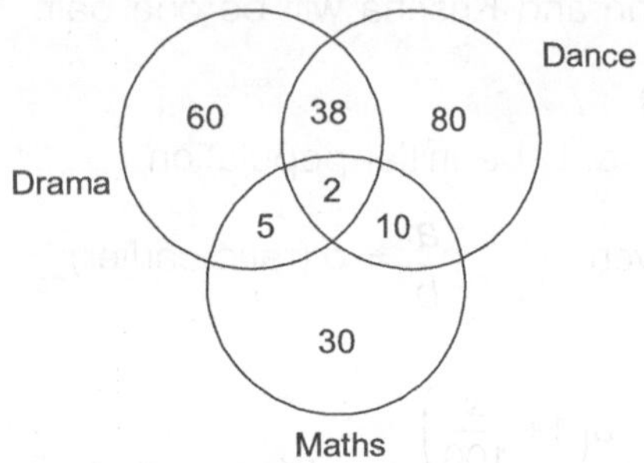

Total number of students = 60 + 80 + 30 + 38 + 5 + 10 + 2 = 225

25% = 225

$$\Rightarrow \quad 100\% = \frac{225}{25} \times 100 = 900$$

316. (a)

Suffix C as criminal, NC as not criminal.

Case I:

Criminals	P	Q	R	S
Assumption	T	F	F	F
Result	Q_C	S_{NC}	R_C	S_C

Q and R are criminal in the result is impossible because only one person committed the crime.

Case II:

Criminals	P	Q	R	S
Assumption	F	T	F	F
Result	Q_{NC}	S_C	R_C	S_C

S and R are criminal in the result is impossible because only one person committed the crime.

Case III:

Criminals	P	Q	R	S
Assumption	F	F	T	F
Result	Q_{NC}	S_{NC}	R_{NC}	S_C

S_{NC} and S_C in the result which is contradiction. [S committed crime and same time not committed crime which is contradiction]

Case IV:

Criminals	P	Q	R	S
Assumption	F	F	F	T
Result	Q_{NC}	S_{NC}	R_C	S_{NC}

R is criminal in the result.

Hence this case satisfies only one person committed the crime.

So option (a) is correct.

317. (c)

Percentage of Administrators

$$= \frac{\text{Administrators}}{\text{Total}} \times 100$$

$$= \frac{50}{160} \times 100 = 31.25$$

319. (d)

Machine X = 4 hours

Machine Y = 2 hours

In 1 hour X can work = $\frac{1}{4}$

In 1 hour Y can work = $\frac{1}{2}$

In 1 hour $(X + Y)$ can work = $\frac{1}{4} + \frac{1}{2} = \frac{3}{4}$

$(X + Y)$ together = $\frac{4}{3}$ hours

320. (d)

Here R and S are fixed in 1st and 2nd places from right.

P Q T [S] [R]

Therefore arrangement of

P, Q, T = $^3P_3$ = 3! = 6

321. (c)

Let total population

in 2001 = 100

in 2011 = 100

In 2001, males = 60, females = 40

illiterates,

50% of 60 = 30

60% of 40= $\underline{24}$

$\underline{54}$

literates = 100 − 54 = 46

In 2011, males = 50, females = 50

illiterates,

40% of 50= 20

40% of 50= $\underline{20}$

$\underline{40}$

literates = 100 − 40 = 60

Percentage of increase literates from 2001 to

$2011 = \frac{60-46}{46} \times 100 = 30.43$

322. (d)

P and *Q* started 8 am

1 hour = 30

$$P = \frac{210}{30} = 7 \text{ hours}$$

$$Q = \frac{240}{30} = 8 \text{ hours}$$

After break time,

P = 7:30 min = 6 hours 30 min

After break time,

Q = 8:20 min = 7 hours40 min

Total time = 6 hours 30 min + 7 hour 40 min

= 14 hours 10 min

$$= 14 \text{ hour} + \frac{10}{16} \text{ hour}$$

$$= 14 \text{ hour} + \frac{1}{6} \text{ hour} = \frac{85}{6} \text{ hours}$$

In every 1 hour USD 200

$$\therefore \frac{85}{6} \text{ hours} = \frac{85}{6} \times 200 = 2833.33$$

Remaining= 3000 − 2833.33

= 166.666666 = 166.67

323. (c)

Only option (c) shows P and Q in same office, R and T in same office with S being in the adjacent office. Thus, option (c) is correct.

324. (d)

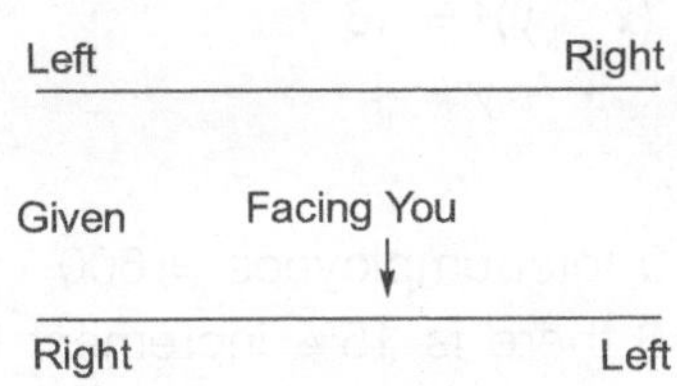

According to given conditions sequence is

Lohit, Seema, Rahul, Mathew,

↓ ↓

DoctorEngineer

∴Engineer= Mathew

325. (d)

7^3, 11^3, ____ , 17^3

∵ 7, 11, 13 and 17 are all prime numbers.

The series is cube of these numbers.

326. (c)

Work done by *X* in 1 hour = $\frac{1}{2}$

Work done by *Y* in 1 hour = $\frac{1}{4}$

Work done by X and *Y* together in 1 hour

$$= \frac{1}{2} + \frac{1}{4} = \frac{3}{4}$$

∴ Total time to complete mowing of lawn

$$= \frac{4}{3} \text{hour} = \frac{4}{3} \times 60 = 80 \text{ min}$$

327. (c)

$$X = \{1, 2, 3\}$$

$$Y = \{2, 3, 4\}$$

$$Z = \left\{\frac{1}{2}, \frac{1}{3}, \frac{1}{4}, \frac{2}{3}, \frac{2}{4}, \frac{3}{2}, \frac{3}{4}\right\}$$

Minimum value in $\{Z\} = \frac{1}{4}$

Maximum value in $\{Z\} = \frac{3}{2}$

$$\text{Product} = \frac{1}{4} \times \frac{3}{2} = \frac{3}{8}$$

328. (c)

All numbers between 100 and 1000 can be 3 digit numbers only

For units and tens digits = 5 integers can be filled (0, 2, 4, 6, 8)

For hundreds digit = 0 can't be filled only 4 digits can be filled (2, 4, 6, 8)

∴ Total choices = 4 × 5 × 5

= 100 numbers

329. (a)

Let, Number of boys = $4x$

and Number of girls = $3x$

Total passed candidates = $\frac{80}{100} \times 7x = \frac{28}{5}x$

Number of girls candidates who passed

$$= \frac{90}{100} \times 3x = \frac{27}{10}x$$

Now total number of candidates passed = Number of girls who passed + Number of boys who passed

⇒ Number of boys who passed

$$= \left(\frac{28}{5} - \frac{27}{10}\right)x = \frac{56-27}{10}x = \frac{29}{10}x$$

% of boys = $\frac{29}{10 \times 4x} x \times 100 = 72.5\%$

Option (a) is correct.

330. (c)

According to given data the only possible arrangement is 10 5 4 7 2

So, second from right will be 7.

331. (c)

Given statements: Some students were not involved in the strike.

Venn diagram:

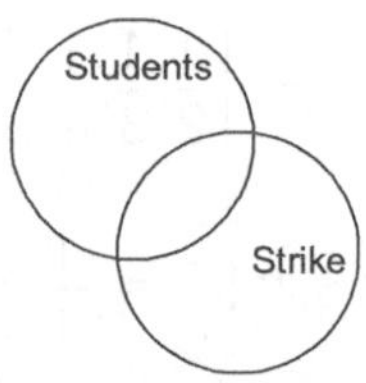

Only conclusion 4 is logically necessary.

Hence, option (c) is correct.

332. (c)

We know formula for volume of a (right circular) cone is $\frac{1}{3}\pi r^2 h$

Original volume, $V_0 = \frac{1}{3}\pi r_1^2 h_1$ ... (i)

Now we know radius and height both are increased by 10%. So after increase the new volume is

$$V_n = \frac{1}{3}\pi(1.1r_1)^2(1.1)h_1$$

$$= 1.331\left(\frac{1}{3}\pi r_1^2 h_1\right) = 1.331\ (V_0)$$

% change in volume

$$= \frac{V_n - V_0}{V_0} \times 100$$

$$= \frac{1.331V_0 - V_0}{V_0} \times 100 = 33.1\%$$

333. (a)

Total population = 1400 million

According to the question 70% of mobile phone users data is an unnecessary data. Its not necessary to calculate tabing the relevant data into consideration,

Number of internet users = 294 million

Half of which (internet) use e-commerce portal

$$\frac{294}{2} = 147$$

Percentage of these (e-commerce portals) in the country (out of 1400 million)

$$= \frac{147}{1400} \times 100 = 10.5\%$$

334. (b)

According to the concept of relative speed in opposite direction speeds should be added

Time of activity = $\frac{\text{Sum of distance}}{\text{Sum of speeds}}$

$$= \frac{540}{100+80} = \frac{540}{180}$$

= 3 hours from 7 am = 10 am

335. (b)

We know that watch hour hand complete one rotation (360°) in12 hours.

So, for 225°, corresponding time

$$= \left(\frac{225}{360}\right) \times 12 = 7.5 \text{ hours}$$

336. (d)

Let two numbers are x and y

Given, $(x + y) = 26$

$xy = 165$

We know that,

$$(x - y)^2 = (x + y)^2 - 4xy$$

$$= (26)^2 - 4 \times 165$$

$$(x - y)^2 = 16$$

$$x - y = 4$$

337. (d)

In 2010 total employees = 600

In 2016 there is 15% increment in number of employees.

In 2016 total employees = 1.15 × 600 = 690

Given, there is no change in P, Q and R type employees.

There is 40% increment in 'S' type employees.

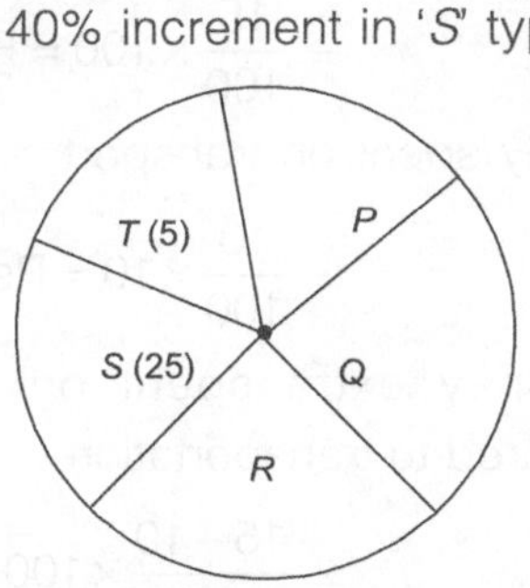

Now,Total employees in 2016

$$= \left(600 \times \frac{25}{100}\right) \times 1.4 = 150 \times 1.4$$

=210 employees

Increase in 'S' type employees

$$= 210 - \left(600 \times \frac{25}{100}\right) = 60$$

Initially, 'T' type employees = $\frac{5}{100} \times 600 = 30$

In 2016, increase in 'T' type employees = (690 − 600 − 60) = 30

In 2016, total 'T' type employees = (30 + 30) = 60 employees

338. (d)

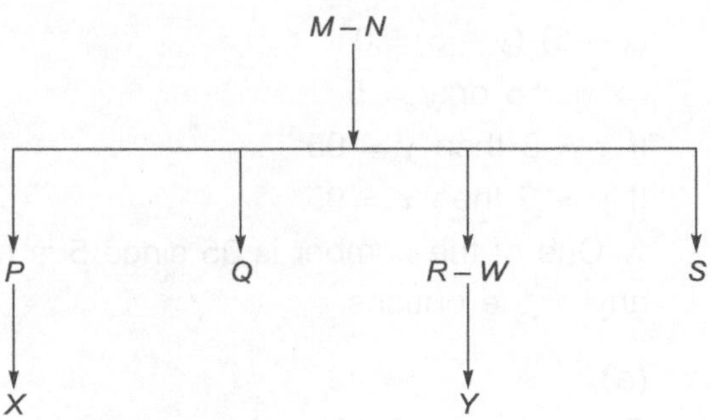

W can't be wife of P.

W can be husband or wife of R.

R can be father or mother of Y.

M can be grandfather of Y.

339. (c)

1st condition

Total amount = 100000

Assume, A has invested Rs.x at 10% rate and Rs (100000 − x) at 12% interest.

$$\text{Total interest} = 0.1x + (100000 - x) \times 0.12$$

$$(TI)_I = 12000 - 0.02x$$

2nd condition

Assume, A has invested Rs x at 12% rate and Rs (100000 − x) at 10% interest.

$$\text{Total interest} = 0.12x + (100000 - x) \times 0.1$$

$$(TI)_{II} = 10000 + 0.02x$$

Given,

$$(TI)_I - (TI)_{II} = 120$$

$$(12000 - 0.02x) - (10000 + 0.02x) = 120$$

$$2000 - 0.04x = 120$$

$$x = \frac{1880}{0.04} = \text{Rs. } 47000$$

Now,Another part

$$= 100000 - x$$

$$= 100000 - 47000$$

$$= \text{Rs. } 53000$$

$$\frac{x}{(100000 - x)} = \frac{47}{53}$$

340. (b)

So,

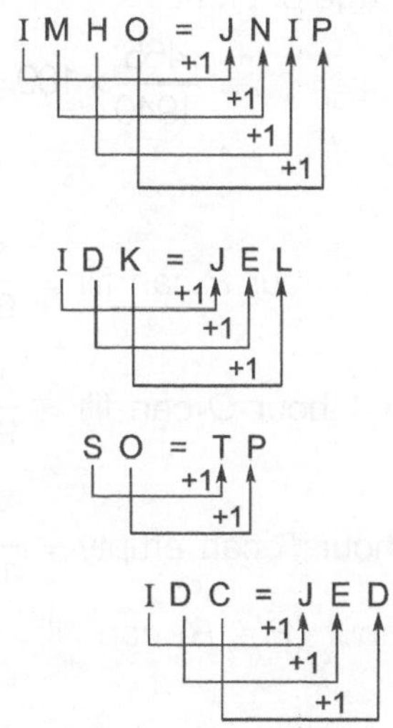

341. (b)

Given, XYZ = 192, Z = 4

Now, XYZ = 192

$$XY = \frac{192}{4} = 48$$

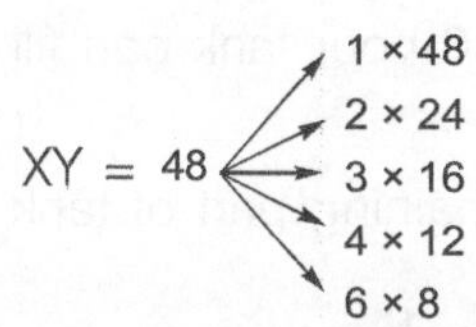

Hence, minimum possible value of P

$$= \frac{X+Y}{2} = \frac{6+8}{2} = 7$$

342. (d)

Let GDP in 2015 = 100

GDP in 2016 = 100 + 10 = 110

Deficit in 2015 = 4

Deficit in 2016 = 5% of 110 = 5.5

Percentage increase in fiscal deficit

$$= \frac{5.5-4}{4} \times 100$$

$$= \frac{1.5}{4} \times 100 = 37.5$$

343. (b)

Revenue by pool

= 170 + 320 + 215 + 190 = 895

Revenue by mini

= 110 + 220 + 180 + 70 = 580

Revenue by prime

= 75 + 180 + 120 + 90 = 465

Total revenue by Mola

= 895 + 580 + 465 = 1940

Percentage of prime

$$= \frac{465}{1940} \times 100 = 23.97\%$$

344. (c)

$$\text{In 1 hour } P \text{ can fill} = \frac{1}{6}$$

$$\text{In 1 hour } Q \text{ can fill} = \frac{1}{9}$$

$$\text{In 1 hour } R \text{ can empty} = \frac{1}{12}$$

In 4 hour $(P + R)$ can fill

$$= 4 \times \left(\frac{1}{6} - \frac{1}{12}\right) = \frac{1}{3}$$

In 6 hour $(Q + R)$ can fill

$$= 6\left(\frac{1}{9} - \frac{1}{12}\right) = \frac{1}{6}$$

$$\text{In 10 hour tank can fill} = \frac{1}{3} + \frac{1}{6} = \frac{1}{2}$$

$$\text{Remaining part of tank} = 1 - \frac{1}{2} = \frac{1}{2}$$

$$\frac{1}{2} \text{ tank } Q \text{ can fill} = \frac{1}{2} \times 9 = 4.5 \text{ hour}$$

Total time = 10 + 4.5

= 14.5 hours

345. (b)

Let total monthly earning = Rs. 100

Monthly spent on education

$$= \frac{15}{100} \times 100 = \text{Rs.}15$$

Monthly spent on transport

$$= \frac{10}{100} \times 10 = \text{Rs. } 10$$

% money extra spent on education as compared to transportation

$$= \frac{15-10}{10} \times 100 = 50\%$$

346. (b)

$\simeq 26591749^{(110016)}$

$\simeq$ Unit place of $9^{even} = 1$

∴ Cyclicity of 9 is (9, 1) (9, 1), (9, 1)

So answer will be 1.

347. (a)

If the product of two positive numbers should be zero, one of the number must be zero. After subtracting 5 if a positive number should become zero, that number should be 5. If one number is 5 and the sum is 100 then the other number must be 95.

Let the two positive numbers be x and y.

∴ $x + y = 100$ ...(i)

$(x - 5)(y - 5) = 0$ ...(ii)

$\Rightarrow x = 5$ or $y = 5$

If $x = 5$ then $y = 95$

If $y = 5$ then $x = 95$

∴ One of the number is 95 since 5 is not in any of the options.

348. (a)

Option (a) satisfies the given conditions in the paragraph.

349. (c)

2, a, $(a + d)$, $(a + 2d)$, ... $(a + 6d)$, 34

∴ Total number of terms of $AP(n)$ = 9

Let sum of seven inserted numbers = S

$$\therefore \quad S = \frac{7}{2}[a + (a + 6d)] = 7[a + 3d]$$

$$T_n = 34$$

Also, $a - 2 = (a + d - a)$

$\Rightarrow \quad a - d = 2$

Similarly $a - 2 = 34 - (a + 6d)$

$\Rightarrow \quad a - 2 = 34 - a - 6d$

$\Rightarrow \quad 2a = 36 - 6d = 36 - 6(a - 2)$

$\Rightarrow \quad 2a = 36 - 6a + 12$

$\Rightarrow \quad 8a = 48$

$\Rightarrow \quad a = 6$

$\therefore \quad d = a - 2 = 6 - 2 = 4$

$\therefore \quad S = 7(a + 3d)$

$= 7(6 + 3 \times 4) = 126$

350. (a)

0,	1,	2,	3,	4,	5,	6,	7,	8,	9
↓	↓	↓	↓	↓	↓	↓	↓	↓	↓
O	P	Q	R	S	T	U	V	W	X

$\therefore$ 45 is coded as 'ST'.

351. (*)

LED bulbs sold

By $\quad Y_{April} = \frac{2}{5} \times 15 = 6\%$ ...(A)

$Y_{May} = \frac{4}{5} \times 10 = 8\%$

$Y_{June} = \frac{10}{20} \times 11 = 5.5\%$

LED bulb sold by Y during April-June

A+ B + C = (6 + 8 + 5.5)% = 19.5%

Total LED bulb sold by X and Y 35% this value = 50000

So, LED bulb sold by Y (April to June)

$\frac{50000}{35} \times 19.5 = 27857.142$

[As 35% of total = 5000

$\text{Total} = \frac{50000}{0.35} = 142857.142$

Y (April- June)= 19.5% of total 0.195 × 1442857.14 = 27857.142]

No option is matching.

Mistake in paper was that examiner intended to give total as 50000 and the options were place accordingly as 19.5% of 50000 = 9750 which is (d).

But this will be wrong as total is NOT 50000 as per language. So correct answer is 27857.142.

Which matches with none of options.

353. (a)

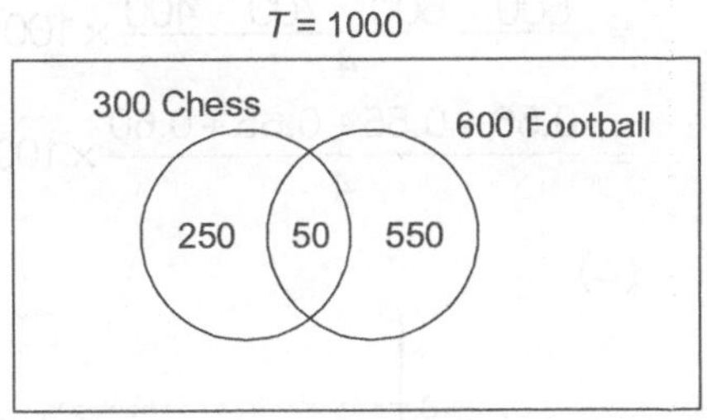

Total number of students playing sports = 850

Total number of students not playing sports = 1000 – 850 = 150

354. (b)

Year	Number of odd days
2013	1
2014	1
2015	1
2016	2
2017	1
2018	1
Total	**7**

Number of odd days in 2019 = 1, so 2013 calendar is same as 2019.

355. (d)

$$f(x) = x^2$$

$$f(f(f(x))) = f(f(x^2)) = f(x^4) = (x^4)^2 = x^8$$

So, $$\frac{f(f(f(x)))}{f(x)} = \frac{x^8}{x^2} = x^6 = (x^2)^3 = (f(x))^3$$

356. (a)

$$\frac{\text{Sum of old numbers from 1 to 100}}{\text{Sum of even numbers from 150 to 200}}$$

$$= \frac{50^2}{175 \times 26} = \frac{50 \times 50}{175 \times 26} = \frac{2500}{4550}$$

Ratio = 50 : 91

From 1 to 100 = 50 odd number

From 150 to 200 = 26 even number

357. (b)

Average percentage of passing students

$$= \frac{\frac{\text{Total no. of passed students}}{\text{Total no. of students}}}{4} \times 100$$

$$= \frac{\frac{280}{500}+\frac{330}{600}+\frac{455}{700}+\frac{240}{400}}{4} \times 100$$

$$= \frac{0.56+0.55+0.65+0.60}{4} \times 100 = 59\%$$

359. (c)

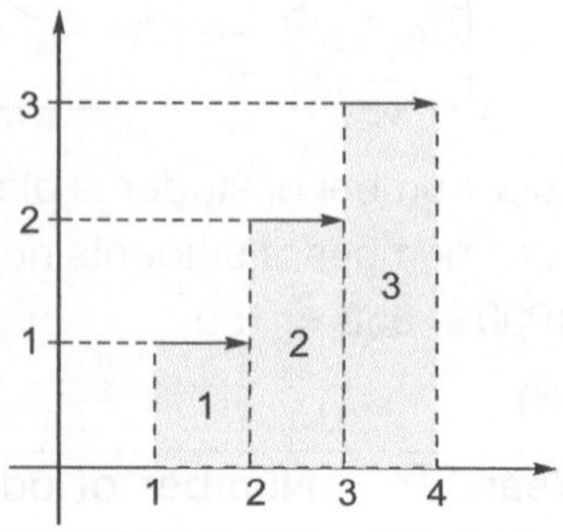

Area = 1 × 1 + 1 × 2 + 1 × 3
= 1 + 2 + 3 = 6

360. (a)

βα and ββ

Given, code of P = αα

code of Q = αβ

these code of R = βα

and code of S = ββ

361. (d)

Using throw options put n = 1 and n = 2

Actual method:

8 + 88 + 888 + ...

8[1 + 11 + 111 + ...]

$$\frac{8}{9}[9+99+999+......]$$

$$\frac{8}{9}[(10-1)+(10^2-1)+(10^3-1)+.....]$$

$$\frac{8}{9}[(10+10^2+10^3+.....)-(1+1+1+......+n)]$$

$$\frac{8}{9}[(10+10^2+10^3+.....)-n]$$

$$S_n = \frac{80}{81}(10^n-1)-\frac{8n}{9}$$

362. (a)

52 men can do in 10 days.

Since 12 men were sent out

Remaining men left = 52 − 12 = 40

We know $M_1D_1 = M_2D_2$

$52 \times 12 = 40 \times x$

$$x = \frac{52\times10}{40} = 13 \text{ days}$$

Total number of days taken = 13 days

3 days more than the original estimate.

364. (d)

P → Q → R → S → T → Customer

↓

120 at each level increased 25% price paid by customer

$$=120\times\frac{125}{100}\times\frac{125}{100}\times\frac{125}{100}\times\frac{125}{100}\times\frac{125}{100}=366.21$$

365. (d)

$$\frac{(\text{Boys})_{\text{Arts}}}{(\text{Girls})_{\text{Management}}}$$

$$= \frac{20\% \text{ of } 5000 - 30\% \text{ of } 1500}{15\% \text{ of } 1500}$$

$$= \frac{1000-450}{225} = \frac{550}{225}$$

$$= \frac{110}{45} = \frac{22}{9}$$

= 22 : 9

366. (a)

This is the log i.e. (n = 4)

5 + 4 = 9

t = 20, x = 24

20 + 4 = 24

Similarly, h = 8

Case-1: 8 + 4 = 12(l)

? + 4 = 8

? = 4 (d)

367. (c)

P = 3, R = 27, T = 243

P	Q	R	S	T
3^1	3^2	3^3	3^4	3^5

Hence Q + S = 90

369. (d)

Unpainted area = Area of annular ring − Area of n blue color circles

Area of annular ring = $\pi b^2 - \pi a^2$

i.e. $\pi(b^2 - a^2)$

$$\text{Area of 1 blue circle} = \pi\left(\frac{b-a}{2}\right)^2$$

$$\text{Hence, area of } n \text{ blue circles} = n\pi\left(\frac{b-a}{2}\right)^2$$

$$\Rightarrow \text{Unpainted area} = \pi\left[(b^2-a^2)-n\frac{\pi}{4}(b-a)^2\right]$$

370. (c)

- Cost of 1-a-C-2 : 200 + 200 + 100 + 100 + 100 = 700
- Cost of 1-b-2 : 300 + 100 + 200 = 600
- Cost of 1-f-b-2 : 100 + 100 + 100 + 200 = 500
- Cost of 1-f-e-2 :100 + 100 + 100 + 200 + 200 = 700

Hence, 1-f-b-2 is having minimum cost.

371. (d)

Total expenditure = 2500 million

Total revenue = 3000 million

So, $\text{profit }\% = \dfrac{500}{2500}\times 100 = 20\%$

372. (c)

Verify with options

Option (a):

$$\frac{1}{x_1+x_2} > \frac{1}{x_1}+\frac{1}{x_2}$$

$$\frac{1}{2+3} > \frac{1}{2}+\frac{1}{3}$$

$$\frac{1}{5} > \frac{5}{6}$$

0.2 > 0.8 which is wrong.

Option (b):

$$\sqrt{x_1+x_2} > \sqrt{x_1}+\sqrt{x_2}$$

$$\sqrt{2+3} > \sqrt{2}+\sqrt{3}$$

$$\sqrt{5} > 1.414 + 1.732$$

2.23 > 3.146 which is wrong.

Option (c):

$$e^{x_1+x_2} > e^{x_1}+e^{x_2}$$

$$e^{1+2} > e^1 + e^2$$

20.085 > 2.718 + 7.389 Satisfying.

373. (b)

Average number of students in school,

$$P = \frac{3+5+5+6+4}{5} = \frac{23}{5}$$

Average number of students in school,

$$Q = \frac{4+7+8+7+5}{5} = \frac{31}{5}$$

Difference of the number of students enrolled in school P and Q

$$= \frac{31-23}{5} = \frac{8}{5}$$

Ratio of the average number of the students enrolled in school P to the average of the difference of the number of students enrolled in schools P and Q is = 23 : 8

376. (a)

Area of shaded portion

= Area of circle – area of rectangle

Maximum possible area of rectangle inscribed in the circle = $2a^2$

So,Required shaded area = $\pi a^2 - 2a^2$

377. (b)

Z,	M V,	R Q P	K J I H
26,	23,22	18,17,16	11,10,9,8

378. (d)

${}^3C_1 + {}^3C_2 + {}^3C_3$

3 + 3 + 1 = 7

Possible combinations,

PQ, RQ, SQ, PRQ, PSQ, RSQ, PQRS

379. (c)

10 × 10 + 9 × 10 + 9 × 10 = 280

Question is asking 37 appears how many times and not how many numbers.

380. (c)

As given in question holding period was reduced by one quarter each time.

Therefore, after third division holding period remains 90 days.

381. (a)

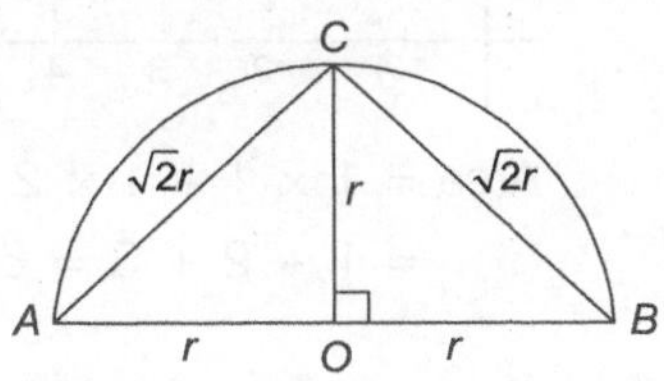

The ratio,

$$\frac{\overline{AC}+\overline{CB}}{\overline{AB}} = \frac{\sqrt{2}r+\sqrt{2}r}{2r} = \frac{2\sqrt{2}}{2} = \sqrt{2}$$

382. (a)

Let the revenue of company *Q* in 2010 = x in million rupees.

Then the revenue of company *Q* in 2015 = $1.2x$ in million rupees.

Given in graph,

$1.2x = 45$

$x = 37.5$ in million rupees

As given in question Q had earned a profit of 10%.

$\because$ Expenditure of company Q in 2014

$$= \frac{37.5}{1.1} = 34.09 \approx 34.1$$

383. (c)

Given: Code of $P = \alpha\alpha$

Code of $Q = \alpha\beta$

Then, Code of $R = \beta\alpha$

and Code of $S = \beta\beta$

384. (c)

Average percentage of passing students

$$= \frac{\frac{\text{Total no. of passed students}}{\text{Total no. of students}}}{4} \times 100$$

$$= \frac{\frac{280}{500}+\frac{330}{600}+\frac{455}{700}+\frac{240}{400}}{4} \times 100$$

$$= \frac{0.56+0.55+0.65+0.60}{4} \times 100 = 59\%$$

385. (c)

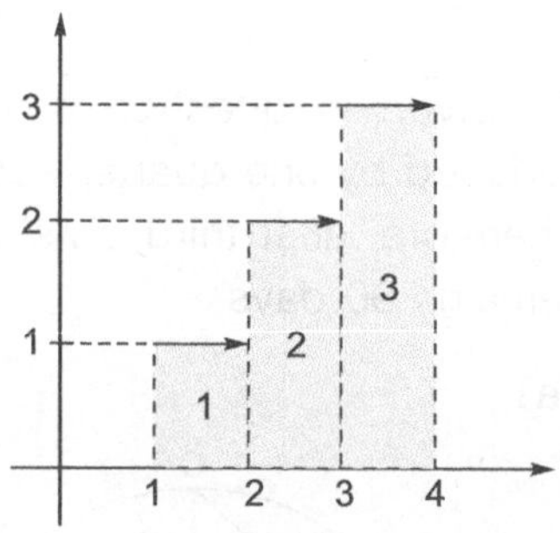

Area $= 1 \times 1 + 1 \times 2 + 1 \times 3$

$= 1 + 2 + 3 = 6$

386. (a)

Put n = 1 in all options.

$$\frac{80}{81}\left[10^1 - 1\right] - \frac{8}{9}[1] = \frac{80}{9} - \frac{8}{9} = \frac{72}{9} = 8$$

387. (c)

$y = x^m$ and $y = x^{1/m}$

For $m = 2$ in $0 \le x \le 1$

$y = x^2$ and $y = \sqrt{x}$

$\therefore$ Graphs are shown as

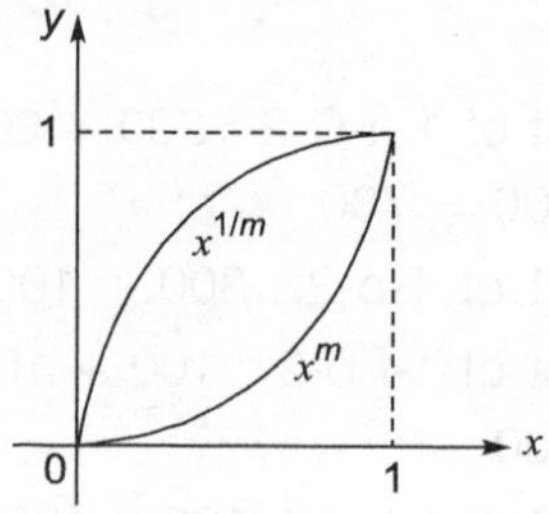

388. (d)

52 men can do in 10 days.

Since 12 men were sent out

Remaining men left = 52 − 12 = 40

We know $M_1D_1 = M_2D_2$

$52 \times 10 = 40 \times x$

$$x = \frac{52 \times 10}{40} = 13 \text{ days}$$

Total number of days taken = 13 days

3 days more than the original estimate.

390. (d)

This is the log i.e. (n = 4)

5 + 4 = 9

t = 20, x = 24

20 + 4 = 24

Similarly, $h = 8$

Case-1: 8 + 4 = 12(l)

? + 4 = 8

? = 4 (d)

391. (a)

P → Q → R → S → T → Customer

↓

120 at each level increased 25% price paid by customer

$$= 120 \times \frac{125}{100} \times \frac{125}{100} \times \frac{125}{100} \times \frac{125}{100} \times \frac{125}{100} = 366.21$$

392. (c)

$$\frac{(\text{Boys})_{\text{Arts}}}{(\text{Girls})_{\text{Management}}}$$

$$= \frac{20\% \text{ of } 5000 - 30\% \text{ of } 1500}{15\% \text{ of } 1500}$$

$$= \frac{1000-450}{225} = \frac{550}{225}$$

$$= \frac{110}{45} = \frac{22}{9}$$

$= 22 : 9$

393. (b)

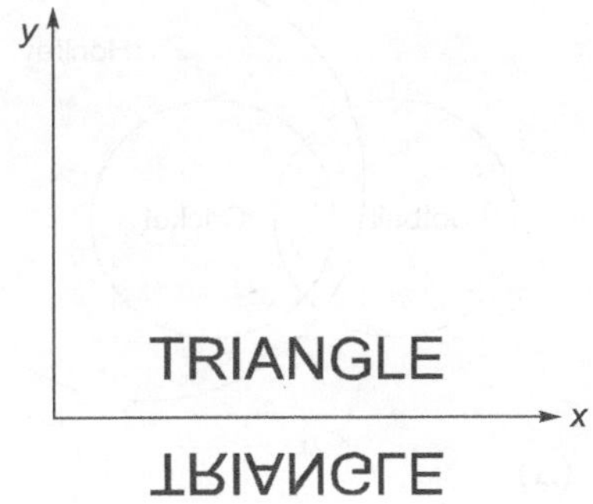

394. (c)

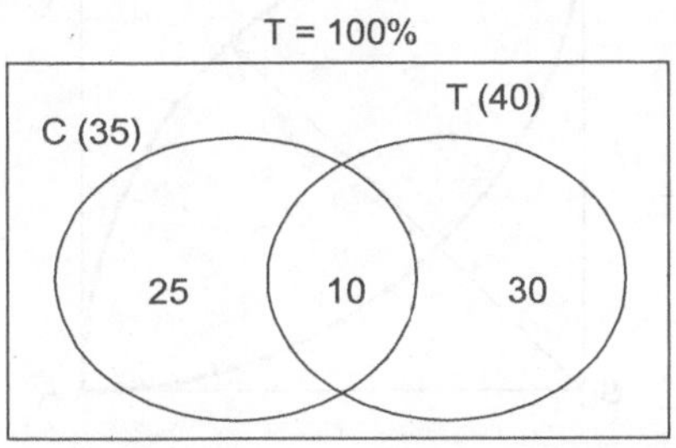

Percentage of employees who drink neither tea nor coffee = 100 − 25 − 10 − 35 = 35

395. (d)
Following cases can be
PQSR, RQSP, QPSR, QRSP, RQPS and PQRS
and $\begin{bmatrix} P & R \\ R & P \end{bmatrix}$ not possible.

396. (b)

$$\frac{\lambda(-(-3+2),(-2+3))}{(-(-2+1))} = \lambda\frac{(1,1)}{1} = \lambda(1,1)$$

So, 1st definition will be applicable as $p = q$.
Hence, $\lambda\,(1, 1) = (1 - 1)^2 = 0$

397. (a)

$$x \oplus y = 2 \odot 2$$

$$\frac{x^2+y^2}{xy} = \frac{2^2}{2}$$

$$x^2 + y^2 = 2xy$$

$$(x - y)^2 = 0$$

$$x = y$$

398. (c)

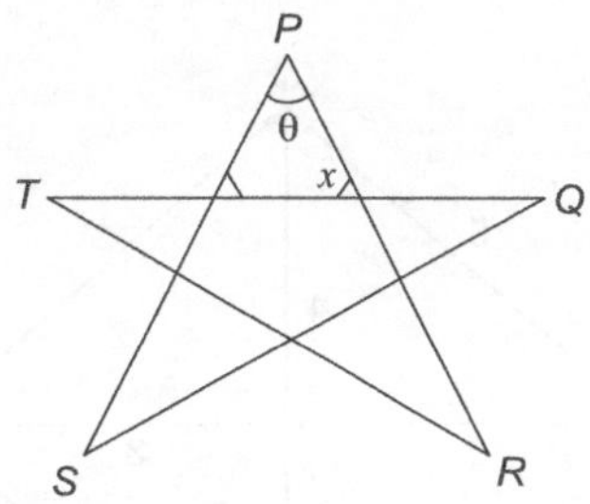

Sum of angle formed at the pentagon = 540°

$$\text{Each angle of pentagon} = \frac{540}{5} = 108°$$

$$\angle x = 180 - 108 = 72°$$

Sum of angle of triangle
$$= 180°$$
$$72° + 72° + \theta = 180°$$
$$\theta = 36°$$

399. (a)

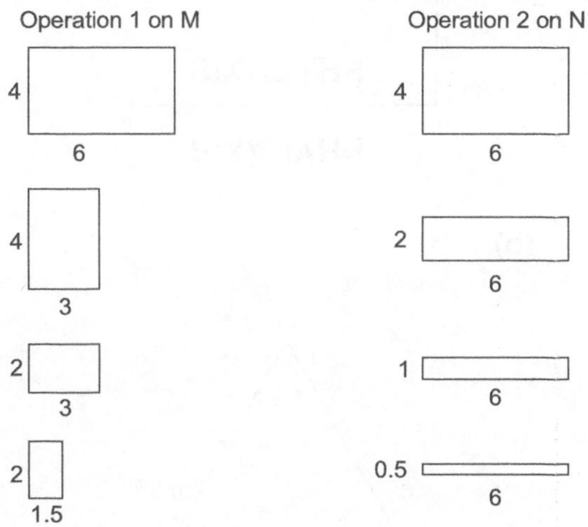

$$(\text{Perimeter})_M = 2\,(2 + 1.5) = 7$$
$$(\text{Perimeter})_N = 2\,(0.5 + 6) = 13$$

$$\text{Required ratio} = \frac{13}{7}$$

400. (c)
$$[9 - (6 \times 7)] - [7 \times 1] = -33 - 7$$
$$= -40$$

401. (b)

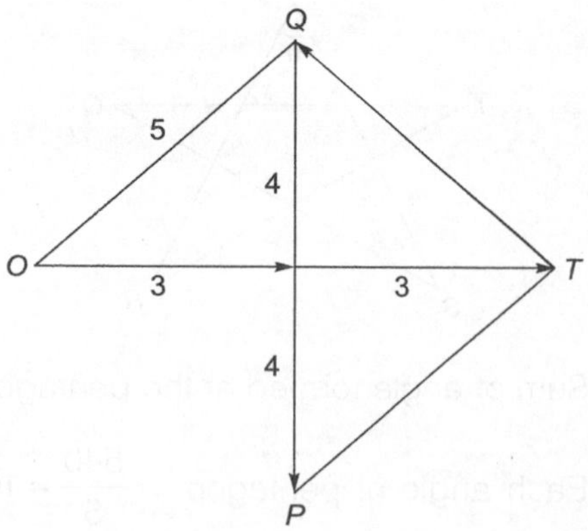

$$OQ = \sqrt{3^2 + 4^2} = 5$$

402. (c)

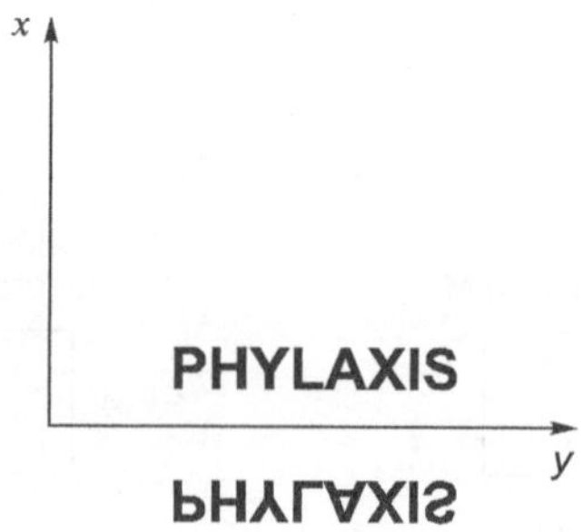

403. (b)

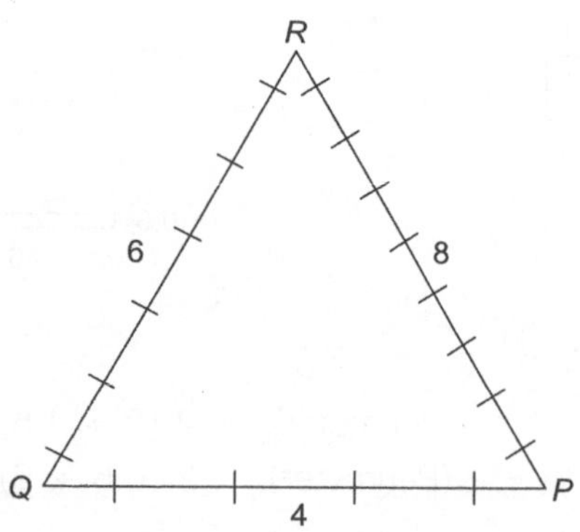

For $\left(\frac{a}{4}, \frac{a}{6}, \frac{a}{8}\right)$ to be integer, *'a'* must be LCM of 4, 6 and 8.

So $a = 24$

$$\text{Area} = \frac{\sqrt{3}}{4}a^2 = \frac{\sqrt{3}}{4} \times 24^2 = 144\sqrt{3}$$

404. (b)

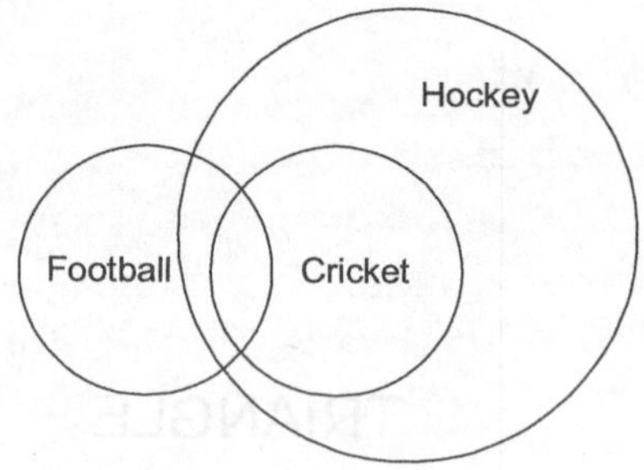

405. (a)

Let feasible area be fA, then total area is TA

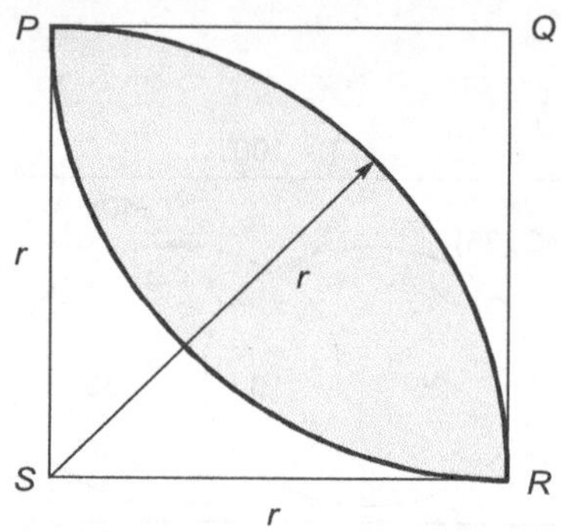

So, Probability $= \frac{fA}{TA}$

$$fA = \left(\frac{\pi r^2}{4} - \frac{r^2}{2}\right) \times 2$$

$$\frac{fA}{TA} = \frac{\left(\frac{\pi r^2}{4} - \frac{r^2}{2}\right) \times 2}{r^2} = \left(\frac{\pi}{2} - 1\right)$$

406. (d)

Number of arrangements = 3 × 3! = 18

407. (c)

Probability of getting even number on a dice

$$= \frac{3}{6} = \frac{1}{2}$$

∵ Two dice are rolled simultaneously,

Hence, probability of an even number rolled out on each dice $= \frac{1}{2} \times \frac{1}{2} = \frac{1}{4}$

409. (b)

Number of employees in C2 company

$$= \frac{5}{100} \times 10000 = 500$$

Number of management degree holder employees in C2 = $\frac{1}{5} \times 500 = 100$

Number of employees in C5 company

$= \frac{20}{100} \times 10000 = 2000$

Number of management degree holder employees in C5 = $\frac{9}{10} \times 2000 = 1800$

Total management degree holder employees = 100 + 1800 = 1900

410. (d)

In farm *P*,

Hens = 65, Ducks = 91, Goats = 169

In farm *Q*,

Hens : Ducks : Goats

5 : 14 : 13

Number of Hens in farm *Q*

$$= \frac{5}{32} \times 416 = 65$$

Number of Ducks in firm *Q*

$$= \frac{14}{32} \times 416 = 182$$

Number of Goats in farm *Q*

$$= \frac{13}{32} \times 416 = 169$$

$\because$ From farm *Q*, hens, ducks and goats are sent to farm P,

$\therefore$ Total hens = 65 + 65 = 130

Total ducks = 91 + 182 = 273

Total goats = 169 + 169 = 338

New ratio = 130 : 273 : 338

= 10 : 21 : 26

411. (b)

$\because$ Area of $\Delta MON = 50 \text{ cm}^2$

$$\frac{1}{2} \times r \times r = 50$$

$$r^2 = 100$$

Area of circle = πr^2

$= \pi \times 100$

$= 100\pi$

412. (c)

The possible seating arrangements as per the given condition are

QPRST, QTRSP

QPRTS, QTRPS

PQTRS, TQPRS

SPQTR, STQPR

RPQTS, RTQPS

SRPQT, SRTQP

SPRTQ, STRPQ

PSRTQ, TSRPQ

Hence, total seating arrangements are 16.

413. (7)

$$\Delta 2 \oplus 3\Delta((4 \otimes 2)\nabla 4) = +2 - 3 + ((4 \div 2) \times 4)$$
$$= +2 - 3 + (2 \times 4)$$
$$= +2 - 3 + 8$$
$$= 7$$

414. (a)

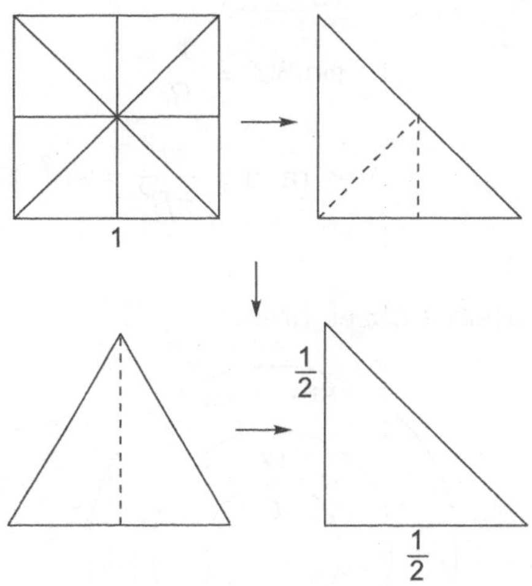

$$\text{Area} = \frac{1}{2} \times \frac{1}{2} \times \frac{1}{2} = \frac{1}{8}$$

415. (b)

The possible distinct arrangement are

S R P T Q, Q R P T S, S R T P Q

Hence, number of distinct seating arrangement. = 3

416. (b)

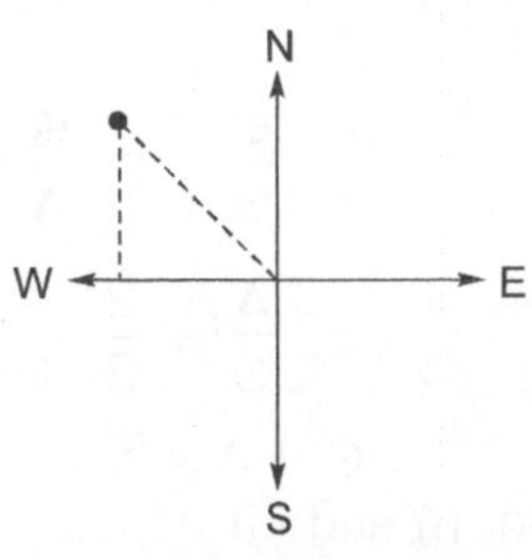

417. (a)

The probability of first ball is blue and second ball is black is given as,

$$P = \frac{15}{60} \times \frac{45}{59} = \frac{45}{236}$$

418. (b)

LCM of (30 and 32) is 480

480 seconds = 8 minutes

Hence, time will be 10.08 pm when they will beep together.

419. (a)

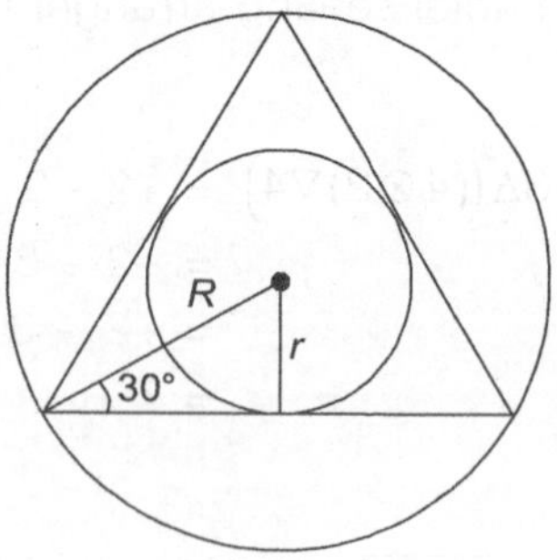

$$\sin 30° = \frac{r}{R}$$

$$\text{Area ratio} = \frac{\pi r^2}{\pi R^2} = \sin^2 30 = \frac{1}{4}$$

420. (c)

Possible cases are:

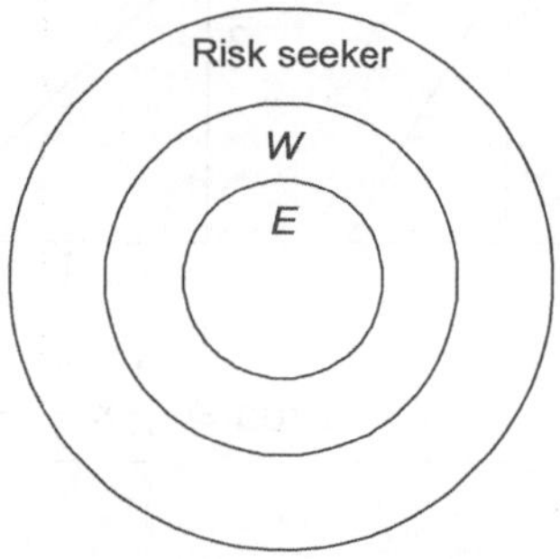

Conclusion-I is incorrect because some risk seeker are wealthy.

Conclusion-II is also incorrect because all the entrepreneurs are risk seeker as well as wealthy.

421. (c)

$$\frac{\oplus}{\odot} = 2, \frac{\oplus}{\Delta} = 3$$

$$\therefore \quad \frac{\Delta}{\odot} = \frac{2}{3} \quad \text{...(i)}$$

$$\odot + \Delta = 5 \quad \text{...(ii)}$$

From eq. (i) and (ii)

$$\Delta = 2, \odot = 3$$

And $\oplus = 6,\ 2 \times \otimes = 10$

$$\otimes = 5$$

$$\Rightarrow (\otimes - \oplus)^2 = (5-6)^2 = 1$$

422. (d)

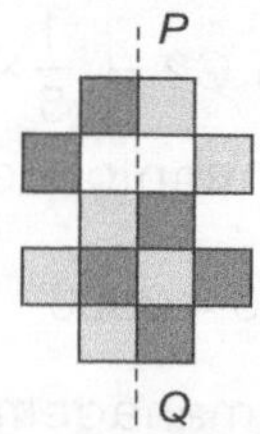

To make *P-Q* as symmetric line, minimum number of ■ square added = 6.

423. (b)

Given, $\frac{p}{q} + \frac{q}{p} = 3$

Squaring both sides,

$$\left(\frac{p}{q} + \frac{q}{p}\right)^2 = 3^2$$

$$\Rightarrow \frac{p^2}{q^2} + \frac{q^2}{p^2} + 2 = 9$$

$$\Rightarrow \frac{p^2}{q^2} + \frac{q^2}{p^2} = 9 - 2 = 7$$

424. (c)

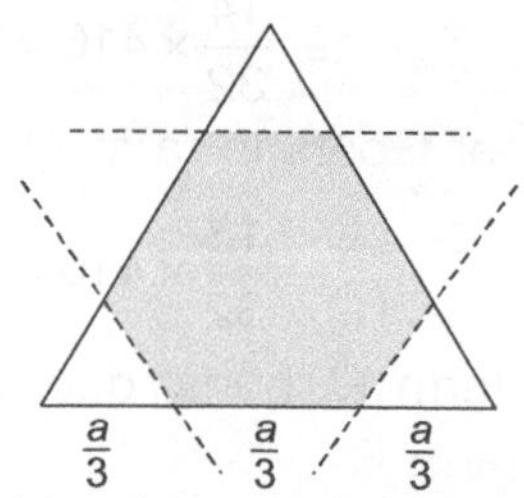

Let the side of the two larger equilateral triangle be *a*

Then, side of regular hexagon will be $\left(\frac{a}{3}\right)$

$$\text{Area of regular Hexagon} = 6 \times \frac{\sqrt{3}}{4}\left(\frac{a}{3}\right)^2$$

$$\text{Area of triangle} = \frac{\sqrt{3}}{4}(a)^2$$

$$\text{Required ratio} = \frac{6\sqrt{3}}{4} \times \frac{a^2}{9} : \frac{\sqrt{3}}{4}a^2$$

$$= 6 : 9 = 2 : 3$$

425. (c)

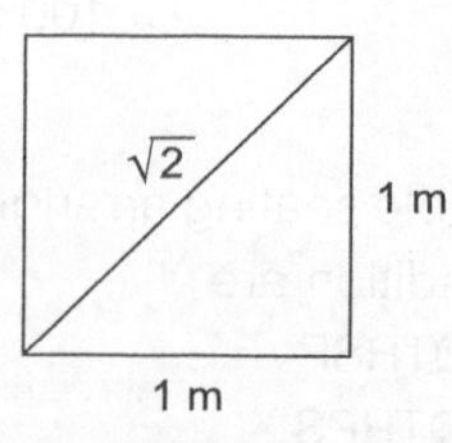

One of the triangle is revolved about its short side, resulting a cone.

Hence, $r = 1, \quad H = 1$

Volume of cone, $V = \frac{1}{3}\pi r^2 H = \frac{\pi}{3}(1)^2\,(1)$

$$V = \frac{\pi}{3}$$

426. (c)

Possible cases can be,

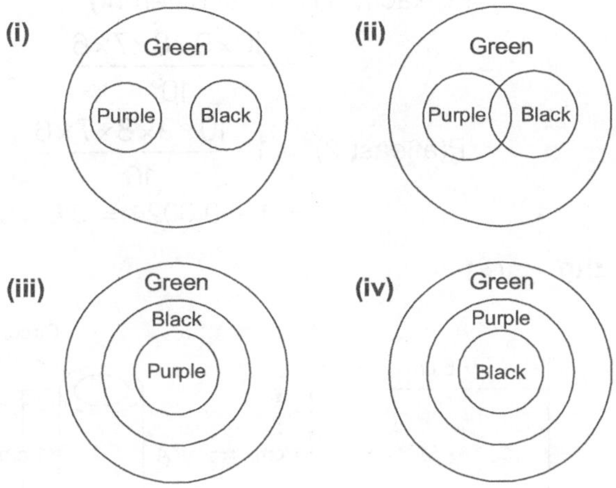

427. (b)

On Sunday, the percentage

$$= \frac{65-55}{55}\times 100 = 18.18\%$$

On Saturday, the percentage

$$= \frac{60-50}{50}\times 100 = 20\%$$

On Friday, the percentage

$$= \frac{35-20}{20}\times 100 = 75\%$$

On Thursday, the percentage

$$= \frac{60-55}{55}\times 100 = 9.09\%$$

On Wednesday, the percentage

$$= \frac{60-50}{50}\times 100 = 20\%$$

On Tuesday, the percentage

$$= \frac{65-55}{55}\times 100 = 18.18\%$$

On Monday, the percentage

$$= \frac{70-45}{45}\times 100 = 55.56\%$$

Total six days are there when one of the students spent a minimum of 10% more than the other student.

428. (b)

Let the population years ago be x.

$$1102500 = x\left(1+\frac{5}{100}\right)^2$$

$$1102500 = x\left(\frac{21}{20}\right)^2$$

$$x = 10{,}00{,}000$$

429. (a)

∵ S and V must be parked together and Q and S also must be parked together.

These two condition possible parkings are:

Q S V – – – –

V S Q – – – –

We can observe that in all cases only one car can be parked in between Q and V which is car S.

Hence option (a) is incorrect.

430. (a)

Side of outer first square = 10 cm

Side of outer second square = $\frac{10}{\sqrt{2}}$ cm

Likewise,

Side of outer third square = $\left(\frac{10}{(\sqrt{2})^2}\right)$

Hence,

Side of smallest square = $\frac{10}{(\sqrt{2})^5}$

Area of smallest square

$$= \left(\frac{10}{(\sqrt{2})^5}\right)^2 = \frac{100}{2^5} = 3.125$$

431. (b)

Consider expression $\frac{x^n - a^n}{x+a}$.

Going by options and trying to convert it in the above format seeing option (b)

$11^{52} - 1 = (11^{13})^4 - (1)^4$

$$\frac{(11^{13})^4 - (1)^4}{11^{13}+1} = \frac{((11)^{13})^2 - (1)^2)((11^{13})^2 + (1)^2)}{11^{13}+1}$$

$$= [(11^{13} - 1)][(11^{13})^2 + (1)^2]$$

Hence, $11^{52} - 1$ is divisible by $11^{13} + 1$.

432. (b)

n is a continuous random variable for continuous random variable

$$\therefore \quad f(x) = \begin{cases} \dfrac{1}{\beta - \alpha} & \alpha \le x \le \beta \\ 0 & \text{otherwise} \end{cases}$$

From here, $\alpha = 0,\ \beta = 100$

For continuous probability function mean

$$= \frac{\alpha + \beta}{2} = \frac{0 + 100}{2} = 50$$

Alternate Solution:

Given: pdf $f(x) = 10^{-2},\ 0 < x < 100$

$$E(x) = \int_0^{100} x f(x) dx = \int_0^{100} x 10^{-2} dx$$

$$= \frac{1}{100} \left. \frac{x^2}{2} \right]_0^{100} = \frac{100}{2} = 50$$

433. (b)

Sum of the interior angle

$$= (n - 2) \times 180° = 1440°$$

The interior angle between two sides of polygon is $\dfrac{1440°}{10} = 144°$

434. (b)

Upon rotation, we get the below figure.

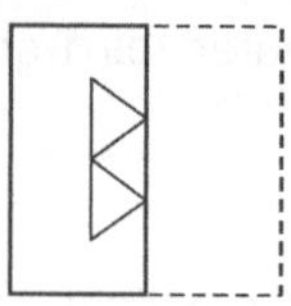

435. (d)

Total number of students in the year-2

$$= 60 + 5 = 65$$

Students who failed in the year-1 and appeared in year-2 = 10

So the students who appeared first time in year-2 = 65 – 10 = 55

Similarly, total number of students in year-3 = 53

Students who failed in the year-2 and appeared in year-3 = 5

So, the students who appeared first time in year-3 = 53 – 5 = 48

436. (d)

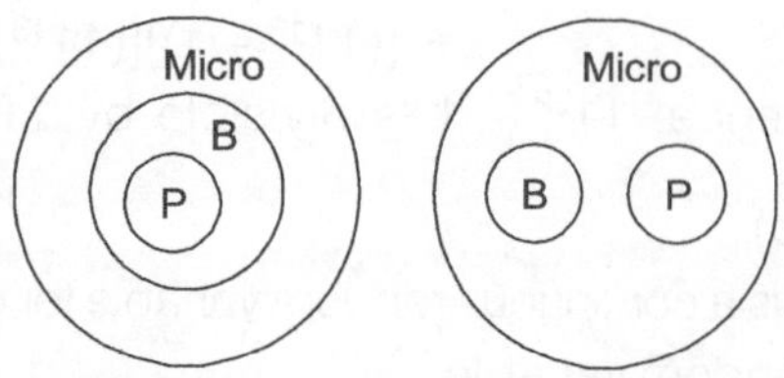

Case 1 Case 2

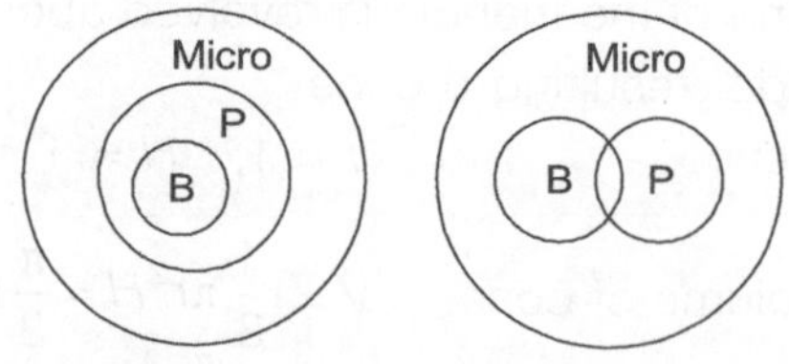

Case 3 Case 4

None of the two conclusions will satisfy all the 4 cases.

437. (a)

$$P(\text{atleast } 2) = 1 - P(\text{exactly } 1)$$

$$P(\text{exactly } 1) = P(\text{all different})$$

$$= \frac{10 \times 9 \times 8 \times 7 \times 6}{10^5}$$

$$P(\text{atleast } 2) = 1 - \frac{10 \times 9 \times 8 \times 7 \times 6}{10^5}$$

$$= 1 - 0.3024 = 0.6976$$

439. (a)

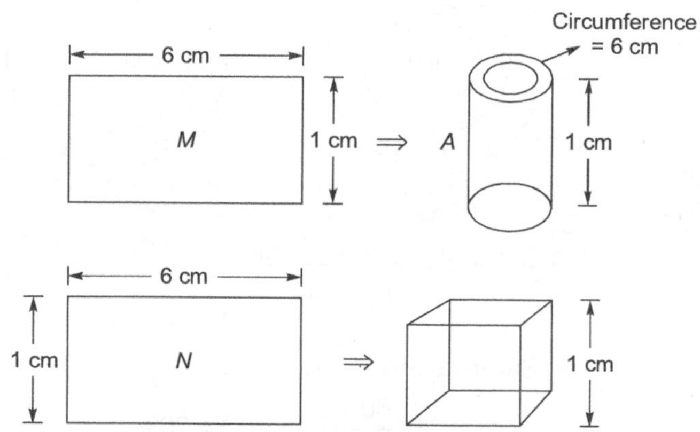

Volume of cylinder $= \pi r^2 h$

Now, $2\pi r = 6$ (figure A)

$$r = \frac{3}{\pi}$$

$$\text{Volume of cylinder} = \pi \times \frac{3}{\pi} \times \frac{3}{\pi} \times 1 = \frac{9}{\pi}$$

$$\text{Volume of cube} = (1)^3$$

$$\text{Ratio} = \frac{\frac{9}{\pi}}{1} = \frac{9}{\pi}$$

441. (b)

Given: CP_P = Rs. 5400

MP_P = Rs. 5860

$$\frac{5400 \times 4}{5} = CP_Q$$

CP_Q = Rs. 7200

SP_Q = ?

$$\text{Profit \%} = \frac{\text{Profit}}{\text{CP}} \times 100 \text{ [for Q]}$$

$$25 = \left(\frac{\text{SP} - 7200}{7200}\right)100$$

$$\text{SP}_Q = \text{Rs. } 9000$$

$$\text{Discount Q} = \text{MP}_Q - \text{SP}_Q = \text{Rs.}1000$$

$$\text{Discount \%} = \frac{1000}{10000} \times 100 = 10\%$$

442. (b)

Boys= $7x$

Girls = $3x$

Total number of students = $10x$

Now, $7x$ and $3x$ should be integers.

Hence, x should be integer.

So $10x$ should be an integer.

Only 50 satisfies the above constraints.

443. (c)

$$\left(x - \frac{1}{2}\right)^2 - \left(x - \frac{3}{2}\right)^2 = x + 2$$

Using $a^2 - b^2 = (a + b)(a - b)$

$$\left(x - \frac{1}{2} + x - \frac{3}{2}\right)\left(x - \frac{1}{2} - x + \frac{3}{2}\right) = x + 2$$

$$(2x - 2) = x + 2$$

$$2x - x = 2 + 2$$

$$x = 4$$

445. (a)

Let angle of longest diagonal of cube with an edge of cube be θ, then

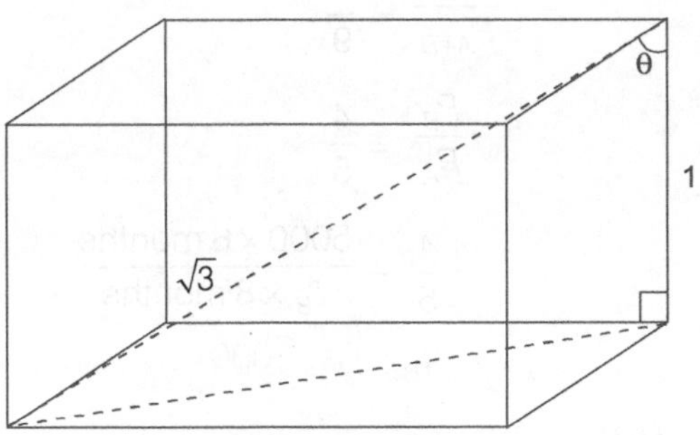

(Assuming side of curve as 1 unit)

$$\cos\theta = \frac{\text{Base}}{\text{Hypotenuse}} = \frac{1}{\sqrt{3}}$$

446. (c)

S > R ...(i)

Q is shortest and U is taller than only one. ...(ii)

T > S ...(iii)

Hence, possible order is : P > T > S > R > U > Q.

∵ Number of students taller than R = 3.

∴ Number of students shorter than S = 3.

Hence, option (c) is the correct answer.

447. (a)

Let $3k + 13k + 6k = n$ ...(i)

Now, adding 18 students in each class new ratio we get,

15 : 35 : 21

$\therefore 15y + 35y + 21y$

$= 22k + 5y$ [$3k + 13k + 6k = 22k$]

$71y = 22k + y$

For $k = 1$

$71y = 16$

(Answer will not be integer so proceed ahead)

For $k = 2$

$71y = 98$

For $k = 3$

$71y = 120$ ...(ii)

For $k = 4$

$71y = 142$

$y = 2$

So substituting $k = 4$ in eq. (i) we get $n = 88$

448. (d)

Cost/unit in year 3 = Re. 1

Cost/unit in year 2 = Rs. 2

Cost/unit in year 1 = Rs. 3

Net Profit = SP – (Cost + Taxes)

In year-2,

$$296 = \text{SP} - (2 \times 200 + 0.13 \text{ SP})$$

$$\text{SP} = 800$$

Selling price in year-2

= Rs. 800

In year 3,

$$210 = \text{SP} - (300 \times 1 + 0.15 \text{ SP})$$

$$210 = \text{SP} - 300 - 0.15 \text{ SP}$$

$$\text{SP} = \frac{210 + 300}{1 - 0.15}$$

Selling price in year-3

= Rs. 600

Hence, required ratio = 800 : 600 = 4 : 3

451. (d)

Given: $\frac{P}{Q} = \frac{1}{2}, \frac{Q}{R} = \frac{4}{3}$

$$\frac{R}{S} = \frac{4}{5}$$

$$\frac{R}{S} = \frac{4 \times 3}{5 \times 3}$$

$$\frac{Q}{R} = \frac{4 \times 4}{3 \times 4}$$

$$\frac{P}{Q} = \frac{1 \times 8}{2 \times 8}$$

P	:	Q	:	R	:	S
8	:	16	:	12	:	15

$$U = 15 \times 1.5 = 22.5$$

$$\frac{P}{U} = \frac{8}{22.5} = \frac{8 \times 2}{22.5 \times 2} = \frac{16}{45}$$

452. (b)

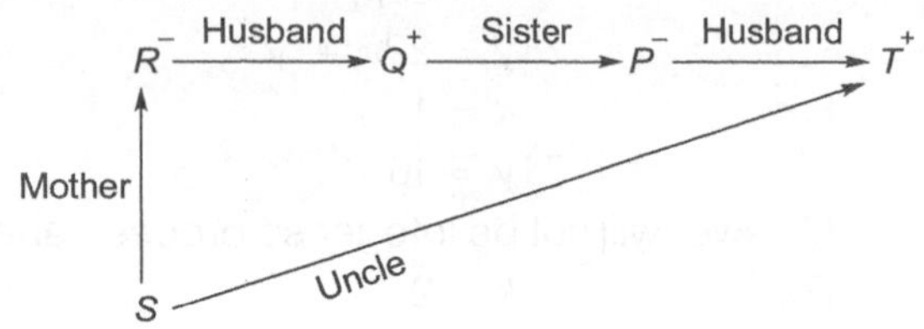

453. (b)

$R = a \cos \theta$

$A_{RUVZ} = r^2$

$= a^2 \cos^2\theta$

$A_{PXWR} = a^2$

$A_{SPQT} = (a \sin\theta)^2$

$= a^2 \sin^2\theta$

$A_{SPQT} = A_{PXWR} - A_{RUVZ}$

$= a^2 - a^2 \cos^2\theta$

$= a^2 (1 - \cos^2\theta)$

$a^2 \sin^2 \theta = a^2 \sin^2 \theta$

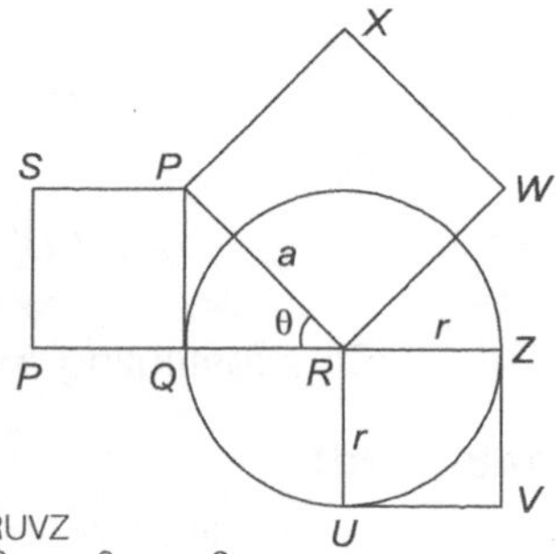

455. (b)

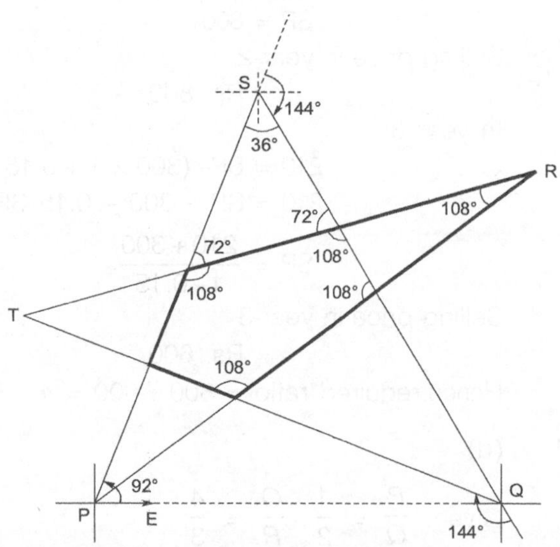

and PS = SQ = QT = TR = RP = 10 cm

456. (a)

$$\frac{(\text{Profit})_A}{(\text{Profit})_B} = \frac{(\text{Investment})_A}{(\text{Investment})_B} \times \frac{t_A}{t_B}$$

Given, ratio of initial investment of friends P, Q, R is 3 : 4 : 5 respectively.

Ratio, after increment in investment of P, Q, R.

P increase the investment by 5%.

Q increase the investment by 10%

R increase the investment by 20%

i.e. 3 + (3 × 1.1) : 4 + (4 × 1.5) : 5 + (5 × 1.2)

= 6.3 : 8.6 : 11

= 63 : 86 : 110

457. (a)

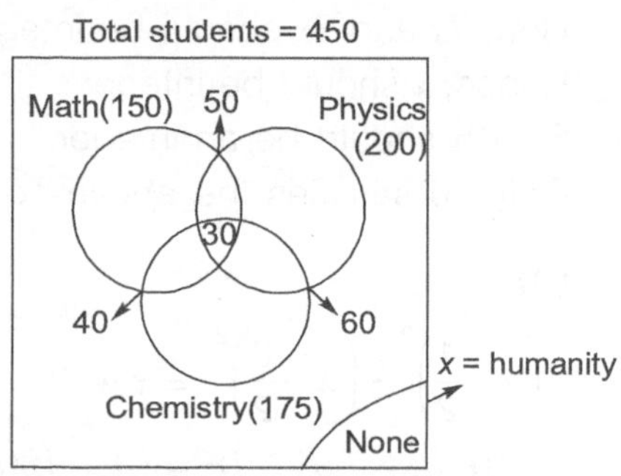

$$T = n(M \cup P \cup C) + x$$

$450 = [(150 + 200 + 175)$

$- (50 +$

458. (b)

$$\frac{(P_R)_A}{(P_R)_B} = \frac{(I_A) \times T_A}{(I_B) \times T_B}$$

$$\frac{P_B}{P_{A+B}} = \frac{4}{9}$$

$$\frac{P_B}{P_A} = \frac{4}{5}$$

$$\frac{4}{5} = \frac{5000 \times 6 \text{ months}}{I_B \times 8 \text{ months}}$$

$$I_B = \text{Rs. } 3000$$

462. (a)

Given: $x : y : z = \frac{1}{2} : \frac{1}{3} : \frac{1}{4} = 6 : 4 : 3$

As we need to find ratio of $(x + y + z)$ to y, so directly x, y, z values can be taken as

$x = 6, \; y = 4, \; z = 3$

∴ The value of

$$= \frac{x + z - y}{y} = \frac{6 + 3 - 4}{4} = \frac{5}{4} = 1.25$$

463. (a)

Let's take the intger as x.

Now as per given statement,

$$\frac{3+x}{4+x} = \frac{15-x}{17-x}$$

By solving for $x = 3$

464. (c)

Line L1 : $2x - 3y = 5$ Line L2 : $3x + 2y = 8$

$$y = \frac{2x-5}{3}$$

$$y = \frac{8-3x}{2}$$

$$\frac{dy}{dx} = \frac{2}{3} \quad \frac{dy}{dx} = \frac{-3}{2}$$

Line L3 : $4x - 6y = 5$ Line L4 : $6x - 9y = 6$

$$y = \frac{4x-5}{6}$$

$$y = \frac{6x-6}{9}$$

$$\frac{dy}{dx} = \frac{2}{3}$$

$$\frac{dy}{dx} = \frac{2}{3}$$

$$\therefore \quad \left(\frac{dy}{dx}\right)_{L1} = \left(\frac{dy}{dx}\right)_{L3} = \left(\frac{dy}{dx}\right)_{L4} = \frac{2}{3}$$

and $\left(\frac{dy}{dx}\right)_{L2} \times \left(\frac{dy}{dx}\right)_{L1}$

$$= \left(\frac{dy}{dx}\right)_{L2} \times \left(\frac{dy}{dx}\right)_{L3} = \left(\frac{dy}{dx}\right)_{L2} \times \left(\frac{dy}{dx}\right)_{L4} = -1$$

$\therefore$ L1, L3, L4 are parallel to each other and perpendicular to L2.

465. (c)

Statement 1: Vane diagram

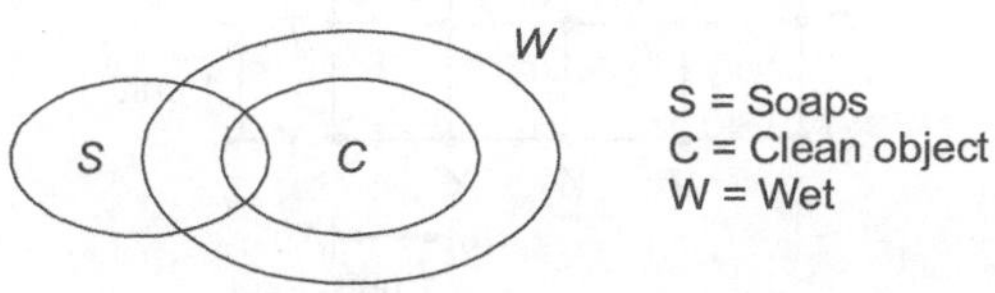

Statement 2: Vane diagram

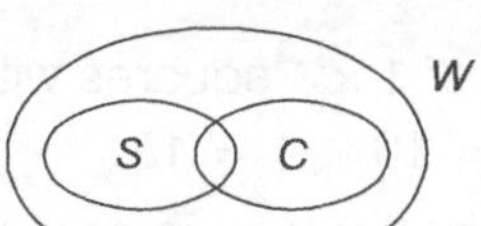

From the above vane diagram we can conculate that conclusion I and conclusion III are correct and conclusion II and IV are in correct.

466. (c)

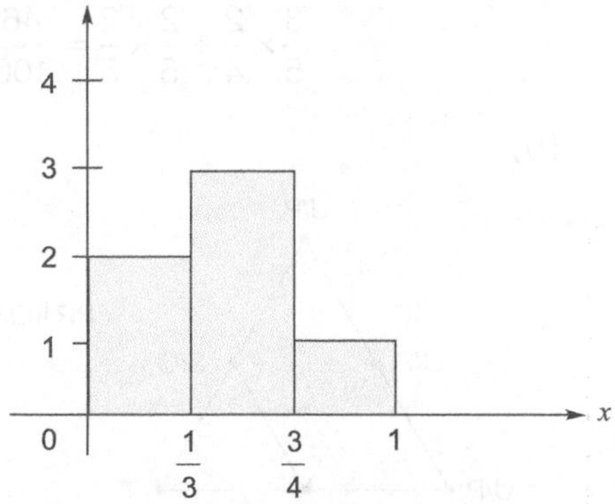

$$A = \frac{1}{3}\times 2 + \left(\frac{3}{4}-\frac{1}{3}\right)\times 3 + \left(1-\frac{3}{4}\right)\times 1 = \frac{13}{6}$$

467. (d)

$$= (r+2)(r+3)(r+4)(r+5)$$
$$= (r^2+5r+6)(r^2+9r+20)$$

Now, $r^2 + 2r = -6$

$$= (3r+6-6)(7r+20-6)$$
$$= 3r(7r+14)$$
$$= 21r(r+2)$$
$$= 21(r^2+2r) = 21 \times (-6) = -126$$

468. (a)

- If all children are inquisitive is true. Then no children are inquisitive is false.
- There is a possibility both (i) and (iii) can be false.

  Assume that some children are inquisitive in that case both (i) and (iii) are false.

470. (d)

$$\text{Appeared} = x$$

$$\text{Boys} = \frac{65x}{100}$$

$$\text{Girls} = \frac{35x}{100}$$

$$\text{Qualified} = y$$

$$\text{Boys} = \frac{40y}{100}$$

$$\text{Girls} = \frac{60y}{100}$$

471. (d)

3 green balls and 2 orange balls are there.
Second ball is orange,

$$P = P(G \cap O) + P(O \cap O)$$

$$= \frac{3}{5} \times \frac{2}{4} + \frac{2}{5} \times \frac{2}{5} = \frac{46}{100} = 0.46$$

472. (b)

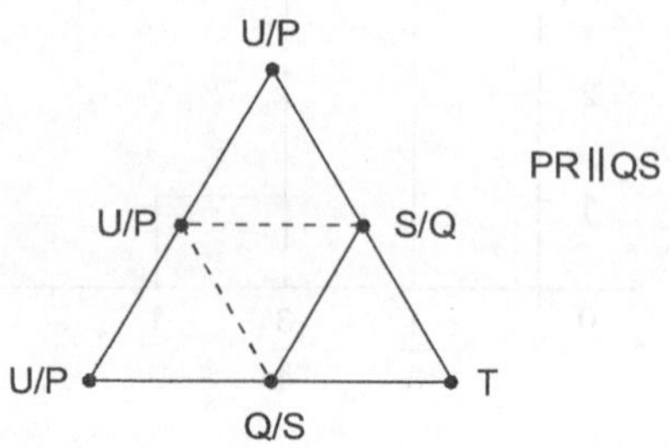

473. (d)

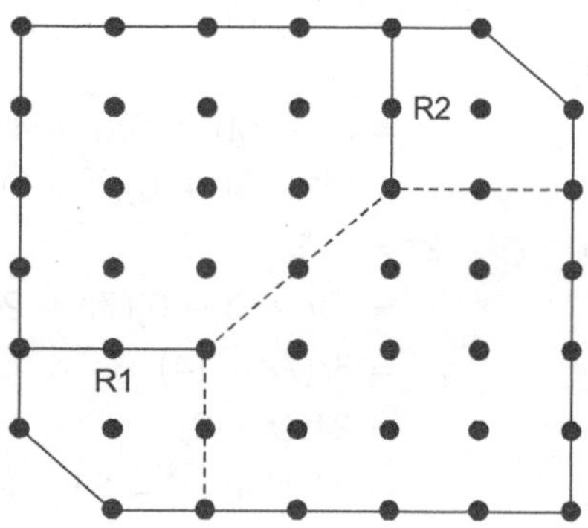

474. (d)

Given : $P : Q : R : S = 5 : 2 : 4 : 3$

Assume $P = 5x$, $Q = 2x$, $R = 4x$, $S = 3x$

Given that R gets Rs. 1000 more than S, therefore,

$$4x - 3x = 1000$$

$$\Rightarrow \quad x = 1000$$

$\therefore$ Share of $Q = 2x = 2 \times 1000 = 2000$

475. (b)

Given :

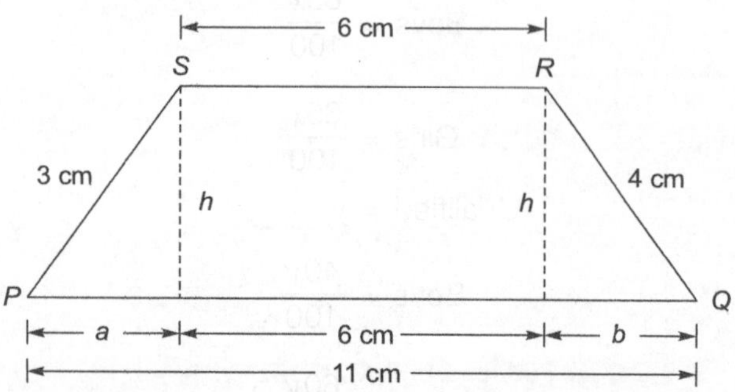

Let h is the shortest (i.e. perpendicular) distance between PQ and SR.

From the trapezium $PQRS$, using Pythagoras theorem,

$$h^2 + a^2 = 9 \quad \text{...(i)}$$

$$h^2 + b^2 = 16 \quad \text{...(ii)}$$

Subtracting equation (i) and (ii), we get,

$$b^2 - a^2 = 16 - 9 = 7$$

$$\Rightarrow (b - a)(b + a) = 7 \quad \text{...(iii)}$$

The length of side PQ,

$$a + 6 + b = 11$$

$$\Rightarrow \quad a + b = 11 - 6 = 5 \quad \text{...(iv)}$$

Using in equation (iii), we get,

$$(b - a) \times 5 = 7$$

$$\Rightarrow \quad b - a = \frac{7}{5} \quad \text{...(v)}$$

Using equation (iv), we get,

$$(5 - a) - a = \frac{7}{5}$$

$$2a = 5 - \frac{7}{5} = \frac{18}{5}$$

$$\Rightarrow \quad a = \frac{9}{5}$$

From equation (i),

$$h^2 = 9 - a^2 = 9 - \left(\frac{9}{5}\right)^2$$

$$h^2 = 9 - \frac{81}{25} = \frac{144}{25}$$

$$h = \frac{12}{5} = 2.40 \text{ cm}$$

476. (b)

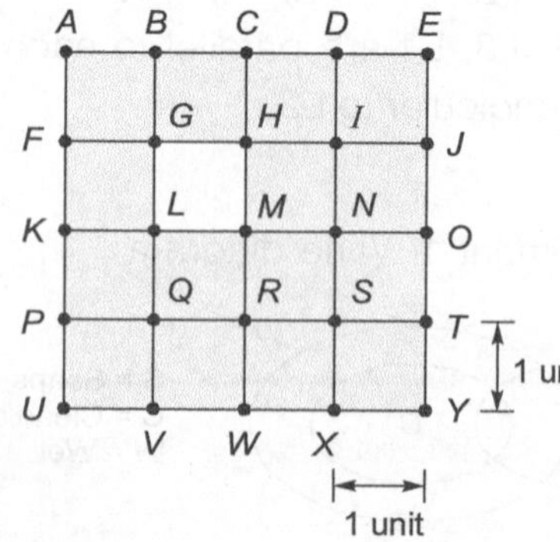

Total number of squares in 4 × 4 grid = 4 × 4 = 16

Number of 1 × 1 squares without a hole in the interior = 16 − 1 = 15

Number of 2 × 2 squares without a hole in the interior

= 5(KMWU, LNXV, MOYW, HJTR, CEOM)

$\therefore$ Maximum number of squares that can be formed without a hole in the interior = 15 + 5 = 20.

477. (c)

Considering the first inequality,

$$2x - 1 > 7$$
$$\Rightarrow \quad 2x > 8$$
$$\Rightarrow \quad x > 4 \quad \ldots\text{(i)}$$

From the second given inequality,

$$2x - 9 < 1$$
$$\Rightarrow \quad x < 5 \quad \ldots\text{(ii)}$$

Combining equations (i) and (ii), we get,

$$4 < x < 5$$

478. (c)

According to the given data, the quadrilateral *PSQR* can be plotted as below:

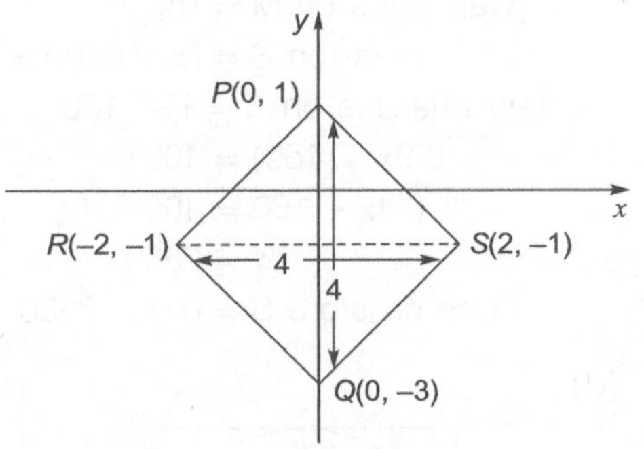

The length of the sides of the given quadrilateral can be calculated as below:

$$PS = \sqrt{(2-0)^2 + (-1-1)^2} = \sqrt{8}$$
$$QS = \sqrt{(2-0)^2 + (-1+3)^2} = \sqrt{8}$$
$$QR = \sqrt{(0+2)^2 + (-3+1)^2} = \sqrt{8}$$
$$PR = \sqrt{(0+2)^2 + (1+1)^2} = \sqrt{8}$$

Hence, all the sides of the quadrilateral are equal.

The length of the diagonals,

$$PQ = \sqrt{(0-0)^2 + (1+3)^2} = 4$$
$$RS = \sqrt{(2+2)^2 + (-1+1)^2} = 4$$
$$\therefore \quad PQ = RS = 4$$

Since, the sides of the given quadrilateral are equal. Also, the length of the diagonals are equal and bisect at right angles. Hence, the given quadrilateral is a square.

$\therefore$ Area of quadrilateral = (side)^2 = $(PS)^2$ = 8 units

479. (b)

The figure obtained after applying the given sequences of operation are as below:

(1) *XYZZ*

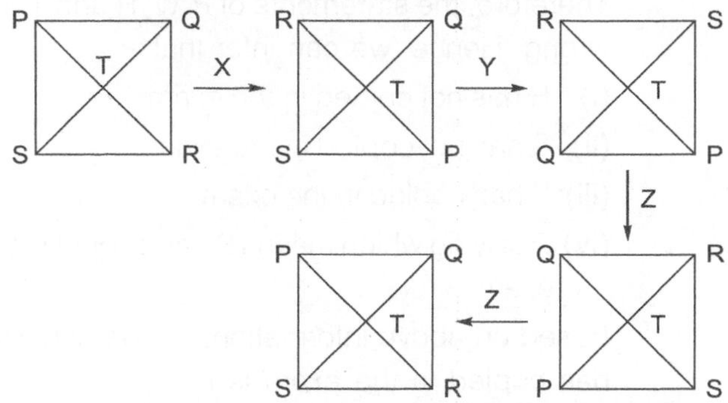

(2) XY

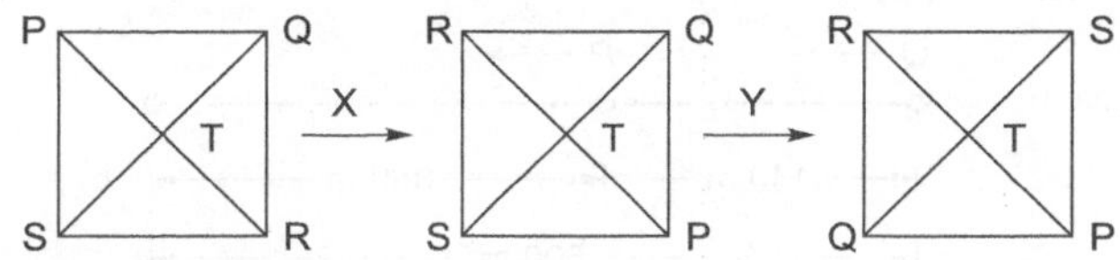

(3) *ZZZZ*

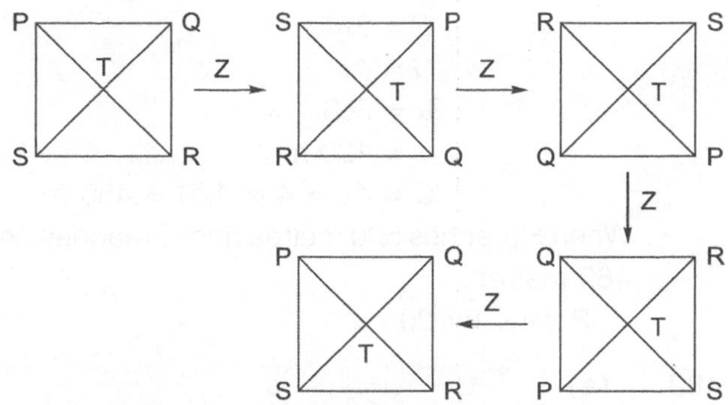

Hence, the sequence of operations (1) and (3) are equivalent.

480. (c)

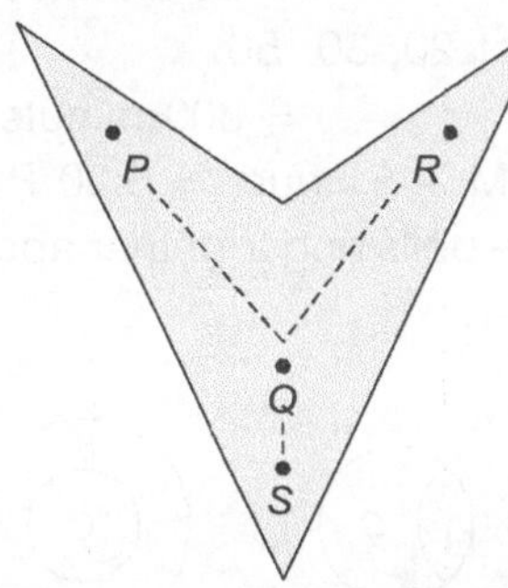

Given that, the plan of the gallery where the boundary walls are opaque, if the security guard will be posted on '*P*' then he doesn't have the visibility of *R* and vice-versa. But if the security guard will be posted at '*Q*' and '*S*', then the security guard can watch over the entire inner space of the gallery.

481. (b)

Since, S never lies, therefore the statements of *S* is correct i.e. only one of us telling the truth, which is S only as S never lies.

Therefore, the statements of P, Q, R and T are wrong. Hence, we can infer that

(i) R has not copied in the exam.

(ii) S has not copied in the exam.

(iii) P has copied in the exam.

(iv) R is lying which means P has copied in the exam.

Based on above information, the person who has copied in the exam is P.

482. (a)

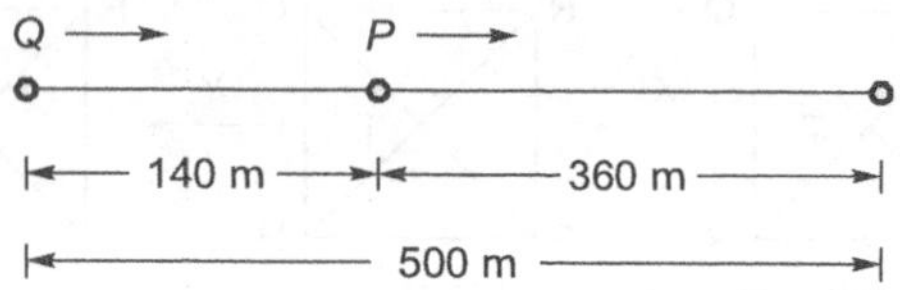

$$P : Q = 3 : 4$$
$$P = 3x$$
$$Q = 4x$$
So, $$3x = 360$$
$$x = 120 \text{ m}$$
$$Q = 4x = 4 \times 120 = 480 \text{ m}$$

When *P* reaches 500 metres and *Q* reaches only 480 metres.

$\therefore$ P wins by 20 m.

483. (a)

Three bells *P*, *Q*, *R*.

$$P = 20 \text{ minutes}$$
$$Q = 30 \text{ minutes}$$
$$R = 50 \text{ minutes}$$

L.C.M. of 20, 30, 50

$$= 300 \text{ minutes} = 5 \text{ hours}$$

12:00 PM + 5 hours = 5:00 P.M.

So, three bells ring together again at 5:00 P.M.

484. (a)

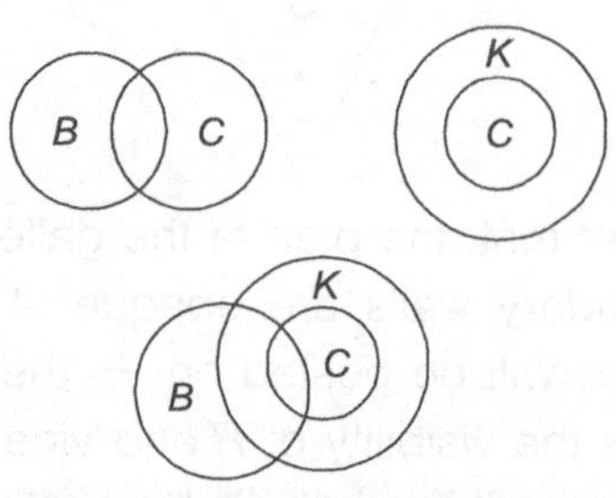

Conclusions I and II are correct.

485. (c)

n = Total = 3

m = Favourable = disc does not retain the same front and rear views after the flipping = 2

$$\text{Probability} = \frac{m}{n} = \frac{2}{3}$$

486. (a)

Given two dice.

$$P(E) = \frac{\text{Favourable}}{\text{Total}}$$

$$P(E) = \frac{3}{6} \times \frac{3}{6} = \frac{1}{4}$$

487. (b)

Given online stores S and M.

Cap price on M = Rs. x

Price on S = $(x - 0.1x) = 0.9x$

Delivery charges on *S* = Rs. 100

$$\therefore \quad x - (0.9x + 150) = 100$$
$$0.1x - 150 = 100$$
$$x = 2500$$

Price on store S = 0.9 × 2500 = 2250

490. (d)

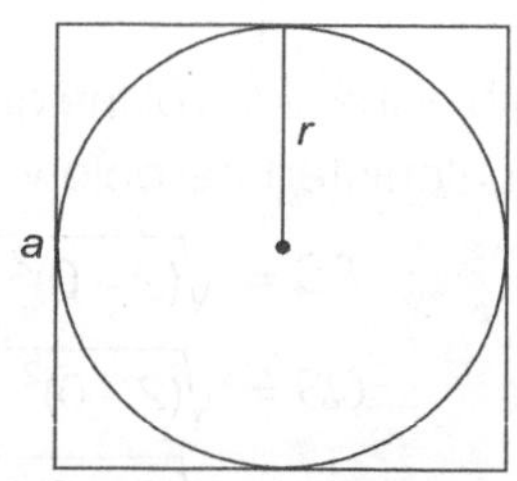

Diameter of sphere = Side of cube

$$a = 2r$$
$$\text{Volume of cube} = (2r)^3$$
$$= 8r^3$$

491. (c)

Let the capacity of storage tank is x liters. Amount of water drawn only by pipe *R* is in one minute.

$$\frac{x}{60} + \left(\frac{x}{6} + \frac{x}{10}\right) = \frac{17x}{60}$$

Total time taken *R* to draw a storage tank

$$= \frac{x}{\frac{17x}{60}} = \frac{60}{17} \text{min}$$

$$\text{Capacity of tank} = \frac{60}{17} \times 34 = 120 \text{ liters}$$

492. (c)

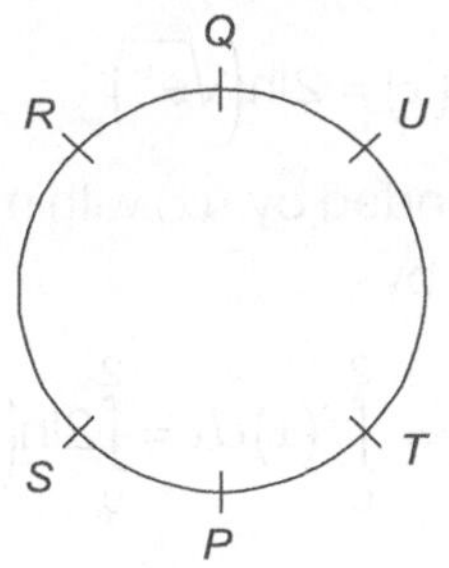

494. (d)

Probability of getting only one number $= \frac{1}{8}$

As they are independent spins probability of getting one number for each disc is

$$P = \frac{1}{8} \times \frac{1}{8}$$

Possible outcomes to the sum of 8

$= (1, 7), (2, 6), (3, 5), (4, 4), (5, 3), (6, 2), (7, 1)$

$$\text{Required probability} = \frac{7}{64}$$

495. (a)

Adding both equation

$$2p + 2q < 16$$
$$p + q < 8$$

496. (c)

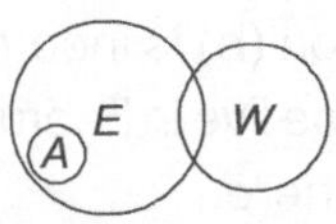

498. (a)

$$M + N + S = 4000 \times 3 = ₹12000 \quad \ldots \text{(i)}$$
$$N + S + P = 5000 \times 3 = ₹15000 \quad \ldots \text{(ii)}$$
$$\because \quad P = ₹6000$$

From (ii),

$$N + S = ₹9000$$

From (i),

$$M + 9000 = 12000$$
$$M = 3000$$
$$\frac{M}{P} = \frac{3}{6} = \frac{1}{2} \approx 50\%$$

499. (c)

Let x be the distance covered with 10 kmph so, $80 - x$ will be covered with 18 kmph.

$\because$ Total time to cover distance is 6 hr.

$$\Rightarrow \quad \left(\frac{x}{10}\right) + \left(\frac{80-x}{18}\right) = 6$$
$$\Rightarrow \quad \left(\frac{9x}{90}\right) + \frac{5(80-x)}{90} = 6$$
$$\Rightarrow \quad 9x + 400 - 5x = 540$$
$$\Rightarrow \quad x = 35$$

So, % Distance covered by 10 kmph

$$= \frac{35}{80} \times 100 = 43.75\%$$

500. (b)

Languages as per difficulty or hardness w.r.t each other

Chinese > Japanese > Dutch > French > Spanish

So, among P, Q, R and S, Q is learning the most difficult languages (i.e. Chinese and Japanese).

501. (a)

If viewed from the direction stated above, the 2-D view will appear as a trapezium placed over a rectangle.

502. (a)

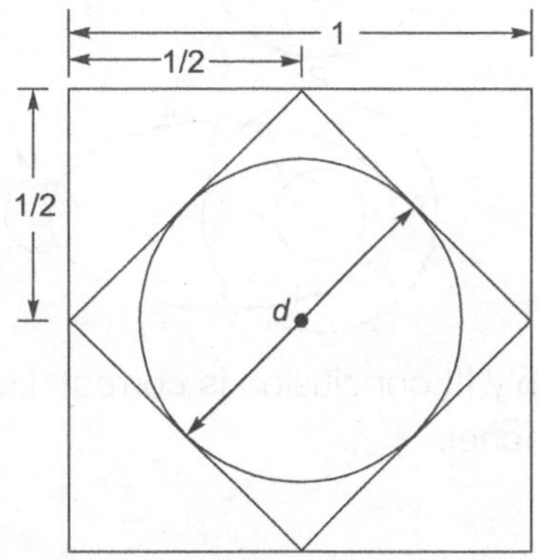

Diameter of circle = Side of rhombus

$$d = \sqrt{\left(\frac{1}{2}\right)^2 + \left(\frac{1}{2}\right)^2} = \frac{\sqrt{2}}{2} = \frac{1}{\sqrt{2}}$$

503. (b)

Let x = Side of equilateral D

a = Side of square

R = Radius of circle

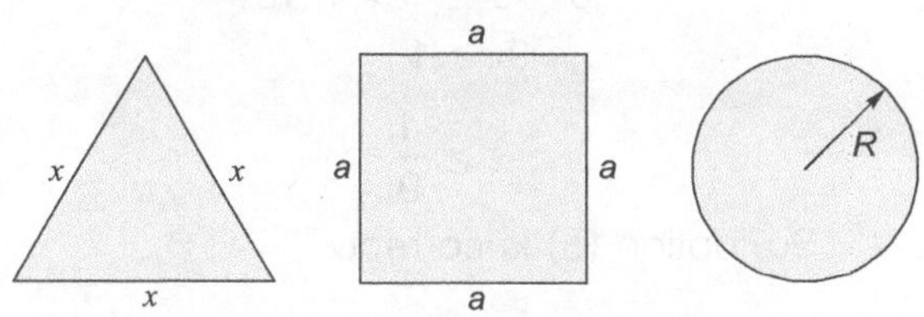

Area:

$$\frac{\sqrt{3}}{4}x^2 = a^2 = \pi R^2$$

or $$\frac{3^{1/4}}{2}x = a = \sqrt{\pi}R = k$$

Perimeter:

$3x : 4a : 2\pi R$

$$\frac{3.2k}{3^{1/4}} : 4k : 2\pi\frac{k}{\sqrt{\pi}}$$

$$\sqrt{(3\sqrt{3})} : 2 : \sqrt{\pi}$$

504. (a)

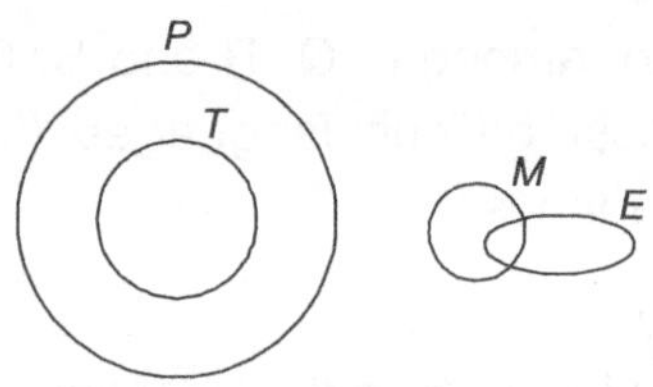

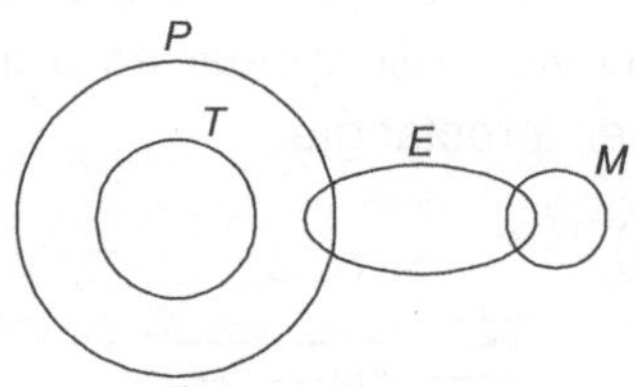

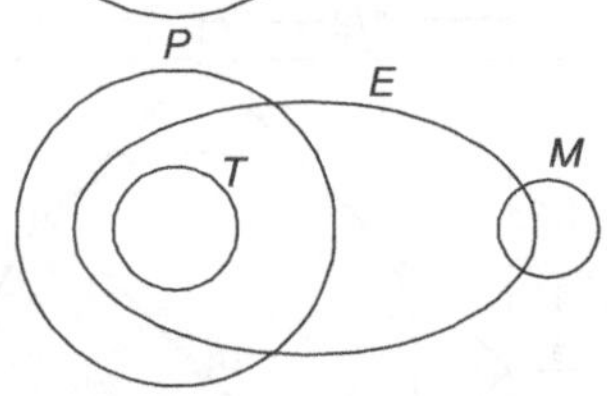

∴ Only III conclusion is correct, i.e. no male is a teacher.

505. (a)

Every 1 hr coincide → 1 time and in 12 h coincides → 11 times because between 11 to 1 they only meet one time.

506. (c)

$$2 - 5x \le -\frac{(6x-5)}{3}$$

$$\Rightarrow \quad 6 - 15x \le -6x + 5$$

$$6 - 5 \le -6x + 15x$$

$$9x \ge 1$$

$$x \ge \frac{1}{9}$$

So, option (c) is correct.

507. (c)

Given: $f(x) = 2\ln\left(\sqrt{e^x}\right)$

Area bounded by $f(x)$ within $x \in [0, 2]$ on x-axis is given by,

$$A = \int_0^2 f(x)dx = \int_0^2 2\ln\left(\sqrt{e^x}\right)dx$$

$$= \int_0^2 2\ln\left(e^{x/2}\right)dx = \int_0^2 2\times\frac{x}{2}\ln e\,dx$$

$$= \int_0^2 x dx = \left[\frac{x^2}{2}\right]_0^2 = 2$$

508. (a)

In a particular year, 5 Mondays can only come in the month of February if that year is a leap year.

So, on 29th February there is a Monday.

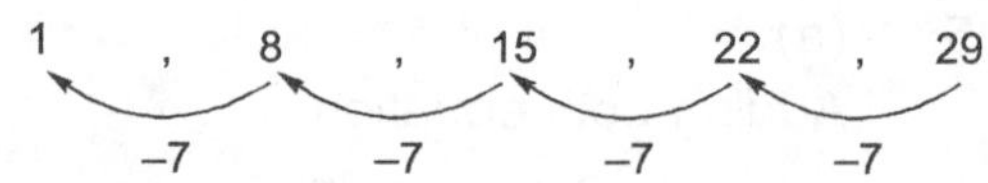

- Option (a) is correct, since 1st February is a Monday and hence 2nd February is a Tuesday.
- Option (b) is incorrect, since only Mondays can be five in February. So, only four Sundays are there.
- Option (c) is incorrect since 1st February is a Monday.
- Option (d) is incorrect as Mondays are on 1th, 8th, 15th, 22nd, 29th February which are all not even dates.

509. (b)

The given shape can be obtained by using 4 triangles (1, 2, 3 and 4) and 1 rectangle (5) as shown below.

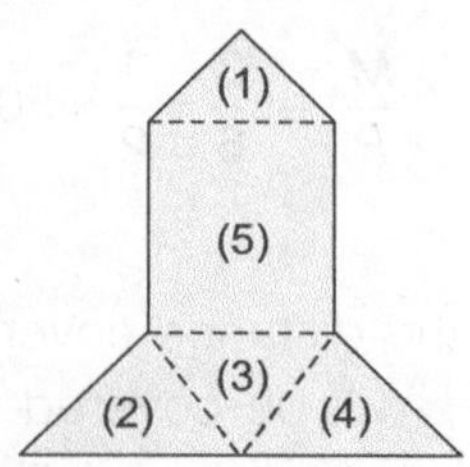

510. (c)
Total production in m days = 100 × m
Total production after today's day i.e. in $(m + 1)$ days = 100 × $(m + 1)$
The average production after todays' production = 110

$$\Rightarrow \frac{100m+180}{(m+1)} = 110$$

$$100m + 180 = 110(m + 1)$$
$$100m + 180 = 110m + 110$$
$$70 = 10m$$
$$\Rightarrow m = 7$$

511. (a)

Given: $f(p, q) = p^q$, $g(p, q) = p^{p^{p^{p^{\cdots}}}}$
Here, $f(2, 2) = 2^2 = 4$
$f(2, 1) = 2^1 = 2$
$f(3, 2) = 3^2 = 9$
$g(2, 2) = 2^2 = 4$
$g(2, 1) = 2 = 2$
$g(3, 2) = 3^3 = 9$
Also, $f(g(2, 2), 2) = f(4, 2) = 4^2 = 16$
$f(2, g(2, 2)) = f(2, 4) = 2^4 = 16$
On checking options, the correct option is (a).

512. (a)
The person starts from R at 7 AM and visits S and then returns to R if he follows the routes $R \to Q \to P \to S \to R$.
Since, travel time between any two connected cities is one hour, the total travel time for the route will be 4 hours as shown below.

$$R \xrightarrow[+1]{} Q \xrightarrow[+1]{} P \xrightarrow[+1]{} S \xrightarrow[+1]{} R$$

Now, the person reaches the city Q at 8 AM but the next train is at 9 AM. At 9 AM, he starts from Q and reaches P at 10 AM. The next train at P is at 11 AM, hence the waiting time at P is 1 hour.
He starts from P at 11 AM and reaches S at 12 AM but the next train at S is at 1:30 PM. Hence, waiting time at R is 1 hour and 30 Minutes.
Thus, the minimum time required is 6 hours 30 minutes.

513. (b)

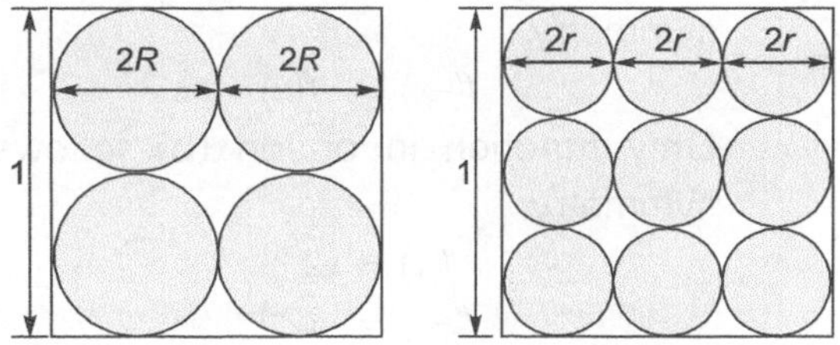

Case M Case N

Here, $6r = 1$ and $4R = 1$

$$\Rightarrow r = \frac{1}{6} \text{ and } R = \frac{1}{4}$$

Area of square in case M = Area of square in case N = 1 cm^2
Total area of all circles in case

$$M = 4(\pi R^2) = 4\pi \times \left(\frac{1}{4}\right)^2 = \frac{\pi}{4} cm^2$$

Total area of all circles in case

$$N = 9(\pi r^2) = 9\pi \times \left(\frac{1}{6}\right)^2 = \frac{\pi}{4} cm^2$$

Thus, area of unshaded region in case M,

$$A_1 = \left(1 - \frac{\pi}{4}\right) cm^2$$

Thus, area of unshaded region in case N,

$$A_2 = \left(1 - \frac{\pi}{4}\right) cm^2$$

Thus, the required ratio is

$$\frac{A_1}{A_2} = \frac{1-\frac{\pi}{4}}{1-\frac{\pi}{4}} = 1:1$$

514. (a)

$$P(1) = \frac{1}{6}$$
$$P(4) = \frac{1}{6}$$

Required probability

$$= P(1) \times P(4) = \frac{1}{6} \times \frac{1}{6} = \frac{1}{36}$$

515. (c)
Total number of triangles is 20.

516. (b)

The function shown in figure follows ODD symmetry.

$$f(-x) = -f(x)$$

Only function in option (b) follows ODD symmetry.

$$f(x) = x2^{-|x|}$$
$$f(-x) = -x2^{-|-x|}$$
$$= -x2^{-|x|}$$
$$f(-x) = -f(x)$$

517. (a)

$$\Rightarrow \quad 3^{999} \times 7^{1000}$$
$$= 3^{4(249)} \times 3^3 \times 7^{4(250)}$$
$$= 3^0 \times 3^3 \times 7^0 = 1 \times 7 \times 1 = 7$$

Unit place = 7

518. (b)

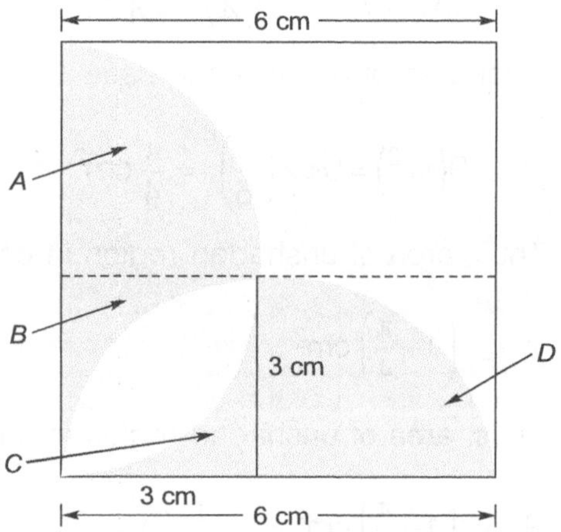

$$A = D = \frac{\pi r^2}{4} = \frac{\pi \times 3^2}{4}$$

$$A = D = \frac{9\pi}{4} \qquad \text{...(i)}$$

$$B = C = r^2 - \frac{\pi r^2}{4} = 9 - \frac{9\pi}{4} \qquad \text{...(ii)}$$

Shaded area = $A + B + C + D$

$$= 2 \times \frac{9\pi}{4} + 2 \times \left(9 - \frac{9\pi}{4}\right)$$
$$= 2 \times 9 = 18$$

519. (a)

Given,

Number of colleges, $C = 25951$

Number of universities, $U = 504$

Number of students with grade 3 in colleges

$$G_3(C) = 0.23 \text{ of } 25951 = 0.23 \times 25951$$
$$G_3(C) = 5968.73$$

Number of students with grade 3 in universities

$$G_3(U) = 0.07 \text{ of } 504 = 0.07 \times 504$$
$$G_3(U) = 35.28$$

Thus, required percentage

$$= \frac{\text{Total number of students with Grade 3}}{\text{Total number of students}}$$

$$= \frac{G_3(C) + G_3(U)}{c + u} = \frac{35.28 + 5968.73}{25951 + 504}$$

$$= 0.2269 \text{ or } 22.7\%$$

520. (a)

The minute hand and second hand will cross each other 1 time every minute after 9:15 i.e. 1 time each in 9:16 AM, 9:17 AM, 9.18 AM and so on upto 9:45 AM. Thus, they will cross each other 30 times.

521. (b)

□	*	○	△
△	○	*	□
○	△	□	*
*	□	△	○

522. (a)

Given inequalities,

$$p^2 - 4q < 4 \qquad \text{...(i)}$$
$$3p + 2q < 6 \qquad \text{...(ii)}$$

Since p and q are positive integers and satisfy the given inequalities, only possibility is

$$p = q = 1$$

This can be easily verified by putting values of p and q in inequality (ii),

For $p = q = 1$; $3(1) + 2(1) < 6$

For $p = 1$ and $q = 2$; $3(1) + 2(2) > 6$

And, similarly any other value of p and q will not satisfy the inequality.

Hence, $p + q = 1 + 1 = 2$

Alternate solution:

Multiplying inequality (ii) with 2,

$$6p + 4q < 12$$

or, $$4q < 12 - 6p \qquad \text{...(iii)}$$

From (i) and (iii),

$$p^2 - 4 < 12 - 6p$$

$$p^2 + 6p - 16 < 0$$
$$(p + 8)(p - 2) < 0$$

Using wavy curve method,

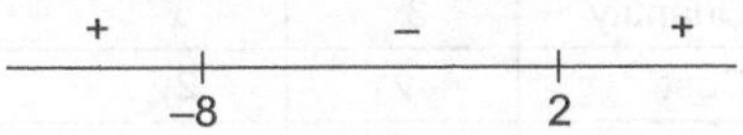

$\therefore \quad p \in (-8, 2)$

But it is given that p is positive integer,

$\therefore \quad p = 1$

From (iii), $\quad 4q < 12 - 6(1)$

$4q < 6$

$\Rightarrow \quad q < \frac{3}{2}$

But again q is positive integer,

$\therefore \quad q = 1$

Hence, $\quad p + q = 1 + 1 = 2$

523. (a)

Let T consists of 5 elements from the given set then the number of subsets possible is

$^6C_5\,[2^5 - 1] = 6 \times 31 = 186$

Let T consists of 4 elements from the given set then the number of subsets possible is

$^6C_2\,[2^4 - 1] = 15 \times 15 = 225$

Similarly,

$^6C_3\,[2^3 - 1] = 20 \times 7 = 140$

$^6C_2\,[2^2 - 1] = 15 \times 3 = 45$

$^6C_1\,[2^1 - 1] = 6 \times 1 = 6$

Thus, total number of subsets = 186 + 225 + 140 + 45 + 6 = 665

525. (d)

As we know that there are 5 odd digits.

So, total frabjous number = 5 × 4 × 4 = 80

527. (a)

Use (1) if Ankita move one stairs up

Use (2) if Ankita move two stairs up

Possible ways in which Ankita can reach the 5th stair.

F(5) ⇒ ① ① ① ① ① | ① ② ① ①
① ② ② | ② ① ① ①
② ② ①
② ① ②
① ① ① ②
① ① ② ①

531. (c)

The digits should be distinct and smallest number is 123456789.

532. (c)

Let students who like Romantic Movies = R.

Students who like Comedy Movies = C.

Given $R = 2C$

Also, 30 students do not like Romantic and Comedy Movies both.

$\therefore \quad 100 - 30 = 70 = n(R \cup C)$

and $\quad n(C \cap R) = 20$

$n(R \cup C) = n(R) + n(C) - n(C \cap R)$

$70 = 2C + C - 20$

$3C = 90$

$C = 30$

$R = 2C = 60$

533. (c)

Number of rectangles = 10 as square is also called as rectangle.

534. (a)

Since, the given function is an even function. Option (d) is only represents the even function.

535. (c)

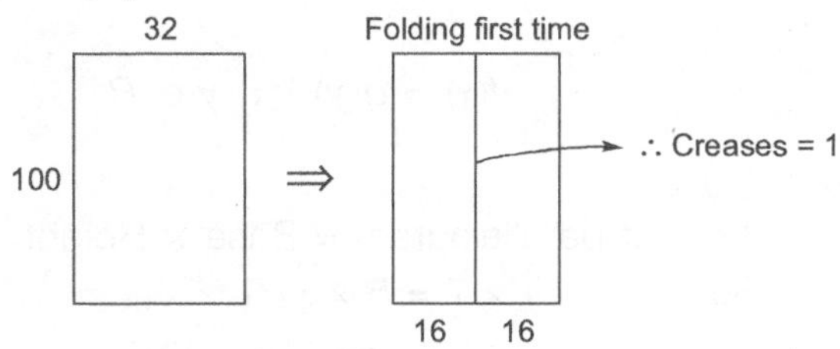

∴ After folding 5 times

$= 1 + 2^1 + 2^2 + 2^3 + 2^4$

$= 1 + 2 + 4 + 8 + 16 = 31$

536. (a)

$F_{n+1} = F_n + F_{n+1} \quad \forall n \geq 2$

$F_{n-1} = F_{n+1} - F_n$

For $n = 6$, $\quad F_5 = F_7 - F_6$

$= 60 - 37$

($\because$ Given that $F_7 = 60$, $F_6 = 37$)

$= 23$

For $n = 5$, $\quad F_4 = F_6 - F_5$

$= 37 - 23 = 14$

For $n = 4$, $\quad F_3 = F_5 - F_4$

$= 23 - 14 = 9$

For $n = 3$, $\quad F_2 = F_4 - F_3$

$= 14 - 9 = 5$

For $n = 2$, $\quad F_1 = F_3 - F_2 = 9 - 5 = 4$

$\therefore \quad F_1 = 4$

538. (c)

$f(t) = 0.01t^2$

$g(t) = 4t$

(i) For some $t > 0$, $g(t) > f(t)$ is true
For example if $t = 1$, $g(t) = 4$, $f(t) = 0.01$
$\therefore$ $g(t) > f(t)$ for some $t = 1$
(ii) There exist $T = 400$ such that
$f(t) > g(t)$ $\forall t > 400$ so it is true

539. **(b)**
Given that $f(x) = g(y)$ for all real value of x and y it is necessary that image of 'x' under f is same as image of 'y' is same as image of 'y' using 'g' for all real values of x and y
i.e. $f(x) = g(y) = \text{Constant}$

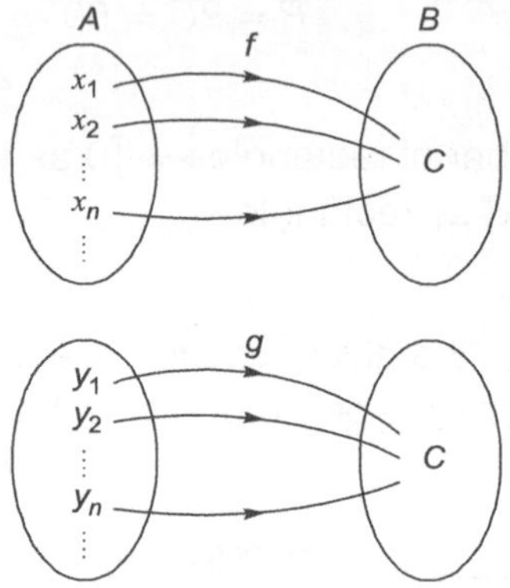

$\therefore$ $f(x) = g(y)\ \forall x, y \in R$

541. **(b)**
Area of parallelogram = Base × Height
So, $4 \times 7 = 5 \times x$

$$x = \frac{28}{5} \text{ cm}$$

Alternate Method:

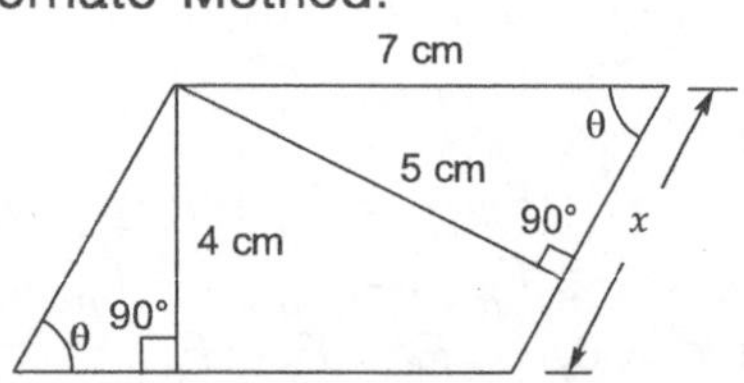

$$\sin\theta = \frac{5}{7} = \frac{4}{x}$$

$$\Rightarrow \quad x = \frac{28}{5} \text{ cm}$$

542. **(c)**

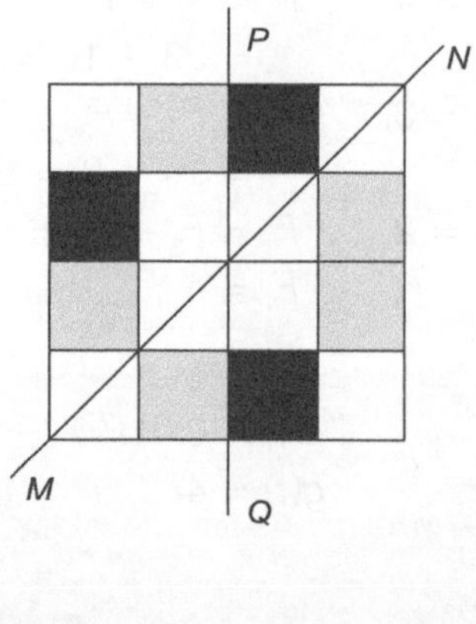

543. **(a)**

	Sand	Cement
Quantity	$3x$	x
Cost	y	$2y$
Total cost	$3xy$	$2xy = 1000$

$$\frac{S}{C} = \frac{3xy}{2xy} = \frac{3}{2} = \frac{600}{400}$$

$$C = \frac{4}{10} \times 1000$$

Cost of cement = 400

544. **(a)**
Given: $(x-1)^3 (x-2)^3$
$[x^3 - 1 - 3x^2 + 3x]\,[x^3 - 8 - 6x^2 + 12x]$ here we consider the coefficient of x^4 only.
$12 + 18 + 3 = 33$

546. **(a)**
For a polygon with n sides,
Interior angle sum $= (n-2) \times 180°$
Each angle of regular Hexagon.

$$\frac{(n-2)\times 180°}{n} = \frac{(6-2)\times 180°}{6} = 120°$$

Area of shaded region

$$= \frac{\theta}{360°}\times\pi\times r^2 = \frac{120°}{360°}\times\pi\times 5^2 = \frac{25\pi}{3}$$

547. **(b)**
If box d or i is coloured black, we get one line of symmetry, i.e., horizontal or vertical, respectively. But for two line of symmetry, we need to atleast colour 3 boxes, i.e., c, d and i.

548. **(d)**

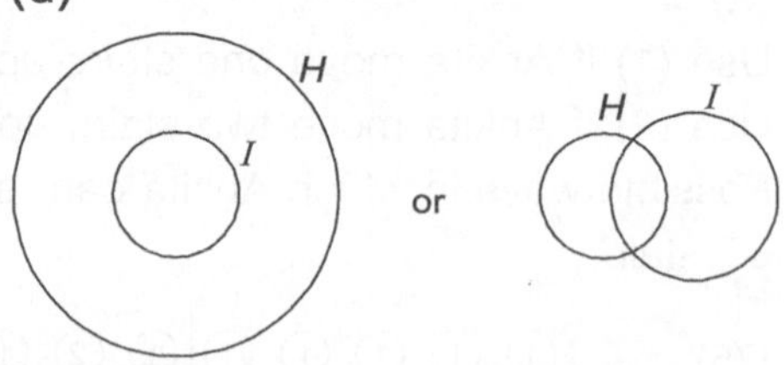

550. **(a)**
Given,
$a = 30!$, $b = 50!$ and $c = 100!$

We know that $\log_n m = \dfrac{\log m}{\log n}$

$\log_a c, \log_c a, \log_a b, \log_b a$

$$\frac{\log c}{\log a}, \frac{\log a}{\log c}, \frac{\log b}{\log a}, \frac{\log a}{\log b}$$

$$\Rightarrow \frac{\log 100!}{\log 30!}, \frac{\log 30!}{\log 100!}, \frac{\log 50!}{\log 30!}, \frac{\log 30!}{\log 50!}$$

$$\therefore \frac{\log 30!}{\log 100!} < \frac{\log 30!}{\log 50!} < \frac{\log 50!}{\log 30!} < \frac{\log 100!}{\log 30!}$$

$$\log_c a < \log_b a < \log_a b < \log_a c$$

551. (a)

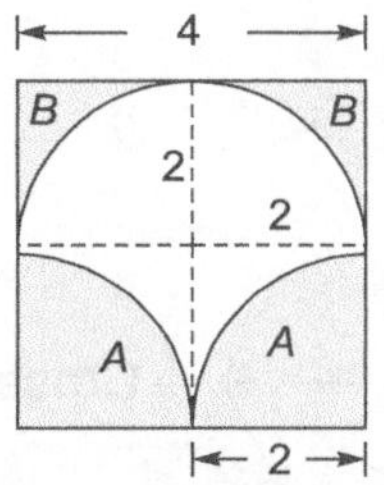

Area of region, $A = \frac{1}{4}\pi(2)^2 = \pi$

Area of region,

B = Area of small square – Area of quadrant of circle

$= 4 - \pi$

So,total shaded area $= 2A + 2B$

$= 2 \times \pi + 2 \times (4 - \pi)$

$= 2\pi + 8 - 2\pi = 8$ unit^2

552. (d)

Since only one cell is to be shaded in each row and each column. So if one cell is shaded so the size of original square (3 × 3) reduces to (2 × 2) grid square. Thus, number of possible ways = 3 × 2 × 1 = 6

553. (a)

Given: 4 Red, 5 Green and 6 Blue balls,

So, in worst case, we can select 3 balls each with different colour, i.e., 1 Red, 1 Green and 1 Blue ball. Now if we select the forth ball, then we found two balls are of same colour. Thus,. smallest number of balls that guarantees we have at least two balls of same colour is 4.

554. (a)

Operation 1:

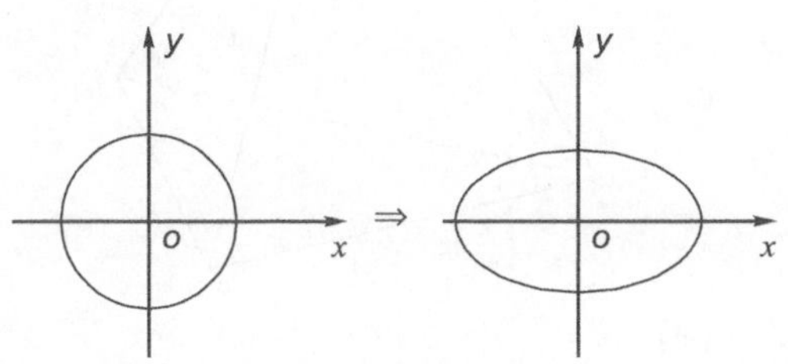

Suppose, $s_{x\text{-axis}} > S_{y\text{-axis}}$

Operation 2:

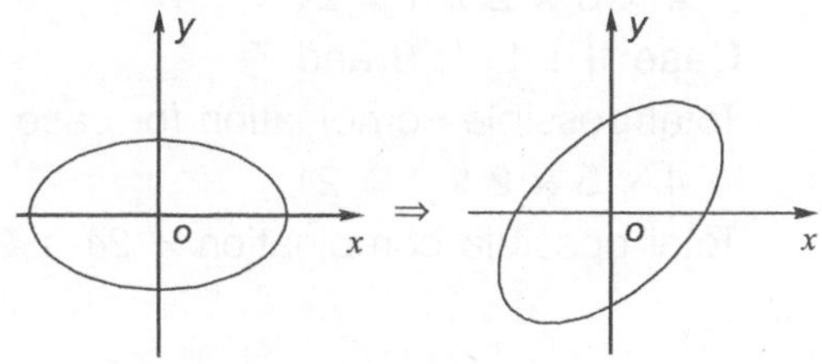

555. (a)

Let the three husband – wife pairs be H_1 W_1, H_2 W_2 and H_3 W_3

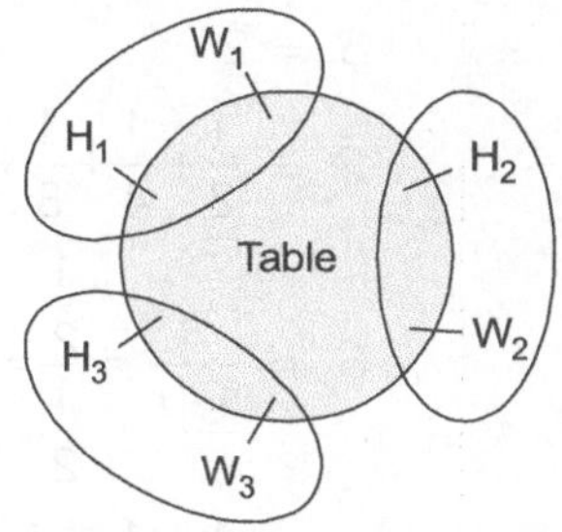

Treat every husband-wife pairs as one block. So, there are three pairs and all three pairs can be arranged by (3 – 1)! ways and in all pairs husband and wife can be arranged by 2! ways. So, required number of ways are

$(3 - 1)! \times 2! \times 2! \times 2! = 2! \times 2^3 = 16$

556. (c)

Given: $4^{8^x} = 256$

$\Rightarrow \quad 4^{8^x} = 4^4$

$\Rightarrow \quad 8^x = 4$

$\Rightarrow \quad (2^3)^x = 2^2$

So, $\quad x = \frac{2}{3}$

558. (a)

The minimum number of colors required is 4.

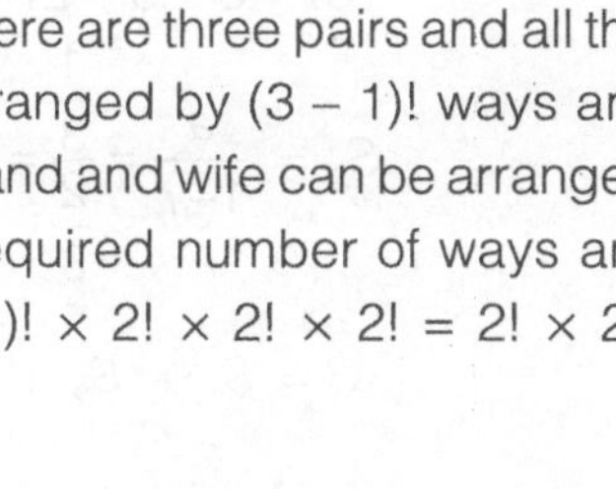

559. (b)

There are two such combination of digit

Case I : 1, 3, 4 and 7

Total possible combination for case I

$4 \times 3 \times 2 \times 1 = 24$

Case II : 1, 4, 6 and 7

Total possible combination for case II

$4 \times 3 \times 2 \times 1 = 24$

Total possible combination = 24 + 24 = 48

560. (b)

Given series

$$2+\frac{1}{2}+\frac{1}{3}+\frac{1}{4}+\frac{1}{8}+\frac{1}{9}+\frac{1}{16}+\frac{1}{27}+....$$

We can break

$$S_1 = 2$$

$$S_2 = \frac{1}{2}+\frac{1}{4}+\frac{1}{8}+\frac{1}{10}+----$$

$$\therefore \quad S_2 = \frac{a}{1-r} = \frac{\frac{1}{2}}{\frac{1}{2}} = 1$$

$$S_3 = \frac{1}{3}+\frac{1}{9}+\frac{1}{27}+----$$

$$\therefore \quad S_3 = \frac{a}{1-r} = \frac{\frac{1}{3}}{\frac{2}{3}} = \frac{1}{2}$$

$$S = S_1 + S_2 + S_3 = \frac{7}{2}$$

561. (b)

Total valid votes = 115000 – 5000

= 110000

Total votes received by *B* and *C* in percentage

= 20 + 25 = 45%

$$\text{Total votes} = 110000\times\frac{45}{100} = 49500$$

562. (a)

There are 8 possible combinations with three pirth events BBB, BBG, BGB, BGG, GBB, GBG, GGB, GGG.

In these possible results there are three that gives exactly two girls BGG, GBG, GGB.

$$\therefore \text{ Required probability} = \frac{3}{8}$$

563. (b)

Let amount rate of return for the mutual funds

$$A = x\%$$

Person 1 earning

$$= 10000\times\frac{x}{100}+20000\times\frac{15}{100}+20000\times\frac{15}{100}$$

$$= 100x + 6000$$

Person 2 earning

$$= 20000\times\frac{x}{100}+15000\times\frac{15}{100}+20000\times\frac{15}{100}$$

$$= 200x + 4500$$

Person 1 earning – Person 2 earning = 500

$$-100x + 1500 = 500$$

$$x = 10\%$$

565. (a)

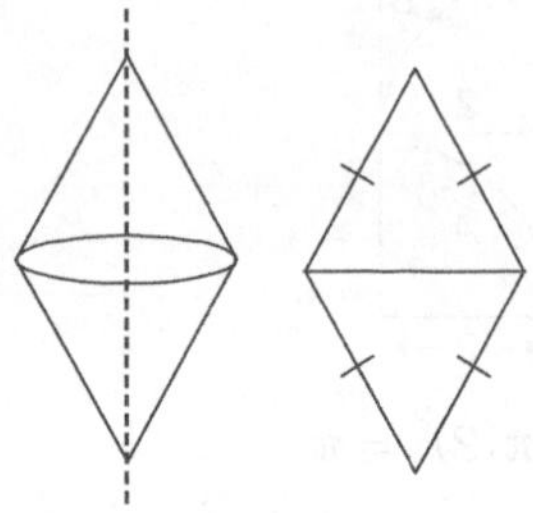

566. (d)

$$\text{Let, } \frac{x}{pq-r^2} = \frac{y}{qr-p^2} = \frac{z}{rp-q^2} = K$$

$$\frac{px}{p^2q-pr^2} = K$$

$$\frac{qy}{q^2r-qp^2} = K$$

$$\frac{rz}{r^2p-rq^2} = K$$

$$px + qy + rz$$

$$= kp^2q - kpr^2 + kq^2r - kqp^2 + kr^2p - krq^2$$

$$= 0$$

567. (*)

Tetrahedron Angles : In a regular tetrahedron, all the faces are equilateral triangles. Therefore, all the interior angles of a tetrahedron are 60° each.

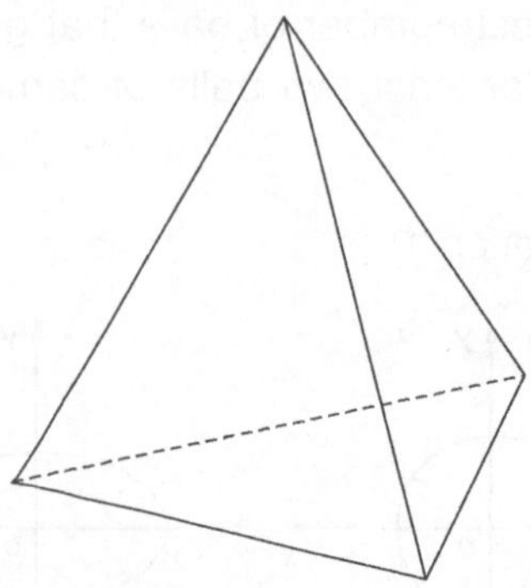

570. (a)

Total outcomes : (2, 2), (2, 3), (2, 5), (2, 7)
(3, 2), (3, 3), (3, 5), (3, 7)

(5, 2), (5, 3), (5, 5), (5, 7)
(7, 2), (7, 3), (7, 5), (7, 7)

Favourable outcomes (sum > 11)
: (5, 7), (7, 5), (7, 7)

$$\text{Probability} = \frac{\text{Number of favourable outcomes}}{\text{Number of total outcomes}} = \frac{3}{16}$$

571. (b)
Except 21 all are prime number.

572. (c)

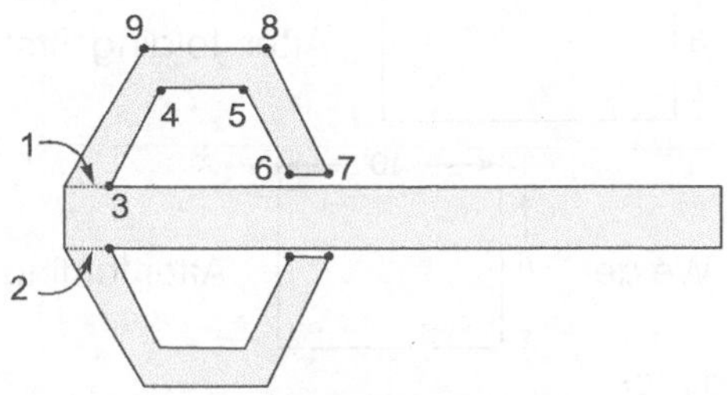

Hence, the total number of folds required to form the above object is 9.

573. (d)
In Fibonacci series, every number is equal to the sum of previous two numbers of the series.
i.e. 0, 1, 1, 2, 3, 5, 8, 13, 21, ...

$$0 + 1 = 1$$
$$1 + 1 = 2$$
$$2 + 1 = 3$$
$$3 + 2 = 5$$
$$5 + 3 = 8$$
$$8 + 5 = 13$$
$$13 + 8 = 21$$

577. (d)

Given, $p > 0$, $q > 0$ and $\frac{p}{q} \neq 1$

$$\left(\frac{p}{q}\right)^{p/q} = p^{\left(\frac{p}{q}-1\right)}$$

$$\left(\frac{p}{q}\right)^{p/q} = p^{\left(\frac{p-q}{q}\right)}$$

$$\Rightarrow \left(\frac{p}{q}\right)^{p} = p^{p-q}$$

$$\Rightarrow \left(\frac{p}{q}\right)^{p} = \frac{p^p}{p^q}$$

$$\Rightarrow q^p = p^q$$

578. (c)
Between 12:05:00 to 12:55:00
There are 50 intervals of 1 minute in which a second hand and minute hand crosses one another time. So during 12:05:00 to 12:55:00, minute and second hand crosses 50 times.

579. (b)

$$3 \times 2 + 1 = 7$$
$$7 \times 2 + 1 = 15$$
$$15 \times 2 + 1 = \boxed{31}$$
$$31 \times 2 + 1 = 63$$
$$63 \times 2 + 1 = 127$$
$$127 \times 2 + 1 = 255$$

581. (d)

301 303 305 ------

302 304 306 ------

Let sum of 301 + 303 + 305 + ... = A
and 302 + 304 + 306 + ... = B
Given, A − B = 27
and 302 − 301 = 1
304 − 303 = 1
As difference of one pair is 1 and total difference is 27.
∴ Number of house in one side is 27.

582. (b)

Conclusion 1 ×
Conclusion 2 ✓
Conclusion 3 ✓

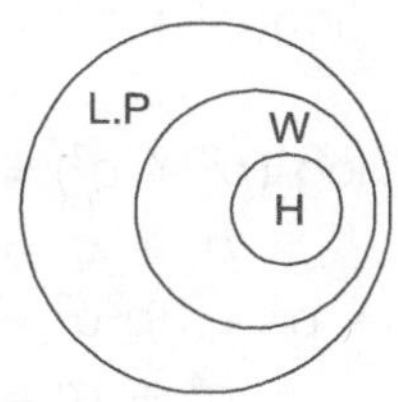

583. (d)

$$\frac{R\times q-\frac{R}{q}}{R\times q} = \frac{80}{100} = \frac{4}{5}$$

$$\frac{Rq-\frac{R}{q}}{Rq} = \frac{4}{5}$$

$$1-\frac{1}{q^2} = \frac{4}{5}$$

$$\frac{1}{q^2} = \frac{1}{5}$$

$$q = \sqrt{5}$$

584. (a)

The sum of first n odd natural number $= n^2$

$\therefore$ Result is $\dfrac{20^2}{20^2} = 1$

585. (b)

	Class VIII	Class IX
Total students	450	360
Number of girls	G	G
Number of boys	450 – G	360 – G

According to question,

$$\frac{G}{450-G} = \frac{360-G}{G}$$

$$G^2 = 360 \times 450 - 450\,G - 360\,G + G^2$$

$$810\,G = 360 - 450$$

$$G = \frac{360 \times 450}{810} = 200$$

586. (b)

Volume of outer cylinder $= \pi R^2 h$

Volume of smaller cylinder $= \pi r^2 h$

$R = 3r$

Ratio of occupied space to empty space

$$= \frac{7 \times \pi r^2 h}{9\pi r^2 h - 7 \times \pi r^2 h} = \frac{7}{2}$$

590. (b)

$$\log(p^2 + q^2) = \log p + \log q + 2 \log 3$$
$$= \log p + \log q + \log 9$$
$$\log(p^2 + q^2) = \log(9pq)$$
$$p^2 + q^2 = 9pq$$
$$p^4 + q^4 + 2p^2q^2 = 81p^2q^2$$
$$p^4 + q^4 = 79p^2q^2$$

The value of $\dfrac{p^4 + q^4}{p^2q^2} = \dfrac{79p^2q^2}{p^2q^2} = 79$

591. (b)

Area of rectangular sheet $= 54 \times 4 = 216$ cm^2

Surface area of cube $= 6\,(\text{side})^2$

$216 = 6\,(\text{side})^2$

Side $= 6$

Volume of cube $= 6^3 = 216$

Radius of cylinder $= \dfrac{4}{2\pi}$

Volume of cylinder $= \pi r^2 h$

$$= \pi\left(\frac{4}{2\pi}\right)^2 \times 54 = \frac{216}{\pi}$$

$$\frac{\text{Volume of cylinder}}{\text{Volume of cube}} = \frac{\frac{216}{\pi}}{216} = \frac{1}{\pi}$$

592. (a)

Given dimension 20×8 folded 3 times symmetrically along perpendicular to its longest edge.

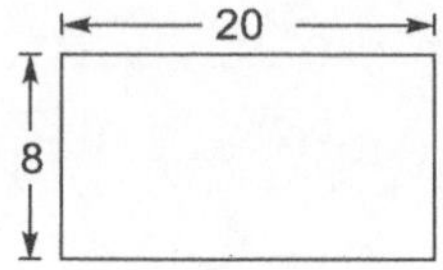

After folding first times

We get

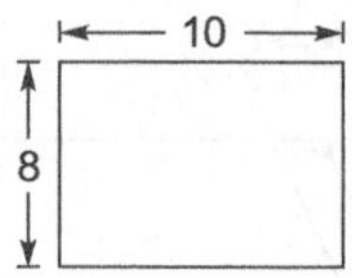

After folding second times

We get

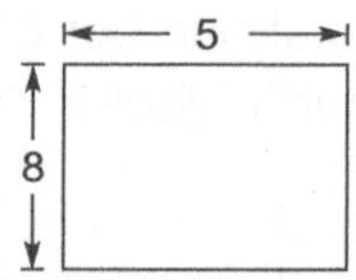

After folding third times

We get

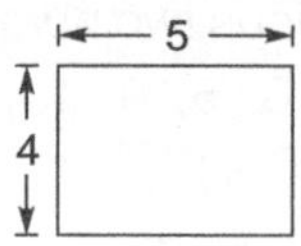

So, the perimeter of final figure $= 2(4 + 5) = 18$ unit

593. (b)

Total Fat = Saturated fat + unsaturated fat

$= 20 + 20 + 5$

$= 45\%$

Total fat in Kcal $= 2000 \times \dfrac{45}{100} = 900$ Kcal

Given all fat energy density = 9 Kcal/g

There of total fat in gram $= \dfrac{900}{9} = 100$ g

594. (d)

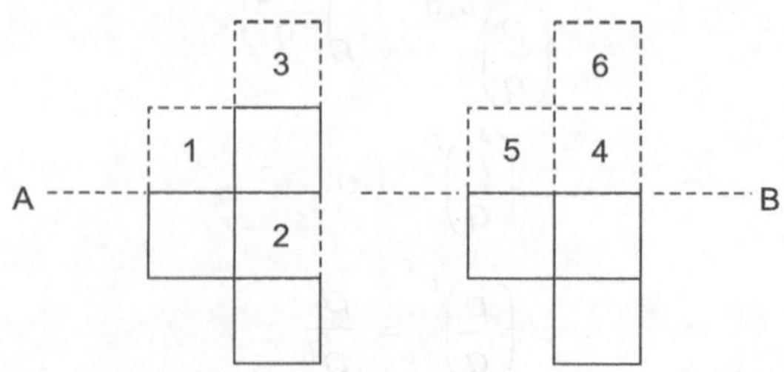

595. (b)

$$\text{Percentage of ₹5} = \frac{5\times3}{1\times5+5\times3+10\times13}\times100$$

$$= \frac{15}{150}\times100 = 10\%$$

596. (a)

Given $(x + y) \propto (x - y)$

$\therefore$ $(x + y) = k(x - y)$

$$x + y = kx - ky$$

$$(k - 1)x = (k + 1)y$$

$$\frac{x}{y} = \frac{k+1}{k-1} = k' \text{ (constant)}$$

597. (c)

Arrange given series in ascending order

9, 10, 11, 11, 13, 14, 15, 17, 18, 69

$$\text{Median of the given data} = \frac{14+13}{2} = 13.5$$

598. (b)

1	X	4	3
X	5	5	4
3	X	6	X
2	4	3	2

Sum of fourth row = 2 + 4 + 3 + 2 = 11

599. (b)

$6 + 3 = 9$

$9 + 5 = 14$

$14 + 7 = \boxed{21}$

$21 + 9 = 30$

$30 + 11 = 41$

600. (b)

Combination of $4 \times 3 \times 2 \times 1 = 24$

601. (b)

$$\text{If S.P. is same than overall loss} = \frac{x^2}{100} = 1\%$$

604. (a)

Number of students = 10000

Number of students not interested in both branches = 1500

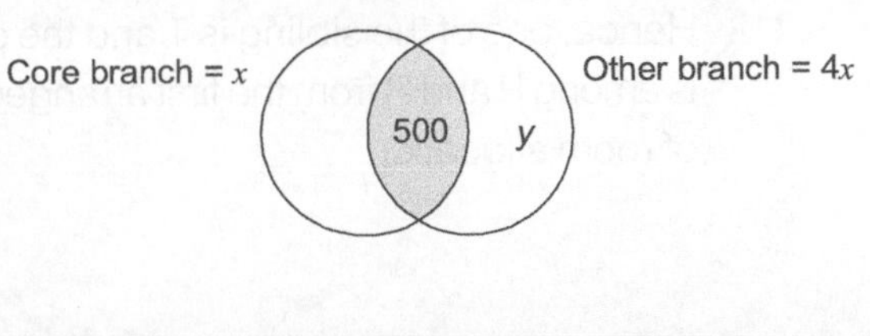

$\therefore$ When diagram represent 10000 − 1500 = 8500

x = Core

y = Other branch (core branch not include) = $4x$

$$x + 4x = 8500 + 500$$

$$5x = 9000$$

$$x = 1800$$

606. (c)

$$\log\frac{(x+y)}{2} = \frac{1}{2}(\log x + \log y)$$

$$\log\left(\frac{x+y}{2}\right) = \log\sqrt{xy}$$

$$\frac{x+y}{2} = \sqrt{xy}$$

$$x^2 + y^2 + 2xy = 4xy$$

$$x^2 + y^2 = 2xy$$

$$\frac{x}{y} + \frac{y}{x} = 2$$

607. (b)

(i) $$\%\left(\frac{\text{Overweight}}{\text{Normal}}\right)_{2003} = \frac{40\%}{30\%} = \frac{4}{3} = 1.33$$

$$\%\left(\frac{\text{Overweight}}{\text{Normal}}\right)_{2023} = \frac{25\%}{50\%} = \frac{1}{2} = 0.5$$

Hence, the ratio of the percentage of population falling into overweight category to the percentage of population falling into normal category has decreased in 20 years. Therefore, option (a) is not correct.

(ii) $$\%\left(\frac{\text{Obese}}{\text{Normal}}\right)_{2003} = \frac{25\%}{30\%} = \frac{5}{6} = 0.83$$

$$\%\left(\frac{\text{Obese}}{\text{Normal}}\right)_{2023} = \frac{15\%}{50\%} = \frac{3}{10} = 0.3$$

Hence, the ratio of the percentage of population falling into obese category to the percentage of population falling into normal category has decreased in 20 years. Therefore, option (b) is correct.

(iii) $$\%\left(\frac{\text{Underweight}}{\text{Normal}}\right)_{2003} = \frac{5\%}{30\%} = \frac{1}{6} = 0.166$$

$$\%\left(\frac{\text{Underweight}}{\text{Normal}}\right)_{2023} = \frac{10\%}{50\%} = \frac{1}{5} = 0.2$$

Hence, the ratio of the percentage of population falling into underweight category to the percentage of population falling into normal category has increased in 20 years. Therefore, option (c) is not correct.

(iv) The percentage of population falling into normal category has increased from 30% to 50% in 20 years. Hence, option (d) is not correct.

608. (d)

Let five years ago,

Son's age = x

Father's age = $4x$

After five years,

Son's age = $x + 10$

and Father's age = $4x + 10$

$$\therefore \quad \frac{x+10}{4x+10} = \frac{2}{5}$$

$$5x + 50 = 8x + 20$$

$$3x = 30$$

$$x = 10$$

Son's age five years ago = 10 years

Father's age five years ago = 40 years

At the time of son's birth, father's age

= 40 − 10 = 30 years

609. (b)

$$3^{199} - 3^{196} = 3^{196}(3^3 - 1) = 3^{196}(27 - 1)$$

$$= 3^{196} \times 26 = 2 \times 3^{196} \times 13$$

The highest prime factor is 13.

610. (c)

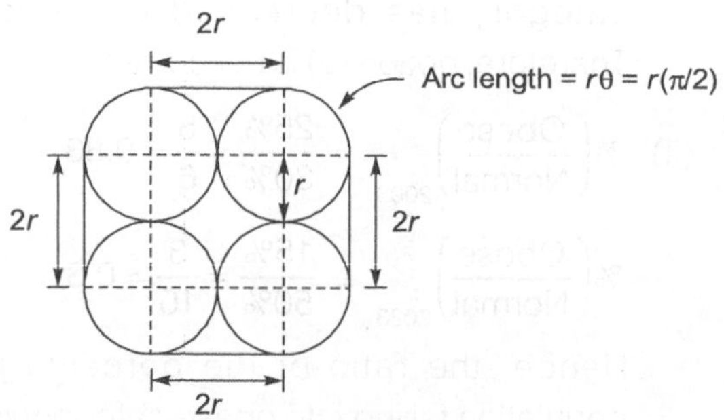

Width of the duct-tape, $W = l = 10$ cm

Length of the duct-tape required,

$$L = 4 \times (2r) + \left(4r \times \frac{\pi}{2}\right)$$

$$L = 8 \times 0.5 + 2 \times 0.5 \times \pi$$

$$L = (4 + \pi) \text{ cm}$$

Hence, the area of duct-tape required,

$$A = W \cdot L = 10(4 + \pi) \text{ cm}^2$$

611. (d)

FO1 operation on sheet A:

Folding first time

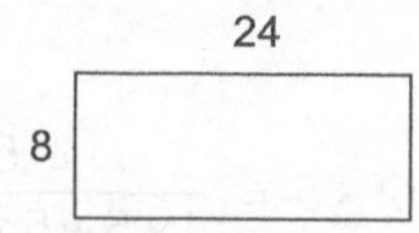

Folding second time

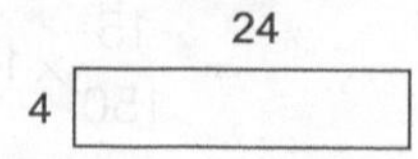

Perimeter = 2(24 + 4) = 56 cm

FO2 operation on sheet B:

Folding first time

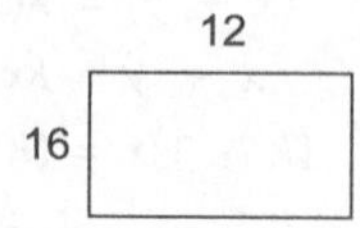

Folding second time

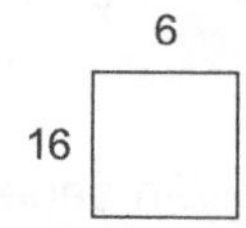

Perimeter = 2(16 + 6) = 44 cm

Required ratio = 56 : 44 = 14 : 11

612. (a)

$$\frac{1}{\log_2 x} + \frac{1}{\log_3 x} + \frac{1}{\log_4 x} = 1$$

$$\Rightarrow \log_x 2 + \log_x 3 + \log_x 4 = 1$$

$$\Rightarrow \log_x(2 \times 3 \times 4) = 1 \quad \Rightarrow \log_x 24 = 1$$

$$\Rightarrow \quad x = 24$$

613. (b)

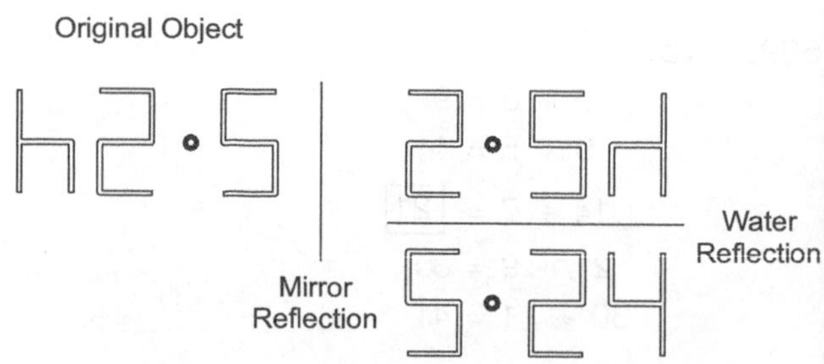

Hence, option (b) is the correct answer.

614. (d)

From the given acceptable room allocations, we can infer that:

(i) R and T are not siblings as they share the same room.

(ii) Since, S and Q are single children, the siblings are sharing the rooms with different person and are not allocated single room. Hence, one of the sibling is T and the other is among P and R from the first arrangement of room allocation.

(iii) From the second arrangement, it is clear that R and T can't be siblings. Hence, the only possible option is P and T.

615. (c)

Object A : Right (→)

Object P : Clockwise (↻)

Object Q : Anti-clockwise (↺)

Object R : Clockwise (↻)

Object B : Left (←)

Object B translates towards the left direction.

616. (c)

$$x : y = (a+1) : (a-1)$$

$$\frac{x}{y} = \frac{(a+1)}{(a-1)}$$

$$\frac{x^2-y^2}{x^2+y^2} = \frac{\left(\frac{x}{y}\right)^2-1}{\left(\frac{x}{y}\right)^2+1} = \frac{\left(\frac{(a+1)}{(a-1)}\right)^2-1}{\left(\frac{(a+1)}{(a-1)}\right)^2+1}$$

$$= \frac{(a+1)^2-(a-1)^2}{(a+1)^2+(a-1)^2} = \frac{4a}{2a^2+2} = \frac{2a}{a^2+1}$$

617. (d)

Divisor	Dividend	Remainder
8	108	
8	13	4
8	1	5
8	0	1

(LSB at top, MSB at bottom)

$$(108)_{10} = (154)_8$$

618. (*)

∵ AFCD is a rhombus

∴ $AD = CD = 5$ cm

∴ $\angle DAE = 90°$

∴ $AD^2 + AE^2 = DE^2$

$25 + 144 = DE^2 \Rightarrow DE = 13$ cm

$DB + BE = DE = 13$ cm

$DB = BF = 13 - 10 = 3$ cm

There is data inconsistency because the values given in question are not practical and such figures can not be made.

620. (c)

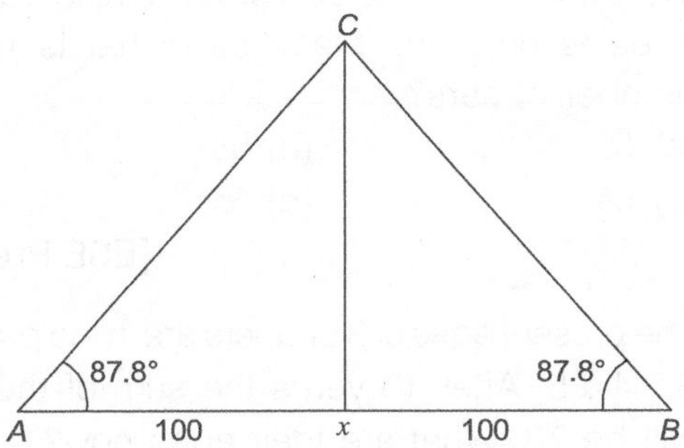

$$\tan\theta = \frac{Cx}{Ax}$$

$$\tan 87.8 = \frac{Cx}{100}$$

$$Cx = 100 \times \tan 87.8 = 2603$$

622. (a)

Let Cost price = x

$$x \times \frac{120}{100} \times \frac{118}{100} = 3186$$

$$x = \frac{100 \times 100 \times 3186}{120 \times 118} = 2250$$

■■■■

Previous ESE Prelims Solved Questions
(General Aptitude)

1. Three hundred passengers are travelling in white, silver and black cars; each of these cars is carrying 6, 5 and 3 passengers, respectively. If the number of white and silver cars is equal and three is only one black car, what is the total number of cars?
(a) 52 (b) 53
(c) 54 (d) 55
[ESE Pre-2017]

2. The present ages of 3 brothers are in the proportion 3 : 4 : 5. After 10 years the sum of their ages will be 78. What are their ages now?
(a) 12, 16 and 20 (b) 15, 20 and 25
(c) 21, 28 and 35 (d) 24, 32 and 40
[ESE Pre-2017]

3. A total of 324 notes comprising of ₹ 20 and ₹ 50 denominations make a sum of ₹ 12,450. The number of ₹ 20 notes is
(a) 200 (b) 144
(c) 125 (d) 110
[ESE Pre-2017]

4. Five Men can paint a building in 20 days, 8 Women can paint the same building in 25 days and 10 Boys can paint it in 30 days. If a team has 2 Men, 6 Women and 5 Boys, how long will it take to paint the building?
(a) 12 days (b) 13 days
(c) 14 days (d) 15 days
[ESE Pre-2017]

5. Rajiv spends 40% of his salary on food, 20% on house rent, 10% on entertainment and 10% on conveyance. If his savings at the month end are ₹ 2,000, then his monthly salary is:
(a) ₹ 6,000 (b) ₹ 8,000
(c) ₹ 10,000 (d) ₹ 12,000
[ESE Pre-2017]

6. A group of workers estimate to finish a work in 10 days, but 5 workers could not join the work. If the rest of them finished the work in 12 days, the number of members present in the team originally is
(a) 50 (b) 45
(c) 35 (d) 30
[ESE Pre-2017]

7. The sum of squares of successive integers 8 to 16, both inclusive, will be
(a) 1126 (b) 1174
(c) 1292 (d) 1356
[ESE Pre-2018]

8. Given that 0.8 is one root of the equation, $x^3 - 0.6x^2 - 1.84x + 1.344 = 0$. The other roots of this equation will be
(a) 1.1 and –1.4 (b) –1.2 and 1.4
(c) 1.2 and –1.4 (d) –1.1 and 1.4
[ESE Pre-2018]

9. The equation, $x^3 - 8x^2 + 37x - 50 = 0$ is factored and it has $(3 + 4i)$ as one of its roots. What is the real root of this equation?
(a) 2 (b) 4
(c) 6.5 (d) 13
[ESE Pre-2018]

10. Circle *A* is 4 cm in diameter; circle *B* is 5 cm in diameter. Circle *C* has its circumference equal to the sum of the circumferences of both *A* and *B* together. What will be the ratio of the area of circle *C*, with respect to the area of circle *A* and circle *B* respectively?
(a) 5.0625 and 1.84 (b) 3.875 and 1.84
(c) 5.0625 and 3.24 (d) 3.875 and 3.24
[ESE Pre-2018]

11. The 12 digits on the face of a clock are to be represented employing contributions of only the number 9 as either 9 or $\sqrt{9}$. The other prescribed conditions are (i) the least number of uses alone are permitted; and (ii) when alternates are possible,

use of 9 will be preferred over use of $\sqrt{9}$, which should be used minimally. How many times would $\sqrt{9}$ have to be used?

(a) 6 (b) 5
(c) 4 (d) 3

[ESE Pre-2018]

12. In a particular test, the marks scored by 4 candidates – *A*, *B*, C and *D* are as follows:
- Marks obtained by *A* and *B* add to 100;
- Marks obtained by *C* and *D* add up to those scored by *A*;
- B scores 4 times of *D*;
- D scores 10 marks less than *C*.

The marks obtained by *C* will be

(a) 30 (b) 15
(c) 20 (d) 25

[ESE Pre-2018]

13. What is the form of the function $f(x)$ for the following data?

x	0	1	2	3
$f(x)$	3	6	11	18

(a) $x^2 + 2x + 3$ (b) $x^2 - 2x + 3$
(c) $x^2 + 2x - 3$ (d) $x^2 - 2x - 3$

14. What is the maximum value of z, if $z = 10x + 6y$ subject to the constraints $3x + y \le 12$, $2x + 5y \le 34$, $x \ge 0$, $y \ge 0$?

(a) 56 (b) 52
(c) 50 (d) 40

[ESE Pre-2018]

15. Let the sum of the squares of successive integers 0, 1, 2, ..., $n - 1$, n be denoted by *S*. Let the sum of the cubes of the same integers be denoted by *C*. It is desirable that $\frac{C}{S}$, as n increases in steps of 'unity' from 'zero', is given by the series:

$\frac{0}{1}, \frac{3}{3}, \frac{9}{5}, \frac{18}{7}, \frac{30}{9}, \ldots$ (for $n = 0, 1, 2, 3, 4 \ldots$). What will this ratio be for $n = 8$?

(a) $\frac{108}{17}$ (b) $\frac{103}{17}$
(c) $\frac{103}{15}$ (d) $\frac{100}{15}$

[ESE Pre-2018]

16. A wall, rectangular in shape, has a perimeter of 72 m. If the length of its diagonal is 18 m, what is the area of the wall?

(a) 224 m² (b) 486 m²
(c) 572 m² (d) 606 m²

[ESE Pre-2018]

17. To isolate an enclosed area for conservation, an open traverse is run keeping close to (but outside of) the exterior boundary of the area through ground points $A \to B \to C \to D \to E \to F \to G \to$ towards *H* (to be eventually located). *AB* is 80° to the East of the North line at *A*. Deflection/ Interior angles at *B*, *C*, *D*, *E*, *F* are indicated. What would be the magnitude of the deflection angle at *G* (as marked) so that *GH* many run parallel to *BA*? (Lengths are immaterial in this case.)

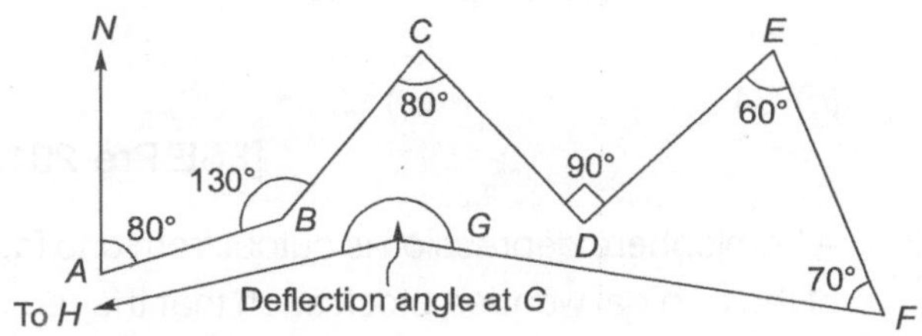

(a) 190° (b) 210°
(c) 200° (d) 230°

[ESE Pre-2018]

18. The objective function $z = 3x_1 + 5x_2$ is to be maximized subjected to constraints.

$x_1 + 2x_2 \le 200$
$x_1 + x_2 \le 150$
$x_1, x_2 \ge 0$

The values of x_1 and x_2 in this context are, respectively

(a) 100 and 75 (b) 125 and 75
(c) 100 and 50 (d) 125 and 50

[ESE Pre-2018]

19. Consider the length of a room is 15 m and width is 10 m. If the sum of the areas of the floor and ceiling is equal to the sum of the areas of the four walls, then the volume of the room is

(a) 900 m³ (b) 1000 m³
(c) 1200 m³ (d) 1500 m³

[ESE Pre-2018]

20. A small production unit now works 6 days per week with $3\frac{1}{3}$ hours of first shift every one of the 6 days and 3 hours of second shift for each of the first 5 days. Wage negotiations led to an agreement to work on 5 days a week with both shifts together clocking $7\frac{1}{2}$ hours per day with an 8% increase in weekly wages. How much change in the hourly production would mean parity in the agreement for both management and employees?
(a) 3.68% (b) 2.15%
(c) 1.82% (d) 1.33%
[ESE Pre-2018]

21. A student purchases some books for Rs. 1600. If he had bought 8 more books for the same amount, each book would cost Rs. 10 less. The number of books he buys is
(a) 30 (b) 32
(c) 34 (d) 36
[ESE Pre-2019]

22. A hemisphere depression is cutout from one face of the cubical wooden block such that the radius r of the hemisphere is equal to half of the edge of the cube. What will be the surface area of the remaining solid?
(a) $2r^2(\pi + 24)$ (b) $r^2(\pi + 24)$
(c) $2r^2(\pi + 36)$ (d) $r^2(\pi + 36)$
[ESE Pre-2019]

23. A rod of length l is to be divided into two parts, such that if 5 times the smaller portion is added to half of the larger portion, it will always be less than l. This can be achieved by taking length of the larger portion more than
(a) $\frac{9}{10}l$ (b) $\frac{7}{8}l$
(c) $\frac{6}{7}l$ (d) $\frac{5}{6}l$
[ESE Pre-2019]

24. Which of the following conditions hold good for a train which crosses the bridge of length l in time t_1 and crosses another bridge of length $\frac{l}{2}$ in time t_2?
1. $t_2 = \frac{t_1}{2}$
2. $2t_2 > t_1$
3. $t_2 < \frac{t_1}{2}$
4. Speed of train is $\frac{l}{10}$ if $t_1 - t_2 = 5$

Select the correct answer using the codes given below:
(a) 1 and 4 only (b) 2 and 4 only
(c) 1 and 3 only (d) 2 and 3 only
[ESE Pre-2019]

25. A tourist covers half of his journey by train at 60 km/h, half of the remainder by bus at 30 km/h and the rest by cycle at 10 km/h. Average speed of the tourist during the journey is
(a) 36 km/h (b) 33 km/h
(c) 24 km/h (d) 18 km/h
[ESE Pre-2019]

26. In a lake, the tip of a bud of lotus is seen 10 cm above the surface of water. Forced by the wind, it gradually moved, and just submerged at a distance of 30 cm. The depth of water at the root of the lotus plant will be
(a) 40 cm (b) 50 cm
(c) 60 cm (d) 70 cm
[ESE Pre-2019]

27. A man sold a chair and a table together for Rs. 7,600, thereby making a profit of 25% on the chair and 10% on the table. By selling them together for Rs. 7,500 he would make a profit of 10% on the chair and 20% on the table. Then the cost price of chair and table will be
(a) Rs. 3000 and Rs. 4000
(b) Rs. 3500 and Rs. 4000
(c) Rs. 3000 and Rs. 3500
(d) Rs. 3500 and Rs. 3500
[ESE Pre-2019]

28. In two concentric circles, a chord length 80 cm of larger circle becomes a tangent to the smaller circle whose radius is 9 cm. The radius of the larger circle will be
(a) 13 cm (b) 41 cm
(c) 52 cm (d) 75 cm
[ESE Pre-2019]

29. Consider following diagram: AC is a diameter of the large circle and $AB = BC$.

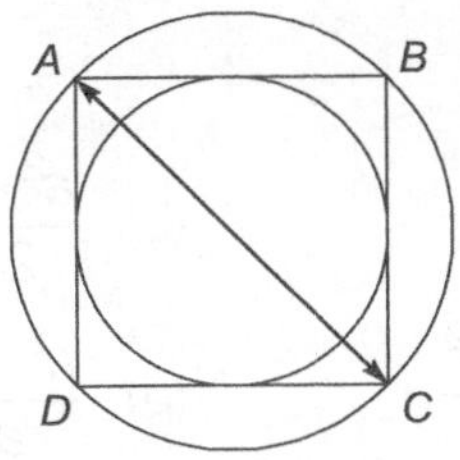

The ratio of areas of the large circle to the small circle of a square is

(a) 4 : 1 (b) 1 : 4
(c) 2 : 1 (d) 1 : 2

[ESE Pre-2019]

30. Ten years ago father was 12 times as old as his son and after 10 years father will be 2 times older than his son. The present ages of father and son respectively are

(a) 32 years and 14 years
(b) 34 years and 14 years
(c) 32 years and 12 years
(d) 34 years and 12 years

[ESE Pre-2020]

31. A number of friends decided to go on a picnic and planned to spend Rs. 96 on eatables. Four of them, however, did not turn up. As a consequence, the remaining ones had to contribute Rs. 4 each extra. The number of those friends who attended the picnic is

(a) 8 (b) 12
(c) 16 (d) 20

[ESE Pre-2020]

32. Consider the following gold articles P, Q, R, S and T with different weights:

- P weighs twice as much as Q
- Q weighs four and a half times as much as R
- R weighs half as much as S
- S weighs half as much as T
- T weighs less than P but more than R

Article T will be lighter in weight than

(a) P and S (b) P and R
(c) P and Q (d) Q and R

[ESE Pre-2020]

33. Consider the rectangle ABCD with $DE = \frac{1}{3}DC$ in the figure:

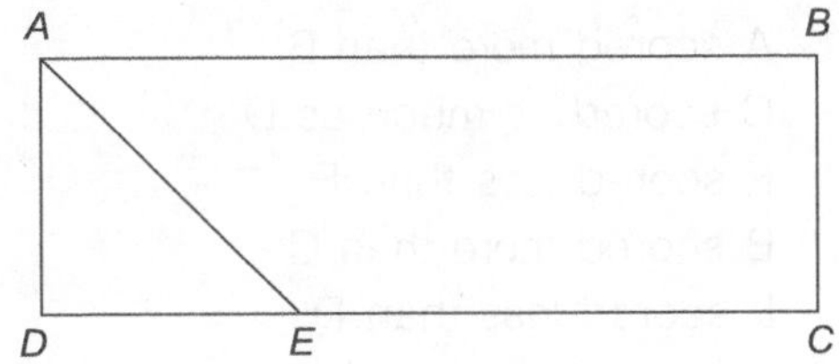

When the area of the triangle ADE is 20 cm^2, the area of the rectangle ABCD will be

(a) 60 cm^2 (b) 80 cm^2
(c) 100 cm^2 (d) 120 cm^2

[ESE Pre-2020]

34. Four metal rods of lengths 78 cm, 104 cm, 117 cm and 169 cm are to be cut into parts of equal length. Each part must be as long as possible. The maximum number of pieces that can be cut will be

(a) 27 (b) 36
(c) 43 (d) 52

[ESE Pre-2020]

35. A man walked 3 km towards East, then 5 km towards North-East, then 8 km towards South and finally 5 km towards North-East direction. The distance of his present location from the starting point will be

(a) 9 km (b) 11 km
(c) 15 km (d) 21 km

[ESE Pre-2020]

36. A clock strikes once at 1 o'clock, twice at 2 o'clock, thrice at 3 o'clock and so on. The number of times it strikes in 24 hours will be

(a) 116 (b) 136
(c) 156 (d) 196

[ESE Pre-2020]

37. The sum of all the natural numbers between 1 and 101 which are divisible by 5 is

(a) 1000 (b) 1050
(c) 1500 (d) 2550

[ESE Pre-2020]

38. In a group of 1000 people, 750 speak Hindi and 400 speak English. The number of only Hindi speaking people is

(a) 150 (b) 350
(c) 600 (d) 750

[ESE Pre-2020]

39. Consider the following students in an examination:
- A scored more than B
- C scored as much as D
- E scored less than F
- B scored more than C
- F scored less than D

Who scored the lowest?
(a) E (b) C
(c) D (d) F
[ESE Pre-2020]

40. What is the angle between the hour hand and minute hand of a clock at 3 : 30?
(a) 105° (b) 180°
(c) 75° (d) 90°
[ESE Pre-2021]

41. Sum of the series $2^2 + 4^2 + 6^2 + ... + 20^2$ is
(a) 1040 (b) 1540
(c) 2540 (d) 3080
[ESE Pre-2021]

42. If A X B means $(A^2 + B^2)$, then the value of 5 X (4 X 3) is
(a) 60 (b) 300
(c) 650 (d) 710
[ESE Pre-2021]

43. The number of triangles in the given figure is

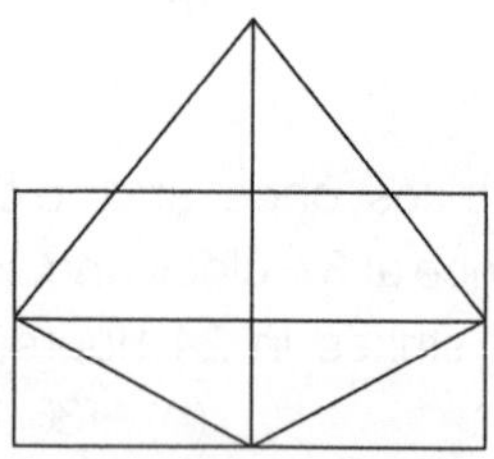

(a) 11 (b) 13
(c) 15 (d) 17
[ESE Pre-2021]

44. Statement 1 : A has more coins than B.
Statement 2 : B has fewer coins than C.
Statement 3 : C has fewer coins than A.
If the statement 1 is true and statement 2 is false, then the statement 3 is
(a) True (b) False
(c) Uncertain (d) Insufficient data
[ESE Pre-2021]

45. If $\frac{x+1}{x-1} = \frac{a}{b}$ and $\frac{1-y}{1+y} = \frac{b}{a}$, then the value of $\frac{x-y}{1+xy}$ is
(a) $\frac{2ab}{a^2-b^2}$ (b) $\frac{a^2-b^2}{2ab}$
(c) $\frac{a^2+b^2}{2ab}$ (d) $\frac{a^2-b^2}{ab}$
[ESE Pre-2021]

46. If $(2x + 3y) : (3x + 5y) = 18 : 29$, then the value of $x : y$ is
(a) 4 : 1 (b) 4 : 5
(c) 3 : 4 (d) 3 : 1
[ESE Pre-2021]

47. A is twice as good a workman as B and together, they finish a piece of work in 18 days. In how many days will A alone finish the work?
(a) 28 days (b) 30 days
(c) 27 days (d) 29 days
[ESE Pre-2021]

48. In how many years will a sum ₹800 at 10% per annum compounded semi-annually become ₹926.10?
(a) $1\frac{1}{3}$ years (b) $1\frac{1}{2}$ years
(c) $2\frac{1}{3}$ years (d) $2\frac{1}{2}$ years
[ESE Pre-2021]

49. The diagonal of a rectangle is $\sqrt{41}$ cm and its area is 20 sq.cm. The perimeter of the rectangle is
(a) 9 cm (b) 18 cm
(c) 20 cm (d) 41 cm
[ESE Pre-2021]

50. Four persons are chosen at random from a group of 3 men, 2 women and 4 children. The chance that exactly 2 of them are children, is
(a) $\frac{2}{9}$ (b) $\frac{4}{5}$
(c) $\frac{7}{12}$ (d) $\frac{10}{21}$
[ESE Pre-2021]

51. A man walks 1 km to East and then he turns to South and walks 5 km. Again, he turns to East and walks 2 km. After, he turns to North and walks 9 km. Now, how far he is from his station point?
(a) 3 km (b) 4 km
(c) 5 km (d) 7 km
[ESE Pre-2021]

52. The population of a village is 5500. If the number of males increases by 11% and the number of females increases by 20% then the population becomes 6330. The population of female in the village is
(a) 2000 (b) 2500
(c) 3000 (d) 3500
[ESE Pre-2021]

53. *A*, *B*, *C*, *D* and *E* are five different integers. When written in the ascending order of values, the difference between any two adjacent integers is 4. *D* is the greatest and *A* is the least. *B* is greater than *E* but less than *C*. The sum of the integers is equal to *E*. What is the positive difference between the lowest and the highest integers?
(a) 8 (b) 6
(c) 16 (d) 18
[ESE Pre-2021]

54. Mary introduces Jack as the son of the only daughter of my father's wife. How is Jack related to Mary?
(a) Brother (b) Son
(c) Husband (d) Father
[ESE Pre-2021]

55. Some environmentalists question the prudence of exploiting features of the environment, arguing that there are no economic benefits to be gained from forests, mountains, or wetlands that no longer exist. Many environmentalists claim that because nature has intrinsic value it would be wrong to destroy such features of the environment, even if the economic costs of doing so were outweighed by the economic costs of not doing so.
Which one of the following can be logically inferred from the passage?
(a) It is economically imprudent to exploit features of the environment.
(b) Some environmentalists appeal to a noneconomic justification in questioning the defensibility of exploiting features of the environment.
(c) Most environmentalists appeal to economic reasons in questioning the defensibility of exploiting features of the environment.
(d) Many environmentalists provide only a noneconomic justification in questioning the defensibility of exploiting features of the environment.
[ESE Pre-2022]

56. Fact 1 : Jessica has four children.
Fact 2 : Two of the children have blue eyes and two of the children have brown eyes.
Fact 3 : Half of the children are girls.
If the first three statements are facts, which of the following statements must also be a fact?
I. At least one girl has blue eyes.
II. Two of the children are boys.
III. The boys have brown eyes.
Select the correct answer using the code given below:
(a) II only
(b) I and III only
(c) II and III only
(d) None of the statements is a known fact
[ESE Pre-2022]

57. Children are in pursuit of a dog whose leash has broken. James is directly behind the dog. Ruby is behind James. Rachel is behind Ruby. Max is ahead of the dog walking down the street in the opposite direction. As the children and dog pass, Max turns around and joins the pursuit. He runs in behind Ruby. James runs faster and is alongside the dog on the left. Ruby runs faster and is alongside the dog on the right. Which child is directly behind the dog?
(a) James (b) Ruby
(c) Rachel (d) Max
[ESE Pre-2022]

58. At the baseball game, Henry was sitting in seat 253. Marla was sitting to the right of Henry in seat 254. In the seat to the left of Henry was George. Inez was sitting to the left of George. Which seat is Inez sitting in?
(a) 251 (b) 254
(c) 255 (d) 256
[ESE Pre-2022]

59. The difference between simple interest and compound interest on a sum for 2 years at 8% when the interest is compounded annually is ₹16. If the interest were compounded half yearly the difference in two interests would be nearly
(a) ₹21.35 (b) ₹24.64
(c) ₹27.85 (d) ₹29.94
[ESE Pre-2022]

60. A library has two books each having three copies and three other books each having two copies. In how many ways can all these books be arranged in a shelf so that copies of the same book are not separated?
(a) 80 (b) 100
(c) 120 (d) 140
[ESE Pre-2022]

61. 21 mango trees, 42 apple trees and 56 orange trees have to be planted in rows such that each row contains the same number of trees of one variety only. Minimum number of rows in which the above trees may be planted is
(a) 9 (b) 12
(c) 14 (d) 17
[ESE Pre-2022]

62. A general wishes to draw up his 36562 soldiers in the form of a solid square. After arranging them, he found that some of them are left over. How many are left?
(a) 81 (b) 75
(c) 61 (d) 52
[ESE Pre-2022]

63. Most serious students are happy students, and most serious students go to graduate school. Furthermore, all students who go to graduate school are overworked.
Which one of the following can be properly inferred from the statements above?
(a) Most overworked students are happy students
(b) Some happy students are overworked
(c) All overworked students are serious students
(d) Some unhappy students go to graduate school
[ESE Pre-2022]

64. A tank can be filled by 20 buckets each of capacity 13.5 litres. If the capacity of each bucket be 9 litres, how many buckets will fill the same tank?
(a) 30 (b) 25
(c) 20 (d) 15
[ESE Pre-2022]

65. One side of a rhombus is 10 cm and one of its diagonals is 12 cm. The area of the rhombus is
(a) 24 sq. cm (b) 48 sq. cm
(c) 72 sq. cm (d) 96 sq. cm
[ESE Pre-2022]

66. Two boys begin together to write out a booklet containing 817 lines. The first boy starts with first line, writing at the rate of 200 lines an hour and the second boy starts with the last line. He writes line 817 and so on backwards proceeding at the rate of 150 lines an hour. At what line will they meet?
(a) 469th (b) 467th
(c) 465th (d) 463th
[ESE Pre-2022]

67. Rohith spends 40% of his monthly income on food items and 50% of the remaining on clothes and conveyance. He saves one-third of the remaining amount after spending on food, clothes and conveyance. If he saves ₹19,200 every year, what is his monthly income?
(a) ₹32,000 (b) ₹16,000
(c) ₹12,000 (d) ₹6,000
[ESE Pre-2022]

68. Of the five boys A, B, C, D and E two are good, one is poor and two are average in studies. Two of them study in post-graduate classes and three in under graduate classes. One comes from a rich family two from middle class families and two from poor families. One of them is interested in music, two in acting and one in sports. Of those studying in under graduate classes, two are average and one is poor in studies. Of the two boys interested in acting, one is a post-graduate student. The one who is interested in music comes from a middle class family. Both of the boys interested in acting are not industrious, good in studies come from middle class families, are average in studies and one of them is interested in acting. The boy interested in sports comes from a poor family, while the one interested in music in industrious. E is industrious, good in studies comes from a poor family and is not interested in acting, music or sports. Cis poor in studies in spite of being industrious. A comes from a rich family, is not industrious but good in studies. B is industrious and comes from a middle class family. Name the boy who is not industrious and is average in studies.
(a) A (b) B
(c) C (d) D
[ESE Pre-2023]

69. At an electric Data Processing Unit five out of the eight program sets, P, Q, R, S, T, U, V and W are to be operated daily. On any one day except for the first day of the month only three of the program sets must be the ones that were operated on the previous day. The program operating must also satisfy the following conditions:
   1. If program P is to be operated on a day, V cannot be operated on that day.
   2. If Q is to be operated on a day, T must be one of the programs to be operated after Q.
   3. If R is to be operated on a day, V must be one of the programs to be operated after R.
   4. The last program to be operated on any day must be either S or U.

   If the program sets R and W are to be operated on the first day which of the following could be the other programs on that day?

   (a) Q, V, S (b) Q, T, V
   (c) T, S, V (d) T, S, V

   [ESE Pre-2023]

70. Read the following information carefully and answer the question give below it:
   1. Eight doctors *P, Q, R, S, T, U, V* and *W* visit charitable dispensary every day except on a holiday i.e., Monday.
   2. Each doctor visits for one hour from Tuesday to Sunday except Saturday. The timings are 9 A.M. to 1 P.M. and 2 P.M. to 6 P.M., 1 P.M. to 2 P.M. is lunch break.
   3. On Saturday it is opened only in the morning i.e. 9 A.M. to 1 P.M. and each doctor visits for only half an hour.
   4. No other doctor visits the dispensary, before doctor *Q* and after *U*.
   5. Doctor *W* comes immediately after the lunch break is followed by *R*.
   6. *S* comes in the same order as *P* in the afternoon session.

   If the lunch break and subsequent visiting hours are reduced by 15 minutes at what time doctor *U* is expected to attend the dispensary?

   (a) 3.15 P.M. (b) 4 P.M.
   (c) 4.15 P.M. (d) 4.45 P.M.

   [ESE Pre-2023]

71. Study the following information carefully and answer the question given below it:
   1. *P, Q, R, S, T* and *U* are six members in a family in which there are two married couples.
   2. *T*, a teacher, is married to the doctor who is mother or *R* and *U*.
   3. *Q*, the lawyer, is married to *P*.
   4. *P* has one son and one grandson.
   5. Of the two married ladies on is housewife.
   6. There is one student and one male engineer in the family.

   How is *R* related to *U*?

   (a) Brother only (b) Sister only
   (c) Brother or Sister (d) Mother

   [ESE Pre-2023]

72. Read the following information carefully and answer the question that follow:
   1. Madhu and Shobha are good in Dramatics and Computer Science.
   2. Anjali and Madhu are good in Computer Science and Physics.
   3. Anjali, Poonam and Nisha are good in physics and History.
   4. Nisha and Anjali are good in Physics and Mathematics.
   5. Poonam and Shobha are good in History and Dramatics.

   Who is good in History, Physics, Computer Science and Mathematics?

   (a) Poonam (b) Nisha
   (c) Madhu (d) Anjali

   [ESE Pre-2023]

73. The question given below the three statements followed by three conclusions numbered I, II and III. You have to take the given statements to be true even if they seem to be at variance from commonly known facts. Read all the conclusions and then decide which of the given conclusions logically follows from the given statements disregarding commonly known facts.

   Statements:
   All lions are tigers
   All tigers are leopards
   Some leopards are wolves

   Conclusions:
   I. No elephant is lion.
   II. Some wolves are lions.
   III. Some leopards are lions.

   (a) Only I follows (b) Only II follows
   (c) Only III follows (d) Only I and III follows

   [ESE Pre-2023]

74. Rohith went 15 km to the west from his house, then he turned left and walked 20 km. He then turned east and walked 25 km and finally turning left covered 20 km. How far is he from his house?

   (a) 5 km (b) 10 km
   (c) 40 km (d) 80 km

   [ESE Pre-2023]

75. Cryptic language is popular since ages, mostly in the field of espionage and sending classified messages. If 'I LOVE YOU' is coded as 7, then how would you code 'GO TO HELL' in the same language?
(a) 1 (b) 4
(c) 3 (d) 5
[ESE Pre-2023]

76. What letter should replace the question mark?

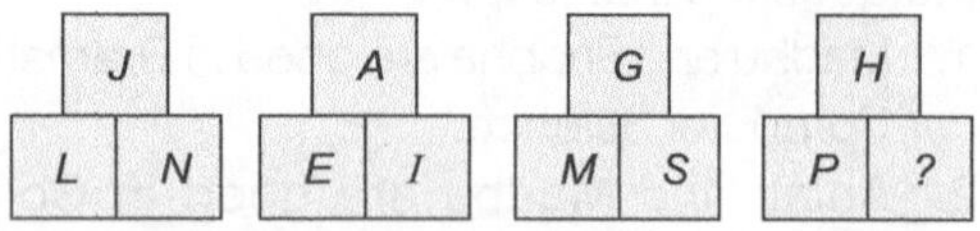

(a) Z (b) Y
(c) X (d) W
[ESE Pre-2023]

77. In the first two circles, the number inside the circle is written according to a particular relation. What is the number inside the third circle which follows the same relation as that of the first two circles?

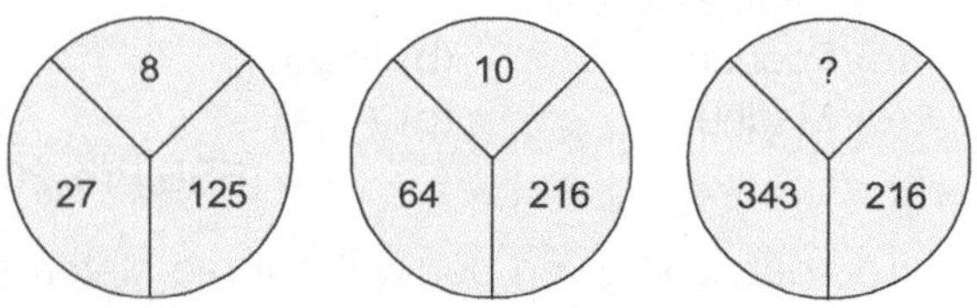

(a) 12 (b) 13
(c) 9 (d) 14
[ESE Pre-2023]

78. Deepthi is playing a treasure hunt game. At the first stage, Deepthi needs to choose a five-digit code to unlock the vault which contains the treasure. She gets the following codes to choose from

15342	26540	35415
23105	15320	13402
35047	71024	28305

The following clues are given to her to help her to find the code
P. The code number is not an even number.
Q. The product of the first two digits is odd.
R. The sum of the first four digits is 12.
S. The code number is not a multiple of 5.

If Deepthi had the option of selecting only one clue, which of the four clues will give her the best chance of finding out the five digit code?
(a) S (b) R
(c) Q (d) P
[ESE Pre-2023]

79. Suppose you enter an elevator at a certain floor. Then the elevator moves up 5 floors, down 3 floors, and up 2 floors. If you are then at the 8th floor, on what floor did you first enter the elevator?
(a) 8 (b) 7
(c) 6 (d) 9
[ESE Pre-2023]

80. A number series is given with one term missing. Choose the correct alternative from options.
0.5, 0.55, 0.65, 0.8, ?
(a) 0.9 (b) 0.95
(c) 0.82 (d) 1
[ESE Pre-2023]

81. The soccer club is putting together a mural using lightly colored transparent paper. This paper then is cut into squares of different sizes that are placed next to each other to make the designs for the mural. Of course, the club wants to save money, so its members are trying to buy the minimum number of sheets of colored paper. Below is one of the designs they are going to use. What is the minimum number of squares they will need to make this design?

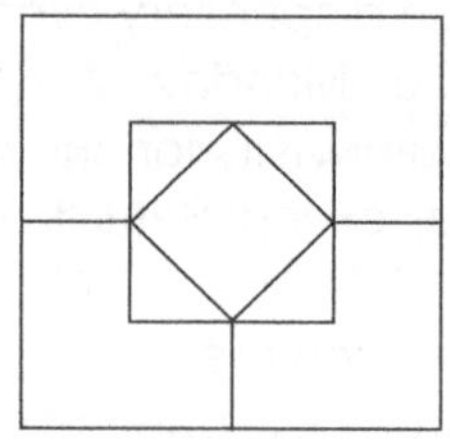

(a) 4 (b) 5
(c) 6 (d) 7
[ESE Pre-2023]

82. What is the correct alternative for the question mark?
2, 3, 8, 63, ?
(a) 1038 (b) 1998
(c) 3008 (d) 3968
[ESE Pre-2023]

83. Brianna, Ryan, Tyler, and Ashley were recently elected as the new class officers (president, vice president, secretary, treasurer) of the sophomore class at Summit College. From the following clues, determine which position each holds.
1. Ashley is younger than the president but older than the treasurer.
2. Brianna and the secretary are both the same age, and they are the youngest members of the group.

3. Tyler and the secretary are next door neighbors.

(a) Tyler is the president, Ashley is the vice president, Ryan is the secretary, and Brianna is the treasurer.
(b) Tyler is the president, Ashley is the vice president, Brianna is the secretary, and Ryan is the treasurer.
(c) Tyler is the president, Ryan is the vice president, Ashley is the secretary, and Brianna is the treasurer.
(d) Tyler is the president, Ryan is the vice president, Brianna is the secretary, and Ashley is the treasurer.

[ESE Prelims-2024]

84. You need to buy groceries at the supermarket, deposit a cheque at credit union, and purchase a book at the bookstore. You can complete the errands in any order, however, you must start and end at your home. The driving time, in minutes, between each of these locations is given in the following figure.

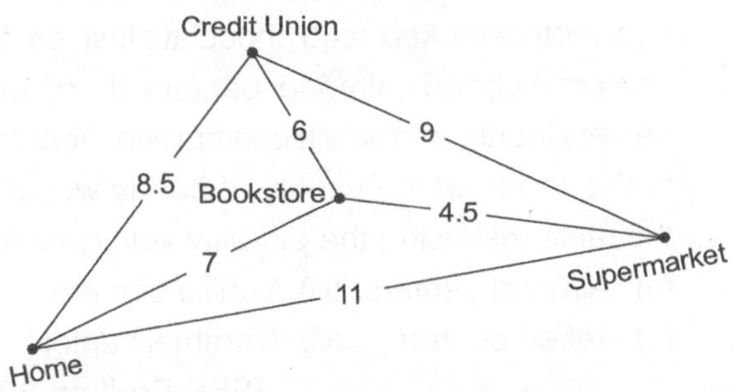

What is the route whose total driving time is less than 30 minutes?
(a) home, bookstore, credit union, supermarket, home
(b) home, supermarket, bookstore, credit union, home
(c) home, bookstore, supermarket, credit union, home
(d) home, supermarket, credit union, bookstore, home

[ESE Prelims-2024]

85. Each of four siblings (Anita, Tony, Maria and Jose) is given ₹5000 to invest in the stock market. Each chooses a different stock. One chooses a utility stock, another an automotive stock, another a technology stock, and the other an oil stock.

1. Anita and the owner of the utility stock purchased their shares through an online brokerage, whereas Tony and the owner of the automotive stock did not.
2. The gain in value of Maria's stock is twice the gain in value of the automotive stock.
3. The technology stock is traded on NASDAQ, whereas the stock that Tony bought is traded on the New York Stock Exchange.

From the above clues, match the name of the sibling and stock bought.
(a) Maria: the utility stock; Jose: the automotive stock; Anita: the technology stock; Tony: the oil stock
(b) Maria: the utility stock; Anita: the automotive stock; Jose: the technology stock; Tony: the oil stock
(c) Maria: the utility stock; Tony: the automotive stock; Anita: the technology stock; Jose: the oil stock
(d) Jose: the utility stock; Maria: the automotive stock; Anita: the technology stock; Tony: the oil stock

[ESE Prelims-2024]

86. If six people greet each other at a meeting by shaking hands with one another, how many handshakes will take place?

(a) 14	(b) 16
(c) 15	(d) 18

[ESE Prelims-2024]

87. Anuhya picks a number. She doubles the number, squares the result, divides the square by 3, subtracts 30 from the quotient, and gets 18. What are the possible numbers that Anuhya could have picked?

(a) 6 or −6	(b) 16 or −16
(c) 26 or −26	(d) 36 or −36

[ESE Prelims-2024]

88. Nothing is known about the personal life of the ancient Green Mathematician Diophantus except for the information in the following:

"Diophantus passed $\frac{1}{6}$ of his life in childhood, $\frac{1}{12}$ in youth, and $\frac{1}{7}$ more as a bechelor. Five years after his marriage was born a son who died four years before his father, at $\frac{1}{2}$ his father's (final) age." How old was Diophantus when he died?

(a) 64	(b) 54
(c) 74	(d) 84

[ESE Prelims-2024]

89. Select a two-digit number between 50 and 100. Add 83 to your number. From this number form a new number by adding the digit in the hundreds place to the number formed by the other two digits (the digits in the tens place and the ones place). Now subtract this newly formed number from your original number, to arrive at the final result. What is the final result?

(a) 16 (b) 26
(c) 36 (d) 46

[ESE Prelims-2024]

90. An activities director for a cruise ship has surveyed 240 passengers. Of the 240 passengers, 135 like swimming, 80 like swimming and dancing, 150 like dancing, 40 like swimming and games, 65 like games, 25 like dancing and games, 15 like all three activities. How many passengers like exactly two of the three types of activities?

(a) 220 (b) 20
(c) 30 (d) 100

[ESE Prelims-2024]

91. The following map shows the 10 bridges and 3 islands between the suburbs of North Bay and South Bay. During your morning workout, you decide to jog over each bridge exactly once. Which one of the following statements is correct?

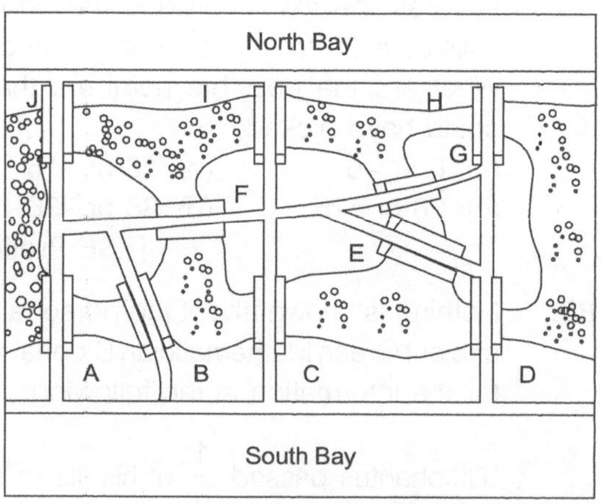

(a) You want to start from North Bay and that your workout concludes after you jog over the 'D' bridge.
(b) You want to start from North Bay and that your workout concludes after you jog over the 'E' bridge.
(c) You want to start from North Bay and that your workout concludes after you jog over the 'H' bridge.
(d) You want to start from North Bay and that your workout concludes after you jog over the 'G' bridge.

[ESE Prelims-2024]

92. Fifty people were asked to rank their preferences of five varieties of chocolate candy, using 1 for their favorite and 5 for their least favorite. The results are shown in the table below.

	Rankings					
Caramel center	5	4	4	4	2	4
Vanilla center	1	5	5	5	5	5
Almond center	2	3	2	1	3	3
Toffee center	4	1	1	3	4	2
Solid chocolate	3	2	3	2	1	1
Number of voters	17	11	9	8	3	2

According to the table (see the column in grey), three voters ranked solid chocolate first, caramel centers second, almond centers third, toffee centers fourth, and vanila centers fifth. According to this table, which variety of candy would win the taste test using the plurality voting system?

(a) Almond centers (b) Vanilla centers
(c) Toffee centers (d) Caramel centers

[ESE Prelims-2024]

93. The members of a club are going to elect a president from four nominees. In each first-place vote receives 4 points, each second-place vote receives 3 points, each third-place vote receives 2 points, and each last place vote receives 1 point. If the 100 members of the club mark their ballots as shown in the table below, who will be elected president?

	Rankings					
Avalon	2	2	2	2	3	2
Branson	1	4	4	4	2	1
Columbus	3	3	1	3	1	3
Dunkrirk	4	1	3	1	4	4
Number of voters	30	24	18	12	10	6

(a) Avalon (b) Branson
(c) Columbus (d) Dunkirk

[ESE Prelims-2024]

94. Study the given information carefully and answer the question :

There are seven books, one each of Psychology, Hindi, English, Sociology, Economics, Education and Accountancy lying on a table one above the other. Sociology is on the top of all books. Accountancy is immediately below Education which is immediate below Sociology. Economics is immediately above Psychology but not in the middle. Hindi is immediately below Psychology.

Which three books are between Accountancy and Hindi?

(a) English, Economics and Psychology
(b) Economics, Psychology and Education
(c) Economics, Psychology and Hindi
(d) Cannot be determined

[ESE Prelims-2024]

95. Read the information given below and answer the questions :

There is a group of five girls. Hasini is second in height but younger than Madhavi. Pooja is taller than Pranati but younger in age. Madhavi and Pranati are of the same age but Madhavi is tallest among them. Neelam is taller than Pooja and elder to Madhavi.

If they are arranged in the descending order of their ages who will be in fourth position?

(a) Neelam (b) Hasini
(c) Pranati (d) Data inadequate

[ESE Prelims-2024]

96. Read the following information and answer the question :

Seven students P, Q, R, S, T, U and V take a series of tests. No two students get similar marks. V always scores more than P. P always scores more than Q. Each time either R scores the highest and T gets the least, or alternatively S scores the highest and U or Q scores the least.

If V is ranked fifth, which one of the following is correct?

(a) S scores the highest
(b) R is ranked second
(c) T is ranked third
(d) Q is ranked fourth

[ESE Prelims-2024]

97. A man has a certain number of small boxes to pack into parcels. If he packs 3, 4, 5 or 6 in a parcel, he is left one, if he packs 7 in a parcel, none is left over. What is the number of boxes, he may have to pack?

(a) 106 (b) 301
(c) 309 (d) 400

[ESE Prelims-2024]

■■■■

ANSWER KEY

1. (d)	2. (a)	3. (c)	4. (d)	5. (c)	6. (d)	7. (d)	8. (c)	9. (a)
10. (c)	11. (b)	12. (d)	13. (a)	14. (a)	15. (a)	16. (b)	17. (a)	18. (c)
19. (a)	20. (a)	21. (b)	22. (b)	23. (a)	24. (b)	25. (c)	26. (a)	27. (c)
28. (b)	29. (c)	30. (d)	31. (a)	32. (c)	33. (d)	34. (b)	35. (b)	36. (c)
37. (b)	38. (c)	39. (a)	40. (c)	41. (b)	42. (c)	43. (c)	44. (a)	45. (a)
46. (c)	47. (c)	48. (b)	49. (b)	50. (d)	51. (c)	52. (b)	53. (c)	54. (b)
55. (d)	56. (a)	57. (d)	58. (a)	59. (b)	60. (c)	61. (d)	62. (a)	63. (b)
64. (a)	65. (d)	66. (b)	67. (b)	68. (d)	69. (d)	70. (b)	71. (c)	72. (d)
73. (c)	74. (b)	75. (b)	76. (c)	77. (b)	78. (b)	79. (*)	80. (d)	81. (b)
82. (d)	83. (a)	84. (c)	85. (a)	86. (c)	87. (a)	88. (d)	89. (a)	90. (d)
91. (b)	92. (c)	93. (a)	94. (a)	95. (d)	96. (a)	97. (b)		

EXPLANATIONS

1. (d)
Since the number of passengers in Black car is 3, it leaves 300 − 3 = 297 passengers to be accommodated in White and Silver coloured cars which are equal in number ($W = S$).
We can form a linear equation:
$(6 + 5)W = 297$ giving $W = 27$ as also $S = 27$
Total number of cars being 27 + 27 + 1 = 55

2. (a)
Let $3x$, $4x$, $5x$ be the present age of 3 brothers. After 10 years, sum of their age is $12x + 30 = 78$ which gives $x = 4$
Hence the present age of three brothers is 12, 16 and 20 years.
Alt: Can be directly solved using the given options since the sum of their present age = 78 – 30 = 48 which is satisfied by option 'A' alone.

3. (c)
We can form 2 linear equations taking the number of Rs. 20 and Rs. 50 notes as T and F

$$T + F = 324 \quad ...(1)$$

$$20T + 50F = 12450 \quad ...(2)$$

Solving the 2 equations, we get $T = 125$

Alt 1: If all the notes are Rs. 50 notes, total amount = 324 × 50 = Rs. 16200 which is 16200 − 12450 = 3750

3750/(50 − 20) = 125 notes of Rs. 20.

Alt 2: This question can also be solved by putting options.

4. (d)
5 men in 20 days = 100 man days
8 women in 25 days = 200 woman days
10 boys in 30 days = 300 boy days
which means 100 MD = 200 WD = 300 BD or 1 man day = 3 boy days; 1 woman day = 1.5 Boy days 2 men, 6 women and 5 boys = 2 × 3 + 6 × 1.5 + 5 = 20 boys deployed to complete the task which will get completed in 300/20 = 15 days
Alt: $[2(1/100) + 6(1/200) + 5(1/300)]x = 1$
Which gives $x = 15$

5. (c)
The remaining percentage after expenditure 20% which equals Rs. 2000 leading to total monthly income = Rs. 10000.

6. (d)
If 'W' is the original number of workers estimated to finish the work, we can form an equation as: 10 W = (W – 5) × 12 which gives W = 30.

7. (d)
Apply formula for Σn^2 to solve this question where

$$\Sigma n^2 = \frac{n(n+1)(2n+1)}{6}$$

$$\Sigma_{n=1}^{n=16} n^2 - \Sigma_{n=1}^{n=7} n^2$$

$$= \frac{16(16+1)(32+1)}{6} - \frac{7(7+1)(14+1)}{6}$$

= 1496 – 140 = 1356

8. (c)
Use the formulae for sum and product of roots
Given equation is $x^3 - 0.6x^2 - 1.84x + 1.344 = 0$
Let the roots of this equation be α, β, γ

$$\alpha + \beta + \gamma = -\frac{b}{a} = 0.6 \quad \text{...(i)}$$

$$\alpha\beta + \beta\gamma + \gamma\alpha = \frac{c}{a} = 1.84 \quad \text{...(ii)}$$

$$\alpha\beta\gamma = -\frac{d}{a} = 1.334 \quad \text{...(iii)}$$

If α = 0.8 then β + γ = 0.6 – 0.8 = –0.2 which can be picked up as option since numerically greater value should have negative sign.

9. (a)
Use the formulae for sum and product of roots
Given equation is $x^3 - 8x^2 + 37x - 50 = 0$
Let the roots of this equation be α, β, γ
With one root α as 3 + 4i; the other root β is its conjugate i.e. 3 – 4i

$$\alpha + \beta + \gamma = -\frac{b}{a} = 8 \quad \text{.....(i)}$$

this gives the third root = 8 – 6 = 2

10. (c)
Circumference of circle A = 4π as its diameter = 4 cm; Area of circle A = 4π or $\frac{16}{4}\pi$

Circumference of circle B = 5π as its diameter = 5 cm; Area of circle B = $\left(\frac{5}{2}\right)^2 \pi$ or $\frac{25}{4}\pi$

Circumference of circle C = Circumference of circle A + Circumference of circle B = 4π + 5π = 9π
Diameter of circle C = 9 cm

Area of circle C = $\left(\frac{9}{2}\right)^2 \pi = \frac{81}{4}\pi$

Ratio of are of circle C w.r.t. area of circle A and that of circle B will be $\frac{81}{16}, \frac{81}{25}$ or 5.0625 and 3.24.

11. (b)
Care is to be taken to meet the given conditions i.e. least number of 9s and or √9 to be used and preference to be given to use of 9.

$1 = \frac{9}{9}$; $2 = \frac{(9+9)}{9}$; $3 = \sqrt{9}$; $4 = \sqrt{9} + \frac{9}{9}$;

$5 = 9 - \sqrt{9} - \frac{9}{9}$; $6 = 9 - \sqrt{9}$; $7 = 9 - \frac{9+9}{9}$; $8 = 9 - \frac{9}{9}$;

$9 = 9$; $10 = 9 + \frac{9}{9}$; $11 = \frac{99}{9}$; $12 = 9 - \sqrt{9}$;

12. (d)

$$A + B = 100$$
$$C + D = A$$
$$B = 4D$$
$$D = C - 10 \text{ or } C = D + 10$$
$$A + 4D = 100$$

and $D + D + 10 = A$

or $A = 2D + 10$

Solving these 2 equations given D = 15 and C = 25

13. (a)
This question can be solved by just putting values given in various options.

14. (a)
This question can be solved by finding point of intersection of the 2 linear equations representing the constraints.

$$3x + y \le 12$$

and $2x + 5y \le 34$

which is x = 2 and y = 6
Putting this value in the given objective function i.e. z = $10x + 6y$, we get maximum value of z = 56.

15. (a)

Ignoring 1st term, $\frac{c}{s} = \frac{\Sigma n^3}{\Sigma n^2} = \frac{\frac{\{n(n+1)\}^2}{2}}{\frac{n(n+1)(2n+1)}{6}}$

Putting, $n = 8$ we get $\frac{c}{s} = \frac{108}{17}$

16. (b)

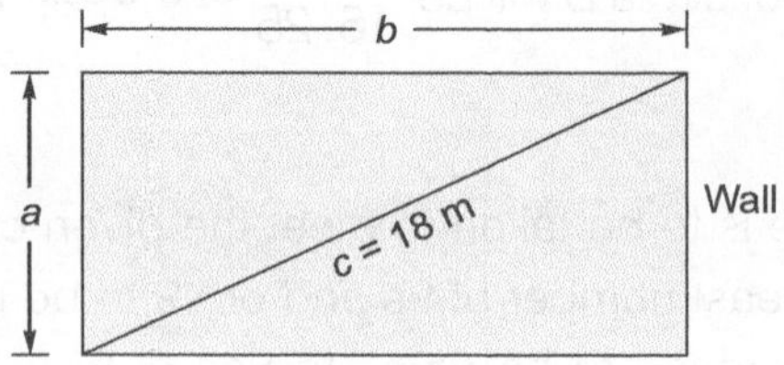

$$\text{Perimeter} = 72 \text{ m}$$

$$\Rightarrow \quad (a + b) \times 2 = 72 \text{ m} \quad \text{... (i)}$$

$$a + b = 36 \text{ m}$$

$$a^2 + b^2 = c^2$$

$$\Rightarrow \quad a^2 + b^2 = 18^2 = 324 \quad \text{... (ii)}$$

$$\text{Area} = a \times b \quad \text{... (iii)}$$

we know, $(a + b)^2 = a^2 + b^2 + 2ab$

$$\Rightarrow \quad 36^2 = 324 + 2 \times (\text{Area})$$

$$\Rightarrow \quad \text{Area} = 486 \text{ m}^2$$

17. (a)

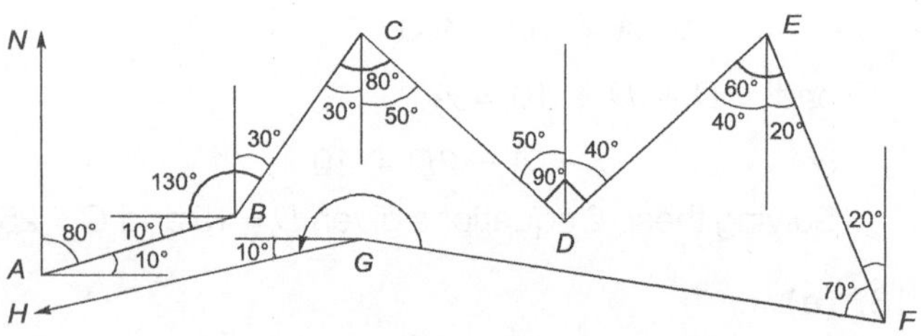

$$\angle FGH = 190°$$

Note: Line FG will be horizontal and not inclined as given in question.

18. (c)

This question can be solved by finding point of intersection of the 2 linear equations representing the constraints

$$x_1 + 2x_2 \leq 200$$

and $\quad x_1 + x_2 \leq 150$

which is $\quad x_1 = 100$ and $x_2 = 50$

19. (a)

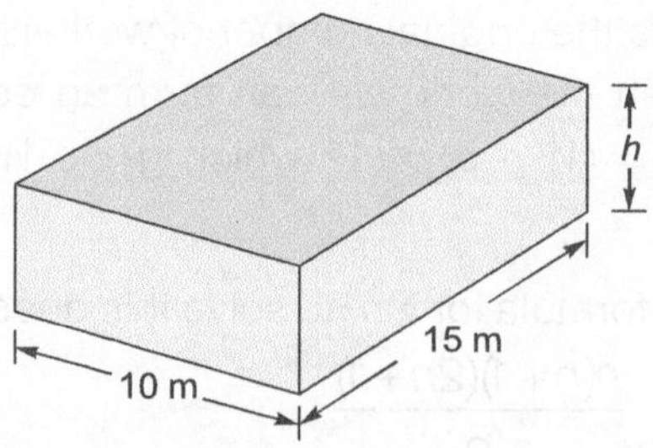

Given information is

Area of floor + Area of ceiling = Sum of area of four walls

$$\Rightarrow (10\times15)+(10 \times 15) = (15\times h\times2) + (10 \times h \times 2)$$

$$\Rightarrow \quad 300 = 50h$$

$$\Rightarrow \quad h = 6 \text{ m}$$

$$\text{Volume of room} = 10 \text{ m} \times 15 \text{ m} \times 6 \text{ m}$$

$$= 900 \text{ m}^3$$

20. (a)

In the first case, total number of hours per week = 3.5 × 6 + 3 × 5 = 36 hours

In the second case, total number of hours = 7.5 ×5 = 37.5 hours

With increased wages, the effective hourly rate is

$\frac{(36 \times 1.08)}{37.5} = 1.0368$ implying 3.68% as the increase in expected productivity.

21. (b)

Let x be the number of books

Using options

$$\frac{1600}{x} - \frac{1600}{x+8} = 10$$

$$x = 32$$

Alternatively

$$32 \times 50 = 40 \times 40$$

22. (b)

Required area = Trough area of hemisphere [A]

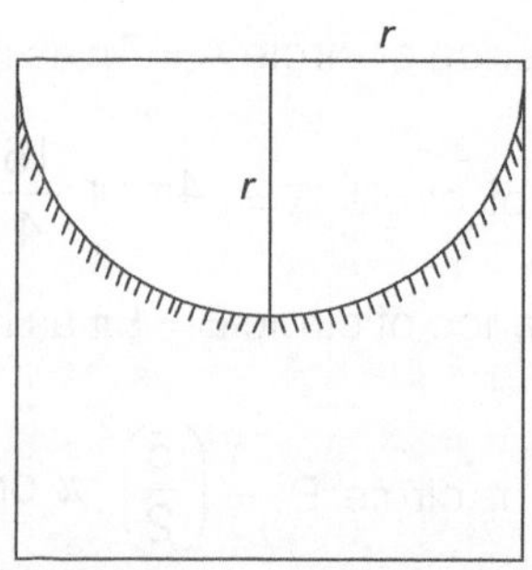

+ Area of five remaining faces of cube [B]

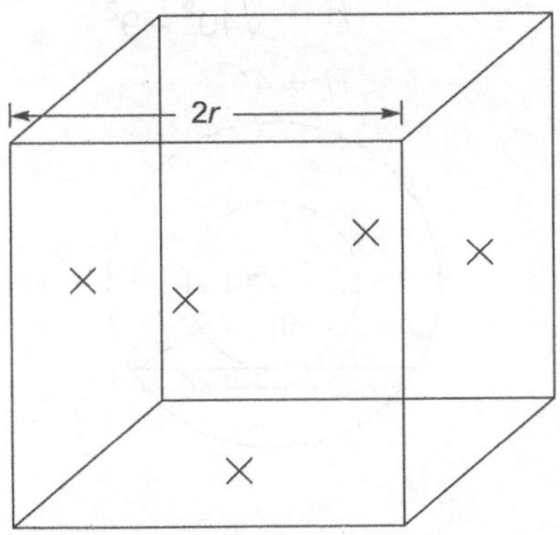

+ Left out area on the face of cube after taking out hemisphere [C]

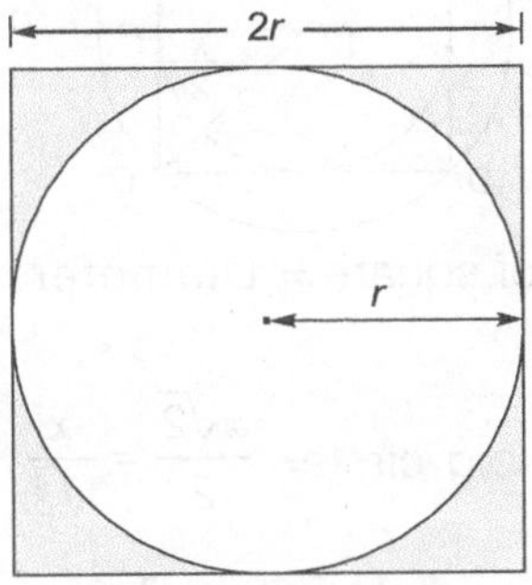

$$A = 2\pi r^2$$
$$B = 5 \times (2r)^2 = 20r^2$$
$$C = 4r^2 - \pi r^2$$
$$\text{Required area} = A + B + C$$
$$= \pi r^2 + 24r^2$$
$$= r^2 (\pi + 24)$$

23. (a)

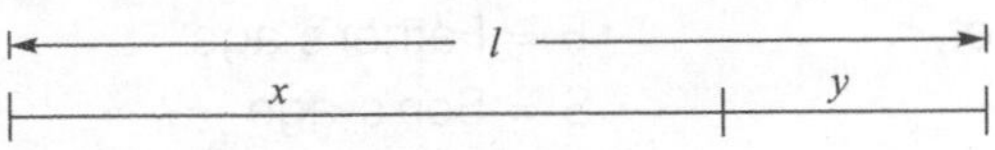

$$x + y = l$$
$$y = l - x$$
$$5y + \frac{x}{2} < l$$
$$5l - 5x + \frac{x}{2} < l$$
$$-\frac{9}{2}x < -4l$$
$$x > \frac{8}{9}l$$

Only option greater than $\frac{8}{9}l$ is option (a)

Alternatively

In option (a), let length of rod = 10

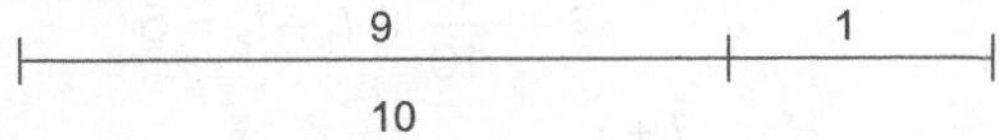

$$5 + 4.5 = 9.5 < 10$$

In option (b) let length of rod = 8 and likewise.

Only option (a) satisfies.

24. (b)

Let the length of train = x

Length of the bridge = l

Speed of train = s

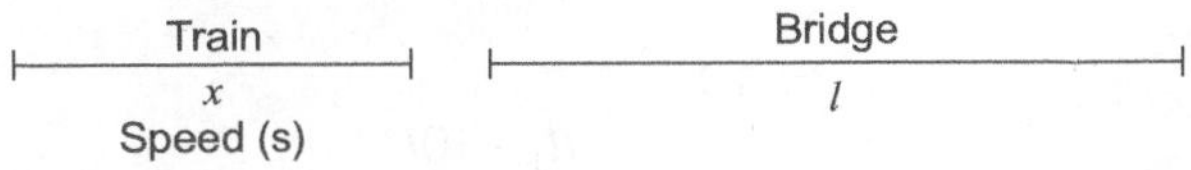

$$\frac{l+x}{s} = t_1 \quad \text{...(i)}$$

$$\frac{\frac{l}{2}+x}{s} = t_2 \quad \text{..(ii)}$$

Option (1)

If $\quad t_2 = \frac{t_1}{2}$

Replacing eq. (i) and (ii)

$$\frac{\frac{l}{2}+x}{s} = \frac{1}{2}\left(\frac{l+x}{s}\right)$$

$$2\left(\frac{l}{2}+x\right) = (l+x)$$

$$x = 0 \quad \text{(Not possible)}$$

Length of train must be greater than zero.

Therefore option (1) is false.

Option (2)

$$2(t_2) > t_1$$

$$\frac{2\left(\frac{l}{2}+x\right)}{s} > \frac{l+x}{s}$$

$$l + 2x > l + x$$

$$x > 0$$

Length of train has to be positive.

Therefore option (2) is true.

Option (3)

$$t_2 < \frac{t_1}{2}$$

$$\frac{\frac{l}{2}+x}{s} < \frac{1}{2}\left(\frac{l+x}{s}\right)$$

$$2\left(\frac{l}{2}+x\right) < l + x$$

$$x < 0$$

Length of train cannot be negative.

Therefore option (3) is false.

Option (4)

$$s = \frac{l}{10} \text{ if } t_1 - t_2 = 5$$

$$\frac{l + x}{s} = t_1 \quad \text{...(i)}$$

$$\frac{\frac{l}{2} + x}{s} = t_2 \quad \text{...(ii)}$$

Replacing $s = \frac{l}{10}$ in eq. (i) and (ii)

$$\frac{l + x}{\frac{l}{10}} = t_1 \quad \text{...(iii)}$$

$$\Rightarrow \quad x = \frac{lt_1 - 10l}{10} \quad \text{...(iv)}$$

$$\frac{\frac{l}{2} + x}{\frac{l}{10}} = t_2 \quad \text{...(v)}$$

$$\Rightarrow \quad x = \frac{lt_2 - 5l}{10} \quad \text{...(vi)}$$

Eq. (iv) = eq. (vi) solve $t_1 - t_2 = 5$

Option (2) and (4) are true.

25. (c)

Let total distance = D

$$\text{Average speed} = \frac{\text{Total distance}}{\text{Total time}}$$

$$= \left(\frac{D}{\frac{\frac{D}{2}}{60} + \frac{\frac{D}{4}}{30} + \frac{\frac{D}{4}}{10}} \right) = 24 \text{ kmph}$$

26. (a)

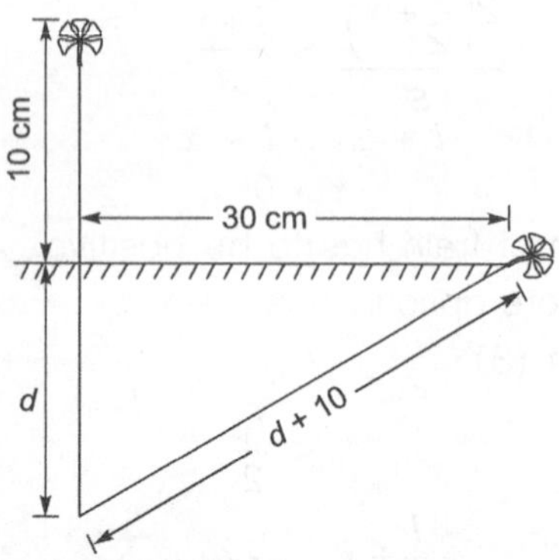

Applying Pythagoras Theorem

$$d^2 + 30^2 = (d + 10)^2$$

Using options, d = 40 (Pythagorean Triplet)

27. (c)

Using options

3000 × 1.25 + 3500 × 1.1 = 7600

3000 × 1.10 + 3500 × 1.20 = 7500

28. (b)

$$R = \sqrt{40^2 + 9^2}$$

$$R = 41$$

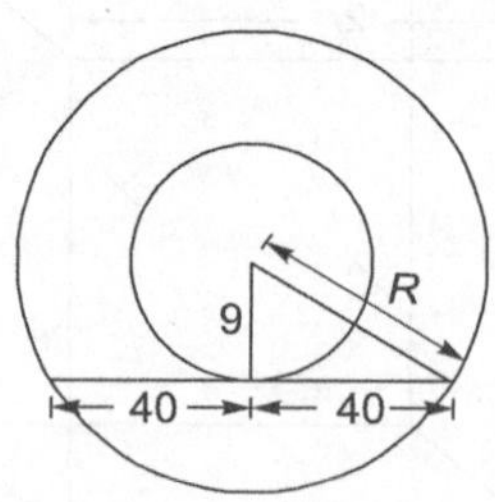

29. (c)

Let the length of side of square = x

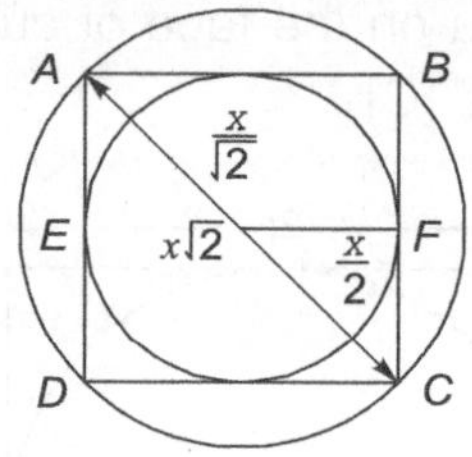

Diagonal of square = Diameter of big circle $= x\sqrt{2}$

$$\text{Radius of big circle} = \frac{x\sqrt{2}}{2} = \frac{x}{\sqrt{2}}$$

$$\text{Radius of small circle} = \frac{1}{2}(\text{side of square}) = \frac{x}{2}$$

∴ Ratio of the areas of big circle and small circle

$$= \frac{\pi \frac{x^2}{2}}{\pi \frac{x^2}{4}} = \frac{2}{1} = 2:1$$

30. (d)

Let, F = Father's age

S = Son's age

$$\frac{F - 10}{S - 10} = 12 \quad \text{...(i)}$$

$$\frac{F + 10}{S + 10} = 2 \quad \text{...(ii)}$$

Solving equations (i) and (ii),

F = 34; S = 12

Alternative solution:

Using option (d)

Father's age (F) = 34 years

Son' age (S) = 12 years

$$\frac{F - 10}{S - 10} = \frac{34 - 10}{12 - 10} = \frac{24}{2} = 12$$

$$\frac{F + 10}{S + 10} = \frac{34 + 10}{12 + 10} = \frac{44}{22} = 2$$

31. (a)

$$x \times y = 96 \quad \text{...(i)}$$
$$(x - 4)(y + 4) = 96 \quad \text{...(ii)}$$

Solving eq. (i) and (ii) we get,

$$x = 12,\ y = 8$$
$$\therefore \quad 12 - 4 = 8$$

Alternative Solution:

Using option (a)

$$8 \times 12 = 96$$
$$(12 - 4) \times (8 + 4) = 8 \times 12 = 96$$

32. (c)

$$\begin{array}{ccccccccc} P & : & Q & : & R & : & S & : & T \\ 2x & : & x & : & \frac{x}{4.5} & : & \frac{2x}{4.5} & : & \frac{4x}{4.5} \\ 9x & : & 4.5x & : & x & : & 2x & : & 4x \end{array}$$

R < S < T < Q < P.

Alternative Solution:

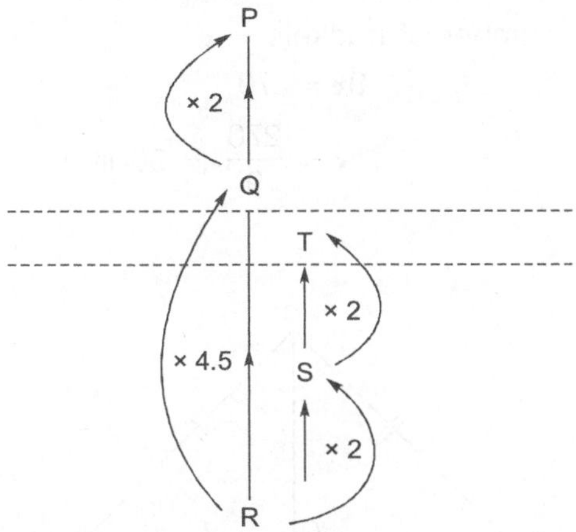

33. (d)

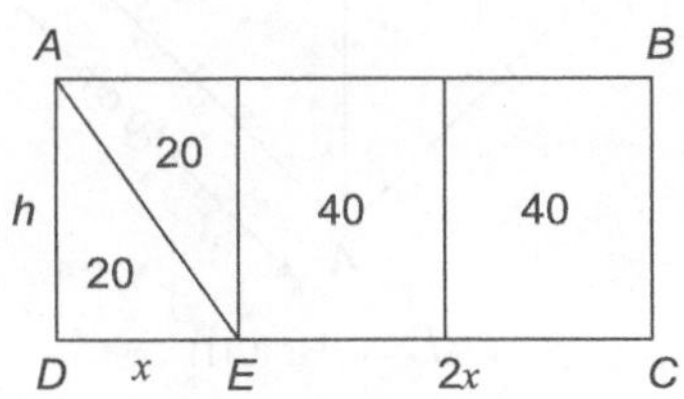

$$\Delta_{ADE} = \frac{1}{2} \times x \times h = 20 \text{ (given)}$$
$$x \times h = 40$$

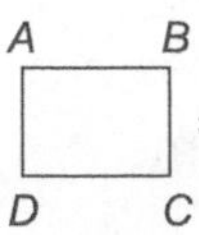

$$\Rightarrow \quad 3x \times h = 3(xh) = 3(40) = 120$$

34. (b)

HCF of 78, 104, 117, 169 is 13 (i.e. maximum length).

$$\text{Total pieces} = \frac{78}{13} + \frac{104}{13} + \frac{117}{13} + \frac{169}{13}$$
$$= 6 + 8 + 9 + 13 = 36$$

35. (b)

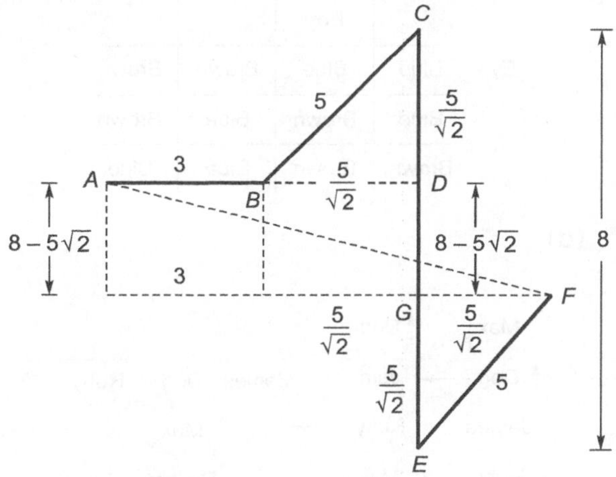

$$AF = \sqrt{(8 - 5\sqrt{2})^2 + (\sqrt{3} + 5\sqrt{2})^2} = \sqrt{102} = 10.1$$

Best case answer is option (b).

36. (c)

Time		No. of strikes
1	→	1
2	→	2
⋮		⋮
12	→	12

Time		No. of strikes
13	→	1
14		2
⋮		⋮
24	→	12

$$2[1 + 2 + 3 \ldots + 12] = 2 \times \frac{12 \times 13}{2} = 156$$

37. (b)

$$\text{Sum} = 5 + 10 + 15 + \ldots + 100$$
$$= 5(\Sigma 20) = 5 \times \frac{20 \times 21}{2} = 1050$$

38. (c)

$$1000 = 750 + 400 - n(H \cap E)$$
$$\Rightarrow \quad n(H \cap E) = 150$$

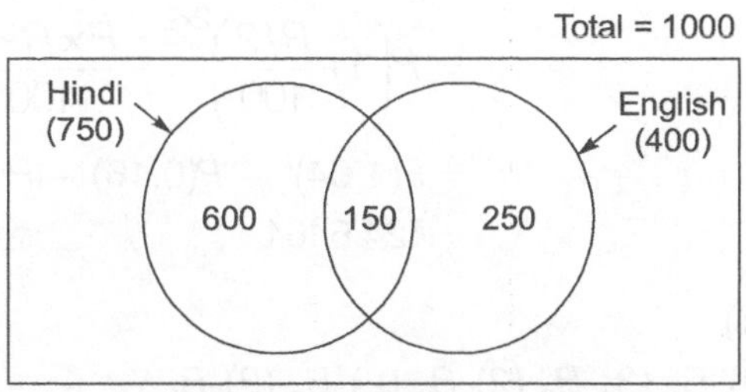

Only Hindi = 600

39. (a)

$$A > B > C = D > F > E$$

56. (a)

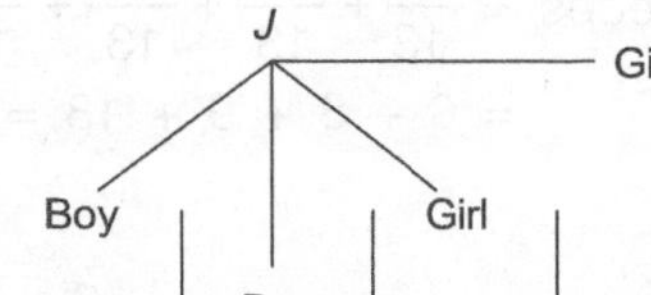

	Boy	Boy	Girl	
Eye	Blue	Blue	Brown	Brown
	Blue	Brown	Blue	Brown
	Brown	Brown	Blue	Blue

57. (d)

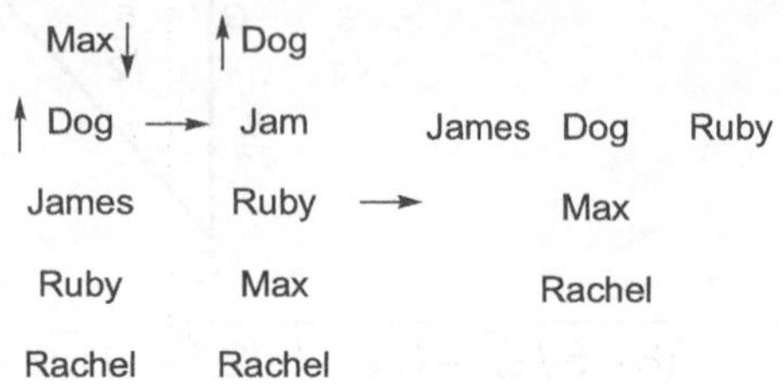

58. (a)

251	252	253	254
I	G	H	M

59. (b)

For interest compounded annually

$$\left[P\left(1+\frac{R}{100}\right)^2 - P\right] - \frac{P\times R\times 2}{100} = 16$$

$$\Rightarrow \quad P(1.08)^2 - P - P(0.16) = 16$$

$$\Rightarrow \quad P(1.1664) - P - P(0.16) = 16$$

$$\Rightarrow \quad P(0.0064) = 16$$

$$P = \frac{16}{0.0064} = 2500$$

For compounded half yearly

$$= P\left(1+\frac{R/2}{100}\right)^{2\times 2} - \frac{P\times R\times 2}{100} - P$$

$$= P(1.04)^4 - P(0.16) - P$$

$$= ₹24.6464$$

60. (c)

(3) B_1 (3) B_2 (2) B_3 (2) B_4 (2) B_5

Arrange in shelf so that copies of the same book are not separates

$$= 5! = 120$$

61. (d)

Minimum number of row contain same number of tree of one variety

$$= \frac{\text{Total}}{\text{HCF}(21,42,56)} = \frac{21+42+56}{7}$$

$$= 3 + 6 + 8 = 17$$

62. (a)

$$(191)^2 + 81 = 36562$$

81 soldiers are left.

	$\overline{3}\,\overline{65}\,\overline{62}$
1	3 65 62
+1	1
29	265
9	261
381	462
1	381
	81

64. (a)

Total volume of the tank = 13.5 × 20 = 270 litres

Volume of the tank will remain same.

Let number of buckets = x

$$\therefore \quad 9x = 270$$

$$\therefore \quad x = \frac{270}{9} = 30 \text{ litres.}$$

65. (d)

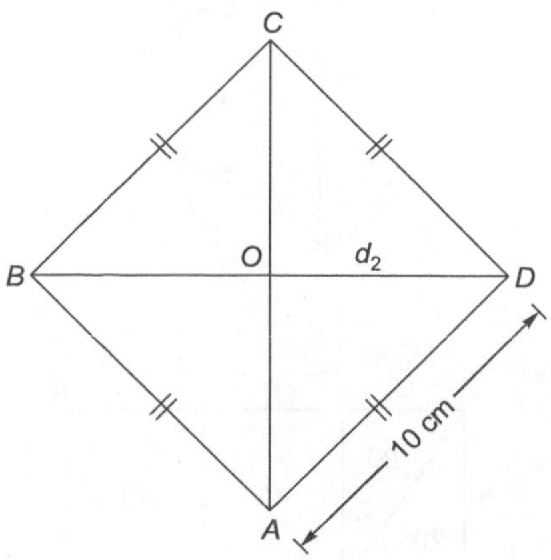

$$d_1 = 12 \text{ cm}$$

For ΔODA, $OA^2 + OD^2 = AD^2$

$$\left(\frac{d_1}{2}\right)^2 + OD^2 = AD^2$$

$$= 6^2 + OD^2 + 10^2$$

$$\Rightarrow \quad OD = 8 \text{ cm}$$

$$\Rightarrow \quad d_2 = 2(OD) = 16 \text{ cm}$$

$$\therefore \quad \text{Area of } ABCD = \frac{1}{2}d_1\times d_2 = \frac{1}{2}\times 12\times 18$$

$$= 96 \text{ sq. cm.}$$

66. (b)

Total lines 817 litres.

Time to complete by both

$$= \frac{817}{200+150} = \frac{817}{350} \text{ hours}$$

$$\text{Both meet at line} = \frac{817}{350}\times 200$$

or

$$817 - \frac{817}{350}\times 150$$

$$= 466.857 \approx 467^{th} \text{ line}$$

67. (b)

$$\text{Let Income} = 100k \text{ per month}$$

$$\text{Spend on food} = \frac{40}{100}\times 100k = 40k$$

Spend on cloth and conveyance

$$= \frac{60}{100}\times\frac{50}{100}\times 100k = 30k$$

$$\text{Saving} = \left(\frac{1}{3}\right)\times(100-40k-30k)$$

$$= \frac{1}{3}\times 30k = 10k$$

$$\text{Monthly saving} = \frac{19200}{12} = ₹\ 1600 = 10k$$

$$\Rightarrow \quad k = 160$$

$$\therefore \text{Monthly income} = 100k = 100\times 160$$

$$= ₹\ 16{,}000$$

68. (d)

Forward difference table is

A = Rich, E = Poor

A, E = good in studies and C = poor in studies

B, D - Average in studies

B, D, C - middle class family

A, E in PG

A, D - Interested in acting

B, D - middle class family

D - Music

C - Sports

B, C, E - Industrious

A and D - Not industrious

So, D is average in studies and NOT industrious.

69. (d)

(PV) not together on any day.

$(\overrightarrow{R\ V})$ as V after R

Last S or U

$(\overrightarrow{Q\ T})$ Q followed by T

T, S, V only correct (d).

70. (b)

B = 4 PM

Q visit first and 4 visit last

After break there are foru people W, R, P, U and total time reduced by 15 minute for break and 15 mins each for Dr.'s visiting after break total reduction of 75 min.

6 PM – 75 min = 4 : 45 PM

So, U came @ 4 : 00 - 4 : 45

As break is from 1 to 2 reduced by 15 mins.

So, break only from1 to 1:45

Next	1:45 to 2:30
Next	2:30 to 3:15
Next	3:15 to 4:00
Next	4:00 to 4:45

U attended from 4 to 4:45

So, he began @ 4 PM = B

U is expected to attend from 4 to 4:45.

72. (d)

Following table can be plotted

	Dramatics	C.S.	Physics	History	Maths
Madhu	✓	✓	✓		
Shobha	✓	✓		✓	
Anjali	✓	✓	✓	✓	✓
Poonam	✓		✓	✓	
Nisha			✓	✓	✓

Only Anjali good in History, Physics, C.S. and Maths.

73. (c)

(i) No mention of elephant

(ii) May or may not

(iii) Always follows

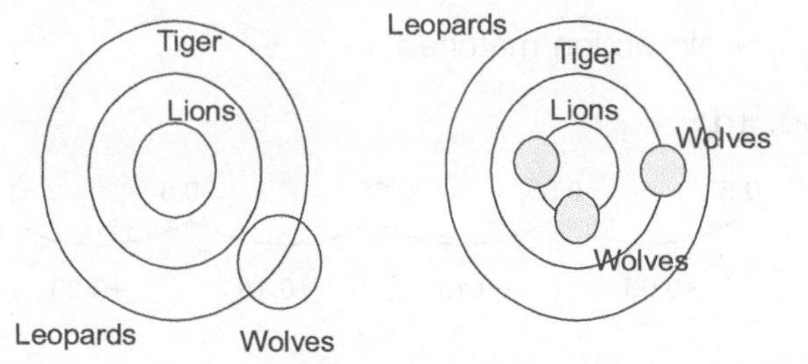

74. (b)

25 − 15 = 10

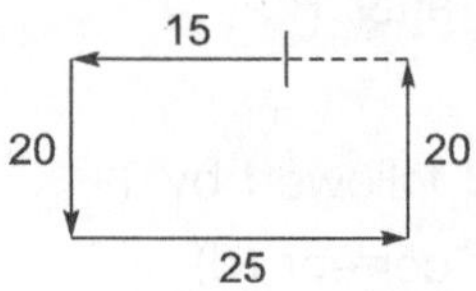

75. (b)

I L O V E Y O U

9 12 15 22 5 25 15 21 = 124 = 1 + 2 + 4 = 7

7 15 20 15 8 5 12 12

G O T O H E L L = 94 = 13 = 4

76. (c)

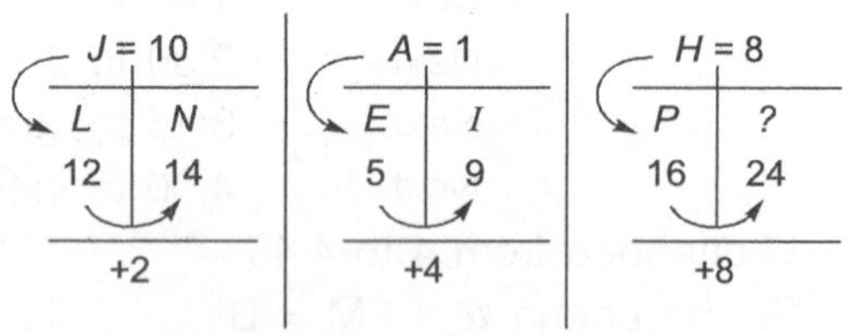

24 = X

77. (b)

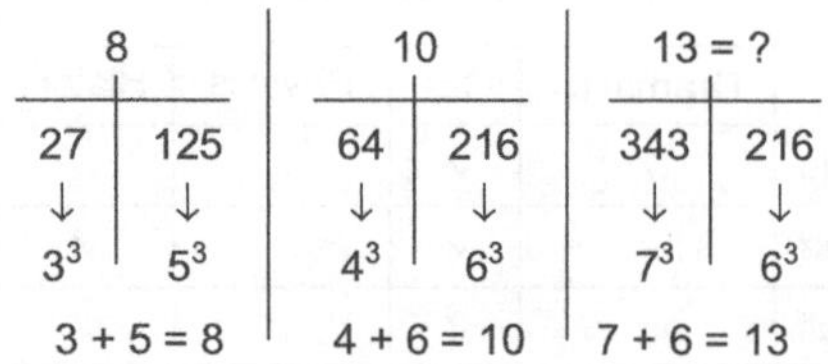

78. (b)

As 3504

$R \to$ first four digits

3504 = 3 + 5 + 0 + 4 = 12

Only one number (code) 35047 satisfies it

79. (*)

$x + 5 - 3 + 2 = 8$

$x = 4$

* No option matched

80. (d)

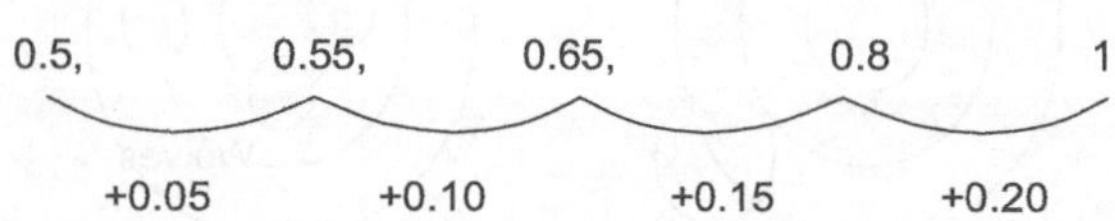

81. (b)

$PQRS \to 1$

$ABCD \to 2$

$\alpha\beta\gamma\delta \to 3$

$EFGS \to 4$

$iFGR \to 5$

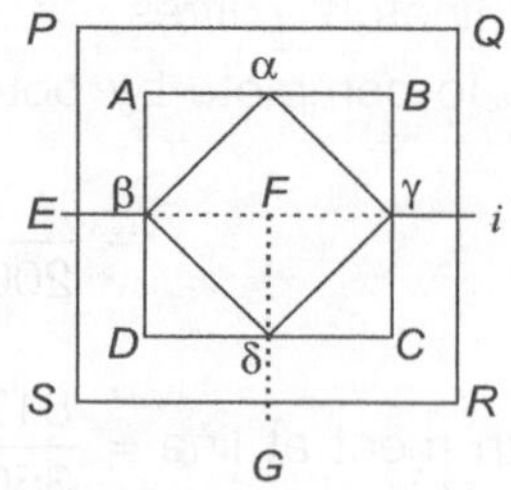

82. (d)

$2^2 - 1 = 3$

$3^2 - 1 = 8$

$8^2 - 1 = 63$

$63^2 - 1 = 3968$

Ans. = D = 3968

83. (a)

The given information can be compiled as below:

	President	Vice President	Secretary	Treasurer
Brianna	✘ (S2)	✘ (S2)	✘ (S2)	✓
Ryan	✘ (S3)	✘ (S2)	✓	✘ (S2)
Tayler	✓	✘ (S2)	✘ (S3)	✘ (S2)
Ashley	✘ (S1)	✓	✘ (S1)	✘ (S1)

84. (c)

The driving time for the different route locations is as given below:

Option	Route	Driving time
(a)	home, bookstore, credit union, supermarket, home	7+6+9+11 = 33 mins
(b)	home, supermarket, bookstore, credit union, home	11 + 4.5 + 6 + 8.5 = 30 mins
(c)	home, bookstore, supermarket, credit union, home	7 + 4.5 + 9 + 8.5 = 29 mins
(d)	home, supermarket, credit union, bookstore, home	11 + 9 + 6 + 7 = 33 mins

Hence, the drive time of route (c) has driving time less than 30 minutes.

85. (a)

The given information can be compiled as below:

	Utility stock	Automotive Stock	Technology Stock	Oil stock
Anita	✘ (S1)	✘ (S1)	✓	✘
Tony	✘ (S1)	✘ (S1)	✘ (S3)	✓
Marita	✓	✘ (S2)	✘	✘
Jose	✘	✓	✘	✘

86. (c)
In the handshake between two people, the order doesn't matter. If 6 people shakes hands with each other,
Total number of handshakes

$$= {}^6C_2 = \frac{6!}{2!4!} = \frac{5\times6}{2} = 15$$

87. (a)
Assuming anuhya picks the number 'n', we have

$$\frac{(2n)^2}{3} - 30 = 18$$

$$\Rightarrow \frac{4n^2}{3} = 48$$

$$\Rightarrow n^2 = 36$$

$$\Rightarrow n = +6 \text{ or } -6$$

88. (d)
Let the age at which Diophantus died be 'x'. From the given information, the age of Diophantus at the time of marriage,

$$y = \frac{x}{6} + \frac{x}{12} + \frac{x}{7}$$

$$= \frac{(14+12+7)x}{84} = \frac{33x}{84} = \frac{11x}{28}$$

We can write,

$$(x-4) - \left(\frac{11x}{25} + 5\right) = \frac{x}{2}$$

$$\Rightarrow \frac{x}{2} - \frac{11x}{28} = 9$$

$$\Rightarrow 3x = 28 \times 9$$

$$\Rightarrow x = 84 \text{ years}$$

89. (a)
Select a random number between 50 and 100, say 90. Adding 83 to this number, we get

83 + 90 = 173

Forming a new number by adding the digit in the hundreds place to the number formed by the other two digits, we get

1 + 73 = 74

Subtracting from the original number, the final result is obtained as:

90 – 74 = 16

90. (d)
The given information can be represented in the Venn diagram as below:

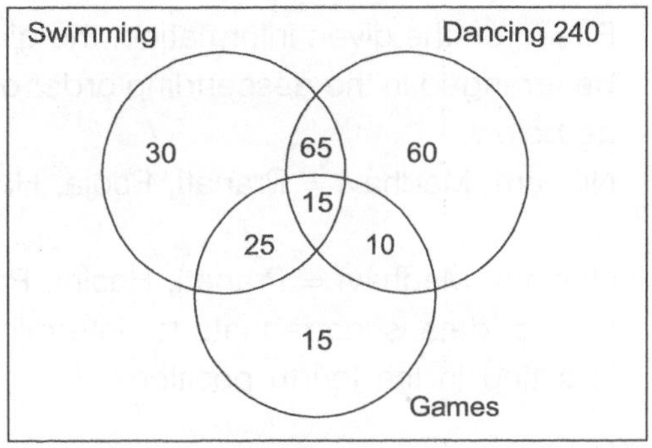

Number of passengers who like exactly two of the three types of activities.

= 65 + 25 + 10 = 100

91. (b)
Starting from North Bay and given that each bridge can be jogged exactly once, the path for jogging would be I → F → B → A → J → H → G → C → D → E.
Hence, the workout concludes over the 'E' bridge.

92. (c)
In plurality voting system, one who gets the most votes wins. Considering the total voters, the number of votes ranking '1' received by different varieties of chocolate candy are as detailed below:

Caramel center 0
Vanilla center 17
Almond center 8
Toffee center 20
Solid chocolate 2

Hence, Toffee center candy would win the taste test using the plurality voting system.

93. (a)
Total points received by different members:
Avalon: (5 × 3) + 2 = 17 points
Branson: (2 × 4) + 3 + (3 × 1) = 14 points
Columbus: (2 × 4) + (4 × 2) = 16 points
Dunkrirk: (2 × 4) + 2 + (3 × 1) = 13 points
Hence, Avalon will be the elected president.

94. (a)
Based on the given information, the order in which the books are arranged can be determined as below:

Sociology Education Accountancy English Economics Psychology Hindi

95. (d)

Based on the given information, the girls can be arranged in the descending order of ages as below:

Neelam, Madhavi = Pranati, Pooja, Hasini

or

Neelam, Madhavi = Pranati, Hasini, Pooja

Hence, data is inadequate to determine who is sitting in the fourth position.

96. (a)

In terms of scores, we have : V > P, P > Q i.e. V > P > Q.

If R scores the highest, we have R > .....> T,

If S scores the highest, we have S > .....> Q or S > .....> U.

If V ranks fifth, P and Q coming after it will occupy sixth and seventh places respectively i.e. Q ranks least. So, S will score the highest.

97. (b)

The required number should be such that it leads a remainder of 1 when divided by 3, 4, 5 or 6 and no remainder when divided by 7.

LCM (3, 4, 5, 6) = 60

So, number in the form = $60x + 1$

The minimum value of x for which the number leaves no remainder when divided by 7 is 5.

Hence,

Required number = 60 × 5 + 1 = 301

■■■■